STUDIES ON THE
STRUCTURE & DEVELOPMENT
OF VERTEBRATES

STUDIES ON THE STRUCTURE & DEVELOPMENT OF VERTEBRATES

EDWIN S. GOODRICH

THE UNIVERSITY OF CHICAGO PRESS

This is a facsimile of the 1930 edition,
published by Macmillan, London.
Pages 507, 770–74, and 776–77 have been
reproduced from the Dover edition of 1958.

THE UNIVERSITY OF CHICAGO PRESS, CHICAGO 60637

Originally published 1930
Foreword © 1986 by The University of Chicago
All rights reserved
University of Chicago Press edition 1986
Printed in the United States of America

95 94 93 92 91 90 89 88 87 86   5 4 3 2 1

Library of Congress Cataloging-in-Publication Data

Goodrich, Edwin S. (Edwin Stephen), 1868–1946.
    Studies on the structure & development of
vertebrates.

    Reprint. Originally published: London : Macmillan,
1930.
    Bibliography: p.
    Includes index.
    1. Vertebrates—Anatomy.   2. Vertebrates—Evolution.
I. Title.   II. Title: Studies on the structure and
development of vertebrates.
QL605.G6   1986   596′.04   86–7002
ISBN 0–226–30354–3 (pbk.)

FOREWORD

E. S. GOODRICH'S *Studies on the Structure and Development of Vertebrates,* universally known as "Goodrich," is an indispensable part of any zoologist's library, not just as a reference work charting a course through an older and often bafflingly diverse literature, but also as a useful guide to work in progress. Goodrich is a book that successfully draws from several traditions, particularly European descriptive morphology and embryology, and combines them with the evolutionary approach more popular, then as now, in Britain and North America. It is a comparative morphology set out in explicitly evolutionary terms, dealing with the grand sweep of vertebrate evolution at the major group level and proceeding organ group by organ group rather than taxonomically. It is no exaggeration to say that it opened up this sort of zoology for the English-speaking audience. The approach and style are also highly individual, and one has to learn one's way around the book in order to get the most out of it. But once that is done a complex world opens up to the reader, one based upon the simple premise that a study of formal (and we would now add experimental) embryology stands as the foundation of any study of evolutionary homology of vertebrate structures. Throughout his whole career, as the biography written by Hardy makes plain,[1] E. S. Goodrich showed that by teasing out the details of fine structure and developmental history of organ systems one can derive a wealth of phylogenetically useful data. Well-known examples of his method are his papers on the scales or archaic fishes or on the excretory organs of invertebrates.

The republication of Goodrich is particularly timely because the sort of work that it illuminates is currently enjoying a renaissance. It is probably not original to observe that fields within biological science tend to proceed in a cyclical pattern, with periods of greater interest in comparative studies alternating with periods when the science charges ahead on a very narrow front with little regard either

[1] A. C. Hardy, "Edwin Stephen Goodrich, 1868–1946." *Quarterly Journal of Microscopical Science* 87, part 4 (1946); reprinted in the 1958 Dover edition of Goodrich.

v

for comparison or history. We now seem to be entering a new phase
of vertebrate morphology when technological advances in developmental biology, especially dealing with later stage mor-
phogenesis, have brought forth so many exciting new data that they
are being eagerly applied across a broad comparative front with
hopes of great value obtaining for evolutionary studies. Most of
these results come from work using avian embryos, which have
replaced the amphibian embryos of the nineteen thirties as experimental organisms of choice. Questions concerning the homology of
a whole range of features, the tetrapod limb, vertebral column,
visceral skeleton (and particularly the mammalian ear ossicles), the
heart and aortic arches, the neural crest and the whole subject of
head segmentation, have all been opened up for restudy in the last
few years. And for all such questions the original terms of discussion are set out in Goodrich, and I delight in demonstrating to
students that one can open the book at random and still find a
doctoral dissertation problem on every page.

At the same time, tremendous advances made in recent years in
systematic theory, in evolutionary theory (and particularly the cru-
cial area of intersection between developmental and evolutionary
theory) and in general theory and philosophy of biology have set
Goodrich within a wholly new context, so that in our own researches, while Goodrich may have laid a foundation, we need to
examine carefully the premises of both the arguments and the
method itself. While the embryo may be neatly laid out in the form
of apparently discrete rudiments, the morphogenetic processes that
turn these rudiments into the adult phenotype are complex and we
need to know the nature of these processes before we can be sure that
the structure of embryos, which is surely in great part an embryonic
adaptation, is also a faithful guide to the relationships of adults. As
has been stated many times, Darwinian evolutionary theory transformed the *Bauplan* of older theoretical morphology into the "common ancestor." Modern comparative zoologists have to be wary of
slavishly adhering to either notion when interpreting the embryo
and they also need to be alert to the problems of deciphering the
conceptual bases from which older contributions may have been
written.

There are difficulties in working with this book. An organ system
approach requires an excellent index, but the one provided is woe-
fully inadequate (a good project for someone to take on as a service to
the science). Goodrich tended to focus his arguments on subjects

that interested him, leaving gaps. There is nothing on the central nervous system, for example. In general we may find that his schemes of homology are a little forced (as in the aortic arch story perhaps) so that, as it is fashionable and often disingenuous to say, his patterns must be seen as hypotheses to test, along with the methodology itself.

One of the book's enduring strengths is the way in which it is illustrated. Vertebrate morphology, like so much of biology, is a fundamentally visual subject. E. S. Goodrich had a special gift in that, having trained as an artist, he could present to the reader just the right drawings to match his text. The 785 pages of text included 754 figures, a very high proportion of which one sees reproduced time an time again in other texts, because of the way in which Goodrich crafted them to help make a point clear.

On a personal note, while I am frankly an admirer of Goodrich, both the book and the man, I have to confess that this was not a book that I took to instinctively as a student. I wrestled with its internal structure, chafed over the gaps, and fought out paragraph by paragraph ways in which modern data agreed with or contradicted his schemes of homology. But once we had come to terms, the book became a favorite guide, one that I tend to read in parallel with E. S. Russell's *Form and Function* (1916) and A. S. Romer's *The Vertebrate Body* (latest edition, 1977, edited by T. S. Parsons) before delving into the original literature. Russell provides the historical background to theory; Romer's book (which Romer originally began in an attempt to revise Goodrich until he realised he had to write his own) makes most things clear; Goodrich gives the wealth of detail that makes the subject complicated. At a time when a whole new generation of zoologists with diverse backgrounds in developmental and cell biology are turning to examine questions of vertebrate morphology, the republication of E. S. Goodrich's classic, yet still current, text will make available once again a masterpiece of comparative biology.

KEITH S. THOMSON

CONTENTS

ix

CONTENTS

CHAPTER III

CHAPTER IV

CHAPTER V

CONTENTS

CHAPTER VI

CHAPTER VII

CHAPTER VIII

CHAPTER IX

CHAPTER X

CONTENTS

CHAPTER XI

CHAPTER XII

CHAPTER XIII

CHAPTER XIV

CONTENTS

STUDIES ON THE STRUCTURE AND DEVELOPMENT OF VERTEBRATES

ERRATA ET CORRIGENDA

P. xx: line 31 from top, for 'Climatias' read 'Climatius'.

P. xxi: line 9 from top, for 'Tarassius' read 'Tarrasius'.

P. xxix: from the Order **Ciconiiformes** remove to the Order **Pelecaniformes** the genera Fregata, Odontopteryx*, Pelecanus, Phaethon, Phalacrocorax; add Anhinga and Sula to the latter Order.

P. xxix: line 25, for 'Scops' read 'Scopus'; line 31, for 'Serpentaria' read 'Serpentarius'; line 34, for 'Laphophorus' read 'Lophophorus'.

P. xxx: line 19 from bottom, for 'Cinolus' read 'Cinclus'; in Order **Amphitheria*** insert 'Amphitherium*'.

P. 4: 5 lines from top, delete 'other'.

P. 68: 5 lines from top, for 'axial' read 'atlas'.

P. 221: in 2nd line of legend and in figure above, 'vr' should be 'cr'.

P. 233: in Fig. 245 A, for 'ac' read 'acv'; in legend, insert 'acv, anterior cardinal vein'.

P. 285: line 3 of legend of Fig. 290, for 'Er' read 'Fr'.

P. 295: line 9 from bottom, for 'prefrontal' read 'postfrontal'.

P. 377: line 2 from top, for 'prefrontal' read 'postfrontal'.

P. 464: legend of Fig. 490, insert 'ty, tympanic'.

P. 481: line 10 from bottom, for 'Fig. 570 A' read 'Fig. 507 A'.

P. 588: line 3 from bottom, for '**849**' read '**949**'.

P. 592: legend of Fig. 600, insert 'sa, sinus endolymphaticus'.

P. 617: legend of Fig. 619, and p. 619 legend of Fig. 621, insert 'ptc, splanchnocoele'.

P. 658: legend of Fig. 655, insert 'sc, splanchnocoele'.

P. 659: line 3 from top, for 'Fig. 665' read 'Fig. 655'.

P. 683: legend of Fig. 680, for 'cardinal view' read 'cardinal vein'.

P. 722: line 4 from bottom, for 'Fig. 747' read 'Figs. 745 and 750'.

P. 750: line 2 from bottom, for 'ophthalmicus' read 'ophthalmica'.

P. 795: line 14 from bottom, for 'Geol. Mag. : . 1914' read 'Anat. Anz. v. 44, 1913'.

P. 837: in Index, insert 'Weber's apparatus, 590, 599 f., 600 f.'.

PREFACE

THIS book has been written in the hope that it may help advanced students and others engaged in teaching and research. It is not a complete treatise, but deals with certain subjects and problems of special interest and importance, some of which receive but scant notice in current text-books. My original intention was to cover the whole range of vertebrate morphology; but the preparation of this volume has taken so many years, that I thought it better to publish what is ready than to wait for the remainder which might possibly never be completed. The literature dealing with the Morphology of the Vertebrata is so vast, the accumulation of known facts so large, that students are apt to feel discouraged from the start, and to turn perhaps to some newer branch of zoological science. On the one hand, they may think that little remains to be done in so ancient a study; or, on the other hand, that its conclusions, for instance in Phylogeny, are so insecure that they afford little trustworthy evidence concerning the process of Evolution. It has, therefore, been my endeavour not only to give an account within reasonable compass of the facts already known and to discuss their significance, but also to point out where our knowledge is deficient, and where further research is desirable. During the last fifty years or so much has been accomplished, many old theories have been overthrown, some new conclusions have been firmly established; yet a great deal remains to be done, and new fields for research are continually being opened up.

The triumph of the doctrine of Evolution has owed much in the past to the study of the structure and development of the Vertebrates, and the correct interpretation of their morphology still plays an important part in the elucidation of the evolutionary process. No other group of animals presents us with so complete a record of the divergent phylogenetic lines along which they have evolved. Although this book is written mainly from the morphological point of view, function has not been lost sight of. For it must not be forgotten that structure and function go hand in hand and evolve *pari passu*.

The bibliography at the end of this volume contains, of course, only a selection of the more important and more recent works; a complete

xvii

list would take at least a volume to itself. Further references the reader
may find in well-known text-books. For Fishes in particular he may be
referred to Dr. Bashford Dean's excellent " Bibliography ", New York,
1916–23. The numbers printed in heavy type after authors' names in
the text refer to the bibliography, in which ' Bibl.' has been added to those
works containing good lists of references on special subjects. Space has not
allowed me to discuss the views of individual authors, except in special
cases, and the citation of a name in brackets does not necessarily imply
that the author holds the view set forth in this volume. So far as possible
I have tried to verify statements of fact by personal observation, and
sometimes have denoted this by adding the initials E.S.G.

The names of the genera figured will be found in the index ; and the
systematic position of the genera mentioned in the text can be seen in
the scheme of classification at the beginning of the volume. Much care
has been devoted to the illustrations. The new figures, over 300 in
number, have been drawn by myself or from my sketches, and many
have been skilfully touched up for reproduction by Mr. A. K. Marshall.

To many authors and the following publishers I am indebted for leave
to reproduce figures : Messrs. G. Allen & Unwin, Ltd., Messrs. G. Bell &
Sons, Ltd., Messrs. A. & C. Black, Ltd., Messrs. Blakiston's, Sons, & Co.,
Messrs. Constable & Co., Ltd., Messrs. Henry Holt & Co., Messrs. Long-
mans, Green & Co., Ltd., Messrs. Macmillan & Co., Ltd., The Macmillan
Company, The W. B. Saunders Company, Messrs. H. F. & G. Witherby,
The Wistar Institute, The Univ. of Chicago Press, The Cambridge Uni-
versity Press, The Clarendon Press, Oxford, The Akademische Verlagsgesell-
schaft, Messrs. Alb. Bonnier, Stockholm, Messrs. W. Engelmann, Leipzig,
Messrs. G. Fischer, Jena, Messrs. W. de Gruyter & Co., Berlin, Messrs.
Hirzel, Leipzig, Messrs. J. Springer, Berlin ; also the Zoological Society,
the Geological Society, and Linnean Society of London.

It is with pleasure that I gratefully acknowledge the help received
from various friends and colleagues. More particularly must I thank
Prof. F. H. Edgeworth for the loan of sections, Prof. D. M. S. Watson for
much useful information, Mr. G. R. de Beer for co-operation in the
preparation of several figures in Chapters VI. and VII., and the use of his
accurate wax reconstructions, and particularly to Prof. J. P. Hill for
constant help and the loan of many series of sections from his magnificent
collections at University College, London.

E. S. GOODRICH,
Department of Zoology and Comparative Anatomy,
University Museum, Oxford.

February 6, 1930.

CLASSIFICATION

(The extinct Orders or lesser divisions down to Genera are marked by an asterisk.)

PHYLUM VERTEBRATA (Chordata).

SUBPHYLUM ACRANIA (Cephalochorda): Amphioxus (Branchiostoma), Asymmetron.

SUBPHYLUM CRANIATA.

Branch **MONORHINA.**[1]

Class **Cyclostomata** (Marsipobranchii).

Subclass **Myxinoidea** : Bdellostoma, Myxine, Paramyxine.
Subclass **Petromyzontia** : Petromyzon.
Incertae sedis : Palaeospondylus*.

Branch and Class **Ostracodermi.**
Order **Anaspida***, Birkenia, Lasanius, Pterolepis, Rhyncholepis.
Order **Cephalaspidomorphi*** (Osteostraci): Ateleaspis, Auchenaspis (Thyestes), Cephalaspis, Kiaeraspis, Tremataspis.
Order **Pteraspidomorphi*** (Heterostraci): Coelolepis, Cyathaspis, Drepanaspis, Pteraspis, Thelodus.
Order **Pterychthyomorphi*** (Antiarchi): Asterolepis, Bothriolepis, Pterychthys.

---

[1] Much evidence has recently been brought forward by Stensiö (**135**) and Kiaer (**126**) for the view that some of the Ostracodermi are closely related to the Cyclostomata, and possessed like the latter a single median opening for the hypophysis and nasal organs. If confirmed, the Myxinoidea, Petromyzontia, Cephalaspidomorphi, and Anaspida may be united in one Branch Monorhina.

Branch **GNATHOSTOMATA** (Amphirhina).

Grade **Ichthyopterygii.**

Class **Pisces.**

Subgrade **Chondrichthyes.**

Subclass **Elasmobranchii.**

Order **Selachii.**

Group 1. **Notidani** : Chlamydoselachus, Heptanchus, Hexanchus, Notidanus.

Group 2. Division A. Suborder **Heterodonti** : Acrodus*, Cochliodus*, Edestus*, Helodus*, Heterodontus (Cestracion), Hybodus*.

Division B. Subdivision *a*. Suborder **Scyllioidei** : Carcharias, Carcharodon, Cetorhinus (Selache), Lamna, Mitsukurina, Odontaspis, Pristiurus, Rhinodon, Scyllium (Scylliorhinus), Stegostoma, Zygaena.

Subdivision *b*. Suborder **Squaliformes** : Echinorhinus, Laemargus, Pliotrema, Pristiophorus, Squalus (Acanthias).

Suborder **Rajiformes** (Hypotremata).

Section Squatinoidei : Rhina, Squatina.

Section Rhinoraji : Pristis, Raja, Rhinobatus, Rhynchobatus.

Section Centrobatoidei : Myliobatis, Psammodus*, Pteroplatea, Ptychodus*, Rhinoptera, Trygon.

Section Torpedinoidei : Narcine, Torpedo.

Order **Holocephali** : Callorhynchus, Chimaera, Ganodus*, Harriottïa, Myriacanthus*, Pyctodus*, Rhinochimaera, Squaloraja*.

Order **Pleuracanthodei*** (Ichthyotomi) : Chondrenchelys, Cratoselache (?), Pleuracanthus, Xenacanthus.

Subclass **Cladoselachii** (Pleuropterygii)*[1] : Cladodus, Cladoselachus, Symmorium.

Subclass **Acanthodii*** : Acanthodes, Cheiracanthus, Climatias, Diplacanthus, Gyracanthus (?).

Subclass **Coccosteomorphi.**[2]

Order **Anarthrodira*** : Macropetalichthys.

Order **Arthrodira*** : Coccosteus, Dinichthys, Homosteus, Mylostoma, Titanichthys.

Incertae sedis : Jagorina*, Rhamphodus*.

---

[1] If Jaekel's description of claspers on the pelvic fins were confirmed, the Cladoselachii could be placed as an Order of the Elasmobranchii.

[2] A group of uncertain affinities, at one time considered to be allied to the Dipnoi. Although superficially resembling the Pterychthyomorphi they are probably true Gnathostomes and derived from an early group of Elasmobranchii provided with bony armour.

Class **Pisces**—*Continued*

Subgrade **Osteichthyes.**

Subclass **Dipnoi**: Ceratodus (Neoceratodus), Ctenodus*, Dipterus*, Lepidosiren, Phaneropleuron*, Protopterus, Sagenodus*, Scaumenacia*, Uronemus*.

Subclass **Teleostomi.**

Division **Crossopterygii.**[1]

Order **Osteolepidoti*.**

Suborder **Haplistia** : Tarassius.

Suborder **Rhipidistia**: Dictyonosteus, Diplopterus, Eusthenopteron, Glyptopomus, Gyroptychius, Holoptychius, Megalichthys, Onychodus, Osteolepis, Rhizodopsis, Sauripterus, Tristichopterus.

Order **Coelacanthini*** : Axelia, Coelacanthus, Macropoma, Undina, Wimania

Division **Actinopterygii.**

Subdivision A.

Order **Chondrostei.**

Suborder **Palaeoniscoidei***: Acrorhabdus, Birgeria, Boreosomus, Catopterus, Cheirodus, Cheirolepis, Coccolepis, Dictyopyge, Elonichthys, Eurynotus, Gonatodus, Gyrolepis, Oxygnathus, Palaeoniscus, Perleidus, Phanerosteon, Platysomus, Pygopterus, Trissolepis.

Suborder **Acipenseroidei**: Acipenser, Chondrosteus*, Gyrosteus*, Pholidurus*, Polyodon (Spatularia), Psephurus, Scaphirhynchus.

Suborder **Saurichthyoidei*** : Belonorhynchus, Saurichthys.

Order **Polypterini** : Calamoichthys, Polypterus.

Subdivision B. **Holostei.**

Group *a.*[2]

Order **Amioidei** (Protospondyli): Acentrophorus*, Amia, Archaeonemus*, Callopterus*, Caturus*, Dapedius*, Eurycormus*, Euthynotus*, Gyrodus*, Hypsocormus*, Lepidotus*, Macrosemius*, Megalurus*, Mesodon*, Oligopleurus*, Osteorhachis*, Pachycormus*, Pholidophorus*, Pycnodus*, Semionotus*, Spathiurus*.

Order **Lepidosteoidei** (Aetheospondyli): Aspidorhynchus (?)*, Lepidosteus.

Group *b.*

Order **Teleostei.**

Division A.

Suborder **Leptolepiformes*** : Leptolepis, Thrissops.

[1] Used in a restricted sense, excluding the Polypterini.
[2] Group *a* is distinguished from Group *b* by the 'lepidosteoid' histological structure of the scales and bony skeleton (Goodrich, 36).

Class **Pisces,** Order **Teleostei**—*Continued*

Division B.

Group *a* : **Ostariophysi.**

Suborder **Cypriniformes.**

Tribe Characinoidei (Cyprinoidei) : Abramis, Barbus, Catostomus, Citharinus, Cobitis, Cyprinus, Erythrinus, Gymnotus, Ichthyoborus, Lebiasina, Leucissus, Rhodeus, Tinca.

Tribe Siluroidei : Aspredo, Auchenoglanis, Callichthys, Clarias, Loricaria, Macrones, Malapterurus, Pimelodus, Silurus, Synodontis.

Group *b.*

Subgroup 1.

Suborder **Clupeiformes** (Isospondyli, Malacopterygii) : Albula, Alepocephalus, Arapaima, Argyropelecus, Chanos, Chauliodus, Chirocentrus, Clupea, Coregonus, Cromeria, Crossognathus, Ctenothrissa, Elops, Gonorhynchus, Hyodon, Idiacanthus, Mormyrus, Notopterus, Osmerus, Osteoglossum, Pantodon, Phractolaemus, Salmo, Saurodon, Stomias.

Subgroup 2.

Series 1.

Suborder **Esociformes** (Haplomi) : Dallia, Enchodus, Esox, Galaxias, Umbra.

Suborder **Scopeliformes** (Iniomi) : Alepidosaurus, Ateleopus, Aulopus, Ipnops, Sardinoides, Scopelus, Sudis, Synodus.

Suborder **Lyomeri** : Eurypharynx, Saccopharynx,

Series 2.

Suborder **Anguilliformes** (Apodes) : Anguilla, Anguillarus, Conger, Derichthys, Muraena, Nemichthys, Urenchelys.

Suborder **Amblyopsiformes** (Microcyprini): Amblyopsis, Anableps, Cyprinodon, Fundulus, Gambusia, Girardinus, Poecilia, Typhlichthys.

Suborder **Scombresociformes** (Synentognathi) : Belone, Exocoetus, Hemirhamphus, Scombresox.

Suborder **Notacanthiformes** (Heteromi) : Dercetes, Halosaurus, Notacanthus.

Suborder **Symbranchiformes** : Amphipnous, Symbranchus.

Suborder **Gasterosteiformes** (Catosteomi).

Tribe 1. **Gasterosteoidei** : Aulorhynchus, Gasterosteus, Spinachia.

Tribe 2. **Hemibranchii** : Amphisile, Aulostoma, Centriscus, Fistularia, Protosyngnathus.

Class **Pisces,** Order **Teleostei**—*Continued*

Tribe 3. **Lophobranchii** : Hippocampus, Phyllopteryx, Solenostomus, Syngnathus.

Tribe 4. **Hypostomides** : Pegasus.

Suborder **Mugiliformes** (Percesoces) : Anabas, Atherina, Chiasmodon, Mugil, Ophiocephalus, Osphronemus, Sphyraena, Tetragonurus.

Division **Lampridiformes** (Allotriognathi).

Subdivision **Lamproidei** (Selenichthyes) : Lampris.

Subdivision **Veliferoidei** (Histichthyes) : Velifer.

Subdivision **Taeniosomi** : Lophotis, Stylephorus, Trachypterus.

Suborder **Percopsiformes** (Salmopercae) : Aphredoderus, Percopsis.

Suborder **Acanthopterygii.**

Section **Berycoidei.**

Division **Beryciformes** : Beryx, Holocentrum, Hoplopteryx, Monocentris.

Division **Zeiformes** : Antigonia, Capros, Zeus.

Division **Rhombiformes.**

Subdivision **Amphistioidea*** : Amphistium.

Subdivision **Heterosomata** : Arnoglossus, Bothus, Cynoglossus, Limanda, Pleuronectes, Psettodes, Solea.

Section **Percoidei.**

Division **Perciformes** : Brama, Caranx, Cepola, Centrarchus, Cichla, Haplognathus, Mullus, Pagellus, Perca, Psettus, Rhacicentrum, Sciaenus, Serranus, Sparus, Toxotes.

Division **Chaetodontiformes.**

Subdivision **Squammipennes** : Chaetodon, Drepane, Platax, Pomacanthus.

Subdivision **Plectognathi.**

Section *a.* Acanthurus, Siganus, Teuthis.

Section *b.* Subsection 1. **Sclerodermi** : Balistes, Monacanthus, Ostracion, Triacanthus.

Subsection 2. Triodon.

Subsection 3. Diodon, Orthagoriscus, Tetrodon.

Division **Cirrhitiformes** : Chilodactylus, Haplodactylus, Latris.

Division **Pomacentriformes** : Pomacentrus.

Class **Pisces,** Order **Teleostei**—*Continued*

Division **Labriformes** : Labrichthys, Labrus. Scarus.

Division **Embiotociformes** : Embiotoca.

Division **Gadopsiformes** : Gadopsis.

Division **Trichodontiformes** : Trichodon.

Division **Ammodytiformes** : Ammodytes.

Division **Champsodontiformes** : Champsodon.

Division **Trachiniformes** : Percophis, Trachinus, Uranoscopus.

Division **Gobiesociformes** (Xenopteri) : Gobiesox.

Division **Nototheniiformes** : Notothenia.

Division **Callionymiformes** : Callionymus.

Section **Gobiiformes** : Gobius, Periophthalmus.

Section **Echeneiformes** (Discocephali): Echeneis, Opisthomyzon, Remora.

Section **Scorpaeniformes** (Scleroparei) : Agonus, Comephorus, Cottus, Cyclopterus, Dactylopterus, Hexagramma, Platycephalus, Scorpaena, Sebastes, Trigla.

Section **Blenniiformes** (Jugulares) : Anarhicas, Blennius, Clinus, Congrogadus, Dactyloscopus, Fierasfer, Ophidium, Pholis, Stichaeus, Xiphidion, Zoarces.

Section **Batrachiformes** (Pediculati).

Subsection **Batrachoidea** : Batrachoides, Opsanus (Batrachus).

Subsection **Lophioidea** : Antennarius, Ceratias, Chaunax, Lophius, Melanocetus, Anchocephalus (Malthe).

Section **Scombriformes** : Blochius, Histiophorus, Luvarus, Palaeorhynchus, Scomber, Thunnus, Trichiurus, Xiphias.

Section **Kurtiformes** : Kurtus.

Section **Mastacembeliformes** (Opisthomi) : Mastacembelus.

Suborder **Gadiformes** (Anacanthini).

Division 1. Bathygadus, Macrurus.

Division 2. Gadus, Lota, Molva, Motella, Pseudophycis.

Branch **GNATHOSTOMATA**—*Continued*

Grade **Tetrapoda** (Cheiropterygii).

Subgrade **Anamnia.**

Class **Amphibia** (Batrachia).

Subclass **Stegocephalia** (Labyrinthodontia).

Order **Embolomeri*** [1] : Baphetes, Cricotus, Diplovertebron (Gephyrostegus), Eogyrinus (Anthracosaurus, Pteroplax), Orthosaurus (Loxomma), Palaeogyrinus, Pholidogaster, Pholiderpeton.

Order **Rhachitomi*** (Temnospondyli) : Acheloma, Actinodon, Archegosaurus, Aspidosaurus, Cacops, Chelydosaurus, Dasyceps, Dwinasaurus, Eryops, Lydekkerina, Micropholis, Rhinesuchus, Trematops, Trimerorhachis, Zatrachis.

Order **Stereospondyli*** : Anachisma, Bothriceps, Capitosaurus, Cyclotosaurus, Lyrocephalus, Mastodonsaurus, Metaposaurus, Trematosaurus.

Order **Branchiosauria*** (Phyllospondyli) : Branchiosaurus, Eugyrinus, Leptorophus, Melanerpeton, Micrerpeton, Micromelerpeton, Pelosaurus.

Order **Ceraterpetomorpha*** (Nectridea) : Batrachiderpeton, Ceraterpeton, Diceratosaurus, Diplocaulus, Urocordylus, Ptyonius.

Order **Aistopoda*** : Dolichosoma, Ophiderpeton.

Subclass and Order **Anura** (Salientia).

Suborder **Phaneroglossa.**

Section **Arcifera** : Bufo, Calyptocephalus, Ceratophrys, Cystignathus, Discoglossus, Hyla, Pelobates.

Section **Firmisternia** : Engystoma, Palaeobatrachus*, Rana.

Suborder **Aglossa** : Hymenochirus, Pipa, Xenopus.

Subclass and Order **Urodela** (Caudata).

Suborder **Lysorophida*** : Lysorophus.

Suborder **Hylaeobatrachia*** : Hylaeobatrachus.

Suborder **Caducibranchiata** : Amphiuma, Amblystoma (Siredon), Cryptobranchus, Megalobatrachus, Molge (Triton), Plethodon, Salamandra, Spelerpes, Tylototriton.

Suborder **Perennibranchiata** : Necturus (Menobranchus), Proteus.

Suborder **Sirenoidea** (Meantes) : Siren.

Subclass and Order **Apoda** (Gymnophiones, Coeciliae) : Dermophis, Hypogeophis, Ichthyophis, Siphonops.

Subgrade **Amniota.**

Class **Reptilia.**

Group **Anapsida** or **Prosauria.**

Subclass and Order **Microsauria*** : Dawsonia, Hylonomus, Petrobates, Seeleya.

Subclass and Order **Seymouriamorpha*** : Conodectes (Seymouria), Karpinskiosaurus, Kotlassia.

---

[1] This Order may include some forms which should be placed in a basal group of Tetrapoda ancestral to both Anamnia and Amniota.

Grade **Tetrapoda,** Class **Reptilia**—*Continued*

    Subclass **Anapsidosauria.**

        Order **Cotylosauria*** : Captorhinus, Diadectes, Eunotosaurus (?), Labidosaurus, Limnoscelis, Pantylus, Pariotichus, Sauravus.

        Order **Pareiasauria*** : Anthodon, Bradysaurus, Elginia, Pareiasaurus, Propappus.

        Order **Procolophonia*** : Koiloskiosaurus, Procolophon, Sclerosaurus Thelegnathus.

    Group **Synapsida,** or **Reptilia Theropsida.**

      Subclass **Theromorpha.**

        Superorder **Theraptosauria.**

          Order **Pelycosauria*** : Clepsydrops, Dimetrodon, Sphenacodon.

          Order **Edaphosauria*** : Edaphosaurus, Naosaurus.

          Order **Poliosauria*** : Ophiacodon, Poliosaurus, Varanops, Varanosaurus.

          Order **Caseasauria*** : Casea.

        Superorder **Therapsida.**

          Order **Anningiamorpha*** : Anningia, Glaucosaurus (?), Mycterosaurus (?).

          Order **Dromasauria*** : Galechirus, Galepus.

          Order **Dinocephalia*.**

            Suborder **Tapinocephalia** : Delphinognathus, Deuterosaurus (?), Mormosaurus, Moschops, Rhopalodon (?), Tapinocephalus.

            Suborder **Titanosuchia** : Titanosuchus.

          Order **Dicynodontia** (Anomodontia)***** : Dicynodon (Oudenodon ?), Endothiodon, Geikia, Gordonia, Placerias, Pristerodon, Procynodon.

          Order **Theriodontia*.**

            Suborder **Therocephalia** : Aelurosaurus, Alopecodon, Ictidosuchus, Lycosaurus, Lycosuchus, Pristerognathus, Scylacosaurus, Scymnosaurus.

            Suborder **Gorgonopsida** : Arctops, Gorgonops, Galesuchus, Scymnognathus, Theriodesmus (?).

            Suborder **Burnetiomorpha** : Burnetia.

            Suborder **Bauriamorpha** : Bauria, Microgomphodon, Sesamodon.

            Suborder **Cynodontia** : Cynognathus, Diademodon, Galesaurus, Gomphognathus, Nythosaurus, Thrinaxodon, Trirhacodon.

    Group **Parapsida.**

      Subclass and Order **Mesosauria** (Proganosauria)***** : Mesosaurus, Stereosternum.

      Subclass **Sauropterygia.**[1]

        Order **Nothosauria*** : Lariosaurus, Neusticosaurus, Nothosaurus, Simosaurus.

        Order **Plesiosauria*** : Cryptocleidus, Elasmosaurus, Muraenosaurus, Peloneustes, Plesiosaurus, Pliosaurus, Polycotylus, Thaumatosaurus.

---

[1] The Sauropterygia are possibly related to the Chelonia (Broom).

Grade **Tetrapoda,** Class **Reptilia**—*Continued*

Subclass and Order **Placodontia*** : Cyamodus, Placochelys, Placodus.

Subclass and Order **Pleurosauria** (Protorosauria)* : Araeoscelis, Palaeohatteria (?), Pleurosaurus, Sauranodon.

Subclass and Order **Ichthyosauria*** : Cymbospondylus, Ichthyosaurus, Mixosaurus, Ophthalmosaurus, Shastosaurus, Toretocnemus.

Group **Eusauria** or **Reptilia Sauropsida.**

Subclass **Testudinata.**

Order **Chelonia.**

Suborder **Triassochelydia*** : Triassochelys.

Suborder **Amphichelydia*** : Baëna, Glyptops, Platychelys, Pleurosternum, Probaëna.

Suborder **Pleurodira** : Bothremys*, Chelodina, Hydraspis, Miolania*, Pelomedusa, Podocnemis, Sternothaerus.

Suborder **Cryptodira** : Archelon*, Chelone, Chelonides*, Chelydra, Dermochelys (Sphargis), Desmatochelys*, Emys, Eurysternum*, Platysternum, Protostega*, Psephophorus*, Testudo, Thalassemys.

Suborder **Trionychoidea** : Conchochelys*, Plastomenus*, Trionyx.

Subclass **Diapsida.**

Superorder **Diaptosauria.**

Order **Rhynchocephalia.**

Suborder **Rhynchosauria*** : Howesia, Hyperodapedon, Mesosuchus (?), Rhynchosaurus.

Suborder **Sphenodontia** : Homoeosaurus*, Sphenodon (Hatteria).

Suborder **Champsosauria*** (Choristodera) : Champsosaurus, Simaedosaurus.

Suborder **Thalattosauria** (?)* : Nectosaurus, Thalattosaurus.

Suborder **Protorosauria** (?)* : Protorosaurus.

Superorder **Lepidosauria** (Streptostylica).[1]

Order **Lacertilia.**

Suborder **Hydrosauria.**

Tribe **Varanomorpha** : Megalania*, Palaeovaranus*, Varanus.

Tribe **Dolichosauria*** : Acteosaurus, Dolichosaurus, Pontosaurus.

Tribe **Aigialosauria*** : Aigialosaurus.

Tribe **Mosasauria*** (Pythonomorpha) : Clidastes, Hainosaurus, Lestosaurus, Mosasaurus, Platecarpus.

Suborder **Lacertae** : Agama, Amblyrhynchus, Anguis, Aniella, Dibama, Draco, Euposaurus*, Gerrhosaurus, Heloderma, Iguana, Lacerta, Scincus, Tejus, Zonurus.

Suborder **Amphisbaenia** : Amphisbaena, Chirotes, Cremastosaurus*.

[1] The Lepidosauria are included in the Diaptosauria on the assumption that their monapsid skull has been derived from the diapsid type.

Grade **Tetrapoda,** Class **Reptilia**—*Continued*

Suborder **Geckones :** Eublepharus, Gecko, Ptychozoon, Uroplates.

Suborder **Chamaeleontes** (Rhiptoglossa) : Chamaeleo.

Order **Ophidia :** Boa, Coluber, Crotalus, Dipsas, Elaps, Eryx, Glauconia, Hydrophis, Ilysia, Naja, Python, Tropidonotus, Typhlops, Uropeltis, Vipera, Xenopeltis.

Superorder **Archosauria** (Thecodontia).

Order **Eosuchia* :** Palaeagama, Paliguana (?), Youngina.

Order **Proterosuchia* :** Proterosuchus.

Incertae sedis, Howesia, Mesosuchus.

Order **Parasuchia*.**

Suborder **Pseudosuchia :** Aëtosaurus, Erpetosuchus (?), Euparkeria, Notochampsa, Ornithosuchus, Scleromochlus, Sphenosuchus.

Suborder **Erythrosuchia** (Pelycosimia): Erythrosuchus.

Suborder **Phytosauria :** Machaeroprosopus, Mesorhinus, Mystriosuchus, Parasuchus, Phytosaurus (Belodon), Rhytidodon, Stagonolepis.

Order **Crocodilia.**

Suborder **Mesosuchia* :** Atoposaurus, Bernissartia, Geosaurus, Goniopholis, Notosuchus, Pelagosaurus, Pholidosaurus, Steneosaurus, Teleosaurus.

Suborder **Eusuchia :** Alligator, Caiman, Crocodilus, Gavialis, Osteolaemus, Thoracosaurus,* Tomistoma.

Suborder **Thalattosuchia* :** Dacosaurus, Metriorhynchus.

Order **Saurischia*** (Dinosauria Theropoda and Sauropoda).

Suborder **Megalosauria** (Theropoda) : Ammosaurus, Anchisaurus, Ceratosaurus, Coelurus, Compsognathus, Creosaurus, Megalosaurus, Plateosaurus, Thecodontosaurus, Zanclodon.

Suborder **Cetiosauria** (Sauropoda, Opisthocoelia): Apatosaurus (Brontosaurus), Atlantosaurus, Camarosaurus (Morosaurus), Cetiosaurus, Diplodocus, Titanosaurus.

Order **Ornithischia** (Dinosauria Orthopoda, Predentata)*.

Suborder **Iguanodontia** (Ornithopoda) : Camptosaurus, Claosaurus, Hadrosaurus, Hypsilophodon, Iguanodon, Laosaurus, Nanosaurus, Trachodon.

Suborder **Stegosauria :** Polacanthus, Scelidosaurus, Stegosaurus.

Suborder **Ceratopsia :** Ceratops, Diceratops, Triceratops.

Order **Pterosauria*.**

Suborder **Tribelesodontia :** Tribelesodon.

Suborder **Rhamphorhynchoidea :** Dimorphodon, Rhamphorhynchus, Scaphognathus.

Suborder **Pterodactyloidea :** Nyctosaurus, Ornithodesmus, Pteranodon, Pterodactylus.

Grade **Tetrapoda**—*Continued*

Class **Aves.**

Subclass **Archaeornithes** (Saururae) : Archaeopteryx*, Archaeornis*.

Subclass **Neornithes** (Ornithurae).

Section **Odontormae.**

Order **Ichthyornithes*** : Ichthyornis.

Section **Odontolcae*.**[1]

Order **Hesperornithes** : Hesperornis.

Section **Ratitae.**[2]

Order **Casuarii** : Casuarius, Dromaeus.

Order **Struthiones** : Struthio.

Order **Rheae** : Rhea.

Order **Dinornithes*** : Dinornis, Palapteryx

Order **Aepyornithes*** : Aepyornis.

Order **Apteryges** : Apteryx.

Section **Carinatae.**

Group *a.* **Palaeognathae.**

Order **Tinamiformes** : Cryptura, Rhynchotus, Tinamus.

Group *b.* **Neognathae.**

Order **Colymbiformes** : Colymbus, Podiceps.

Order **Sphenisciformes** : Aptenodytes, Eudiptes, Palaeeudyptes*.

Order **Procellariiformes** : Diomedea, Fulmarus, Procellaria, Puffinus.

Order **Ciconiiformes** : Ardea, Balaeniceps, Ciconia, Eudocimus, Fregata, Ibis, Nycticorax, Odontopteryx*, Palaelodus*, Pelecanus, Phaethon, Phalacrocorax, Phoenicopterus, Scops.

Order **Pelecaniformes** :

Order **Anseriformes** : Anas, Anser, Chauna, Cygnus, Fuligula, Mergus, Nesonetta, Palamedea, Somateria.

Order **Falconiformes** : Accipiter, Aquila, Buteo, Cathartes, Circus, Falco, Gyparchus, Gypaëtus, Gyps, Haliaëtus, Milvus, Pandion, Polyborus, Sarcorhamphus, Serpentaria, Vultur.

Order **Opisthocomi** : Opisthocomus.

Order **Galliformes** : Argusianus, Crax, Gallinuloides*, Gallus, Lagopus, Laphophorus, Megapodius, Meleagris, Numida, Ortalis, Palaeortix*, Pavo, Pedionomus, Perdix, Phasianus, Tetrao, Turnix (Hemipodius).

Order **Gruiformes** : Aphanapteryx*, Balearica, Cariama, Crex, Erythromachus*, Eurypyga, Fulica, Gallinula, Grus, Heliornis, Mesites, Notornis*, Ocydromus, Otis, Parra, Phororacus*, Rallus, Rhinochetus.

Order **Charadriiformes.**

Suborder **Limicolae** : Aegialitis, Charadrius, Chionis, Dromas, Gallinago (Capella), Glareola, Haematopus, Machetes, Numenius, Oedicnemus, Pluvianus, Scolopax, Thinocorys, Totanus, Tringa.

[1] Should perhaps be placed with the Colymbitormes.

[2] The true affinities of these flightless and ' palaeognathous ' Orders are still doubtful.

Grade **Tetrapoda,** Class **Aves**—*Continued*

Suborder **Lari** : Alca, Anous, Larus, Megalestris, Mergulus, Rissa, Rhynchops, Sterna, Uria.

Suborder **Pterocles** : Pterocles, Syrrhaptes.

Suborder **Columbae** : Columba, Didunculus, Didus*, Ectopistes, Goura, Pezophaps*, Turtur.

Order **Cuculiformes.**

Suborder **Cuculi** : Coccystes, Crotophaga, Cuculus, Musophaga, Necrornis*, Turacus.

Suborder **Psittaci** : Cacatua, Eos, Loriculus, Lorius, Melopsittacus, Nestor, Psittacula, Psittacus, Stringops.

Order **Coraciiformes.**

Suborder **Coraciae** : Alcedo, Buceros, Bucorvus, Coracias, Dacelo, Eurystomus, Merops, Momotus, Upupa.

Suborder **Striges** : Asio, Bubo, Strix, Syrnium.

Suborder **Caprimulgi** : Caprimulgus, Podargus, Steatornis.

Suborder **Cypseli** : Collocalia, Cypselus, Lophornis, Panyptila, Patagona, Trochilus.

Suborder **Colii** : Colius.

Suborder **Trogones** : Haploderma, Harpactes, Trogon.

Suborder **Pici** : Bucco, Capito, Dendrocopus, Galbula, Gecinus Indicator, Iynx, Picus, Pteroglossus, Rhamphastus.

Order **Passeriformes.**

Suborder **Anysomyodi** (Clamatores) : Cotinga, Dendrocolaptes, Eurylaemus, Formicaria, Philepitta, Pipra, Pitta, Rupicola, Xenicus.

Suborder **Diacromyodi** (Oscines) : Ampelis, Atrichornis, Cardinalis, Certhia, Chelidon, Cinolus, Corvus, Hirundo, Fringilla, Icterus, Lanius, Loria, Loxia, Meliphaga, Menura, Mimus, Mniotilta, Muscicapa, Nectarinia, Oriolus, Paradisea, Parus, Passer, Pica, Pratincola, Prunella, Pycnonotus, Ruticilla, Serinus, Sitta, Sturnus, Sylvia, Tanagra, Troglodytes, Turdus, Vidua, Zosterops.

Class **Mammalia.**

Subclass and Order **Multituberculata*** (Allotheria).

Suborder **Tritylodontoidea** : Oligokyphus, Stereognathus, Tritylodon.

Suborder **Plagiaulacoidea** : Bolodon, Ctenacodon, Hypsprymnopsis (Microlestes), Loxaulax, Microcleptes (Microlestes, in part), Neoplagiaulax, Plagiaulax, Ptilodus, Taeniolabis (Polymastodon).

Subclass and Order **Triconodonta*** : Amphilestes, Phascolotherium, Triconodon, Trioracodon.

Subclass **Pantotheria.**

Order **Spalacotheria** (Symmetrodonta)* : Peralestes, Spalacotherium.

Order **Amphitheria***: Amblotherium (Dryolestes), Kurtodon, Phascolestes.

Subclass and Order **Monotremata** (Ornithodelphia, Prototheria), Echidna, Ornithorhynchus.

Grade **Tetrapoda,** Class **Mammalia**—*Continued*
   Subclass **Ditremata** (Theria).
     Grade **Marsupialia** (Didelphia, Metatheria).
       Order **Polyprotodontia.**
         Suborder **Didelphoidea** : Amphiproviverra*, Borhyaena*, Dasyurus, Didelphys, Eodelphys*, Marmosa, Myrmecobius, Prothylacinus*, Thlaeodon*, Thylacinus.
         Suborder **Notoryctoidea** : Notoryctes.
         Suborder **Parameloidea** [1] : Perameles.
       Order **Caenolestoidea** (Paucituberculata): Caenolestes, Epanorthus, Garzonia*.
       Order **Diprotodontia** : Bettongia, Diprotodon*, Hypsiprymnus, Macropus, Petaurus, Phalanger, Phascolarctos, Phascolomys, Tarsipes, Trichosurus, Thylacoleo*.
     Grade **Placentalia** (Monodelphia).
       Order **Insectivora.**
         Group *a.*
         Suborder **Deltatheroidea*** : Deltatheridium, Deltatheroides.
         Group *b.* **Zalambdodonta.**
         Suborder **Centetoidea** : Centetes, Chrysochloris, Microgale, Necrolestes*, Potamogale, Solenodon, Xenotherium*, Zalambdalestes*.
         Group *c.* **Dilambdodonta.**
         Suborder **Erinacoidea** : Diacodon*, Erinaceus, Gymnura, Leptictis*, Palaeolestes*, Palaeoryctes*.
         Suborder **Soricoidea** : Blarina, Crocidura, Myogale, Nectogale, Scalops, Sorex, Talpa.
       Order **Tillodontia*** : Esthonyx, Tillotherium.
       Order **Pholidota** (Nomarthra in part) : Manis.
       Order **Xenarthra** (Edentata in part).
         Suborder **Palaeanodonta*** : Metacheiromys, Palaeanodon.
         Suborder **Loricata** (Hicanodonta) : Chlamydophorus, Dasypus, Doedicurus*, Glyptodon*, Panochthus*, Peltephilus*, Priodon, Tatusia, Tolypeutes.
         Suborder **Pilosa** (Anicanodonta) : Bradypus, Cholaepus, Cycloturus, Glossotherium (Neomylodon)*, Megalonyx*, Megatherium*, Mylodon*, Myrmecophaga, Scelidotherium*, Tamandua.
       Order **Taeniodontia*** (Ganodonta) : Conoryctes, Psittacotherium, Stylinodon.
       Order **Rodentia.**
         Suborder **Duplicidentata** : Lagomys, Lepus.
         Suborder **Simplicidentata** : Arctomys, Bathyergus, Castor, Cavia, Cephalomys*, Chinchilla, Dasyprocta, Dipus, Dolichotis, Erethizon, Geomys, Gerbillus, Hydrochaerus, Hystrix, Mus, Myoxus, Octodon, Pedetes, Sciurus, Spalax, Spermophilus.

[1] Perhaps more closely allied to the Diprotodontia.

Grade **Tetrapoda,** Class **Mammalia**—*Continued*
    Order **Carnivora.**
        Suborder **Creodonta*.**
            Tribe **Procreodi**: Arctocyon, Chriacus, Claenodon, Deltatherium, Oxyclaenus.
            Tribe **Acreodi** : Andrewsarchus, Coconodon, Dissacus, Mesonyx, Pachyaena.
            Tribe **Pseudocreodi** : Hyaenodon, Limnocyon, Oxyaena, Proviverra, Palaeonictis, Patriofelis, Sinopa.
            Tribe **Eucreodi** [1] : Didymictis, Miacis, Viverravus, Vulpavus.
        Suborder **Fissipedia.**
            Section *a*. Ælurictis, Acinonyx* (Cynaelurus), Eusmilus*, Felis, Lynx, Machairodus*, Nimravus*.
            Section *b*. Arctogale, Cryptoprocta, Cynictis, Eupleres, Fossa, Genetta, Gulo, Hyaena, Latax, Lutra, Mephitis, Mellivora, Mustela, Palaeogale*, Paradoxurus, Plesictis*, Proteles, Viverra.
            Section *c*. Aeluropus, Aelurus, Bassaris, Canis, Cercoleptes, Cynodictis*, Daphaenus*, Hemicyon*, Hyaenarctos*, Lycaon, Melursus, Nasua, Procyon, Otocyon, Ursus.
        Suborder **Pinnipedia** : Cystophora, Halichaerus, Monachus, Otaria, Phoca, Trichechus.

    Order **Artiodactyla.**
        Suborder **Protartiodactyla*** : Diacodexis (Pantolestes, Trigonolestes).
        Suborder **Bunoselenodontia*** : Ancodon, Anoplotherium, Anthracotherium, Dichobune (?), Merycopotamus.
        Suborder **Bunodontia** (Suina) : Achaenodon*, Chaeropotamus*, Dicotyles, Elotherium*, Entelodon*, Hyotherium*, Palaeochoerus*, Phacochoerus, Platygonus*, Potamochoerus, Sus.
        Suborder **Selenodontia** (Ruminantia).
            Section *a*. Dichodon*, Caenotherium*, Xiphodon*.
            Section *b*. **Oreodonta*** : Agriochoerus, Merycochoerus, Mesoreodon, Oreodon, Protoreodon.
            Section *c*. **Tylopoda** : Auchenia, Camelus, Poebrotherium*, Protolabis*, Stenomylus*.
            Section *d*. **Tragulina** : Gelocus*, Blastomeryx*, Cryptomeryx*, Dorcatherium, Hyaemoschus, Hypertragulus*, Leptomeryx*, Lophiomeryx*, Tragulus.
            Section *e*. **Pecora** : Alces, Antilocapra, Bos, Capra, Cephalophus, Cervus, Elaphodus, Gazella, Giraffa, Moschus, Okapia, Ovibos, Ovis, Rangifer, Sivatherium*, Protoceras*.

    Order **Cetacea.**
        Suborder **Archaeoceti*** : Agorophius, Pappocetus, Patriocetus, Protocetus, Prozeuglodon, Zeuglodon.
         Suborder **Mystacoceti** : Balaena, Balaenoptera, Megaptera, Rhachianectes.

---

[1] Should probably be placed in the Fissipedia as ancestral forms.

Grade **Tetrapoda,** Class **Mammalia**—*Continued*

Suborder **Odontoceti** : Cogia, Delphinus, Grampus, Globiocephalus, Hyperoodon, Mesoplodon, Monodon, Orca, Phocaena, Physeter, Platanista, Pontoporia, Squalodon*, Steno, Ziphius.

Order **Amblypoda***.

Suborder **Taligrada** : Ectoconus, Mioclaenus, Pantolambda, Periptychus.

Suborder **Dinocerata** : Coryphodon, Eudinoceras, Prodinoceras, Uintatherium (Dinoceras).

Order **Condylarthra*** : Hyopsodus (?), Meniscotherium (?), Phenacodus Tetraclaenodon.

Order **Subungulata.**

Suborder **Hyracoidea** : Dendrohyrax, Hyrax (Procavia), Megalohyrax*, Sagatherium*.

Suborder **Embrithopoda*** : Arsinoitherium.

Suborder **Proboscidea** : Baritherium*, Dinotherium*, Elephas, Mastodon*, Moeritherium*, Palaeomastodon*, Stegodon*, Tetrabelodon*.

Suborder **Sirenia** : Desmostylus*, Elotherium*, Eosiren*, Halicore, Halitherium*, Manatus, Prorastomus*, Protosiren*, Rhytina.

Order **Notoungulata***.

Suborder **Typotheria** : Archaeohyrax, Archaeopithecus, Interotherium, Hegetotherium, Typotherium.

Suborder **Entelonychia** : Arctostylops, Homalodontotherium, Isotemnus, Leontinia, Notostylops, Palaeostylops.

Suborder **Astrapotherioidea** : Astrapotherium, Trigonostylops.

Suborder **Toxodontia** : Adinotherium, Nesodon, Notohippus, Toxodon.

Order **Pyrotheria*** : Pyrotherium.

Order **Perissodactyla.**

Suborder **Rhinocerotoidea** : Aceratherium*, Amynodon*, Aphelops*, Atelodus*, Baluchitherium*, Caenopus*, Ceratorhinus*, Coelodonta*, Diceratherium*, Elasmotherium*, Hyrachyus*, Hyracodon*, Metamynodon*, Rhinoceros, Teleoceros*.

Suborder **Tapiroidea** : Colodon, Helaletes*, Heptodon*, Homogalax*, Megatapirus*, Protapirus*, Tapirus.

Suborder **Hippoidea** : Anchilophus*, Anchitherium*, Eohippus (Proterohippus)*, Epihippus*, Equus, Hipparion*, Hippidium*, Hyracotherium* (Pliolophus), Merychippus*, Mesohippus*, Miohippus*, Orohippus*, Palaeotherium*, Paloplotherium*, Parahippus*, Plesippus*, Pliohippus*, Propalaeotherium*, Protohippus*.

Suborder **Chalicotheria*** (Ancylopoda) : Chalicotherium, Eomoropus, Macrotherium, Moropus, Schizotherium.

Suborder **Titanotheria*** : Brontotherium (Titanotherium), Eotitanops, Megacerops (Brontops), Palaeosyops, Protitanotherium, Telmatotherium.

Order **Litopterna*** : Adiantus, Diadiaphorus, Didolodus, Epitherium, Macrauchenia, Thoatherium.

Grade **Tetrapoda,** Class **Mammalia**—*Continued*

    Order **Tubulidentata** : Orycteropus.

    Order **Menotyphla** : Adapisorex (?)*, Macroscelides, Mixodectes (?)*, Plesiadapis* (Nothodectes) (?), Ptilocercus, Tupaia.

    Order **Dermoptera** : Galaeopithecus.

    Order **Chiroptera.**

      Suborder **Megachiroptera** : Archaeopteropus* (?), Cephalotes, Epomophorus, Harpyia, Notopteris, Pteropus, Xantharpyia.

      Suborder **Microchiroptera** : Desmodus, Emballonura, Glossophaga, Hipposiderus, Megaderma, Miniopterus, Molossus, Mormops, Mystacops, Natalus, Noctilio, Nycteris, Phyllostoma, Rhinolophus, Scotophilus, Thyroptera, Vespertilio, Vesperugo, Zanycteris*.

    Order **Primates.**

      Suborder **Lemuroidea.**

        Series **Lemuriformes** : Adapis*, Avahis, Cheiromys, Chirogale, Hapalemur, Indris, Lemur, Lepidolemur, Megaladapis*, Microcoebus, Nesopithecus*, Notharctus*, Opolemur, Pelycodus*, Propithecus, Protoadapis*.

        Series **Lorisiformes** : Galago, Loris, Nycticebus, Perodicticus.

        Series **Tarsiiformes** : Anaptomorphus*, Hemiacodon*, Microchoerus*, Necrolemur*, Omomys*, Tarsius, Tetonius*.

      Suborder **Anthropoidea.**

        Section **Parapithecoidea*** : Moeripithecus, Parapithecus.

        Section **Platyrrhina** : Ateles, Callithrix, Cebus, Hapale, Homunculus*, Lagothrix, Midas, Mycetes, Nyctipithecus, Pithecia.

        Section **Catarrhina** : Australopithecus*, Anthropopithecus, Cercocoebus, Cercopithecus, Colobus, Cynocephalus, Cynopithecus, Dryopithecus*, Gorilla, Hylobates, Macacus, Mesopithecus*, Nasalis, Pliopithecus*, Propliopithecus*, Semnopithecus, Simia Simopithecus*, Sivapithecus*, Theropithecus*.

        Section **Bimana** : Eoanthropus*, Homo, Pithecanthropus*, Sinanthropus*.

CHAPTER I

THE ENDOSKELETON IN GENERAL

THE entire body of a vertebrate is supported by a framework of connective tissue which packs and binds the various parts together, delimits spaces, and serves for the attachment of muscles. Doubtless primitively the vertebrates had an elongated body stiffened by the notochordal rod, and moved by a side-to-side bending more especially of the caudal region. Correlated with this mode of progression is the segmentation of the somatic or body-wall muscles, entailing the corresponding segmentation

1

B

of the peripheral nervous system and the skeleton.   The parts of the endoskeleton of cartilage or bone may be looked upon as local specialisations of the general connective tissue system developed in those regions where the stresses are most pronounced and where the muscular attachments need most support.   Although this primitive segmentation may be much modified in the adult, especially of the higher forms, and even scarcely recognisable, yet it is always distinctly shown in the embryo and persists more or less completely in lower forms.   Now, in such

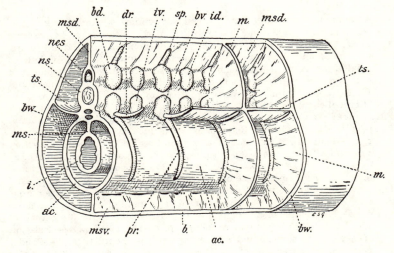

Fig. 1.

Diagram of the connective-tissue system in the trunk of a Craniate Vertebrate, showing the relation borne by the axial skeleton to the transverse and longitudinal septa. *ac,* wall of splanchnocoel ; *bd,* basidorsal ; *bv,* basiventral ; *bw,* cut body-wall ; *dr,* dorsal rib ; *i,* intestine hanging in the coelom ; *iv,* interventral ; *m,* transverse septum (myocomma) ; *ms,* mesentery ; *msd,* median dorsal septum ; *msv,* median ventral septum ; *nes,* neural tube ; *ns,* notochordal sheath ; *pr,* ventral or pleural rib ; *sp,* supraneural spine ; *ts,* horizontal septum. Oblique view of left side, from which the septa have been partially removed. (From Goodrich, *Vert. Craniata,* 1909.)

primitive forms, the connective tissue surrounds the somatic muscle segments or myomeres, forming not only closed boxes in which they lie, but also a lining to the skin outside and to the body-cavity or coelom within.   Connective tissue sheaths also surround the notochord, the neural canal enclosing the central nervous system, and the alimentary canal hanging in the coelom.   Moreover, since the mesoblastic segments and the coelomic cavities are of paired origin, the body is divided into right and left halves by a longitudinal vertical median septum continuous with the sheaths enclosing the nervous system, notochord, and gut ; this septum remains as a median dorsal and median ventral septum separating the myomeres, and as a median mesentery suspending the gut.   The ventral mesentery below the gut disappears almost completely (p. 620), but the dorsal mesentery usually remains. Thus is formed a system of longitudinal septa and tubular coverings, and of transverse septa (myosepta, myocommata) intersegmental in position between the myomeres, Fig. 1.

Before dealing with the endoskeleton in further detail, something

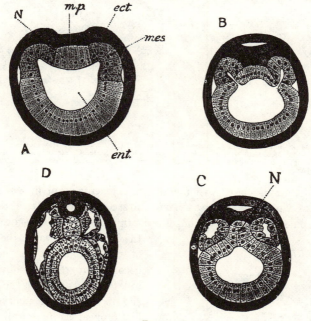

Fig. 2.

Transverse sections of young *Amphioxus* illustrating the origin of the mesoderm. (After Hatschek.) *ect*, Ectoderm; *ent*, enteric cavity; *m.p*, medullary plate; *mes*, mesoderm; *N*, notochordal rudiment. The dark tone indicates ectoderm, the pale tone endoderm, and the medium tone mesoderm. (From J. G. Kerr, *Zoology*, 1921.)

must be said about the embryonic origin of the connective tissue and the development of the mesoblast in Vertebrata. In general terms the middle or mesoblastic germ layer may be described as arising from paired outgrowths of the roof of the archenteron near the blastopore. These outgrowths develop on either side of a median longitudinal band which gives rise to the notochord, and when the notochordal band and the right and left mesoblastic rudiments become separated off, the archenteric walls meet below them to complete the hypoblastic lining of the gut.[1] In

[1] As the embryo elongates backwards the three germ-layers, epiblast, hypoblast, and mesoblast (ecto-, endo-, mesoderm) become differentiated from the indifferent tissue proliferating forwards from the lip of the blastopore.

Cephalochorda the lateral outgrowths destined to give rise to all the mesodermal tissues develop as a series of paired segmental pouches from before backwards, Fig. 2. They soon become nipped off from the archen-

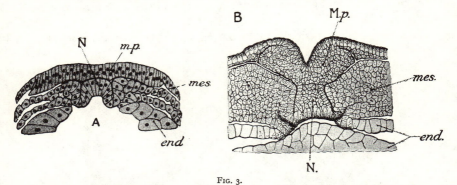

FIG. 3.

Transverse sections through embryos of (A) *Triton* and (B) *Rana temporaria* showing continuity of endoderm and mesoderm on each side of the notochord. (After O. Hertwig, 1882 and 1883.) *end*, Endoderm ; *m.p*, medullary plate ; *mes*, mesoderm ; *N*, notochordal rudiment. (From J. G. Kerr, *Embryology*, 1919.)

teric wall forming closed coelomic sacs or segments ; but in *Amphioxus* the first, and in other Craniates the first and second pairs tend to be delayed in development and to remain longer in continuity with the fore-gut. From

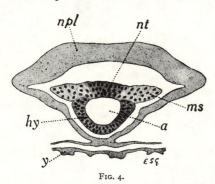

FIG. 4.

Scyllium canicula, Stage B. Transverse section through head. *a*, Archenteron ; *hy*, hypoblast ; *ms*, first outgrowing mesoblastic somite ; *npl*, neural plate ; *nt*, notochordal plate ; *y*, hypoblastic layer of yolk sac.

the first few backwards, the origin of the sacs as hollow pouches becomes progressively obscured ; in later stages towards the hinder end of the growing embryo they first appear as practically solid outgrowths. Each coelomic sac becomes subdivided into a dorsal somite (future myotome and sclerotome) and a ventro-lateral chamber. The series of ventro-lateral cavities of *Amphioxus* soon combine to form a continuous longitudinal splanchnic coelom, with an outer wall (somatopleure) and an inner wall (splanchnopleure), by the breaking down of the transverse partitions. In the Gnathostomata the mesoderm is of essentially similar origin, but distinct hollow pouches are not clearly formed ; paired grooves, however, of the archenteric wall near the blastopore have been found by Hertwig and G. Schwink in Amphibia, Fig. 3, and by others in Selachii, indicating a folding off of the mesoblast. Similarly there is distinct evidence of paired segmental outgrowths in the first two or three segments of Selachian embryos, Fig. 4, and still more obvious pouches occur in this region in *Petromyzon* where, consequently, the mesoblast is at first completely segmented as in Cephalochorda (Kupffer, 55 ; Hatschek, 43 ; Hatta, 44 ; Koltzoff, 361).

It is very improbable that the formation of coelomic sacs as hollow pouches is of any phylogenetic significance and represents an adult ancestral condition ; nevertheless this mode of separating off the mesoblastic building material seems to have been established at an early stage in the history of the Vertebrate phylum. Since it occurs also in the development of related groups such as Echinodermata and Enteropneusta the origin of the mesoblast from paired pouches is probably an embryonic device inherited from the ontogeny of the remote common ancestor of all these coelomate phyla. It may, therefore, be considered as an ancient and so far primitive mode of development which has become modified and obscured in the higher Vertebrates ; for in them not only do the coelomic sacs develop by the secondary hollowing out and subdivision of continuous mesoblastic bands, but the ventral lateral-plate region and its contained splanchnocoel never show distinct signs of segmentation except to a slight extent in Cyclostomes.

In *Amphioxus* all the coelomic segments except the first pair become subdivided into dorsal and ventral portions. The adjacent walls of the latter soon fuse and break down and there is so formed a continuous splanchnocoel as mentioned above. From the dorsal portions or somites are formed the hollow myotomes and sclerotomes. Most of the adult myomere is developed from the inner wall of the myotome, but its outer wall also contributes muscle cells (Sunier, 86, E.S.G.). The sclerotome arises by outfolding from the inner wall of the base of the somite, and grows up between the myomere and axial organs. While the cavity of the myotome or myocoel is soon obliterated, that of the sclerotome (sclerocoel) persists in the adult, Fig. 707. The splanchnocoel forms the adult perivisceral coelom behind, and the right and left suprapharyngeal coelom in front; from which extend coelomic canals down the primary gill-bars to a median subpharyngeal or endostylar coelom. Segmental coelomic canals also pass down the inner side of the metapleural folds to the genital coelomic chambers (Legros, 844). The thin layers of connective tissue seem to develop from the surface of all these mesoblastic structures derived from the original coelomic sacs. The mesoblastic bands of the Craniata become differentiated into dorsal segmental somites and ventral unsegmented lateral plate, Figs. 5 and 12. The segmental somite becomes further differentiated into

myotome, sclerotome, and nephrotome or stalk connecting it with the lateral
plate (Chapter XIII.). A cavity (myocoel), continuous through the nephrotome with the splanchnocoel, extends at first into the myotome, but disappears later. The definitive myomere or muscle segment ('somatic'
muscle) is usually developed from the inner wall of the myotome, though in
Dipnoi the outer so-called cutis layer may also contribute muscle (Kerr,
840). The bulk of the axial connective tissue is derived from the sclerotome,
an outgrowth or proliferation from the ventral inner region of the myotome.
Mesenchyme cells may also be proliferated from the nephrotome, the

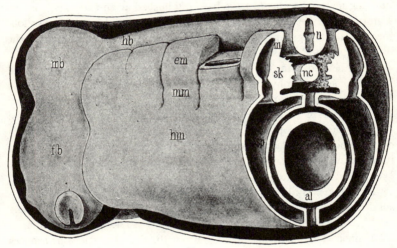

FIG. 5.

Stereogram of head region of craniate embryo, showing segmentation of mesoblast (from J. S.
Kingsley, *Comp. Anat. of Vertebrates*, 1926). *al*, Alimentary canal ; *c*, coelom ; *em*, epimere (myotome) ; *fb*, fore-brain ; *hb*, hind-brain ; *hm*, hypomere (lateral-plate) ; *m*, myomere ; *mb*, mid-brain ;
mm, mesomere (nephrotome) ; *n*, spinal cord ; *nc*, notochord ; *s*, stomodaeal region ; *sk*, sclerotome ;
so, *sp*, somatic and splanchnic walls of coelom approaching above and below *al* to form mesenteries.

outer cutis layer of the myotome, and the outer surface of the somatopleure. All these cells scatter and multiply, filling the spaces between the
various layers and organs. During this process the original segmentation of
the sclerotomes is lost, the later segmentation displayed by the skeleton
being, so to speak, reimposed upon the mesenchyme by the myomeres.

The Acrania and Cyclostomata illustrate the segmental structure
in its least modified condition. In the Gnathostomata the myomeres
become subdivided into dorsal (epiaxonic) and ventral (hypoaxonic)
portions by an additional longitudinal horizontal septum stretching from
the base of the notochordal sheath to the body wall surface,[1] completing

[1] In Pisces and aquatic Amphibia this horizontal septum reaches the
surface at the level of the main 'lateral line'.

the system of chief membranes in which the various parts of the skeleton arise, Fig. 1.

The axial skeleton of the trunk first develops in connexion with the covering of the notochordal axis and the transverse septa. Speaking generally, where these septa meet the longitudinal sheaths neural arches develop above and haemal arches below, while vertebral centra appear round the notochord itself. Thus, since the myomeres mark the segments of the body, these skeletal segments of the vertebral column alternate with them, and become intersegmental in position, Fig. 1.

The right and left neural arches tend to meet and fuse above the neural canal, and the haemal arches to meet and fuse below the caudal artery and vein in the tail region. Interneural and interhaemal elements may also appear. Dorsal ribs develop where the transverse septa meet the horizontal septum, and ventral ribs where they meet the coelomic wall. Such is the fundamental plan of the axial skeleton in the Gnathostomes.

THE NOTOCHORD AND ITS SHEATHS

The notochord is perhaps the most constant and characteristic feature of the Vertebrata, or Chordata, as they are often called. Essentially a rather stiff but flexible rod lying below the central nervous system and between the paired series of muscle segments, it not only forms the chief skeletal axis in the more primitive Cephalochorda and Cyclostomata, but always serves as the foundation of the vertebral column in higher Craniata. Indeed it must surely have been one of the chief factors in the development of the fundamental plan of structure of early vertebrates as free-swimming segmented animals moving by a lateral bending of the body.

The presence of the notochord not only in all the Vertebrata, but also in the degenerate Ascidians, shows that it is an extremely ancient organ; yet we have little real evidence as to its origin. We may, however, conjecture that it arose simply as a longitudinal thickening of the wall of the gut. The interesting suggestion made by Bateson, that an anterior diverticulum of the oesophagus in the Enteropneusta represents the notochord, rests on a very slender foundation, and in any case *Balanoglossus* differs too widely from the true Chordata for much help to be derived from this comparison.

The notochord invariably develops from the dorsal wall of the archenteron as a thickening or upfolding which (except at its extreme anterior end) becomes nipped off from before backwards, and continues to grow at its posterior end as the embryo lengthens. The growing point is situated just in front of the neurenteric canal at a point representing the original dorsal lip of the blastopore; thus, when a primitive streak is established the notochordal cells may no longer arise from the wall of the gut itself, but directly from the undifferentiated tissue of the streak. In front the

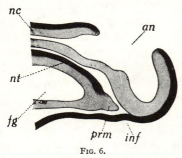

FIG. 6.

Median sagittal section through head-region of embryo *Torpedo marmorata*, 2·8 mm. long (after Dohrn, 1904). *an*, Anterior neuropore ; *fg*, foregut ; *inf*, beginning of hypophysial invagination ; *nc*, nerve-cord ; *nt*, notochord ; *prm*, premandibular mesoblast joining paired somites.

notochord always extends to near the extremity of the archenteron, to the wall of which it may remain for some time attached at a place immediately behind the hypophysis and below and behind the infundibulum, Figs. 6, 7, and 234. Never, in Craniates, does the notochord pass beyond this region. Here, then, is a meeting-point of these three structures, hypophysis, infundibulum, and notochord, constant throughout the Craniata from Cyclostomes to man (p. 214). This disposition is probably fundamental and primitive, and the growth forward of the notochord beyond the brain and mouth in *Amphioxus* would then be a secondary

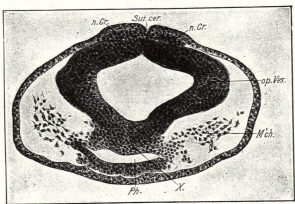

FIG. 7.

Transverse section of fore-brain and optic vesicles of chick embryo of 7 somites (from F. R. Lillie, *Develt. Chick*, 1919). *M'ch*, Mesenchyme ; *n.Cr*, neural crest ; *op.Ves*, optic vesicles ; *Ph*, pharynx ; *Sut.cer*, anterior cerebral suture ; *X*, mass of cells in which anterior end of intestine and notochord fuse.

adaptation to its well-known habit of burrowing rapidly in the sand (Willey, 94).

Soon after the separation of the notochord from the wall of the archenteron the cells become flattened and arranged in a single row like a pile of coins (Boeke, 8). In *Amphioxus*, although the nuclei may divide and become distributed, the flattened cuticularised cells retain this original disposition; there is no special covering epithelium, but a dorsal and ventral strand of superficial cells. But in the Craniata, after repeated division of the nuclei, the cells multiply, become cuticularised and vacuolated, forming the characteristic notochordal tissue of polygonal cells. At the periphery, however, an epithelium is formed of cells rich in protoplasm, which apparently secrete the covering sheaths.

The history of the two sheaths present in the Craniata, and secreted by the notochordal epithelium, is of considerable interest. Their origin and structure has been worked out chiefly by Kölliker (54), Hasse (39-42), v. Ebner (20), Klaatsch (53), Schneider (81), Schauinsland (78), and others, and is now well understood. At an early stage the notochordal epithelium secretes a thin covering membrane in which intercrossing elastic fibres become differentiated; this is the elastica externa. Next is secreted, also by the notochordal epithelium, an inner and usually thicker fibrous sheath.[1] The fate of these true notochordal sheaths varies in different groups, and their structural importance is inversely proportional to the development of the mesoblastic vertebral column outside.

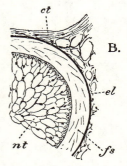

FIG. 8.

Ammocoete larva of *Petromyzon fluviatilis*, L. Portion of a transverse section of the notochord, enlarged. *ct*, Connective tissue; *el*, elastica externa; *fs*, fibrous sheath; *nt*, notochord. (From Goodrich, *Vert. Craniata*, 1909.)

In the Cyclostomata, where the large notochord forms the main axial support in the adult without a trace of centra, and remains as a continuous unconstricted rod stretching from the infundibular region to the tip of the tail, the fibrous sheath is quite thick, Figs. 8 and 30. But in most of the living Gnathostomata (all Teleostomi with the exception of the Chondrostei, and especially in the Tetrapoda) it remains comparatively thin and frequently disappears. In the Tetrapods, indeed, the sheaths together with the notochord are reduced to a mere vestige in the adult.

In the Elasmobranchii mesoblastic cells, from the skeletogenous layer originally outside the sheaths, make their way through the thin elastica externa and invade the underlying fibrous sheath in every segment at four points in its circumference corresponding to the bases of the dorsal and ventral arches, Figs. 9 and 22. These cells penetrate in large numbers, arrange themselves in concentric layers, and contribute matrix to the ever-widening fibrous sheath in which the centra are thus eventually

---

[1] There is still some doubt as to the share taken by the chordal epithelium in the formation of the sheaths; Tretjakoff, one of the most recent writers on this subject, believes the elastica externa to be of mesoblastic origin (87-89).

formed (p. 21).   Next to the notochord is generally seen a clear cell-less layer, often called the 'elastica interna'; but apparently it is usually only the last-formed layer of the fibrous sheath, and the true elastica interna, when present, is an extremely thin layer of fibres near the

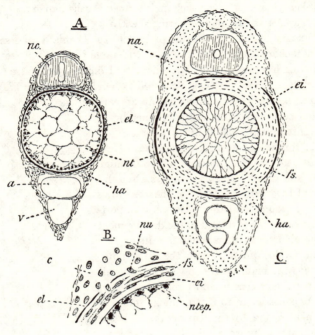

FIG. 9.

Scyllium canicula, L. A, Transverse section of the vertebral column of the tail of a young embryo before invasion; B, the base of the neural arch in an older embryo showing invasion; C, of a later stage: all magnified. *a*, caudal artery; *c*, cartilage; *ei*, 'elastica interna' or inner layer of the fibrous sheath; *el*, elastica externa; *fs*, fibrous sheath which becomes invaded by mesoblastic cells; *ha*, haemal arch; *na*, neural arch; *nc*, nerve-cord; *nt*, notochord; *ntep*, notochordal epithelium; *nu*, nuclei of mesoblastic cells passing through the broken elastica externa; *v*, caudal vein. (From Goodrich, *Vert. Craniata*, 1909.)

notochord.[1]   The centrum may develop as a cartilaginous ring in continuity with the arches, and carrying on its outer surface the remains of the ruptured elastica.   Or cartilage may spread from the base of the arches both inside and outside; the elastica then becomes embedded in the centrum and eventually disappears as a rule.   Such centra, in which

[1] There has been much controversy concerning the presence of a distinct elastica interna inside the fibrous sheath of the Craniata.   O. Schneider has recently shown that such a thin layer of elastic fibres exists in the Elasmobranchs, in Acipenser, and in some but not in all Teleosts, and in the form of scattered fibres in the Cyclostomes.   It has been described by many observers in fishes.

the invaded fibrous sheath takes a share, are called chordal centra, as distinguished from the usual perichordal centra developed outside the unbroken elastica externa.

A similar invasion of the thick fibrous sheath through the elastica takes place in the Dipnoi ; but the invasion is less thorough, and although a small amount of cartilage is formed at the base of the arches by the

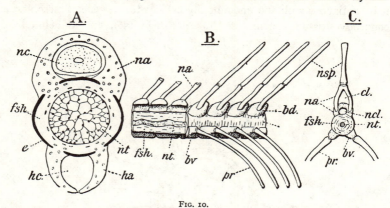

Fig. 10.

A, Transverse section of vertebral column of a young *Protopterus*, showing the invasion of the fibrous notochordal sheath by the mesoblastic cells ; B, left-side view of a portion of the vertebral column (abdominal region) of *Ceratodus Forsteri*, Krefft., of which the anterior half has been cut longitudinally ; C, view of the same cut across. *bd*, Basidorsal cartilage ; *bv*, basiventral cartilage ; *cl*, canal for ligament ; *e*, elastica externa ; *fsh*, fibrous sheath ; *ha*, haemal arch (basiventral) ; *hc*, haemal canal ; *na*, neural arch ; *nc*, nerve-cord ; *ncl*, neural canal ; *nsp*, supraneural spine ; *nt*, notochord ; *pr*, pleural rib. (From Goodrich, *Vert. Craniata*, 1909.)

mesoblastic cells, true complete centra are not developed at all events in living genera, Fig. 10.

True centra are not formed in the living Holocephali, although the fibrous sheath is invaded by cells. The cells gather into a middle zone in the sheath, and a series of calcified rings are here developed which are more numerous than the segments, Fig. 31. The extinct *Squaloraja*, from the Lower Lias, had vertebral rings much better developed and more like centra, Fig. 32.

The Chondrostei also have a well-developed fibrous sheath, Fig. 34. No centra are formed in living genera ; but, as first discovered by Schneider, the fibrous sheath is invaded through the ruptured elastica, in spite of many statements to the contrary (E.S.G.).

The concurrence of this invasion in such diverse groups of fish naturally suggests the question whether it is a primitive condition in Pisces, or a secondary modification of no phylogenetic significance, and connected with the thickening of the sheath. No definite answer can as yet be given ; but whichever it should be, there can be little doubt that the centra themselves were originally formed outside the elastica externa. In this connexion it should not be forgotten that the occipital region of the skull is always so developed, even in the Selachian, and that where, as in Holocephali, Dipnoi, and Chondrostei, several anterior vertebral rings may merge into the occipital region, these are perichordal in structure.

The notochord in *Amphioxus* is surrounded by a strong sheath; but it is continuous with and undoubtedly belongs to the general mesoblastic connective tissue system, as Lankester held (56), and therefore not

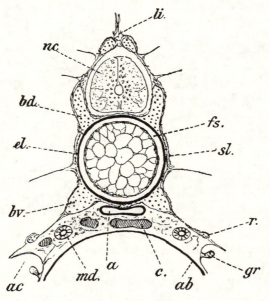

Fig. 11.

Transverse section of the vertebral column in the trunk of a very young trout (*Salmo*), enlarged. *ab*, Dorsal wall of air-bladder; *ac*, abdominal coelom; *bd*, basidorsal cartilage; *bv*, basiventral cartilage; *c*, posterior cardinal vein; *el*, elastica externa; *fs*, fibrous sheath surrounding notochord; *gr*, genital ridge; *li*, dorsal longitudinal ligament; *md*, mesonephric duct; *nc*, nerve-cord; *r*, rib; *sl*, skeletogenous tissue. (From Goodrich, *Vert. Craniata*, 1909.)

homologous with either of the true sheaths described above. Whether there is really a very thin separate sheath between it and the notochord, as described by v. Ebner (20) and Klaatsch (53), is doubtful. If present it probably corresponds to the elastica externa.

The vertebral centra in Teleostomes and Tetrapods are formed from the mesoblastic skeletogenous layer surrounding the notochordal sheaths, and belong to the perichordal type, Fig. 11. In these forms, then, the centra typically lie outside the notochord and its sheaths; the latter remain intact, although they often become much constricted and even obliterated in the adult.[1] Probably all the Osteichthyes, as well as the Tetrapods, have vertebral columns really belonging to this type ; for though the fibrous sheath is invaded in modern Dipnoi and Acipenseroidei, as mentioned above, yet the cartilaginous elements representing the centra in these fish are placed outside the sheaths.

A remarkable alteration of the notochord takes place in the middle of the vertebra in Amphibia, *Sphenodon*, and many Lacertilia, Figs. 56, 58, 268. It consists in the late appearance of a zone of cartilage-like tissue, which may constrict the ordinary notochordal tissue to a mere thread (Gegenbaur, **29**, in Urodela, Apoda, and Lacertilia ; Goette, 1875, in Anura). About the origin of this ' chordal cartilage ' there has been much controversy, some believing it to be produced by invading mesoblastic cells (Lwoff, **56** ; Zykoff, **99** ; Gadow, **26** ; Tretjakoff, **89**), others holding that it is formed from modified notochordal cells which secrete a matrix resembling that of true cartilage (Field, **21** ; v. Ebner, **20** ; Klaatsch, **53** ; Kapelkin, **52** ; Schauinsland, **78**). There can be little doubt that the latter is the correct interpretation. There is good evidence that the notochordal sheaths remain unbroken, that mesoblastic cells do not pierce them, and that the notochordal tissue becomes converted into the chordal ' cartilage '.

THE VERTEBRAL COLUMN

The structure and development of the vertebral column in the Craniata may now be considered. Great diversity occurs in the various groups not only in the number of elements serving to arch over the neural canal and to enclose the haemal canal, but the vertebral centra themselves may vary greatly. They may be of more or less complex build, may be well developed, vestigial, or altogether absent, and even may be more numerous than the segments in which they lie. Yet it is probable that the vertebrae are fundamentally homologous throughout the Craniata. The older authors roughly identified arches and centra in fishes and land vertebrates ; but recently attempts have been made to compare in detail different types of vertebral column and refer them to a single scheme of homologous parts. It is doubtful whether such a proceeding is altogether justifiable, since at least some of the types may have been independently evolved. Moreover, there is danger of adopting too uniform and artificial a scheme for the whole length of the column, forgetting that the arrangement of the dorsal elements may never have

[1] According to Tretjakoff, the fibrous sheath may possibly be slightly invaded in Amphibia (**89**).

corresponded exactly to that of the ventral elements, and that the anterior region may never have exactly resembled the posterior. Nevertheless

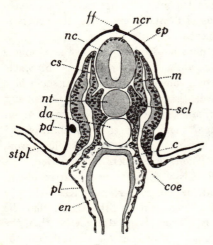

considerable success has already been achieved, notably by Cope (13), Hay (45), Goette (31-4), and more especially by Gadow and Abbott (27), and Schauinsland (78).

Selachii.—It is convenient to begin with an account of the vertebral column of the Selachii, and its structure is best understood by describing its development. The sclerotomes which yield the mesoblastic tissue of which it is composed are of course strictly segmental in origin (p. 6), and grow so as to surround both the notochord in its sheaths (p. 9) and the neural canal, Figs. 12, 14, 15, 16. Between successive sclerotomes extend upwards small segmental

Fig. 12.

Scyllium canicula; transverse section through trunk of embryo, stage H. *c*, Coelomic canal between ventral splanchnocoele, *coe*, and dorsal myocoele, *cs*; *da*, dorsal aorta; *ep*, epidermis; *en*, endodermal gut; *ff*, ectodermal fin fold; *m*, myotome; *nc*, nerve cord; *ncr*, neural crest; *nt*, notochord; *pd*, pronephric duct; *pl*, splanchnopleure; *scl*, sclerotome; *stpl*, somatopleure.

blood-vessels, arteries from the dorsal aorta, and veins from the cardinals. These primitive vessels are, therefore, intersegmental in

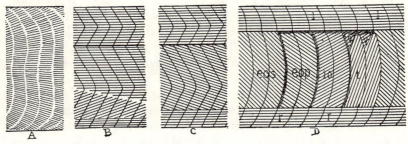

Fig. 13.

Scheme of arrangement of *myomeres* and muscle fibres in left-side view of trunk of: A *Petromyzon*, B *Selachian*, C *Teleost*, D Urodele (from J. S. Kingsley, *Comp. Anat. of Vertebrates*, 1926). *eop*, *eos*, Deep and superficial external oblique; *io*, internal oblique; *l*, longissimus dorsi; *r*, rectus abdominis; *t*, transversus; *v*, subvertebral.

position; and when, as soon happens, the limits between consecutive sclerotomes become obscured as the cells become distributed along the column to form a continuous covering of skeletogenous tissue or membrana reuniens, the blood-vessels even in the adult serve to show the original limit between the segments (Rabl, 71-2; Schauinsland, 78). Outside this sclerotomal tissue develop the segmental myotomes, giving rise to myomeres which soon become bent first in < shape, then in ⋛ shape, so that their disposition no longer corresponds to the original vertical intersegmental divisions still indicated by the intersegmental vessels, Figs. 13 and 15. Moreover, the myocommata are set obliquely to the long axis, each myomere being partly covered by the one in front.

In the course of development the posterior half of each sclerotome becomes denser than the anterior, and through the anterior half of looser tissue pass the ventral and dorsal nerve-roots, the latter bearing a ganglion. The anterior half may be called the cranial half, the posterior the caudal half-sclerotome. Primitively, no doubt, the dorsal roots are intersegmental and the ventral roots segmental in position; but, owing to relative shifting of parts and obliquity of septa, the ganglia appear to move forward and come to occupy a position opposite the middle of the segment to which they belong, Figs. 15, 16.

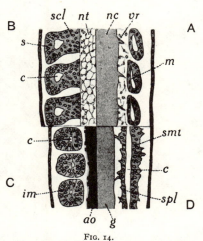

FIG. 14.

Scyllium canicula. Longitudinal frontal sections of trunk of embryo, stage I. A, Most dorsal, D, most ventral; B, through sclerotomes, *scl,* and scleromyocoeles, *c*; C, through nephrotomes and nephrocoeles, *c*; D, through ventral lateral plate and splanchnocoele, *c*; *ao,* dorsal aorta; *g,* gut wall; *im,* intermediate mass (nephrotome); *nc,* nerve cord; *smt,* somatopleure; *spl,* splanchnopleure. Other letters as in Figs. 12 and 15.

The main branch of the mixed spinal nerve passes out in the transverse septum posterior to its myomere. Into this septum also grows outwards a prolongation of the ventral region of the posterior dense half of the sclerotome to form the rudiment of the rib (p. 71). Dorsally to the notochord now appear two condensations of the sclerotomal mesenchyme on each side of the neural canal in each segment. The posterior and larger rests on the elastica externa and is the rudiment of the cartilaginous neural arch which takes up an approximately intersegmental position, between the intersegmental vessels behind and the dorsal root in front. The anterior and smaller rudiment does not usually reach the elastica, is situated behind the intersegmental vessels between the ventral and the dorsal nerve-roots, and gives rise to the interneural or 'intercalary' cartilaginous arch.[1] Ventrally to the notochord on each side in each segment appear a pair of rudiments corresponding to those above. The posterior and larger gives rise to the haemal or ' transverse process ' bearing the rib in the trunk or the haemal arch in the tail. The anterior rudiment is far less regularly developed and in many forms vestigial or absent ; in some Selachii and in certain regions of the column it may be well developed and persist

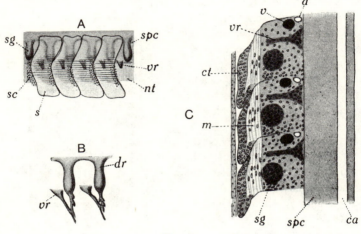

FIG. 15.

Scyllium canicula. A, Left side view of myotomes, etc., of trunk of embryo stage J. B, Reconstruction of spinal nerve roots of later stage. C, Longitudinal frontal section of trunk of embryo, stage K, left side only complete ; a, segmental artery ; ca, neural canal ; ct, outer ' cutis ' layer of myotome ; dr, dorsal root ; m, myomere ; nt, notochord ; s, segmental myotome ; sc, scleromere ; sg, spinal ganglion ; spc, spinal cord ; v, segmental vein ; vr, ventral root.

as a separate ' intercalary ' cartilage. It is in connexion with the posterior dorsal and ventral paired elements that the centrum of the vertebra is developed. These four cartilages spread at their bases over the elastica externa, Fig. 22, and it is at these four points in the circumference, chiefly if not entirely, that the piercing of the elastica takes place, allowing the mesoblastic cells to invade and spread throughout the fibrous sheath (p. 9). Chondrification now extends in the sheath forming rings of cartilage in continuity with the neural and haemal elements outside ; thus are formed the ' chordal ' centra constricting the notochord inter-

[1] Van Wijhe has recently shown that at an early stage of the development of the column of Acanthias the arcualia are represented by four continuous bands of cartilage along the notochordal sheath, which subsequently break up into separate basidorsal and basiventral elements (v. Wijhe, 93). Similar bands of early cartilage have been described by de Beer in Heterodontus (7) ; it would seem, however, that the continuity is better ascribed to the temporary fusion of rudiments set very close together than to the arcualia having been evolved from an originally continuous cartilaginous band.

segmentally, but allowing it to continue growing segmentally. The fibrous sheath between the consecutive cartilaginous centra remains to form the intervertebral fibrous rings, Figs. 17-21.

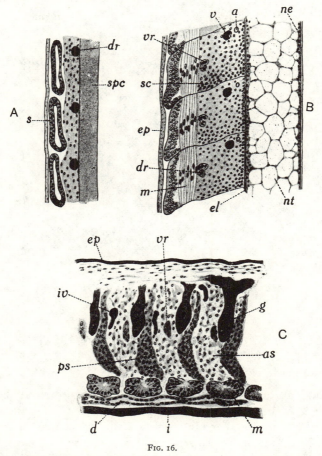

FIG. 16.

Scyllium canicula, embryo stage K ; longitudinal frontal section more dorsal in A, more ventral in B. *dr*, Dorsal root ; *el*, elastica externa ; *ep*, epidermis ; *m*, myomere ; *ne*, notochordal epithelium ; *s*, dorsal region of myotome ; *sc*, denser posterior region of sclerotome ; *spc*, spinal cord ; *vr*, ventral root. Other letters as in Fig. 17. C, Sagittal section to left of middle plane, stage J. *d*, Primary (mesonephric) duct ; *g*, ganglion ; *i*, intestinal wall ; *iv*, intersegmental vein ; *m*, mesonephric tubule rudiment ; *vr*, ventral root ; *as* and *ps*, anterior and posterior halves of sclerotome ; much denser latter half will give rise to basalia.

The important conclusion is reached with regard to the vertebral column in Selachii that four paired elements in each segment contribute to its development. To these may be applied the terms introduced by Gadow and Abbott: the paired elements above and below the notochord

C

are called the dorsal and ventral arcualia respectively ; the larger posterior
arcualia in each segment are the basalia (basidorsalia and basiventralia);
the smaller anterior arcualia are the interbasalia (interdorsalia and inter-

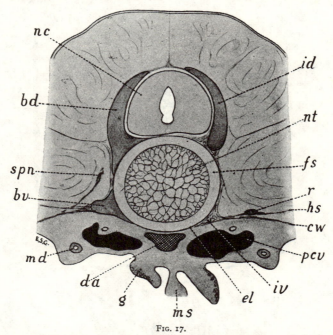

FIG. 17.

Transverse section through the anterior trunk region of an embryo *Scyllium canicula* about 32 mm.
long. Being slightly oblique the section cuts basidorsal, *bd*, and basiventral, *bv*, on left, and interdorsal,
id, and interventral, *iv*, on right. *cw*, Connective tissue of splanchnocoelic wall ; *da*, dorsal aorta ;
el, elastica externa ; *fs*, fibrous sheath ; *g*, gonad ; *hs*, horizontal septum ; *md*, Müllerian duct ;
ms, mesentery ; *nc*, nerve cord ; *nt*, notochord ; *pcv*, posterior cardinal vein ; *r*, dorsal rib ; *spn*,
spinal nerve.

ventralia). The basidorsal (neural arch) and the basiventral (haemal
arch) are derived from the ' caudal ' or posterior denser half of the sclerotome, and the interdorsal and interventral from the ' cranial ' or anterior
half of the sclerotome.[1] The segmentation of the sclerotomes is early

[1] For the theory of Gadow and Abbott (**27**), that each scleromere or
vertebral segment is formed on each side of the ventral half of one sclerotome
combined with the dorsal half of the sclerotome of the segment next following
(each sclerotome being obliquely and not vertically divided), there seems to be
no good evidence. These authors were apparently misled by the secondary
bending of the myomeres causing the septa to pass obliquely across the column ;
while the original segmentation is better indicated by the intersegmental
vessels as explained above. The evidence seems to be clear that the basidorsal and basiventral of each vertebral segment are derived from the posterior
region of the same sclerotome. Nevertheless Marcus and Blume (**62**) have
revived Gadow's theory in dealing with the Apoda.

lost, but the original limits between them are shown by the intersegmental blood-vessels. The bases of the basidorsals and basiventrals spread over the elastica externa but do not as a rule meet round it, and the chordal centra are developed in continuity with them as complete rings. Since the basalia (and centra) take up a position between consecutive myomeres, they become connected with the intersegmental myocommata, while the interbasalia become segmentally placed. The

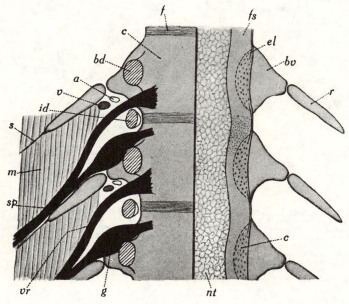

Fig. 18.

Scyllium canicula. Reconstruction of vertebral column, etc., of advanced embryo; dorsal view. On left, arcualia cut through; on right, deeper cut through notochord. *a*, Intersegmental artery; *bd*, basidorsal; *bv*, basiventral; *c*, centrum developing in fibrous sheath; *el*, interrupted elastica externa; *f*, intervertebral ligament; *fs*, notochordal fibrous sheath; *g*, ganglion on dorsal root; *id*, interdorsal; *m*, myomere; *nt*, notochord; *r*, rib; *s*, intersegmental septum; *sp*, spinal mixed nerve; *v*, intersegmental vein; *vr*, ventral root.

finished column thus consists of vertebral centra alternating with muscle segments, two myomeres being connected to each vertebra, an adaptation for the bending of the vertebral column seen in Gnathostomes generally.

The structure of the vertebral column in adult Selachians has been studied by Kölliker (54), Goette, Hasse (38), and others. Lately Ridewood has extended and corrected earlier observations, and provided a sounder interpretation of results (76). The column forms a flexible skeletal covering not only to the notochord but also to the neural canal above, and the haemal canal below in the caudal region. There are no

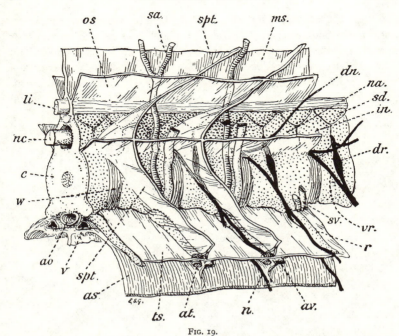

FIG. 19.

Scyllium canicula, L. Left-side view, enlarged, of a portion of the vertebral column and ribs with the connective tissue septa, to show their relation to the blood-vessels and nerves. *ao*, Dorsal aorta ; *as*, abdominal wall ; *at*, artery ; *av*, vein ; *c*, centrum ; *dn*, dorsal nerve ; *dr*, ganglion of dorsal root ; *in*, interdorsal ; *li*, dorsal ligament ; *ms*, median dorsal septum ; *n*, ventral branch of spinal nerve ; *na*, basidorsal ; *n.c*, nerve-cord ; *r*, rib ; *s.a*, segmental dorsal artery ; *sd*, supradorsal ; *spt*, vertical transverse septum passing between successive myotomes ; *sv*, segmental dorsal vein ; *ts*, chief transverse horizontal septum in which lie the ribs ; *os*, oblique upper longitudinal septum, a similar lower septum occurs between it and the transverse horizontal septum ; *v*, posterior cardinal vein ; *vr*, ventral spinal root ; *w*, intervertebral ligament. (From Goodrich, *Vert. Craniata*, 1909.)

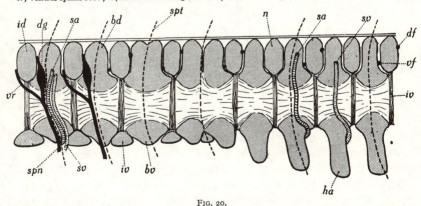

FIG. 20.

Diagram illustrating structure of vertebral column and transition to diplospondylous condition in caudal region of a *Selachian*. Cartilage stippled. *bd*, basidorsal ; *bv*, basiventral ; *df*, foramen for dorsal root ; *dg*, spinal ganglion on dorsal root ; *ha*, haemal arch, prolongation of basiventral ; *id*, interdorsal ; *iv*, interventral ; *n*, secondary intercalated interneural ; *sa*, segmental artery ; *spn*, mixed spinal nerve ; *spt*, position of intersegmental septum ; *sv*, segmental vein ; *vf*, foramen for ventral root ; *vr*, ventral root.

articular joints between the centra and no articular processes between the arches, the consecutive centra being firmly united by fibrous rings and the arcualia by fibrous tissue. The dorsal and ventral nerve-roots escape through foramina situated respectively either in front and behind each interdorsal (Scyllioidei) or through the basidorsal and interdorsal, which have grown backwards so as to enclose them (Goodrich, 35; v. Wijhe, 93; de Beer, 7). Above the column runs the strong longitudinal elastic ligament found in all Craniates. Below this there is usually a series of small median supradorsal cartilages (sometimes wrongly called neural spines, p. 87) wedged between the tips of the basidorsals and inter-dorsals, and forming keystones to the arches; or the basidorsals and interdorsals fuse above the neural canal as in *Squalus*. In the caudal region the interventrals are reduced or gener-ally lost, and the right and left basiventrals fuse below the caudal vessels, and are pro-longed as broad median spines supporting the ventral caudal fin. The basiventral may some-

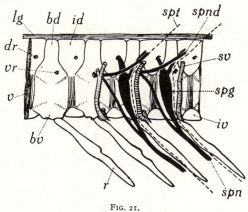

FIG. 21.

Squalus acanthias. Left-side view of portion of vertebral column of mid-trunk. *bd*, Basidorsal; *bv*, basiventral; *dr*, dorsal root foramen; *id*, interdorsal; *iv*, interventral; *lg*, dorsal ligament; *r*, dorsal rib; *spg*, spinal ganglion; *spn*, mixed spinal nerve; *spnd*, dorsal branch of spinal nerve; *spt*, course of intersomitic septum; intersegmental vein (corresponding artery shown in next segment); *v*, intervertebral disc; *vr*, ventral root foramen.

times show a ventral prolongation at the side of the aorta in addition to the more dorsal process supporting the dorsal rib (*Cetorhinus, Laemargus*, etc.). In such a form as *Lamna* the basiventrals and interventrals together with cartilages apparently representing the ribs form in the trunk con-tinuous outstanding flattened flanges in the horizontal septum, Fig. 25. As a rule the interventrals are irregularly developed, vestigial, or absent.

The most important modifications occur in the structure and develop-ment of the centra. To what may be termed the chordal centrum, formed in the fibrous sheath and growing by expansion into typically biconcave or amphicoelous cartilages in continuity with the basalia, may be added later on cartilage developed at four points between the bases of these arches from skeletogenous tissue outside the elastica externa. These are the four intermedialia (dorsal, ventral, and two lateral) of Ridewood, which grow by addition of new layers peripherally, forming wedges between the arches and burying the remains of the elastica deeper and deeper in the centrum. Eventually the elastica may be absorbed and the limits between the components of the cartilaginous centrum be lost, Figs. 28, 29.

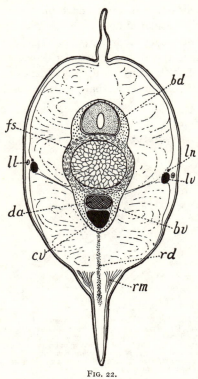

FIG. 22.

Transverse section of the tail of *Scyllium canicula*, late embryo. *bd*, Basidorsal; *bv*, basiventral; *cv*, caudal vein; *da*, caudal artery or dorsal aorta; *fs*, fibrous sheath of notochord invaded by mesoblastic cells; *ll*, lateral-line canal; *ln*, lateral-line nerve; *lv*, lateral cutaneous vein; *rd*, procartilaginous rudiment of radial of anal fin; *rm*, radial muscles.

Besides the usual crust of calcification extending over the surface of the cartilages, special much denser calcifications are deposited in the centra, except in degenerate forms, giving rise in transverse section to characteristic patterns. Hasse attempted to classify the Selachii into three main groups according to these patterns : Cyclospondyli with a simple cylinder, Tectospondyli with concentric cylinders, and Asterospondyli with radiating lamellae, Fig. 23. But a rigid adherence to his definitions leads to a very artificial grouping, and the types are by no means so distinct as he supposed. The calcifications increase with age, and radiating lamellae occur in nearly all the families.

The primary cartilaginous centrum becomes differentiated into outer, middle, and inner zones, and calcification starts early in the middle zone forming a simple cylinder or double cone expanded at both ends. Such a primary cylinder with an outer strengthening of calcification derived from the outer zone is found in the Squalidae, Figs. 24 and 29. To the primary cone may be added successive concentric cylinders developed in the outer zone as in *Squatina* (tectospondylous type), Fig. 26. The outer zone in *Pristis* and *Rhynchobatus* is occupied by a solid calcification. To the primary cones are frequently added longitudinal radiating lamellae spreading in the outer zone. Four such diagonal lamellae occur in most Scyllioidei (not, however, in the common Dog-fish, *Scyllium*,

[*Scylliorhinus*] *canicula*). The lamellae may become more numerous, subdividing from their outer ends as in Notidani and Heterodonti.

Another set of calcifications may develop either as radiating lamellae (*Stegostoma, Lamna*) or as concentric lamellae (*Cetorhinus*) in the four intermedialia. They extend inwards and in some forms reach to near the primary cylinder, Figs. 25, 28.

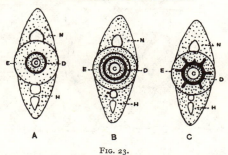

FIG. 23.

Diagrammatic transverse sections of vertebrae of Selachians. A, Cyclospondylous ; B, tectospondylous ; C, asterospondylous condition (after Hasse, from A. Sedgwick, *Zoology*, 1905). C, Notochord ; D, central calcareous ring ; E, elastica externa ; H, haemal arch ; N, neural arch.

Degeneration of the vertebral column occurs in several groups of the Selachii. It is usually more pronounced anteriorly, the caudal region retaining a more primitive structure. In the Notidani, *Echinorhinus,* and *Laemargus* (*Somniosus*), for instance, the centra become narrower

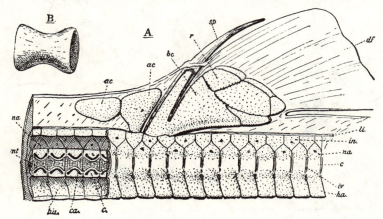

FIG. 24.

Squalus vulgaris, Risso. A, Dorsal fin and portion of the vertebral column, which has been cut through longitudinally in front. B, Calcified cylinder of a centrum. *ac*, Anterior cartilages (either modified radials or neural spines) ; *bc*, basal cartilage ; *ca*, calcified cylinder ; *df*, dorsal fin ; *ha*, haemal arch ; *in*, interdorsal ; *iv*, intervertebral ligament ; *li*, dorsal ligament ; *na*, neural arch ; *nt*, notochord ; *sp*, fin spine, with base cut away to expose cartilage core ; *r*, radial. The cartilage is dotted. (From Goodrich, *Vert. Craniata*, 1909.)

and the intervertebral rings wider, until the vertebral constrictions are reduced to septa separating large blocks of swollen partially liquefied notochord. In *Chlamydoselachus* the notochord is no longer constricted anteriorly, Fig. 27.

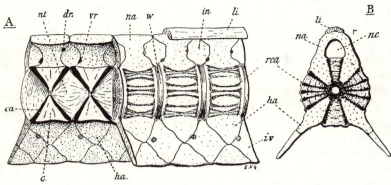

FIG. 25.

Lamna cornubica, Gm. A, Portion of the vertebral column of the trunk, partly cut longitudinally (right-side view); B, transverse section of the same through the middle of a centrum. *c*, Centrum; *ca*, calcareous constricted ring; *dr*, foramen for dorsal nerve-root; *ha*, basiventral; *in*, interdorsal; *iv*, interventral; *li*, ligament; *na*, basidorsal; *nc*, neural canal; *nt*, notochord; *rca*, radial calcifications; *vr*, foramen for ventral root; *w*, intervertebral ligament. (From Goodrich, *Vert. Craniata*, 1909.)

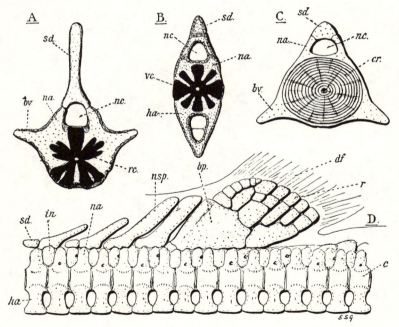

FIG. 26.

Transverse section through the centre of A, a trunk vertebra of *Raja*. B, A caudal vertebra of *Rhinobatus granulatus*, Cuv.; C, a trunk vertebra of *Rhina squatina*, L.; D, left-side view of a portion of the vertebral column, and of the skeleton of the first dorsal fin of *Rhina squatina*, L. *bp*, Basal; *bv*, basiventral (haemal arch); *c*, centrum; *cr*, calcareous ring; *df*, dorsal fin; *ha*, haemal arch; *in*, interdorsal; *na*, basidorsal (neural arch); *nc*, neural canal; *nsp*, supraneural spine (or anterior radial); *r*, distal end of radial; *rc* and *vc*, radiating calcification (black); *sd*, supradorsal. (From Goodrich, *Vert. Craniata*, 1909.)

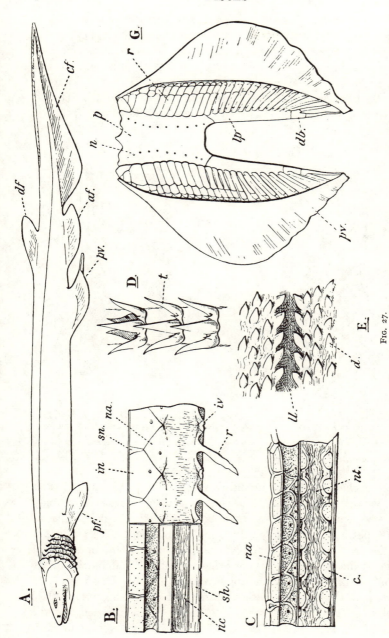

FIG. 27.

Chlamydoselachus anguineus, Garman. A, Left-side view of male; B, portion of vertebral column of trunk, part of which has been cut longitudinally (right-side view); C, longitudinal section of more anterior region; D, three teeth; E, lateral line; G, pelvic fins and girdle (ventral view). B, C, D, and E are modified from Garman. *af*, Anal fin; *c*, centrum; *cf* caudal fin; *d*, denticle; *db*, distal segment of basipterygium; *df*, dorsal fin; *in*, interdorsal; *iv*, interventral; *ll*, lateral-line groove; *lp*, proximal segment of basipterygium; *n*, nerve foramen; *na*, neural arch; *nc*, notochord; *nt*, notochord; *p*, pelvic girdle; *pf*, pectoral fin; *pv*, pelvic fin; *r*, radial (in G), rib (in B); *sh*, notochordal sheath; *sn*, supraneural; *t*, tooth. (From Goodrich, *Vert. Craniata*, 1909.)

A strange modification of the normal structure occurs in the caudal region of the Elasmobranchs. It is known as diplospondyly, for here the number of vertebral segments is double that of the myomeres and their nerves; that is to say, two centra and two sets of dorsal and ventral arcualia correspond to one pair of myomeres and spinal nerves. Naturally this exceptional structure has attracted much attention, and it has been studied most successfully of late by Mayer (128), Ridewood (75), and Sečerov (84). The regular doubling of the arcualia and the centra takes

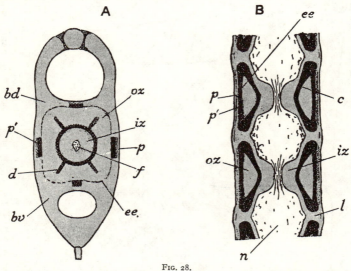

FIG. 28.

A, Transverse section of caudal vertebra from region of second dorsal fin of late embryo of *Mustelus vulgaris*, 286 mm. long; B, horizontal section of vertebrae from same region (from W. G. Ridewood, *Tr. Roy. Soc.* 1921). *bd*, Basidorsal; *bv*, basiventral; *d*, diagonal calcified lamella; *ee*, membrana elastica externa, outer limit of notochordal sheath; *f*, funiculus chordae; *iz*, inner-zone cartilage of sheath; *l*, intervertebral ligament; *n*, remains of notochord; *ox*, and *oz*, outer-zone cartilage of sheath; *p*, cartilaginous part of intermediale of perichondrial origin outside sheath; *p'*, calcified part of same.

place throughout the caudal region, except at the tip, where irregularities occur, and the transition zone in the pelvic region, Figs. 20, 24, 26. A study of development shows conclusively that diplospondyly is due neither to a mere splitting of the ready-formed rudiment of the vertebra nor to a reduction by fusion of successive pairs of myotomes and disappearance of alternate pairs of nerves as held by Schauinsland. The duplication of the skeletal parts is related to the lengthening of the caudal segments, and judging from the structure of the transitional region seems to take place as follows. The first step is the lengthening of the centrum and the basidorsal and basiventral; these arcualia are then transversely divided, and later a new interdorsal appears between the basidorsals. Lastly a centrum is formed in relation to each set of basalia. As the caudal segment lengthens the originally intersegmental vessels become separated. The artery remains in front, and the vein is shifted to the middle of the segment; the new interdorsal lies between them (E.S.G.). The transition may be abrupt as in *Squalus,* short but gradual as in *Scyllium,* or long as in *Hexanchus,* where several centra bear double sets of arches. The same phenomenon occurs in the Holocephali; but in these fish towards the tip of the tail the arcualia become small and very numerous, leading to an obviously secondary condition misleadingly called poly-spondyly. It is by lashing the tail from side to side that fish propel themselves, and ' diplo-spondyly ' seems to secure greater flexibility for this purpose.

The extreme tip of the column is much specialised. Here the arcualia become irregular and replaced by blocks of cartilage which tend to fuse to a continuous rod completely en-veloping and replacing the posterior extremity of the notochord. A similar modification is found to some extent in all Pisces, excepting those specialised Teleostei in which the verte-bral column is truncated in the adult (p. 113).

Cyclostomata.—We may now review the structure and development of the vertebral column in the Craniata generally. Dealing first with the Cyclostomata we find a per-sistent unconstricted notochord enclosed in the typical sheaths: elastica externa, and thick fibrous sheath. The mesoblastic cells outside

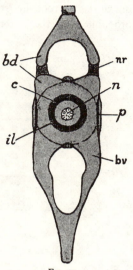

Fig. 29.

Transverse section of caudal vertebra of *Squalus acanthias* (from W. G. Ridewood, *Tr. Roy. Soc.* 1921). *c,* Primary double cone, calcification of middle zone of noto-chordal sheath; *il,* investing layer calcification; *nr,* nerve foramen; other letters as in Fig. 28.

form a connective tissue covering to the notochord and enclosing the neural canal as well as a longitudinal median dorsal space filled with fatty tissue above, and a haemal canal below in the tail, Fig. 30. There is no trace of centra;[1] but in the Petromyzontia the axial skeleton is represented by paired rods or arches resting on the notochordal sheath on either side of the neural canal (Parker, **67**; Schneider, **81**; Schauinsland, **78**; Tretja-koff, **87-8**).

Throughout the greater length of the trunk there are two pairs of such arches in every segment not meeting above, but passing vertically

---

[1] The doubtful traces described by Schauinsland in anterior segments scarcely deserve the name.

upwards in the anterior trunk region. They alternate regularly with the dorsal and ventral root nerves ; the posterior arch being approximately intersegmental, just behind the dorsal nerve and about on a level with the intersegmental blood-vessels, and the anterior arch just behind the ventral nerve. Schauinsland compares the posterior pair to the basi-dorsals and the anterior pair to the interdorsals of Gnathostomes. But in the Selachian it is the basidorsal or neural arch which lies between a dorsal nerve root in front and a ventral root behind, so it is more probable that the anterior cartilage (cranial of Schauinsland) is the

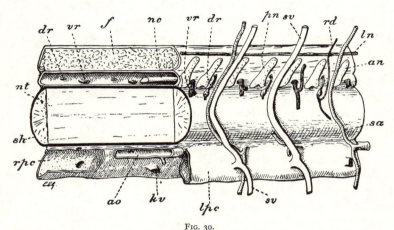

FIG. 30.

Petromyzon marinus, L. Left-side view of a portion of the notochord and neighbouring organs ; the left half has been removed by a median longitudinal section in the anterior region. *an*, neural arch (basidorsal ?) ; *ao*, dorsal aorta ; *dr*, dorsal nerve-root ; *f*, fatty tissue ; *kv*, kidney vein ; *ln*, lateral-line nerve ; *lpc*, left posterior cardinal vein ; *nc*, nerve-chord ; *nt*, notochord ; *pn*, interneural arch (interdorsal ?) ; *rd*, ramus dorsalis ; *rpc*, right posterior cardinal ; *sa*, segmental artery ; *sh*, notochordal sheath ; *sv*, segmental vein ; *vr*, ventral nerve-root. (From Goodrich, *Vert. Craniata*, 1909.)

homologue of the Gnathostome basidorsal ; moreover, it is the larger and more important of the two arches in *Petromyzon*. In the anterior region the anterior arches tend to fork at their base over the ventral nerves which they come to surround. (The first arch, perhaps formed of two fused arches, encloses the motor nerve to the fourth metaotic myomere and also the dorsal nerve in front of it in *P. fluviatilis*.) In the caudal region the arches dwindle in size, become irregular, tend to fuse to a continuous covering of the tip of the notochord, and come into connexion with the dorsal median fin supports. These are slender carti-laginous rods, Fig. 99, which in the hinder region of the caudal fin fork at their bases over the neural canal like neural arches and fuse to a longi-tudinal plate on each side. Similar median fin-supports occur in Myxinoidea (p. 97), but there are no true vertebral elements, Fig. 100. In this respect these Cyclostomes are possibly degenerate.

Chondrichthyes.— The diagram given in Fig. 20 illustrates the general morphology of the vertebral column in Selachians. Each vertebral segment is composed typically of paired basidorsals and interdorsals above and paired basiventrals and interventrals below, the basidorsals and basiventrals (intersegmental or vertebral in position) alone rest on the notochordal sheath, and the chordal centrum is formed intersegmentally in continuity with them as a complete ring. Typically, also, the basalia do not meet round the notochord, but perichordal cartilage may be added to the chordal centrum from skeletogenous tissue between their bases. Nevertheless, although the possession of chordal centra is so characteristic of the Selachii, it is not impossible that this condition is secondary and that they have been derived from ancestors with perichordal centra derived chiefly from the bases of the arcualia as in other Pisces. In this connexion it should be noticed that near the skull the basidorsals tend to meet the basiventrals forming a perichordal investment, and that in such an archaic form as Heterodontus they do so throughout the column.

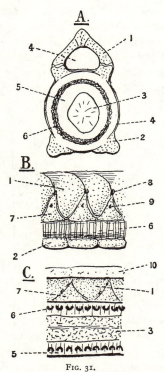

FIG. 31.

Chimaera monstrosa, L. Vertebral column : A, transverse section ; B, side view ; and C, longitudinal median section. (All after Hasse, slightly modified.) 1, Interdorsal ? ; 2, basiventral ; 3, notochord ; 4, elastica externa and connective tissue ; 5, fibrous sheath ; 6, calcified ring ; 7, basidorsal ? ; 8, dorsal nerve-root foramen ; 9, ventral nerve-root foramen ; 10, supradorsal. (From Goodrich, *Vert. Craniata*, 1909.)

The Holocephali (Hasse, 38 ; Klaatsch, 53 ; Schauinsland, 78), closely related to the Selachii, have a persistent and unconstricted notochord, Fig. 31. The thick fibrous sheath is invaded by mesoblastic cells through the ruptured elastica externa. Typical chordal centra are not developed ; but the cells spreading throughout the sheath form complete rings, which acquire a calcified bone-like structure in *Chimaera*, and are much more numerous than the segments. Such rings do not occur in *Callorhynchus* ; while in the extinct *Squaloraja* of the Lower Lias they are, on the contrary, very strong and closely packed, Fig. 32. It is possible that the Holocephali are descended from early ancestors with chordal centra like those of the Selachii, and that the rings have been formed by their duplication or subdivision. Their reduced condition or absence in modern forms is doubtless due to degeneration in fish which live in deep waters and use their paired fins more than their tails for swimming. Though somewhat irregular, the arcualia are developed, much as in Selachians, as four paired elements in each trunk segment. The basiventrals may fuse with the interventrals, and there are no ribs. As interpreted by Schauinsland it is the interdorsal only which reaches and spreads over

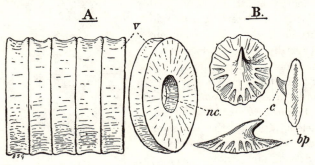

FIG. 32.

Squaloraja polyspondyla, Ag. Enlarged views of, A, vertebral rings ; B, denticles. *bp*, Basal plate ; *c*, projecting spine ; *nc*, cavity for notochord ; *v*, vertebral ring. (From Goodrich, *Vert. Craniata*, 1909.)

the notochordal sheath, and surrounds from behind the ventral nerve-root foramen. Supradorsals close the neural tube above. Immediately behind the skull some dozen or more vertebral segments are fused below the anterior dorsal fin ; here, by the fusion of the dorsal with the ventral arcualia round the notochord and of the consecutive cartilages with each other, is formed a rigid continuous perichordal cartilage enclosing both nerve-cord and notochord, and stretching upwards into a process articulating with the cartilage of the fin-base. Numerous nerve apertures betray its compound nature, and in the extinct *Squaloraja* and *Myriacanthus* other traces of segmentation appear in it (Dean, **17**). The same doubling of the skeletal elements takes place in the segments of the tail region as in Selachians.

The structure of the vertebral column in the extinct groups of Chondrichthyes is but imperfectly known. The Pleuracanthodii, Cladoselachii, and Acanthodii seem all to have had a persistent unconstricted notochord without centra. Basidorsals and basiventrals bearing ribs are well developed in *Pleuracanthus*, but the interdorsals and interventrals are small (Brongniart).   Basidorsals and basiventrals have been described in *Cladoselache* (Dean, 154) and in Acanthodians.

Dipnoi.—Modern Dipnoi have a persistent notochord provided with an elastica externa and a thick fibrous sheath.   The latter is invaded by mesoblastic cells (p. 11) to a considerable extent, yet complete chordal centra are never formed so as to constrict the notochord as in Selachians, Figs. 10 and 33.   But in *Ceratodus* it is pushed alternately from above and below by the cartilaginous masses growing inwards from the basidorsals and basiventrals (Goodrich, 35).   Well-developed somewhat irregular interdorsals appear in the tail of *Ceratodus*, but elsewhere they are probably fused

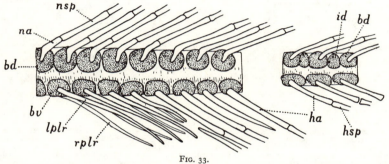

FIG. 33.

Ceratodus Forsteri.   Left-side view of vertebral column of posterior trunk, anterior and middle caudal regions.   *bd*, Basidorsal cartilage ; *bv*, basiventral cartilage ; *ha*, haemal arch ; *hsp*, infrahaemal spine ; *lplr*, and *rplr*, left and right pleural ribs ; *na*, neural arch ; *nsp*, supraneural spine.

with the bases of the large basidorsals.   Similarly the large basiventrals probably include the interventrals.   The basidorsals meet above and below the longitudinal ligament ; and since, except at the extreme ends of the column, they do not meet the basiventrals the notochordal sheath remains exposed at the sides.   The enlarged bases of basidorsals and basiventrals remain cartilaginous, but the slender neural and haemal arches are ossified.   The vertebral column of the modern genera is probably degenerate.

Teleostomi.—The centra of the Teleostomi are of the perichordal type (p. 12).   Usually well developed in the higher forms, they strongly constrict the notochord and its sheaths.   No centra, however, occur in living Chondrostei, where the notochord is persistent and unconstricted, passes uninterruptedly into the base of the skull, and is surrounded by a thick fibrous sheath (slightly invaded, p. 11).   An absence of centra is characteristic of some of the earlier and extinct groups of Teleostomes which apparently retained an unconstricted notochord (Coelacanthini, Pycnodontidae).   In many of the extinct Osteolepidoti centra seem to have been absent, though they occur as ossified rings in *Megalichthys*, and as paired wedge-shaped pieces in Eusthenopteron (Bryant, 465). In many of the earlier Amioidei the centra are absent (*Hypsocormus, Caturus* (A. S. Woodward, 663)) or but slightly developed as thin rings. They rarely occur in Palaeoniscoidei, where, however, traces of centra have been described in some genera (*Pygopterus, Phanerosteon* (Traquair 616,

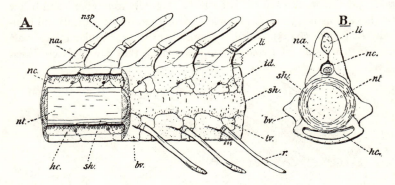

FIG. 34.

Vertebral column of *Acipenser sturio*, L., from the trunk region. A, Left-side view of a piece partly cut through longitudinally; B, the same cut transversely. *bv*, Basiventral; *h.c*, haemal canal; *id*, interdorsal; *iv*, interventral; *li*, longitudinal ligament; *na*, basidorsal; *n.c*, neural canal; *n.sp*, supraneural spine; *nt*, notochord; *r*, pleural rib; *sh*, notochordal fibrous sheath bounded outside by the elastica externa. (From Goodrich, *Vert. Craniata*, 1909.)

Fritsch, 23)). The absence or slight development of centra in these various groups is probably due to degeneration.

The vertebrae of the adult Polypterini are very thoroughly ossified. The larva of *Polypterus* shows cartilaginous basidorsals and basiventrals, but no interbasals. Anteriorly the basiventrals seem to be subdivided into separate elements bearing the dorsal and the pleural ribs (Budgett, 10, p. 73).

The arcualia, on the other hand, are usually well formed and occur as ossified basidorsals (neural arches) and basiventrals (haemal arches) in Coelacanthini, some Osteolepidoti, and the Actinopterygii generally. The modern Chondrostei have basals and interbasals well represented. The large and partly ossified basidorsals meet below and above the dorsal ligament, and may bear supraneural spines. Interdorsals are present, wedged between their bases. Below basiventrals alternate with interventrals and their haemal processes enclose the dorsal aorta, Figs. 34 and 132*a*. As a sign of degeneration it may be noted that these cartilages are often irregular and tend to be secondarily subdivided.[1]

[1] In the Belonorhynchidae, which are probably Chondrostei related to the Acipenseridae (Stensiö, 134), there are no centra; but the early Triassic

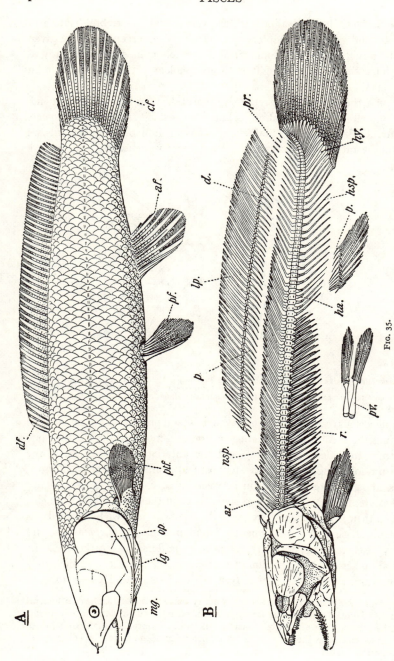

FIG. 35.

Amia calva, L. A after Brown Goode, slightly altered; B after Franque, slightly altered. *af*, Anal fin; *ar*, anterior dorsal radials (?); *cf*, caudal fin (hypochordal); *d*, distal segment of dorsal radial; *ha*, haemal arch; *hsp*, haemal spine; *hy*, hypural spine; *lg*, lateral gulars; *lp*, lepidotrichia; *mg*, median gular; *nsp*, neural spine; *op*, operculum; *p*, proximal segment of radial; *pf*, pelvic fin; *plf*, pectoral fin; *pv*, pelvic bone; *r*, pleural rib. (From Goodrich, *Vert. Craniata*, 1909.)

Holostei.—Before dealing with the very variable vertebral column in the remainder of the Actinopterygii, the Holostei, an account must be given of the structure and development of the vertebrae in *Amia*, the study of which has shed important light on the morphology of the vertebral column in general.

Amia, the only surviving representative of the large Sub-order Amioidei, has a well-ossified vertebral column remarkable in several respects (Franque, 1847; Zittel, 98; Shuffelt, 85). Passing from before backwards, we find in the anterior trunk region vertebral segments consisting of biconcave

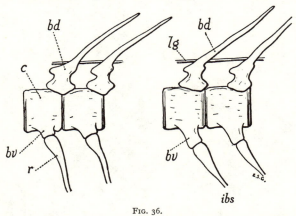

FIG. 36.

Amia calva. Left side view of vertebrae 7 and 8 on left, and 30 and 31 on right. In the former the pleural rib, *r*, is attached to the lateral process of the basiventral, *bv*. In the latter the basiventrals join to enclose a haemal canal and bear a median infrahaemal, *ibs*. *c*, Centrum; *bd*, basidorsal; *lg*, longitudinal ligament.

(amphicoelous) centra with paired ventro-lateral processes or parapophyses bearing the pleural ribs. Resting on the posterior dorsal surface of one centrum to which they belong and the anterior dorsal surface of the centrum following are the separately ossified neural arches or basidorsals, thus to some extent wedged in between successive centra. The neural arches are prolonged into paired spines beyond the longitudinal ligament, and on their inner surface, just below the ligament, are two little cartilages of doubtful significance, frequently found in Holostei (*Lepidosteus*, and many Teleostei) and perhaps representing supradorsals. Farther back the neural arches come to rest almost entirely on the centrum to which they belong, and the spinal processes fuse to a median spine, Figs. 35-40.

---

form *Saurichthys ornatus* has twice as many ossified dorsal arches as ventral arches. Stensio believes the latter to be fused basiventrals and interventrals and the former to be alternately basidorsals and interdorsals equally developed. This view that the interdorsals form arches with spinous processes, etc., requires confirmation.

In the caudal region below each pair of neural arches is a pair of haemal arches, separately ossified and at first bearing a separate median

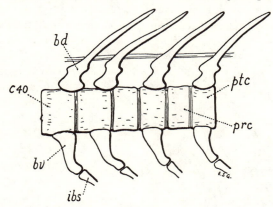

FIG. 37.

Amia calva : left side view of caudal vertebrae. *c40*, centrum of fortieth vertebra ; *ibs*, median infrahaemal ; *prc*, precentrum'; *ptc*, postcentrum.

haemal spine, which soon becomes continuous with the fused arches farther back. Finally, in the posterior upturned region the haemal

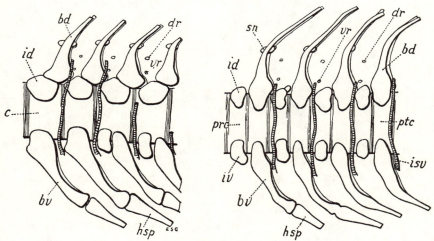

FIG. 38.

Amia calva, about 45 mm. long, cartilaginous stage ; left side view of anterior (on left) and posterior caudal regions of vertebral column. *bv*, basiventral ; *hsp*, infrahaemal spine. Other lettering as in Figs. 37, 39.

arches themselves fuse with their centra, Fig. 35. Throughout the column dorsal and ventral spinal nerve-roots issue through the gap between successive pairs of neural arches which are intersegmental in position and have the transverse septa attached to them, Fig. 39.

At about the 6th or 7th caudal segment there is a remarkable and sudden change, for from this point backward to near the extreme tip

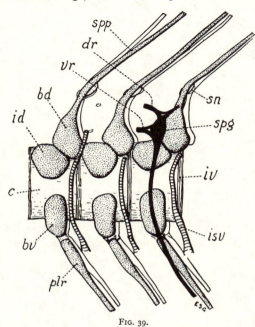

there are two similar biconcave bony centra to each segment of the body, as shown by the myomeres and spinal nerves. The neural and haemal arches, therefore, occur on every alternate centrum.[1] The arch-bearing and the arch-less centrum have been called respectively centrum and intercentrum by Schmidt (80), pleuro- and hypo-centrum by Hay (45), pre- and post-centrum by Gadow and Abbott (27), Figs. 35, 37, 38.

Fig. 39.

Amia calva, about 45 mm. long ; left-side view of 3 anterior trunk vertebrae. *bd*, Basidorsal ; *bv*, basiventral ; *c*, bony centrum ; *dr*, dorsal nerve-root ; *id*, interdorsal ; *isv*, intersegmental artery ; *iv*, intervertebral ligament ; *plr*, left pleural rib ; *sn*, supradorsal ; *spg*, spinal ganglion ; *spp*, paired spinal process of neural arch ; *vr*, ventral nerve-root. Cartilage dotted.

To understand the morphology of these vertebrae it is essential to study their development (Goette, Gadow and Abbott, and especially Hay, and Schauinsland). In the young, before extensive ossification, there are seen in every caudal segment four pairs of cartilages : two basidorsals meeting above the dorsal ligament and two basiventrals meeting below the caudal vein just in front of the intersegmental vessels, and two interdorsals and two interventrals behind these vessels and the spinal nerve-roots, Figs. 38, 39. These cartilages are developed at an earlier stage from four pairs of accumulations of cells derived from the skeletogenous layer. The interventrals disappear as separate elements in the anterior caudal region, becoming fused with the basiventrals from this

---

[1] Many irregularities may occur in the fusion of the elements of the vertebral column. In the tail an arch-bearing centrum may occasionally fuse with the arch-less centrum in front, or that behind, or with both.

point forwards. Passing towards the head the interdorsals become wedged from in front below the basidorsals which succeed them. Ossification begins, according to Hay, as a thin layer on the neural arches, and on the bases of the interdorsals, basiventrals, and interventrals of the tail. Later the separate ossifications on the interdorsals and basi-

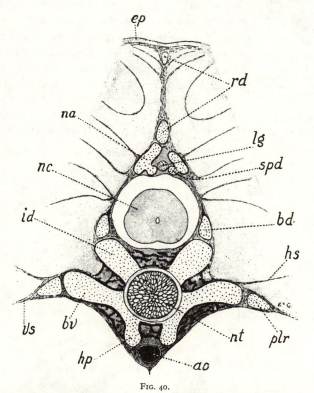

FIG. 40.

Transverse section of *Amia calva*, about 60 mm. long, anterior trunk region. *ao*, Dorsal aorta; *bd*, basidorsal; *bv*, basiventral; *ep*, epidermis; *hp*, haemal process; *hs*, horizontal septum; *id*, interdorsal; *lg*, longitudinal ligament; *na*, dorsal end of neural arch (basidorsal); *nc*, nerve cord; *nt*, notochord surrounded by fibrous sheath and elastica externa; *plr*, pleural rib; *rd*, radial; *spd*, supradorsal; *vs*, wall of splanchnocoele. Cartilage dotted, bone in black.

ventrals of the trunk region spread and join to a complete cylindrical centrum. In that region of the tail where two 'centra' exist in each segment, separate ossifications appear at the base of each of the four paired cartilaginous elements, and also on the projecting neural and haemal arches. The 'post-centrum' is formed by the fusion to a complete ring of the basal ossifications of the basidorsals and basi-ventrals; the arch-less 'pre-centrum' by the fusion to a complete ring of the basal ossifications of the interdorsals and interventrals. Thus the adult bony centra are of compound origin.

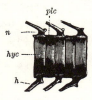

As the bony centra thicken outwards the bases of the cartilaginous elements remain embedded in the bone, and show in transverse section as cartilaginous wedges reaching to near the notochordal sheath, Fig. 40. The basiventrals in the trunk tend to spread upwards, developing a strong outgrowth (parapophysis) to bear the ribs and a smaller ventral outgrowth at the side of the aorta ('aortal support', haemal process). Later these two processes become separated, and the cartilaginous haemal process remains as a ventral projection of the centrum in the adult. From the account given above it is clear that the parts of the vertebral column of the

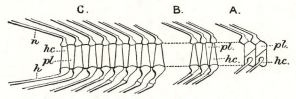

FIG. 42.

Eurycormus speciosus : A, Foremost abdominal vertebrae ; B, posterior abdominal vertebrae ; C, caudal vertebrae. U. Jurassic : Bavaria. (From A. S. Woodward, *Vertebrate Palaeontology,* 1898.) *h,* Haemal arches ; *hc,* 'hypocentrum' ; *n,* neural arches ; *pl,* 'pleurocentra'.

adult *Amia* correspond very imperfectly in number and disposition to the fundamental four paired elements which appear as separate

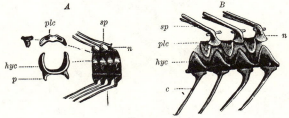

FIG. 43.

Vertebrae of *Euthynotus* (*A*), and *Caturus furcatus,* Ag. (*B*). *c,* Rib ; *hyc,* hypocentrum ; *n,* neural arch ; *p,* parapophysis ; *plc,* pleurocentrum ; *sp,* cleft neural spine. (From K. Zittel, *Palaeontology.*)

cartilages in the young. In the formation of a complete single vertebral segment the basal elements become associated and fused with the interbasal elements lying anteriorly to them. If this association holds good in other Teleostomes (which has not yet been satisfactorily proved), there is a fundamental difference between the vertebra of a Teleostome and that of a Tetrapod, since, in the latter, the basal elements combine with the interbasal elements lying behind them.

Certain Jurassic Amioids have, as in *Amia*, two complete centra in each caudal segment, but farther forward these become reduced to alternate dorsal and ventral crescents which remain distinct even near the skull although combining to form one vertebral segment, Fig. 44 (Zittel, 98 ; Woodward, 663, 665 ; Schmidt, 80 ; Goette, 32 ; Hay, 45). *Caturus* shows perhaps the most primitive condition (Zittel) ; here each vertebral

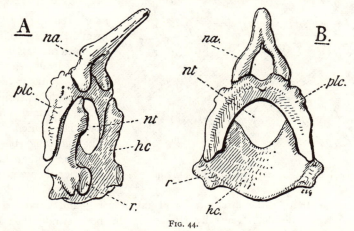

Fig. 44.

Vertebra of *Osteorhachis leedsi*, A. S. W. ; Oxford Clay, Peterborough. A, Oblique side view from behind ; B, front view. *hc,* hypocentrum ; *na,* neural arch ; *nt,* aperture for notochord ; *plc,* pleurocentrum ; *r,* parapophysis for rib. (From Goodrich, *Vert. Craniata,* 1909.)

segment of both trunk and tail shows an anterior dorsal and a posterior ventral crescent embracing the notochordal space, Figs. 43 B, 44. Between the former interdorsal crescents are wedged the neural arches ; while the ventral (basiventral) crescents bear parapophyses in the trunk and haemal arches in the tail. But in the mid-caudal region of *Caturus heterurus* the four typical elements are separately ossified in each segment (E.S.G.).[1] *Eurycormus* has in the tail alternate arch-bearing and arch-less centra round the notochord ; but in the trunk the interdorsal anterior elements remain crescentic, and the neural arches tend to shift on to them, Figs. 41 and 42. The fusion of the interdorsal anterior crescent with the posterior crescent or ring would give rise to the condition seen in the trunk of *Amia*. In *Hypsocormus* only the arches are ossified, Figs. 45, 46.

[1] In the posterior caudal region of this fish the interdorsal and interventral elements are no longer ossified, and each vertebral segment has a dorsal crescent formed by the basidorsals and a ventral crescent by the basiventrals, recalling the structure of the vertebral column in Pycnodontidae.

In conclusion it may be said that the bony vertebral segment of the Amioidei seems to have been laid down on the four original paired elements already recognised in the Selachian; that from the basidorsals are developed the neural arches and from the basiventrals the haemal arches and parapophyses, while the cartilaginous interdorsals and interventrals remain small (the interventral disappearing in the trunk) and become buried in the growing bony centrum; that an anterior bony crescent develops dorsally in relation to the interdorsal and interventral, and a posterior bony crescent develops ventrally related to the basiventral; that in the tail the two crescents form complete rings in the more advanced

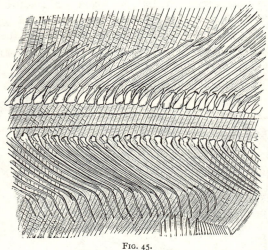

Fig. 45.

Hypsocormus insignis, Wagn. Portion of trunk. Upper Jurassic (Lithographic Stone); Eichstädt, Bavaria. (From K. Zittel, *Palaeontology*.)

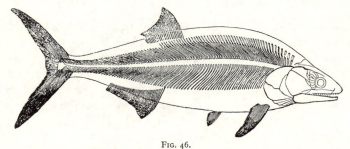

Fig. 46.

Hypsocormus insignis: restoration, scales omitted. U. Jurassic: Bavaria. (From A. S. Woodward, *Vertebrate Palaeontology*, 1898.)

Amioids; and lastly that in the anterior tail and trunk-region of *Amia* they are represented by a single centrum (showing, however, no signs of a compound origin in the adult). In the Aspidorhynchidae, according to Zittel (**98**), annular centra occur bearing neural and haemal arches.

The vertebrae of *Lepidosteus* are unique among fishes in being opisthocoelous, the centrum having a convex anterior face fitting into a concavity of the centrum in front (Gegenbaur, 30; Balfour and Parker, 2; Gadow and Abbott, 27; Schauinsland, 78). The notochord becomes completely constricted intervertebrally. The neural arches are continuous with the centra, and in the trunk separate neural spines rest on the dorsal ligament below which are supradorsal cartilages. The development has been studied by Balfour and Parker, and Gadow and Abbott. The notochordal sheaths are thin and in later stages the notochord is almost obliterated. In the young separate basidorsals and basiventrals appear, the latter meeting below the haemal canal in the caudal region. The slight vertebral constriction of the notochord due to these basalia is soon obliterated by the pronounced intervertebral constriction brought about by the development of a wide cartilaginous ring formed apparently by the coalesced interdorsals and interventrals. As usual in the Holostei, supradorsal cartilages appear at the sides of the dorsal ligament

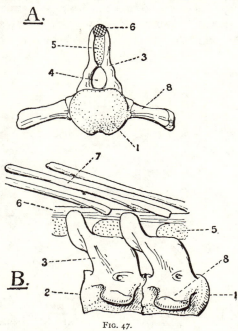

FIG. 47.

Lepidos eus osseus, L. A, Vertebra from in front; B, vertebral column of trunk, right-side view. (After Balfour and Parker.) 1, Convex anterior surface of centrum; 2, concave posterior surface of centrum; 3, neural arch (basidorsal); 4, neural canal; 5, supradorsal cartilage; 6, longitudinal ligament; 7, neural spine; 8, transverse process of centrum (parapophysis). (From Goodrich, *Vert. Craniata*, 1909.)

between the neural arches. The opisthocoelous joint is carved out of the intervertebral ring which becomes divided into anterior and posterior portions; these are ossified in continuity with the outer vertebral cylinder, Fig. 47.

The Teleostei usually have well-developed amphicoelous bony centra strongly constricting the notochord and its sheaths, Fig. 11. These remain thin vertebrally but thicken intervertebrally, contributing to the usual strong intervertebral ligament an inner layer or ligamentum intervertebrale internum of Kölliker (1864). The vertebrae show great variety in

detail within the group, but the centra are always simple and intersegmental in the adult.

Special intervertebral articulations are formed by anterior processes
from the neural or haemal arches which rest on corresponding processes
in front, and these may be supplemented or replaced by dorsal and ventral
processes from the centra themselves, analogous to the pre- and postzygapophyses of Tetrapods (Gegenbaur, 270 ; Goette, 32 ; v. Ebner, 20 ;
Ussow, 90 ; Bruch, 1862 ; Lotz, 57 ; Grassi, 37 ; Scheel, 79 ; Gadow and
Abbott ; Schauinsland). Basidorsal and basiventral cartilages rest on the

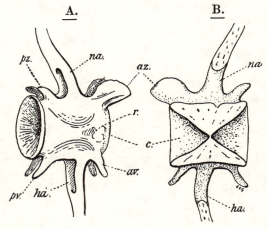

FIG. 48.

Caudal vertebrae of *Thynnus vulgaris* (Cuv. and Val.) A, Right-side view ; B, left-side view of
the same cut in half. *av*, Anterior ventral process ; *az*, anterior " zygapophysis " ; *c*, centrum ;
ha, haemal arch ; *na*, neural arch ; *pv*, posterior ventral process ; *pz*, posterior " zygapophysis " ;
r, place of attachment of rib. (From Goodrich, *Vert. Craniata*, 1909.)

notochordal sheath in the young, Fig. 11 ; but interbasals do not usually
appear at all, though traces of interventrals occasionally have been found
(*Salmo fario*, Schauinsland). Supradorsal cartilages usually appear
between the neural arches as in other Holostei, Figs. 48, 49, 50.

According to Schauinsland the vertebra is really formed by the
coalescence of an anterior half (developed in relation to the basalia) and
a posterior half (developed in relation to the interbasalia) derived from
the sclerotome of the segment behind. He claims that in *Fistularia*
where the nerves issue through the neural arches these two halves develop
similar arches both dorsal and ventral. Though this interpretation of
the bisegmental origin of the Teleostean vertebra is possibly correct,
the absence of clearly developed interbasals in the types hitherto studied
makes it difficult to verify.

The Teleostean bony centrum itself usually shows no trace of compound origin, but develops as a cylinder embracing the arcualia. The bases of the basalia do not, as a rule, join round the notochord to form a complete vertebral ring, and may remain as four conspicuous radiating cartilages in the adult centrum (as in *Esox*, Fig. 49 B). Following the general tendency towards a reduction of cartilage in the Holostei, and the direct development of bony tissue without a preceding cartilaginous

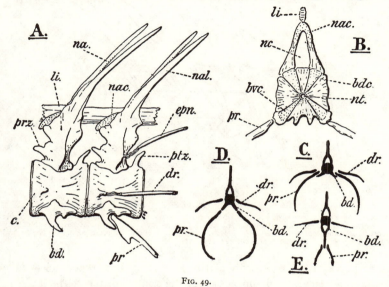

FIG. 49.

A, Left-side view of two trunk vertebrae of *Esox lucius*, L. B, Vertical median section of the same; D, C, and E, *Thynnus vulgaris*, Cuv.; C, anterior trunk region; D, posterior trunk region; E, caudal region. *bd*, Basiventral (haemal arch); *bdc*, basidorsal cartilage; *bvc*, basiventral cartilage; *c*, centrum; *dr*, dorsal rib (epipleural); *epn*, epineural; *li*, longitudinal ligament; *na*, basidorsal (neural arch); *nac*, supradorsal cartilage; *nal*, left neural arch; *nc*, neural canal; *nt*, notochord; *pr*, pleural rib; *prz*, anterior articulating process; *ptz*, posterior articulating process. (From Goodrich, *Vert. Craniata*, 1909.)

stage, the bony centrum appears early as a thin cylinder which thickens rapidly outwards. Later on a thin layer of cartilage may extend from the basalia between the bone and the elastica externa as in *Amia*—a vestige perhaps of an ancestral cartilaginous centrum. In *Rhodeus*, however, the basidorsals join the basiventrals at the sides, forming a cartilaginous ring of considerable thickness interrupted above and below (Scheel, **79**).

Ossification may extend from the centrum over the basalia, the neural and haemal arches being then continuous with the centrum as in most of the higher Teleostei and the tail region of even the lower forms. But in the trunk of some of the Cypriniformes, Esociformes, and Clupeiformes the basiventrals and even the basidorsals (*Esox*) may be separately ossified and remain distinct in the adult—presumably a more primitive but less efficient condition, Fig 49.

Concerning the general morphology of the vertebrae in the Teleostomi it may be said that, although the four fundamental paired elements may

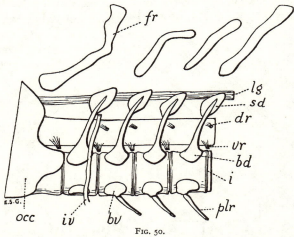

FIG. 50.

Anterior region of vertebral column of *Salmo truta*, 35 mm. long. Left-side view. *bd*, Basidorsal; *bv*, basiventral; *dr*, dorsal nerve-root; *fr*, radial (?) ; *i*, intervertebral ligament between thin bony centra ; *iv*, intersegmental artery ; *lg*, longitudinal ligament ; *occ*, occipital cartilage ; *plr*, pleural rib ; *sd*, supradorsal cartilage ; *vr*, ventral nerve-root.

often be made out as separate cartilages in the young, they tend to take less and less share in the formation of the adult centrum. The latter in the form of an intersegmental biconcave bony cylinder develops chiefly in the skeletogenous layer surrounding the notochordal sheaths ; this is in accordance with the general tendency throughout the endoskeleton of these fishes for the cartilage to be reduced even in development, and for the bone to be formed more and more directly and precociously from skeletogenous tissue.

Reviewing the structure of the vertebrae in Pisces generally, it appears that although the four fundamental paired arcualia may be traced with considerable certainty in all the large groups, yet the part played by the individual elements varies greatly. For instance, in the Selachii the interdorsals take no share in the formation of the centrum, but in the trunk of *Amia* they contribute greatly to its structure while the basidorsals are excluded. Again, in the Teleostei, it is the basidorsal which together with the basiventral is the important element. The interdorsal, extending primarily between the ventral and dorsal roots of a spinal nerve are liable to be interrupted and hindered in development; hence the main element of the neural arch is always the basidorsal. It will be gathered from the above account that a good deal more work is required before the history of the centrum in the various groups can be satisfactorily described.

THE VERTEBRAL COLUMN OF TETRAPODA

The morphology of the vertebral column of the Tetrapoda may be studied from two points of view : with the help of comparative anatomy and palaeontology we may attempt to reconstruct its phylogenetic history in the adult, or we may try to determine the homology of its parts by following their development in existing forms. The interpretation of the embryological facts is, however, difficult, since most modern Tetrapods are specialised and have departed along divergent lines from the primitive structure.

In adaptation to progression by means of walking limbs there is an increasing tendency for the body of the terrestrial vertebrates to become differentiated into well-defined head, neck, trunk, and tail regions, mainly determined by the position of the limbs and their girdles. The slender neck lifts and moves the head, the trunk contains the coelomic cavities and viscera, and the tail becomes narrow and tapering. Corresponding variations of structure occur in the vertebral column which becomes differentiated into regions whose limits are arbitrarily defined for descriptive purposes. The cervical vertebrae are typically very movable and bear small ribs. Between pectoral and pelvic regions are the trunk vertebrae ('dorsal', or thoracico-lumbar vertebrae) with ribs well developed anteriorly where they reach the sternum, but dwindling posteriorly. The pelvic girdle becomes attached to the sacral vertebrae by means of stout, short, sacral ribs. In Mammalia and some of the higher Reptilia a few of the trunk vertebrae in front of the sacral may lose their ribs more or less completely, and are then distinguished as lumbar from the more anterior thoracic vertebrae. In the tail the ribs tend to disappear and the vertebrae become simplified and reduced towards the tip. In no region of the column is there any trace of separate supraspinal elements above the neural arches, or of infraspinal elements below the haemal arches, such as occur in Pisces.

In general build a typical trunk vertebra consists of a body or centrum and a neural arch, with which are associated a pair of ribs (p. 75). The two halves of the arch join above the longitudinal ligament to a median dorsal spine, and bear paired pre- and postzygapophyses with which they articulate with neighbouring arches,[1] and paired lateral transverse processes (diapophyses). The capitulum of the rib articulates at the anterior intercentral region, and the tuberculum at the end of the diapophysis. The arch and ribs are essentially intersegmental; the fibrous intervertebral discs uniting consecutive centra are segmental. The spinal nerves issue behind the neural arches.[2] Muscular segments

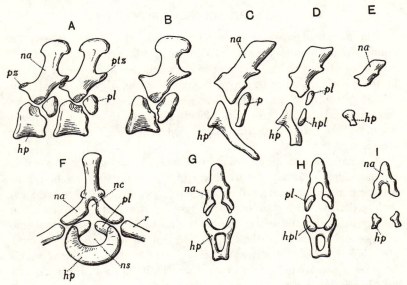

FIG. 51.

Diagram of rhachitomous vertebrae of *Archegosaurus* (chiefly from Jaekel, 1896). Left-side view of anterior thoracic A, posterior trunk B, anterior caudal C, and more posterior caudal vertebrae D and E. F anterior view of A. G anterior and H posterior view of D. I anterior view of E. *hp*, Hypocentrum (basiventral); *hpl*, hypocentrum pleurale (interventral); *na*, neural arch (basidorsal); *nc*, neural canal; *ns*, space for notochord; *pl*, pleurocentrum (interdorsal); *ptz*, postzygapophysis; *pz*, prezygapophysis; *r*, rib.

and vertebrae alternate, and as always in vertebrates one muscular segment is connected to two vertebral segments.

Amphibia.—Important evidence regarding the morphology of the

[1] Additional articular processes have been developed in various groups with long flexible vertebral column. A projecting zygosphene above the prezygapophysis fitting into a zygantrum above the postzygapophysis occurs not only in all Ophidia, but also in certain Lacertilia (Iguanidae, Mosasauridae), and in some Stegocephalia (*Urocordylus, Diplocaulus*, etc.). Similar hyposphenes and hypantra occur below the zygapophyses in the Cotylosaurian *Diadectes*. The more specialised Pterodactyls develop accessory articulating processes below the centrum.

[2] Although the spinal nerves issue between the cartilaginous neural arches, in the adult they may pierce the arches in many cases owing to secondary overgrowth of bone.

Tetrapod vertebra has been obtained from the fossil Stegocephalia (Owen; v. Meyer; Cope, 12, 13; Fritsch, 23; Gaudry, 28; Zittel, 98; Jaekel, 51; Schwarz, 83; Williston, 95; Watson, 631, 643, 644, and others). The vertebrae of the earlier and more primitive forms are found to consist of several elements contributing to the formation of amphicoelous centra constricting the notochordal space. Such vertebrae have been called temnospondy-lous by Cope, who divides them into rhachitomous and embolomerous types. *Archegosaurus* affords a good example of the rhachitomous structure, Fig. 51. Each typical trunk vertebra consists of a neural arch of two halves fused above to a median neural spine, a large crescentic anterior wedge partially surrounding the space for the notochord from below (hypocentrum of Gaudry, intercentrum of Cope), and a pair of posterior dorsal elements (pleurocentra of Gaudry) which together par-

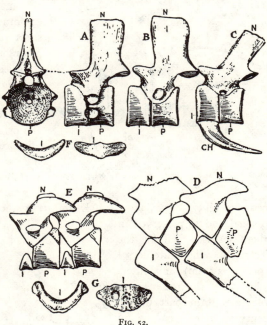

FIG. 52.

Vertebrae of *Cricotus*, A, anterior and left-side views of dorsal (thoracic); B and C, left side views of anterior and median caudals. D, Left-side view of caudals of *Eryops*. E, Left-side view of thoracics of *Conodectes*. F, Dorsal intercentrum (hypocentrum) of *Dimetrodon*, and G of *Trimerorhachis* from in front and below. (From Williston, *Osteology of Reptiles*, 1925.) N, Neural arch; I, hypocentrum; P, pleurocentrum; CH, chevron.

tially surround the notochordal space from above. The intercentrum and pleurocentra, together with the bases of the arches wedged in between them, make up the amphicoelous body of the vertebra. Passing forwards the hypocentrum is seen to increase somewhat in size. Passing backwards to the tail region the hypocentrum is seen to diminish in size and to become incompletely divided into two wedges which send down long processes meeting below, enclosing the haemal canal and fusing to a median haemal spine. Still farther back the neural arch and the hypocentrum are represented by small paired elements. The pleuro-

centra are elongated in the pelvic region almost surrounding the notochord, and in the tail become divided into paired dorsal pleurocentra

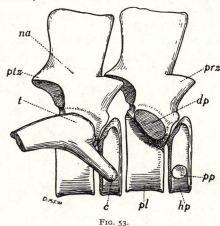

FIG. 53.

Trunk vertebrae of *Eogyrinus*, right-side view, drawn
by D. M. S. Watson. *hp*, Hypocentrum; *prz*, prezygapophysis.

and paired ventral hypocentra
pleuralia (Fritsch). Both these
disappear towards the end
of the tail. Most of the rhachitomous vertebral columns
conform to this structure ; but
separately ossified hypocentra
pleuralia have only been
recorded in *Archegosaurus*,
Chelydosaurus, and *Spheno-
saurus*.

It would appear, then, that
the original four paired arcualia already described as entering into the composition of the
vertebra of most fish can be
identified in that of the Stego-
cephalia (Hay, 45; Gadow, 26)[1], and that they may be separately

ossified and so preserved in
these fossils. But in the
Tetrapod the interdorsal
(pleurocentrum) and interventral (hypocentrum
pleurale) form part of the
centrum only, the neural
arch being formed from the
basidorsals alone above and
the haemal arch from the
basiventrals alone below.
There are no separate arches
developed from the inter-
basals; see, however, p. 51.

In the embolomerous type
each vertebra consists of the
neural arch and two complete
amphicoelous discs surround-

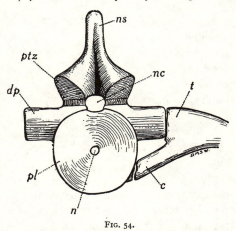

FIG. 54.

Trunk vertebra of *Eogyrinus*, posterior view, drawn by
D. M. S. Watson. *c*, Capitulum ; *dp*, transverse process
of neural arch ; *nc*, neural canal ; *n*, canal for notochord;
ns, neural spine ; *pl*, pleurocentrum; *ptz*, postzygapophysis;
t, tuberculum of rib.

ing the notochordal space, Figs. 52 A, 53, 54. The hypocentrum has

[1] The interpretation of the constitution of the vertebral column of Tetrapods adopted in this volume differs from that of Gadow.

grown round dorsally to form the anterior disc, and the pleurocentra have fused and grown round ventrally (or combined with the hypocentra pleuralia) to form the posterior disc. In the anterior segments the hypocentral disc is somewhat wedge-shaped, and in the tail it is prolonged ventrally to form the haemal arch. Since the embolomerous vertebra occurs only in primitive Carboniferous and Lower Permian forms (*Cricotus*, Cope ; *Diplovertebron*, Fritsch ; *Orthosaurus*, *Eogyrinus*, Watson), Williston and Watson believe it to be the most primitive type. Its exact phylogenetic relationship to the rhachitomous type is, however, still obscure ; and on general grounds we should expect to find the four paired elements separately and more equally developed in the ancestral vertebral column.

The later and more specialised large Stegocephalia such as *Mastodonsaurus* have vertebrae belonging to the so-called stereospondylous type. Here the neural arch rests on an amphicoelous centrum notched or pierced for the notochord and composed of a single bone. Apparently the stereospondylous has been derived from the rhachitomous type by the enlargement of the hypocentrum to form the whole body, while the other elements, if present, remain small and unossified.

The Branchiosauria, a specialised group of small Stegocephalia with a somewhat degenerate and weakly ossified endoskeleton, have so-called

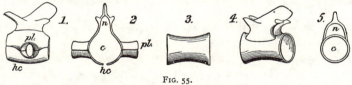

FIG. 55.

Diagrams of Branchiosaurian (1, 2) and Microsaurian (3, 4, 5) vertebrae. 1, Vertebra of *Branchiosaurus*, lateral aspect. 2, The same, end view. 3, Vertebra of *Hyloplesion*, inferior aspect. 4, The same, lateral aspect. 5, The same, end view (after A. S. Woodward, *Vertebrate Palaeontology*, 1898). *c*, Notochordal canal ; *hc*, hypocentrum ; *n*, neural canal ; *pl*, transverse process.

phyllospondylous vertebrae, Fig. 55. The centrum is preserved in the form of a thin bony cylinder composed of four pieces belonging to the neural arches above and hypocentra below. These basidorsal and basiventral ossifications meet half-way, sharing in the formation of a large ' transverse process ', and enclosing a wide space in which doubtless lay cartilage and notochord. Another and highly specialised group of extinct Amphibia, the snake-like Aistopoda or Dolichosomatidae (Huxley, 1871 ; Fritsch, **23** ; Schwarz, **83**) have numerous amphicoelous vertebrae ossified in one piece and closely resembling those of modern Apoda, to which this family is probably related, Fig. 323*a*. The more normal Urocordylidae (Huxley ; Fritsch) have vertebrae also ossified in one piece and very similar to those of modern Urodela.

E

Coming now to the modern Amphibia we find that their bony vertebrae, whether of the trunk or the tail, are all of one piece, the centrum being continuous with the neural arch and in the caudal region with the haemal arch as well. There are never any separate 'chevron bones', an important fact distinguishing them from the Amniota. Thus there are in the adult no separate bony elements corresponding to those seen in primitive Stegocephalia, nor is it easy to recognise the elements in the

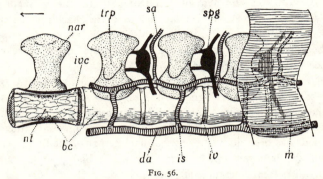

FIG. 56.

Larva of *Amblystoma*. Diagram of reconstructed left-side view of four vertebrae (3-6), showing relation to nerves (black), arteries (cross-lined), and myomere, *m*. First vertebra cut to remove left half. *bc*, Bony cylinder of centrum; *da*, median dorsal aorta; *is*, intersegmental artery to longitudinal anastomosis; *iv*, intervertebral disc; *ivc*, intervertebral cartilage; *nar*, neural arch; *nt*, notochord with notochordal 'cartilage'; *sa*, dorsal segmental artery; *spg*, spinal ganglion; *trp*, transverse process.

embryo since the cartilage is much reduced and ossification sets in very early somewhat as in the higher Teleostomes.

The development has been studied by many authors, including Gegenbaur (29), Goette (32-3), Field (21), Gadow (26), and Schauinsland (78). In the Urodela, as in other Craniates, the notochord secretes an elastica externa and an inner fibrous sheath; these remain thin, are not invaded (p. 9), and are of little structural importance in the adult. The cartilaginous elements develop in the continuous mesoblastic skeletogenous layer derived from the sclerotomes which soon become confluent. This layer, surrounding neural canal, notochord, and caudal haemal canal, passes laterally into the transverse intersegmental septa, and here (in the intersegmental or vertebral zones) appear a pair of cartilaginous basidorsals above and a pair of basiventrals below. The former join to a neural arch dorsally,[1] and the latter to a haemal arch ventrally in the tail. The base of the basidorsal does not join that of the basiventral at the side; but soon these arcualia become embedded in

[1] The spinal nerve-roots issue behind these arches; but in the adult, owing to their enclosure by backwardly growing bone, they may pass through the neural arches.

the bony cylindrical centrum which develops round the notochord and extends over the arches in each vertebra ; it is wider at each open end than in the middle, where it constricts the notochord. This cylinder first appears as a thin layer of cell-less bone and was mistaken by Hasse (39) for a chordal sheath ('cuticula sceleti'). Unossified connective tissue unites consecutive centra in the intervertebral (segmental) zones ; and here the skeletogenous tissue thickens to form intervertebral cartilaginous rings between the notochordal sheath and the expanded ends of the bony cylinders, Figs. 56, 57. The intervertebral cartilage spreads and thickens, usually constricting the notochord intervertebrally. The modern Urodela

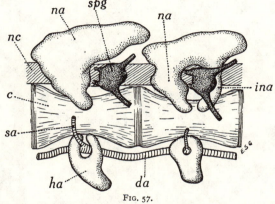

Fig. 57.

Reconstruction of two caudal vertebrae of the larva of *Amblystoma tigrinum. c*, Bony centrum ; *da*, dorsal aorta ; *ha*, cartilaginous haemal arch ; *ina*, cartilaginous interneural, and *na*, neural arch ; *nc*, nerve-cord ; *sa*, segmental artery ; *spg*, spinal ganglion.

have become more or less readapted to an aquatic life, and their vertebral column tends to remain correspondingly undeveloped. Thus in such forms as *Proteus, Ranidens,* and *Necturus,* the intervertebral cartilaginous ring is little developed, remains undivided, and the notochord is continuous ; but in more terrestrial Salamandridae the ring enlarges, strongly constricts or even obliterates the notochord, and becomes divided transversely so as to form an opisthocoelous joint between consecutive vertebrae, Fig. 58 (Wiedersheim, 311 ; Gadow, 26).[1]

Gadow's description of the development of the intervertebral ring from four primary rudiments (interdorsals and interventrals) has not

[1] Schauinsland's claim to have found vestigial interdorsal arches in the caudal region of *Amblystoma* needs confirmation (small vestiges have been described by Marcus and Blume in *Hypogeophis*, 62). Neither interdorsal nor interventral arches persist separately in any known Tetrapod, where these elements if represented at all form part of the centrum and its arches.

been confirmed ; nevertheless his comparison of the ring with the inter-basals and the cartilages of the neural and haemal arches with the basals is probably well founded.

The Apoda possess well-ossified amphicoelous vertebrae which differ

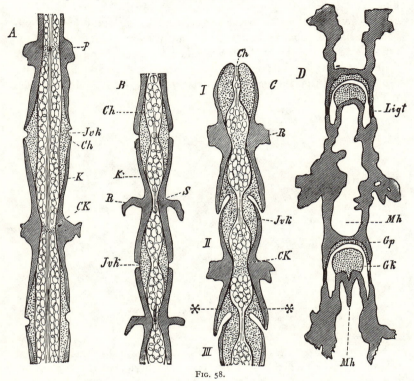

Fig. 58.

Longitudinal section through the vertebral centra of various Urodeles. A, *Ranodon sibericus* ; B, *Amblystoma tigrinum* ; C, *Gyrinophilus porphyriticus* (*I, II, III*, the three anterior vertebrae) ; D, *Salamandrina perspicillata*. *Ch*, Notochord ; *CK*, intravertebral cartilage and fat-cells ; *Gp*, concave posterior face, and *Gk*, convex anterior face of centrum with articular socket and head ; *Jvk*, invertebral cartilage ; *K*, superficial bone of centrum ; *Ligt*, intervertebral ligament ; *Mh*, marrow cavity ; *R*, transverse process ; *S*, intravertebral constriction of notochord in *Amblystoma*, without cartilage and fat-cells ; **, intervertebral cartilage. (From Wiedersheim, *Comp. Anatomy.*)

little in development from those of Urodela, except that the intervertebral cartilage is almost if not quite lost (Peter, 68 ; Marcus and Blume, 62).

Of all vertebrates the Anura have the shortest vertebral column ; it consists of not more than nine free vertebrae and a rigid bony rod, the urostyle or os coccygeum, representing the postsacral vertebrae. The well-ossified free vertebrae have a low neural spine and usually large 'transverse processes' (much dilated on the sacral vertebra of Disco-glossidae and Pelobatidae, and other Arcifera). Their articular surfaces

are either opisthocoelous as in Cystignathidae and Discoglossidae, or
procoelous as in Bufonidae, Ranidae, etc., while the last free vertebra
may have two convexities articulating with the urostyle. The development of the vertebral column is very specialised, and varies from the
'perichordal' type seen in *Rana* and *Bufo* to
the 'epichordal' type of *Pelobates Bombinator*
and *Pipa* (Gegenbaur, **29**; Goette, **32**; Gadow,
26; Ridewood, **74**; Schauinsland, **78**). In the
former, cartilage first appears as a series of
intersegmental basidorsals which fuse at their
bases to two continuous longitudinal rods
above the notochord. Ingrowths of the cartilage now take place intervertebrally, probably
representing interdorsals, and constricting the
notochord. Later a continuous median rod of
cartilage, sometimes indistinctly paired, appears below the notochord, and eventually becomes subdivided into vertebral pieces; it is
taken to represent the basi- and interventrals.
Ossification spreads round the notochord and
arches at each vertebra connecting the dorsal
with the ventral cartilages which become included in the finished vertebra. The intervertebral cartilage surrounds and completely
constricts the notochord, a remnant of which
may sometimes remain vertebrally as in
Rana. This intervertebral cartilage becomes
ossified, and according as it attaches itself
chiefly to the centrum in front or behind
gives rise to a pro- or opisthocoelous joint.
In the 'epichordal' type the ventral cartilaginous elements are still further reduced,
and the notochord in extreme cases is
only enclosed below by the overgrowth of the bony cylinder of the
centrum.

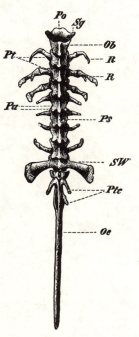

Fig. 59.

Vertebral column of *Discoglossus
pictus*. *Ob*, Upper arch of first vertebra; *Pa*, articular processes; *Po*,
anterior process of first vertebra;
Ps, neural spine; *Pt*, transverse
processes of trunk vertebrae; *Ptc*,
transverse processes of caudal vertebrae (urostyle, *Oc*); *R*, ribs; *Sg*,
condylar facets of first vertebra;
SW, sacral vertebra. (From Wiedersheim, *Comp. Anatomy*.)

In the swimming tail of the tadpole larva no vertebrae are developed,
but only the notochord with its sheaths and an enveloping skeletogenous
layer. The whole structure degenerates and is absorbed at metamorphosis.

That the urostyle is formed of fused post-sacral vertebrae is indicated
in the adult by nerve apertures and sometimes, as in *Discoglossus*, by

'transverse processes' and vestiges of neural arches anteriorly, Fig. 59. It develops from two dorsal longitudinal cartilaginous rods, which later join, and a similar ventral rod ; these eventually surround and obliterate the notochord, and enclose in front a narrow vestigial dorsal spinal canal. Traces of segmentation indicate that the urostyle is formed by the fusion of some twelve vertebrae (Gadow, 26).

Amniota.—In spite of various modifications in details in adaptation to different modes of life which need not be described here, the vertebrae of Reptiles, Birds, and Mammals conform in essentials to the fundamental plan of structure described above (p. 45). It is characteristic of the Amniote vertebra that in addition to the neural arch and body or centrum, which ossify separately, there is an anterior ventral element, the hypocentrum or intercentrum as it is often called, Fig. 61. In the caudal region it gives rise to the 'chevron' bones enclosing the haemal canal. The hypocentrum is particularly large in the very primitive reptile *Conodectes* (*Seymouria*), where the vertebra approaches the embolomerous type, but it remains unossified dorsally, never completely surrounding the notochord (Williston, 95 ; Watson, 643).

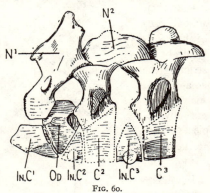

FIG. 60.

Seymouria (*Conodectes*) *bayloriensis*, left-side view of first three vertebrae (from D. M. S. Watson, *P.Z.S.*, 1918). C^2, C^3, centra (pleurocentra) of 2nd and 3rd vertebrae ; $In.C^1$, ^{2}, ^{3}, intercentra (hypocentra) of three vertebrae ; Od, odontoid process = pleurocentrum of atlas ; N^1 and N^2, neural arches of atlas and axis.

Three important points remain to be discussed bearing on the general morphology of the vertebral column of Tetrapods. They are : (1) the structure of the atlas and axis and significance of its parts ; (2) the possible identification of the primitive four paired elements seen in Pisces, and the share these take in the building up of the vertebra ; (3) the relation of these elements to the embryonic sclerotomes and body segments.

Taking the last point first, we find, owing to the fact that in Amniotes the sclerotomes are more definite in structure and retain their segmentation longer than in most other Craniates, their relation to the developing vertebrae can be more easily traced, Fig. 62.

The sclerotome is large and arises ventrally from the inner wall of the primitive hollow somite (whose outer wall forms the cutis layer), while only a portion of the inner wall thickens to form the myotome. The

sclerotomes enlarge, forming blocks closely packed together and reaching
to the notochord. Each is developed from the somite as a saccular outgrowth, whose thickened anterior and posterior walls together with
immigrating cells reduce the cavity or sclerocoel to a narrow space
dividing the block into an anterior and a posterior half, except for the
thin original inner wall applied to the notochord (Corning, 15, Männer, 60,

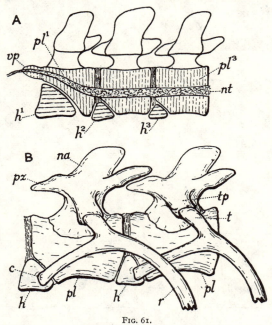

FIG. 61.

A, Diagrammatic longitudinal section of atlas, axis, and third cervical vertebra of a primitive
Tetrapod to show composition of centra and of odontoid process. *h*, Hypocentrum; *nt*, notochord;
pl, pleurocentrum; *vp*, vestigial pleurocentrum at end of odontoid process, and belonging to last
occipital segment. B, Diagrammatic left-side view of two thoracic vertebrae of a primitive *Tetrapod*,
showing relation of rib to vertebra. *c*, Capitulum; *na*, neural arch; *pz*, prezygapophysis; *t*,
tuberculum; *tp*, transverse process.

Schauinsland, 78, Higgins, 48, in Reptiles; O. Schultze, 82, in Birds; O.
Schultze, 82, Bardeen, 3, in Mammals). The narrow space just mentioned
may be called the sclerocoelic cleft; it is the 'Intervertebralspalte' of
v. Ebner, who first pointed out its significance as marking the position
of the future intervertebral joint. In Reptiles it is clearly developed,
is at first in communication with the myocoel, and may persist for a
considerable time. But, according to O. Schultze, in Birds it develops
later and only secondarily opens into the myocoel, while in Mammals,
though the division of the sclerotome is visible, the cavity is only virtual.
In Amniotes generally it is of course cut off from the myocoel when the

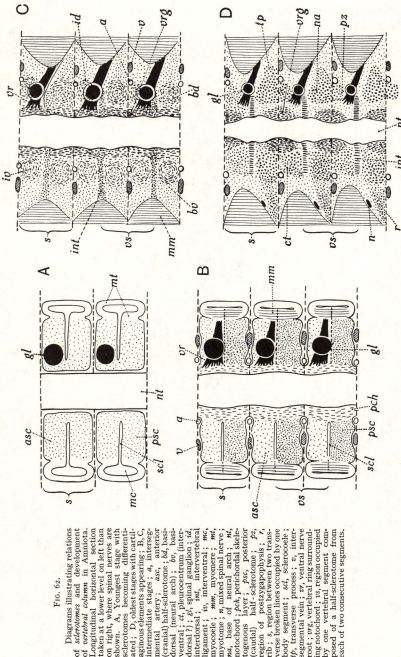

FIG. 62.

Diagrams illustrating relations of *sclerotomes* and development of *vertebral column* in Amniota. Longitudinal horizontal section taken at a lower level on left than on right, where spinal nerves are shown. A, Youngest stage with sclerotomes becoming differentiated; B, C, intermediate stages; D, oldest stages with cartilaginous elements appearing; *a*, intersegmental artery; *asc*, anterior (cranial) half-sclerotome; *bd*, basidorsal (neural arch); *bv*, basiventral; *ct*, pleurocentrum (interdorsal?); *gl*, spinal ganglion; *id*, interdorsal; *iv*, interventral; *mc*, myocoele; *mm*, myomere; *mt*, myotome; *n*, mixed spinal nerve; *na*, base of neural arch; *nt*, notochord; *pch*, perichordal skeletogenous layer; *psc*, posterior (caudal) half-sclerotome; *pz*, region of postzygapophysis; *r*, rib; *s*, region between two transverse broken lines occupied by one body segment; *scl*, sclerocoele; *tp*, transverse process; *v*, interventral vein; *vr*, ventral nerve root; *vrg*, vertebral ring surrounding notochord; *vs*, region occupied by one vertebral segment composed of a half-sclerotome from each of two consecutive segments.

myotome is developed, and ultimately disappears. The posterior half of the sclerotome usually becomes denser than the anterior half, which is to a considerable extent invaded dorsally by the developing spinal ganglion. Thus, owing to the sclerotomes retaining their individuality for a considerable time, stained longitudinal sections of embryos show alternating light and dark half-sclerotomes. During subsequent development

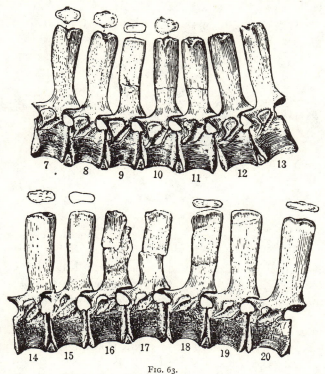

Fig. 63.

Ophiacodon mirus; left-side view of seventh to twentieth vertebrae showing pleurocentra and hypocentra. (From Williston, *Osteology of Reptiles*, 1925.)

each light anterior half-sclerotome of segment B fuses with the dark posterior half-sclerotome of segment A in front of it to give rise to a complete vertebral segment. The vertebrae then come to alternate with the original segmental somites. The sclerocoelic clefts are intervertebral, the intersegmental vessels vertebral in position.

Meanwhile the adjacent sclerotomes early fuse at their inner ends next to the notochord, where a continuous skeletogenous perichordal layer is formed which spreads over the neural canal above and the caudal haemal canal below. The myotome grows inwards as a wedge between the two halves of a sclerotome, and the denser half-sclerotome grows outwards in the intermuscular septum. Cartilages then appear : the neural arch develops in the dorsal region and the hypocentrum (intercentrum, or haemal arch in the tail) in the ventral region of the posterior half-sclerotome of segment A, while the centrum (pleurocentrum) develops in the anterior half-sclerotome of segment B. These three cartilaginous elements may ossify separately, but may later become fused. In the completed vertebra, then, the neural arch and hypocentrum belong to

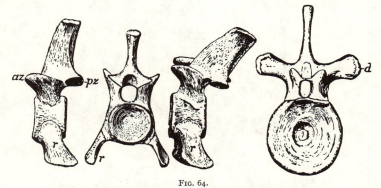

FIG. 64.

Plesiosaur vertebrae : *Polycotylus.* Left-side and posterior views of cervical and thoracic vertebrae. *az*, Prezygapophysis ; *pz*, postzygapophysis ; *d*, articulation for tuberculum on transverse process ; *r*, rib. (From Williston, *Osteology of Reptiles*, 1925.)

one body segment and the centrum to the segment behind it.[1] Such is the general course of development of a vertebra in an Amniote.

We may now attack point (2), and examine more in detail the morphology of the vertebral column in various groups.

Reptilia.—In the Reptilia the neurocentral suture may persist as in many Crocodilia and Chelonia, and especially in aquatic forms such as the Ichthyosauria ; but it may disappear early as in Lacertilia and Ophidia. A hypocentrum is always present in the atlas (p. 66) and in primitive forms on all or nearly all vertebrae throughout the column, as in *Sphenodon* and Geckones among living forms. But usually they disappear as separate bones more or less completely in the more specialised and modern Reptiles, except in the tail, where they persist as originally paired elements prolonged ventrally to form haemal arches and spines (chevron bones). Rarely they are paired in the cervical region (Chelonia, Plesiosauria, *Procolophon*). Although the hypocentrum belongs to the centrum behind it, it is usually wedged in between two consecutive centra in the region of the intervertebral connective tissue disc. Secondarily it may

[1] See, however, p. 61.

fuse either with its own centrum or that in front, and even be carried at the end of the hypapophysis of the anterior centrum (Osborn, 65).

The early primitive Reptiles, such as the Cotylosauria, possess amphicoelous centra (pleurocentra) pierced by the constricted but continuous notochord. Such 'notochordal' amphicoelous centra are present in *Sphenodon* and the Geckones among modern forms, Fig. 65.

The Geckones (Gadow, 27) seem to combine primitive with degenerate characters : hypocentra are present, the centra are amphicoelous cylinders constricting the notochord vertebrally ; but the cartilage is little developed, forms an almost continuous tube, and there are no distinct intervertebral joints.

FIG. 65.

Vertebra of *Sphenodon*, showing the amphicoelous centrum (*C*). (After Headley.)

But from the middle Permian onwards the Reptilian hollow centra tend to be replaced by solid centra with shallow excavations (platycoelous) or flat faces (amphiplatyan). The typical reptilian vertebra is procoelous, with a concavity in front and a convexity behind, Figs. 66, 67. Such centra appeared about the middle of the Jurassic epoch and occur in Lacertilia, Ophidia, Crocodilia, and Pterosauria. Rarely opisthocoelous centra are found, with an anterior convexity and posterior concavity, as

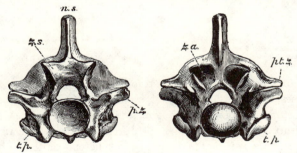

FIG. 66.

Vertebra of *Python*, anterior and posterior views. *n.s*, Neural spine ; *p.z*, prezygapophysis ; *pt.z*, postzygapophysis ; *t.p*, transverse processes ; *z.a*, zygantrum ; *z.s*, zygosphene. (After Huxley.) (From Parker and Haswell, *Zoology*.)

in the cervical region of Dinosaurs. The first caudal centrum is biconvex in modern Crocodilia. More or less amphicoelous vertebrae persist among modern Chelonia, but in this order the vertebral articulations become very specialised and all types occur. In the trunk their neural arches are intercentral in position, resting partly on their own and partly on the centrum in front. In the Edaphosauria the neural arch may be immensely developed, apparently to support a membranous fold of skin, Fig. 68.

The early development of the reptilian vertebra and its relation to the sclerotomes has been studied by Corning and Brünauer in Snakes, by Männer in Snakes and Lizards, by Higgins in the Alligator, and by Schauinsland in *Sphenodon* and other forms. Although their observations do not agree in every detail, and some points are not yet quite established, their chief results may be summarised as follows. The neural arch rudiment arises in the denser posterior (caudal) half of the original sclerotome (see p. 57), chondrifies separately, and grows out laterally to form the transverse process. The right and left cartilages meet dorsally, while their basal regions expand and clasp the developing centrum. The neurocentral division may be lost in the cartilage, but is plain, at all events for a period, as a neurocentral suture when ossification has set in. The

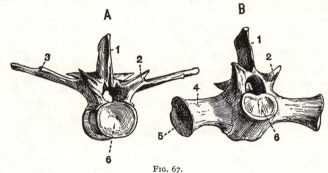

FIG. 67.

Anterior view of A, a late thoracic, and B, first sacral vertebra of a young *Crocodilus palustris* (from S. H. Reynolds, *Vertebrate Skeleton*, 1913). 1, Neural spine; 2, prezygapophysis; 3, facet for articulation with capitulum of rib; 4, sacral rib; 5, surface united with ilium; 6, concave anterior face of centrum.

exact history of the hypocentral structures, their relation to the half-sclerotomes, is less easily made out, owing perhaps to their taking up an intercentral position. They form a 'hypochordal bridge' from the united ventral regions of the same half-sclerotomes as give rise to the neural arches, are paired at first, but as a rule become united by bone into a single wedge below the intervertebral membranous joint. In the tail the originally paired cartilages unite below the haemal canal, and sometimes are joined by bone above it. The centrum itself appears as a chondrification in the now thickened perichordal layer, apparently developed from the anterior cranial half-sclerotomes of the segment behind (p. 57). Generally it appears ventrally at first (Brünauer, 9), but soon grows round the notochord to a complete ring slightly posterior to the neural arch. These rings spread along the perichordal layer, and slightly constrict the notochord vertebrally. The regions of the perichordal layer between the cartilaginous centra form the intervertebral ligaments.

In Reptiles with procoelous vertebrae the central cartilage develops greatly posteriorly, and eventually completely obliterates the notochord intervertebrally. The bulk of this posterior region of the centrum forms

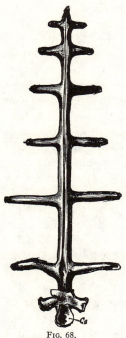

FIG. 68.

Naosaurus claviger: restoration of dorsal vertebra, anterior aspect. Permian, Texas. (After A. S. Woodward, *Vertebrate Palaeontology*, 1898.) *ce*, Centrum.

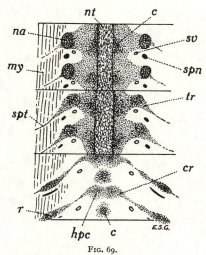

FIG. 69.

Longitudinal horizontal (frontal) sections of the mid-trunk region of an embryo *Lacerta* sp., taken at three successive levels, the upper being most dorsal and the lower most ventral. They show the cartilaginous elements beginning to appear. *c*, Body of vertebra (pleurocentrum); *cr*, capitulum of rib; *hpc*, intercentral vestige (hypocentrum); *my*, myomere; *na*, base of neural arch; *nt*, notochord with its sheaths; *r*, rib; *spn*, spinal nerve; *spt*, intersegmental septum; *sv*, segmental blood-vessel; *tr*, transverse process.

the convexity of the adult vertebra. A vestige of the notochord may sometimes remain near the middle of the centrum.

An important question remains to be considered concerning the origin of the neural arches—namely, whether only one or both half-sclerotomes share in their formation, or in other words whether in the Amniote there is any trace of an interneural arch (interdorsal arch). Although the bulk of the material forming the neural arch in Reptiles is undoubtedly derived, as described above, from the posterior caudal half-sclerotomes of segment A, there is reason to believe that the anterior cranial half-sclerotomes of segment B (see p. 58) also contribute tissue to build up the posterior dorsal region of the arch with its postzygapophyses (Goette, Schauinsland in *Sphenodon*, Higgins in *Alligator*, Piiper in Birds).

The structure and development of the caudal vertebrae in Rhynchocephalia and the Lacertilia throws some light on this question. It is
well known that in *Sphenodon* and most modern Lacertilia the tail can be
easily broken off. The break takes place at definite planes of autotomy
corresponding to the myosepta and penetrating the vertebral column
itself (Hyrtl, '52 ; Gegenbaur, '62 ; H. Müller, '64 ; Goette, 34 ; Gadow,
26 ; detailed description by Woodland, 97). Throughout a considerable
region of the tail each vertebra is transversely divided by a zone of soft
tissue passing approximately through the middle of the centrum and neural
arch and even the transverse process (*Anguis*). Now this plane corresponds
to the original division between the two half-sclerotomes which built
up the vertebra and have not become as firmly coalesced as in the trunk.

As Männer (60) and Schauinsland (78) have shown, not only does the
main neural arch develop in the anterior half, but a second weaker arch
in the posterior half (original cranial half of a sclerotome) ; these two
arches combine to form the foundation of the bony neural arch of the
caudal vertebrae, the plane of autotomy being between them. Thus in
the tail of these Reptiles there would seem to be present interneural
(interdorsal) arches comparable to those of fishes (Schauinsland), and of
which traces are also occasionally found in Amphibia (p. 51). But this
structure of the caudal vertebrae is quite peculiar to *Sphenodon* and
certain Lacertilia, and can hardly be truly primitive since it certainly does
not occur in early unspecialised Reptiles and Amphibia. Rather would
it seem to be a comparatively recent specialisation in the course of which
possibly the interdorsal arch has been reinstated. Nevertheless it is
possible that the portion of the neural arch bearing the posterior zygapophyses is derived in the Amniota from the half-sclerotome behind that
which produces the main part of the arch (see below, in Birds). Further
research is necessary to decide this difficult point.

Aves.—The avian vertebral column has been greatly modified in adaptation to flight. The thoracic vertebrae show a strong tendency to become
rigidly ankylosed in certain regions ; the last thoracic, lumbar, and caudal
vertebrae become assimilated to the sacral vertebrae between the expanded ilia. Most remarkable is the shortening of the tail. The Jurassic
Saururae (*Archaeopteryx* and *Archaeornis*) still have an elongated tail
with free vertebrae (only the anterior four or five have transverse processes) ; but in the Ornithurae, including all modern birds, it is reduced
to a vestige with but a few free vertebrae, the remainder being fused into
a characteristic pygostyle (urostyle) (W. Marshall, 63). In the embryo
this bone is seen to correspond to a varying number of vertebrae at first
laid down as separate cartilages.

Archaeopteryx and *Archaeornis* have retained the amphicoelous centra of their reptilian ancestor. Similar articulations occur only in Ichthyornis, the toothed Carinate of the Cretaceous (Marsh), Fig. 70. Opisthocoelous vertebrae occur in some Carinates such as Penguins, Auks, Parrots, Darters, and Cormorants; and procoelous vertebrae more rarely in the tail (Parker). But the characteristic and usual articulation is saddle-shaped (heterocoelous), Figs. 71, 72.

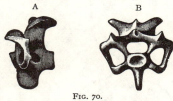

Separate hypocentra (intercentra) are found only in the tail of birds, where they may be paired and produce chevron bones '.

FIG. 70.

Ichthyornis dispar, Marsh. Lateral (A) and anterior (B) aspect of cervical vertebra, ⅞. (After Marsh.) (From K. Zittel, *Palaeontology*.)

The development of the avian vertebrae has been well described by Froriep (24), and more recently by T. J. Parker (66), Schauinsland (78), Lillie (845), Sonies (385), and Piiper (70). The posterior half-sclerotomes, distinguished in early stages by their denser structure, give rise to skeletogenous tissue in which appear a prochondral neural arch and a ventral subchordal crescent ('hypochordalen Spange' of Froriep). A hypocentral cartilage develops in the crescent, and a neural arch cartilage dorsally.

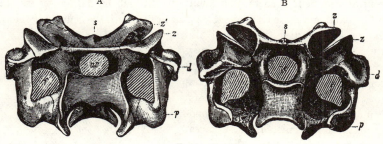

FIG. 71.

Hesperornis regalis, Marsh. Upper Cretaceous: Kansas. Anterior (A) and posterior (B) aspect of thirteenth cervical vertebra, ¼. *d*, Transverse process; *f*, costal canal for vertebral artery; *nc*, neural canal; *p*, parapophysis; *s*, rudimentary neural spine; *z*, *z'*, anterior and posterior zygapophyses. (After Marsh.)   (From K. Zittel, *Palaeontology*.)

In the perichordal layer derived from the anterior half of the sclerotome behind appears the cartilaginous centrum. At first ventral it soon surrounds the notochord and grows forwards. Later the four cartilages of a vertebra join and fuse, and the notochord is obliterated by the centrum. Each half-sclerotome does not develop a dorsal arch and ringlike centrum which later combine to form the adult vertebra, as described by Schauinsland (Sonies, 385, E.S.G.). The composition of the avian vertebra, according to the latest researches of Piiper, is shown in Fig. 73, and according to this author both the half-sclerotomes contribute to the neural arch.

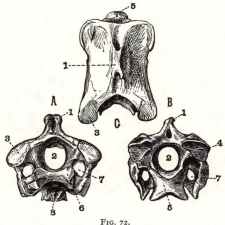

FIG. 72.

Third cervical vertebra of *Struthio camelus*. A, Anterior; B, posterior; and C, dorsal view. (From S. H. Reynolds, *Vertebrate Skeleton*, 1913.) 1, Neural spine; 2, neural canal; 3, prezygapophysis; 4, postzygapophysis; 5, posterior articular surface of centrum; 6, anterior articular surface of centrum; 7, vertebrarterial canal; 8, hypapophysis.

Mammalia.—The development of the mammalian vertebrae has been described by Froriep (24), Macalister (59), Weiss (91), Bardeen (3), and others; it resembles that in Birds, and agrees closely with that shown in the diagram above, Fig. 62, p. 56. The denser posterior half-sclerotome surrounds as usual neural canal and notochord (and haemal canal in the tail), and spreads outwards in the myoseptum.

Paired neural arch rudiments appear in the dorsal region, and a ventral hypocentral rudiment (generally median) in the subchordal crescent. In most mammals these crescents disappear or fuse with the centrum behind, except in the atlas (p. 66) and also the caudal region, where they may give rise to the 'chevrons' of paired origin. Such Y-shaped haemal arches (or sometimes median crescents) commonly occur in the tails of Monotremata, Marsupialia, Sirenia, Cetacea, Xenarthra, and Nomarthra; but hypocentra rarely survive elsewhere in the adult except in the lumbar region of Insectivora. The centrum arises from usually paired chondrifications in the perichordal region of the anterior half-sclerotome of the next segment behind. These soon fuse round the notochord, which is

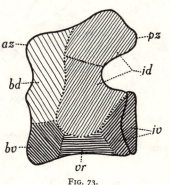

FIG. 73.

Diagram of the constitution of the *vertebra of a bird* (after Piiper, 1928). *az*, Anterior zygapophysis; *bd*, basidorsal; *bv*, basiventral; *id*, interdorsal; *iv*, interventral including intervertebral disc; *pz*, posterior zygapophysis; *vr*, vertebral ring.

completely obliterated, excepting for a small vestige (nucleus pulposus) in the middle of the intervertebral ligament, Fig. 74. In some mammals this ligament is formed secondarily after a temporary fusion of the centra into a continuous cartilaginous rod (O. Schultze, 82; Schauinsland, 78).

From the account given above of the structure and development of the vertebrae in the various groups of Tetrapoda, it appears that they may all be derived from an ancestral form and made up as follows : paired anterior dorsal elements combine to a neural arch (p. 47) ; paired anterior hypocentra remain separate, combine to a median hypocentrum, or join

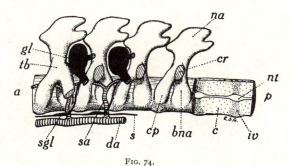

FIG. 74.

Trichosurus vulpecula, embryo 13·5 mm. long. Reconstruction of developing vertebral column in thoracic region, left-side view. *a*, Anterior ; *bna*, base of neural arch ; *c*, pleurocentrum ; *cp*, capitulum of rib ; *cr*, cut shaft of rib ; *da*, dorsal aorta ; *gl*, spinal ganglion ; *iv*, intervertebral procartilage (hypocentral region) ; *na*, neural arch ; *nt*, notochord ; *p*, posterior ; *s*, longitudinal sympathetic nerve ; *sgl*, sympathetic ganglion ; *tb*, tubercular region of rib continuous with transverse process of arch.

to a haemal arch in the tail ; and paired posterior pleurocentra (p. 47) tend to surround and combine to a ' centrum ' behind. The relation of these elements to the sclerotomes and body-segments has already been sufficiently explained above (p. 54), and the homology of the anterior dorsal elements with the basidorsals, and anterior ventral elements with the basiventrals of fishes may be considered as established. But the nature of the ' centrum ' of the Amniote is not so well understood. Gadow considers it to be formed by the interventrals (26). On the other hand, Cope and others would derive it from the pleurocentra of temnospondylous Stegocephalia, in which case it would represent the interdorsals. This interpretation is more plausible since interventrals are known for certain only in the caudal region of certain Stegocephalia among Tetrapods (p. 48). Possibly it is a combination of both interventral and interdorsal elements. However this may be, there can be little doubt that in all Tetrapods with well-developed vertebrae the basidorsals and basiventrals of one segment join the pleurocentra of the segment next behind to form a complete vertebra.

This result brings us to an important conclusion, first indicated by

F

Cope, that there has been a divergence between the Amphibia on the one hand and the Amniota on the other in the building up of the vertebra. For whereas in the former the hypocentrum gains in importance (at all events in most Stegocephalians) and the pleurocentra become relatively small, in the Amniote on the contrary the pleurocentra more and more form the bulk of the centrum or body of the vertebra, and the hypocentra take up an intercentral position or disappear.

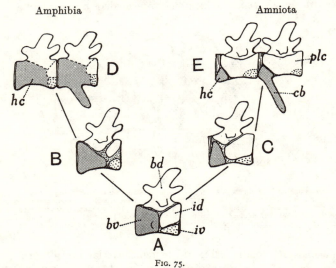

Fig. 75.

Diagram illustrating supposed divergence in development of vertebral elements leading from a primitive ancestral Tetrapod A to a typical Amphibian D, and a typical Amniote E. In D and E a caudal vertebra has been added to the trunk vertebra. *bd*, Basidorsal (neural arch) ; *bv*, basiventral = *hc*, hypocentrum ; *cb*, chevron = intercentrum (shaded) ; *id*, interdorsal = *plc*, pleurocentrum ; *iv*, interventral (coarsely dotted).

The structure of the vertebra, then, is a valuable guide to phylogeny, Fig. 75.

Atlas and Axis.—Another striking difference distinguishes the Amphibia from the Amniota. It is the specialisation of the first two cervical vertebrae in the Amniota to form a characteristic atlas and axis complex concerned with the articulation of the skull, Figs. 61 A, 76. The Amphibian second cervical differs in no special respect from the third and succeeding vertebrae, and the skull articulates with the anterior face of the first vertebral body expanded to receive it. When the typical paired amphibian condyles develop they fit into corresponding depressions in the centrum of this 'atlas' which in modern forms usually has a projection between them (see further, p. 68), Figs. 85, 267.

It is a significant fact that in the Amniota the atlas is permanently temnospondylous (p. 47). Its three elements, neural arch, hypocentrum, and pleurocentrum (all of paired origin), are separate in primitive early Reptiles ; its neural arch differs little from the succeeding ones in *Seymouria*, Fig. 60. Even in living Reptiles, Birds, and Mammals the pleurocentrum is separate from the other two elements (Gadow, **26** ; Osborn, **65**). This pleurocentrum becomes more or less closely attached to or fuses with the pleurocentrum (centrum) of the axis vertebra to form its odontoid process (dens epistrophei), Fig. 61. Thus, in Amniotes, the neural arch and hypocentrum of the atlas tend to combine to a ring supporting the

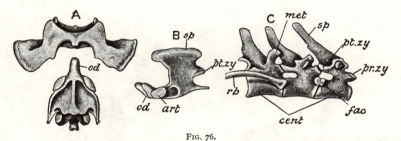

Fig. 76.

Lepus cuniculus. A, Atlas and axis, ventral aspect. *od*, Odontoid process of axis. B, Lateral view of axis ; *art*, articular facet for atlas ; *od*, odontoid process ; *pt.zy*, postzygapophysis ; *sp*, neural spine. C, Thoracic vertebrae, lateral view. *cent*, Centrum ; *fac*, facet for rib ; *met*, metapophysis ; *pr.zy*, prezygapophysis ; *pt.zy*, postzygapophysis ; *rb*, rib ; *sp*, spinous process. (From Parker and Haswell, *Zoology*.)

skull and movably articulated to the compound second cervical or axis vertebra (epistropheus). For the reception of the occipital condyle a depression is formed by the hypocentrum below, the expanded bases of the neural arch at the sides and the tip of the odontoid process between them. In the case of the mammal the hypocentral articulation with the basioccipital is reduced or lost, and the articulation becomes bicondylar.

It may be noted that in a primitive Reptile like *Ophiacodon* the pleurocentrum of the atlas is large and in the form of a disc pierced by the notochord much as in an embolomerous Stegocephalian (Williston, **95**) ; but in specialised forms it tends to become reduced to a mere process of the axis centrum concealed from without, Figs. 78, 79, 80. The hypocentrum of the atlas tends to become more closely united with its neural arch, and finally fuses with it to a complete ring, as in Birds and Mammals ; though in *Thylacinus* a separate bony hypocentrum persists in the adult.

It would be interesting to determine the exact position of the occipital joint in relation to the sclerotomes and original segmentation. According to Barge (**4**) it corresponds in Amniotes (Sheep) to the limit between the last occipital and first cervical segments. That is to say, the hind-most region of the skull cartilage is derived from the posterior (caudal) half of the last sclerotome of the head ; while the neural arch and hypo-centrum of the atlas is from the posterior half of the next sclerotome (first cervical). The axial ring comes to articulate directly with the occipital condyles since the cranial half-sclerotome between them fails to develop a pleurocentrum of any size, Fig. 77.

In the Urodela the occipital joint is probably not strictly intersegmental, but appears to correspond to the articulation formed between vertebrae

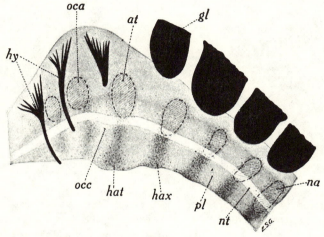

FIG. 77.

Trichosurus vulpecula. Embryo 7·25 mm. long. Partially reconstructed thick longitudinal sagittal section of occipital and anterior cervical region, left-side view. Two hypoglossal, *hy*, and the first cervical ventral nerve roots are shown ; also part of the cervical ganglia, *gl* ; *at*, base of arch of atlas vertebra ; *hat*, procartilage rudiment of hypocentrum of atlas ; *hax*, same of axis ; *na*, neural arch ; *nt*, notochord ; *oca*, occipital arch ; *occ*, posterior pleurocentral cartilaginous rudiment belonging to occipital region ; *pl*, pleurocentrum of third cervical vertebra.

in the intervertebral cartilage. The cartilage in which the occipital joint arises belongs to the anterior (cranial) half-sclerotome of the first cervical segment (Fortman, **22**).

The Proatlas.—There occurs in many Amniotes, overlying the neural canal between the occipital region of the skull and the arch of the atlas, a pair of small bones (or a single bone) having the appearance of a vestigial neural arch, Figs. 78, 79, 80. They have been found in many Cotylosauria, Theromorpha, Rhynchocephalia, Dinosauria, and Crocodilia, and even in the Mammal *Erinaceus*, and may articulate with the skull in front and the atlas arch behind (Albrecht, **1** ; Dollo, **18** ; Hayek, **46, 47** ; Howes and Swinnerton, **49** ; Schauinsland, **78** ; Gadow, **26** ; Williston, **95**).

Whether paired or single, the bones develop from paired cartilages, and were named ' proatlas ' by Albrecht [1] under the impression that they are the vestiges of a vertebra more or less completely crushed out between the skull and the atlas. Albrecht's theory, at first accepted by many (Baur, 6 ; Dollo, and others), is now discredited. It has been suggested that the proatlas is a ' neomorph ' of no special morphological significance, or that it is the vestige of the neural arch of a vertebral segment, of which the remainder is included in the occipital region. Barge (4) considers it to be the arcual of the anterior (cranial) half of the sclerotome which produces the arch of the atlas. On the whole this seems the most plausible explanation of its presence.

FIG. 78.

Ophiacodon; left-side view of proatlas, atlas, axis, and ribs. (From Williston, *Osteology of Reptiles*, 1925.)

In connexion with the proatlas, Schauinsland points out that the tip of the odontoid process chondrifies

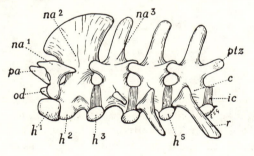

FIG. 79.

Sphenodon punctatum ; left-side view of anterior vertebrae, diagrammatic. *c*, Body of vertebra (pleurocentrum) ; *h*, hypocentrum ; *ic*, intervertebral disc ; *na*, arch ; *od*, odontoid process or first pleurocentrum ; *pa*, ' proatlas ' ; *ptz*, postzygapophysis ; *r*, rib.

separately in Sphenodon (78). Gaupp (512) believes the odontoid process in Echidna to contain the elements of two centra, and much the same

[1] Misled by the fact that the first spinal nerve (the ' suboccipital ' nerve between skull and atlas) has disappeared in the adult Anura, Albrecht erroneously concluded that the first or atlas vertebra in Amniotes corresponds to the second in Amphibia. But a ' suboccipital ' nerve exists in other Amphibia, and embryology shows that no segment has been lost in this region. The question is further complicated by the fact that in Urodela the first spinal nerve may secondarily become overgrown by the bone of the neural arch, and so pass through it.

result has been reached by de Burlet in Bradypus (11), by Hayek in Reptiles, Birds, and Mammals (46-7). There is reason to believe that in all Amniotes an anterior (cranial) half-sclerotome is present between the posterior (caudal) half-sclerotome, which gives rise to the cartilage enclosing the last hypoglossal nerve-root and completes the occipital region of the skull behind, and that which gives rise to the atlas arch. From this anterior half-sclerotome the 'proatlas' arch may

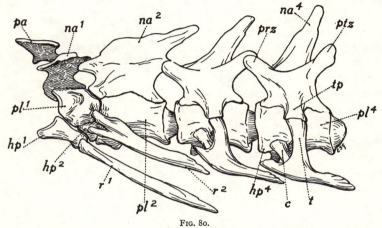

FIG. 80.

Anterior four cervical vertebrae of *Crocodilus niloticus*; left-side view. *c*, Capitulum; *hp*, hypocentrum; *na*, neural arch; *pa*, proatlas; *pl*, pleurocentrum; *prz*, prezygapophysis; *ptz*, postzygapophysis; *r*, rib; *t*, tuberculum; *tp*, transverse process; neural arch 1 and proatlas of left side removed to expose pleurocentrum 1 = odontoid process.

perhaps develop, but at all events its vestigial pleurocentrum appears to be added to the tip of the odontoid process, Figs. 61, 77.

General Review.—In this study of the morphology of the vertebral column we have seen that there is considerable evidence that throughout the Gnathostomes four paired elements or arcualia are concerned in the building up of a vertebral segment. Of these the basidorsals and basiventrals are the more important, originate in the posterior (caudal) half of the sclerotome, and give rise to the bulk of neural and haemal arch respectively. The interdorsal and interventral elements develop in the anterior (cranial) half of the sclerotome. But, while the morphological position of the basalia (especially basidorsals) near the middle of the vertebra, where comes the septum between two consecutive myomeres, is constant from Selachians upwards, that of the interarcuals varies considerably, and the homology of the parts supposed to represent them is by no means so clearly established. For instance, in Petromyzon and the Selachii the interdorsal is primarily between the dorsal and ventral nerve-root of a segment; but in other groups, where the interdorsals appear to be represented only by their bases closely applied to the notochord and the arches proper are little or not at all developed, the relation to the roots seems to vary. Moreover, if, as Schauinsland supposes, the interdorsal arch is occasionally developed in Urodeles (p. 51, and Fig. 57), then it passes behind both roots of the spinal nerve, which may become enclosed between the basidorsal and interdorsal arches of one vertebra, Fig. 57. On the other hand, if both these elements can combine to form the adult neural arch in Amniota, as held by Schauinsland and others (see above, p. 62), then in these Tetrapods both roots of the spinal nerve pass posteriorly to the interdorsal (p. 64, and Fig. 73).[1] Not only is the relation of the nerves to the cartilages different in these various groups, but the relation of the interdorsal to the intervertebral joint is also variable : it lies above it in Selachians, behind it in some Teleostomes, in front of it in Amniotes. In the modern Amphibia it would appear that the joint forms across the interarcuals. These are some of the obscurities which future research may clear up.

THE RIBS

Ribs first make their appearance in the Gnathostomata, and it is now recognised that, as already mentioned above (p. 7, and Fig. 1), they are of two kinds[2]: upper d o r s a l   r i b s extending outwards in the horizontal septum at its intersection with the transverse myosepta, and lower ventral or p l e u r a l   r i b s extending downwards where these myosepta join the coelomic wall (A. Müller, 1853; Goette, 31-3; Rabl, 71; Baur, 100-2; Hatschek, 109). Both kinds are originally connected proximally with the basiventral, but while the former passes between the epiaxonic and the hypoaxonic muscles, the pleural rib passes below (internal to) the myomeres. Several important questions arise in connexion with the morphology of ribs : (1) whether they were derived phylogenetically from the axial skeleton as held by Gegenbaur (1870) or are independent structures which have become secondarily connected with it (Bruch, '67 ; Kölliker, '79 ; Hasse and Born, '79) ; (2) what relation have the two kinds to each other and to the haemal arches ; (3) whether the bicipital ribs of Tetrapods are strictly homologous with one kind of rib or are a combination of both.

[1] These variations in relative position may be due to the cutting through, so to speak, of the interdorsal arch by the nerve-roots.

[2] The earlier view of Gegenbaur and others, that all ribs and haemal arches are homologous structures, was abandoned when it was shown that dorsal and pleural ribs may coexist in the same segment of the trunk, and dorsal ribs and haemal arches in the same segment of the tail.

Before attempting to answer these questions we must examine the structure and development of the ribs in the various groups, beginning with the Pisces.

Pisces.—The Selachii have well-developed ribs in the trunk which dwindle in size anteriorly, being usually absent in the first few segments. They also dwindle posteriorly, and are absent in the tail region except for vestiges in the most anterior caudal segments. These ribs are of the dorsal type, and are attached to the basiventrals provided with outer processes (' parapophyses ') for the purpose. In development (Balfour, 1878; Göppert, 107; Schauinsland, 78) they arise in the distal region of the tissue which grows out into the myoseptum from the posterior (caudal) half of each sclerotome, Fig. 16. At first basiventral and rib rudiments are in perfect blastematous and even procartilaginous continuity. In some cases true cartilage may appear independently in the rib ; but often it appears to develop continuously from basiventral to rib, and then the rib becomes secondarily detached, a fibrous joint appearing, which separates it from the basiventral or ' basal stump '. In the caudal region the haemal arches are developed in continuity with the basiventrals. Dorsal ribs and haemal arches may be considered as independent extensions of the basiventrals, the former developed in the trunk, the latter in the tail. Both may be present in the intermediate region (Balfour, 2). As pointed out by Schauinsland in *Laemargus,* a haemal process may be present on the inner side of the haemal arch in Selachians supporting the membrane

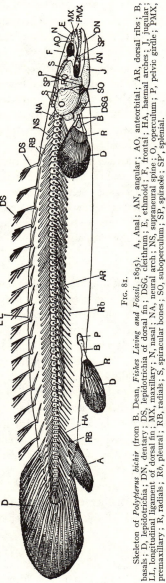

FIG. 81.

Skeleton of *Polypterus bichir* (from B. Dean, *Fishes Living and Fossil,* 1895). A, Anal ; AN, angular ; AO, anteorbital ; AR, dorsal ribs ; B, basals ; D, lepidotrichia ; DN, dentary ; DS, lepidotrichia of dorsal fin ; DSG, cleithrum ; E, ethmoid ; F, frontal ; HA, haemal arches ; J, jugular ; LL, longitudinal ligament of dorsal fin ; MX, maxillary ; N, nasal ; NA, neural arch ; NS, supraneural spine ; O, operculum ; P, pelvic girdle ; PMX, premaxillary ; R, radials ; Rb, pleural ; RB, radials ; S, spiracular bones ; SO, suboperculum ; SP, spiracle ; SP', splenial.

which separates the aorta from the caudal vein. On passing forwards towards the trunk the closed arch opens out, the basiventrals spread dorsalwards and tend to separate into a pair of haemal processes protecting the aorta, and a more dorsal outstanding part supporting the rib (see also p. 21).

The condition of the ribs in the Polypterini is very instructive (Dollo, 104; Hatschek, 109; Göppert, 107; Budgett, 10). In *Polypterus* each segment of the vertebral column in the mid-trunk bears two almost equally developed pairs of bony ribs. The upper dorsal pair is articulated to the middle of the centrum; the lower ventral or pleural pair is secondarily attached farther forward and even intercentrally. The dorsal ribs increase in length forwards and disappear in the tail; the pleural ribs increase in length backwards, bend downwards, and, coming together ventrally, pass into the haemal arches in the tail, Fig. 81. Budgett has shown that in the young larva the upper ribs appear as prolongations of a series of lateral cartilages at the side of the notochord, and the pleural ribs as prolongations of a similar series of more ventral cartilages. Both lateral and ventral cartilages are situated vertebrally (in the same transverse plane as the basidorsals), and are no doubt of basiventral origin. It is clear that in *Polypterus* the haemal arches are serially homologous with the pleural ribs and basiventral bases of support. This conclusion is amply borne out by observations on the structure and development of these parts in certain other Teleostomes and in the Dipnoi.

The ribs in the lower Teleostomes (Chondrostei, Amioidei, Lepidosteoidei) appear to be all of the pleural type, and passing backwards the basiventrals and ribs of the trunk are seen to correspond to the basiventrals and haemal arches of the tail (Goette, 31-3; Balfour, 2, and others).[1] But the majority of the Teleostei, in addition to these pleural ribs, have upper ribs in the horizontal septum. The pleural ribs are usually very well developed and may almost meet ventrally. Proximally they are articulated to rib-bearing processes of the basiventrals which may be separately ossified as in *Esox* and *Cyprinus*. The upper ribs ('epipleurals') are usually attached by ligament to the sides of the centra, but sometimes to the pleural ribs themselves, in the anterior region.

---

[1] In spite of what has been said above, it is not impossible that the upper dorsal ribs of Selachians are really homologous with the pleural ribs of Teleostomes, as held by Balfour. In *Lepidosteus* the free ends of the ribs are embedded in the muscles and therefore in a somewhat intermediate condition. If, in phylogeny, pleural ribs have been derived from 'dorsal ribs', the dorsal ribs of the Teleostomes must have appeared as a new set of extensions in the horizontal septum. A closer study of the more primitive fossils might throw light on this difficult question.

Frequently they persist in the caudal region.   Passing backwards to the tail the pleural ribs are seen to persist as such in its anterior region. The haemal arch is formed by the prolongation and meeting of two processes of the basiventrals which may be continued into a median spine. This condition of the posterior pleural ribs, very conspicuous in many

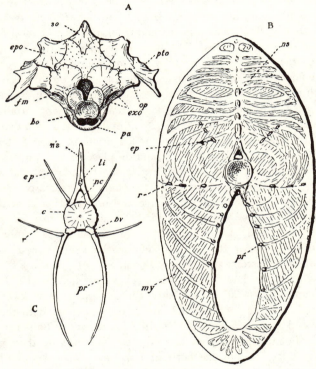

Fig. 82.

Salmo salar, L.   (After Bruch.)   A, Skull from behind ;   B, transverse section of the trunk ; C, transverse section of a trunk vertebra.   *bo*, Basioccipital ;   *bv*, basiventral ;   *c*, centrum ;   *ep*, epipleural ;   *epo*, epiotic ;   *exo*, exoccipital ;   *fm*, foramen magnum ;   *li*, longitudinal ligament ;   *my*, myotome ;   *nc*, neural canal ;   *ns*, neural spine ;   *op*, opisthotic ;   *pa*, parasphenoid below the myodome ;   *pr*, pleural rib ;   *pto*, pterotic ;   *r*, rib ;   *so*, supraoccipital.   (From Goodrich, *Vert. Craniata*, 1909.)

of the higher Teleosts, may be related to the secondary shortening of the abdominal cavity and shifting forwards of the anus.   On the whole, the evidence is against the conclusion of Goette and Balfour that the pleural rib enters into the formation of the haemal arch in Teleostei, Figs. 49, 141.

It may be added that pleural ribs develop in Teleostomes as extensions of the basiventral rudiment, continuous with it in the procartilaginous stage. The rib and basiventral often chondrify separately, but may become continuous. Later a fibrous joint is formed separating the rib from the process which bears it (Goette, Göppert). The upper rib cartilage arises separately, and is only connected with the basiventral by ligament.

The slender 'intermuscular bones' frequently present in the myosepta of Teleosts, extending from the centrum ('epicentrals') or neural arch ('epineurals'), were distinguished by J. Müller as ossified tendons from the true ribs preformed in cartilage; but in Salmonidae the epineurals are at first cartilaginous, Fig. 82.

The Dipnoi, which in so many respects approach the Tetrapoda, differ from them markedly in the absence of upper or dorsal ribs and the presence of pleural ribs in the trunk, Figs. 10, 33. They articulate with the cartilaginous basal stumps (basiventrals) and are separately ossified. A very characteristic feature, seen even in Devonian forms, is the persistence and enlargement of the anterior pleural ribs giving rise to the so-called 'cranial ribs' attached to the skull and belonging to those vertebral segments which have been fused to its occipital region, Figs. 315-17.

Concerning the phylogenetic derivation of the ribs, since they are connected with the column, and always arise at least in blastematous continuity with it, we may conclude that they belong to the axial skeleton and are derivatives of the basiventral element. This conclusion is also borne out by what we know of the structure and development of the ribs in Tetrapods.

Tetrapoda.—The ribs of the Tetrapoda, in spite of the fact that they extend ventralwards to enclose the thoracic cavity and may even meet below, are of the upper or dorsal type, situated primarily in the septum between epiaxonic and hypoaxonic muscles. It is unnecessary, now, to enter in detail into the controversies carried on in the past as to their homology. Claus, 1876, pointed out that the coexistence of ribs and haemal arches in the tail of Amphibia and Reptilia disposed of the view that dorsal ribs, pleural ribs, and haemal arches are all homologous structures. Later Hoffmann urged that the ribs of Amniota are connected intervertebrally with the column (110). As indicated above (p. 15, and Fig. 62), it is now generally accepted that the ribs develop in the intersegmental septa from tissue derived from the posterior (caudal) half-sclerotome, and therefore from the same half-sclerotome as the hypocentrum and neural arch to which they belong.

In a typical primitive Tetrapod the rib is bicipital, the ventral head or capitulum being articulated to the hypocentrum below, the dorsal head or tuberculum to the extremity of the transverse process (diapophysis) of the neural arch above, Fig. 61. The shaft of the rib passes outward in the myoseptum. Between capitulum, tuberculum, and vertebra is enclosed a space, the ' vertebrarterial canal ', through which runs a longitudinal vertebral artery joining successive segmental arteries.

The origin of the bicipital rib is an interesting problem which has not yet been satisfactorily solved.[1] If the upper rib is an outgrowth of the basiventral, or is at all events primitively related to it, its connexion with the transverse process of the neural arch must be secondary, and the tubercular head must have been added to secure better articulation to the vertebral column. It is not easy to see how the intermediate stages could have been useful. But the double head might also have been evolved by the spreading upwards of the original basal articulation from the hypocentrum to the arch and its subsequent subdivision into two accompanied by the forking of the proximal end of the rib itself. Such, for instance, is the view of Williston (95), who points out that apparently primitive single-headed (holocephalous) ribs occur in Temnospondyli among the Stegocephali and in some early Reptilia, Figs. 52, 84. This theory, however, is difficult to reconcile with the fact that typical bicipital ribs already occur throughout the vertebral column of *Seymouria*, perhaps the most primitive of all known reptiles, and in the earliest known Embolomeri, such as *Eogyrinus* (Watson, 644), Figs. 53, 54, 83. The ribs of this carboniferous Stegocephalian are of peculiar interest since even the sacral ribs are of considerable length, being scarcely yet modified to support the ilium, and bicipital. There can be little doubt that they represent an early stage in the evolution of the Tetrapoda before the typical short one-headed sacral rib had become developed.

This evidence, combined with that of the widespread occurrence of

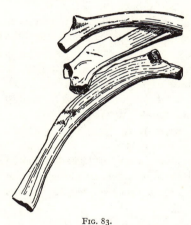

FIG. 83.

Eogyrinus Attheyi; typical dorsal, last dorsal, and first sacral rib. (From D. M. S. Watson, *Tr. Roy. Soc.*, 1926.)

[1] The theory put forward by A. Müller, '53, and supported by Dollo (104), that bicipital ribs are formed by the coalescence of dorsal and ventral ribs, is not supported by the study of their development, and seems to have been finally disposed of by the discovery of the coexistence with haemal arches of typical bicipital ribs in the anterior caudal segments of *Conodectes* (*Seymouria*) (Williston, 95; Watson, 643).

bicipital ribs among Reptiles, Birds, and Mammals, strongly favours the view that separate capitular and tubercular heads were present on the ribs of the ancestral Tetrapod. The holocephalous condition present always in the sacrum (except in certain Embolomeri mentioned above), and sometimes throughout the length of the column (Temnospondyli, Cotylosauria, many Theromorpha, Rhynchocephalia, Lacertilia, Ophidia, etc.), would then be secondary. Sometimes it appears to be due to the loss of the tubercular articulation as in Monotremes and some other Mammals; or to the fusion of the two heads to one as in *Sphenodon* (Howes and Swinnerton, 49), Ophidia, and most Lacertilia. But a

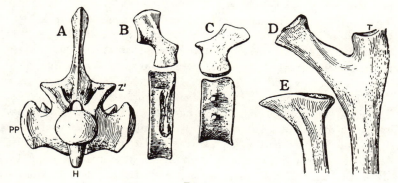

FIG. 84.

Vertebra : A, of *Clidastes*, posterior cervical from behind ; B, of *Cymbospondylus*, anterior dorsal from right (after Merriam) ; C, of *Ichthyosaurus*, middle dorsal from left (after Broili). D, anterior dorsal rib of *Dimetrodon*, and E, of *Diadectes*. (From Williston, *Osteology of Reptiles*, 1925.)

detailed study of the articulation of the ribs in Reptiles remains to be made.

The Tetrapod rib chondrifies separately, and although its capitular region is always developed in blastematous continuity with the hypocentral blastema (hypochordal Spange of Froriep), yet its point of attachment may shift in the adult (Hoffmann, 110; Howes and Swinnerton, 49; Schauinsland, 78, in Reptiles; Froriep, 24, in Birds; Schultze, 82; Bardeen, 3, in Mammals). In primitive Stegocephalia and Reptilia, where the hypocentrum is large, the capitulum is articulated to it; but when the hypocentrum is reduced or disappears as such, the rib comes to articulate either in the intercentral region on facets belonging to two consecutive centra (as in some Mammals), Fig. 76, or on the centrum belonging to its arch (as in the majority of Reptiles and in Birds). A parapophysis may be developed on the centrum to support it. The tubercular articulation may also shift downwards on to the centrum, as in *Ichthyosaurus*, and a second more dorsal parapophysis be formed for it on the centrum, Fig. 84.

In the Archosauria there is a tendency for the capitulum of the trunk ribs to shift upwards on to the transverse process of the arch, at the tip of which it finally meets the tuberculum in the posterior segments. This peculiar mode of articulation is typically developed in Crocodilia, Fig. 67.

The rib of the atlas is generally lost ; but it remains in Archosauria (including modern Crocodilia) and some early Reptiles (Theromorpha such as *Ophiacodon*), Figs. 78, 80. As the neck region becomes more fully differentiated the cervical ribs tend to become shortened and reduced. In most Amniotes, however, they retain two heads and often become fused to the centrum below and the arch above, thus completely enclosing the vertebrarterial canal in bone (Birds and Mammals), Figs. 70, 71, 72. In the sacral and caudal regions the single head of the rib becomes, as a rule, fused to the arch above and the centrum below, and in the tail so-called ' transverse processes ' are thus formed which are not strictly homologous with the diapophyses of the neural arches farther forward.[1]

It is in the anterior trunk region that the ribs (thoracic ribs) are best developed. Typically a considerable number encircle the visceral cavity and reach the sternum below (p. 81). These ribs become divided into a dorsal and a ventral portion (sternal rib), the latter articulating with the dorsal rib above and the sternum ventrally. Except in Birds and a few Mammals the sternal ribs remain cartilaginous. The jointing allows for the expansion and contraction of the thoracic cavity for respiratory purposes (p. 84). The ribs of the Chelonia are highly modified. The cervical ribs have been lost except in the primitive *Triassochelys*, where they are small and bicipital. The ribs of the trunk in Chelonia are greatly modified owing to the formation of a dorsal carapace. The sternum is lost, the ribs stretch outwards, and eight pairs contribute usually to the carapace. The capitulum is suturally attached to the arch and two consecutive centra. In Ophidia, where there is no sternum, and also in the snake-like Lacertilia (where the sternum is reduced or absent), the ribs are well developed and tend to become uniform along the whole elongated trunk.

No modern Amphibian has typical ribs (Mivart, 1870 ; Göppert, 106, 108, and others). They are absent on the atlas, but usually

---

[1] Ossified, or more usually cartilaginous, lateral projections known as uncinnate processes occur on the trunk ribs of some Temnospondyli among Stegocephalia, and possibly are represented by the distal dorsal process of the ribs of Urodela. Uncinnate processes occur also in fossil and living Archosauria (Sphenodon, Crocodilia), and in Birds. They chondrify separately, and do not unite with the ribs in Reptiles ; but in Birds, where they ossify, they fuse with the ribs.

present on all the presacral, sacral, and even anterior caudal vertebrae. Much reduced in all Anura, they may disappear altogether in the more specialised forms such as *Rana*. Even in Urodeles they are scarcely bent downwards to encircle the body cavity, never join the sternum, and often end in the horizontal septum at the level of the lateral line. The ribs in Urodeles sometimes fork distally, and usually are provided proximally with two heads attached to the bifurcated extremity of a 'transverse process' of peculiar structure and doubtful homology, Fig. 85. This rib-bearing process in the adult has a strong dorsal limb continuous with the neural

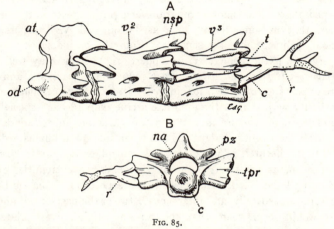

Fig. 85.

Necturus maculatus. A, Left-side view of first three vertebrae; B, posterior view of fourth cervical vertebra. *at*, Neural arch of first cervical vertebra; *c*, ventral process of rib (capitulum?); *na*, neural arch; *nsp*, neural spine; *od*, anterior process of first centrum; *pz*, postzygapophysis; *r*, rib; *t*, dorsal process of rib (tuberculum?); *tpr*, transverse process or 'rib-bearer', pierced at base for artery; $v^{2,3}$, second and third vertebrae.

arch, and a slender ventral limb continuous with the centrum below (the ventral limb may be reduced to a slender bone or even to a ligament in some genera). The two limbs meet distally, thus enclosing the vertebral artery; and the rib having no connexion with the vertebrarterial canal, is carried away distally from it. The development of Amphibian ribs has been studied in detail by Göppert (108), Mayerhofer (112), Gamble (105), and others. As in other Tetrapods it arises in blastematous continuity with the basiventral or 'basal stump', but at a higher level than usual owing to the more dorsal attachment of the horizontal septum. In *Necturus* the cartilaginous basiventral stretches outwards and upwards below the vertebral artery and may form the main portion of that part of the 'transverse process' which bears the capitulum of the rib. In others, such as *Salamandra*, this basiventral bar is scarcely preformed in cartilage and is represented by a ligament which ossifies directly. Göppert concluded that the original connexion of the rib was by means of its ventral capitular head with an outgrowth of the basiventral passing below the vertebral artery, that this parapophysis tends to become reduced and that an upgrowth of the basiventral, distal to the artery, joins the neural arch above, forming the main part of the ribbearer (Rippenträger) and affording further support to the rib. The upper branch of the rib would then be secondary. This theory of a special ' rib-bearer ' has been generally accepted, but is of very doubtful value. The view that the dorsal limb of the rib-bearer is only secondarily attached to the neural arch and is not a true transverse process rests on the slender evidence that in some forms, such as Triton and *Necturus*, its cartilaginous rudiment is early separated from the basidorsal by a thin layer of bone (Knickmeyer, 111 ; Göppert, 106). This may well be a secondary separation due to the great reduction of cartilage and premature appearance of bone ; for, in many Urodeles, the cartilage occurs in continuity with that of the arch like a true diapophysis, especially in the caudal region (*Amphiuma*, Davison, 16). Gamble, indeed, finds that in *Necturus* the rib-bearer develops not as an upgrowth but as a downgrowth from the arch which meets the ventral parapophysis. In the anterior segments the rib is borne by the combined extremities of the rib-bearer and the basal stump ; while more posteriorly the rib-bearer enlarges, grows ventrally, and intervenes between the basiventral and the rib. It would seem, then, that the rib was originally articulated as usual by its capitulum to the basiventral below and by its tuberculum to the basidorsal above, that it has been carried outwards at the extremity of a process formed by the combination of the parapophysis with the diapophysis; and that, as the portion of this ' transverse process ' derived from the diapophysis increased more and more at the expense of the parapophysis, the attachment of the capitulum shifted on to it, Figs. 85, 626.

This conclusion is borne out by observations on the structure and development of the bicipital ribs of the Apoda, where the tuberculum abuts against an outgrowth of the neural arch, and the capitulum against a more ventral cartilage representing the basiventral (Göppert, Marcus and Blume, 62).

The reduced single-headed rib of Anura is attached to a ' transverse process ', apparently a lateral outgrowth of the neural arch, but passing below the vertebral artery. It is doubtful, however, whether the vertebral artery in the various groups is strictly homologous and constant in position (Schöne, 114), and the evidence does not allow us to decide whether the process represents a parapophysis which has shifted dorsally, or a true diapophysis.

In conclusion it may be pointed out that the ribs of Gnathostomes are extensions outwards in the myosepta of the basiventrals ; that in the Chondrichthyes the ribs are of the upper or dorsal type, situated in the horizontal septum ; that in Tetrapods the ribs are of the same type although they may bend ventrally and surround the coelomic cavity of the trunk ; that the Osteichthyes are provided with ribs of the lower or pleural type situated internally to the musculature in the wall of the coelom ; and that in some (Polypterini and Teleostei) both types of ribs may apparently coexist. That the haemal arches are derived from the basiventrals seems certain ; but it is not clear that the pleural ribs always share in their formation (Teleostei), although this seems to be their fate in the lower Teleostomes and Tetrapods.

Further research is necessary to determine whether the pleural ribs of Osteichthyes have or have not been phylogenetically derived from ancestral dorsal ribs.

THE STERNUM

A typical sternum is found among living Gnathostomes only in Amniota, where it occurs as a median ventral endoskeletal plate in the thoracic region. It serves for protection of the organs in the thoracic cavity, for the attachment of pectoral limb muscles, and in respiratory movements, the first and a varying number of more posterior thoracic ribs being articulated to it ventrally. In Reptiles it is usually in the form of a shield-shaped cartilage narrowing behind and prolonged into paired slender processes joined to ribs. In flying Birds the plate develops a median keel affording greater surface of attachment for the powerful wing muscles, Figs. 86, 192. An analogous but smaller keel may arise on the sternum of Pterodactyles and Chiroptera. Paired ossifications may occur in the sternum of Ornithischian Dinosaurs ; and in Aves the whole sternum becomes completely ossified from similar paired centres, Figs. 86, 188.

The mammalian sternum is in the form of a comparatively narrow longitudinal bar of cartilage which becomes more or less completely segmented and ossified into median sternebrae alternating and articulating with the ventral ends of the sternal ribs. The anterior segment (manubrium sterni), however, is longer and projects beyond the articulation with the first rib. Behind the sternum usually expands into a xiphisternal cartilaginous plate or is continued into two diverging

G

processes.[1] When a dermal interclavicle is present it is applied to the
ventral surface of the sternum.

In all Amniotes a close relation is early established between the
sternum and anterior ribs which together come to encircle the thoracic
cavity. But in no modern Amphibian does such a connexion exist. The

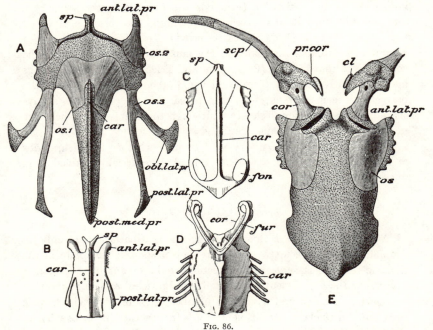

FIG. 86.

Sterna of various birds. A, *Gallus* (common fowl, young); B, *Turdus* (thrush); C, *Vultur* (vulture);
D, *Procellaria* (petrel); E, *Casuarius* (cassowary). *ant.lat.pr*, Anterior lateral process; *car*, carina;
cl, clavicle; *cor*, coracoid; *fon*, fontanelle; *fur*, furcula; *obl.lat.pr*, oblique lateral process; *os*,
paired ossification of sternum in E; *os*. 1, carinal ossification in A; *os*. 2, *os*. 3, lateral ossifications;
post.med.pr, posterior median process; *post.lat.pr*, posterior lateral process; *pr.cor*, procoracoid;
scp, scapula; *sp*, spina sterni. (A and E after W. K. Parker; B, C, and D from Bronn's *Thierreich*.)

sternum is absent in Apoda, where it has no doubt been lost, and in Urodela
is represented by a median plate of cartilage wedged in between the
posterior ventral ends of the coracoids, and widely separated from the short
ribs, Fig. 87. A somewhat similar plate occurs in arciferous Anura; and in
Firmisternia there may be, in addition to this posterior sternum (ossified in
part), a similar median plate and ossification ('omosternum') attached to
the anterior border of the girdle, Fig. 87. Whether this anterior 'prezonal'
element is of the same nature as the 'postzonal' sternum of these and

[1] The Mammalian sternum is sometimes described as composed of anterior
presternum (manubrium), middle mesosternum (several sternebrae), and
posterior xiphisternum (behind last sternal rib).

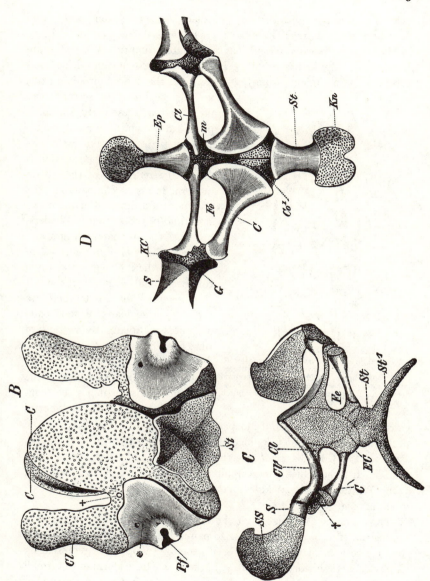

FIG. 87.

Pectoral arch of various amphibians. From the ventral side. B, Axolotl (*Amblystoma*); C, *Bombinator igneus*; D, *Rana esculenta*. C, coracoid; Cl, procoracoid; Cl¹ (Cl in D), clavicle; EC, Co¹, epicoracoid; Ep, omosternum; Fe, fenestra between procoracoid and coracoid bars; Kn, cartilaginous xiphisternum; †, Pf, G, glenoid cavity for the humerus; S, scapula; SS, supra-scapula St, St¹, sternum. *, † (in B) indicate nerve-apertures. (From Wiedersheim, *Comp. Anatomy*.)

other forms is not quite certain, but there seems to be no good reason to deny it (Parker, **113**; Gegenbaur, **170**; Wiedersheim, **230**, and others).

It has sometimes been held that the isolated condition of the sternum in modern Amphibia is primitive, and that the connexion with ribs is secondary; but judging from the structure of the dermal girdle and ribs in Stegocephalia, it seems more probable that these early Amphibia possessed a typical sternal cartilaginous plate connected with anterior ribs, and that the isolation is due to the reduction of the ribs in modern forms accompanying the adoption of a different mode of respiration (p. 598). Indeed Wiedersheim has shown that in Anura and Urodela the sternum develops in relation to the intersegmental myocommata, and even in the adult (*Necturus*) may have extending into these septa so many as four pairs of processes, apparently vestiges of ribs, Fig. 88.

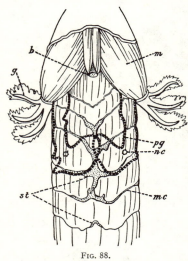

Fig. 88.

Ventral view of *Necturus maculatus* (after R. Wiedersheim, 1892), showing sternal cartilages, *st*, extending up myocommata, *mc*, and pectoral girdle drawn in thick broken line, *pg*; *b*, basihyal; *g*, external gill; *m*, muscle; *mc*, transverse septum (myocomma); *nc*, nerve foramen.

There has been much controversy about the phylogenetic history of the sternum. By some it has been considered as derived from the ventral ends of ribs, by others as of independent origin or as a derivative of the pectoral girdle. Its development has been studied by Dugés, 1835; Parker, **113**; Götte, **254**; Wiedersheim, **230**; T. J. Parker, **113**; Braus, **238**; Fuchs, **249**, chiefly in Amphibia; and by Ruge, **112b**; Paterson, **112a**; Whitehead and Waddell, **115**; Krawetz, **273**; and Hanson, **257**, in Mammalia.

The conclusion of Ruge, that the mammalian sternum develops from the ventral ends of ribs which fuse to paired longitudinal bands and later combine to a median plate, has been generally adopted as representing its phylogenetic history. But T. J. Parker traced the sternum to a small median cartilage, first described by Haswell, between the ventral ends of the coracoids in *Notidanus* (**113**), and Howes distinguished an anterior 'coracoidal' from a posterior 'costal' element (**263-4**). No true sternum has been found in any Dipnoan or Teleostome; but infracoracoid cartilages sometimes occur in Pisces (pp. 166, **171**), and may fuse to a median piece. There is no convincing evidence, however, that such coracoidal derivatives have given rise to even a part of the sternum of Tetrapods, as held by these authors and recently by Hanson.

On the other hand, from the work of Paterson, Krawetz, Whitehead and Waddell, and Hanson, it appears that in mammals the paired sternal bands first appear as procartilaginous rudiments independently of ribs with which they later become connected, and that Ruge missed the early stages,

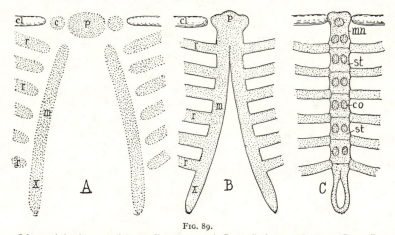

Fig. 89.

Scheme of development of *mammalian sternum.* A, Precartilaginous, early stage ; B, cartilage, halves beginning to unite ; C, beginning of ossification. *c* (?) Coracoid procartilage ; *cl*, clavicle ; *co*, centres of ossification ; *m*, longitudinal mesosternal element ; *mn*, manubrium ; *p*, pro- or presternum ; *r*, ribs ; *st*, sternebrae ; *x*, xiphisternum. (From J. S. Kingsley, *Vertebrate Skeleton*, 1925.)

Fig. 89. In front of the bands appears a separate median rudiment of the 'prosternum', an element of doubtful origin which contributes to the formation of the manubrium sterni. It is this prosternal rudiment which has been claimed as of coracoidal origin ; but it possibly represents a vestige of the interclavicle. In addition, paired cartilaginous rudiments may combine with it as in Marsupials, or remain separate as in many lower Mammals, and give rise to small cartilages or bones in adult Insectivores and Rodents (omosternal elements of Parker ; praeclavium). The origin and significance of these paired structures, however, are insufficiently known, and worthy of further study.

In Reptiles and Birds the sternum also develops from paired longi-tudinal bands, at first widely separated but later fusing from before backwards. These rudiments are continuous with the ribs even in early procartilaginous stages (Juhn, **271** ; Bogoljubsky, **273**).

It may be concluded that the true sternum of the Amniota is of paired origin, and is closely associated with the ventral end of ribs if not actually derived from them ; that it belongs to the axial rather than to the appendicular skeleton ; that the sternum of modern Amphibia is probably of the same nature, and has become secondarily isolated owing to the reduction of the ribs ; and, finally, that the pectoral girdle has probably contributed little or nothing to its development.

CHAPTER II

MEDIAN FINS

THE MEDIAN ENDOSKELETAL FIN-SUPPORTS AND NEURAL SPINES

An important question relating to the morphology of the skeleton of the median fins of Fishes concerns the significance of the so-called 'neural spines'. There has been some confusion about the use of this term and the parts to which it has been applied. Doubtless, strictly speaking, the

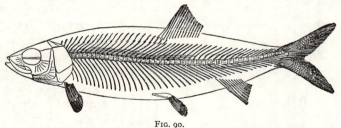

FIG. 90.

Leptolepis dubius : restoration, scales omitted. U. Jurassic : Bavaria. (From A. S. Woodward, *Vertebrate Palaeontology*, 1898.)

name neural spine should only be given to the median dorsal process projecting above the longitudinal ligament and formed by the fusion of the right and left half of the neural arch. Such a true neural spine is found in Amniotes and Amphibia, but does not occur either in Chondrichthyes or in Dipnoi. The neural arches of Teleostomes, however, are often prolonged dorsally far beyond the ligament as slender rods, which

87

remain separate as a rule in the anterior segments but fuse to a median rod in the caudal region. This kind of neural spine is little developed in Chondrostei, but more so in Holostei, and most of all in the higher Teleostei. Probably such spines are merely analogous to the true neural spine of Tetrapods, and may have been independently developed in the Pisces, Figs. 35-9, 49.

Corresponding to the dorsal neural spines similar haemal spines may be developed in the caudal region of Pisces (Chondrichthyes and Teleostomes) from the haemal arches; to determine their exact homology is a difficult task (see pp. 101, 110).

Examining more closely the neural 'spines' of Teleostomes we find that, whereas in the caudal region they are continuous with the arches in Chondrostei and *Lepidosteus*, in the more anterior segments they become separated off above the longitudinal ligament from the true neural arches below— these separate pieces may be called supraneural

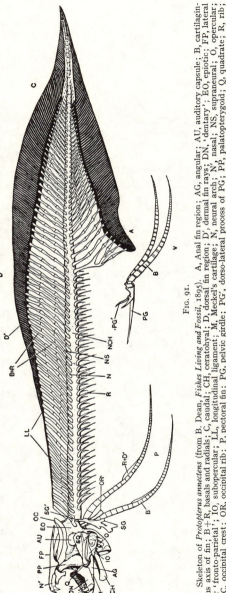

FIG. 91.

Skeleton of *Protopterus annectens* (from B. Dean, *Fishes Living and Fossil*, 1895). A, Anal fin region; AG, angular; AU, auditory capsule; B, cartilaginous axis of fin; B+R, basals and radials; C, caudal; CH, ceratohyal; D, dorsal fin region; D', dermal fin rays; DN, 'dentary'; EO, epiotic; FP, lateral or 'fronto-parietal'; IO, subopercular; LL, longitudinal ligament; M, Meckel's cartilage; N, neural arch; N', nasal; NS, supraneural; O, opercular; OC, occipital crest; OR, occipital rib; P, pectoral fin; PG, pelvic girdle; PG', dorso-lateral process of PG; PP, palatopterygoid; Q, quadrate; R, rib; R+D', radial and dermal fin elements; SG, shoulder girdle; SG', dorsal process of shoulder girdle; SQ, squamosal; V, pelvic fin.

spines or supraneurals. Such separate supraneurals occur possibly in some Selachii, especially near the base of the dorsal fins. Separate supraneurals are well developed in the lower Amioidei (Eugnathidae, Pachycormidae) and lower Teleostei (Leptolepidae, Fig. 90), but tend to disappear in the higher forms. Rods persisting near the head in modern forms (*Amia*, many Teleosts, Figs. 35, 50) are probably radials.

Similarly the haemal arches meet ventrally to form median haemal spines, and in the anterior caudal region become separated off to form what may be called infrahaemal spines or infrahaemals. In Teleostomes, such as *Amia*, Figs. 35-8, the gradual transition from haemal spine to infrahaemal spine can easily be seen. There can be no doubt, then, that there is no fundamental difference between the fused and the free pieces ; neural and supraneural spines are homologous elements, and the same may be said of haemal and infrahaemal spines.

But in the Dipnoi, where supraneurals and infrahaemals are most regularly developed, they form the basal piece of a segmented skeletal rod extending and supporting the base of the median dorsal and median ventral fins respectively. Towards the head, where the dorsal fin is reduced, this rod becomes two-jointed, and finally is formed of a single piece. A similar reduction occurs towards the tip of the tail, Figs. 33, 91.

The study of the ' neural spines ' in living and fossil fishes brings us to the consideration of an important question regarding the general morphology of the median fin-supports or radials and their relation to the axial skeleton : whether they are only secondarily connected with the vertebral column, or were primarily derived from it.

Thacher (219) and others considered that the radials are special structures developed to support the median fins, comparable to the radials of the paired fins, therefore forming part of the appendicular skeleton and only coming into secondary connexion with the axial skeleton.[1] On the contrary, Gegenbaur, 1870, and Cope (152) regarded the median fin radials as derivatives of the axial skeleton which may become secondarily separated off and specialised. Gegenbaur, indeed, considered them to be merely extensions of the neural and haemal spines. According to Cope the several pieces of each ray were simultaneously developed in lines of

---

[1] Schmalhausen (133), after a careful study of the development and structure of the median fins of fishes, adopts Thacher's view that the radials have secondarily become connected with the axial skeleton, especially in the tail. He bases his conclusion chiefly on embryological grounds. But here, as so often when dealing with ontogeny, the evidence may be read either way. The fact that the distal ends of the caudal haemal arches may sometimes chondrify separately and later become fused, may be considered either as a repetition of phylogeny, or as an indication that they are tending to become free and would fail to fuse if the process were carried further.

maximum strain, extended originally from neural arch to fin-base, and became differentiated into proximal neural spine, middle 'axonost', and distal 'baseost'. The axonost afterwards being separated from

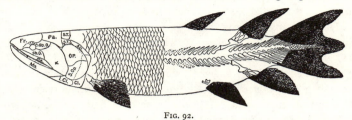

FIG. 92.

Eusthenopteron foordi. Restoration by J. F. Whiteaves, scales omitted in caudal region to expose axial skeleton and bases of median fin. U. Devonian: Scaumenac Bay, Canada. (From A. S. Woodward, *Vertebrate Palaeontology*, 1898.) *Cl*, Cleithrum; *Fr*, frontal; *I.Cl*, infracleithrum; *Mn*, mandible; *Mx*, maxillary; *OP*, operculum; *Pa*, parietal; *S.Cl*, supraclavicle; *S.Op*, suboperculum; *S.T*, supratemporal; *Sb.O*, suborbital; *x*, cheek-plate.

the spine, became the 'interspinal' of Cuvier, which together with the baseost supports the fin-base.

In support of this theory of the derivation of the median fin-supports from the axial skeleton it may be pointed out that the separation of the radials from the vertebral column is clearly related to the need for the

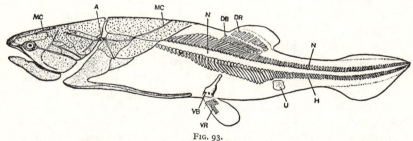

FIG. 93.

Coccosteus decipiens. Left-side view, restored (after A. S. Woodward, from B. Dean, *Fishes Living and Fossil*, 1895). A, Articulation of head with trunk; DB, cartilaginous basals; DR, cartilaginous radials of dorsal fin; H, haemal arch and spine; MC, lateral line canals; N, neural arch and spine; U, median plate; VB, pelvic girdle; VR, radials of pelvic fin.

independent action of well-developed and isolated median fins in the more active fishes. It generally accompanies the necessary concentration of the radials at the base of dorsal and anal fins. Originally they were of necessity strictly segmental as in Dipnoi; but, when concentration occurs, some if not all of the radials must become detached from their corresponding neural arches (p. 118). Yet even in Selachians, where the fin-skeleton is usually well separated from the column, it may rest directly on the dorsal ligament in some sharks (as in *Squalus*, Fig. 24), and especially in Rhina and other Rajiformes (*Raja*, Fig. 136). Moreover, in the practically continuous dorsal fin of the Pleuracanthodii the radials are closely connected with the neural arches much as in Dipnoi, Fig. 94 (Fritsch, **23** ; Brongniart, **149**, etc.).   Little is known about the median fin-supports of the Osteolepidoti ; in *Eusthenopteron*, however, not only are they articulated to the arches in the dorsal and ventral caudal fins, but also some of them in the dorsal and anal, Fig. 92. The Coelacanthini also have caudal radials resting on the neural and haemal spines ; but those of the dorsal and anal fins are extremely concentrated to single pieces separated from the axial skeleton, Fig. 95.

The radials of the Actinopterygii seem never to be so closely connected with the distal ends of the spines as those of the Dipnoi, though they may reach them and often even overlap them (interspinous bones, Fig. 141).

Generally, and probably primitively, the radials are subdivided into three pieces (dorsal fin of Pleuracanthodii, etc.). In Selachii these may be of approximately equal length, Fig. 132 ; but specialisation leads to the further subdivision of the elements or to their fusion (Mivart, and p. 119). The radials of the Holocephali remain unsegmented, and those of the Dipnoi have segments.   The Teleostomi also have their radials usually subdivided into three pieces (Thacher, **219** ; Bridge, **118** ; Schmalhausen, **133**) ; but while the proximal piece becomes the main support (interspinal, axonost), the middle piece is shortened, and the distal element further reduced in size. In the higher forms the bases of the paired lepidotrichia, which come to correspond to them in number, are firmly fixed to these rounded distal pieces, Fig. 97.

In the paired fins there are typically two radials to each segment (p. 125), and in the median fins also there may be two to each body-segment,

Pleuracanthus Decheni. Lower Permian of Bohemia ; skeleton and outline restored.   (After A. Fritsch, modified ; from *Brit. Mus. Guide.*)

FIG. 94.

one opposite the neural arch and the other between two arches. Such a double series of radials occurs in the median fin of Pleuracanthodii and in a large number of Teleostomes (Chondrostei, and many Teleostei, Figs. 94, 96). Owing, however, to concentration it is not always easy to distinguish between true duplication and mere crowding together of originally segmental radials such as occurs, for instance, in *Scyllium* (Goodrich, **172**) and *Salmo* (Harrison, **179**). Both processes may be combined in the same fin. Whether the double number is due to secondary duplication of the original radial, or to the development of a radial in

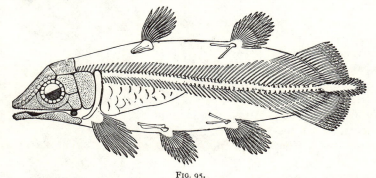

FIG. 95.

Undina gulo. Restoration, omitting scales and supraclavicle. L. Jurassic : Dorsetshire. (From A. S. Woodward, *Vertebrate Palaeontology*, 1898.) The extent of the ossified air-bladder is indicated beneath the notochordal axis in the abdominal region.

each half of a segment (corresponding to basalia and interbasalia), is uncertain.

Another question to be considered is how can a true radial be distinguished from a distal extension of a neural spine. If the former is derived phylogenetically from the latter no hard and fast line can be drawn between them ; but it may be held that a supporting rod is a true radial when it is related to a pair of radial muscles (p. 114). These right and left special fin-muscles, whose fibres run parallel to the rods and not to the longitudinal axis of the body as in the myomeres, are present in the median fins of *Petromyzon* both on the trunk and tail,[1] Fig. 98, and occur regularly in all Pisces, Figs. 102, 130.[2] Their distal ends are attached

[1] Their absence in Myxinoidea and at the extreme end of the tail of *Petromyzon* is probably due to reduction and degeneration.

[2] They tend to become modified or reduced in the caudal fin. Schmalhausen's theory that the median fins of the Dipnoi are not true dorsal and ventral fins, but caudal fins which have grown forward, is manifestly contrary to palaeontological, anatomical, and embryological evidence. His contention that the Dipnoan fin-muscles are not true radial muscles because they do not develop from muscle-buds cannot be accepted. Median fin-muscles like

chiefly to the skin and dermal rays. In the Selachii each muscle becomes subdivided into inner and outer regions, and they are further specialised in Teleostomes. Those of the Holostei become differentiated

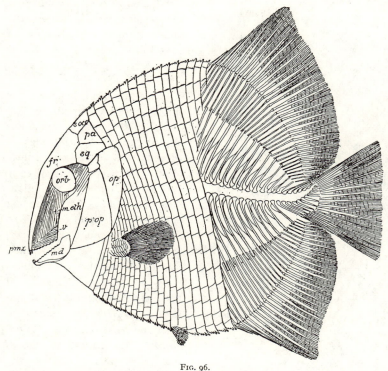

Fig. 96.

Mesodon macropterus. Restoration, with cheek-plates removed. U. Jurassic: Bavaria. (From A. S. Woodward, *Vertebrate Palaeontology*, 1898.) *fr*, Frontal ; *m.eth*, mesethmoid ; *md*, mandible, showing narrow dentary in front ; *op*, operculum ; *orb*, orbit ; *p.op*, preoperculum ; *pa*, parietal ; *pmx*, premaxillary ; *s.occ*, supraoccipital ; *sq*, squamosal ; *v*, prevomer. The caudal region is destitute of scales.

into an elaborate system of mm. inclinatores, erectores, and depressors serving to move and fold the fins (Harrison, **179** ; Schmalhausen, **133**).

In conclusion, it would appear that the evidence is not yet sufficient to enable us to assert that the radials of the median fins were derived from the axial skeleton. Should future evidence prove the correctness of this theory we may be compelled to adopt some such view as that

paired limb-muscles may arise either from typical muscle-buds (p. 114) or from proliferations of the myotomes (anal fin of Selachians, etc.). Ceratodus has typical radial muscles in its median fins (Goodrich, **122**) ; but they are reduced and degenerate in the fins of the more sluggish Protopterus and Lepidosiren.

foreshadowed by Owen in describing his ' archetype ' : namely, that the skeleton of the paired limbs was also originally related to the axial skeleton by means of the ribs. For there can be little doubt that the median radials are of the same nature as the radials of the paired fins (p. 132).

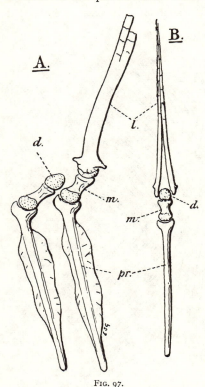

FIG. 97.

Esox lucius, L. A, two radials of the dorsal fin, left-side view ; B, radial and dermal ray from in front. *d*, Distal cartilage ; *m*, median segment, and *pr*, proximal segment of radial ; *l*, lepidotrich, broken short in A. (From Goodrich, *Vert. Craniata*, 1909.)

THE MEDIAN FINS

Longitudinal median fins occur throughout the aquatic vertebrates in the form of outstanding folds covered with epidermis without and containing mesoblastic tissue within. Already in Cephalochorda a continuous median fin-fold is found running dorsally from head to tail. In front it passes round the anterior end into a ventral ' subrostral ' fin, which becomes continuous with the right fold of the oral hood as in *Amphioxus* (*Branchiostoma*) and *Heteropleuron*, or with both oral hood folds as in *Asymmetron*. Posteriorly it passes round the tip of the tail and runs forward ventrally to the atriopore, passing to the right of the anus. In the postanal region the median fins are usually expanded to a caudal fin ; but in *Asymmetron* the pronounced ventral expansion is farther forward. In both *Heteropleuron* and *Asymmetron* the right metapleural fold is continued into the median fin behind the atriopore (Lankester, Andrews, Willey, Kirkaldy, **127**). The dorsal fin-fold is strengthened by a series of ' fin-rays ' (some 250 in *Amphioxus*), more numerous than the segments (about 5 per segment). They dwindle and disappear before reaching the anterior and posterior ends of the body. Each ' fin-ray ' consists of a mass of gelatinous tissue projecting into a box formed of a double layer of epithelium apparently derived from the sclerotomes (Stieda, 1873 ; Hatschek, **124**). A similar series of rays sup-

ports the ventral fin behind the anus; but between anus and atriopore
the rays are for the most part in pairs, suggesting that in this region the
fin is formed by the coalescence of paired folds.

Speaking generally, the median fins of the Craniata are similar longitudinal folds into which extends the median connective tissue septum
dividing the body into right and left halves.[1]   In this septum are developed
supporting endoskeletal rods or radials (fin-supports, pterygiophores,

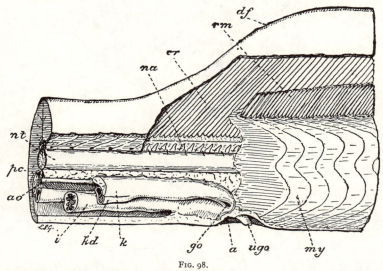

FIG. 98.

Petromyzon marinus, L.   Left-side view of the trunk region near the base of the dorsal fin; the
skin and muscles have been partially removed. *a*, Anus; *ao*, dorsal aorta; *cr*, cartilage rays
supporting fin; *df*, dorsal fin; *go*, left genital aperture into urinogenital sinus; *i*, intestine; *k*,
mesonephros, and *kd*, its duct; *my*, myotome; *na*, neural arch; *nt*, notochord; *pc*, posterior
cardinal vein; *rm*, radial muscles of fin; *ugo*, urinogenital opening on papilla. (From Goodrich,
Vert. Craniata, 1909.)

somactidia), primarily segmental, and the nature of which is discussed
above (p. 89).   Special paired radial muscles, parallel to and corresponding to the radials, serve to move the fin.   Although the median fins
appear in the young as continuous dorsal and ventral folds, these usually
become subdivided and specialised into separate dorsal, caudal, and postanal fins.   As a rule the subdivision is most complete in the most actively
swimming forms, and the fins vary much in shape and extent according
to the habits of the fish (Osburn, **202**; Gregory, **123**; Breder, **117**).

Cyclostomata.—The Cyclostomes possess median fins, better developed

[1] The median fin-folds of larval Anura and of aquatic Urodela have no
special muscles, endoskeletal radials, or dermal rays.   These fins are possibly
new structures reacquired in adaptation to their aquatic habits, and not
directly derived from the median fins of fishes.

in the active Petromyzontia than in Myxinoidea, where they form con-

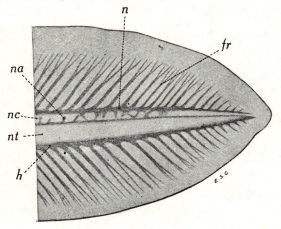

FIG. 99.

Petromyzon marinus. Left-side view of end of tail showing the cartilaginous endoskeleton. *fr*, Fin-ray (radial); *h*, fused bases of ventral fin rays forming a haemal canal; *n*, fused bases of dorsal radials in the membranous wall of the neural canal, *nc*; *na*, vestigial neural arch; *nt*, notochord. (Cp. Fig. 101.)

tinuous dorsal and ventral folds meeting round the tip of the tail. The ventral fold reaches to the cloacal pit, and in Myxinoids divides into two

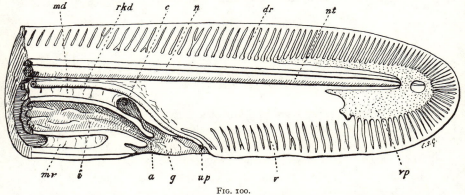

FIG. 100.

Tail of *Myxine glutinosa*, L., cut so as to show the skeleton and the opening of the intestine, etc.; left-side view. *a*, Anus; *c*, gap behind mesentery leading from right to left coelomic cavities; *dr*, cartilage radials of dorsal median fin; *g*, median opening through which the genital cells escape; *i*, intestine; *md*, dorsal mesentery; *mv*, ventral mesentery; *n*, nerve-cord; *nt*, notochord; *rkd*, left kidney duct; *up*, urinary papilla; *v*, cartilage radials of ventral median fin; *vp*, cartilaginous plate. (From Goodrich, *Vert. Craniata*, 1909.)

ridges passing on either side and meeting again in front to a small median fold which extends forwards to near the branchial apertures.

There is a continuous median fin in the Ammocoete larva, but in the adult Petromyzontia the dorsal fold is subdivided and expanded into two dorsal fins separate from the caudal. These dorsal fins are supported by cartilaginous unsegmented median rods, with tapering and branching distal ends reaching to near the margin, and with bases resting on the axial membranous covering of the neural tube. They are about three to

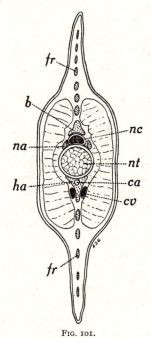

Fig. 101.

Petromyzon merinus. Transverse section of tail. *b*, Base of dorsal fin-radial; *ca*, caudal artery; *cv*, caudal vein ; *fr*, cartilaginous fin - ray or radial; *ha*, base of ventral fin-radial; *na*, neural arch; *nc*, nerve cord ; *nt*, notochord with sheaths.

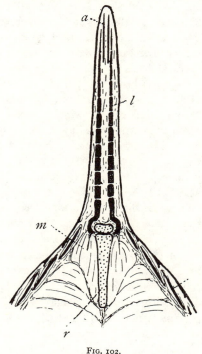

Fig. 102.

Diagram of a section through the dorsal fin of a Teleost. *a*, Actinotrich ; *l*, lepidotrich ; *m*, radial muscle; *r*, endoskeletal fin-radial; *s*, scale. (From E. S. Goodrich, *Quart. Jour. Micr. Sci.*, v. 47, 1903.)

four times as numerous as the segments, Fig. 98. In both Myxinoidea and Petromyzontia similar cartilaginous rays support the dorsal and ventral lobes of the caudal fin extending round the tip of the notochord ; they tend to fuse at their base, forming plates partially enclosing both spinal cord and notochord, Figs. 99, 100, 101. Since these rods are unsegmented, no distinction can be drawn in Cyclostomes between true fin radials and axial rods (p. 92), but radial muscles occur in the dorsal fins of Petromyzontia. All these cartilaginous structures are differentiated in the skeletogenous tissue surrounding the axial nerve chord and noto-chord and extending into the fin-folds in the embryo (Schaffer, **132**).

H

Very similar unsegmented branching median rods springing from the axial skeleton seem to have been present in the Devonian *Palaeospondylus* (Traquair, 137-8), and afford evidence of its relationship with the Cyclostomes.

Pisces.—Both the median and the paired fins of Pisces are provided not only with endoskeletal radials, but also with dermal fin-rays disposed on both sides of the fins and to which the radial muscles become attached. There are two chief kinds of these dermotrichia (Goodrich, 122) : the unbranched and unjointed horny fin-rays, or ceratotrichia, of the Chondrichthyes, and the bony lepidotrichia of Teleostomes, usually branching and jointed. Delicate horny rays (actinotrichia) develop in the embryonic fins of all Teleostomes, and are doubtless homologous with the ceratotrichia. They persist at the growing margin of the fins ; but they are functionally replaced in later stages by the more superficial lepidotrichia formed from modified scales, Fig. 102. The Dipnoi also have jointed bony rays (camptotrichia), possibly formed by the combination of outer lepidotrichia with deeper ceratotrichia. While in the more primitive Pisces the radials extend far into the fin-folds, they tend in the higher forms, and especially in the higher Teleostomes, to become more restricted to their base, the web of the fins being more and more supported by the dermotrichia.

Median fins occur in Ostracodermi, usually in the form of a short dorsal (*Thelodus*, among Pteraspidomorphi ; Cephalaspidomorphi, Fig. 151 ; Pterichthyomorphi) or an anal fin (Anaspida). They may be covered with denticles, or bony plates in rows somewhat resembling lepidotrichia, or with large scales (whether deeper ceratotrichia were present is not known, but there appear to have been endoskeletal rods, at all events in the caudal fin of Anaspida (Kiaer, 126)).

THE CAUDAL FIN

The study of the general homology of the skeleton of the median fins has a bearing on our interpretation of the structure of the caudal fin in fishes, a subject of very considerable interest from the point of view not only of morphology, but also of evolution in general. For the development of the caudal fin in the higher fishes affords a most striking instance of so-called recapitulation ; as in other cases, however, it may be interpreted as a repetition of the developmental stages of the ancestor and not of adult phylogenetic stages.

The structure and development of the caudal fin of aquatic Vertebrates has been studied by many authors. Though L. Agassiz, in his great work on Fossil Fishes, 1833, introduced the terms heterocercal and homocercal to indicate externally asymmetrical and symmetrical caudal fins, and McCoy later the term diphicercal for the truly symmetrical form, it was not till Huxley, 1859, studied the development of the tail in *Gasterosteus* that the full significance of the false or secondary symmetry of the homocercal type became apparent. Kölliker then described the anatomy of the tail in ' Ganoids ', and Lotz in Teleosts, while A. Agassiz (1878) and Huxley (**125**) completed our knowledge of the development, and Ryder (**130**), Emery (**121**). Gregory (**123**), and Dollo (**119-120**) have since made important contributions. Recently Whitehouse (**139-140**), Totton (**136**), Regan (**129**), and Schmalhausen have studied the question in further detail. Three main types have been distinguished : (1) Protocercal (diphycercal)[1] : primitively symmetrical externally and internally, with continuous dorsal epichordal and ventral hypochordal fin-folds equally developed respectively above and below the notochordal axis which is straight. (2) Heterocercal : the posterior end of the notochord is bent upwards, the ventral hypochordal lobe of the fin being more developed than the epichordal, and the caudal fin consequently asymmetrical both externally and internally. (3) Homocercal : the caudal fin is externally

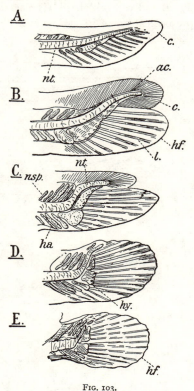

Fig. 103.

Successive stages in the development of the homocercal tail of the Flounder, *Pleuronectes flesus*, L., showing the disappearance of the axial lobe, *c*, and growth of the hypochordal fin, *hf*. *ac*, Actinotrichia ; *ha*, haemal arch ; *hy*, hypural cartilage ; *l*, dermal ray ; *nsp*, neural spine ; *nt*, notochord. (After A. Agassiz.) (From Goodrich, *Vert. Craniata*, 1909.)

symmetrical but internally asymmetrical, and the notochordal axis upturned and shortened, Fig. 105. As the later types have been derived from the earlier, intermediate forms occur which bridge over the gaps between them and do not quite fit these definitions.

---

[1] The term diphycercal having been somewhat loosely applied to secondarily symmetrical as well as to primitively symmetrical tails, it is better to use the term protocercal only for the latter (Whitehouse).

A protocercal fin is found in the Acrania (Cephalochorda), the extinct *Palaeospondylus*, and Cyclostomata, Fig. 98. It is adapted for propulsion by an undulating lateral motion of the body. When radials are developed, as in Cyclostomes, they are symmetrically disposed in the dorsal and ventral lobes.[1] The caudal fin of all Pisces passes through a protocercal stage in early development (before the appearance of the skeleton), Figs. 103, 104; but this structure is probably not retained

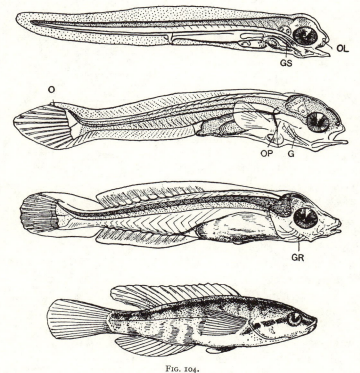

FIG. 104.

Larvae of Teleost, *Ctenolabrus* (after A. Agassiz, from B. Dean, *Fishes Living and Fossil*, 1895). G, Gill-arches; GR, branchiostegal rays; GS, gill-slit; O, region of upturned notochord; OP, opercular fold; OL, olfactory pit.

unmodified in any living fish, although persisting in aquatic Amphibia. The tail of the majority of the lower Pisces is frankly heterocercal. In the typical heterocercal tail of a shark (or Teleostome) the centra remain separate on the upturned axis, except at its extreme tip where they may be represented by a cartilaginous rod (p. 27). Dermal fin-rays are

[1] Even in Cyclostomes the dorsal and ventral lobes are not quite equal. The structure of these fins has already been described above (p. 95).

present in both lobes; especially well developed in the hypochordal lobe, they may become highly modified and reduced in the epichordal lobe. This lobe is supported by a series of closely set radials resting on but mostly separate from the neural arches, Figs. 105, 132 A. Usually, except near the tip, the radials are twice as numerous as the centra, thus corresponding to interbasals as well as basals. The larger hypochordal lobe is strengthened by what appear to be broad prolonga-

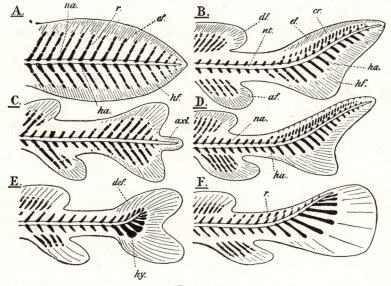

FIG. 105.

Diagrams showing the modifications of caudal fins, and the relations of the endoskeletal radials of median fins to the axial skeleton. A, Protocercal (diphycercal) type, with equal dorsal and ventral lobes (Dipnoi); B, heterocercal type (Selachii); C, modified diphycercal (Coelacanthini); D, heterocercal (Chondrostei); E, homocercal type (Teleostei); F, abbreviate heterocercal type (Amioidei). af, Anal fin; axl, axial lobe; cr, epichordal radial; def, dorsal lobe partly formed by epichordal lobe; df, dorsal fin; ef, epichordal lobe; ha, haemal arch; hf, hypochordal lobe; hy, hypural arch; na, neural arch; nt, notochord; r, radial. The endoskeleton is black. (From Goodrich, Vert. Craniata, 1909.)

tions of the haemal spines, known as hypurals. There is good evidence that these represent combined true radials and haemal spines (p. 89), and separate radials often persist anteriorly (Whitehouse, Schmalhausen). Special radial muscles are absent in the dorsal but present in the ventral lobe. Heterocercy is related to fast swimming and active motion, and useful perhaps to counteract the heaviness of the head. The caudal fin, the chief organ of propulsion, is separated from the more anterior dorsal and anal fins, and the enlarged hypochordal lobe enables the fish to rise rapidly. But in many Selachians heterocercy seems to have been secondarily reduced, the ventral lobe diminishes, and the axis tends to become straight (*Scyllium, Chlamydoselachus*). The Holocephali may be typically heterocercal (Callorhynchus), but in many genera inhabiting the deeper waters the caudal fin is much elongated and may almost return to a symmetrical condition (*Harriotta*). In Pleuracanthodii also heterocercy is but feebly developed; in Acanthodii, however, and especially in Cladoselachii, it is pronounced, Fig. 106. Among Ostracodermi, the Pteraspidomorphi, Cephalaspidomorphi, and Pterichthyomorphi have heterocercal caudals. Heterocercy was not only prevalent in Devonian, but already established in Silurian times (*Pteraspis*, etc.).[1]

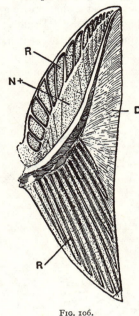

FIG. 106.

Heterocercal caudal fin of *Cladoselache fyleri* (from B. Dean, *Fishes Living and Fossil*, 1895). D, Dermal fin-rays; N+, neural arches; R, radials.

The Teleostomes show every gradation between heterocercy and homocercy. Among Osteolepidoti, some like *Osteolepis* have heterocercal tails, while in others like *Glyptopomus* and *Tristichopterus* the axis is almost straight and the epichordal lobe large. In *Eusthenopteron* the dorsal and ventral lobes are about equally developed above and below the straight vertebral column, and a middle axial lobe, or opisthure, is borne by the projecting extremity of the notochord. A very similar structure is seen in the Coelacanthini, where the notochord is straight and the opisthure distinct, Fig. 95. It is generally held that these more or less symmetrical tails have been secondarily derived from a more primitive heterocercal type; but the evidence is uncertain, Figs. 107-109.

Still more difficult to interpret is the symmetrical tail of modern Dipnoi. Here neither in ontogeny nor in the adult is there any evidence of heterocercy. The posterior extremity of the notochord remains straight, though it, as in many other fishes, undergoes degeneration, becoming enclosed in a cartilaginous rod representing the modified vertebral elements, Figs. 110, 91. Balfour, indeed, considered that the true caudal had degenerated and been replaced by the extended dorsal

---

[1] According to Kiaer's recent account (126) the tail of the Anaspida was of the reversed heterocercal type, with the axis bent downwards, a very small hypochordal and a large epichordal lobe. Such a 'hypocercal' tail is unique among fishes, but is paralleled by the analogous caudal fin of Ichthyosauria.

and anal fins (116), while Schmalhausen on the contrary believes the median fins of modern Dipnoi to be formed by the forward growth of the

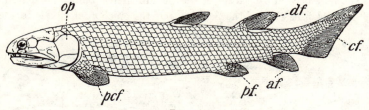

FIG. 107.

Osteolepis macrolepidotus, Ag.; restored. (After Traquair.) *af*, Anal fin; *cf*, caudal fin; *df*, second dorsal fin; *op*, opercular; *pcf*, pectoral fin; *pf*, pelvic fin. (From Goodrich, *Vert. Craniata*, 1909.)

caudal; neither of these views has good evidence in its favour. Dollo (119) derives the continuous median fins of modern Dipnoi from the

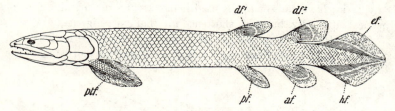

FIG. 108.

Restoration of *Glyptopomus Kinnairdi*, Huxley; Devonian. (After Huxley, modified.) *af*, Anal, *df*, dorsal; *ef*, epichordal; *hf*, hypochordal; *pf*, pelvic, and *ptf*, pectoral fin. (From Goodrich, *Vert. Craniata*, 1909.)

condition seen in *Dipterus*, where the dorsal and anal fins are separate and short, and the caudal is heterocercal, through such intermediate forms as *Uronemus, Phaneropleuron,* and *Scaumenacia* described by Traquair, in which the median fins become more and more elongated until they meet and finally fuse into continuous fins as in modern Dipnoi, Figs. 111, 112. This view is now generally accepted; but although *Dipterus* is the earliest known Dipnoan, yet the occurrence of *Scaumenacia* already in the Devonian suggests that the continuous fins may after all be primitive. What-

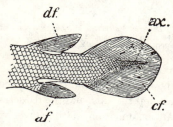

FIG. 109.

Tail of *Diplopterus Agassigii*, Traill. (After Traquair.) *ax*, Extremity of axis; *af*, anal fin; *cf*, hypochordal fin; *df*, dorsal fin. (From Goodrich, *Vert. Craniata*, 1909.)

ever conclusion is reached with regard to the primitive type of caudal fin in Dipnoi and Teleostomi, it is clear that it must be admitted that either the symmetrical or the asymmetrical form has been independently acquired in various groups.

The heterocercal caudal fin of the Chondrostei is built on essentially the same plan as that of the Selachii, but the epichordal lobe is further

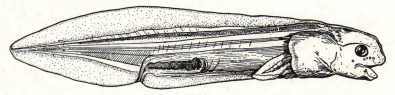

FIG. 110.

Larva of *Ceratodus forsteri*.   (After R. Semon, '93, from B. Dean, *Fishes Living and Fossil*, 1895.)

reduced and the lepidotrichia generally remain only as a double series of pointed scales often fused to ∧ shaped fulcra, Figs. 113, 132*a*.   The Polypterini have tails of a disguised heterocercal structure almost symmetrical outwardly, but with shortened internal axis, Fig. 114.

In the Holostei a progressive modification takes place leading from the heterocercal to the homocercal type, by the relative shortening of the

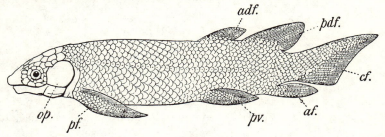

FIG. 111.

Dipterus Valenciennesii, Sedgw., restored. (After Traquair, slightly modified.)   *adf*, Anterior dorsal fin;   *af*, anal fin;   *cf*, caudal fin;   *op*, operculum;   *pf*, pectoral fin;   *pv*, pelvic fin. (From Goodrich, *Vert. Craniata*, 1909.)

axis which is withdrawn to the base of the fin, the great development of the hypochordal fin (more especially its anterior lobe) which projects far beyond the axis, and the reduction of the epichordal fin to a mere vestige. An examination of the stages in the development of these tails shows that the original axial lobe or opisthure takes little or no share in the formation of the adult fin, the dermal rays of which are entirely derived from the hypochordal fin, and in the higher types from its anterior portion chiefly.   Thus, in the ordinary forked homocercal caudal, the upper part is derived from the more posterior and the lower part from the more anterior region of the same hypochordal fin.   A few of the dermal rays of the epichordal fin may persist dorsally near the base. Accompanying this reduction of the axis is the gradual establishment of the outward

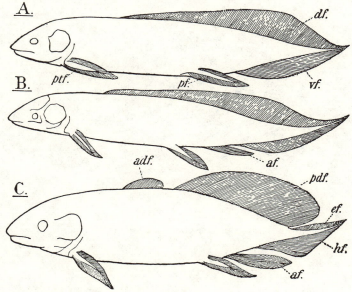

FIG. 112.

Restorations from Traquair of A, *Uronemus lobatus*, Ag., Lower Carboniferous ; B, *Phanero-pleuron Andersoni*, Huxley, Upper Devonian ; C, *Scaumenacia curta*, Whiteaves, Upper Devonian. *adf*, Anterior dorsal fin ; *af*, anal fin ; *df*, dorsal fin ; *ef*, epichordal lobe, and *hf*, hypochordal lobe, of caudal fin ; *pdf*, posterior dorsal fin ; *pf*, pelvic fin ; *ptf*, pectoral fin ; *vf*, ventral fin. (From Goodrich, *Vert. Craniata*, 1909.)

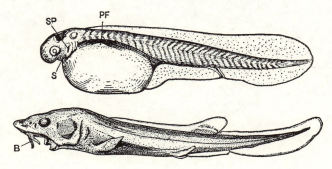

FIG. 113.

Larvae of *Acipenser* (from B. Dean, *Fishes Living and Fossil*, 1895). B, Barbel ; PF, pectoral fin ; S, mouth ; SP, spiracle.

symmetry carried to a wonderful state of perfection in the higher Teleostei ; it effects not only the scaling on the tail, but also the disposition of the lepidotrichia in the fin, Figs. 103, 116, 121-2.

The homocercal form is assumed in development not suddenly but gradually ; and tails belonging to the higher grades pass through lower

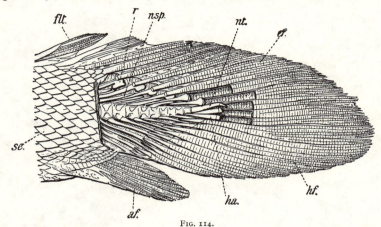

FIG. 114.

Dissected tail of *Polypterus bichir*, Geoffr. (After Kölliker.) *af*, Anal fin ; *ef*, epichordal fin ; *flt*, dorsal finlet ; *ha*, haemal arch ; *hf*, hypochordal fin ; *nsp*, neural spine ; *nt*, slightly upturned tip of the notochord ; *r*, endoskeletal radial ; *sc*, scales. The proximal ends of the dermal rays have been cut off to expose the radials and tip of the notochord. This tail is probably secondarily almost diphycercal. (From Goodrich, *Vert. Craniata*, 1909.)

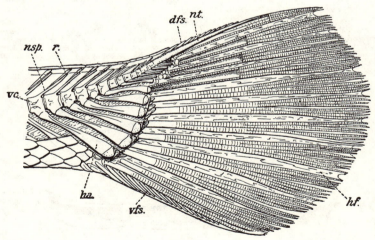

FIG. 115.

Dissected tail of *Lepidosteus*. (After Kölliker.) *dfs*, Dorsal fulcral scales ; *ha*, expanded haemal arch ; *hf*, hypochordal fin ; *nsp*, neural spine ; *nt*, upturned tip of the notochord ; *r*, dorsal radial ; *vc*, vertebral centrum ; *vfs*, ventral fulcral scales. The proximal ends of the dermal rays have been cut off to expose the endoskeleton. (From Goodrich, *Vert. Craniata*, 1909.)

grades in regular sequence. Moreover, intermediate forms occur, as for instance in *Lepidosteus* and *Amia*, between the heterocercal and homo-

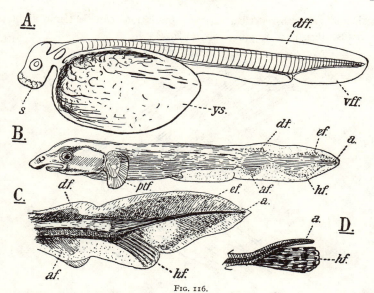

FIG. 116.

The development of the caudal fin of *Lepidosteus*. (After A. Agassiz.) A, Young larva with anterior sucker, *s*; yolk-sac, *ys*; continuous dorsal and ventral fin folds, *dff* and *vff*; and straight notochord. The later stages, B, C, and D, show the upbending of the notochord, the dwindling of the axial lobe, *a*, which disappears in the adult (cp. Fig. 115), and the great development of the hypochordal fin, *hf*. *af*, Anal, *df*, dorsal, *ef*, epichordal, and *ptf*, pectoral fin. (From Goodrich, *Vert. Craniata*, 1909.)

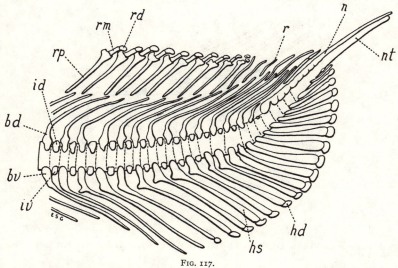

FIG. 117.

Left-side view of skeleton of tail of young *Amia calva*, about 45 mm. long. Endoskeleton cartilaginous; limits of future bony centra just visible. *bd*, Basidorsal; *bv*, basiventral; *hd* and *hs*, distal and proximal elements of haemal spine; *id*, interdorsal; *iv*, interventral; *n*, covering of neural canal composed of fused neural arches; *nt*, tip of notochord; *r*, radial; *rd*, *rm*, and *rp*, distal, middle, and proximal elements of dorsal fin radial.

cercal, and an almost perfect series of gradations can be found among the fossils, Figs. 115-17, 35.

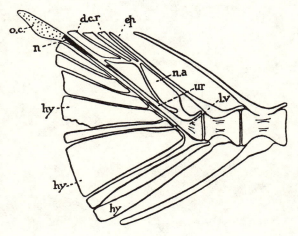

FIG. 118.

Skeleton of caudal fin of *Clupea pilchardus* (from R. H. Whitehouse, *P.R.S.*, 1910).   *d.c.r*, Dorsal radials ; *ep*, epural ; *hy*, hypural ; *l.v*, last vertebra ; *n*, notochord ; *n.a*, neural arch ; *o.c*, opisthural cartilage ; *ur*, urostyle.

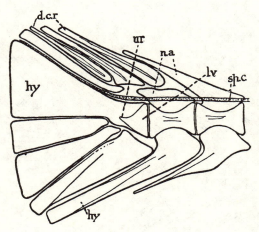

FIG. 119.

Skeleton of caudal fin of *Trigla lineata* (from R. H. Whitehouse, *P.R.S.*, 1910).   *sp.c*, Spinal cord, other letters as in Fig. 118.

In the most advanced or homocercal type, occurring only in the Teleostei, the notochord is much shortened and withdrawn ; though in some primitive forms (*Clupea*) it may still project surrounded by cartilage a little beyond the vertebral elements. The upturned region of the noto-chord is enclosed in a urostyle, a process of the last vertebral centrum which may represent the fused centra of this posterior region. This urostyle, at first independent (*Clupea*, Fig. 118), becomes in more specialised forms fused with the last hypural and reduced to a mere vestige. The neural arch of the penultimate centrum may become modified into a

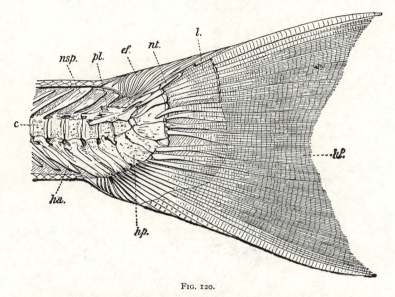

FIG. 120.

Dissected tail of *Salmo*. (After Kölliker.) *c*, Centrum of caudal vertebra ; *ef*, epichordal dermal ray ; *ha*, haemal arch ; *hf*, hypochordal fin ; *hp*, expanded haemal arch or hypural ; *l*, dermal ray of opposite (right) side ; *nsp*, neural spine ; *nt*, upturned extremity of the notochord ; *pl*, covering bony plate (modified neural arch ?). (From Goodrich, *Vert. Craniata*, 1909.)

pair of bones embracing the urostylar region. A few dorsal radials may remain free bearing dermal rays in this region, but tend to disappear in the more specialised tails. One of these radials may fuse with a dorsal arch to form a true ' epural '. But the most characteristic modification is the great development of the hypurals. More numerous and free along the urostyle in primitive forms (Clupeiformes) they tend to become reduced in number, increased in size, and the hindmost fuses with the last centrum. Finally, in the most specialised tails the hypurals are two in number, fused to the last centrum, and symmetrically disposed above and below the longitudinal axis, Figs. 119-23, 125.

With regard to the morphology of the hypurals it may be pointed out that there is good reason to believe that, as in the Selachians so in the Holostei, they are typically formed by the fusion of a radial with the spine of a haemal arch. Their structure in *Amia*, for instance, with a

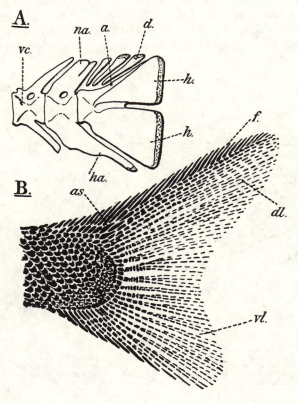

FIG. 121.

A, Endoskeleton of the tail of *Cottus gobio*, L. (After Lotz.) B, Tail of *Pachycormus heterurus*, Ag., showing the last external trace of the heterocercal structure. *a*, Bony sheath of tip of notochord ; *as*, scales covering the upturned tip of notochord ; *d*, dorsal radials (?) ; *dl*, dorsal lobe of hypochordal fin ; *f*, fulcra ; *h*, hypural bone ; *ha*, haemal arch ; *na*, neural arch and spine ; *vc*, vertebral centrum ; *vl*, ventral lobe of hypochordal fin. (From Goodrich, *Vert. Craniata*, 1909.)

distal or terminal cartilage resembling that of the radial, strongly suggests this origin, Fig. 117. In Teleosts, also, separate terminal cartilages may appear ; but in the higher forms they no longer occur, and possibly the hypurals are there formed entirely from the haemal spine.

Although many Teleostei have more or less tapering and apparently symmetrical tails, they are never truly protocercal.[1] The assumption of a tapering form is an adaptation to sluggish habits either in bottom living or deep sea fishes, which leads to the reduction in size of the hypochordal fin, and the elongation of the dorsal and anal fins until finally a con-

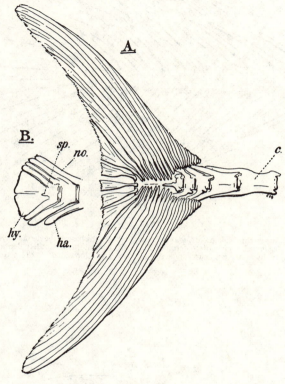

FIG. 122.

A, Skeleton of the tail of *Thynnus vulgaris*, Cuv. and Val., Tunny fish ; B, terminal caudal vertebrae of *Scomber scomber*, L., Mackerel. *c*, Centrum ; *ha*, haemal arch ; *hy*, hypural compound bone, partially concealed in A by the dermal rays ; *no*, neural arch ; *sp*, spine. (From Goodrich, *Vert. Craniata*, 1909.)

tinuous fin-fold is re-established (Dollo, **120**). Such isocercal tails, as they may be called, have been repeatedly acquired in various Teleostean families though with differences of detail (Anguilliformes, Notopteridae, Gymnarchidae, Macruridae, Zoarcidae, etc.) ; but their true nature is

[1] The Gadidae have a peculiar tail-fin in which either the true caudal has extended forwards, or it has combined with the posterior part of the median dorsal and ventral fins (Figs. 123-4).

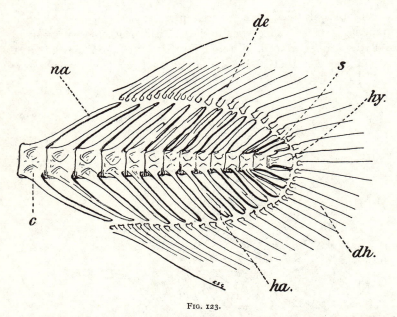

FIG. 123.

Tail end of the vertebral column of *Gadus morrhua*, L. Only the base of the dermal rays is indicated. *c*, Centrum ; *de*, dorsal (epichordal) dermotrichia ; *dh*, ventral (hypochordal) dermotrichia ; *ha*, haemal spine ; *hy*, hypural ; *na*, neural spine ; *s*, detached spine (or radial). (From Goodrich, *Vert. Craniata*, 1909.)

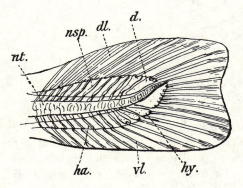

FIG. 124.

Tail of a young cod. (After A. Agassiz.) *d*, Dorsal cartilage (neural arch ?) ; *dl*, lepidotrichia of dorsal lobe of caudal fin ; *ha*, haemal arch ; *hy*, hypural ; *nsp*, neural spine ; *nt*, notochord. (From Goodrich, *Vert. Craniata*, 1909.)

always betrayed by their internal asymmetry, at all events in young stages, the extremity of the notochord being sharply bent upwards, and hypurals

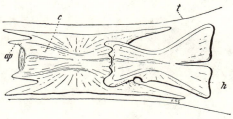

FIG. 125.

Callionymus lyra, L.   Left-side view of the two last caudal vertebrae, enlarged.   *ap*, Anterior articulating process ;   *c*, centrum ;   *h*, hypural expansion ;   *t*, outline of tail incomplete.   (From Goodrich, *Vert. Craniata*, 1909.)

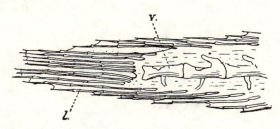

FIG. 126.

Skeleton of the extremity of the tail of *Fierasfer dentatus*, Cuv.   (After Emery.)   *l*, Lepidotrich ; *v*, last vertebra.   (From Goodrich, *Vert. Craniata*, 1909.)

being present.   They are cases of disguised homocercy.   More complete disappearance of the caudal fin with truncation of the axis leads to yet another type of fin, called gephyrocercal, seen in *Fierasfer*, Fig. 126, and *Orthagoriscus* (Ryder, **130** ; Emery, **121**).

DEVELOPMENT OF THE MEDIAN FINS OF FISHES

The development of the median fins is best known in the Selachians where it has been studied by Balfour (**142**), Dohrn (**159**), Mayer (**128**), Goodrich (**172**), and Schmalhausen (**133**).   They are first indicated by a narrow ridge or fold of epidermis, the precursor of a wider longitudinal fold into which penetrates mesoblastic mesenchyme, Fig. 12.   A continuous fold is thus formed running dorsally along the trunk to the tip of the tail, where it meets a similar ventral fold running forwards to just behind the cloaca, Figs. 104, 110.   At an early stage the first indications of the definitive fins appear as local thickenings of the mesoblastic tissue.   Meanwhile the whole median fold continues to grow and usually forms a

I

provisional embryonic fin of considerable size. Soon, however, the local thickenings become enlarged, these regions alone developing into the adult

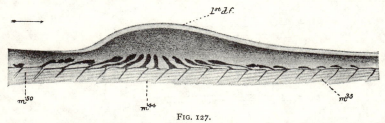

FIG. 127.

Scyllium canicula. Embryo 24 mm. long; right-side view of reconstruction of first dorsal fin. A single muscle-bud is shown derived from each myotome, *m.* (E. S. Goodrich, *Quart. Jour. Micr. Sci.*, 1906.)

fins while the intervening parts of the fold disappear. In each fin rudiment is formed a thick longitudinal median plate of mesenchyme in which the fin-skeleton later develops (Balfour, **142**). On either side of this plate, in the case of the dorsal fins, the myotomes give off into the fold small epithelial outgrowths or ' muscle-buds ', the rudiments of the

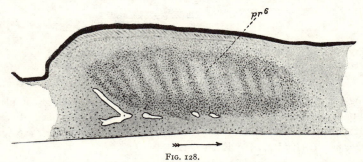

FIG. 128.

Scyllium canicula. Longitudinal section of first dorsal fin of embryo 33 mm. long, showing originally separate mesenchymatous rudiments of radials spreading and joining above and below. (E. S. Goodrich, *Quart. Jour. Micr. Sci.*, 1906.)

radial muscles (Dohrn, **159**; Mayer, **128**). Each myotome gives off one muscle-bud from its dorsal edge (Goodrich, **172**); but since, as will be explained later, the dorsal fins of Selachians are considerably shortened or ' concentrated ' (the body growing faster in length than the base of the fin), the more anterior and posterior buds appear to move away from their myotomes to reach the base of the fin; they also decrease in size at the two ends. The buds become detached and give rise to the radial muscles, Figs. 127-9. Thus each adult radial muscle corresponds to and is (in the main) derived from one bud and therefore one segment. This primary strict segmentation is, however, to some extent disturbed in later development, partly owing to increasing concentration at both ends of the fin where the buds may be crowded and fuse to form compound muscles. Moreover, throughout the length of the fin the buds may be joined by anastomosing bridges allowing some tissue to pass from a segment to its neighbours, Fig. 129. The right and left buds are regularly disposed in pairs, and between each pair develops a skeletal radial, Figs. 130-31. The first indication of the radials is

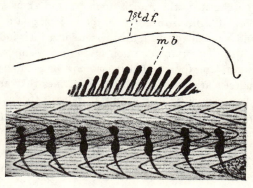

FIG. 129.

Scyllium canicula. Embryo 28 mm. long; left-side view of reconstruction of region of first dorsal fin. *mb*, Muscle-buds, becoming converted proximally into muscle fibres, and united by bridges at their base. (E. S. Goodrich, *Quart. Jour. Micr. Sci.*, 1906.)

seen as a series of separate streaks of denser mesenchyme in the median plate. These become defined as procartilaginous rods merging into each other above and below, Figs. 128, 130-31. The whole skeleton of the fin is so preformed and the various pieces of the adult structure

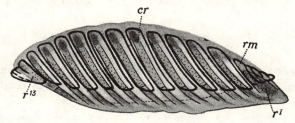

FIG. 130.

Reconstruction of first dorsal fin of embryo *Scyllium canicula*, 37 mm. long. Right-side view; blastema and procartilage grey, cartilage dotted, radial muscles in black outline. *cr*, Radial; *rm*, radial muscle; r^1-r^{13}, radials. (E. S. Goodrich, *Quart. Jour. Micr. Sci.*, 1905.)

chondrify separately in it. The anal fin of Selachians develops in essentially the same way; but here (as also in the caudal fin) distinct muscle-buds do not occur, the radial muscles being derived from proliferations of cells from the ventral border of the myotomes.

In all important points the development of the median fins of the Osteichthyes resembles that of the Selachii. Segmental muscle-buds appear to be absent in the Dipnoi but have been described in *Acipenser*, *Amia*, and *Lepidosteus* (Salensky, 131; Schmalhausen, 133), and in *Salmo*

(Harrison, **179**), where they occur not only in the dorsal but also in the anal fin. In the median fins of the Osteichthyes, which are usually little concentrated, the radials tend to appear as more distinctly separate procartilaginous rods than in the Selachii ; later they become chondrified and generally ossified, usually retaining their independence.

The important question of the nerve-supply of the fins will be discussed later (p. **133**), but it may now be pointed out that since the radial muscles are derived from the myotomes they are innervated by efferent fibres coming from the ventral roots of the spinal nerves. Each radial muscle primarily is supplied from the nerve of that segment which gave rise to the bud from which it developed. This occurs in regular order

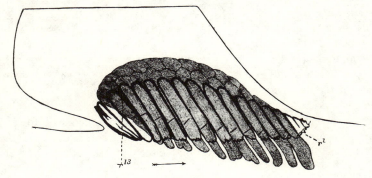

Fig. 131.

Scyllium canicula. Right-side view of skeleton and radial muscles (black outline) of first dorsal fin of adult. r^{1-13}, Radials. (E. S. Goodrich, *Quart. Jour. Micr. Sci.*, 1906.)

from before backwards. The number of spinal nerves supplying motor fibres to the fin, therefore, agrees with the number of buds contributing to its musculature, Figs. 143, 144.

We may now turn to the general morphology of the median fins. In many fish which in the adult state have discontinuous dorsal, caudal, and anal fins, these separate fins develop as differentiations in a continuous embryonic fin-fold, which becomes subdivided by the obliteration of certain regions. Moreover, traces of the fin-skeleton and musculature are found between the separate fins of many such fish, as for instance in *Squalus, Rhina, Pristis* (Thacher, **219** ; Mivart, **196**), and Teleosts, Figs. 26, **141**. In Cephalochorda and in Myxinoidea the folds remain continuous. It is natural, then, to suppose that such has been the phylogenetic history of median fins, and that discontinuity is secondary. In support of this view it may be pointed out that in the Pleuracanthodii, an extinct group of Chondrichthyes, Carboniferous forms occur with a complete or almost complete dorsal fin-fold. Nevertheless, it is a notable fact that in most groups of fishes the most ancient known representatives are already provided with separate short dorsal or anal fins. This is the case not only with the Ostracodermi, which date from the Silurian, the Clado-selachii, and the Acanthodii, but also with the early Teleostomes such as the Osteolepidoti, Chondrostei, Amioidei, and Teleostei. Indeed greatly

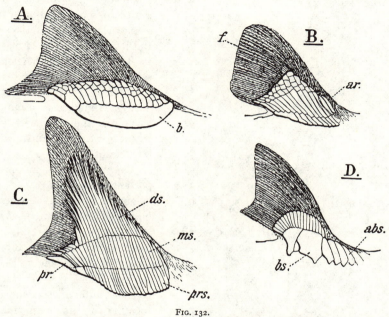

FIG. 132.

Dorsal fins, with the endoskeleton exposed, of : A, *Notidanus* (*Heptanchus*) *cinereus*, Gm. ; B, *Ginglymostoma cirratum*, Gm. ; C, *Zygaena malleus*, Risso ; D, *Rhynchobatus djeddensis*, Forsk. (After Mivart.) *abs*, Anterior radials lying on vertebral column ; *ar*, anterior radial ; *b*, longitudinal basal ; *bs*, basal ; *ds*, distal segment of radial ; *f*, fin web supported by ceratotrichia ; *ms*, median segment of radial ; *pr*, posterior radial ; *prs*, proximal segment of radial. These figures illustrate the formation of basals by the concrescence of radials. (From Goodrich, *Vert. Craniata*, 1909.)

extended dorsal and anal fins are characteristic of the more specialised forms, such as the Platysomidae among Chondrostei, Amiidae and Pycnodontidae among Amioidei. The primitive extinct Leptolepidae among Teleostei possess short dorsal and anal fins ; but these are greatly extended or even continuous folds in many modern families such as the Gymnotidae among Cypriniformes, the Lampridiformes, Mastacembelli-formes, the Zeorhombiformes, and the Anguilliformes. It also seems to hold good for the Dipnoi, where two dorsal and an anal occur in the Devonian *Dipterus*, but continuous fin-folds are present in the modern genera (p. 102).

The ontogenic history of the median fins, however, clearly shows that they are of fundamentally segmental structure as regards their nervous, muscular, and skeletal elements even if this primary segmentation agreeing with that of the body is obscured or lost owing to later specialisations. But it must be admitted that for the present the question remains open as to whether continuity is primitive or not. One thing seems certain, that discontinuity is generally if not always accompanied by concentration.

Concentration is an important process in the development of the fins of fishes both median and paired. As indicated above (p. 114), it is due to differential growth of fin and body, leading to the relative shortening of the base of the fin and crowding together of its segmental elements, especially at its anterior and posterior ends. Thus a dorsal fin, which in an embryo *Scyllium* is developed from some fourteen segments, comes to occupy only about six segments in the adult (Goodrich, **172**). As a rule more muscle-buds arise in the embryo than come to full development in the adult, some being suppressed at each end, Figs. 127, 143. It is owing to concentration that, increasingly towards each end of the fin, the radial

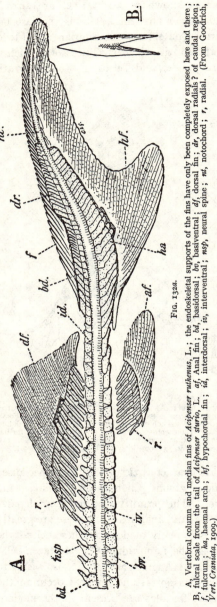

FIG. 132*a*.

A, Vertebral column and median fins of *Acipenser ruthenus*, L.; the endoskeletal supports of the fins have only been completely exposed here and there; B, fulcral scale from the tail of *Acipenser sturio*, L. *af*, Anal fin; *bd*, basidorsal; *bv*, basiventral; *df*, dorsal fin; *dr*, dorsal radials ? of caudal region; *f*, fulcrum; *ha*, haemal arch; *hf*, hypochordal fin; *id*, interdorsal; *iv*, interventral; *nsp*, neural spine; *nt*, notochord; *r*, radial. (From Goodrich, *Vert. Craniata*, 1909.)

muscles become separated from the myotomes which gave rise to them, and the nerves are made to converge towards the narrowed base of the fin to supply them. Thus are formed longitudinal 'collector' nerves, especially in front of the fin; for here concentration is usually more pronounced.

The skeletal elements are likewise affected, and concrescence of the radials is the second important factor in the modification of the skeleton of fins [1] (Thacher, Mivart). It plays but a small part in the median fins of Actinopterygii, but may be seen in such Crossopterygii (Osteolepidoti) as *Eusthenopteron*, or *Glyptolepis*, where the numerous radials of the dorsal fin fuse at their base to form a longitudinal axis of several elements, Figs. 92, 155 A. A bony plate at the base of the very short median fins of Coelacanthini seems to represent the base of fused radials, Fig. 95. Concrescence is clearly shown in *Pleuracanthus* (Fritsch, 23), where two short anals are formed containing a number of radials more or less completely combining proximally into jointed axis and basal pieces, Fig. 133. But it is in the Elasmobranchii that the

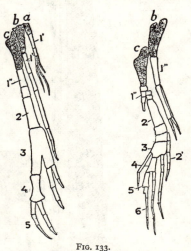

FIG. 133.

Skeleton of anal fins of *Pleuracanthus (Xenacanthus) Decheni*, Goldf. *a, b, c,* Haemal arches; 1-6, segments of fin-radials. Right border preaxial, and left postaxial. (After Fritsch, from A. S. Woodward.)

effects of concrescence are best seen. In the Holocephalan *Chimaera* the posterior dorsal fin is scarcely if at all concentrated, and here the parallel radials are separate and regularly distributed, Fig. 134. But in Holocephali generally the anterior dorsal becomes much concentrated behind the anterior spine; in *Myriacanthus* of the L. Lias, the radials are here represented by two or three pieces, while in the modern genera they fuse to a single cartilage, the whole fin-spine and all being erectile.

Many of the modern sharks still preserve separate segmental radials in the dorsal and anal fins, as in *Scyllium* and *Sphyrna*; even in these, however, a few of the radials may be fused, especially at their base. In others concrescence is very pronounced, as in the Notidanidae, where nearly all

[1] As noted above, it may also affect the muscles. A detailed study of the anatomy and development of unconcentrated fins is greatly needed.

the radials contribute to the formation of a continuous basal piece, Fig. 132.

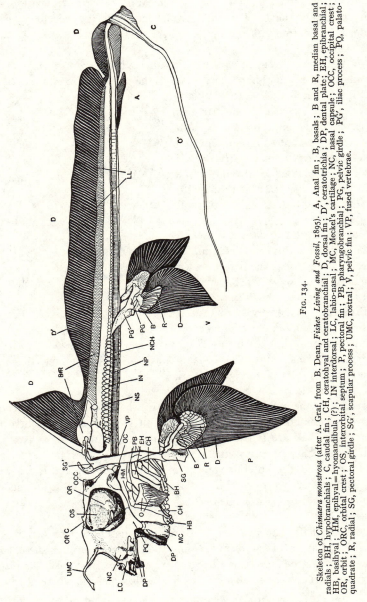

Fig. 134.

Skeleton of *Chimaera monstrosa* (after A. Graf, from B. Dean, *Fishes Living and Fossil*, 1895). A, Anal fin; B, basals; B and R, median basal and radials; BH, hypobranchials; C, caudal fin; CH, ceratohyal and ceratobranchial; D, dorsal fin; D′, dorsal fin; D″, ceratotrichia; DP, dental plate; EH, epibranchial; HB, basihyal; HM, epihyal = hyomandibula (?); IN interdorsal; LC, labio-nasal; MC, Meckel's cartilage; NC, nasal capsule; OCC, occipital crest; OR, orbit; ORC, orbital crest; OS, interorbital septum; P, pectoral fin; PB, pharyngobranchial; PG, pelvic girdle; PG′, iliac process; PQ, palato-quadrate; R, radial; SG, pectoral girdle; SG′, scapular process; UMC, rostral; V, pelvic fin; VP, fused vertebrae.

Those forms in which the fin is armed with an anterior spine tend to have the radials much fused sometimes on both sides of it (Hetero-

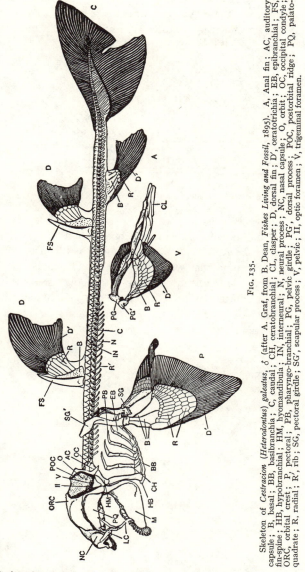

FIG. 135.

Skeleton of *Cestracion (Heterodontus) galeatus*, ♂ (after A. Graf, from B. Dean, *Fishes Living and Fossil*, 1895). A, Anal fin; AC, auditory capsule; B, basal; BB, basibranchia; C, caudal; CH, ceratobranchial; CL, clasper; D, dorsal fin; D', ceratotrichia; EB, epibranchial; FS, fin-spine; HB, hypobranchial; HM, hyomandibula; IN, interneural; N, neural process; NC, nasal capsule; O, orbit; OC, occipital condyle; ORC, orbital crest; P, pectoral; PB, pharyngo-branchial; PG, pelvic girdle; PG', dorsal process; POC, postorbital ridge; PQ, palato-quadrate; R, radial; R', rib; SG, pectoral girdle; SG', scapular process; V, pelvic; II, optic foramen; V, trigeminal foramen.

donti, Spinacidae); and in the Rajiformes the radials may combine to form an oblique basal jointed axis, bearing separate distal radials, Fig. 136. In such fins the fin-skeleton usually rests on the vertebral column and the primitive metamery is almost completely lost (*Acanthias, Squatina, Rhinobatus, Raja*).

Finally it may be pointed out in favour of the theory that the median fins were probably primitively continuous ; and that, if we trace the origin

FIG. 136.

Left-side view of a portion of the tail of *Raja*. The vertebral column has been exposed in front, also the endoskeleton of the two dorsal fins, showing the concrescence of the radials at their base to form a posterior axis. (From Goodrich, *Vert. Craniata*, 1909.)

of the muscular and nervous elements which go to build up the dorsal fins of such a form as *Scyllium*, and spread them out as if concentration had not taken place, we find that whereas in the adult they occupy quite a short space, when ' deconcentrated ' they extend over some thirty segments forming a continuous fin-fold along a considerable length of the trunk and tail (see diagrams, Fig. 143). Discontinuity in such a case is due to concentration.

CHAPTER III

PAIRED LIMBS

THE ORIGIN OF THE PAIRED LIMBS

Theories of the Origin of the Paired Fins.—The consideration of concentration and concrescence in median fins brings us to the problem of the origin of the paired limbs. Few questions concerning the general morphology of Vertebrates have aroused greater interest. Since it is generally recognised that the pectoral and pelvic limbs of Tetrapods must have been derived phylogenetically from the paired fins of their fish-like ancestors, the discussion of the general problem may be confined to the origin of the paired fins or 'ichthyopterygia' of fishes of which there are typically two pairs, each strengthened by an endoskeleton composed of radials projecting into the outstanding fin-lobe, and supported by an endoskeletal girdle in the body-wall. The pectoral fins are placed immediately behind the gill-arches, and the pelvic fins (often called the ventrals), typically, just in front of the anus.

Two main rival and incompatible theories have been held of the origin of paired fins. According to that put forth by Gegenbaur (**168-9**) the paired fins are modified gill structures, the girdles representing gill-arches, and the fin-folds with their contained fin-skeleton representing gill-flaps or septa with their branchial rays. The position of the pelvic fins far back is explained as due to the shifting backwards or migration of these posterior arches which have lost their branchial function. This may be called the 'gill-arch theory', Fig. 137.

The modern version of the second theory, put forth almost simultaneously by Balfour (**142**) and Thacher (**219**), and later by Mivart (**196**), holds that the paired fins are of essentially the same nature as the unpaired fins, and have been derived from paired longitudinal fin-folds. The endoskeletal radials (somactidia, pterygiophores) would in both kinds

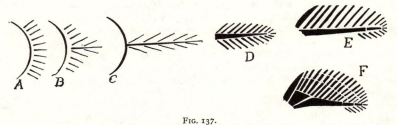

FIG. 137.

Diagram illustrating Gegenbaur's theory of origin of fin-skeleton from gill-arch and gill rays A-D ; D, biserial archipterygium ; D, E, F, origin of Selachian pectoral fin-skeleton from archipterygium. (From J. S. Kingsley, *Comp. Anat. Vertebrates*, 1926.)

have arisen for the stiffening of the fin-folds ; and the girdles would have been developed by the extension inwards of the base of the radials so as to afford firm support for the outstanding fin. This, the now most generally accepted view, is known as the 'lateral-fold theory'.

Each of these theories may claim to have had among its supporters some of the most eminent exponents of vertebrate morphology. V. Davidoff, Fürbringer, Braus (**144-8**), and others have followed Gegenbaur ; while Dohrn (**159**), Mayer (**128**), Wiedersheim (**230**), Haswell (**180**), Rabl (**206**), Mollier (**197**), Dean (**154-6**), A. S. Woodward (**231**), Harrison (**179**), Regan (**208**), Sewertzoff (**215-17**), and others have written in favour of the lateral-fold theory ; Osburn (**202**) and Goodrich (**172**) have recently discussed these theories in detail, and brought forward much evidence in favour of the view that the paired fins were developed from lateral folds, and against the rival theory.

An examination of the development of the paired fins shows that in every important particular it resembles that of the median fins. First usually indicated by a narrow ridge of folded epidermis, they subsequently grow out as longitudinal folds of the body-wall into which penetrates mesenchymatous mesoblast. Epithelial muscle-buds soon push their way into the fin-fold ; they spring from the ventral ends of the neighbouring myotomes and give rise to the radial muscles, Figs. 138, 143. The distribution and development of these buds is of great interest, all the phenomena characteristic of concentration in median fins being clearly shown in the less specialised forms, and more especially in Elasmobranchs. To begin with, the lateral fin-folds extend longitudinally over many more segments in the embryo than they do in the adult ; the shortening of the base of the fin as it becomes more clearly marked off from the body leads to the formation of a distinct notch at its hinder end, Figs. 113, 139.

Muscle-buds may develop from the segments along the whole of the embryonic fold, and even beyond. They gradually become vestigial and disappear beyond the fin areas. Two such buds arise from each myotome in Selachians, except near the extreme ends of the series where they may be reduced to one. A large number reach the base of the fin, separate from their myotomes, divide into upper (dorsal) and lower (ventral) halves, and spreading outwards develop into the radial muscles of the adult fin, Figs. 139, 620. In Osteichthyes only one primary bud, which divides, or two buds are given off from each myotome (Teleostei :

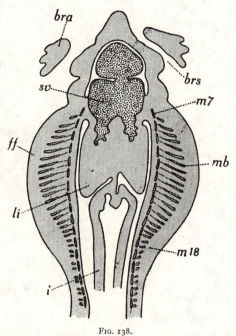

FIG. 138.
Scyllium canicula, embryo stage *N*. Reconstructed horizontal section. *bra*, Branchial bar ; *brs*, branchial slit ; *ff*, pectoral fin-fold : *i*, intestine ; *li*, liver ; m^{7-18}, ventral ends of metaotic myotomes ; *mb*, muscle-buds; those from myotomes 7 to 18 contribute to pectoral-fin musculature, behind them are vestigial buds.

Harrison, 179; Guitel, 175; Derjugin, 157. Chondrostei: Sewertzoff, 216-17; Ducret, 160; Kryžanovsky, 193. Dipnoi: Salensky, 213; Semon, 214; Agar, 141). Typically in all fishes a radial is differentiated between each pair of upper and lower muscle-buds (Rabl, Mollier, Derjugin). The radial muscles of the paired fins being thus derived from the myotomes naturally receive their motor nerve-supply from the ventral roots of the corresponding spinal nerves. Owing to concentration the nerves converge towards the base of the fins. In front and behind they may be gathered together so as to form a compound or 'collector' nerve, Fig. 143.

As in median so in paired fins the endoskeleton is differentiated in
the continuous plate of dense mesenchyme situated in the fin-fold between
the dorsal and ventral muscle rudiments.   In Selachians the radials first
appear as denser streaks, separate distally, but more or less combined
proximally into a plate which extends inwards in the body-wall as the
rudiment of the girdle, Fig. 620. Thus the separate limb girdle, basals, and
peripheral radials appear in a continuous procartilaginous rudiment. Later
on the individual elements seen in the adult arise *in situ* as separate
chondrifications, leaving non-cartilaginous joints (Balfour, 142 ; Mollier,
197 ; Ruge, 211, and others).   Since the girdle is primarily outside the

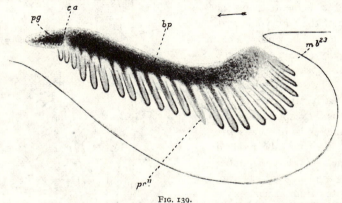

FIG. 139.

Scyllium canicula. Reconstruction of left pelvic fin of embryo 28 mm. long. Procartilaginous
skeleton drawn complete ; outer epithelial ends of muscle-buds overlying radials shown in outline
except over preaxial radial 11.   *bp*, Basipterygium ; *ca*, nerve canal ; *mb*[23] twenty-third muscle-bud ;
pg, pelvic girdle.   (E. S. Goodrich, *Quart. Jour. Micr. Sci.*, 1906.)

myotomes, and is an extension inwards from the base of the fin-skeleton,
it is natural that some of the nerves should become surrounded by it as
they pass from the inner surface of the myotomes round their ventral
ends and outwards to supply the fin.   Thus certain nerves, called diazonal,
may be found in the adult to perforate the girdles, while other nerves
pass before or behind them, Fig. 142.

These essential facts in the ontogeny of the paired fins have been well
established in Selachians which are in many respects primitive ; but most
of them have been confirmed by observations on other fish such as the
Chondrostei and higher Teleostomes, including the Teleostei.

On the other hand, it is difficult to find any facts which actually support
the gill-arch theory, and much evidence may be brought against it.
Indeed the theory could hardly have been conceived except at a time
when the ontogeny of fins was scarcely known at all.   If developed from
gill-septa situated transversely to the long axis of the body, paired fin-

folds would rather hinder than favour progression ; moreover, the two pairs would presumably be at first close together behind the other gills and in a position mechanically very disadvantageous. Now, in ontogeny, a paired fin never makes its appearance as a dorso-ventral fold ; but on the contrary always as a more or less extended longitudinal ridge.

The position of the pelvic fins was attributed by Gegenbaur to their backward migration. But neither in primitive fishes generally nor in their early fossil representatives are the pelvic fins more anterior. When, as in some of the specialised Teleostei, the pelvic fins are far forwards, there is good evidence that this position is secondary, Figs. 140-41 (p. 138).

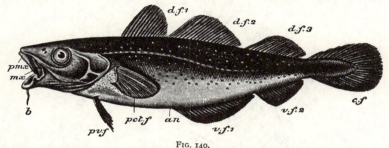

FIG. 140.

Gadus morrhua (Cod). *an*, Anus ; *c.f*, caudal fin ; *d.f* 1-3, dorsal fins ; *mx*, maxilla ; *pct.f*, pectoral fin ; *pmx*, premaxilla ; *pv.f*, pelvic fin ; *v.f* 1 and 2, anal fins. (After Cuvier.) (From Parker and Haswell, *Zoology*.)

The presence in ontogeny of vestigial muscle-buds in front of the pelvic fin-folds has been supposed to indicate backward migration. This, however, can hardly be so, since such buds are also found behind these same fins, and on both sides of the pectoral fins also. It has been urged that the presence of a ' nerve-plexus ' or compound collector nerve at and in front of the base of the pelvic fins (v. Davidoff, **153**), and that the greater extension of the collector in the young than in the adult (Punnett, 1900), are evidence of backward migration. But, again, such a collector and greater extension are found at the posterior end of these same fins, and are similarly developed on both sides of the pectoral fins. The attempt to explain the posterior position of the pelvics on the supposition that ancestral Gnathostomes had very numerous gill-arches extending farther back than in any known forms (Fürbringer, Braus) has no evidence to support it. That the paired limbs occupy very different relative positions on the trunk in various forms is an obvious and striking, fact which will be dealt with below (p. 136).

The gill-arch theory gives no intelligible explanation of the participation of a large number of segments in the formation of the musculature and nerve-supply of the paired fins. Yet a considerable and sometimes a very large number of spinal nerves and myotomes always contribute

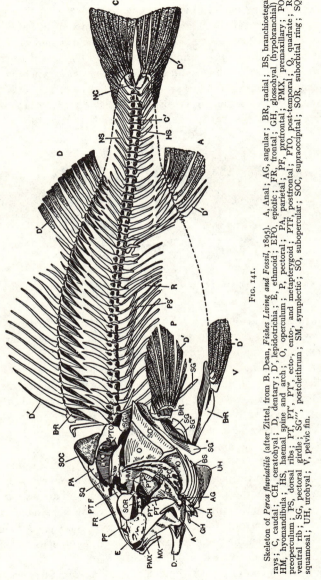

FIG. 141.

Skeleton of *Perca fluviatilis* (after Zittel, from B. Dean, *Fishes Living and Fossil*, 1895). A, Anal; AG, angular; BR, radial; BS, branchiostegal rays; C, caudal; CH, ceratohyal; D, dentary; D', lepidotrichia; E, ethmoid; EPO, epiotic; FR, frontal; GH, glossohyal (hypobranchial); HM, hyomandibula; HS, haemal spine and arch; O, operculum; P, pectoral; PA, parietal; PF, prefrontal; PMX, premaxillary; PO, preoperculum; PS, dorsal ribs; PT', PT'', PT''', ecto-, ento-, and metapterygoid; PTF, postfrontal; PTO, post-temporal; Q, quadrate; R, ventral rib; SG, pectoral girdle; SG''', postcleithrum; SM, symplectic; SO, suboccipital; SOC, supraoccipital; SOR, suborbital ring; SQ, squamosal; UH, urohyal; V, pelvic fin.

towards them; the area from which they are derived is wider than the actual base of the fin. This also applies to the paired limbs of Tetrapods, and, speaking quite generally, the lower the class of vertebrate concerned, the more segments take part in the formation of its paired limbs.

If the radials of the paired fins were derived from gill-rays and the girdles from gill-arches, we should expect their muscle-supply to be drawn, not from the myotomes, but from the unsegmental lateral-plate musculature innervated from the dorsal roots (p. 218). In the head region, although epi- and hypobranchial muscles of myotome origin may be associated with the visceral arches, yet the great bulk of the musculature of these arches is developed from the lateral-plate mesoblast and innervated

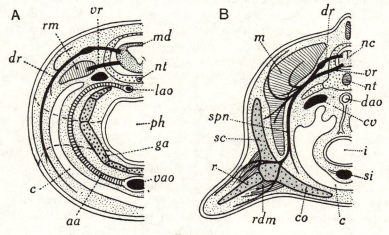

Fig. 142.

Diagrams to illustrate different morphological relations of *skeletal gill-arch* and *limb-girdle*. A, Transverse section of pharyngeal region; B, of anterior trunk region. *aa*, Arterial arch; *c*, coelom; *co*, coracoid region; *cv*, cardinal vein; *dao*, dorsal aorta; *dr*, dorsal root nerve; *ga*, skeletal gill-arch; *i*, gut; *lao*, lateral dorsal aorta; *md*, medulla; *m*, myomere; *nc*, nerve cord; *nt*, notochord; *ph*, pharynx; *r*, radial; *rdm*, radial muscle; *rm*, reduced myomere; *sc*, scapular region; *spn*, mixed spinal nerve; *si*, subintestinal vein; *vr*, ventral root nerve; *vao*, ventral aorta.

by cranial dorsal root nerves. It is true that in many forms there is a lateral-plate trapezius muscle attached to the dorsal end of the scapula and supplied from the vagus; but it may be pointed out that it only becomes so attached at a late stage in development, that it does not penetrate into the fin-fold, and that in any case there is no trace of such a muscle in the pelvic region.

But perhaps the most fatal of all objections to Gegenbaur's theory is the fact that gill-arches and limb-girdles differ radically from each other as regards their position relative to the coelom, nerves, blood-vessels, etc.; for whereas the gill-arches are morphologically internal, the girdles are external to these various structures. In fact, the girdles lie in the outer body-wall, while the visceral arches lie in the wall of the

K

alimentary canal. Hence the paired fins and their girdles are supplied with blood-vessels from the segmental somatic system, with muscles from the myotomes, and with motor nerve-fibres from the ventral spinal roots. These various relations are illustrated in diagrams, Fig. 142.

Finally, the gill-arch theory offers no explanation whatever of the striking resemblance borne by the paired fins to the median fins, even in specialised forms, and in spite of their many necessary divergencies due to adaptation to different functions. The resemblance in development and structure is so close that we are forced to the conclusion that they are organs of essentially the same nature. Not only is the fundamentally segmental character of the parts of both median and paired fins clearly established, but they agree in the minutest histological details. Especially remarkable is the identity in structure of the dermal fin-rays : the ceratotrichia of the Chondrichthyes, the camptotrichia of the Dipnoi, the lepidotrichia of the Teleostomes are faithfully reproduced in all the fins.[1]

Turning now to the rival fin-fold theory, we find that although some difficulties in its application may occur, and some points still remain obscure, yet the evidence of embryology and comparative anatomy is overwhelmingly in its favour. The palaeontological evidence, though unfortunately still very incomplete, is also favourable.

As already mentioned, paired fins first appear as longitudinal folds of the body-wall. Thacher and Balfour, indeed, believed that the pectoral and pelvic folds were differentiated from an originally continuous fold. This supposed primary continuity from pectoral to pelvic region is not, however, an essential part of the theory.[2] Possibly from the first the paired fins, and indeed the unpaired fins also, were discontinuous. In modern forms the paired folds vary much in extent, and are not known to be truly continuous except in the case of Rajiformes, where the pectoral fins

---

[1] A variant of the gill-arch theory has been put forward by Kerr, 840, according to whom paired fins have been derived from external gills. This theory is open to all the objections urged against Gegenbaur's theory and to others besides. Prominent true external gills occur but rarely, and in rather advanced forms, among Osteichthyes and Amphibia. It is true that the pelvic fin (not the pectoral) of *Lepidosiren* may be provided with vascular filaments ; but these are only known to occur at a certain time and for a special purpose in the male of that one genus. Except for a very superficial resemblance to the highly specialised and reduced limbs of modern Dipneumones, external gills differ from paired limbs in every important respect.

[2] The palaeozoic Ostracodermi possessed lateral folds which in Cephalaspidae grew into outstanding lobes resembling the pectoral fins of higher fish (see Goodrich, 35, Fig. 173). It is possible that the paired pectoral folds developed first, and the pelvic folds only later in the Gnathostomata.

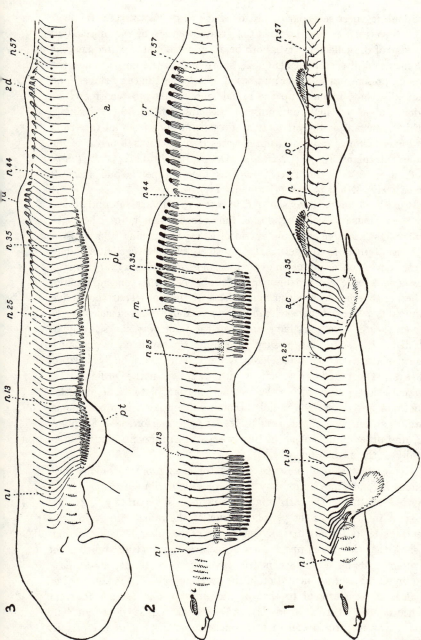

FIG. 143.

1, Diagram of an adult *Scyllium canicula*, showing the nerve-supply of the fins. The nervous, muscular, and skeletal segmental elements are distributed as if concentration had not taken place. **2,** Diagram of an adult *Scyllium canicula*; the fins are expanded, and their shaded oval areas; the girdles themselves are not shown. **3,** Diagram of an embryo *Scyllium canicula* about 19 mm. long, in which are shown the ganglia, the myotomes, and the muscle-buds. *a,* Anal fin; *ac,* anterior collector of first dorsal fin; *cr,* cartilaginous radial projecting beyond the radial muscles; *n* 1-57, spinal nerves and ganglia; *pc,* collector nerve of second dorsal fin; *pl,* pelvic fin; *pt,* pectoral fin; *rm,* radial muscle; *1d* and *2d,* first and second dorsal fins. (From E. S. Goodrich, *Quart. Jour. Micr. Sci.,* v. 50, 1906.)

have doubtless been secondarily extended backwards, Figs. 145, 146 (Mollier, **197**). Nevertheless, such facts as the great extension of the muscle-bud areas and of the collector nerves both before and behind the paired fins, the concentration of the fin-elements and shortening of the fin-base, and the frequent presence of a greater number of muscle-buds in early stages than eventually come to build up the radial muscles, may be considered as evidence of an original continuity of the lateral folds. Very striking, in this connexion, is the production of a pair of buds from all the trunk myotomes forming an uninterrupted series from pectoral to pelvic region in some Selachians, Fig. 143 (*Pristiurus*; Dohrn, **159**; Braus, **145**. *Scyllium*; Goodrich, **172**). In the Teleost *Lophius* the first pelvic bud succeeds immediately the last pectoral (Derjugin, **157**).

It has been convincingly shown (Thacher, **219**; Mivart, **196**, and others) that the various types of the endoskeleton of the median fins of fishes, with more or less extensive basal plates and often rays branching from an axis, have been formed from a series of primitively discrete segmental radials by the concrescence and fusion of their basal regions, Figs. 132, 133, 135-6. This concrescence accompanies and is probably due to concentration or gathering together at the shortening base of the fin. That the similar manifold types of the endoskeleton of the paired fins, where concentration is even greater as a rule, have arisen in the same way can scarcely be doubted. Indeed it is necessary to assume this even on Gegenbaur's theory, since there the fin-skeleton is derived from originally separate gill-rays. The radials arise, not as outgrowths from the girdle, but as differentiations *in situ* within the mesenchymatous plate. The objection sometimes raised that the radials become differentiated in a continuous plate (as shown by Balfour, **142**; Ruge, **211**; Swinnerton, **300**; Derjugin, **157**, and others) and not as distinctly separate rudiments would apply equally to the median fins.

This early fusion of mesenchymatous or even procartilaginous rudiments with indefinite limits may be reasonably attributed to their being crowded together. The cartilaginous elements arise separately, Figs. 128, 130-31. In Selachians the peripheral ends of the radials first appear as separate streaks (Goodrich, **172**). Supporters of the gill-arch theory, anxious to show that the paired fins are not of metameric structure and that the myotome-buds have only secondarily invaded them (Braus, **144-8**), have also urged that the segmentation of the radial muscles of the adult paired fin does not correspond to that of the embryo, and further that the 'concordance' of radial muscles and skeletal radials does not hold good. Now it is true that in greatly concentrated fins, and more especially at the narrow base of such fins, fusion may take place between neighbouring muscle-buds and also between the rudiments of radials, that the metamerism may be greatly obscured, and that discrepancies may occur. But peripherally these disturbances occur chiefly at the two ends of the fin-fold. Usually in the middle region the concordance between muscle and radial is perfect ; and usually, even in highly concentrated fins, it is manifest in the peripheral parts, Figs. 131, 139.

Although adjacent muscle-buds anastomose at their base (Mollier, 197 ; E. Müller, 199-200), and the substance of an adult radial muscle may not be derived entirely from one bud, yet it has been shown to be derived mainly from that bud the position of which it continues to occupy throughout development, and the radial muscles, as in median fins, correspond in number and relative position to the buds from which they were formed, Fig. 144. This holds good, excepting for the above-mentioned discrepancies, in the paired fins of Selachians (Mollier, 197 ; Goodrich, 172). Since in these fish two primary buds are given off from each myotome, the number of adult radials and radial muscles is double that of the contributing segments (excepting always for reductions and disturbances at the anterior and posterior ends).

Nerve Plexus.—The consideration of the partial loss of the primary metamerism of radial muscles brings us to the question of the nature of the nervous plexus. That in a series of segments the motor component of each spinal nerve remains faithful to its myotome throughout the changes which may take place in ontogeny, if not actually proved, is in the highest degree probable from the evidence of comparative anatomy and embryology. That a motor nerve does not forsake the muscle in connexion with which it was originally developed to supply the muscle of some other segment is a conclusion in harmony with our knowledge of the nerve-supply of muscles in general, and in particular of those of the fins. We have seen above that the fin-muscles are primarily innervated from the segments from which they arose in ontogeny, the dorsal fins from branches of the dorsal rami, and the ventral and paired fins from branches of the ventral rami of spinal nerves. Now dissection reveals that these rami divide near the base of the fin and branch repeatedly, mingling to form a basal plexus whence arises a network of extraordinary complexity which spreads throughout the fin. Such a plexus of connected nerves is related not only to the median and paired fins of fishes, but also to the paired limbs of all Tetrapods (Mayer, 128, Harrison, 179, Goodrich, 172, in median fins; Mollier, 197, Braus, 144, 147, Hammarsten, 178, E. Müller, 199-200, in paired fins ; Fürbringer, 162-3, Sewertzoff, 215, and others in Tetrapods).

The number of spinal nerves sending motor fibres to the plexus

indicates the number of segments which have contributed to the musculature of the fin. But in the peripheral network the nerves would seem to
be inextricably mixed, and the original segmental order would seem to
be entirely lost. However, the branches of the spinal nerves are of mixed
character (sensory and motor), and the unsegmental appearance of the
network is chiefly due to the intermingling of sensory fibres ; the motor
fibres pass through it to their destination, where they may be distributed
to the muscles with little over-
lapping (E. Müller, **200-1**) or
none at all (as in the ventral
lobe of the caudal fin of *Scyllium*, Goodrich, **172**). It has
been shown by experimental
stimulation that in the median
fins of Selachians the radial
muscles are supplied by the
spinal nerves in regular sequence
from before backwards (Goodrich, **172**), and the same is the
case in the paired fins (Goodrich, **173** ; Braus, **148a** ; E.
Müller, **200-1**). In the main,
then, segmental order is preserved. Even in the limbs of
Tetrapods, where all obvious
signs of segmentation are lost
in the adult muscles, this rule
has been shown to hold good
by dissection and experiment.
Here also plexus formation does
not entirely destroy segmental

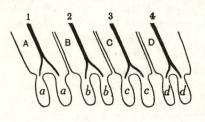

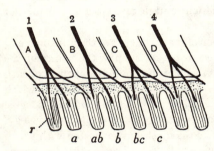

FIG. 144.

Diagram illustrating formation of adult *radial
muscles* from embryonic muscle-buds, and their motor
nerve-supply. Above, embryonic stage with a pair of
buds to each segment ; below, adult with radial muscles
compounded of material from adjacent buds. 1-4, Four
spinal nerves ; A-D, four myomeres ; *a-d*, muscle buds ;
r, radial muscle.

order (Herringham, **182** ; Patterson, **203-4**, and others). How then are
we to explain the fact that muscles of median fins and paired limbs may
be innervated by more than one spinal nerve ? The answer is, in all
probability, that this is due to the combination of the muscular tissue
of two or more segments. We have already mentioned above that the
muscle-buds of fins do not as a rule retain their original independence
but become connected by anastomosing bridges with their neighbours,
Figs. 129, 144. By such means myoblasts of adjacent segments may
combine to form compound radial muscles, whose segmental purity is
to that extent lost—to what extent varies, no doubt, in different fishes

and in different regions of the same fin. Experiments on paired fins in Selachians have yielded somewhat discordant results. While Braus concluded the overlap to be extensive, and that a single spinal nerve could supply some six or seven radial muscles, others have found that throughout the greater length of the fin each nerve supplies, as a rule, only one complete and two adjacent half-muscles (see diagram, Fig. 144).[1] In Tetrapods the combination and mixing of muscle substance derived from several segments is even more pronounced. But, strictly speaking, even here the nerves probably remain faithful to the muscle-fibres derived from their own segments; for it has been proved that each motor root supplies

[1] It would be interesting to determine the nerve-supply in median unconcentrated fins such as occur in Teleostei and Dipnoi.

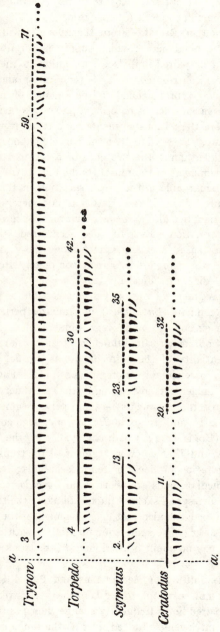

Fig. 145.

Diagrams, drawn to the same scale, indicating the nerve-supply of the paired fins in four fish. The spinal nerves are represented by the series of dots and strokes, the latter being those nerves which share in the formation of the 'limb plexus'. The thick horizontal line extends over the region supplying the pectoral fin, and the broken line over that supplying the pelvic fin. The line a-a shows the limit between the head and the trunk, and the numbers indicate the order of the true spinal nerves (from the results of H. Braus). (E. S. Goodrich, *Vert. Craniata*, 1909.)

its own special fibres, which are merely assembled and bound together in the same adult muscle (Sherrington).

Variation in Extent.—From the above it will be gathered that the number of spinal nerves which supply motor fibres to the muscles of median fin or paired limb is a sure guide to the number of segments which have contributed to its musculature; and, since the size of a motor nerve is proportional to the amount of muscle it supplies, the size of the various nerves entering a plexus may be taken as proportional to the share the various segments have taken in the composition of the musculature. So we find that the nerve components are usually stouter in the middle than at the two ends of a plexus, in agreement with the usual development of the muscle-buds, Fig. 143.

Very remarkable is the great variation in the extent of the paired and unpaired fins. This variation is not merely due to differences in concentration, but also to the fact that more or fewer segments contribute to their formation. Even if 'deconcentrated' the median fins would not in most cases meet and stretch along the whole length of the body, nor would the paired fins spread completely from pectoral to pelvic region (see diagrams, Figs. 143, 145).

Every segment of the body seems capable of producing median fin-elements (compare *Amphioxus*; in Craniates perhaps the head segments should be excepted); and every segment of at least the trunk seems capable of producing paired fin-elements, muscular, nervous, and skeletal. This potentiality of the segments is, as a rule, actually expressed or called into force along restricted regions only. For instance, with regard to paired fin-elements in Selachians, the spinal nerves entering the plexus belong to pectoral segments 2-13 and pelvic segments 25-35 in *Scyllium*, segments 2-19 and 29-50 in *Heptanchus*, segments 4-30 and 31-42 in *Torpedo* (Braus, **144**; Goodrich, **172**). In the embryo of *Scyllium*, *Pristiurus*, and *Torpedo* every segment of the trunk from the pectoral to the pelvic region produces paired muscle-buds; but while in the sharks the intermediate buds are abortive, in *Torpedo* they form radial muscles to which correspond skeletal radials all along the trunk.

It may be concluded that every segment of the trunk is equipotential in this respect. The same equipotentiality with regard to paired limb-elements may no doubt be attributed to the trunk segments of Tetrapods (see below).

On the Shifting of Median Fins and Paired Limbs in Phylogeny.—We must now attempt to explain the very striking fact that median fins and paired limbs frequently change their position on the body in the course of phylogeny. Indeed, just as the limbs rarely are derived from exactly the same number of segments, so also they rarely occupy exactly the same place in the series of segments. This remarkable variation is seen not only in comparing closely related species, but even individuals of the same species, and occasionally the two sides of the same individual. Moreover, the variation affects both their position with regard to each other and to the body as a whole, and the shifting may take place in either direction.

First of all it may be noticed that, even if the ancestral Gnathostomes had possessed continuous fin-folds, the position of the median fins and

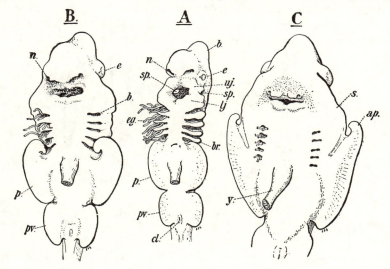

FIG. 146.

A, B, and C, Three successive stages in the development of *Torpedo ocellata*, Raf. The external gill-filaments have been removed on the left side. *ap*, Anterior region of pectoral fin growing forward; *b*, region of fore-brain; *br*, fourth branchial arch; *cl*, cloacal aperture; *e*, eye; *eg*, external gills; *lj*, lower jaw; *n*, opening of nasal sac; *p*, pectoral fin; *pv*, pelvic fin; *s*, ridge along which the pectoral fin will grow; *sp*, spiracle; *uj*, upper jaw; *y*, stalk of yolk-sac. (From Goodrich, *Vert. Craniata*, 1909).

paired limbs cannot be accounted for merely by their persistence in some regions and suppression in others. For in all classes there has been a perpetual alteration of position in the course of phylogeny both up and down the body.

The supporters of the gill-arch theory (Gegenbaur, **169**; Braus, **145**) held that these changes of position are due to the actual migration of the limb rudiment from one place to another. This view, put forward more especially to explain the posterior position of pelvic fins, is not borne out by embryology. There is little or no migration of the whole fin in onto-geny. A fin is not first formed in one place and then moved to another

later. Considerable apparent motion may be brought about by processes
of concentration, growth, and reduction. But the fin as a whole retains
its position throughout ontogeny (with the single exception noted below
of certain Teleosts, which does not affect the argument). This will be
readily understood on comparing the diagrams, Fig. 143.

That there is no motion of the whole fin-rudiment is proved by the
examination of the nerve-plexus. As already explained, the nerve-supply
of an adult limb is a sure guide to the identification of the segments from
which its elements have been derived, and the position of the plexus
always corresponds to that of the limb.

Gegenbaur's view that the pelvic fins were originally anterior is borne
out neither by palaeontology nor by comparative anatomy. For there

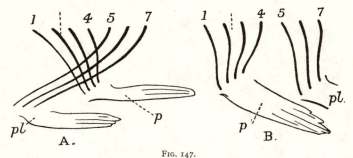

FIG. 147.

Diagram of motor nerve-supply of paired fins of *Gadus merlangus* (from E. S. Goodrich, *Quart.
Jour. Micr. Sci.*, 1913). A, In natural position ; B, with pelvic brought back to its place of origin.
p, Pectoral fin ; *pl*, pelvic fin ; 1-7, spinal nerves.

is no evidence of the pelvic being further forwards either in primitive
fishes generally or in their early fossil representatives. When, as in some
Teleostei, the pelvic fins are placed far forwards, their position is on good
evidence considered to be secondary. In the Acanthopterygii they tend
to shift from an abdominal to a ' thoracic ' and finally to a ' jugular '
position, and in such cases their nerve-plexus and the muscle-buds which
give rise to their radial muscles are transposed with them as far as they
can go (see Fig. 147, and footnote p. 140).

But if the shifting of limbs is not due to the actual migration of ready-
formed embryonic rudiments, nevertheless such migration may occur
to a very limited extent. For instance, in the frog the straining backwards
of the pelvic plexus shows that the base of the hind-limb has moved
backwards with the elongation of the pelvic girdle. In such fish as the
Gadidae, where the pelvic fins have attained a jugular position actually
in front of the pectorals, their real place of origin is betrayed by the
nerves which are derived from the spinal cord from segments behind

the origin of the pectoral plexus, crossing those which supply the pectoral fins, Fig. 147. Certain variations in the attachment of the pelvic girdle to the sacral vertebrae in Tetrapods may also be explained in the same way. For instance, the pelvic girdle of *Salamandra*, supplied by the 16th, 17th, and 18th spinal nerves, is occasionally attached to the 17th instead of to the 16th vertebra (v. Jhering, **187**). Only such small variations are due to migration, and the course of the nerves always indicates the real place of origin of the limb.

Three other explanations have been offered of the shifting of a limb together with its plexus : the theory of intercalation and excalation, the theory of redivision, and the theory of progressive modification or transposition.

According to the theory of intercalation (v. Jhering, **187**) the relative shifting of parts along a series of segments is due to segments having been dropped out or new segments having been added. To account for the extension of a limb or plexus over more or fewer segments, as well as for its shifting up or down the series, it is necessary to suppose that segments may be intercalated or excalated at any point in front of, within, or behind the limb and its plexus. At first sight this theory seems to afford a plausible explanation of simple cases of segmental variation, but it fails when applied to more complicated cases.[1] To account for the vast changes in the extent and position of the limbs which occur in most classes of Gnathostomes in the course of phylogeny, it would be necessary to assume that large numbers of segments have repeatedly been added to or dropped out of the series.[2] If such were the case we should expect to find zones of growth or zones of reduction where

---

[1] Taking the pelvic plexus of Mammals as an example, we find that v. Jhering first deals with such simple cases as *Sorex*, where a whole segment seems to have been added in front of it, some individuals having 13 and others 14 thoracic vertebrae, and the whole plexus and sacrum in the latter being situated further back. More difficult is it to explain variation in the Rabbit, where the lumbo-sacral plexus may include either nerves 25-30 or 24-30, the first sacral vertebra being the 27th in both cases. Here a nerve seems to have been intercalated in the middle of the plexus, since the crural moves one segment forward, while the more posterior ischiadic retains its position; yet the vertebrae are not changed. It has then to be further assumed that nervous and mesoblastic series of elements can move independently and fit on in new places, an assumption for which there is no evidence.

[2] A few examples of the variation in extent and of the shifting in position of the limb-plexus in Tetrapods may here be given (Fürbringer, **162-3**; Gadow, **165**, and others) :

The pectoral or brachial plexus of Anura usually is formed by spinal nerves 2, 3, 4 ; of *Proteus* by nerves 3, 4, 5 ; of *Megalobatrachus* by nerves 2, 3, 4, 5, 6. In Reptilia by nerves 4, 5, 6 in *Pseudopus* ; 6, 7, 8 in *Trionyx* ; 6, 7, 8, 9 in

new segments appear or old ones vanish. But no such zones occur, and
segments grow only at the posterior end of the series. Moreover, the
theory leads to absurd conclusions when applied to the paired fins of
fishes.[1] These and other objections have been successfully urged against
the theory of intercalation by Fürbringer and others. But if it is difficult
on this theory to explain such ordinary cases, it becomes impossible to
do so in the case of the relative shifting of median and paired fins in
Elasmobranchs. The interesting and quite conclusive fact is that the
two sets of fins shift independently. Thus the first dorsal is opposite
the pectoral in *Lamna*, between the pectoral and pelvic in *Alopecias*,
opposite the pelvic in *Scyllium*, and behind the pelvic in *Raja*. If it is
granted that the fins are homologous in these four genera, no addition or
suppression of segments can possibly account for their disposition (Goodrich, 173a).

The evidence against the theory of intercalation being overwhelming,
we may examine the theory of redivision (Welcker, 228; Bateson, 143),
which may be stated as follows : if one individual or organ is composed
of, say, twenty segments and another of twenty-one or nineteen, the
difference is due not to the addition or subtraction of a segment, but to
the subdivision of the several individuals or organs into twenty, twenty-
one, or nineteen segments respectively. Therefore, no segment of one
case strictly corresponds to any one segment in the other two. If the
number of segments were sufficiently increased or diminished by renewed
subdivision, the number of segments between parts or organs might thus
be altered. But their relative position to each other could only be altered
if the redivision was unequal along the series. Clearly, such a statement
would be no explanation at all, but merely a restatement in different
words of the problem we set out to solve. No better than the theory of
intercalation can it be applied to such cases as the apparent suppression

---

Phrynosoma ; 7, 8, 9, 10, 11 in *Crocodilus* ; 3, 4, 5, 6, 7 in *Chamaeleo*. In
Aves by nerves 10-15 in *Columba* ; 15-19 in *Anser* ; 22-26 in *Cygnus atratus*.

The pelvic (lumbo-sacral) plexus includes nerves 8, 9, 10 in *Rana* ; 19-24
in *Phrynosoma* ; 22-26 in *Crocodilus*. In Man the brachial plexus is formed
chiefly from nerves 5-9, and the lumbar plexus chiefly from nerves 21-25.

[1] For instance, in the smelt, *Osmerus eperlanus*, the pectoral fin is supplied
by spinal nerves 1-4 and the pelvic by nerves 18-29 (Hammarsten, 178) ; while
in the whiting, *Gadus merlangus*, nerves 1-4 supply the pectoral and nerves
5-6 the pelvic. Again, in *Scymnus* nerves 2-13 supply the pectoral and 23-45
the pelvic ; while in *Torpedo* the pectoral receives nerves 4-30 and the pelvic
nerves 31-42. If the approximation of pectoral to pelvic fin were due to
excalation, we should have to assume that the whole mid-trunk region had
been suppressed in *Gadus* and *Torpedo*, and a new trunk region intercalated
behind the pelvic, for instance, in *Gadus*.

of the mid-trunk in *Gadus* and *Torpedo*, or the independent shifting of the fins mentioned above.

There remains only the theory of transposition based on the 'Umformungstheorie' of Rosenberg (**209-10**), and first used to explain the varying extension of the different regions of the vertebral column. In his important works on anatomy, Fürbringer has shown how a limb-plexus may shift backwards or forwards like the limb it supplies. Obviously the nerves cannot actually move through the vertebral segments; it is, therefore, by progressive growth in one direction and by corresponding reduction in the other direction that change of position takes place. There is no transference of the limb-elements from one segment to another, no actual translocation or duplication of nerves, no intercalation or excalation, but a gradual assimilation of

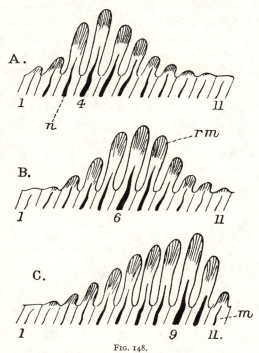

FIG. 148.

Diagram illustrating principle of shifting by *transposition* as seen in development of median fins. 1-11, Motor branches of spinal nerves; *rm*, muscle-buds forming radial muscles. In A, segment 4 contributes most; in B, segment 6; in C, segment 9.

neighbouring segments. Nerves at the anterior or posterior end may increase in size, and new nerves from adjoining segments may enter into the plexus. This, of course, accompanies the participation of new segments in the supply of muscle. So by gradual growth a limb supplied by, say, nerves 4, 5, and 6 may come to be supplied by nerves 3-7 or 1-8, and so on. Or, on the contrary, by a similar but reverse process of reduction, a limb supplied by nerves 1-8 may come to be supplied by nerves 3-7, or 4-6, Fig. 148.

Further, a limb-plexus may shift its position, without necessarily altering its structure from one region to another, by such a process of growth at one end accompanied by reduction at the other end. New segments being assimilated at one end, others may drop out, ceasing to contribute to the plexus, at the other. So we find in a plexus the more important and stouter nerves towards the middle, and the slenderer nerves towards either end. This explanation is in complete harmony with all the findings of comparative anatomy and embryology, and what has been said of the nerves applies equally well to the muscular and skeletal elements of fins and limbs (Goodrich, **173a**). Since the various fins can thus be transposed independently, it is the only theory consistent with the changes of distribution of the fins in Selachians mentioned above (p. 140). The con-

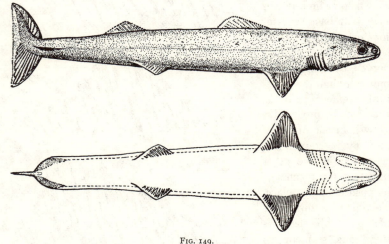

Fig. 149.

Right side and ventral views of *Cladoselache fyleri*, Cleveland Shales, Ohio ; restored (from B. Dean, *Fishes Living and Fossil*, 1895).

clusion is that when structures appear to move up or down a segmental series, the shifting is due to a change in the incidence of the formative stimuli which determine the development of and differentiate the equipotential segments (Ruge, **212** ; Goodrich, **173a** ; Lebedinsky, **195**).

Conclusion.—In conclusion it may be said that the lateral-fold theory alone agrees with the facts revealed by comparative anatomy and embryology. It was pointed out above (p. 130) that, although a primitive continuity of the lateral folds from pectoral to pelvic region is not an essential part of this theory, yet there is considerable embryological evidence in favour of this assumption. Whether much weight should be attached to Thacher's comparison with the metapleural folds of *Amphioxus*, and whether it may further be supposed that the two folds originally joined the median ventral fold behind the anus, is very doubtful. But it may be pointed out that in two genera of the Cephalochorda the right metapleural fold is continued into the preanal fold, and that the latter bears signs of a paired structure.[1]

Palaeontology affords some evidence of the existence of paired fin-folds. The primitive shark-like Palaeozoic Cladoselachii have paired fins in the form of longitudinal folds passing gradually without notch at either end into the body-wall, Figs. 149-50. The Acanthodii also possess paired fins of similar shape, but with the anterior border armed with a powerful spine probably formed of modified scales. As A. S. Woodward pointed out, the presence of a series of smaller spines and finlets between the pectoral

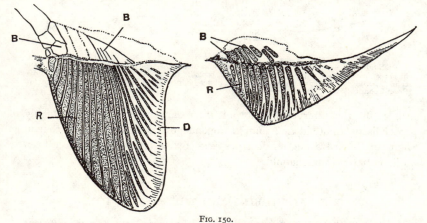

FIG. 150.

Paired fins of *Cladoselache fyleri* (from B. Dean, *Fishes Living and Fossil*, 1895). B, Basal cartilages; D, dermal fin-rays of pectoral; R, distal radials of fins. Pectoral fin on left, pelvic on right.

and pelvic in *Climatius* suggests that these were originally joined by a continuous fold.

The absence of any paired fins in Cyclostomes has always been a difficulty in adopting the view that the ancestral Craniates were provided with paired folds. But, if the conclusion be adopted that the modern Cyclostomes are the degenerate descendants of ancient Ostracoderms (Cope; A. S. Woodward; Stensiö, 135; Kiaer, 190), it appears not improbable that Lampreys and Hagfishes have lost the paired fin-folds, for indications of such folds have been found in most Ostracoderms (Lankester, 194; Traquair, 223-5; A. S. Woodward, 231; Kiaer, 190).

They appear to have been little developed in Anaspida, being chiefly represented by rows of spine-like scales specially developed in the pectoral

[1] It is uncertain what significance should be attached to the preanal fold of Teleostomes (*Amia, Lepidosteus*, and many Teleosts). Probably it is a true continuation of the median fin-fold, since it usually contains actinotrichia; but it may be a secondary extension of it, Figs. 113, 116.

region. In Pteraspidomorphi they form flattened pectoral ridges as in *Drepanaspis*, or more outstanding flaps as in Coelolepidae. The lateral folds become well marked in the pectoral region of Cephalaspidomorphi, and may be developed into prominent fin-like paddles, Fig. 151 ;

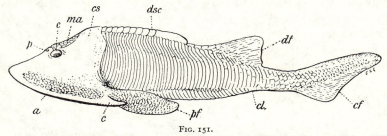

FIG. 151.

Restoration of *Cephalaspis Lyelli*. *a*, Lateral area with polygonal plates ; *c*, cornu of cephalic shield ; *cf*, caudal fin ; *cl*, position of cloaca ; *cs*, cephalic shield ; *df*, dorsal fin ; *dsc*, dorsal ridge scales ; *e*, orbit ; *ma*, median area with polygonal plates ; *p*, position of median pore ; *pf*, pectoral fin. (From Goodrich, *Vert. Craniata*, 1909.)

while in the Pterychthyomorphi these are represented by highly specialised armoured and jointed appendages. It is to be noted that in none of the Ostracodermi does a special pelvic fin appear to have been differentiated from the fold.

THE PAIRED FINS AND THEIR SKELETON

The endoskeleton of the paired fins of fishes varies greatly in the different groups ; but, from what has been said above, it will be gathered that the various types of fin-skeleton, together with the girdles, have probably been derived phylogenetically from originally separate segmentally arranged radials with the help of concentration and concrescence. These radials appear to have been jointed in the ancestral Pisces. Very generally the radials of the median fins are subdivided into three pieces : a proximal deeply embedded (axonost), a middle, and a distal marginal piece (p. 91). Those of the paired fins were probably originally similarly jointed (pelvics of Chondrostei) ; but often they become further subdivided (Elasmobranchii, Fig. 159), or on the contrary simplified and reduced to one piece (Teleostei, etc., Figs. 173, 176, 205).

Although we shall not attempt to describe the different types of the endoskeleton of the paired fins in detail, we may discuss their possible phylogenetic origin, though neither anatomy nor embryology enables us to determine this for certain, and the evidence of palaeontology is still inconclusive.

According to Gegenbaur (**167-9**) the original type resembled the skeleton of the pectoral and pelvic fins of *Ceratodus*. This 'archipterygium' had a median jointed tapering axis articulating with the girdle, and was provided with an anterior preaxial and a posterior postaxial series of rather fewer radials, Figs. 152, 154. The radials were arranged in pairs on the segments of the axis, and diminished in size towards the distal extremity of the leaf-like fin (Günther, 1871; Huxley, 535; Howes, 183; Braus, 147). Such a ' biserial archipterygium ' may be described as ' mesorachic ' and ' rachiostichous ' (Lankester). That it is a very ancient type can hardly be doubted. For, though at the present day it occurs only in the archaic *Ceratodus*,[1] it seems, judging from the outward appearance of the fins, to have been present in the pectoral and pelvic fins of primitive Dipnoi down to their earliest known representatives in the Devonian. It was also possessed by the Devonian Crossopterygii, but here the axis is usually shorter with fewer segments and more reduced postaxial radials in the pectoral fin (*Eusthenopteron*, Fig. 155; Whiteaves, 229; Traquair, 220; A. S. Woodward, 231; Goodrich, 171; Petronievics, 205); while in the only pelvic fin-skeleton known there are no separate postaxial radials at all (*Eusthenopteron*, Fig. 204; Goodrich, 171). Moreover, an almost perfect archipterygial type of endoskeleton is present in the pectoral fin of

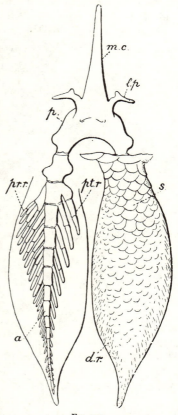

FIG. 152.

Ventral view of the pelvic girdle and fins of *Ceratodus Forsteri*, Kr. (Partly from Davidoff.) *a*, Jointed axis; *d.r*, dermal rays; *l.p*, lateral prepubic process; *m.c*, median epipubic process; *p*, pelvic girdle; *pr.r*, postaxial radial; *pt.r*, preaxial radial; *s*, scale covering axial region. (From E. S. Goodrich, *Quart. Jour. Micr. Sci.*, v. 45, 1901.)

the Carboniferous Pleuracanthodii (Brongniart, 149; Fritsch, 23; Döderlein, 158; Jaekel, 186), though the pelvic fin shows only preaxial

---

[1] The fin-skeleton is entirely cartilaginous in all living Dipnoi (but in the extinct *Ceratodus sturi* its basal segment is ossified). The postaxial radials are not only more slender but also more numerous than the preaxial; consequently more than one may be attached to an axial element. They tend

L

radials attached to the axis. The biserial or archipterygial type is associated with a projecting ' acutely lobate ' shape of fin, with a well-developed muscular lobe, a narrow base and dermal fin-rays set all round; and in the Osteichthyes the lobe is covered with scales.

Gegenbaur further maintained that the various types of paired fin-skeleton found in the Teleostomi may be derived from the archipterygium

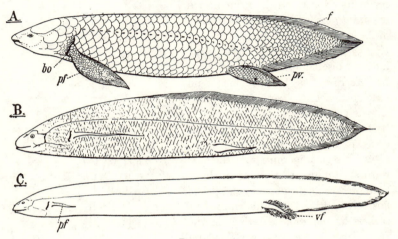

FIG. 152a.

A, *Ceratodus Forsteri*, Krefft ; B, *Protopterus annectens*, Owen (after Lankester) ; C, *Lepidosiren paradoxa*, Fitz. (after Lankester, modified). *bo*, Branchial opening ; *f*, median fin ; *pf*, pectoral fin ; *pv*, pelvic fin ; *vf*, vascular villi present on the male. (From E. S. Goodrich, *Vert. Craniata*, 1909.)

on the supposition that the axis became shortened and finally lost, and that the radials were reduced in number, especially the postaxial radials which soon entirely disappeared. The ' rhipidostichous ' type of skeleton of the pectoral fins of Elasmobranchs, in which the radials have a fan-like arrangement, he also deduced from the archipterygium on the supposition that the axis has been reduced and is represented chiefly by the posterior basal metapterygium, there being a large number of preaxial radials and at most only few vestigial postaxial radials, Fig. 156.

But on this theory the origin of the archipterygium itself still remains to be explained. We have already seen that Gegenbaur's view that it was derived from the branchial skeleton is untenable (p. 130). It may, however,

---

also to fuse at their base. A true joint occurs between the single basal and girdle.

In *Protopterus* and *Lepidosiren* the paired fins are much reduced and filamentous, with a slender jointed axis. The latter genus has lost all trace of lateral radials, and only vestigial preaxial radials remain in *Protopterus*, Figs. 91 and 152a.

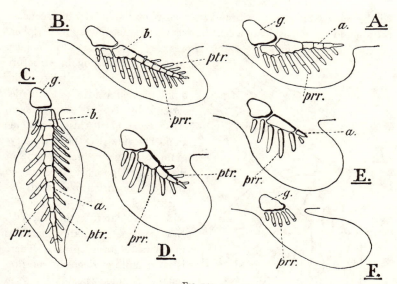

FIG. 153.

Diagrams showing the possible derivation from each other of the various types of pectoral fin skeleton in the Osteichthyes. A, Pleurorachic type (*Cladodus*) ; B, hypothetical stage leading to the mesorachic type C (*Ceratodus*) ; D, hypothetical type leading to E (*Acipenser, Amia*) ; F, teleostean type, reached either from A through E, or from C through D and E. *a*, Segment of axis ; *b*, basal of axis ; *g*, pectoral girdle ; *prr*, preaxial radial ; *ptr*, postaxial radial. (After Goodrich, *Vert. Craniata*, 1909.)

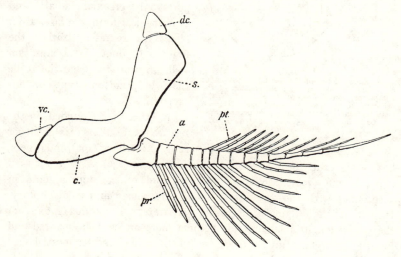

FIG. 154.

Left pectoral girdle and fin-skeleton of *Pleuracanthus Gaudryi*, Fr. *a*, Segmented axis of fin ; *c*, coracoid region ; *dc*, dorsal cartilage ; *pr*, preaxial radial ; *pt*, postaxial radial ; *s*, scapular region ; *vc*, ventral cartilage. (From E. S. Goodrich, *Vert. Craniata*, 1909.)

still be held that it was early formed from segmental radials by their concentration and fusion basally to a central axis with radials diverging peripherally on either side, Figs. 153-4, 156 (Howes, **183**; Haswell, **180-81**; Mollier,

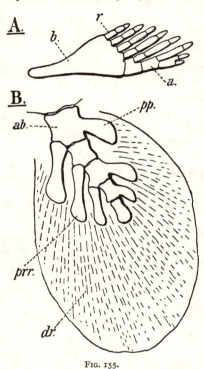

FIG. 155.

A, Endoskeleton of the second dorsal fin of *Glyptolepis leptopterus*, Ag.; B, skeleton of the left pectoral fin of *Eusthenopteron Foordi*, Wht.; restored. *a*, Segment of longitudinal axis; *ab*, basal segment of axis; *b*, basal; *dr*, dermal rays; *pp*, postaxial process (radial?); *prr*, preaxial radial; *r*, radial. (From Goodrich, *Vert. Craniata*, 1909.)

197). The variation in structure and the occasional splitting of the axial elements in *Ceratodus* described by Howes and Haswell strongly support this interpretation. The view that the archipterygium is a very ancient type of fin-skeleton, possibly ancestral to that of Chondrichthyes and Osteichthyes, is quite reconcilable with the 'lateral-fold theory'. Nevertheless, there are difficulties in the way of its acceptance. There is no definite evidence of the presence of distinctly postaxial rays in the pelvic fin of any Elasmobranch or Teleostomé; even in the Pleuracanthodii and Crossopterygii (Osteolepidoti) the jointed axis supports only preaxial radials, and in its typical form the biserial fin seems to represent rather the finished product of specialisation well differentiated from the girdle than a primitive type of fin-skeleton.

Another view, founded on the researches of Balfour (**142**), Thacher (**219**) and Mivart (**196**), and others, has more recently been developed by Wiedersheim (**230**), A. S. Woodward (**231**), Regan (**208**), Sewertzoff (**216-17**), and others, and is more in accordance with the general results of embryology and palaeontology. It is that the originally segmental separate and parallel radials of the paired longitudinal fin-folds become more or less concentrated and fused at their base, giving rise to a 'pleurorachic' and 'monostichous' or uniserial type of skeleton. In such a type the concrescence was greater at the anterior than at the posterior end of the fin, the longitudinal axis lay in the body-wall, and bore a single row of radials along its outer edge. The basal elements of the radials fused, especially anteriorly, to larger pieces of which the most anterior extended inwards and gave rise to the limb girdle, Figs. 153, 157. If this view is correct we should expect to find in primitive forms traces of a segmented longitudinal axis posteriorly, of progressive fusion of the basal elements anteriorly, of peripheral radials still attached to the girdle in front of the axial element,

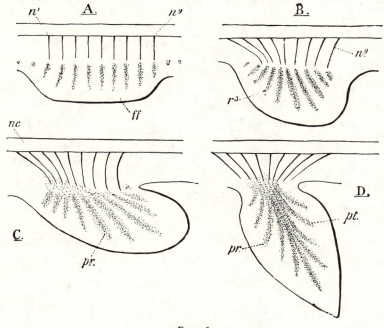

FIG. 156.

Diagrams to show the result of concentration on the skeleton and nerve-supply of a paired fin. A, B, C lead towards the Selachian type of fin ; A, B, D towards the Dipnoan type. n^{1-9}, Nine spinal nerves supplying the fin ; *nc*, nerve-cord ; *r*, radials represented as seen in an early embryonic stage ; *pr*, preaxial, and *pt*, postaxial radials ; *ff*, fin-fold. (From Goodrich, *Vert. Craniata*, 1909.)

and of the segmental origin of the girdle itself. And this is just what we find in most early or primitive fishes, especially in the pelvic fin, for it is a remarkable fact that in many respects the pelvic limb lags behind the pectoral in development, except perhaps in Dipnoi and Tetrapods. From the primitive form, with the axis running parallel with the body-wall, the other types in which it projects outwards at an angle are supposed to have been derived by the freeing of its posterior end, and extension of the lobe at a greater and greater angle to the body. This is due, perhaps, not so much to the motion of the axis, as to its being formed more and more towards the centre of the fin-lobe. This change of position of the axis of concrescence would be accompanied by the shifting of an increasing number of the free ends of the radials to a postaxial position. The typical biserial mesorachic fin would represent, then, the final product of this phylogenetic process. This second theory is more completely in agreement with embryology and palaeontology than is that of Gegenbaur.

Most striking of all are the fins of the ancient Cladoselachii first described by Dean (**154**), already mentioned in connexion with the lateral-

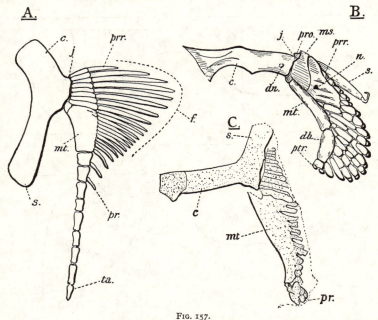

Fig. 157.

Skeleton of the pectoral arch and fin of A, *Cladodus Neilsoni*, Traq. ; B, *Chlamydoselachus anguineus*, Garman ; and C, *Symmorium reniforme*, Cope. (A restored from Traquair's figure, B and C after Braus.) *c*, Coracoid region ; *db*, distal basal or 3rd segment of metapterygial axis ; *dn*, diazonal nerve foramen ; *f*, problematical fin outline ; *j*, joint between girdle and fin ; *ms*, mesopterygium ; *mt*, metapterygium ; *n*, nerve foramen ; *pr*, posterior preaxial radial ; *pro*, propterygium ; *prr*, anterior preaxial radial ; *ptr*, possibly postaxial radial ; *s*, scapular region ; *ta*, distal segment of metapterygial axis. (From E. S. Goodrich, *Vert. Craniata*, 1909.)

fold theory (p. 143). In *Cladoselache* not only are the peripheral radials almost parallel and unfused, but their basal segments are combined to form a longitudinal axial region in which the elements, especially in the pelvic fin, are still for the most part separate and but imperfectly differentiated from the girdle, Fig. 150.

The pectoral fins of the Devonian and Carboniferous genera *Cladodus* and *Symmorium* have a somewhat similar skeleton, but with the axis more definitely formed, especially in *Cladodus* where it is long, tapering, and many-jointed (Traquair, **222**; Cope, **152**). From a somewhat similar

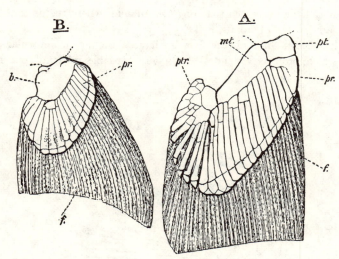

FIG. 158.

Callorhynchus antarcticus, Lac. The web of the fin, and the ceratotrichia, have been cut across. (After Mivart.) *b*, Basipterygium ; *f*, fin-web ; *mt*, metapterygium ; *pr*, preaxial radials ; *pt*, propterygium ; *ptr*, cartilages representing postaxial radials. (From Goodrich, *Vert. Craniata*, 1909.)

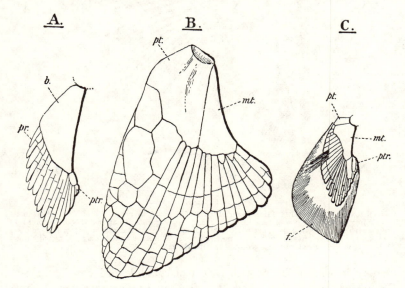

FIG. 159.

Skeleton of the pectoral fin of A, *Scymnus lichia*, Cuv. (after Gegenbaur) ; B, *Heterodontus* (*Cestracion*) *Philippi*, Lac. (after Gegenbaur) ; C, *Centrophorus calceus*, Gthr. (after Woodland). In the latter the web of the fin is represented. *b*, Basipterygium ; *f*, fin-web ; *mt*, metapterygium ; *pr*, preaxial radials ; *pt*, propterygium ; *ptr*, postaxial radials. (From E. S. Goodrich, *Vert. Craniata*, 1909.)

type may have developed that of *Pleuracanthus*, where the pectoral axis has many postaxial radials, Figs. 154, 157.

The pectoral fins of the Selachii have a prominent muscular lobe into which projects an endoskeleton very variable in detail, but built

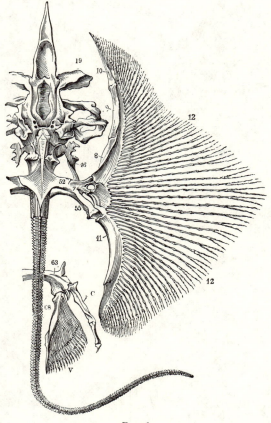

FIG. 160.

Skeleton of *Raja batis*, L. (From Owen, *Anat. of Vertebrates*, by permission of Messrs. Longmans and Co.) 7, Scapular region of pectoral girdle ; 8, 9, 10, segments of anterior axis of pectoral fin ; 11, posterior axis ; 12, radials of expanded pectoral fin ; 19, preorbital process ; 52, cartilage plate joining scapula to vertebral column ; 55, posterior outgrowth of pectoral girdle ; 63, pelvic girdle ; 68, basipterygium ; *c*, anterior enlarged radial ; *v*, pelvic fin radials.

on the rhipidostichous plan with a shortened or ill-defined axis, Figs. 158-9. Gegenbaur attempted to show that the ground plan consisted of three basal pieces, the pro-, meso-, and metapterygium articulated to the girdle, and each bearing radials. The pro- and mesopterygium were considered to be formed by the fusion of the basal joints of preaxial radials, while the metapterygium, with sometimes some posterior terminal elements, was supposed to represent the original axis not formed by concrescence. Some few small radials occasionally found on the posterior border were taken to represent remains of the postaxial series, and the whole was compared to a reduced and modified archipterygium. However, the distinction drawn between the posterior and largest basal (metapterygium) and the others is not justified by comparative anatomy or embryology (p. 144). Indeed, Huxley (535) identified the original axis in the meso-

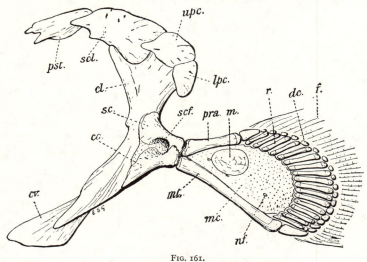

Fig. 161.

Skeleton of the right half of the pectoral girdle and of the right fin of *Polypterus bichir*, Geoffr. Inner view. *cl*, Cleithrum; *co*, coracoid; *cv*, clavicle; *dc*, distal radial cartilage; *f*, dermal rays; *lpc*, lower postcleithrum; *m*, mesopterygial bony plate; *mc*, mesopterygial cartilage; *mt*, metapterygium, or postaxial radial; *nf*, nerve foramen; *pra*, preaxial radial; *pst*, post-temporal; *r*, radial; *sc*, scapula; *scf*, scapular foramen; *scl*, supracleithrum; *upc*, upper postcleithrum. (From Goodrich, *Vert. Craniata*, 1909.)

pterygium. All the basals are doubtless formed by concrescence, traces of which can still be seen even in the adult fin, and an endless variety of detail in the shape and composition of the elements is presented by the various families and genera. There may be a single basal, as in *Scymnus*; two basals as in *Heterodontus* and *Chimaera*; three, as in *Scyllium* and *Squalus*; or five, as in *Myliobatis*. It is also important to note that, in Rajidae, an anterior axis is formed by concrescence, similar to the original posterior metapterygial axis, Figs. 146, 160 (Bunge, 151; Howes, 184).

The endoskeleton of the pectoral fin of the Teleostomi has already been dealt with in the preceding discussions, and it need here only be pointed out with regard to the Crossopterygii (Osteolepidoti) that while in the Osteolepidae it may have been typically mesorachic and biserial,

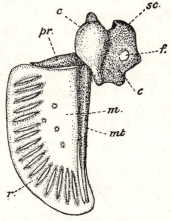

FIG. 162.

Reconstruction of the pectoral girdle and fin-skeleton of a larval *Polypterus*, enlarged. (After Budgett.) *c*, Coracoid region ; *f*, foramen ; *m*, mesopterygial cartilage plate ; *mt*, metapterygium ; *pr*, propterygium ; *r*, radial ; *sc*, scapular region. (From Goodrich, *Vert. Craniata*, 1909.)

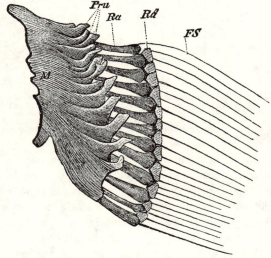

FIG. 163.

Right pelvic fin of a young *Polyodon folium*. From the dorsal side. *FS*, Bony dermal rays ; *M*, metapterygium ; *Pru*, uncinate ('iliac') processes ; *Ra*, *Ra¹*, radii of the first and second orders. (From Wiedersheim, *Comp. Anatomy.*)

in the Rhizodontidae the jointed axis is relatively short and the postaxial radials much reduced, Fig. 155 (Whiteaves, **229** ; A. S. Woodward, **231** ;

Goodrich, **171** ; Broom, **150**). On the other hand, the fin-skeleton of the Actinopterygii shows no definite signs of a mesorachic and biserial structure. It resembles rather that of the Selachii. In the modern Chondrostei the pectoral fin-skeleton has a posterior axis and a consider-

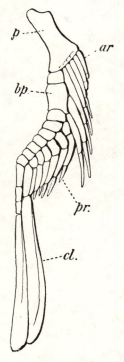

FIG. 164.

Pelvic girdle and fin-skeleton of a male *Pleuracanthus Oelbergensis*, Fr. (After Fritsch.) *ar*, Anterior preaxial radial resting on girdle ; *bp*, segmented basipterygial axis ; *cl*, modified radials of clasper ; *p*, pelvic girdle (left half) ; *pr*, preaxial radials. (From Goodrich, *Vert. Craniata*, 1909.)

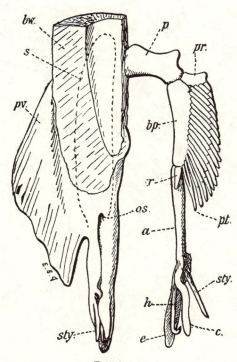

FIG. 165.

Dorsal view of the pelvic girdle and fins of a male *Squalus acanthias* ; the skeleton has been exposed on the right side. *a*, Axial cartilage of clasper ; *bp*, basipterygium ; *bw*, cut body-wall ; *c*, dorsal covering plate ; *e*, ventral plate ; *h*, hook ; *os*, opening of glandular sac ; *p*, pelvic girdle ; *pr*, propterygial, or anterior basal ; *pt*, posterior radial ; *pv*, pelvic fin ; *r*, modified radial ; *s*, outline of glandular sac embedded in body-wall dorsal to girdle ; *sty*, hard style. (From Goodrich, *Vert. Craniata*, 1909.)

able number of preaxial radials, several of which are articulated in front to the girdle (Gegenbaur, **166** ; Wiedersheim, **230** ; Hamburger, **177** ; Rautenfeld, **207** ; Regan, **208** ; Sewertzoff, **217**). That of *Polypterus*, though highly specialised in the adult (Gegenbaur, **166** and **168** ; Klaatsch, **192**), with two basal pieces reaching the girdle, an expanded central piece, and numerous peripheral free radials, has been shown by

Budgett (**10**) to develop in the young much as in the Selachii, with a posterior axis and preaxial radials only, Figs. 161-2.

In *Amia* the pectoral fin-skeleton more closely resembles that of *Acipenser*, but the axis is shorter and the radii ossified ; while in *Lepi-*

FIG. 166.

Raja blanda, Holt. Dorsal view ; the cartilaginous skeleton has been completely exposed on the right side. *a, b, c, d, f,* Cartilages of clasper ; *al,* anterior lobe of pelvic fin ; *bp,* basipterygium ; *cl,* clasper ; *cp,* covering-plate ; *ip,* iliac process ; *os,* opening of sac ; *p,* pelvic girdle ; *pp,* prepubic process ; *prr,* enlarged preaxial radial ; *pt,* posterior preaxial radial ; *s,* dotted line indicating ventral glandular sac ; *st,* second segment of basipterygial axis ; *sty,* styliform cartilage. (From Goodrich, *Vert. Craniata*, 1909.)

dosteus reduction has gone still further, Fig. 173. Finally, in the Teleostei the axis has disappeared and all the radii come to articulate on the girdle. Usually there are not more than five, of which the first is small and closely bound to the enlarged anterior dermal ray. Occasionally there are more, as in *Malapterurus, Muraenolepis,* and Anguilliformes (eight in *Anguilla*),

Fig. 176, a peculiarity which may be due to the retention of a more primitive and larger number (Sagemehl, 378; Derjugin, 157). In connexion

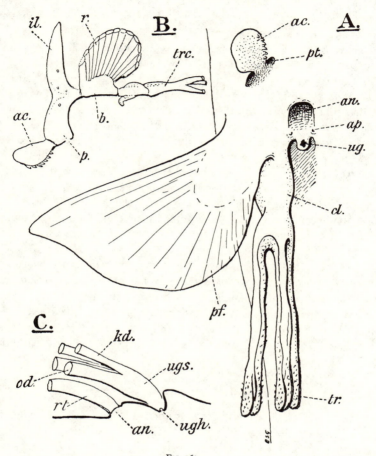

FIG. 167.

Chimaera monstrosa, L. A, Ventral view of the right pelvic fin of a male; B, ventral view of the left half of the pelvic girdle and pelvic fin of a male (after Davidoff); C, diagram showing the opening of rectum and urinogenital sinus in the female. *ac*, Anterior clasper armed with denticles; *an*, anus; *ap*, abdominal pore; *b*, basipterygium; *cl*, posterior clasper; *il*, iliac process; *kd*, kidney duct; *od*, oviduct; *p*, pelvic cartilage; *pf*, pelvic fin; *pt*, pocket into which the anterior clasper can be withdrawn; *r*, radial; *rt*, rectum; *tr*, trifid extremity of clasper; *trc*, its cartilage cut short; *ug*, urinogenital papilla; *ugh*, its opening; *ugs*, urinogenital sinus. (From Goodrich, *Vert. Craniata*, 1909.)

with the pectoral skeleton of the Teleostei it should be mentioned that there is generally developed in the young from the hinder margin of the plate of mesenchyme from which arise the girdle and the radials (p. 132) a long process passing backwards along the body-wall; later it remains attached as a cartilaginous process to the girdle, but always disappears more or less completely in the adult (Emery, **121**; Swinnerton, **300**; Haller, **176**; Derjugin, **157**; Vogel, **226**). By some it has been considered to be a new larval structure; but it may possibly represent a vestige of the longitudinal axis, Figs. 177-8.

Leaving aside the Dipnoi, as a specialised group in which both pectoral and pelvic fins have become biserial, we find in all the other Pisces a pelvic fin conforming to the uniserial type with the axis along the body-wall, and freed from it only at its posterior end. Although the fins of the modern Chondrostei are perhaps somewhat degenerate, they are particularly instructive, since they show clearly the manner in which not only the axial elements but also the girdle itself is formed by the fusion of basal pieces, Figs. 163, 209 (Thacher, **219**; Regan, **208**; Mollier, **198**; Sewertzoff, **216-17**). In *Pleuracanthus* the basal pieces are still for the most part separate, and the girdle still bears many radials, Fig. 164. The pelvic fin of the Selachians, on the other hand, has the axial elements much more completely fused to form the characteristic basipterygium; except in *Chlamydoselachus*, only a few anterior radials rest directly on the girdle,[1] Figs. 27, 165-6.

The pelvic fins of the higher Teleostomes are very much shortened and show little or no trace of an axis, the few remaining radials articulating with the pelvic girdle. *Polypterus* has only four such radials; small vestiges of radials and possibly of the axis are found in *Lepidosteus, Amia,* and the lower Teleostei, but in the higher Teleosts they seem to disappear altogether (?), and the dermal rays then articulate directly with the girdle, Figs. 205-7.

THE PAIRED LIMBS OF THE TETRAPODA

The name Ichthyopterygium is applied to the fin-like paired limbs of Pisces, while the limb of Tetrapods, adapted for walking on land, has been given the name Cheiropterygium. It is also called the pentadactyle limb, because both the fore and the hind limb are primitively provided with five well-developed functional digits.[2]

---

[1] In the male Pleuracanthodii and Selachii (but apparently not in Cladoselachii) the posterior end of the pelvic fin, with its contained endoskeleton consisting of the apex of the axis and about three radials, is converted into a specialised intromittent organ, the so-called clasper or myxopterygium (Gegenbaur, **167**; Jungersen, **188-9**; Huber, **185**; Leigh-Sharpe, **195a**). The Holocephali have in addition a smaller anterior clasper developed from the front end of the fin (Davidoff, **153**). See Figs. 164-7.

[2] The number of digits may be reduced in specialised forms, but the number five was doubtless established in the ancestral Tetrapod. The majority of

It is important to notice that, although the fore and hind limbs may in the course of evolution and in adaptation to various modes of life come to differ greatly from each other, yet there is good evidence that they are built on essentially the same plan and were primitively alike. The pentadactyle limb is differentiated into three regions (stylopodium, zeugopodium, autopodium), bent at an angle to each other.   The first or proximal region (upper-arm or brachium of fore-limb ; thigh or femur of hind-limb) projects outwards from the body ; the second region (fore-arm or antebrachium ; shank or crus) extends downwards ; the third region (hand or manus ; foot or pes) rests on the ground and bears the five separate digits.   The endoskeleton consists of a single element in the proximal region (humerus ; femur), two elements in the middle region (preaxial radius in fore-limb, and tibia in hind-limb ; postaxial ulna in fore-limb, and fibula in hind-limb), and several elements in the distal region.   The elements of these three regions articulate with each other by well-defined joints.   The endoskeletal elements of the autopodium or distal region are further differentiated into proximal, middle, and distal sections (carpus, metacarpus, and phalanges of digits in the fore-limb ; tarsus, metatarsus, and phalanges in the hind-limb).   Well-defined joints may also become differentiated between these various elements, Fig. 168.

In spite of the great variety in the number and disposition of the elements of the carpus and tarsus in the different groups, a generalised primitive plan may be made out consisting of a row of three proximal elements (proximal carpalia or tarsalia), five distal elements (distal carpalia or tarsalia), and about three central elements between them (centralia). The metacarpus and metatarsus consist of five elements each (meta-carpalia, metatarsalia), and the free digits have a varying number of phalanges of which the primitive formula may have been 2, 3, 4, 5, 3.

In general structure and development the cheiropterygium resembles the ichthyopterygium ; there is the same contribution from many spinal nerves, the same sort of nerve-plexus at the base of the limb, the same contribution from many myotomes to form the musculature.

In the Amphibia and many of the higher Amniota the latter arises from the ventral ends of the myotomes as proliferations which combine to a continuous mass ; but in others (Lacertilia) each myotome of the limb region produces one epithelial muscle-bud, and the segmental

---

extinct and all living Amphibia have no more than four digits in the manus, yet some of the earliest and most primitive known representatives of the class have the usual number five (*Diplovertebron*, Watson, 644 ; *Eryops*, Cope, 12). Traces of a preaxial digit in front of the first (pollex) and of a postaxial digit behind the fifth (minimus) have been described.

structure is at first well defined (Mollier, 197 ; Sewertzoff, 215).　As in
fishes the muscle-buds may dwindle at each end of the series, and some
may disappear in the course of development.　For instance, in *Ascalobotes*
(Sewertzoff, 215) spinal nerves 6-10 enter the pectoral plexus, while in
the embryo the plexus includes nerves 4 and 5 ; muscle-buds arise from

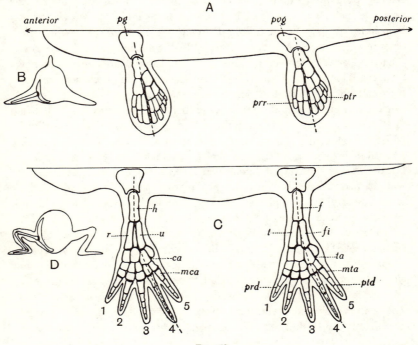

FIG. 168.

Diagrams illustrating comparison between *ichthyopterygium* and *cheiropterygium*, and possible
derivation of latter from former.　A, Primitive aquatic fish-like form with similar pectoral and pelvic
fins provided with a skeleton approximating that described in Sauripterus ; B, transverse section of
same in pectoral region ; C, primitive Tetrapod with similar pectoral and pelvic walking limbs ; main
axis of limb taken through 4th digit ; D, transverse section of same in pectoral region. *ca*, Carpus ;
f, proximal element, femur ; *fi*, fibula ; *h*, proximal element, humerus ; *mca*, metacarpus ; *mta*, metatarsus ; *pg*, pectoral, and *pvg* pelvic girdle ; *prd, prr*, preaxial digit and radial ; *ptd, ptr*, postaxial
digit and radial ; *r*, radius ; *t*, tibia ; *ta*, tarsus ; *u*, ulna.

the myotomes supplied by spinal nerves 2-10 ; but the first four do not
contribute to the adult limb muscles.　The buds grow into the limb-fold,
become mesenchymatous, and fuse to a continuous mass from which
develops a dorsal and a ventral plate.　Between these two the mesenchyme
gives rise to the endoskeleton, spreading inwards to form the girdle.　The
original segmentation of the muscles and nerves is obscured.　Later the
two plates form not only the muscles of the outstanding limb, but grow
inwards to give rise to those muscles passing from limb to girdle which

are supplied by branches of the plexus. New muscles to the girdle may be added later from the myotomes, and are innervated from the thoracic spinal nerves. The development of the cheiropterygium is in full agreement with the lateral-fold theory.

The limbs first appear as longitudinal folds. In some Amniotes the pectoral and pelvic folds are local enlargements of a continuous lateral ridge (Wolffian ridge) which later disappears. It is improbable, however, that this represents an ancestral continuous fin-fold. The limb rudiment grows into an outstanding lobe with pre- and postaxial edges and a narrowing base. Soon the distal region becomes expanded, and from it grow out the free digits. Meanwhile the limb rudiment has become twisted so that its main axis is bent backwards, its original dorsal surface turned outwards, its original ventral surface turned inwards. Its preaxial edge thus becomes ventral and its postaxial edge dorsal. Later the manus and pes are so disposed that the preaxial digit is turned forwards and inwards and the postaxial digit backwards and outwards.

Many attempts have been made to derive the cheiropterygium from the ichthyopterygium, but none has proved convincing. Gegenbaur (169-170) believed it to have arisen from an ' archipterygium ' like that of *Ceratodus*, the base of which articulates with the girdle by a single element comparable to the humerus or femur. However, the comparison of the more distal elements of the two types is not easy, and his view that the axis is represented in the anterior pentadactyle limb by the humerus, radius, and first digit (and the corresponding elements in the hind-limb), and that the other four digits are postaxial, has not been generally accepted. Huxley (535), on the other hand (who identified the main axis of the Selachian fin in the mesopterygium), drew the main axis through the humerus, intermedium centrale, third distal carpal, and third digit (digits 1 and 2 being preaxial, 4 and 5 postaxial, and the radius and ulna being compared to the pro- and metapterygium respectively). The attempt of Klaatsch (192) to derive the cheiropterygium from such a fin as the pectoral of *Polypterus* is still less satisfactory, since this fin appears to be highly specialised. Moreover, the pelvic fin of *Polypterus* is obviously far too much reduced and specialised to have given origin to the pentadactyle limb. More promising is the comparison with the pectoral fin of the Rhizodont *Eusthenopteron* made by Watson (227), who considers the main axis to pass through the humerus, ulna, and fourth digit. But here again the fin (especially the pelvic fin) is too specialised towards the Teleostome type, with short endoskeleton, large fin-web, and powerful dermal rays, to be truly primitive. Lately Broom has drawn attention to the pectoral fin of the related Upper Devonian *Sauri-*

M

pterus, which seems to have possessed a better developed and more primitive endoskeleton. It has a single proximal segment (humerus), two more distal elements (radius and ulna), and about a dozen marginal radials (Broom, 150 ; Gregory, Miner and Noble, 174). From such a fin-skeleton that of a pentadactyle limb might conceivably have been developed, Fig. 168 ; but nothing is known of the pelvic fin of this fish.

We may conclude that as yet nothing for certain is known of the origin of the cheiropterygium. It may have been derived from a fin not unlike that of the Devonian Dipnoi or Crossopterygii. At all events it is clear that we should expect the fish-like ancestor of the Tetrapoda to have possessed pectoral and pelvic fins alike in structure, with outstanding muscular lobe, extensive endoskeleton with at least five radials, small web, and few if any dermal rays.

CHAPTER IV

LIMB GIRDLES

THE PECTORAL GIRDLE

THE pectoral limb is supported by an endoskeletal girdle [1] consisting of right and left halves extending in the body-wall above and below the articulation of the limb ; the scapular region is dorsal and the coracoid region ventral. Primitively the girdle lies not only outside the vertebral column and ribs, but also outside the longitudinal muscles of the body (myomeres) and spinal nerves, which latter pass before, behind, or through the girdle to reach the limb (see p. 129 and Fig. 142). As the girdle sinks below the surface close behind the last gill-arch the myomeres are disturbed and interrupted ventrally and to a less extent dorsally. The dorsal portions of the muscle segments are continued to the skull, only superficial fibres becoming specialised and inserted on the scapula as anterior protractors, levators, and posterior retractors. To this scapular region is also attached the superficial trapezius, a lateral plate muscle supplied by

[1] The first origin and development of the limb girdles are discussed in Chapter III.

163

a branch of the vagus (p. 129). The coracoid region interrupts the ventral portions of the lateral muscle segments, separating the ' hypoglossal musculature ' inserted on its anterior border from the muscles of the ventral body-wall inserted on its posterior border. Primitively (in Elasmobranchii) the limb-muscles are only inserted on the girdle near the limb articulation, but they gradually extend more and more on the girdle in higher forms.

To the original endoskeletal girdle is added in Osteichthyes and Tetrapoda a set of protective and strengthening dermal bones to which many of the muscles become attached.

Pisces.—The endoskeletal pectoral girdle is well developed in the Chon-

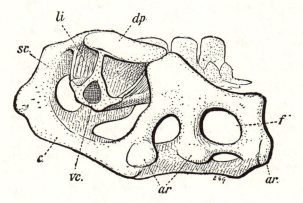

FIG. 169.

Pectoral girdle and portion of the vertebral column of *Raja blanda*, Holt ; oblique left-side view. *ar*, Articular facets for pectoral fin ; *c*, coracoid region ; *dp*, dorsal plate ; *f*, foramen ; *li*, ligament ; *sc*, scapular region ; *vc*, vertebral column. (From Goodrich, *Vert. Craniata*, 1909.)

drichthyes. The two scapulocoracoids remain separate in the Pleuracanthodii, where each is provided with a separate ventral infra-coracoidal and dorsal suprascapular element, Fig. 154 (Fritsch, **23** ; Döderlein, **158** ; Jaekel, **186**). In the Elasmobranchii the two halves are firmly united by ligament (Notidani) or more usually fused in the middle line ventrally, Fig. 169. A small median ventral element has been described in *Heptanchus* and *Hexanchus* (Haswell, **181** ; T. J. Parker, **113** ; Hanson, **257**). It is probably a derivative of the coracoids, and its comparison to the sternum of the Tetrapoda is of doubtful value (see above, p. 84). The scapular region tapers dorsally, and sometimes ends in a separate suprascapula (*Squalus*). The coracoid region expands immediately below the pericardium. Diazonal nerves usually pierce the girdle and divide into dorsal and ventral branches supplying dorsal and ventral fin-muscles. These nerves enter usually by a single internal foramen leading to a canal which divides and opens by external supraglenoid and coracoid foramina above and below the glenoid articulation of the fin. In the Rajiformes, where the girdle has to support the enormously developed pectoral fins, it becomes much enlarged, strengthened, fenestrated, and firmly connected to the vertebral column by a cartilaginous plate probably of suprascapular origin, Fig. 169. The Acanthodii are remarkable in

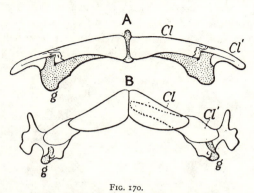

FIG. 170.

Pectoral girdle of A, *Ceratodus*; B, *Polypterus*. Ventral view. (After Gegenbaur, *Vergl. Anat.*, from Goodrich, *Vert. Craniata*, 1909.) *Cl*, Clavicle; *Cl'*, cleithrum; *g*, articular facet for fin-skeleton.

often having calcified dermal plates strengthening the endoskeletal girdle and simulating the clavicle and cleithrum of Osteichthyes (A. S. Woodward, 231).

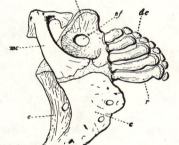

FIG. 171.

Inner view of the right half of the endo-skeletal pectoral girdle and fin of *Salmo salar*, L. *c*, Coracoid; *dc*, distal cartilages of radials; *mc*, mesocoracoid arch; *r*, fifth radial; *s*, scapula; *sf*, scapular foramen. (From Goodrich, *Vert. Craniata*, 1909.)

The girdle of the Osteichthyes is provided with dermal bones forming the posterior boundary of the branchial chambers, and on to which fit the opercular folds (p. 498). This dermal ' secondary ' girdle consists of paired bones : ventral clavicles, lateral cleithra overlying the articular region, supra-cleithra, and dorsal post-temporals or suprascapulars, Fig. 161. It is important to notice that the dermal girdle in all Osteichthyes (except a few specialised forms in this respect degenerate, such as the Anguilliformes, Fig. 176) is connected to the posterior region of the skull by means of the post-temporal, which generally forks, and is attached to the tabular and epiotic regions in Teleostei, Figs. 141, 175.

The endoskeletal ' primary ' girdle of Dipnoi, though unossified, is still well developed, especially in *Ceratodus*, Fig. 170, where it consists of two large scapulocoracoids bearing the articular knobs, and a separate ventral cartilage probably derived from the coracoids. There are no diazonal nerves. These cartilages are firmly connected to the dermal girdle, formed in *Ceratodus* and the earlier fossil forms of clavicle, cleithrum and post-temporal on each side (Gegenbaur, **170, 251**; Gregory, **123**; Watson, **648**). The girdle is more degenerate in the modern *Protopterus*

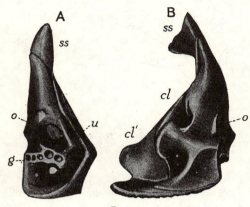

FIG. 172.

Left pectoral girdle of *Acipenser sturio*. A, Oblique view from behind; B, inner view. (After Gegenbaur, *Vergl. Anat.*, from Goodrich, *Vert. Craniata*, 1909.) *cl*, Cleithrum; *cl'*, clavicle; *g*, articular facets for pectoral fin-skeleton; *o*, *u*, open channel; *ss*, suprascapula.

and *Lepidosiren* where the post-temporal is vestigial (Bridge, **433**; Wiedersheim, **230**).

The general tendency in Teleostomes has been for the endoskeletal girdle to become reduced, and for the dermal bones to serve more and more for the support of the fin and the attachment of the muscles. The primitive clavicles, seen in Crossopterygii, still persist in modern Chondrostei, Fig. **174**, and in *Polypterus*, Figs. **170, 161**, and are known to have been present in the extinct Palaeoniscoidei (Traquair, **616**; Stensiö, **218**) and in the Belenorhynchidae (Stensiö, **134**); but they have been lost in all other Actinopterygii, where their place has been taken by forward extensions of the cleithra which meet ventrally, Fig. **174**.[1]

This large dermal bone in the Teleostean girdle was formerly called the clavicle, the bones dorsal to it the supraclavicles, and the bone situated ventral to it in lower fishes the infraclavicle. It was Gegenbaur (**251**) who showed that this ventral bone really corresponds to the clavicle of Stegocephalia and higher Tetrapods, and that the large so-called 'clavicle' of the Teleost is a more dorsal element to which he gave the name cleithrum. The nomenclature of all the bones has, therefore, been altered.

While the Actinopterygian dermal girdle is thus typically much developed, the endoskeletal girdle is usually reduced to shortened scapulo-

[1] Postcleithra are present in *Polypterus* and the Holostei. In Teleostei a postcleithrum is often characteristically lengthened, Fig. **141**.

coracoids at the base of the fin, not meeting in the middle line, and firmly attached to the cleithra alone, Figs. 173-7. Modern Chondrostei, however, still preserve a large cartilaginous scapulocoracoid with a suprascapula, Fig. 172. In these and the lower Holostei (Amioidei, Lepidosteoidei, Cypriniformes, Clupeiformes) the scapulocoracoid has a middle region projecting obliquely, bearing an outer horizontal glenoid articular surface, a scapular region (shortened in Holostei), and a ventral coracoid region bifurcated in front. The scapular region is hollowed out by a muscle canal

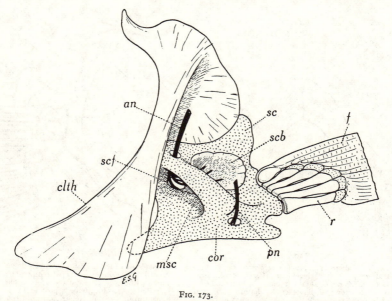

Fig. 173.

Lepidosteus osseus. Inner view of right cleithrum, with endoskeleton of pectoral girdle, and of fin. Nerves black, cartilages dotted. *an,* Anterior compound nerve ; *clth,* cleithrum ; *cor,* coracoid region ; *msc,* mesocoracoid arch ; *pn,* posterior compound nerve ; *r,* sixth radial ; *sc,* scapular region ; *scb,* scapular bone ; *scf,* scapular foramen.

through which pass dorso-medial muscles of the fin and which is closed medially by a mesocoracoid arch, Figs. 171-3 (Spange of Mettenheimer, Spangenstück of Gegenbaur, mesocoracoid of Gill). This arch is a structure of some importance (Parker, 288 ; Gegenbaur, 166 ; Romer, 294). It is present in all the lower Actinopterygii (Chondrostei and Holostei mentioned above), and though absent in Polypterini, is probably primitive and derived from the boundary of the pit for the insertion of fin-muscles and the diazonal nerve canal of Elasmobranchs. In Chondrostei the anterior diazonal nerves enter the canal from in front, and their dorsal branches issue through the posterior opening above the glenoid region, while their ventral branches pass through a foramen in its floor to the ventro-lateral muscles below the glenoid region (Swinnerton, 300).[1]

While the muscle pits and nerve canal are thus confluent in Chondrostei and the Teleostei, in *Amia* and *Lepidosteus*, Fig. 173, the anterior diazonal nerves enter the canal through the mesocoracoid arch. Probably a secondary condition, since in young stages of *Amia* the nerves enter

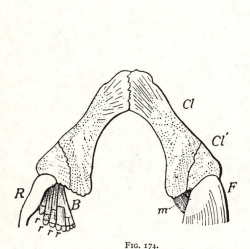

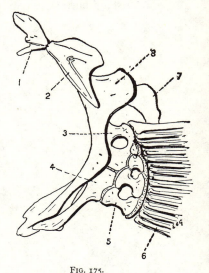

FIG. 174.

Pectoral girdle and fins of *Acipenser sturio*. Ventral view. (After Gegenbaur, *Vergl. Anat.*, from Goodrich, *Vert. Craniata*, 1909.) *B*, Postaxial edge of fin-skeleton; *Cl*, clavicle; *Cl'*, cleithrum; *F*, base of left fin; *m*, muscle; *R*, enlarged preaxial dermal ray; *r*, endoskeletal radials.

FIG. 175.

Skeleton of left half of pectoral girdle and fin of *Pterois volitans*. 1, Post-temporal; 2, supracleithrum; 3, scapular; 4, coracoid; 5, radial; 6, lepidotrich; 7, postcleithrum; 8, cleithrum. (From Goodrich, *Vert. Craniata*, 1909.)

from in front as usual (E.S.G.). The posterior diazonal nerves pierce the coracoid behind. The mesocoracoid arch, already very narrow in many of the Teleostei where it exists, is absent in the higher Teleostei (Boulenger, 426). In these it appears to have been lost owing to the confluence of the openings of the muscle canal now opened up, accompanying the tendency of the base of fin to become more and more vertical (Swinnerton, 300 ; Wasnetzoff, 305).

The endoskeletal girdle remains cartilaginous in *Amia*, develops a dorsal scapular bone in *Lepidosteus*, and a ventral coracoid bone in addition in the Teleostei. These bones share in the formation of the

[1] This foramen enlarges and becomes the scapular foramen present in the scapula of all Holostei.

glenoid articulation, the preaxial radials resting on the scapula and on the coracoid.

Little is known of the phylogenetic history of these bones, nor is it at all certain that they are strictly homologous with the scapula and coracoid of the Tetrapoda (p. 174). Bryant (465) has figured a small scapulocoracoid bone in *Eusthenopteron*, and Stensiö (218) has described the primary girdle of the Palaeoniscid *Acrorhabdus* as formed of a single scapulocoracoid bone, smaller but somewhat like the cartilage in *Acipenser*,

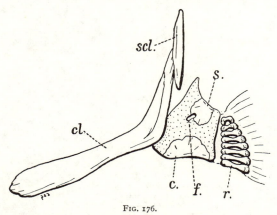

scl.

s.

cl.

c. f. r.

FIG. 176.

Skeleton of the left half of the pectoral girdle and of the fin of *Anguilla vulgaris*, L. c, Coracoid ; cl, cleithrum ; f, foramen ; r, eighth radial ; s, scapula; scl, supracleithrum. The cartilage is dotted. (From Goodrich, *Vert. Craniata*, 1909.)

and apparently with an excavated mesocoracoid arch. Our knowledge of the endoskeletal girdle in the fossil Teleostomes is still very incomplete.

The development of the primary pectoral girdle in Pisces has been studied by many observers, among whom one may mention Balfour (142), Mollier (197), and Braus (145, 148) in Selachii; Mollier (198), Wiedersheim (230), and Sewertzoff (217) in Chondrostei; Swirksi (301), Vogel (226), Derjugin (157), and especially Swinnerton (300) in Teleostei.

The blastematous or procartilaginous rudiment of the girdle in the body-wall is at first continuous with that of the fin-skeleton (p. 126). Later the scapulocoracoid is separately differentiated as a cartilage which spreads so as to enclose the diazonal nerves. In Teleosts the mesocoracoid arch develops late as an outgrowth from the middle region which grows upwards to join a smaller process from the tip of the scapula. The coracoid plate develops a large anterior so-called precoracoid process, representing the main limb of the coracoid, and a slenderer postcoracoid process. The former may meet its fellow in the middle line, and may even fuse with it to a transverse bar in larval Clupeoids (Derjugin, **157**;

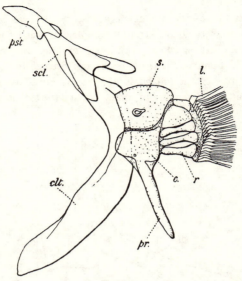

Fig. 177.

Skeleton of the right half of pectoral girdle and right fin of *Fierasfer acus*, L (after Emery). *c*, Coracoid; *clt*, cleithrum; *l*, lepidotrich; *pr*, ventral process; *pst*, post-temporal; *r*, 5th radial; *s*, scapula with small foramen; *scl*, supracleithrum. The cartilage is dotted. (From Goodrich, *Vert. Craniata*, 1909.)

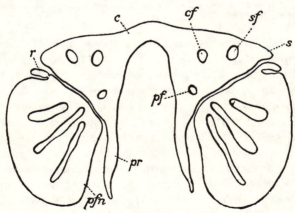

Fig. 178.

Clupea pilchardus, young. Cartilaginous skeleton of pectoral fins and girdle. (Goodrich, *Linn. Soc. J. Zool.*, 1919.) *c*, Coracoid region; *cf*, coracoid foramen; *pf*, postcoracoid foramen; *pfn*, pectoral fin skeleton; *pr*, posterior process; *r*, anterior radial; *s*, scapular process; *sf*, scapular foramen.

Goodrich, **255**); but it is later much reduced and usually shortened, often, however, leaving a small anterior ventral cartilage (Parker, **288**). The postcoracoid process may reach a remarkable length, and run backwards in the body-wall far behind the articulation of the fin, Figs. 177-8; but it also is later reduced and disappears more or less completely. Swinnerton suggests that it represents the original 'metapterygial' axis of the fin-skeleton (p. 157). But this view is difficult to reconcile with the structure of these parts in the lower Teleostomes, and the fact that the postcoracoid process arises from the girdle itself proximally to the glenoid articulation. More probably it is a secondary development of larval Teleosts.

THE PECTORAL GIRDLE IN TETRAPODA

The pectoral girdle of Tetrapods, like that of Osteichthyes, consists of a primary endoskeletal girdle strengthened by dermal bones (Gegenbaur, **166, 170**; Parker, **288**; Fürbringer, **250**). The primary girdle is of two halves, each developed from a continuous scapulocoracoid cartilage in the embryo, which becomes variously modified, fenestrated, and even subdivided in the different groups. A glenoid facet for the humerus is situated on its outer posterior border; dorsally extends the scapular region and ventrally the coracoid region. The coracoid plates ˙meet and even overlap ventrally, and abut against the front edge of the sternum. The secondary dermal

FIG. 179.

Clavicles and interclavicle of *Ophiacodon* (from Williston, *Osteology of Reptiles*, 1925).

girdle in such primitive Tetrapods as the Stegocephalia consists of paired ventral clavicles, and dorsal cleithra applied to the anterior edge of the scapulocoracoid. In addition there is a median interclavicle (episternum), usually lozenge or T-shaped and underlying ventrally the coracoids and sternum, Figs. 179, 180, 182. In Stegocephalia these dermal bones often preserve the external sculpturing characteristic of the covering bones on the head. Thus, excepting for the interclavicle, which is probably an enlarged scale of the median ventral series, the dermal girdle in these early Tetrapods closely resembles that of primitive Osteichthyes, with this important difference, that it is not connected to

the skull. This character is, however, probably secondary, since there is
evidence that in some of the earliest and most primitive Stegocephalia,
such as the Carboniferous genera *Eogyrinus* and *Batrachiderpeton*, a

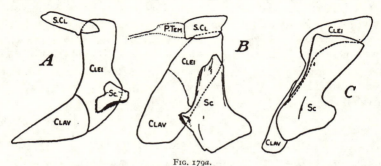

FIG. 179a.

Inner view of right half of pectoral girdle (interclavicle omitted) of : A, *Eusthenopteron* ; B, *Eogyrinus* ; C, *Cacops* (from D. M. S. Watson, *Tr. Roy. Soc.*, 1926). *Clav*, Clavicle ; *Clei*, cleithrum ; *P.Tem*, post-temporal ; *Sc*, scapulocoracoid ; *S.Cl*, supracleithrum.

post-temporal joined the cleithrum to the tabular, Fig. 179a (Watson,
306 ; Romer, 294).

The cleithrum tends to disappear in later forms. It has vanished in
the modern Urodela, together with the other dermal bones, but appears

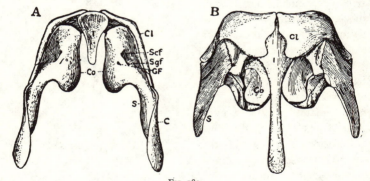

FIG. 180.

Pectoral girdle of A, *Cacops*, dorsal view, and B, *Conodectes* (Seymouria), ventral view. *C*, Cleithrum ; *Cl*, clavicle ; *Co*, coracoid region ; *GF*, glenoid foramen ; *I*, interclavicle ; *S*, scapular region ; *Scf*, supracoracoid foramen ; *Sgf*, supraglenoid foramen. (From Williston, *Osteology of Reptiles*, 1925.)

to persist in the Anura as the ossification which overlies the suprascapula
(Schmalhausen, 296). In a reduced condition it still occurs in many
Cotylosauria (*Diadectes*, *Pareiasaurus*, etc.), in some Theromorpha
(*Dicynodon*, *Clepsydrops*, *Edaphosaurus*, *Moschops*, Williston (95),
Gregory and Camp (256a), Romer (294), Watson (307), Seeley, and
others), and, according to Jaekel (536), in the primitive Chelonian,

Triassochelys; otherwise it is unknown in Reptiles, Birds, and Mammals, unless it is represented as a vestige by an early ossification of the scapular spine described by Broom in Marsupials (**239**).

The clavicle remains in all Tetrapods, excepting highly specialised forms such as the Urodela among Amphibia, certain Lacertilia (Chamele-

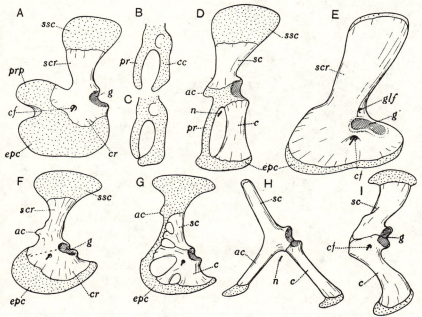

FIG. 181.

Diagrammatic figures of left half of pectoral girdle, outer view. A, *Megalobatrachus*; B, very young, and C, older stage in development of *Rana* (from A. Goette, but mesenchyme should join lower ends of cartilaginous processes); D, later stage, *Rana*; E, *Eryops*; F, *Sphenodon*; G, *Iguana*; H, *Chelone*; I, *Crocodilus*.   Cartilage dotted.   *ac*, Acromial process; *c*, coracoid bone·; *cc*, coracoid cartilage; *cf*, coracoid foramen; *cr*, coracoid region; *epc*, epicoracoid cartilage; *g*, glenoid cavity; *glf*, supraglenoid foramen; *n*, diazonal nerve; *pr*, 'procoracoid' process and bar (part of coracoid plate?); *prp*, 'procoracoid' process; *sc*, scapular bone; *scr*, scapular region; *ssc*, suprascapula.

ontes, etc.), the Ophidia, the higher Crocodilia and Dinosauria among Reptilia. Except in flightless forms in which the clavicles may be reduced or lost (*Struthio*, *Apteryx*) they are fused in Aves to form the characteristic furcula, already present in *Archaeopteryx*, Fig. 194. Clavicles are also lost or vestigial in several groups of the Mammalia (Marsupialia, Ungulata, Cetacea, Carnivora, Rodentia, etc.).

An interclavicle occurs in no modern Amphibian, but persists in modern Reptilia with a well-developed pectoral girdle (Rhynchocephalia, Crocodilia, Chelonia, Lacertilia), and was present in the primitive extinct forms. Alone among Mammalia the Monotremes possess a well-developed interclavicle, an important feature indicating their primitive structure, Fig. 197. A vestige of the interclavicle is probably represented by the so-called omosternum or presternum of Mammalia, which is not really preformed in true cartilage, and sometimes ossifies from two centres (Gegenbaur, **170**; Broom, **239**; Watson, **307**). Vestiges of the interclavicle have been described in Birds.

The primary girdle becomes ossified, but writers by no means agree as to the homology of the elements so formed, and their nomenclature is still in a state of confusion (Howes, **263-4**; Broom, **241**; Williston, **95**; Gregory, **256a**; Watson, **207**; Romer, **290**, **294**; Miner, **285**). The principal facts are as follows: In Stegocephalia (and Urodela) there is a single bone formed in each scapulocoracoid cartilage, primarily occupying the base of the scapu-

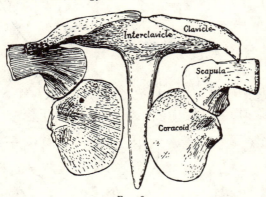

FIG. 182.

Ventral view of pectoral girdle of *Champsosaurus* (after Brown, from Williston, *Osteology of Reptiles*, 1925).

lar region near the glenoid fossa, but spreading in fully ossified forms dorsally as the scapular blade, and ventrally as the expanded coracoid plate, Figs. 180 A, 181 E. The scapulocoracoid bone is typically developed in such a primitive form as *Eryops* (Cope, **243**; Romer, **292**, **294**; Miner, **285**). Its internal supraglenoid buttress represents the mesocoracoid region of the Teleostome scapulocoracoid, and is pierced as in the fish by a supraglenoid foramen, doubtless for the diazonal nerves to the dorsal limb-muscles, and a more ventral supracoracoid or coracoid foramen for nerves to the ventral limb-muscles (supracoracoid or suprascapular nerve).

In primitive Reptilia (*Seymouria, Diadectes,* etc.), on the other hand, two centres of ossification appear—a dorsal scapula and a ventral coracoid, Figs. 180 B, 183 A, separated by a suture which crosses the glenoid cavity. In other Cotylosauria (Captorhinidae, Pareiasauria, Procolophonia) and in Theromorpha three centres of ossification occur; there being, in addition to the scapula, a posterior coracoid and an anterior procoracoid (epicoracoid), Figs. 183 B, C, 184 A, 185 A. The latter bone more or less completely surrounds the coracoid foramen from in front and below, and may penetrate to the glenoid cavity as a wedge between the scapula and the coracoid. The other extinct and all modern Reptilia, and Aves, have only

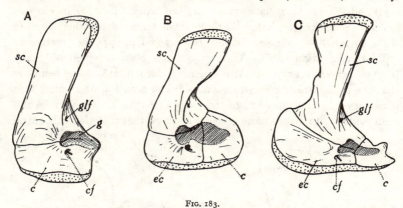

FIG. 183.

Outer view of left half of endoskeletal pectoral girdle with cartilaginous parts restored (dotted). A, *Diadectes*; B, *Labidosaurus*; C, *Dimetrodon*. Lettering as in Fig. 184.

a single coracoid element, sharing as usual in the formation of the glenoid cavity and enclosing as a rule the coracoid foramen, Figs. 181 F-I, 186, 86. The Mammalia, on the other hand, belong to the group with two ventral

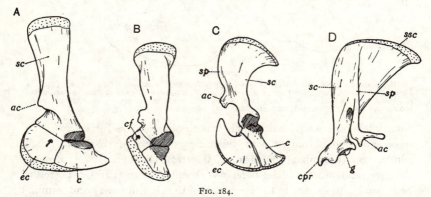

FIG. 184.

Outer view of left half of endoskeletal pectoral girdle of: A, *Dicynodon*; B, *Cynognathus*; C, *Ornithorhynchus*; D, Lepus; cartilage restored (dotted). *ac*, Acromion; *c*, coracoid; *cf*, coracoid foramen; *cpr*, coracoid process; *ec*, procoracoid; *g*, glenoid cavity; *glf*, glenoid foramen; *sc*, scapula; *sp*, spine.

elements in addition to the scapula, as is clearly seen in the Monotremata (Seeley; Cope; Williston, 95; Watson, 207; Romer, 294; Miner, 285; Gregory, 523). Now Williston has shown that in early Pelycosauria there are two ventral elements (as in other Theromorpha), but that of these the posterior bone (his metacoracoid) may be much the smaller, Fig. 183 c, and he believed it to be ossified late in *Ophiacodon* and not at all in the primitive *Varanosauras* and *Varanoops* (Case, 469 ; Williston, 95) ; Broili (436) and Watson (307) have since described a posterior bone in *Varanosaurus*. Williston concluded that the single coracoid of other Reptiles is the enlarged anterior element (our procoracoid, p. 174), the ' metacoracoid ' having disappeared. This opinion has been accepted by many (Broom, 241 ; Gregory, 523 ; Watson, 207 ; Hanson, 258), some believing that the posterior coracoid (metacoracoid) has been lost and others that it has never been acquired in the majority of Reptiles. On very insufficient evidence it must on either view be assumed that the principal ventral

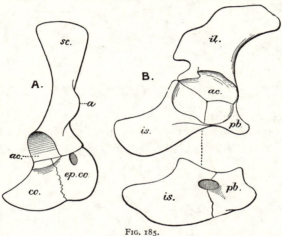

FIG. 185.

Right pectoral arch, outer aspect (A), and right pelvic arch, outer and inferior aspects (B), of a *Dicynodont*. Karoo Formation, Cape Colony. (From A. S. Woodward, *Vertebrate Palaeontology*, 1898.) *a*, Acromial process ; *ac*, glenoid cavity for humerus, and acetabulum for femur ; *co*, coracoid ; *ep.co*, procoracoid ; *il*, ilium ; *is*, ischium ; *pb*, pubis ; *sc*, scapular.

coracoidal elements are not homologous throughout the Tetrapoda. The whole question is fraught with difficulties, and has given rise to much controversy (Gegenbaur, Fürbringer, Goette, Lydekker, Howes, Broom, Gregory, and others). The older and simpler view, that the posterior bone (coracoid) sharing in the formation of the glenoid cavity and forming the posterior edge of the coracoid plate is homologous throughout, would seem to be nearer the truth than the complicated interpretations later authors have tried to substitute for it.

Examining the evidence more in detail we may begin with the Amphibia. Known stages in the development of *Archegosaurus* seem to show that in Stegocephalia the scapulocoracoid arises from one centre of ossification which spreads dorsally and ventrally. Its ventral coracoid plate comes to embrace the coracoid foramen. In modern Urodela also one centre is usually present, and its coracoid region may enclose the foramen. In *Amphiuma* and *Siren*, however, Parker describes separate coracoid and scapular bones. In these and other Amphibia (including doubtless the Stegocephalia) unossified cartilage may remain at the dorsal extremity of the scapula, as a suprascapula, and at the ventral end of the coracoid region, as infracoracoid cartilage (epicoracoid). There is also in Urodela a forwardly directed cartilaginous ' procoracoid process ', membrane closing the notch between it and the coracoid, Figs. 87 B, 181 A.

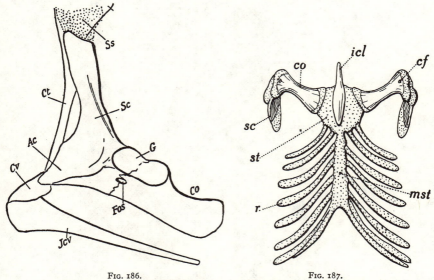

<table>
<tr><td>FIG. 186.</td><td>FIG. 187.</td></tr>
</table>

Pectoral girdle of *Triassochelys* (after Jaekel, from Williston, *Osteology of Reptiles*, 1925). *Ac*, Acromion; *Co*, coracoid; *Ct*, cleithrum; *Cv*, clavicle; *Fos*, coracoid foramen; *G*, glenoid cavity; *Jcv*, interclavicle; *Sc*, scapula; *Ss*, suprascapula.

Ventral view of pectoral girdle, sternum, and sternal ribs of late embryo *Crocodilus acutus* (after W. K. Parker, 1868). *mst*, Median prolongation of sternum. Other letters as in Fig. 190.

The primary girdle of the Anura consists of two halves which overlap ventrally (as in Urodela) in the Arcifera; but meet in the middle line in the Firmisternia, Fig. 87. Each half consists of a scapulocoracoid cartilage with an ossified scapula (with a supraglenoid foramen) and a coracoid. The latter bone forms the hinder limit of a large fenestra in the coracoid plate closed below by the epicoracoid cartilage, and in front by the procoracoid cartilage to which is closely adherent the clavicle (Gegenbaur).[1]   Where

---

[1] Goette (**254**) maintained that this bone is a true ossification of the cartilage; but Gegenbaur (**166, 170**) showed that it is really a membrane bone, separated by a thin layer of tissue from the underlying cartilage. If so, this clavicle must have sunk below the muscles to cling to the procoracoid cartilage.

N

the clavicle abuts on the scapula an 'acromial process' is present, Figs. 87 C, D, 181 B, D.

The 'fenestra' appears to include the coracoid foramen,[1] since it lets through the suprascapular nerve; but it is not closed by membrane.

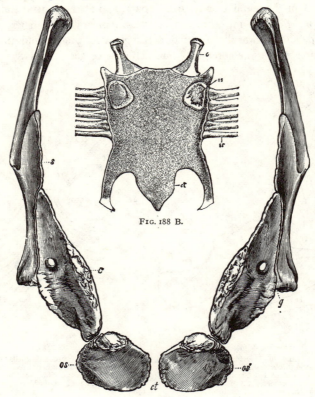

FIG. 188 B.

FIG. 188 A.

A, *Brontosaurus excelsus*. Jurassic: Wyoming. Pectoral arch, anterior aspect, with sternum of young ostrich *Struthio camelus* for comparison (Fig. B). (After A. S. Woodward, *Vertebrate Palae-ontology*, 1898.)   c, Coracoid;   *ct*, sternal cartilage;   *g*, glenoid cavity for humerus;   *os*, sternal bone; *s*, scapula;   *sr*, sternal ribs.

This fact renders it difficult to adopt Gegenbaur's view that the anterior process of Urodela is derived from the procoracoid bar of Anura by the breaking through of a fenestra. The Urodelan process may possibly be

[1] According to Goette (**254**) the fenestra in Rana becomes surrounded in ontogeny by two cartilaginous processes starting from the acetabular region, an observation confirmed by Fuchs (**249**). But it has recently been shown that in other Anura it is a true fenestra originating in a procartilaginous plate, and may be considered as formed by the enlargement of the coracoid foramen (de Villiers, **303**).

derived from the acromial process (Eissler, 246 ; Anthony and Vallois, 235), or more probably is a new development of the ventral plate.

We have already seen that in the most primitive extinct Reptilia there are present separate scapula and coracoid bones. The former bears a small acromial process where the clavicle is attached, and the latter encloses the coracoid foramen. From such a primitive girdle can be derived that of the Sauropsidan Reptiles (including Chelonia and Lacertilia, and all modern Reptiles) and of the Birds. It is typically developed in *Sphenodon* and other Rhynchocephalia, Fig. 181 F (Howes and Swinnerton, 49 ; Schauinsland, 583). The primary girdle is very similar in the Dinosauria ; but they lose the dermal clavicles and interclavicle, Fig. 188. While the Parasuchia still possess a primitive

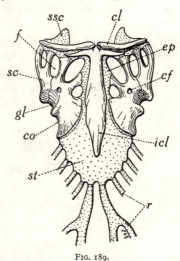

FIG. 189.
Ventral view of pectoral girdle, sternum, and sternal ribs of *Iguana tuberculata*.

pectoral girdle, in the Crocodilia the clavicle is lost and the coracoid much elongated, Fig. 187. There is no evidence in any of these Reptiles

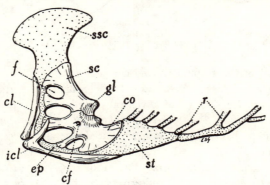

FIG. 190.
Left-side view of pectoral girdle, sternum, and sternal ribs of *Iguana tuberculata*. Cartilage dotted. *cf*, Supracoracoid foramen ; *cl*, clavicle ; *co*, coracoid ; *ep*, epicoracoid cartilage ; *f*, fenestra ; *gl*, glenoid cavity ; *icl*, interclavicle ; *r*, ventral ends of ribs ; *sc*, scapula ; *ssc*, suprascapula ; *st*, sternum.

of the occurrence of a fenestra or of more than one coracoidal bone in the ventral plate.

A highly specialised girdle occurs in the Chelonia. Here the dermal

clavicles and interclavicle are closely associated with more posterior
dermal plates (apparently derived from gastralia) to form a ventral
plastron (Gegenbaur, **170**), while the endoskeletal girdle forms on each
side a separate triradiate support for the limbs, movably connected to
carapace above and plastron below so as to allow for respiratory movements (p. 600). The coracoid (usually without foramen) and the scapula
become slender rods, and the latter bears a long process directed antero-
ventrally, Fig. 181 H. This process has been compared by some to the
procoracoid and by others to the acromion, and the earlier stages of the

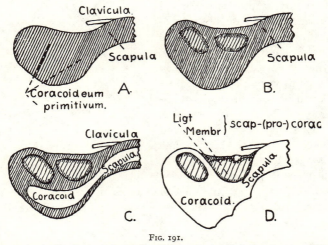

FIG. 191.

Diagrams showing development of coracoid region of pectoral girdle in *Lacerta*; procartilage darkly
shaded. Fenestrae are formed in procartilaginous ventral plate. (From C. van Gelderen, *Anat. Anz.*,
1925.)

development of the scapulocoracoid cartilage have not been sufficiently
described (Goette, **254**; Fuchs, **249**) to determine for certain whether
there has been fenestration of a ventral plate.[1] However, since in the
primitive Chelonian, *Triassochelys* (Jaekel, **536**), the process is much
shorter, it is probably merely an exaggerated acromial process, Fig. 186.

The coracoid and even the scapular regions of the Lacertilia become
pierced by membranous fenestrae, Figs. 189, 190, 191, 181 G, and it has
been shown (Goette, **254**; van Gelderen, **252-3**; Bogoljubsky, **237**) that this
is due to the fenestration during development of an originally continuous
procartilaginous plate (like that which persists in *Sphenodon* as cartilage
and finally as bone). In it appears the coracoid bone with its foramen,
and the fenestra or fenestrae leave behind the clavicle an anterior band of

---

[1] No cartilage can be seen in the embryo joining the ventral ends of the
coracoid and ventral scapular process, nor does a membrane stretch across.

cartilage running from the ventral infracoracoid cartilage to the acromial region. Later this band may be more or less completely replaced by ligaments, Fig. 191.

The Ophidia have lost all trace of a pectoral girdle, and it is very reduced and degenerate in many of the legless Lacertilia (Cope, 244; Boulenger).

The primary girdle of Birds is considerably modified for flight. The scapula in Carinatae is sword-shaped, and lies almost parallel to the vertebral column; it is movably articulated to the elongated coracoid (with

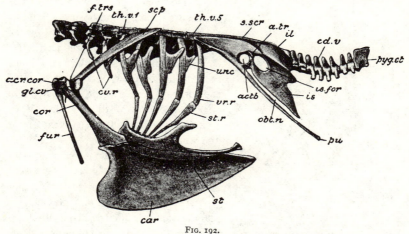

FIG. 192.

Columba livia. The bones of the trunk. *acr.cor*, Acrocoracoid; *a.tr*, anti-trochanter; *actb*, acetabulum; *car*, carina sterni; *cd.v*, caudal vertebrae; *cor*, coracoid; *cv.r*, cervical ribs; *f.trs*, probe passed into foramen triosseum; *fur*, furcula; *gl.cv*, glenoid cavity; *il*, ilium; *is*, ischium; *is.for*, ischiadic foramen; *obt.n*, obturator notch; *pu*, pubis; *pyg.st*, pygostyle; *scp*, scapula; *s.scr*, syn-sacrum; *st*, sternum; *st.r*, sternal ribs; *th.v.1*, first, and *th.v.5*, last thoracic vertebra; *unc*, uncinates; *vr.r*, vertebral ribs. (From Parker and Haswell, *Zoology*.)

a supracoracoid foramen or notch), Fig. 192. The very primitive Jurassic *Archaeopteryx* (Petronievics and Woodward, 289), however, has a scapula and a short coracoid (pierced by the supracoracoid foramen) more closely resembling those of Archosaurian reptiles, Fig. 194 A. The Ratitae have a short broad coracoid, and from the base of the scapular region there arises an anterior process (called 'precoracoid' process), the homology of which is difficult to interpret. The precoracoid process when present in Carinatae is on the coracoid itself, and in *Apteryx* it is known to arise by fenestration of the cartilage (T. J. Parker, 66). An opening is formed, separate in some species from the supracoracoid foramen; later the cartilaginous bar in front becomes ligamentous and may finally be ossified, Figs. 86, 193, 194. The 'precoracoid' or 'prescapular' process of the Struthiones appears to correspond to that of *Apteryx*, but develops as an outgrowth (Broom, **240**), which, in *Struthio* itself, grows down to join the coracoid and enclose a fenestra. The supracoracoid foramen persists in *Casuarius*; in other genera it appears to be confluent with the ' precoracoid ' notch or fenestra. The precoracoid process described above recalls that of the Chelonian scapula, is present in *Archaeopteryx* on the scapula (Petronievics and Woodward, **289**), and is possibly to be regarded as an acromion.

In good fliers the scapula is articulated to the coracoid at an acute angle of less than 90°. In *Archaeopteryx*, however, the two bones, though fused, more nearly resemble those of the Dinosaurs in shape and position; and in those birds which have lost the power of flight the girdle is reduced, the coraco-scapular angle enlarged (Dodo, Solitaire, *Ocydromus*, *Notornis*, etc.), and the coracoid fused to the scapula, Figs. 193, 194. Some of the flightless Moas seem to have lost all trace of the girdle and wing.

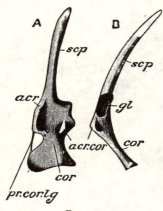

FIG. 193.

Apteryx mantelli. The left shoulder girdle. A, anterior; B, lateral (outer) surface. *acr*, Acromion; *acr.cor*, acrocoracoid; *cor*, coracoid; *gl*, glenoid cavity; *pr.cor.lg*, procoracoid, reduced to a ligament; *scp*, scapula. (After T. J. Parker; from Parker and Haswell, *Zoology*.)

The pectoral girdle of the aquatic Ichthyosauria is but little specialised. It preserves, in addition to the dermal T-shaped interclavicle and clavicles, a scapula and expanded coracoid, Fig. 195. The Sauropterygia, although less completely adapted to aquatic life, acquire a much more modified girdle (Owen, 1865; Conybeare; Hulke, 1892; Seeley, **298**; Boulenger, 1896). Clavicles and interclavicle are still normally developed in Mesosauria (Seeley, **297a**; Broom, **239a**; McGregor, **279a**), and the primary girdle has a scapula with fairly large dorsal blade and an expanded coracoid pierced by a foramen. But in the Nothosauria the dermal bones form a strong arch widely separated from the more posterior coracoid (von Meyer, 1847; Boulenger, 1896). The tendency for the dorsal blade of the scapula to become reduced, and for its ventral extension to enlarge, already seen in this suborder, is carried to an extreme in Plesiosauria (Seeley, **298**; Andrews, **232**; Watson, **308**). Here the coracoid becomes very large; and sends forward a ventral process which in more specialised and later forms (*Microcleidus, Cryptocleidus*, Fig. 196) meets a posterior process of the ventral plate of the scapula, thus surrounding a fenestra. Meanwhile the dermal bones are much reduced (in *Muraenosaurus* the clavicles disappear, and in *Microcleidus* the interclavicle) and come to lie on the inner dorsal side of the scapular plate. These modifications lead to the formation of a very solid ventral endoskeletal girdle which can afford a firm basis of insertion for pectoral muscles and resist the thrust of the powerful swimming paddles. The exact nature of the ventral scapular

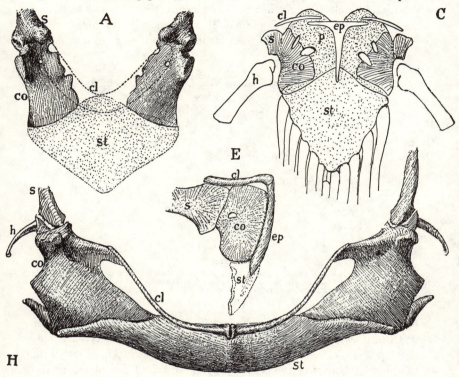

FIG. 194.

Pectoral girdle and sternum of : A, *Archaeopteryx*, Jurassic ; C, *Carsosaurus*, Cretaceous (after Nopsca) ; E, *Euparkeria* (after Broom) ; H, *Hesperornis*, Cretaceous (after Marsh). *cl*, Clavicle ; *co*, coracoid ; *ep*, interclavicle ; *h*, humerus ; *p*, epicoracoid cartilage ; *s*, scapula ; *st*, sternum. (From G. Heilmann, *Origin of Birds*, 1926.)

extension is not quite clear, but it appears to be formed by the growth of the acromial or prescapular bony process ; on the other hand, it may be an ossification of a procoracoid bar due to the fenestration of the ventral plate. It is probable that even in Nothosauria there was an extensive fenestrated cartilaginous plate, ossified only posteriorly, Fig. 196.

It has already been mentioned (p. 174) that in many Cotylosauria there is in addition to the scapula and coracoid a third bone developed in the primary girdle, the epicoracoid of Cuvier (precoracoid or procoracoid), situated in front of the coracoid and more or less completely enclosing the coracoid foramen, Fig. 183. This procoracoid, as it is now generally called, is seen typically developed in *Pareiasaurus* and *Procolophon*, and doubtless together with the coracoid occupied the greater part of a wide originally cartilaginous plate. A similar and quite obviously homologous procoracoid bone is found in the Theromorpha, Fig. 184, and Monotremata.[1] Indeed the pectoral girdle of Monotremes, with its interclavicle, procoracoid, and other points of resemblance, affords the strongest evidence

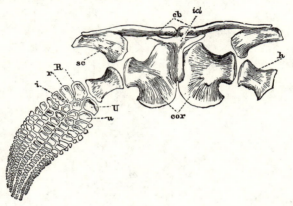

FIG. 195.

Ichthyosaurus communis, Conyb. Lower Lias : England. Pectoral arch and right fore-limb ; ventral aspect, ¼. *cl*, Clavicle ; *cor*, coracoid ; *h*, humerus ; *i*, intermedium ; *icl*, interclavicle (partly covered by clavicles) ; *R*, radius ; *r*, radiale ; *sc*, scapula ; *U*, Ulna ; *u*, ulnare. (From K. Zittel, *Palaeontology*.)

for the derivation of the Mammalia from Theromorph ancestors. *Echidna* and *Ornithorhynchus* preserve a T-shaped expanded interclavicle, well-developed clavicles, a strong acromial process still as in Theromorphs on the anterior border of the scapula, a large coracoid articulated to the sternum, and a procoracoid excluded from the glenoid cavity, Fig. 197.

The scapula of Cynodont Theromorphs already shows in its everted anterior edge the beginning of the formation of the acromial spine and prespinous fossa so characteristic of the Mammalia. The procoracoid may take a share in the formation of the glenoid cavity in Theromorphs.

There is a marked difference between the girdle of Monotrematous and Ditrematous Mammals. In all Ditremata, while the ventral elements have been much reduced and the interclavicle has disappeared, the

[1] The presence of the procoracoid in these forms may be taken as evidence that the order Cotylosauria is polyphyletic, some families being related to the ancestors of the Theromorphs, and others to the Sauropsidan stem.

scapula forms a wide blade expanding in front, so that the acromion and its ridge or spine come to lie on its outer surface separating an anterior from a posterior fossa, Figs. 184 D, 200 (McKay and Wilson). In the

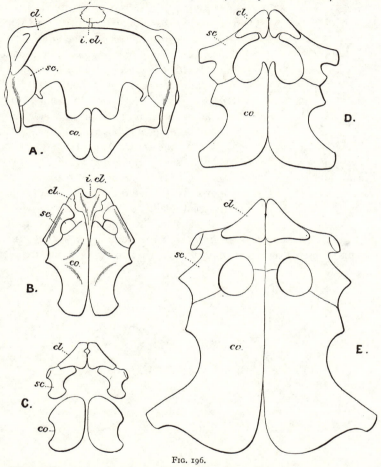

FIG. 196.

Diagram of pectoral arch of Sauropterygia, showing gradual atrophy of clavicular elements. A, *Nothosaurus mirabilis*; dorsal aspect. M. Triassic, Würtemberg. B, *Plesiosaurus*; ventral aspect. L. Jurassic, Lyme Regis. C, D, E, *Cryptocleidus oxoniensis*; dorsal aspect of three successive stages of growth. U. Jurassic, Peterborough. (After A. S. Woodward, *Vertebrate Palaeontology*, 1898.) *cl*, Clavicle; *co*, coracoid; *i.cl*, interclavicle; *sc*, scapula.

adult Marsupial the coracoid region is reduced to merely a ventral bony process of the scapula. But Broom (239) has shown that in such genera as *Trichosurus* and *Dasyurus* the coracoid cartilage in the embryo reaches the sternum, and the acromial process is near the anterior edge of the scapula, Figs. 198-9 (in *Trichosurus* the procoracoid region is still represented in embryonic mesenchyme). These important observations have been confirmed (Watson, **307**), and extended to *Didelphys* (Romer).

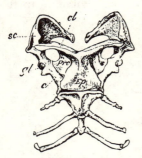

As shown by Howes (**263-4**) the coracoid process may ossify separately in Placentals (*Lepus*) and be retained as a separate bone often of considerable size in the adult (some Edentata). This process has been homologised with the procoracoid by Howes, and with the coracoid by Broom (**241**) and Gregory (**523**). Moreover, in some forms a small ossification occurs on the hinder surface of the glenoid cavity which is considered as the coracoid by Howes, and as an epiphysis by Broom and Gregory. The homology of these two ventral bony elements can hardly be determined on the evidence at present available.

FIG. 197.

Ventral view of pectoral girdle and anterior portion of sternum and sternal ribs of *Ornithorhynchus* (from Flower, *Osteol. of Mammalia*, 1885). *c*, Coracoid; *cl*, clavicle; *Ep*, interclavicle; *gl*, glenoid cavity; *prc*, procoracoid.

Conclusion.—In conclusion, it may be said of the Tetrapod pectoral girdle that the homologies of the dermal elements present no great difficulties. They may be compared to

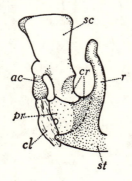

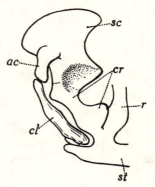

FIG. 198.

Trichosurus vulpecula. Embryo 8·5 mm. long. Reconstruction of shoulder girdle and part of sternum; cartilage white, procartilage dotted; left-side view (from Broom, 1900). *ac*, Acromial process; *cl*, clavicle; *cr*, coracoid region; *pr*, procoracoid region, mesenchymatous; *r*, first sternal rib; *sc*, scapular region; *st*, sternum.

FIG. 199.

Trichosurus vulpecula. Embryo 14·8 mm. long. Left-side view of reconstruction of shoulder girdle (from Broom, 1900). Lettering as in previous figure.

those of a primitive Teleostome, the post-temporal connexion being lost in all but the most primitive Stegocephalia, and a median interclavicle being added ventrally. On the other hand, the history of the endoskeletal elements in the specialised groups is by no means clear.

While in Stegocephalia there is a simple girdle with an expanded ventral plate and a single scapulocoracoid bone, in modern Anura a fenestra develops in the ventral plate (apparently from the coracoid foramen), and a posterior coracoid bone appears separate from the scapula. The Amniota seem to have independently acquired a similar coracoid bone sharing in the formation of the glenoid cavity and strength-

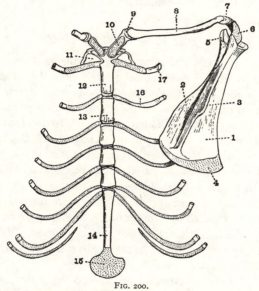

FIG. 200.

Dorsal view of sternum and right half of shoulder girdle of *Mus sylvaticus*. (From S. H. Reynolds, *Vertebrate Skeleton*, 1913.) 1, Postscapular fossa ; 2, prescapular fossa ; 3, spine ; 4, suprascapular border unossified ; 5, coracoid process ; 6, acromion ; 7, secondary cartilage ; 8, clavicle ; 9, secondary cartilage ; 10, ' omosternum ' ; 11 and 12, presternums (manubrium or first sternebra) ; 13, sternebra, first segment of mesosternum ; 14, xiphisternum ; 15, cartilaginous termination of xiphisternum ; 16, second sternal rib ; 17, first vertebral rib.

ening the posterior edge of the ventral plate. Fenestration of this plate occurs in Lacertilia, and probably in Sauropterygia ; but in most Reptiles and in Birds the coracoid bone apparently occupies the whole extent of the ventral plate. In those forms which possess only this coracoid, it typically encloses the nerve foramen. The acromial process of the scapula, already incipient in Amphibia, appears to be secondarily extended ventrally in Chelonia and possibly in Sauropterygia. In some Cotylosauria, the Theromorpha, and the Mammalia, a new ossification, the procoracoid, appears in the ventral plate usually enclosing the coracoid foramen and situated in front of the coracoid. Thus, as far as two bones can ever be said to be homologous with one, the single coracoid of the Diaptosaurian or Sauropsidan Reptiles which lead towards the Birds may be considered homologous with the coracoid and procoracoid of those Synaptosaurian or Theropsidan (and Cotylosaurian) Reptiles which lead towards the Mammals. The pectoral girdle offers evidence of the early divergence of the Reptilia into these two main branches.[1]

Thus, outside the Cotylosauria, Theromorpha, and Mammalia, there is never more than one ventral bone ; and in no Tetrapod are there more than two ossifications in the ventral plate.

THE PELVIC GIRDLE OF PISCES

The general morphology of the pelvic girdle in Pisces is less well understood than that of the pectoral girdle. Like the latter the pelvic girdle is essentially an extension in the body-wall of the fin-skeleton for the support of the pelvic fin and the attachment of its muscles (p. 149). In ontogeny it is differentiated in a blastema continuous with that of the fin-skeleton. Indeed, in the Cladoselachii (Dean, 154-5) it is scarcely differentiated even in the adult from the basal region of the radials and shows distinct signs of segmental origin, Fig. 150 ; and it may be said of primitive fish generally that the consolidation of the girdle from originally segmental elements proceeds from before backwards. Radials are primitively articulated to its outer border, and its posterior end passes gradually into the longitudinal axis of the fin-skeleton. Better defined pelvic plates converging towards the mid-ventral line are found in Pleuracanthodii, Fig. 203 ; Fritsch, 23.

The Selachii have, in front of the cloacal depression, a transverse plate formed by the fusion of paired pelvic cartilages. To its outer border is

---

[1] It may be added that it is not impossible that a fenestration has occurred in the majority of, if not in all, the Amniota, leaving a cartilaginous acromial (or precoracoid) bar extending from the scapula to the infracoracoid cartilage (Gegenbaur, Fürbringer). The excavated anterior edge of the coracoid in such forms as Phytosauria and Sauropterygia suggests this interpretation. If the fenestra grew large and the anterior bar was reduced to an acromion and even finally lost, the coracoid would remain as the only ventral element, as seen in most diving Reptiles and Birds. Likewise the procoracoid and coracoid, in those forms where both these bones occur, may be ossifications in the plate posterior to a fenestra which has broken through in front. The various anterior cartilaginous or bony bars and processes (' precoracoid ', ' acromion', scapular process, etc.) seen in Plesiosauria, Chelonia, Aves, and perhaps in Lacertilia, would then all be derived from the bar originally closing the fenestration in front. Nevertheless, since there is no convincing evidence of the existence of such a complete acromial bar in primitive Amniotes, this view has not been generally adopted.

attached the fin-skeleton, Fig. 201. The pelvic plate interrupts the ventral body-wall muscles inserted on its anterior and posterior edges. Diazonal nerves usually pierce it, and in *Chlamydoselachus*, where the pelvic plate is very long, there are two series of nerve foramina indicating that it is derived from many segments, Fig. 27. Generally a slight elevation occurs above the articulation of the fin, and in Batoidei this may develop into a considerable dorsal process comparable to the ilium of Tetrapods,

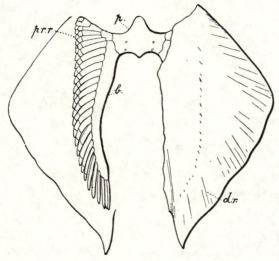

FIG. 201.

Ventral view of the pelvic girdle and fins of *Heptanchus cinereus*, Gm. ; in this and the succeeding figures the complete skeleton is exposed on one side only. *a*, Jointed axis; *b*, basipterygium; *c*, anterior cartilage; *dr*, dermal rays; *p*, pelvic girdle; *prr*, preaxial radial; *s*, scale covering lobe. Lettering for Figs. 204-8. (From E. S. Goodrich, *Quart. Jour. Micr. Sci.*, v. 45, 1901.)

Fig. 166. The Holocephali, in which the two halves of the girdle remain separate, also have a large dorso-lateral ' iliac ' process.

The two halves of the originally cartilaginous girdle of Teleostomes ossify, except in modern Acipenseroidei, in the form of two separate bones lying horizontally in the body-wall, meeting in front of the anus and bearing the fin-skeleton at their divergent posterior ends (Polypterini, Amioidei, Lepidosteoidei, and Teleostei, Figs. 35, 205-7). Small cartilages may be detached at their anterior ends (*Polypterus*, Fig. 205), and rarely the two pelvic plates may meet in a median cartilage as in *Gadus*.

Various observers have considered these pelvic plates to be parts of the fin-skeleton which has shifted inwards, and have hence called them basipterygia or metapterygia (Davidoff, **153**; Gegenbaur, **170**; Wiedersheim, **230**), believing the true girdle to be represented by the small anterior cartilages mentioned above. But, while Wiedersheim held that these

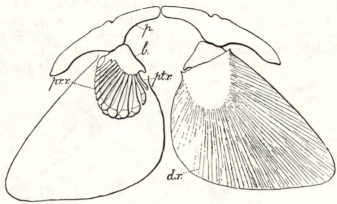

FIG. 202.

Ventral view of the pelvic girdle and fins of *Chimaera monstrosa*, L., ♀. The skeleton is completely exposed on the left side. *b*, Basipterygium ; *d.r*, web of right fin with ceratotrichia ; *p*, pelvic cartilage ; *pr.r*, preaxial radials ; *pt.r*, postaxial radials. (From *Quart. Jour. Micr. Sci.*, v. 45, 1901.)

represent the first rudiments of a girdle, Gegenbaur on the contrary looked upon them as its last vestiges.

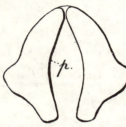

FIG. 203.

Ventral view of the pelvic girdle of *Pleuracanthus Gaudryi*, Brogn. (From E. S. Goodrich, *Quart. Jour. Micr. Sci.*, v. 45, 1901.)

There can, however, be little doubt that these theories are erroneous (Goodrich, **171**). Since pelvic bones are found normally developed in all Teleostomes from the Devonian Osteolepi-doti (*Eusthenopteron*, Fig. 204) to the present day, there is no reason for rejecting the older view that they are the halves of a true girdle, comparable to that, for instance, of the Pleuracanthodii. Moreover, their position in the body-wall and relation to the nerves and musculature clearly show that they do not belong to the fin. The limit and joint between fin-skeleton and girdle remains always near the base of the fin-fold in fishes, and there is no evidence that it has shifted inwards and been carried to the anterior tip of the pelvic bone or cartilage.[1]

It is probably the structure of the girdle in modern Chondrostei which has led to this unlikely interpretation. For, especially in the somewhat degenerate living representatives of the group, the limit between the

[1] Stensiö adopts an intermediate view : that, during the course of phylogeny, to the original pelvic plate of Teleostomes has been added a region belonging to the base of the fin-radials (**134**).

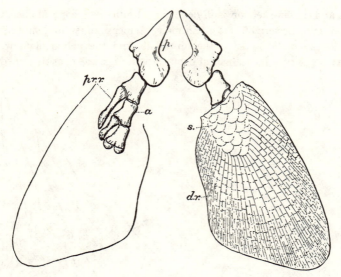

FIG. 204.

Ventral view of the pelvic girdle and fins of *Eusthenopteron Foordi*, Whit., restored. For lettering see Fig. 201. (From E. S. Goodrich, *Quart. Jour. Micr. Sci.*, v. 45, 1901.)

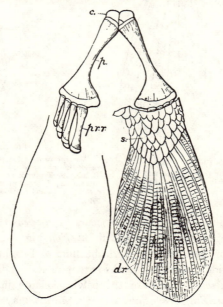

FIG. 205.

Ventral view of the pelvic girdle and fins of *Polypterus bichir*, Geoffr. For lettering see Fig. 201. (From E. S. Goodrich, *Quart. Jour. Micr. Sci.*, v. 45, 1901.)

girdle and the fin-skeleton is ill-defined (v. Rautenfeld, **207**; Mollier, **198**; Stensiö, **134**; Sewertzoff, **216**).   In many Acipenseroidei no definite joint exists between the bases of the fin-radials and the pelvic plate, which preserves traces of its segmental composition, Figs. 208-9, as described by

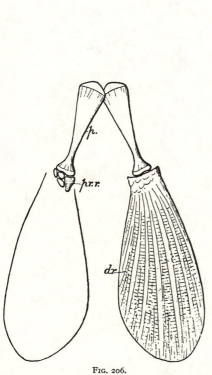

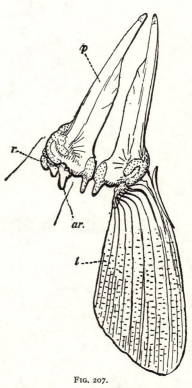

FIG. 207.

FIG. 206.

Ventral view of the pelvic girdle and fins of *Amia clava*, Bon. (Partly from Davidoff; from E. S. Goodrich, *Quart. Jour. Micr. Sci.*, v. 45, 1901.)

Ventral view of the pelvic girdle and left pelvic fin of *Salmo salar*, L. (Modified from Bruch.) *ar*, Posterior radia or remains of basi-pterygium; *l*, lepidotrich; *p*, pelvic bone; *r*, radial. (From E. S. Goodrich, *Vert. Craniata*, 1909.)

Thacher.   An 'iliac' dorsal process occurs in *Acipenser*, and a similar process is developed on each segment in *Polyodon*, Fig. 163 (Thacher, **219**). The homology of these processes with the ilium of Tetrapods is doubtful. The great individual variation in the extent of its subdivision and the almost complete segmentation of the pelvic skeleton in some modern forms (*Acipenser, Scaphirhynchus*, and *Psephurus*) may be due to the more or less complete retention of an embryonic condition, and not to truly primitive structure.   Nevertheless, their pelvic girdle seems to be at about the same stage of differentiation as that of the more primitive

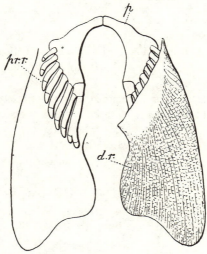

FIG. 208.

Ventral view of pelvic girdle and fins of *Acipenser sturio*, L.   (From E. S. Goodrich, *Quart. Jour. Micr. Sci.*, v. 45, 1901.)

FIG. 209.

Ventral view of the cartilaginous skeleton of the pelvic girdle and fin of *Scaphirhynchus* (after Rautenfeld). *Pm*, median, and *Pd*, dorsal process of girdle; *F*, nerve foramen; 1-9, radials. (From Goodrich, *Vert. Craniata*, 1909.)

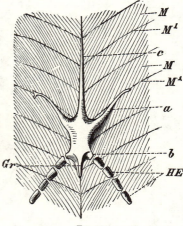

FIG. 210.

Pelvis of *Protopterus*. From the ventral side. *a*, Prepubic process, which may become forked at its distal end ; *b*, process to which the pelvic fin (*HE*) is attached; *c*, epipubic process ; *Gr*, ridge for attachment of muscles ; *M*, myotomes ; *M¹*, intermuscular septa. (From Wiedersheim, *Comp. Anat.*)

Chondrichthyes. The presence of foramina for diazonal nerves is also good evidence that the plate is part of a true girdle.

A well-defined ossified girdle has been found in Crossopterygii in the form of two triangular bones (*Eusthenopteron*; Goodrich, 171) supporting the fin-skeleton, Fig. 204; and in the Coelacanthini the pelvic plates clearly resemble those of the Holostei. The girdle in Palaeoniscoidei and Saurichthyoidei is more like that of *Acipenser* (Stensiö, 134).

The girdle in Dipnoi is unossified and known only in living forms (Günther, 1871; Wiedersheim, 230). The two halves are completely fused to a median cartilage bearing prominent knobs for the articulation of the fin-skeleton, and a long tapering anterior epipubic process. At each side is a slender prepubic process embedded in an intermuscular septum. There are no nerve foramina, Figs. 152, 210. Except for the absence of an ilium the girdle resembles that of Urodela, in particular of *Necturus*.

THE PELVIC GIRDLE OF TETRAPODA

The pelvic girdle of the Tetrapoda consists primitively of two halves, each formed of an originally continuous cartilage bearing an acetabular

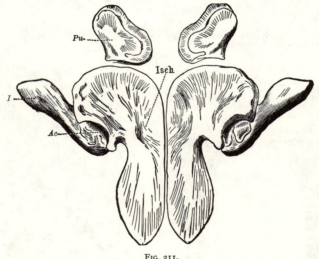

FIG. 211.

Pelvic arch of *Mastodonsaurus giganteus*, Jaeg. (after Fraas). *I*, Ilium; *Isch*, ischium; *Pu*, pubis; *Ac*, acetabulum. (From K. Zittel, *Palaeontology*.)

cavity for the head of the femur. Dorsally rises an iliac process which is typically firmly attached to the distal end of one or more sacral ribs. Ventrally the cartilage expands into a wide pubo-ischiadic plate, which meets that of the opposite side in a long ventral symphysis. The girdle encircles the abdominal cavity in front of the anus, and being more or less rigidly fixed to the vertebral column affords a firm basis for the articulation of the hind-limb, and for the insertion of its muscles. In typical well-ossified terrestrial forms three bones occur, the ilium, pubis, and ischium, all meeting in the acetabulum. The ventral plate interrupts the ventral body-wall muscles, separating the abdominal muscles partially

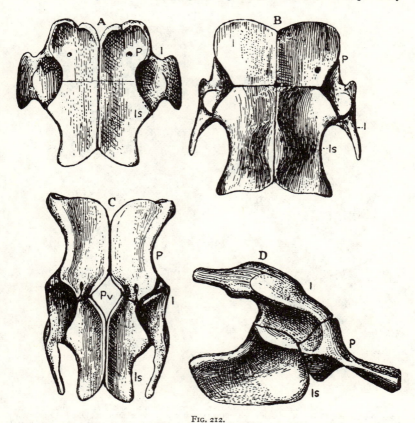

FIG. 212.

Pelvic girdle of A, *Cacops*, ventral view; B, *Conodectes* (*Seymouria*), ventral view; C, D, Varanops, ventral and side view. *I*, ilium; *Is*, ischium; *P*, pubis. (From Williston, *Osteology of Reptiles*, 1925.)

attached to its anterior border from the caudal muscles partially attached to its posterior border, while the outer surface of the plate serves chiefly for the insertion of limb-muscles (Vialleton, 302; Gregory and Camp, 256a; Romer, 391-2, 394-5). To a less extent the ilium interrupts the lateral myomeres more dorsally. In such primitive forms as the Amphibia, muscles derived from them are inserted over the ilium; but in higher Tetrapods these are shifted to the inner surface of the dorsal extremity of the ilium as the insertions of the limb-muscles spread over it.

Amphibia.—The primitive type of girdle is seen in Stegocephalia typically developed (Cope ; Williston, **95** ; Romer). The ilium is almost

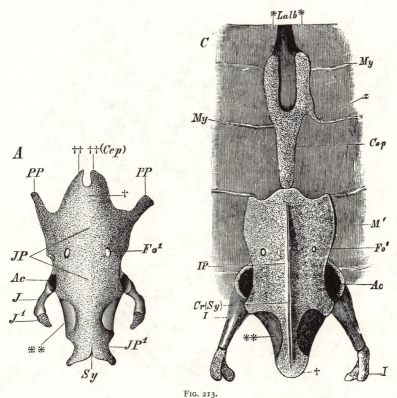

FIG. 213.

Pelvis of (A) *Proteus* and (C) *Cryptobranchus*. From the ventral side. *Ac*, Acetabulum ; *Cep*, ypsiloid cartilage ; *Cr* (*Sy*), muscular ridge on the ventral side of the ischiopubis ; *Fo*, *Fo*[1], obturator foramen ; *J*, *J*[1], ilium ; *JP*, *JP*[1], ventral pelvic plate (ischiopubis) ; *Lalb*, linea alba ; *My*, intermuscular septa ; *PP*, prepubis ; *Sy*, symphysis ; ** (in A), ossified region of the ischium ; *z*, outgrowth from this bifurcation ; † (in C), hypoischiatic process, present in the Derotremata and Necturus ; †† (*Cep*), epipubis. (From Wiedersheim, *Comp. Anatomy of Vertebrates*.)

vertical, and the ventral plate is broad with a large ischium and usually smaller pubis (not ossified in Branchiosauria), both no doubt embedded in a continuous cartilaginous plate in the living. The pubis comes to surround the diazonal obturator nerve in an obturator or pubic foramen, Figs. 211, 212 A. In Amphibia generally the ilium tapers dorsally and meets only one sacral rib, while in Amniota it is expanded as a rule to cover two or more sacral ribs and transverse processes. The Embolomerous

Stegocephalia, however, have an ilium expanded backwards as in *Cricotus* (12) and *Eogyrinus* (Watson, 644), which seems to have been but loosely attached to the long, scarcely modified sacral ribs. The Urodela have a girdle built on the primitive plan, but the ossifications are less developed, and the pubis is not as a rule ossified at all, though rudiments appear in *Salamandra* (Wiedersheim, 230). In *Amphiuma* and *Proteus* with degenerate hind-limbs the ilium does not reach the vertebral column. The cartilaginous ventral plate is in *Necturus* prolonged into a long

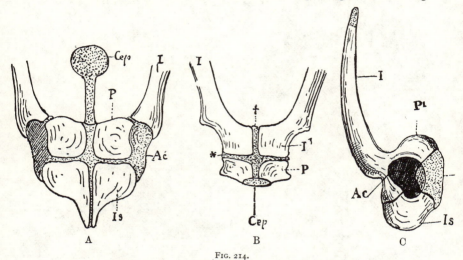

FIG. 214.

Pelvis of Anura. A, *Xenopus*, from below; B, the same from the front; C, *Rana esculenta*, from the right side. *Ac*, Acetabulum; *Cep*, ypsiloid cartilage; *I*, ilium; *I¹* (in *Xenopus*), the proximal end of the ilium, which is separated from its fellow and from the pubis by a +-shaped zone of cartilage, †, *; *Is*, ischium; *P*, pubis (*P¹* in *Rana*, pubic end of ilium). (From Wiedersheim, *Comp. Anat. of Vertebrates*.)

epipubic process recalling that of Dipnoi (p. 194). In other genera there may be a separate Y-shaped ypsiloid cartilage (*Salamandra*, *Cryptobranchus*, Fig. 213). Smaller paired lateral prepubic processes are also generally present. The Anura have a much reduced ventral plate, while the ilium is greatly extended forwards, an adaptation to their leaping mode of progression by means of enlarged hind-limbs. Only in *Xenopus* is the pubis ossified, Fig. 214 A. A median ypsiloid cartilage similar in origin to that of Urodela occurs in this genus.

Believed to be of paired origin, like the plate itself, and to be formed from the pubic cartilage by forward growths, which may fuse more or less completely in the middle line, and may also become secondarily detached from the plate (Wiedersheim, 310), the ypsiloid cartilage has been held to be homologous with the epipubic median process of

Amniotes. But others maintain that it is of separate origin, and primarily a chondrification in the linea alba between the posterior myomeres sometimes extending into the myocommata (Baur, 236; Whipple, 311). It may perhaps be compared to the sternal cartilage further forwards.

Primitive Reptilia.—The pelvic girdle of primitive Reptilia closely re-

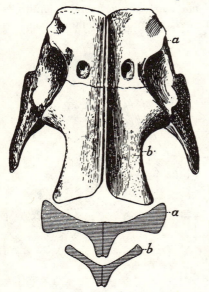

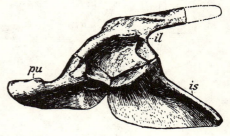

FIG. 215.

Ventral view of pelvic girdle of *Limno-scelis paludis*. Cross-section through pubes at *a*, and through ischia at *b*. (From Williston, *Osteology of Reptiles*, 1925.)

Left-side view of pelvic girdle of *Ophiacodon*. *il*, Ilium; *is*, ischium; *pu*, pubis. (From Williston, *Osteology of Reptiles*, 1925.)

sembles that of the Stegocephalia, with a stout vertical ilium and complete ventral plate composed of an ischium and a pubis pierced by an obturator foramen, Figs. 212, 215. But in some of the Cotylosauria the dorsal region of the ilium is considerably extended both forwards and backwards. The backward extension in these and other reptiles seems to be correlated with the development of a powerful tail. It is often pronounced in Theromorpha such as Dromosauria and Pelycosauria. The obturator foramen tends to become enlarged, and gives rise to a fenestra separating the pubis from the ischium in Dicynodontia and Theriodontia.[1]

[1] The central opening in the mid ventral line between the bony pubes and ischia (pubo-ischiadic of Williston, 95) probably was filled with cartilage, and does not seem to correspond to the true lateral fenestra of either the Synapsidan or Diapsidan pelvic girdles.

The expansion forwards of the ilium with an everted crest, and the

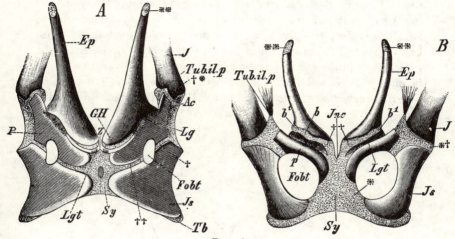

FIG. 216.

Pelvis of (A) *Echidna hystrix* (adult), and (B) *Didelphys azarae* (foetus, 5·5 cm. in length). From the ventral side. *Ep*, Epipubis or prepubis (marsupial bone); *Fobt*, obturator foramen; *J*, ilium; *Js*, ischium; *Lg* and *Lgt*, ligament between the pubis and epipubis; *P*, pubis; *Sy*, ischiopubic symphysis; *Tub.il.p*, iliopectineal tubercle; **, cartilaginous apophysis at the anterior end of the epipubis. In Fig. A, *GH*, articulation between the pubis and epipubis; *Tb*, cartilaginous tuber ischii; *Z*, process on the anterior border of the pubis; †*, †, ††, ilio- and ischio-pubic sutures. In Fig. B, *b*, *b¹*, cartilaginous base of the epipubis, continuous with the interpubic cartilage at †; *, *†, ischio-pubic and ischio-iliac sutures. (From Wiedersheim, *Comp. Anat. of Vertebrates*.)

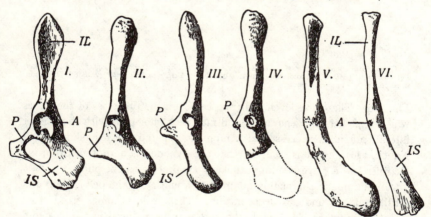

FIG. 217.

Outer view of left pelvic bone of: I. *Eotherium*, Middle Eocene, Egypt; II. *Eosiren*, Upper Eocene, Egypt; III. *Halitherium*, Oligocene, Europe; IV. *Metaxytherium*, Miocene and Pliocene, Europe; V. *Halicore dugong*, recent, Indo-Pacific Ocean; VI. *Halicore tabernaculi*, recent, Red Sea. *A*, Acetabulum; *IL*, ilium; *IS*, ischium; *P*, pubis. (From O. Abel, *Stämme der Wirbeltiere*, 1919.)

growth of the ventral plate backwards in Cynodontia, leads to the mammalian type of girdle.

Mammalia.—The mammalian girdle has the ilium usually greatly extended forwards, the acetabulum being behind the sacrum ; the symphysis somewhat shortened (sometimes very short as in many Insectivora, or absent as in some Insectivora and Chiroptera), and in some Insectivores, Carnivores, and Primates formed by the pubes only. In Edentata the ischia may join the caudal vertebrae behind. The foramen obturatorium becomes a large fenestra. Characteristic epipubic bones ('marsupial' bones) are articulated to the pubes of the Monotremata and Marsupialia,

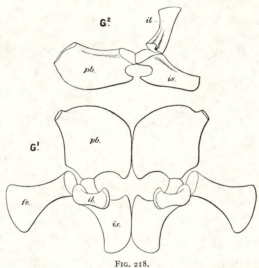

Fig. 218.

G^1, G^2, *Muraenosaurus leedsi*. Pelvis from the superior and left lateral aspect. Oxford Clay, Peterborough. (After C. W. Andrews, from A. S. Woodward, *Vertebrate Palaeontology*, 1898.) *fe*, Femur ; *il*, ilium ; *is*, ischium ; *pb*, pubis.

Fig. 216. These are developed from cartilaginous processes of the pubes which later become separated, and may be homologous with the similar lateral prepubic processes of Amphibia (p. 196) (Wiedersheim, 310 ; Nauck, 286). There is evidence that the Cynodontia and other primitive Reptiles had a corresponding cartilaginous process on the pubis.

In Mammalia thoroughly adapted to aquatic life the pelvic limb and girdle tend to undergo degeneration. The girdle becomes freed from its attachment to the sacral ribs and may be reduced in Cetacea to a small and simple bony rod 'floating' in the body-wall. A similar reduction takes place in Sirenia, Fig. 217.

Sauropterygia and Ichthyosauria.—The Sauropterygia preserve a primitive type of girdle, but the wide ventral plate is pierced by an obturator fenestra between pubis and ischium, Fig. 218. A separate obturator foramen occurs, however, in *Nothosaurus* (Andrews).   While in the more primitive Triassic Ichthyosauria (*Cymbospondylus*, *Toretocnemus* (Merriam,

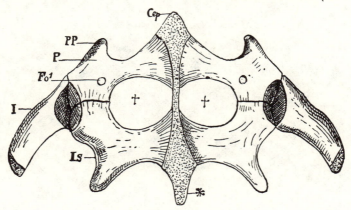

FIG. 219.

Pelvic arch of *Sphenodon*. (After Credner.)   From the ventral side.   *Cep*, Epipubic cartilage ; *Fo¹*, obturator foramen ; *I*, ilium ; *Is*, ischium ; *P*, pubis ; *PP*, prepubis ; ***, hypoischiatic process ; †, †, ischiopubic foramina.   (From Wiedersheim, *Comp. Anatomy.*)

283)) the pelvic girdle is provided with a well-developed ilium connected to a sacral rib, and large expanded pubis and ischium, in later forms,

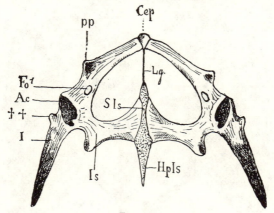

FIG. 220.

Pelvis of *Lacerta vivipara*. From the ventral side.   *Ac*, Acetabulum, in which the three pelvic bones come together ;   *Cep*, epipubis, composed of calcified cartilage ;   *Fo¹*, obturator foramen ; *HpIs*, hypoischium, which becomes segmented off from the hinder ends of the ischia in the embryo as a paired structure ;   *I*, ilium, with its small preacetabular process ††, much more strongly developed in crocodiles, dinosaurians, and birds ;   *Is*, ischium, forming a symphysis at *SIs* ;   *Lg*, fibrous ligament ; *P*, pubis ;   *pp*, prepubic process.   (From Wiedersheim, *Comp. Anatomy.*)

where the hind-limbs are reduced and propulsion was mainly performed by the caudal fin, the girdle becomes more and more degenerate.   The ilium pubis and ischium are small rod-like bones in a 'floating' girdle

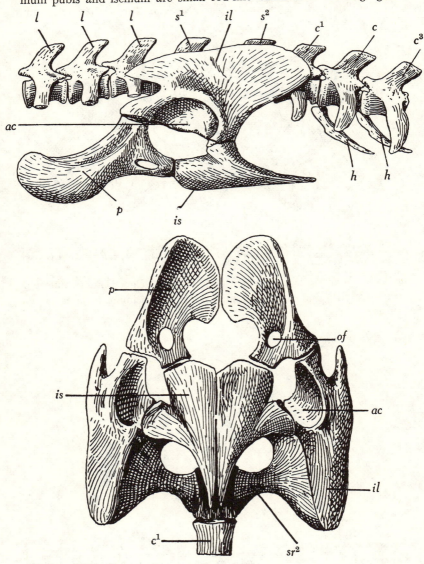

FIG. 221.

Left-side view of pelvic girdle and part of vertebral column of *Aëtosaurus crassicauda* (after C. Fraas, '07, from O. Abel, *Stämme der Wirbeltiere*, 1919). *ac*, Acetabulum ; c^{1-3}, caudal vertebrae ; *h*, chevron bones, haemal arches ; *il*, ilium ; *is*, ischium ; *l*, lumbar vertebrae ; *p*, pubis ; s^{1-2}, sacral vertebrae.

in the Jurassic *Ichthyosaurus*. *Ophthalmosaurus*, which died out in the

Cretaceous, had a vestigial girdle, the small pubis and ischium being fused (Andrews, 232-3).

Reptilia Sauropsida and Aves.—The pelvic girdle of the Sauropsidan Reptiles is no doubt derived from the Stegocephalian type. Typically the ilium is inclined and lengthened backwards, and except for this the

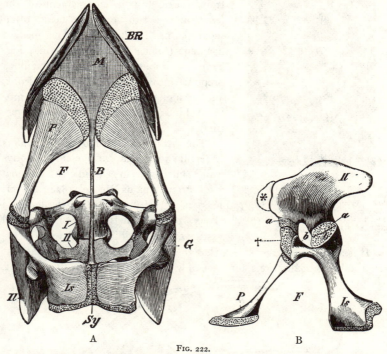

FIG. 222.

Pelvis of a young *Alligator lucius*. A, ventral, and B, lateral view. B, Fibrous band between the pubis and symphysis ischii ; b, foramen in the acetabulum, bounded posteriorly by the two processes, a and b, of the ilium and ischium respectively ; F, ischiopubic foramen ; G, acetabulum ; Il, ilium ; Is, ischium ; M, fibrous membrane extending between the anterior margin of the pubis and the last pair of 'abdominal ribs' (BR) ; P, pubis ; Sy, symphysis of ischium ; *, indication of a forward growth of the ilium, such as is met with in dinosaurians and birds ; †, pars acetabularis, which is interposed between the process a of the ilium and the pubis ; I, II, first and second sacral vertebrae. (From Wiedersheim, *Comp. Anatomy*.)

girdle is little changed in such primitive forms as the Champsosauria and Phytosauria, where the ventral plate is formed of expanded pubes and ischia, meeting no doubt in a long cartilaginous symphysis. But in others a fenestra closed by membrane develops in the ventral plate between the pubis and the ischium, Fig. 219. It is not the enlarged obturator foramen, as it appears to be in the Theropsida (Synapsida), since the obturator nerve passes independently through the pubis (except in those forms where the nerve foramen and the fenestra become secondarily confluent), and it may be called the pubo-ischiadic or ischiopubic fenestra.

The Rhynchocephalia, Lacertilia, and Chelonia usually develop strong prepubic and postischial processes. The ventral plate may be much reduced ventrally owing to the enlargement of the fenestra, the two fenestrae being often separated in the adult merely by a median ligament (some Chelonia, Lacertilia, Fig. 220). A prominent median cartilaginous prepubic process may remain in front, and a similar posterior hypoischial process behind (separate in some adult Lacertilia).

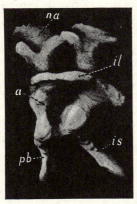

FIG. 223.

Outer view of left half of pelvic girdle and two sacral vertebrae of embryo *Crocodilus*. From wax reconstruction of cartilage made by G. R. de Beer. *a*, Acetabulum; *il*, ilium; *is*, ischium; *na*, neural arch; *pb*, pubis.

The girdle may be reduced in legless Lacertilia to a mere vestige in the body-wall; but even in Ophidia a vestigial girdle often remains (Meckel, 1824; Mayer, 1825–29; J. Müller, 1838; Duerden and Essex, 1923), sometimes with a clawed vestige of the hind-limb (Boidae).

It is among the Archosauria that the pelvic girdle undergoes the most remarkable specialisations, for some of which it is very difficult to account.

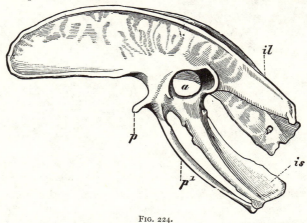

FIG. 224.

Pelvis of *Apteryx australis*. Lateral view. (After Marsh.) *a*, Acetabulum; *il*, ilium; *is*, ischium; *p*, pectineal process from the pars acetabularis; *p*[1], pubis. (From Wiedersheim, *Comp. Anatomy*.)

The Phytosauria (Parasuchia), as already mentioned, still preserve a primitive form of girdle with the ilium expanded chiefly backwards, and large plate-like pubis and ischium (v. Meyer, 1847 ; McGregor, **279**). The ilium extends forwards as well as backwards, especially in Dinosaurs, and the pubo-ischiadic fenestra becomes large between the divergent lengthened pubis and ischium in Pseudosuchia, Fig. 221, and Saurischia. The Theropodous Saurischia have the ventral ends of the pubis and ischium enlarged and flattened at the symphysis, forming apparently surfaces on which the body could be supported at rest. In Saurischia the acetabulum also becomes pierced—the lower border of the ilium being excavated. A similar piercing of the acetabulum occurs in Crocodilia,[1] Pterosauria, Ornithischia, and Aves. Doubtless it was always closed by membrane as it is seen to be in living Crocodiles and Birds. But, whereas in Pseudosuchia

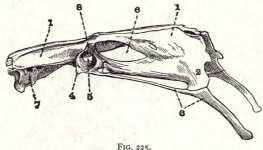

FIG. 225.

Lateral view of pelvis and sacrum of *Anas boschas* (from S. H. Reynolds, *Vertebrate Skeleton*, 1913). 1, Ilium ; 2, ischium ; 3, pubis ; 4, pectineal process ; 5, acetabulum ; 6, ilio-sciatic foramen ; 7, fused vertebrae ; 8, antitrochanter.

and Saurischia the pubis and ischium are normally disposed and form ventral symphyses, in Crocodilia the ilium and ischium surround the acetabulum excluding the pubis ; while the pubis is movably articulated to the ischium, directed forwards and downwards, and forms no true symphysis, Fig. 222. The ischium of Pterosauria is a short broad plate (sometimes pierced by the obturator foramen ?), the pubis being excluded from the acetabulum and articulated to its front edge (the pubes are fused across the mid line in *Rhamphorhynchus*, *Pteranodon*, and *Nyctosaurus*). These facts led Fraas (1878) and Seeley (**297**) to believe that the true pubis has in these reptiles been fused to the ischium, and Seeley maintained that the detached so-called pubis is an epipubis comparable to the marsupial bones of Mammalia. Considering how nearly related the Crocodilia are to other Archosauria with a well-developed normal pubis (Eosuchia, Parasuchia, Pseudosuchia, Saurischia), it is hard to believe that the pubis has been so completely reduced ; yet it must be admitted that a thorough

[1] In the embryo Crocodile the iliac, pubic, and ischiadic cartilages are in continuity and the acetabulum is closed. The piercing of the acetabulum and separation of the pubis take place late, Fig. 223.

investigation of the development of the Crocodilian pubis is urgently needed before its homology can be determined. Lately v. Huene (266) has held that the so-called pubis of Crocodiles and Pterosaurs is a 'prepubis' com-

FIG. 226.

Hesperornis regalis, Marsh. Upper Cretaceous: Kansas. Restoration of skeleton, ⅛. (After Marsh, from K. Zittel, *Palaeontology*.)

parable to the 'prepubic' process of Ornithischia, the structure of whose pelvic girdle we must now consider.

Ever since Huxley insisted on the near relationship of Birds to Dinosaurs, attention has been centred on the interpretation of the homologies of the pelvic bones in these groups. The general resemblance between the girdle of such forms as *Iguanodon* or *Camptosaurus* on the one hand, and the Cassowary, Apteryx, or Tinamou on the other, is striking, Figs. 224, 226, 229; for not only is the ilium greatly expanded in both cases, but the ischium is much lengthened, and ex-tending backward parallel to it is a slender pubic bone provided, in the reptile, with a large anterior 'prepubic' process. Corresponding to it in the bird is a 'pectineal' process of much smaller size. Moreover, the resem-blance is completed by the presence, in many birds, of a process of the ischium similar to the obturator pro-cess overlapping the 'postpubis' in

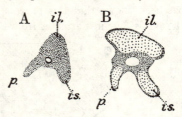

FIG. 227.

Side view of pelvis of bird embryos (after Mehnert, 1888). A, *Podiceps*; B, *Larus*. *il*, Iliac region; *is*, ischium; *p*, pubis.

Ornithischia and incompletely cutting off an obturator space, Figs. 225, 227. Now the older anatomists held that the slender backwardly directed bone is a true pubis, and Huxley compared it to the postpubic process of Ornithischia, believing the pectineal process to be the reduced remnant of their prepubic process. According to this view the main

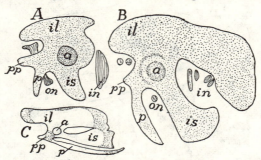

FIG. 228.

Development of pelvic girdle of chick, left-side view (after Johnson, '83, from J. S. Kingsley, *Vertebrate Skeleton*, 1925). A, 6-day chick; B, older stage; C, 20-day chick. Cartilage stippled, bone white. *a*, Acetabulum; *il*, ilium; *in*, ischiadic nerve; *is*, ischium; *on*, obturator nerve; *p*, pubis; *pp*, pectineal or prepubic process.

shaft of the pubis must have been rotated backwards in both Ornith-ischia and Aves. Evidence that rotation has occurred in birds is obtained from their development, since, as shown by Bunge (**242**) and Mehnert (**280**) and corroborated by Johnson (**270**), Parker (**66**), Broom (**240**), Lebedinsky (**274**), and Levin (**276**), the procartilaginous rudiment of the pubis is at first nearly vertical, and later shifts backwards, Figs. 227-8. That the true pubis has been rightly identified in Birds may be considered as established, since in *Archaeopteryx* the pubes, which still

are joined in a symphysis, are elongated slender bones projecting backwards parallel to the ischia, Fig. 233; but the homology of the pectineal
process, to which the ambiens muscle is attached, is far less certain. It
has been shown that this process ossifies, not as Huxley imagined from
the pubis, but in Carinates from the ilium. Marsh and others concluded

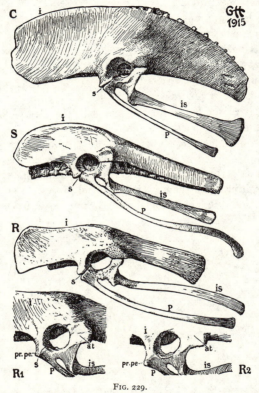

FIG. 229.

Left-side views of pelvic girdles of young : C, *Casuarius* ; S, *Struthio* ; and R, *Rhea* ; R₁, acetabular region of full-grown Rhea ; R₂, of old Rhea. *pr.pe*, Processus pectinealis ; *s*, suture between
ilium and pubis. (From G. Heilmann, *Origin of Birds*, 1926.)

that it is not the homologue of the prepubis but a new formation
(supported since by Lebedinsky and Heilmann). The large prepubic
bone of Ornithischia was then held to be the true pubis from which a
secondary postpubic process developed, absent in Ceratopsia, still small
in *Allosaurus*, fully developed in *Iguanodon*, Figs. 230-31, 232. The
resemblance to Birds on this view is deceptive and due to convergence.
The question, however, was not thus settled, for it was found that the
pectineal process in *Apteryx* and the Ratitae (Baur, **236** ; T. J. Parker,

66) is ossified as much from the pubis as from the ischium ; it could still be interpreted as a stage in the reduction of the prepubic process, Fig. 229. Lebedinsky maintains that in early stages the cartilaginous rudiments of ilium, pubis, and ischium appear separately and that the pectineal process arises from the ilium ; but even this observation is not conclusive, since the separate origin of the cartilages can have no phylogenetic significance. Of more consequence is the fact that the

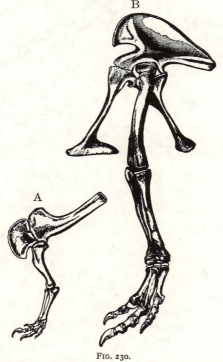

FIG. 230.

Allosaurus agilis, Marsh. Upper Jura : Colorado. Restoration of anterior (A) and posterior (B) limbs, $\frac{1}{20}$. (After Marsh, from K. Zittel, *Palaeontology*.)

pectineal process in *Archaeopteryx* is quite small. Finally, v. Huene (266) and others maintain that the pubis has rotated not only in Birds, but also in Ornithischia, Crocodilia, and Pterosauria (Romer, 291, 293, 295). According to this view the pectineal process and the prepubic process are new formations, and the latter supplants the true pubis. The postpubis of Ornithischia would then represent the dwindling true pubis, small in *Allosaurus*, altogether gone in advanced Ceratopsia (also Crocodilia and Pterosauria), where the prepubic process has taken its place (in somewhat the same position as the original pubis in primitive reptiles). This inter-

P

pretation entails the conclusion that the old Order Dinosauria is composed of two distinct groups : the Saurischia allied to the Phytosauria and

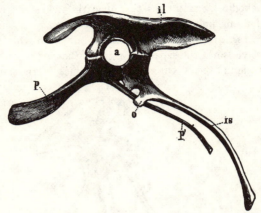

FIG. 231.

Pelvis of *Iguanodon*, ₂₁₀. *a*, Acetabulum ; *il*, ilium ; *is*, ischium ; *o*, obturator process ; *p*, prepubic process ; *p′*, pubis.   (From K. Zittel, *Palaeontology*.)

Pseudosuchia (with normal pubis), and the Ornithischia having a common ancestry with the Crocodilia, Pterosauria, and Aves.   The evidence is still far from complete ; but with regard to the Ornithischia it may be pointed out that a rotation of the pubis has already begun in such

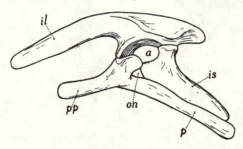

FIG. 232.

Stegosaurus undulatus.   Outer view of left half of pelvic girdle (from Marsh).   *a*, Acetabulum ; *il*, ilium ; *is*, ischium ; *on*, obturator notch ; *p*, pubis ; *pp*, prepubic process.

primitive Archosauria as *Euparkeria* and *Ornithosuchus*, Fig. 233 (Broom, 442 ; Gregory and Camp, 256a), and that the enlargement of the prepubis and dwindling of the pubis can on this view be traced through the newly discovered *Protiguanodon*, and *Protoceratops*, to *Triceratops*, where it appears as a mere vestige.

In Ornithischia the ischia may unite in a symphysis, but not the

postpubes, while the prepubes are widely divergent. The Neornithes
lose the symphysis altogether (except in *Struthio*), the right and left pubes
and ischia being widely divergent. The ilium is also very much enlarged
before and behind as thin plates excavated internally for the kidneys,
and embracing a large number of lumbar and caudal vertebrae in the
sacrum. Usually the ischium is likewise expanded and fuses behind with
the ilium enclosing an ischiadic fenestra, Fig. 225.

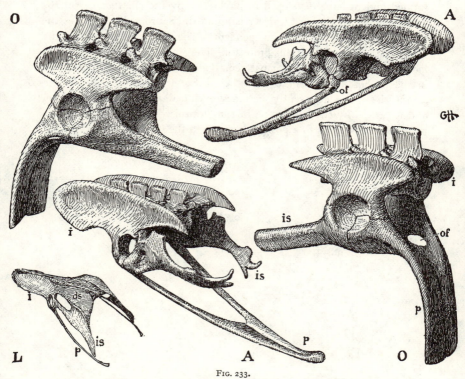

FIG. 233.

Pelvic girdle and sacrum of: A, *Archaeopteryx*, restored; O, *Ornithosuchus woodwardi*, restored
(after E. T. Newton and R. Broom); L, young gull. *ds*, Dorsal expansion of ilium; *i*, ilium; *is*,
ischium; *of*, obturator foramen; *p*, pubis. (From G. Heilmann, *Origin of Birds*, 1926.)

Conclusion.—Of the pelvic girdle in general it may be concluded that it
appears to have been developed at the base of the pelvic limbs from paired
plates in the body-wall, which tend to meet and fuse in the mid-ventral line
in front of the cloacal aperture. In Pisces the girdle may become ossified
as a single bone on each side; but in Tetrapods the ventral plate develops
a dorsal iliac process which articulates with the outer end of one or more
sacral ribs. Whether the 'iliac' processes sometimes found in fish are

the forerunners of the true ilia is doubtful. Of the three typical ossifications in each half of the girdle of Tetrapoda, the pubis (pierced by an
obturator foramen) and ischium combine in primitive forms to make a
complete ventral plate (Amphibia and lowest Reptilia). But in higher
forms a fenestra appears in the plate more or less completely separating
the pubis from the ischium. In Mammalia and Theromorpha (and the
majority of other Synapsidan Reptiles) this fenestra appears to be
formed and to have developed in phylogeny by the enlargement of the
obturator foramen. On the other hand, in the Sauropsidan Reptiles the
two openings are at first independent. The pelvic girdle undergoes
many changes in the various groups in divergent adaptation to different
modes of progression ; the most remarkable of these changes are seen
in the Archosaurian Reptiles and the Birds. In Crocodilia and Pterosauria the pubis apparently becomes excluded from the acetabulum,
and is carried by the ischium ; while in Ornithischia and Aves the pubis
is rotated backwards to a position parallel to the ischium.

CHAPTER V

MORPHOLOGY OF HEAD REGION

GENERAL MORPHOLOGY OF THE HEAD AND SKULL

THE structure of the trunk segments having been described (Chapter I.), that of the head may now be considered. The first question which arises is whether the head is segmented like the rest of the body or represents a primitively unsegmented anterior region. The anatomy of the Craniata would at first sight seem to suggest that the head region differs fundamentally from the trunk; that the skull is not segmented like the vertebral column; or that, if segmentation there exists, it is restricted to the gills and their arches. The muscles show no obvious sign of segmentation; the cranial nerves are some with and others without ganglia, and seem to be distributed in an irregular manner very different from the orderly disposition of the spinal nerves. Closer study of its structure and development, however, soon brings the conviction that the head region of the Craniate is truly segmented, that it is composed of a number of segments essentially similar to those of the trunk, and that segmentation originally extended to the anterior end of the body as it still does in *Amphioxus*. It would appear that owing to the Vertebrates being elongated bilaterally-symmetrical animals the anterior end became progressively differentiated in relation to the presence of the chief organs of sense, the brain, the mouth, and the respiratory gill-slits. Thus head and trunk regions became more and more specialised in divergent directions, and the line of demarcation between them more and more definite. So we find this process of cephalisation more pronounced in higher than in lower forms, in older than in younger stages of development.

In the interpretation of the head region the skeleton has always played the chief part. So early as 1792 Frank compared the skull to a single vertebra, and a few years later Goethe and Oken put forth a celebrated vertebral theory of the skull according to which it is composed of from three to six vertebrae conforming to the same plan as those of the trunk. Rathke and Reichert upheld the theory on embryological grounds, and Owen further elaborated it. The 'vertebral theory' was generally adopted until Huxley shattered its foundations in a famous Croonian lecture (1858). Having traced a fundamental plan common to the skull in all classes of the Craniata, both in development and in adult structure, he showed that the 'skull' is first membranous, then cartilaginous, and that the ossifications which appear later in the higher forms have even less to do with any primary segmentation comparable to that of the vertebral column than had its cartilaginous predecessor. The skull, in fact, arose long before bony vertebrae appeared. Huxley's argument, based on the researches of Rathke and Reichert, led him to the conclusion that the skull is not a modified portion of the bony vertebral column, and that although both started from the same primitive elements they immediately began to diverge.

After the overthrow of the 'vertebral theory' a new interpretation was gradually built up; it may be called the 'segmental theory' of the

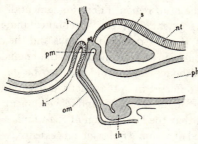

Fig. 234.

Diagram of median sagittal section of embryo *Torpedo* (from Dohrn's figures, 1904) showing region of mouth and hypophysis, *h*. *i*, Infundibular outgrowth of floor of brain; *nt*, notochord; *om*, oral membrane; *ph*, pharynx; *pm*, region of paired outgrowth of premandibular somites; *s*, median vascular sinus cephalicus; *th*, thyroid outgrowth.

head and skull. Huxley had insisted on the importance of the pituitary space as marking a point of comparison in the head of all Craniates. To this place reaches the anterior end of the notochord, above it arises the mesocephalic flexure of the brain, through it pass the internal carotids, and into it grows the hypophysis from below, Figs. 234-5. Behind it extends the basal plate of the cranium, and it is embraced by the divergent posterior ends of the trabeculae cranii. The basal plate behind, the

trabeculae in front, and the auditory optic and olfactory capsules were considered to represent the fundamental elements of the skull which combined into a continuous cranium; to these were added the visceral arches completing the skeleton of the head (Huxley, 1858; Gaupp, 342-3).

The next important step is due to Gegenbaur (1872), who from a study of the skeleton of the Elasmobranchs considered the post-pituitary or vertebral region of the skull into which the notochord extends to be composed of about nine segments, as evidenced by the cranial nerves and visceral arches, comparable to spinal nerves and vertebrae farther back. Specialisation and reduction of the muscles is supposed to have led to the concrescence and solidification of this region of the skull from which the prevertebral region, pierced only by the optic and olfactory nerves, may have grown forwards. Stöhr ('79, '81, '82) held that the hinder limit of the skull extended farther and farther back in the ascending Vertebrate series by the concrescence or assimilation of segments from behind: a con-

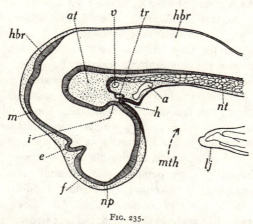

Fig. 235.

Median sagittal section of embryo *Scyllium canicula*, Stage J. *at*, Acrochordal mesenchyme; *e*, epiphysis; *f*, fore-brain; *h*, hypophysis; *hbr*, hind-brain; *lj*, lower jaw; *m*, mid-brain; *mth*, mouth; *np*, recessus neuroporicus; *tr*, transverse canal uniting premandibular somites. Other letters as in Fig. 234.

clusion borne out in a general way by Gegenbaur (345-6), Rosenberg (377), and many others since. Sagemehl (378) drew a distinction between the primary cranium, a 'protometameric' region of fused segments, and the 'auximetameric' region of true vertebral elements later added to it. Froriep (339, 498), who analysed the occipital region in development and discovered vestigial dorsal nerve-roots and ganglia corresponding to the hypoglossal nerve compounded of ventral roots, distinguished between that region of the skull without obvious segmentation in front of the vagus group of nerves, which he called the prespinal region, and the post-vagal or spinal region of assimilated vertebral elements.

The next step in the theory of the head and skull is marked by Fürbringer's monograph on the nerves of the occipital region (340). Following Stöhr and Froriep, he maintained that the vagus marks a division between two regions of different value. Anterior to the vagus extends the older ' Palaeocranium ', the only skull found in the Cyclostomes ; while behind the vagus lies a newer region to which segments have been added.[1] This ' Neocranium ' formed by the assimilation of originally free skeletal segments and including corresponding nerves (called spino-occipital by Fürbringer) is supposed to have been built in successive stages. First of all was added to the palaeocranium the protometameric neocranium of seven segments, represented in Elasmobranchs and Amphibia ; and later the auximetameric neocranium of three additional segments found in the Amniota, with even more segments added in certain of the higher Pisces such as *Acipenser*. Fürbringer's views were chiefly founded on the occurrence of a varying number of hypoglossal ventral nerve-roots (first discovered by Jackson and Clarke (356) in the shark *Echinorhinus*) under-lying the vagus. These seemed to indicate that a varying number of segments had been partially compressed and reduced behind the auditory capsule. Believing the anterior hypoglossal roots, without obvious corre-sponding dorsal roots and ganglia, to belong to his seven protometameric segments, he held that the hind limit of the protometameric neocranium remained constant throughout the Gnathostomes and could be identified in the various groups. To these protometameric segments he applied the letters $t \, u \, v \, w \, x \, y \, z$, and assumed that they were progressively re-duced from before backwards, t being the first to disappear and z the last to remain. Their corresponding spino-occipital nerves were distin-guished as ' occipital ' from the more posterior ' occipito-spinal ' nerves included by the auximetameric neocranium. To these occipito-spinal nerves the letters a, b, and c were applied, the first free spinal nerve being in that case the 4th. Although Sewertzoff and others maintained that the chief distinction is between a ' spinal ' and a ' prespinal ' region, yet Fürbringer's conclusions have been far too generally accepted. The limit between occipital and occipito-spinal segments assumed by him to occur at a constant and recognisable point throughout the Gnathostomes as a matter of fact can only arbitrarily be determined ; nor is there any clear evidence of the postulated disappearance of a large number of protometameric segments. Only definite anatomical and embryological evidence can decide what serial number is to be assigned to the last occipital or the first free spinal nerve in any species (see footnote, p. 227).

The Segmentation of the Head.—Whereas in the higher Vertebrates the head shows no obvious signs of segmentation, as we pass to the lower fishes the distinction between the head and the trunk becomes less marked ;

---

[1] The question of the composition of the vagus nerve, which is of great importance in connexion with the views of Gegenbaur and Fürbringer on the phylogenetic development of the skull, is discussed on p. 767.

in Cyclostomes it is difficult to say where one begins and the other ends ; and finally in *Amphioxus* the body is clearly segmented from end to end. Moreover, a study of the development even of the higher forms makes the signs of segmentation increasingly plain. It has therefore been concluded that the head region is the result of a process of cephalisation of the anterior segments in an originally uniform series, leading to their modification and more or less complete fusion, so that finally almost all trace of the primitive segmentation disappears in the finished product. To unravel the constituent elements and determine their composition and number it is necessary to study their development and first origin. This applies not only to the skeleton, but also to the musculature, nervous elements, sense organs, branchial apparatus, and blood-vessels, all of which may supply corroborative information.

Now since segmentation in the Vertebrates is primarily expressed in the mesoblastic tissues, it is from the mesoblast that we must expect to derive the most convincing evidence. We have already seen (p. 6) that in the trunk of the Craniate the more dorsal part of the mesoblast forms segmental somites and sclerotomes from which are derived segmental myotomes (myomeres) and skeletal elements respectively, and to which correspond segmental nerves. A ventral motor root supplies each myotome and a dorsal root (chiefly sensory) passes behind it. In his epoch-making works on the development of the Elasmobranch Fishes (1874–78) Balfour showed that eight ' head-cavities ' or hollow somites are formed whose walls give rise to the muscles of the head. The first is preoral and the seven others correspond to the mandibular, hyoid, and five branchial arches. His tabular statement is given below, and the best and most

TABLE OF THE CEPHALIC SEGMENTS FROM F. M. BALFOUR, 1877.

Segments.	Nerves.	Visceral arches.	Head-cavities or cranial muscle-plates.
Preoral 1	3rd and 4th and ? 6th nerves (perhaps representing more than one segment)	?	1st head-cavity
Postoral 2	5th nerve	Mandibular	2nd head-cavity
,, 3	7th nerve	Hyoid	3rd ,,
,, 4	Glossopharyngeal nerve	1st branchial arch	4th ,,
,, 5	1st branch of vagus	2nd ,, ,,	5th ,,
,, 6	2nd ,, ,,	3rd ,, ,,	6th ,,
,, 7	3rd ,, ,,	4th ,, ,,	7th ,,
,, 8	4th ,, ,,	5th ,, ,,	8th ,,

recent work has fully confirmed his main conclusions (excepting position of 4th and 6th nerves). Of the many attempts since made by

Dohrn (333), Froriep (1905), and others to prove that there are more or fewer mesoblastic segments in this region of the head of Elasmobranchs, none has succeeded. Marshall, following Balfour, emphasised the comparison between the somites and lateral plate of the trunk and the dorsal head-cavities and more ventral mesoblast in the gill-arches of the head. The segmentation of the head-cavities he held to be independent of that of the gill-clefts ; he described the origin of the four eye-muscles supplied by the oculomotor nerve from the premandibular somite, and from the third somite the origin of the posterior rectus supplied by the abducens, which is the ventral root of the facial segment. Next came an important contribution from van Wijhe, in 1882, who described in detail the development and fate of the eight somites discovered by Balfour, and showed that a typical head segment contains on each side a somite (myotome and sclerotome) below which extends lateral-plate mesoblast with a cavity passing down a visceral arch. The dorsal and ventral nerve-roots related to each segment remain separate from each other, as they had been shown by Balfour to be in early stages in the trunk, and as they permanently remain in *Amphioxus* and *Petromyzon*. *The dorsal ganglionated root supplies the visceral muscles derived from the lateral plate, while the ventral root supplies those derived from the myotome.* The auditory sac marks off three pro-otic segments, of which the ophthalmicus profundus, trigeminal, and facial nerves are the dorsal roots ; while the oculomotor, trochlear, and abducens are the ventral roots. Further, van Wijhe distinguished more clearly between the true somites and the lateral plate, and definitely traced the development of the eye-muscles from the corresponding three myotomes, and the origin of the hypoglossal roots from the metaotic segments of which the glossopharyngeal and vagus represent the dorsal roots only. Van Wijhe, however, attributed nine segments to the head ; the tenth, with typical mixed spinal nerve, he called the first of the trunk (396).

Since the pioneer work of Balfour, Marshall, and van Wijhe, numerous investigators have studied the development of the head not only in Elasmobranchii (Dohrn, 333-4 ; Hoffmann, 354 ; Sewertzoff, 384 ; Neal, 368 ; Ziegler, 399 ; Goodrich, 349 ; de Beer, 320) but in other Gnathostomes, such as the Reptilia (Corning, 331 ; Filatoff, 337 ; Johnson, 358), Birds (Rex, 375 ; Matys, 367 ; Adelmann, 313), and Mammals (Fraser, 338). Although their results are not always in agreement, yet, leaving certain controversial points to be dealt with later, the general conclusions may be summarised as follows.

The head in the Gnathostomata is segmented up to a point just behind the hypophysis, where the notochord merges into the original roof of the archenteron from the side walls of which the mesoblast develops. This mesoblast becomes subdivided into dorsal segmented somites and ventral unsegmented lateral plate. The head somites are of the same nature as and form a continuous series with those of the trunk. In so far as the wall of the head somite develops into a myotome with muscle persisting in the adult it is found to be supplied by the ventral motor root of its segment. The corresponding dorsal gang-lionated root passes behind each somite.[1] The gill-slits pierce the lateral plate intersegmentally. Visceral arches are thus formed,

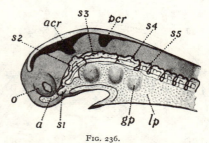

FIG. 236.

Scyllium canicula, left-side view of anterior region of embryo, Stage F. *acr* and *pcr*, Anterior and posterior proliferations of neural crest ; *a*, anterior proliferation of mesoblast ; *gp*, hypoblastic gill-pouch showing through ; *lp*, lateral plate mesoblast ; *o*, optic cup ; *s* 1-5, first to fifth mesoblastic somites.

and down each passes the dorsal nerve-root supplying motor fibres to the visceral muscles derived from the lateral plate (jaw and branchial

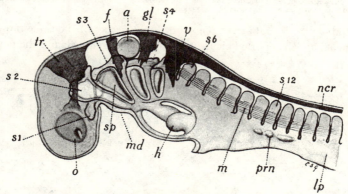

FIG. 237.

Scyllium canicula, embryo 5 mm. long, Stage G. Left-side view of anterior region. *a*, Auditory placode ; *h*, heart in pericardial coelom ; *md*, mandibular arch ; *ncr*, neural crest ; *o*, optic cup ; *prn*, pronephric rudiments ; *s* 1-12, somites ; *lp*, lateral plate of unsegmented mesoblast ; *sp*, first or spir-acular gill-slit. Rudiments of profundus and trigeminal *tr*, facial *f*, glossopharyngeal *gl*, and vagus ganglia *v*, derived from neural crest are shown darkly shaded.

arch musculature). The primitive uniform disposition of the somites is disturbed by the development of the auditory sac and capsule marking off three anterior pre-auditory or pro-otic somites from the

---

[1] The terminal, the olfactory, and the optic nerves are not counted as segmental nerves ; and the auditory is considered to be derived from the facial nerve (see Chapter XIV.).

post-auditory or metaotic somites.   The first somite is the premandibular, the next the mandibular, and the more posterior somites lie

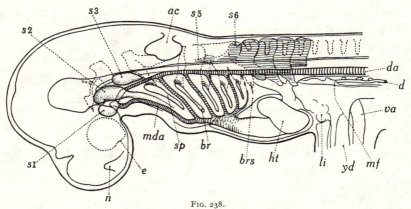

FIG. 238.

Diagrammatic left-side view of anterior region of embryo *Scyllium canicula*, Stage J.   Compare
Fig. 239.   Nerves represented in dotted lines.   *br*, First branchial slit ; *brs*, fifth branchial slit still
closed ;   *d*, Müllerian duct ;   *da*, dorsal aorta ;   *e*, eye ;   *ht*, heart ;   *li*, liver ;   *mda*, mandibular aortic
arch ;   *mf*, Müllerian funnel ;   *n*, nasal sac ;   *s1*, premandibular somite ;   *s2*, mandibular, and *s3*, hyoid
somites ;   *s5*, fifth somite with vestigial myomere ;   *s6*, sixth somite with myomere passing behind vagus ;
sp, spiracular slit ;   *va*, vitelline artery ;   *yd*, yolk duct.

each above a visceral arch.   The dorsal-root nerves of the three pro-otic
segments are the ophthalmicus profundus, the trigeminal, and the facial ;

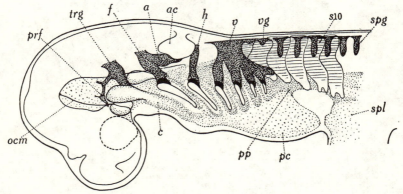

FIG. 239.

Diagram of anterior region of embryo *Scyllium canicula*, Stage J ;   left-side view, showing developing
cranial and spinal nerves darkly shaded, and epibranchial placodes black.   *a*, Auditory ;   *f*, facial ;
h, glossopharyngeal ;   *prf*, profundus ;   *spg*, spinal ;   *trg*, trigeminal ;   *v*, vagus ganglia ;   *ac*, auditory
sac ;   *c*, coelomic canal in mandibular bar ;   *ocm*, oculomotor nerve ;   *pc*, pericardial coelom ;   *pp*,
position of pericardio-peritoneal canal passing above septum transversum ;   *s10*, myotome of tenth
segment ;   *spl*, peritoneal or splanchnic coelom of trunk ;   *vg*, vestigial ganglia of vagus region.

those of the metaotic segments are the glossopharyngeal and the compound vagus, one branch of which corresponds to each somite in this

region (see p. 767). The pro-otic somites are entirely subordinated to the use of the movable optic capsule enclosing the eye, and give rise to the extrinsic eye-muscles. The premandibular forms the rectus superior, rectus anterior and rectus inferior, and obliquus inferior, all supplied by the oculomotor nerve. The mandibular somite, supplied by the trochlear nerve, gives rise to the obliquus superior; while the third or hyoid somite forms the rectus posterior supplied by the abducens nerve (see p. 227). The first metaotic somite crushed by the growing auditory capsule never

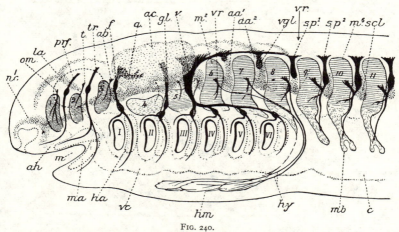

FIG. 240.

Diagram illustrating *segmentation of head* in a Selachian (E. S. Goodrich, *Q.J.M.S.*, 1918, modified). Skeletal visceral arches indicated by dotted outlines. *vr*, Posterior limit of head region; *I-VI*, gill-slits; I-II, somites, pro-otic from 3 forwards, metaotic from 4 backwards; *a*, auditory nerve; *ab*, abducens n.; *ac*, auditory capsule; *ah*, anterior mesoblast (anterior 'head-cavity'); *c*, coelom in lateral plate mesoblast; *f*, facial n.; *gl*, glossopharyngeal n.; *ha*, hyoid arch; *hm*, hypoglossal muscles derived from somites 6, 7, 8; *hy*, hypoglossal nerve compounded from branches of ventral roots of those segments; *la*, orbital cartilage; *m*, mouth; *m²*, *m⁶*, second and sixth metaotic myomeres; *ma*, mandibular arch; *mb*, muscle-bud to pectoral fin; *nc*, nasal capsule; *aa¹*, *aa²*, first and second arches which with third make up occipital region; *om*, oculomotor n.; *prf*, profundus n.; *scl*, sclero-mere of segment 10; *sp¹*, vestigial dorsal root and ganglion of first spinal nerve; *sp²*, second spinal n. complete; *t*, trochlear n.; *tr*, trigeminal n.; *v*, complex root of vagus n.; *vgl*, vestigial dorsal root and ganglion of segment 7; *vc*, ventral coelom extending up each visceral bar; *vr*, ventral nerve-root of segment 6. Myomeres longitudinally striated, nerves black, scleromeres and cartilage dotted.

forms a myomere in the Gnathostomes, and soon breaks down into mesen-chyme; it has no ventral root. The second metaotic somite (fifth of the series) forms a vestigial myomere of a few muscle fibres which may later degenerate together with the ventral root supplying them. The third and succeeding somites persist as more or less complete myo-meres. As in the pro-otic so in the metaotic region the gill-slits interfere with the development of these myotomes which do not extend down the lateral body-wall, but contribute to the formation of the epibranchial and hypobranchial musculature, Fig. 241. The ventral roots of these metaotic segments contribute to the formation of the hypoglossal nerve

supplying the muscles just mentioned.   Although a sclerotome arises from
each somite, fusion between them takes place so early that no definite
trace of segmentation can be detected in the cranium, except in the
occipital region to which sclerotomes in varying number become assimilated and fused (see p. 226).   A skeletal visceral arch develops behind
the mouth and each gill-slit ; vascular arches correspond to them.   Such
in brief are the chief conclusions embodied in the ' Segmental theory of

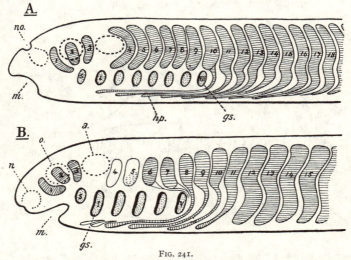

Fig. 241.

Diagrams of anterior region of *Cyclostome*, A, and of *Gnathostome*, B, showing position and development of somites 1-18 ; myomeres are shaded.   In A all somites form myomeres ; in B no myomere
appears in somite 4, and only a vestigial myomere in somite 5.   *n*, Nasal, *o*, optic, and *a*, auditory
capsules indicated by dotted lines ; *gs I-VII*, gill-slits ; *hp*, hypoglossal muscles derived from
myotomes ; *m*, mouth ; *na*, nasal opening ; *s*, spiracular slit.   (From *Vertebrata Craniata*, 1909,
modified.)

the Head '.   Certain special points must now be elaborated, and certain
difficult and controversial subjects discussed.

   While the main conclusions mentioned above have been chiefly based
on work carried out on the Elasmobranchs, very favourable material for
the purpose, they have mostly been corroborated by observations on
other Gnathostomes, as already mentioned.   But it is important also to
notice that they have been strongly supported by the work of Koltzoff
on the development of *Petromyzon* (361).   He finds the same three pro-otic
somites giving rise to the eye-muscles as explained above, and the same
corresponding nerves.   But, possibly owing to the small size of the
auditory organ and the less development of the skull which has not yet
acquired an occipital region (p. 216), all the metaotic somites develop
myotomes which persist in the adult.   Thus in the Cyclostome every

somite from the premandibular backwards is represented in the adult, and there is an uninterrupted series of myotomes from the first metaotic or glossopharyngeal segment to the trunk. Such a condition represents an interesting transition between that of the Gnathostome, in which one or more myotomes are always suppressed behind the auditory sac, and that of the Cephalochorda, in which the myotomes pass evenly from end to end, Figs. 241, 738.

The first mesoblastic segment in *Amphioxus*, however, produces no myotomes, and it may now be asked whether the premandibular somites are really the first of the series in the Craniata. They arise near the extreme anterior end of the archenteron (Hatschek, 352 ; Koltzoff, 361 ; Dohrn, 334 ; Neal, 368-9), but J. B. Platt described in *Squalus* a pair of head-cavities developing still farther forward (373). These she believed to represent the most anterior somites, a view supported by Neal. Since, however, they soon disappear and form no permanent structure, are scarcely or not at all developed in other Elasmobranchs (Dohrn ; de Beer), have not been found at all in other Gnathostomes, and are absent in *Petromyzon*, it seems more probable that they are derivatives of the premandibular segment, as held by Dohrn and van Wijhe. Moreover, there is good reason to believe that the premandibular segment represents the first in *Amphioxus*.[1]

The next point concerns the number and disposition of the somites in the auditory region. While in the scheme outlined above (p. 218) somites, nerves, and gill-slits correspond and follow in regular order, according to van Wijhe the fourth somite, crushed below the auditory capsule, belongs to the facial nerve segment, which would be really double, formed by the combination of the third and fourth somites. In spite of the fact that it would disturb the orderly sequence of these parts of the head, van Wijhe's interpretation has been adopted by a number of authors (Neal, 368 ; Hoffmann, 354 ; Braus, 145 ; Sewertzoff, 384). Nevertheless, it cannot be considered as well founded. It involves the assumption that a gill-slit

---

[1] These first somites acquire an opening to the exterior, either into the oral hood cavity as in *Amphioxus* (on the left side only, Hatschek's pit) or into the hypophysis as in Gnathostomes, for instance Selachians (Ostroumoff ; Dohrn), Birds (Goodrich), and probably Reptiles (Salvi, 379). For this and other reasons they seem to correspond throughout the Vertebrata (Goodrich, 349). The first pair of somites is also remarkable for being joined together in early stages by a transverse connexion, which often becomes hollow (see Figs. 242-3) ; thus the right and left myocoeles may be for a time in communication. The connexion is not secondarily established, but is due to the retention of that portion of the extreme anterior wall of the archenteron from each side of which the first somites grow out. It is not retained in *Amphioxus*, nor apparently in *Petromyzon*.

arch and nerve have disappeared in the hyoid region, for which there is

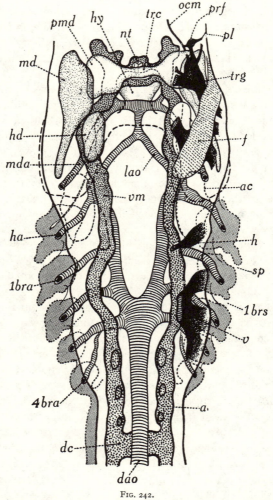

FIG. 242.

Diagrammatic reconstruction of thick frontal section of anterior region of embryo *Scyllium canicula*, Stage K. Arteries cross-lined, veins darkly stippled, nerves on right black. Dorsal outline drawn in continuous line, ventral outline in broken line. *a*, Anterior cardinal ; *ac*, broken line indicating ventral portion of auditory sac ; *bra*, branchial arterial arches ; *dao*, median dorsal aorta ; *dc*, ductus Cuvieri ; *f*, cut surface of acustico-facial ganglion and placode ; *h*, glossopharyngeal ; *ha*, hyoid aortic arch ; *hd*, third or hyoid somite ; *hy*, hypophysis ; *lao*, right lateral aorta ; *md*, second or mandibular somite ; *mda*, mandibular aortic arch ; *nt*, tip of notochord ; *ocm*, oculomotor ; *pl*, placode ; *pmd*, first or premandibular somite ; *prf*, profundus ganglion ; *sp*, spiracular slit ; *trc*, transverse premandibular canal ; *trg*, trigeminal ; *v*, vagus ; *vm*, vena capitis medialis.

no good evidence in any embryo or adult Craniate. A renewed and careful study of the development of several Selachians shows that the

third somite belongs to the facial and the fourth to the glossopharyngeal
segment, and that no segment has disappeared in this region (Ziegler, **399** ;

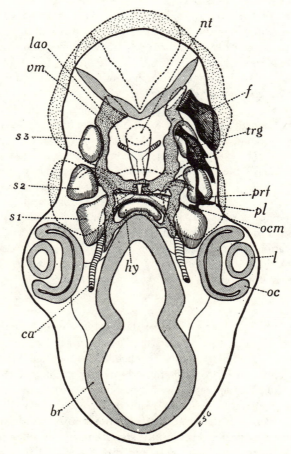

FIG. 243.

Scyllium canicula, embryo Stage K. Reconstructed thick transverse frontal section of head-cutting medulla above and fore-brain below. *br*, Fore-brain ; *ca*, cerebral artery ; *f*, facial nerve with placode ; *hy*, hypophysis ; *l*, lens ; *lao*, lateral aorta (internal carotid) ; *nt*, notochord ; *oc*, optic cup ; *ocm*, oculomotor nerve ; *pl*, placode attachment of profundus ganglion, *prf* ; *s1, s2, s3*, first, second, and third somites or head-cavities ; note transverse connexion between first pair ; *trg*, trigeminal ganglion ; *vm*, vena capitis medialis. Upper anterior section outlined in black line, lower posterior in broken line.

Goodrich, **349** ; de Beer, **320**). Moreover, these results agree with those of
Koltzoff on *Petromyzon*, Figs. 236-40.

The important question as to the number of segments included in
the occipital region of the head involves two separate but related problems:

the first concerns the number of segments suppressed by the auditory capsule overgrowing them from in front ; the second concerns the number of segments added by assimilation from behind.[1] It is clear that in the Gnathostomes both processes have taken place in varying degree in different groups ; but the work on this question has been so much influenced by the theories of Fürbringer that it is difficult to obtain trustworthy evidence on the subject from the literature. It is probable that even in the Selachii the number of post-auditory head segments varies slightly. *Squalus acanthias* has six metaotic somites (van Wijhe, 396 ; Hoffmann, 354 ; Sewertzoff, 383 ; de Beer, 320), of which the last four produce permanent myomeres. *Scyllium* has at least four, and at most five, metaotic somites (Goodrich, 349) ; *Spinax* possibly as many as seven (Braus, 145). The exact number occurring in other fish is less accurately known, owing to the difficulty of making out how many disappear in front. According to Schreiner there would be seven metaotic somites in *Amia* and *Lepidosteus*, the three anterior disappearing, and the last four forming permanent myomeres (381). In *Salmo* occipital somites are found, probably belonging to segments 4, 5, and 6 (Harrison, 179 ; Beccari, 319). According to Greil three metaotic somites develop muscle behind the first, which disappears in *Ceratodus* ; but Sewertzoff describes two more myomeres in the occipital region (592). In *Lepidosiren* and *Protopterus* there appear to be only three metaotic somites, of which the two last persist (Agar, 141).

The Amphibia are remarkable for the shortness of the occipital region. Only three metaotic segments occur in *Amblystoma* ; the first produces no myomere, but even the second is much reduced, its muscle combining with that of the third to form the dorsal temporal muscle (Platt, 374 ; Goodrich, 347 ; Froriep, 499). Marcus describes a third metaotic myomere in Apoda (752), but the Anura seem to have only two (Elliot, 336 ; van Seters, 382). In the Amniota there appear to be always more than three metaotic segments ; though the statement usually made that in the Amniote more segments have been added to the nine already present in the head of Selachians is quite misleading. Froriep (1883), one of the first to attempt the analysis of the occipital region, described four post-vagal myotomes in the chick, and apparently the same number in ruminants. Recently the question has been reinvestigated by Jager (357) in the chick, who finds four occipital myotomes behind the vagus, and one vestigial below it (probably belonging to metaotic somites 2-6 or segments 5-9 of the whole series). The last four have nerves ; but only the last two preserve

---

[1] The most posterior head segment is the last whose myomere is supplied by a ventral nerve-root passing through the skull wall.

their hypoglossal roots in the adult. The researches of Hoffmann (354), van Bemmelen (1889), Chiaruggi (329), and Beccari (319) show that the three hypoglossal roots of adult Amniotes belong to myomeres derived from metaotic segments 3-5 or possibly 4-6 (6-8, or 7-9 of the whole series).

It may be concluded that in the Gnathostomes a varying number of segments may enter into the composition of the head, and that consequently its hinder limit is inconstant in position. The first metaotic segment never produces permanent muscle or nerve. The second and sometimes even the third somite may also disappear. In the Selachii from four to seven metaotic segments share in the formation of the occipital region ; there is a similar variation among the Teleostomes. On the other hand, the Dipnoi appear to possess only from three to possibly five metaotic segments ; thus approaching the Amphibia in which there are rarely if ever more than four. The Amniota have a larger number, for in them five, or possibly six, metaotic segments occur, of which only the last two or three persist. It is important to realise that the head region can only arbitrarily be defined. It may vary in extent according as we take somites, scleromeres, nerves, or gill-slits and arches as our criterion, since these structures do not necessarily involve the same number of segments. This point will be made clearer when we treat of the cranial nerves (Chapter XIV.), and the occipital region of the skull itself (pp. 67, 242).[1]

THE EYE-MUSCLES

Throughout the Craniata the eye-muscles are extraordinarily constant in number, disposition, and innervation. There are six muscles, four musculi recti and two musculi obliqui, adapted for moving the eye-ball in various directions, Fig. 244. The anterior or internal rectus, the superior rectus, the ventral inferior rectus, and the inferior obliquus are supplied by the 3rd nerve ; the posterior or external rectus by the 6th nerve. The 4th nerve supplies the superior obliquus. While the mm. recti are generally attached close together to the basal region of the wall of the orbit postero-ventrally to the optic foramen, the oblique muscles usually spring from the planum antorbitale. Apart from the Cyclostomes, little change occurs except for the addition of a retractor bulbi close to the optic stalk in Reptiles and Mammals (Corning, 332 ; Matys, 367 ; Allis, 402), and two corresponding

[1] From the above account it will be gathered that no line can be drawn between 'protometameric' and 'auximetameric' segments, between 'occipital' and 'occipitospinal' nerves ; that the series of somites and nerves is continuous behind the auditory sac ; and that, although the first few (not more than three) may disappear more or less completely in development, none disappears in the middle of the series. Fürbringer's attempt to identify nerve z or nerve a as fixed points is, therefore, not justified (see further, pp. 216, 242).

muscles in Birds (m. quadratus and m. pyramidalis) used for moving the nictitating membrane (Corning, 332; Slonaker, 1921).

The interesting developmental history of these muscles from the three pro-otic somites has already been described (p. 218). As shown by Balfour (317), Marshall (365-6), van Wijhe (396), Platt (373), Lamb (363a), Gast (341), Neal (369), and others, these somites acquire so-called head-cavities in the Selachii from whose walls the muscles develop; the anterior, superior, and inferior recti, and the inferior obliquus, being derived from the first somite, the superior obliquus from the second somite, and the posterior rectus from the third. To the rudiment of the external rectus is added, in Selachians, some substance from the second somite (muscle E of Platt); but, while Dohrn and Neal maintain that this contributes to the adult muscle, others (Lamb, 363a; Johnson, 358; de Beer, 320-22) hold that it does not do so. This conclusion is supported by the fact that the 6th nerve alone supplies the external rectus, and it does not appear to be a compound nerve.

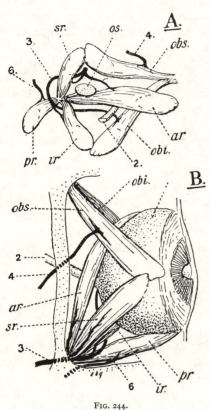

Fig. 244.

Squalus acanthias. A, Outer view of muscles and nerves of right orbit, from which the eye has been removed; B, dorsal view of right eye. *ar,* Anterior, *ir,* inferior, *pr,* posterior, and *sr,* superior rectus muscle; *obi,* inferior, and *obs,* superior oblique muscle; *os,* cartilaginous optic stalk; 2, optic, 3, oculomotor, 4, pathetic, and 6, abducens nerve. (After Goodrich, *Vert. Craniata,* 1909.)

A similar development, derivation of the eye-muscles, from the three pro-otic somites (in which 'head-cavities' usually appear) has been described in Amphibia (Marcus, 752), in Reptilia (Corning, 331; Johnson, 358), in Aves (Rex, 375; Adelmann, 314), and in Mammalia (Fraser, 338). In spite of certain discrepancies in these accounts, it may be considered as firmly established that in all the Gnathostomes the history of the eye-muscles is like that given above for the Selachian. The retractor bulbi of Reptiles and Mammals and the quadratus and pyramidalis of Birds are derived from the third somite and hence are supplied by a branch of the abducens.

The disposition of the eye-muscles in the Petromyzontia differs somewhat from that found in the Gnathostomes (P. Fürbringer, 340a; Corning, 332; Koltzoff, 361; Ducker, 335; Tretjakoff, 1178; Cords, 330; Addens, 312). Koltzoff has described their development in *Petromyzon* from the three pro-otic somites. But, although the muscles conform in general to the Gnathostome plan, the rectus inferior seems to be innervated not only by the oculomotor but also by the abducens (Tretjakoff), or by the abducens alone in most species. According to Addens this is due neither to different development of muscles nor to change of peripheral innervation, but to a rootlet (r. inferior) of oculomotor fibres having shifted backwards and become associated with the abducens root. The two-branched apparent abducens of *Petromyzon* would, then, be a compound nerve; and the general disposition and innervation of the eye-muscles in Cyclostomes would differ in no essential from those of other Craniates. The eye is so degenerate in the Myxinoidea that no muscles are developed.

It is clear from the above account that already in the common ancestor of the Cyclostomes and Gnathostomes the paired eyes must have been well developed, the pro-otic somites specialised, and the chief muscles of the eye-ball differentiated.

THE SKULL AND CHONDROCRANIUM

General Composition.—The skeleton of the head is made up in the Craniata of various elements of diverse origin which become more or less closely combined to form what we loosely call the 'skull' and visceral arches. Many of these constituent elements are more clearly distinguishable in lower than in higher forms, in earlier than in later stages of development. The membranous covering of the central nervous system, becoming strengthened by cartilages to protect the brain and afford attachment to muscles, gives rise to the brain-case or cranium proper. Cartilaginous capsules enclose the three chief paired organs of sense, and of these the olfactory in front and the auditory capsule behind become fused to the brain-case, while the optic capsule between them remains free, thus allowing the eye to move. Out of these parts is built up the primitive skull or chondrocranium (neurocranium of Gaupp), seen in adult Cyclostomes and Chondrichthyes, and the embryo of all the higher Craniates. There are also developed in the head cartilaginous visceral arches: a pair behind the mouth, and a pair behind each pair of gill-slits. They primarily served to bear the gills and to strengthen the wall of the pharynx, while allowing it to contract and expand. In Gnathostomes the first or mandibular arch bends over the corner of the mouth and forms two bars, the upper and lower primitive jaws bearing the lips and teeth and serving to seize the food. The upper bar is called the palato-pterygo-quadrate or palato-quadrate, and the lower the mandibular or Meckel's cartilage. The mandibular and the second or hyoid arch usually become intimately connected with the chondrocranium above, while the more posterior visceral arches, the true branchial arches in Pisces, remain free from the skull, and become much modified and reduced in air-breathing Vertebrates.

In the Osteichthyes and Tetrapoda these primitive cartilaginous elements of the skull may be more or less ossified, and further strengthened by various dermal bony plates added from the outside. A very complex structure is thus built up, the anatomy and development of which will be dealt with later. Our present purpose is to trace out the primary parts of which the skull is composed, to discover whether a fundamental plan can be made out common to all the Craniata, and to ascertain what evidence there may be of a segmental structure.

Setting aside the Cyclostomes in which the skull is in many respects highly specialised, we find in the more primitive Gnathostomes a continuous cartilaginous chondrocranium. For descriptive purposes it may be distinguished into a posterior occipital region connected with the vertebral column and surrounding the foramen magnum; an otic region containing the auditory capsules; an orbito-temporal region; and an anterior ethmoid region with the nasal capsules separated by a median nasal septum. The brain cavity extends from this septum backwards through the three more posterior regions. Primitively the side walls are pierced for the passage of cranial nerves and blood-vessels, and the floor in the orbito-temporal region is pierced by the fenestra hypophyseos for the passage into the cranial cavity of the internal carotid arteries and the hypophysis (pp. 214, 240). The hypophysial ingrowth usually becomes completely nipped off, and lodges as the pituitary body in a special depression, known as the pituitary fossa, on the floor of the brain-case which grows below it. Up to this region extends the notochord in the floor of the embryonic cranium, Fig. 245 A. A posterior ' chordal ' region of the skull can therefore be distinguished from an anterior ' prechordal ' region (Kölliker). A host of observers have studied the structure and development of the chondrocranium, among whom may be mentioned Rathke; Huxley; W. K. Parker (556-74); Gegenbaur; Kölliker; Born, who introduced the method of solid reconstruction of sections ; Stöhr, one of the first to apply it ; van Wijhe, who invented the method of differential staining of whole specimens ; and Gaupp, who contributed much to our knowledge of the skull, and wrote a masterly account of the subject in 1905 (343).

Elements of Chondrocranium of Gnathostome.—Speaking generally, the cartilaginous cranium is better developed in lower than in higher forms, more complete in earlier than in later stages, when it becomes to a great extent replaced by endochondral or dermal bones, and even partially reduced to membrane. To discover the fundamental elements out of which this chondrocranium is built up we must appeal to embryology. Unfortunately, although the later stages are comparatively well known, its first origin has been little studied since the days of Parker. However, the work of later authors, more especially of Sewertzoff (384) and van Wijhe (397) on Selachii, Pehrson on *Amia* (372), Veit on *Lepidosteus* (393), Stöhr (388) on *Salmo*, Platt (374) and Stöhr (386-7) on Amphibia, Sonies on Birds (385), Noordenbos (370), Terry (390), Fawcett on Mammals (489-492), and de Beer (321, 324, 421-2), enables us to conclude that the following elements can be identified throughout the Gnathostomes : the base of the skull in the anterior or prechordal region appears as paired trabeculae cranii, while the side walls in the orbito-temporal region arise from paired orbital or sphenolateral cartilages ; paired parachordals in the chordal region give rise to the basal plate, to which are added in front an acrochordal cartilage and behind more or less distinctly segmental elements completing the occipital region. The nasal septum is formed by the union and upgrowth of the anterior ends of the trabeculae, of which the nasal capsules are also usually an extension, though separate cartilages often help to complete them. The auditory capsules generally appear first of

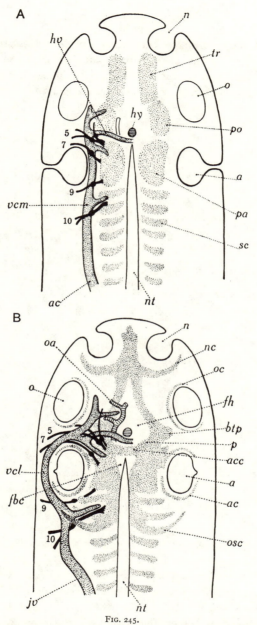

FIG. 245.

Diagrams illustrating development of *chondrocranium in Gnathostome* and relation of basal elements and sense capsules to certain cranial nerves and blood-vessels ; dorsally situated orbital cartilage omitted. A, Younger stage with procartilaginous elements separate ; B, older stage with cartilaginous elements partially fused. *a*, Auditory sac ; *ac*, auditory capsule ; *acc*, acrochordal cartilage ; *btp*, basitrabecular process ; *fbc*, basicranial fenestra ; *fh*, hypophysial foramen ; *hv*, hypophysial vein ; *hy*, hypophysis ; *jv*, jugular vein ; *n*, nasal sac ; *nc*, nasal capsule ; *nt*, notochord ; *o*, optic cup ; *oa*, optic artery from internal carotid ; *oc*, optic capsule ; *osc*, occipital arch = last occipital sclero-mere ; *p*, place of junction with base of pila antotica ; *pa*, parachordal ; *po*, polar cartilage ; *sc*, scleromere ; *tr*, trabecula proper ; *vcl*, vena capitis lateralis ; *vcm*, vena capitis medialis.

all as one or two independent cartilaginous plates, which grow round the
auditory sacs and soon fuse with the parachordals. The notochord is
probably the only truly median element included in the skull.   Its
anterior end, projecting beyond the basal plate, always degenerates, and

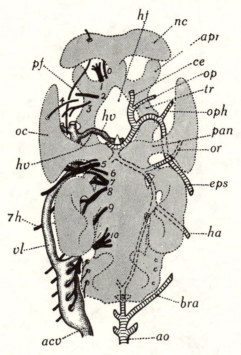

Fig. 246.

Diagram of skull of embryo Selachian; dorsal view. Cranial nerves on left-side black; arteries
cross-lined; veins darkly shaded; cartilage grey.   *acv*, Anterior cardinal; *ao*, dorsal aorta; *apr*,
antorbital process; *bra*, first epibranchial; *ce*, cerebral; *eps*, efferent pseudobranchial; *ha*, efferent
hyoid; *hf*, fenestra hypophyseos; *hv*, hypophysial; *nc*, nasal capsule; *oc*, orbital; *op*, optic;
oph, ophthalmic; *or*, orbital; *pan*, pila antotica; *pf*, profundus; *vl*, vena capitis lateralis; 1-10,
cranial nerves; *o*, nervus terminalis.   Dotted lines indicate the course of arteries below cartilages.

even the portion which becomes enclosed in the plate usually disappears
more or less completely in the adult, Figs. 245-6-7.

   Trabeculae.—It is usual for the trabeculae cranii to develop independently as a pair of rods below the fore-brain, one on either side of the hypophysis (see, however, Mammalia, below, p. 262). They soon fuse in front and
with the parachordals behind. A median hypophysial space (fenestra hypophyseos, or anterior basicranial fenestra) thus becomes enclosed between
the trabeculae and the basal plate, letting through the hypophysis and
internal carotids (p. 214). In Elasmobranchii, Dipnoi, Amphibia, and

Mammalia (p. 240) this fenestra is later closed by cartilage, leaving usually only two carotid foramina.[1] In other Gnathostomes it generally remains open, being only closed ventrally by the dermal roofing bone of the palate, the parasphenoid,[2] Figs. 248, 250-54.

As Gaupp has shown, there are two types of skull : the platybasic and the tropybasic. In the former the trabeculae remain wide apart in the orbito-temporal region, an intertrabecular plate unites them in front, and an extensive brain cavity is continued forward to the nasal capsules, Figs. 251, 259, 260. This possibly more primitive type is found in the Chondrichthyes and lower Osteichthyes (*Acipenser, Amia, Polypterus,* and

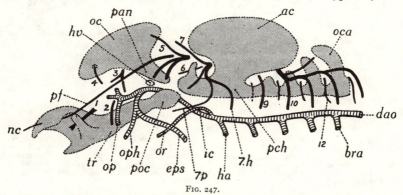

FIG. 247.

Diagram of skull of Selachian embryo before the fusion of main cartilaginous elements ; left-side view.   Cranial nerves black ; arteries cross-lined ; cartilage grey. *ic*, Internal carotid ; *oca*, posterior scleromere of occipital arch ; *poc*, polar cartilage fused to trabecula but not yet to basal (parachordal) plate, *pch* ; *pan*, pila antotica not yet fused to basal plate ; facial foramen for *7p*, palatine, and *7.h*, hyomandibular branch of facial not yet closed by prefacial commissure ; *12*, posterior root of hypoglossal.   Compare with Fig. 246 for later development and lettering.

some Teleostei such as the Cypriniformes among living forms), Dipnoi, and Amphibia.   In the tropybasic type the trabeculae tend to fuse immediately in front of the hypophysis to form the base of a median interorbital septum continuous with the internasal septum farther forward, Figs. 254, 255.

This type occurs in a very pronounced form in the majority of Teleostei

[1] According to Gaupp (504), in the course of development in *Rana* the internal carotid shifts.   The carotid first becomes enclosed in a foramen, then this becomes confluent with the oculomotor foramen, cutting through the trabecula.   The internal carotid and ophthalmic arteries now lie outside the skull, and in the adult the cerebral artery enters the skull by the oculomotor foramen.   *Polypterus* and the Teleost *Amiurus* also show a similar anomalous condition (Allis, 803).

[2] A median hypophysial canal may persist through the parasphenoid region in certain fossil fish (*Pygopterus, Acanthodes* (Jaekel, 1903), and in *Polypterus*).

(p. 256), and also in the Birds and higher Reptiles (p. 392). The interorbital

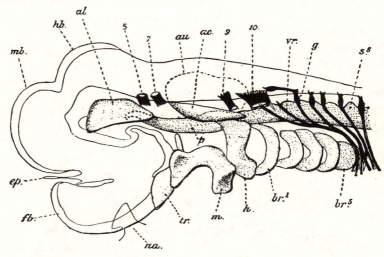

FIG. 248.

Reconstruction of the head of an embryo of *Squalus acanthias*, enlarged. (After Sewertzoff.) *ac*, Cartilage of auditory capsule; *al*, orbital (alisphenoid) cartilage; *au*, auditory capsule; *br*[1-5], first to fifth branchial arches; *ep*, epiphysis; *fb*, fore-brain; *g*, spinal ganglion; *h*, hyoid arch; *hb*, hind-brain; *m*, mandibular arch; *mb*, mid-brain; *na*, nasal pit; *p*, parachordal plate; *s*[8], eighth scleromere; *tr*, trabecula; *vr*, ventral spinal root; 5, 7, 9, 10, roots of the trigeminal, facial, glossopharyngeal, and vagus nerves. (From Goodrich, *Vert. Craniata*, 1909.)

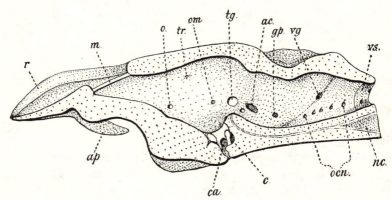

FIG. 249.

Inner view of the right half of the skull of *Hexanchus*. (After Gegenbaur.) *ac*, Foramen for auditory, *gp*, for glossopharyngeal, *o*, for optic, *ocn*, for spino-occipital, *om*, for oculomotor, *tg*, for trigeminal, *tr*, for trochlear, *vg*, for vagus, and *vs*, for occipito-spinal nerve; *ap*, antorbital process; *c*, carotid foramen; *ca*, interorbital canal; *m*, membrane over fontanelle; *r*, rostrum. (From Goodrich, *Vert. Craniata*, 1909.)

septum is less extensive in Mammalia. It was probably slightly developed even in the most primitive Amphibia and Teleostomes (p. 390). In certain specialised reptiles, however, the interorbital septum is no longer formed ; and in the Ophidia the trabeculae remain separate throughout most of the orbito-temporal region (Parker, 563).

When they first develop the trabeculae may be bent downwards at a considerable angle to the basal plate, especially in the Selachii where the cephalic flexure is very pronounced, Fig. 248 (Sewertzoff, 384) ; but later they straighten out. The fusion of the trabecula to the parachordal may take place by means of an at first separate polar cartilage. First described

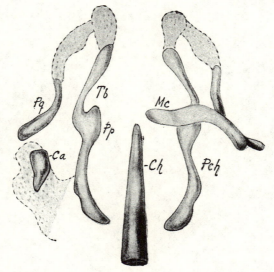

FIG. 250.

Amia calva, Stage 3 ; ventral view of skeleton of head. (From T. Pehrson, 1922.) *Ca*, Auditory capsule ; *Ch*, notochord ; *Mc*, Meckel's cartilage ; *Pch*, parachordal ; *Pp*, polar cartilage ; *Pq*, palatoquadrate ; *Tb*, trabecula.

by van Wijhe in Selachii (*Squalus*, 397), it has since been found in other forms (Veit in *Lepidosteus*, de Beer in *Scymnus*, Pehrson in *Amia*, Sonies in Birds) ('anterior parachordals' of Stöhr). Whether this polar element has any important significance or merely is the hind end of the trabecula separately chondrified remains uncertain. It forms an infrapolar process in Crocodiles and Birds, which may come to surround the internal carotid artery in many Birds.

An outer antorbital process in front and a basitrabecular ('basipterygoid' of authors) process farther back serve for the attachment of the palato-pterygo-quadrate bar in many of those Craniates in which this bar is complete (Chapter VII.).

Of the phylogenetic origin of the trabeculae little is known. That they are not visceral arches, as suggested by Huxley, there can be little doubt.[1] Their position in the cranial wall and relation to neighbouring parts points to the trabeculae belonging to the axial skeleton. They are said to be derived from the sclerotomes of the first or first and second

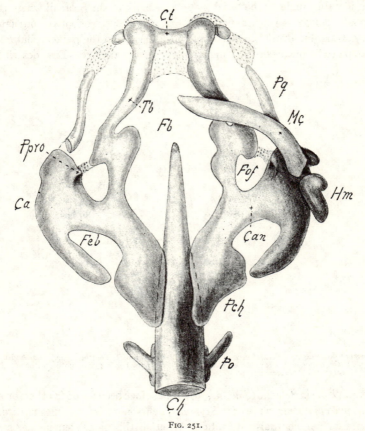

FIG. 251.

Amia calva, Stage 5 ; ventral view of skeleton of head. (From T. Pehrson, 1922.) *Can*, Commissura basicapsularis anterior ; *Ct*, commissura trabecularis ; *Fb*, fenestra basicranialis+fen. hypophyseos ; *Feb*, fenestra basicapsularis ; *Fof*, facial foramen ; *Hm*, hyomandibula ; *Po*, pila occipitalis ; *Ppro*, pro-otic process. Other letters as in Figs. 250 and 258.

segments (Koltzoff, 361 ; Sewertzoff, 384), yet they show no sign of segmentation unless it be the subdivision already mentioned into anterior and polar regions. They can scarcely be compared to a pair of segmental neural

[1] Allis has revived Huxley's view, believing the trabeculae to represent premandibular arches which have swung upwards to fuse with the membranous brain-case, and the polar cartilages to be the dorsal elements of the mandibular arches (1923 and 1925).

arches farther back, since they project forwards beyond the truly segmented region of the head. On the whole the trabeculae are best considered as structures *sui generis* developed to support and protect the
fore-brain and nasal sacs.

Basal Plate.—The notochord, surrounded by its sheaths, extends
primitively throughout the post-hypophysial basal plate region, beyond
which it projects in a hook-like curve bending down below the flexed
brain (p. 214) and just behind the hypophysis, Figs. 235, 245, 249. The
tip of the notochord degenerates early, and the remainder of the free
portion piercing the thick 'acrochordal' connective tissue filling the
space formed by the cerebral flexure
disappears later. Even the part embedded in the basal plate usually
degenerates in still later stages; but
it may persist as a shrunken vestige
in Elasmobranchs, or more completely
in Dipnoi and certain lower Teleostomes (Acipenseridae).

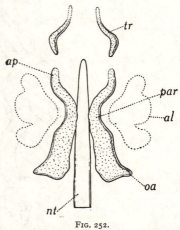

Parachordal cartilages, essentially
paired (see, however, below), give rise
to the greater part of the basal plate,
Figs. 250, 252, 255. They soon join
across, either below, or above, or enclosing the notochord; but often
leaving a membrane between their
diverging front ends. The true relations of the skull elements and the
median fenestrae in the pituitary
region have been considerably mis-

FIG. 252.

Salmo fario, embryo 9·5 mm. long. Skeleton
of head from preparation by G. R. de Beer. *al*,
Auditory labyrinth ; *ap*, 'anterior parachordal'
= polar cartilage (?) ; *nt*, notochord ; *oa*, occipital arch ; *par*, parachordal ; *tr*, trabecula.

understood even by Gaupp (Allis, **406, 412, 413**; de Beer, **421-2**). As
shown in Fig. 256, the anterior end of the notochord is in typical forms
bent upwards following the cerebral flexure. The floor of the cranium
is presumably likewise folded upwards, and here forms a pocket into
which pushes the pituitary body (combined hypophysis and infundibulum). The pituitary vein runs across from left to right orbit in a
space (interorbital canal) below the notochord. This space is either
intramural (excavated in the thickness of the cranial wall) or extra-
cranial, as held by Allis.

As already mentioned, dense mesenchymatous tissue lies in the transverse plica encephali ventralis ; it here surrounds the notochord, and
closes in front the membranous fenestra basicranialis posterior which

remains for a time in development between the divergent anterior ends
of the parachordals. The crista sellaris, a special cartilaginous formation in the acrochordal tissue, forms eventually the dorsum sellae
(Gaupp, 343; Sonies, 385; Noordenbos, 370; Voit, 394-5; Jager, 357;
de Beer, 421). The crista stretches across from the base of one pila antotica
to the other, and lies not at the level of the cranial floor formed by the

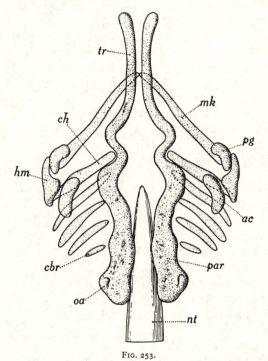

FIG. 253.

Salmo fario, embryo 10·5 mm. long. Skeleton of head from preparation by G. R. de Beer.
ac, Auditory capsule; *ch*, ceratohyal; *hm*, hyomandibula; *mk*, Meckel's cartilage; *pg*, palatoquadrate. Other letters as in Fig. 252.

trabeculae in front and the parachordals behind, but dorsal to these in
the plica, and overhanging the pituitary fossa from behind, Figs. 234-5,
256, 263, 274, etc. It is notched or pierced by the abducens nerve. In
later stages the parachordals join the acrochordal bridge and the fenestra
may be obliterated, Fig. 261.

When the posterior ends of the trabeculae (polar cartilages) come to
fuse somewhat ventrally with the anterior ends of the parachordals there
is enclosed a space, the primitive fenestra hypophyseos, through which
internal carotids and hypophysis reach the cranial cavity. The pituitary

vein runs across dorsally to this fenestra. In later stages cartilage may
extend inwards from the trabeculae (or from special centres in Mammals,
p. 262) so as more or less completely to obliterate the fenestra, leaving,
however, carotid foramina. Thus may be formed a cartilaginous floor
to the pituitary fossa or sella turcica, Fig. 256. From the above
description it is clear that the fenestra hypophyseos and posterior
basicranial fenestra do not lie in the same plane, but that the latter to
some extent overlies the former from behind.

The basal plate is a compound structure, and the first origin of the
parachordals is somewhat complicated and variable. In the Selachii van

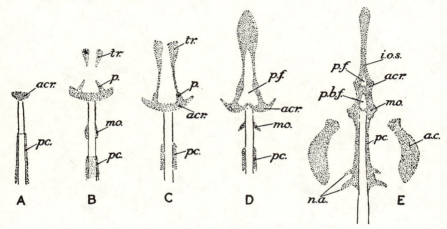

FIG. 254.

Diagrams illustrating the early development of the chondrocranium of birds. (Based on figures
by Sonies, 1907.) A, Chick, 11 mm.; B, duck, 13 mm.; C, duck, 15 mm.; D, duck, 14 mm.;
E, chick, 12 mm.   *a.c*, Auditory capsule; *acr*, acrochordal cartilage; *i.o.s*, interorbital septum;
mo, mesotic cartilage; *n.a*, neural arch; *p*, polar cartilage; *p.f*, pituitary foramen; *p.b.f*, posterior
basicranial fontanelle; *pc*, parachordal; *tr*, trabecula. (From J. G. Kerr, *Embryology*, 1919.)

Wijhe has shown that they appear as plates extending along each side of
and finally enclosing the notochord. No certain signs of segmentation
are found in front ; but in the post-auditory region the myocommata
are attached to the plate, and here upgrowths, passing between and
joining above the roots of the glossopharyngeal and hypoglossal nerves,
may be considered as segmental elements forming the occipital arch or
pila occipitalis. The so-called anterior parachordals and mesotic cartilages, described by Stöhr and others in Teleostomi and Amphibia as
arising separately but soon joining the more posterior true parachordals,
possibly represent the polar elements and really belong to the trabecular
region. In Birds, however, a true anterior 'mesotic' parachordal is
present in early stages, Fig. 254 (Teleostei: Stöhr, **388**; Swinnerton,

R

389; de Beer, 422. *Amia:* Pehrson, 372. Amphibia: Stöhr, 386-7; Platt, 374). In Birds (Parker, 556-7, 568, 572; Sonies, 385) and in Mammals (Noordenbos, 370; Fawcett, 489-92; Terry, 390) the basal plate appears first as a median bilobed cartilage and grows forwards to join the acrochordal cartilage and trabeculae.

The basal plate with its side wings forming occipital arches probably contains rudiments corresponding to the pleuro- and hypocentra

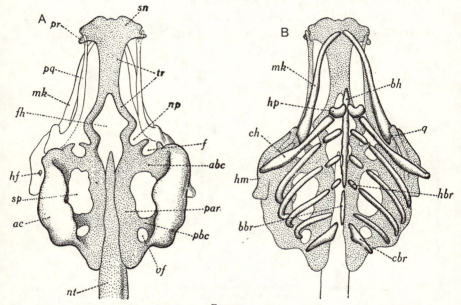

FIG. 255.

Salmo fario, embryo about 14 mm. long. A, Dorsal, and B, ventral view of cartilages of skull and visceral arches (from preparation by de Beer). *abc*, Anterior basicapsular commissure; *ac*, auditory capsule; *bbr*, basibranchial; *bh*, basihyal; *cbr*, ceratobranchial; *ch*, ceratohyal; *f*, facial foramen still open; *fh*, fenestra hypophyseos; *hbr*, hypobranchial; *hf*, foramen for hyomandibular nerve; *hm*, hyomandibular; *hp*, hypohyal; *mk*, mandibular; *np*, notch for palatine nerve; the lateral commissure is seen developing between it and *f*; *nt*, notochord; *par*, parachordal; *pbc*, posterior basicapsular commissure; *pq*, palatoquadrate; *pr*, antorbital process; *q*, quadrate; *sn*, solum nasi (ethmoid plate); *sp*, fissura basicapsularis; *tr*, trabecula; *vf*, foramen for vagus.

and neural arches seen in the vertebral column; segmental elements are usually very indistinct, though just recognisable in the plate of Birds (Froriep, 398; Sonies, 385) and Mammals (Terry, 390). Thus the basal plate is doubtless really formed of paired elements representing the scleromeres of several segments even in its front region.

Side Walls.—Considering now the side wall of the brain-case, we find, as already mentioned, that in the Selachii it shows in the occipital region signs of segmentation, being formed of a number of scleromeres

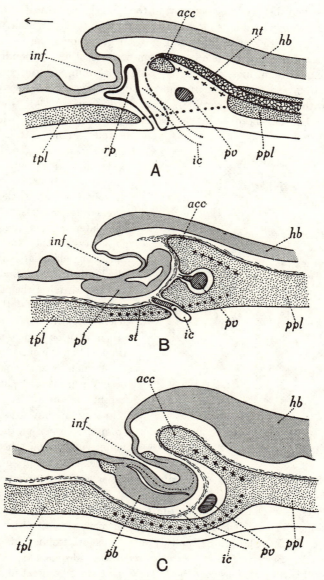

FIG. 256.

Diagrammatic longitudinal sections of *pituitary region* of head of : A, embryo Gnathostome ; B, adult Selachian ; C, late embryo Mammal. Showing relation to surrounding structures of *foramen hypophyseos* (marked by line of black dots) and *fenestra basicranialis posterior* (marked by line of crosses). *acc*, Acrochordal cartilage ; *hb*, floor of hind brain ; *ic*, internal carotid ; *inf*, infundibulum ; *nt*, notochord ; *pb*, pituitary body ; *ppl*, parachordal basal plate ; *pv*, pituitary vein ; *rp*, hypophysis ; *st*, floor of sella turcica ; *tpl*, trabecular plate.

progressively better developed toward the hind limit of the skull, Figs. 247, 248. They combine to form an occipital arch (pila occipitalis) which fuses above with the auditory capsule, and grows over the medulla to join that of the opposite side in a tectum posterius, thus surrounding the foramen magnum, Figs. 259, 263, 274. This tectum posterius generally arises from separate paired supraoccipital cartilages. They may fuse first with the auditory capsules, as in Birds (Sonies, 385), when the roofing bar has been called a tectum synoticum, or first with the occipital arches. It is sometimes held that the tectum posterius and tectum synoticum are distinct structures; they both occur in the early stages of some forms, and combine later. Often there is a gap between the pila occipitalis lying immediately in front of the future occipital joint and the capsule; the pila then at first represents the last scleromere of the head, but others may be added to it in front, filling the gap.

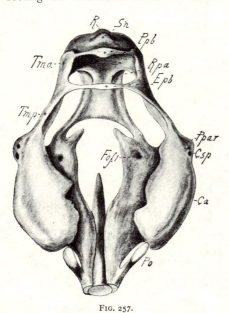

FIG. 257.

Amia calva, Stage 9, dorsal view of chondrocranium (from T. Pehrson, 1922).

A small extent of side wall occurs in front of the auditory capsule as the prefacial commissure, an upward growth from the corner of the basal plate cutting off the facial foramen from the pro-otic foramen, Figs. 263, 271. In Anura and *Acipenser* and Holostei it fails to develop (see p. 259).

The side walls of the cranium in the prechordal or trabecular region are formed as a rule from separate paired elements described by Parker (556-70) in Reptiles, Birds, and Mammals as orbitosphenoids or alisphenoids. Sewertzoff first clearly traced these 'alisphenoid' cartilages in Selachians (384), Figs. 246-7-8. They have since been called sphenolaterals by Gaupp (343) and pleurosphenoids by van Wijhe (397), according to whom they originate from two rudiments: a more anterior dorsal and a more posterior ventral (lamina antotica). Each of these orbital cartilages, as they may be called, soon produces a pila antotica, which, joining the outer corner of the basal plate (or its acrochordal region) behind the hypophysial vein, separates an orbito-nasal fissure in front from a pro-otic fissure behind, Fig. 270.

Later the upper posterior edge of the orbital cartilage grows back as a commissura orbito-parietalis (taenia marginalis posterior) to join the capsule and close the incisura prootica above. Its upper anterior edge grows forward to meet the nasal capsule as a taenia marginalis anterior or commissura spheno-ethmoidalis, thus completing the margin

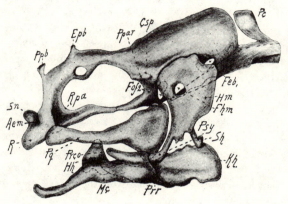

FIG. 258.

Amia calva, Stage 9, left lateral view of chondrocranium, mandibular and hyoid arches (from T. Pehrson, 1922). *Aem*, Anterior eye-muscle canal; *Csp*, spiracular canal; *Epb*, epiphyseal bar; *Feb*, foramen for glossopharyngeal; *Fhm*, foramen for hyomandibular branch of facial; *Fof₁*, palatine for.; *Fof₂*, facial for.; *Hh*, hypohyal; *Kh*, ceratohyal; *Po*, pila occipitalis; *Ppar*, parotic process; *Ppb*, paraphysial bar; *Prco*, coronoid process; *Prr*, retroarticular process; *Psy*, symplectic process; *R*, rostrum; *Rpa*, parethmoid ridge; *Sh*, stylohyal; *Sn*, nasal septum; *Tma, Tmp*, anterior and posterior taenia marginalis. (Lettering for Figs. 251, 257, and 258.)

of the orbit and delimiting an orbito-nasal foramen.[1] By the meeting of downgrowths from the orbital and upgrowths from the trabeculae the optic and other foramina are enclosed and the side wall is completed, Figs. 257-8, 263, 275.

Except in the Elasmobranchii and Amphibia, the side walls are usually very incomplete ; in the latter the orbital cartilages have been described as separate elements only in *Necturus* (Platt, **374**), and the side walls appear to develop in continuity with the trabeculae. It is important to notice that the pila antotica fails to develop in Teleostomes, possibly on account of the formation of a posterior eye-muscle canal (p. 279), Fig. 258. It is absent, also, in Ditrematous Mammals, Fig. 282.

[1] This fissure is lengthened into an orbito-nasal canal in *Amia*, which is enlarged in Teleostei to form the anterior myodome for the oblique eye-muscles.

The orbital cartilages in the Reptiles and Birds spread above the interorbital septum to form the planum supraseptale (Gaupp, 343, 506). In the Mammalia they are represented by the well-known alae orbitales, which meet the trabecular bar here forming a short, narrow, interorbital septum, Fig. 275.

Roof.—The roof of the chondrocranium usually remains very incomplete except in the occipital region, where the foramen magnum becomes surrounded by cartilage, as explained above. Elsewhere the membranous roof is strengthened chiefly by dermal bones. But in the Elasmobranchii, where no such bones exist, the orbito-temporal as well as the occipital region becomes completely roofed by the overgrowth of cartilage from the sides, leaving only a large median epiphysial fontanelle in front and two small fenestrae behind, openings for the endolymphatic ducts, Fig. 249. Paired apertures for nerves supplying the surface of the head pierce the extensions over the orbits.

The chondrocranium is generally well developed in the lower Osteichthyes, and may become thick and massive with advancing age. In such forms as *Ceratodus*, *Acipenser*, *Amia*, and *Salmo*, the dorsal fontanelles may become almost if not quite obliterated. The roof is here formed from a transverse 'epiphysial bar' developed from the supraorbital cartilages, Figs. 257-8, which cuts off the anterior epiphysial from a posterior fontanelle (Swinnerton, 389; Pehrson, 372). By the extension of this bar and its junction with the tectum posterius the roof may be completed. In most of the higher Teleostei, however, the fontanelles remain large, or the bar may even disappear.

No median chondrocranial roof is usually found in the Amniota beyond the tectum synoticum and tectum posterius; but in the Anura, among Amphibia, cartilage extends over the anterior region of the brain-cavity as well, Figs. 450, 506.

Sense Capsules.—The ethmoid region of the chondrocranium of Gnathostomes is developed in relation to the snout, the nostrils, the mouth and upper jaw, and more especially the enclosure of the paired nasal sacs. As the latter sink inwards capsules envelop them, arising mostly on either side from the intertrabecular plate and median septum rising above it, and the posterior and lateral antorbital process of the trabecula.

The nomenclature of Gaupp, devised for the description of this region in Tetrapods, is now generally adopted. A median septum nasi separates the right from the left cavum nasi in which lie the olfactory sacs. The roof of the capsule is called the tectum, the side wall the paries, and the floor the solum nasi. The anterior wall is the cartilago cupularis from which small cartilagines alares may be separated off. The posterior wall, separating the cavum nasi from the orbit, is the planum antorbitale or orbito-nasalis. A fenestra narina, leading to the external nostril, becomes separated from a fenestra basalis serving for the internal nostril or choana by the lamina transversalis anterior of the solum nasi. A foramen epiphaniale remains for the exit of the nervus lateralis nasi at the

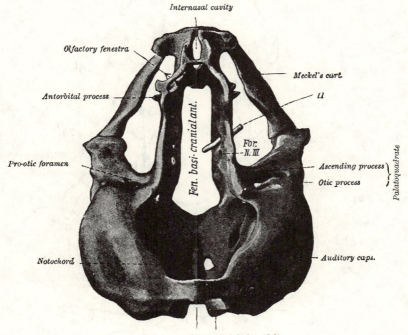

Internasal cavity

Olfactory fenestra

Meckel's cart.

Antorbital process

ll

Fen. basi-cranial.ant.

For. N. III

Pro-otic foramen

Ascending process ⎫
Otic process ⎬ *Palatoquadrate*
                 ⎭

Notochord

Auditory caps.

Tectum synoticum   Occipital condyle

FIG. 259.

Neurocranium (platybasic type) and mandibular arch of larval newt (*Triton taeniatus*), 2 cm. in length, seen from above. ×25. (From a model by E. Gaupp. Wiedersheim, *Comp. Anatomy.*) Fenestra basicranialis anterior = confluent f. basicr. post. and f. hypophyseos.

side, and a foramen apicale in front for the n. medialis nasi, two branches of the profundus nerve, Figs. 255, 257-8, 260, 274.

The nasal capsule in fishes develops in continuity with the front end of the trabeculae. An antorbital process grows out from the trabecula, forms the posterior wall, or planum antorbitale, and joins above with the internasal septum to close off the olfactory foramen. In the Teleostomes, where the original nasal opening becomes divided into two nostrils situated at the side of the head, the tectum is but little developed.

In the Tetrapoda, where the incurrent nostril serves not only to bring air to the nasal sac, but also to the lung, and where the nasal groove has closed over so as to form a canal leading to the buccal cavity by an

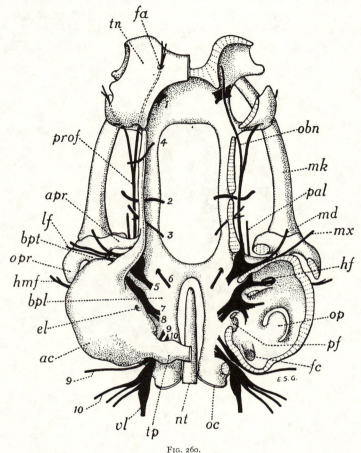

FIG. 260.

Salamandra maculosa, larva. Reconstruction of chondrocranium, etc.; dorsal view. *ac*, Auditory capsule; *apr*, ascending process; *bpl*, parachordal basal plate; *bpt*, basal process; *el*, foramen endolymphaticum; *fa*, foramen apicale; *fc*, fenestra cochleae; *hf*, foramen faciale; *hmf*, hyomandibular branch of facial nerve; *lf*, buccal branch of facial (supraorbital branch removed to expose structures below); *md*, mandibular nerve; *mk*, Meckel's cartilage; *mx*, maxillary nerve; *nt*, notochord in posterior basicranial fenestra, which is closed in front by acrochordal bridge delimiting posterior limit of hypophysial fenestra; *obn*, orbito-nasal branch; *oc*, occipital condyle; *op*, operculum; *opr*, otic process; *pal*, palatine nerve; *pf*, fenestra perilymphatica; *prof*, profundus; *tn*, tectum nasi; *tp*, tectum posterius; *vl*, lateralis branch of vagus; 1-10, cranial nerves in black. On right dorsal parts of cartilages removed by horizontal cut.

internal nostril (a disposition already found in the Dipnoi), the nasal capsule becomes more elaborately developed.

The cavity of the sac tends to be subdivided into a more ventral air

passage and a more dorsal olfactory chamber, whose wall becomes folded
to offer more surface for the olfactory epithelium ; moreover, the ventral
wall becomes strengthened by a transverse extension from the septum
forming a lamina transversalis anterior, separating a fenestra narina from
a fenestra choanalis or basalis for the internal nostril. As this lamina

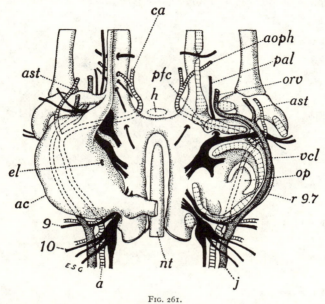

FIG. 261.

Salamandra maculosa, larva. Posterior region of reconstruction shown in Fig. 260, with arteries
(cross-lined) and veins (shaded). *a*, Lateral aorta ; *aoph*, ophthalmic ; *ast*, stapedial ; *ca*, cerebral ;
h, hypophysis ; *j*, jugular ; *orv*, orbital ; *pfc*, prefacial commissure ; *r* 97, ramus from glossopharyngeal to facial ; *vcl*, vena capitis lateralis.

fuses with the outer paries the sac in this region is entirely surrounded
by cartilage.

The nasal capsule in Amphibia develops much as in the Teleostomes
from extensions of the front end of the trabecula and an antorbital
process, Figs. 259, 260. It becomes fully formed after metamorphosis,
though membranous gaps usually remain in the wall ; and, owing to the
great development of large intermaxillary glands passing between the
capsules, the median septum becomes much reduced in the Urodela
(Parker, 558 ; Born, 325 ; Wiedersheim, 651 ; Gaupp, 504 ; Higgins, 353).

In the Amniota the capsules are usually more complicated, and tend
to grow back along the correspondingly elongated septum nasi to accommodate the enlarging nasal chamber (Reptiles: Parker, 564-5 ; Gaupp,
506 ; Rice, 376 ; Kunkel, 362. Aves: Sonies, 385. Mammalia: Noordenbos,

370). The anterior and posterior walls apparently originate as independent cartilages which spread and fuse with each other and the septum. Just behind the fenestra narina the nasal sac becomes surrounded, the lamina transversalis anterior being completed below ; but farther back the floor is usually very incomplete, and represented only medially by a longitudinal strip, the paraseptal cartilage, underlying Jacobson's organ and primitively

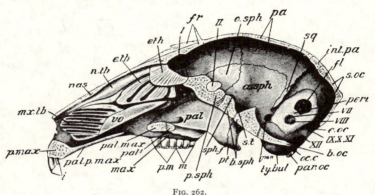

FIG. 262.

Lepus cuniculus. Skull in longitudinal vertical section. The cartilaginous nasal septum is removed. *a.sph,* Alisphenoid ; *e.oc,* exoccipital ; *e.tb,* ethmo-turbinal ; *eth,* ethmoid ; *fl,* fossa for flocculus of brain ; *i,* incisors ; *mx.tb,* maxillary turbinal ; *n.tb,* naso-turbinal ; *pal',* palatine portion of the bony palate ; *peri,* periotic (petrous portion) ; *p.sph,* presphenoid ; *sph.f,* sphenoidal fissure ; *s.t,* sella turcica, or depression in which the pituitary body lies ; I, foramina for olfactory nerves ; II, optic foramen ; *Vmn,* foramen for mandibular division of trigeminal ; VII, for facial nerve ; VIII, for auditory nerve ; IX, X, XI, for glossopharyngeal, vagus, and spinal accessory ; XII, for hypoglossal. (From Parker's *Practical Zoology.*)

joining the lamina transversalis posterior or inturned edge of the planum antorbitale closing the fenestra basalis behind,[1] Figs. 263-4, 274-5.

Although the planum antorbitale appears to be homologous with the part derived from the antorbital process in lower forms, it chondrifies separately in the Amniotes.[2] The paraseptal and antorbital cartilages

[1] It is characteristic of the terrestrial Tetrapoda that an apparatus is developed to keep the surface of the eye moist. Epidermal glands secrete a watery fluid into the orbit which is carried away by a lacrymo-nasal duct from the anterior corner of the orbit to the nasal cavity. This duct develops as a longitudinal groove or thickening of the epidermis which sinks from the orbit to the external nostril, and becoming hollowed out comes to lead into the nasal cavity (Born, 325).

[2] Jacobson's organ is a specialised portion of the sensory region of the nasal sac which tends to become separated off from the main chamber in Tetrapods. Ill-defined and variable in living Amphibia, it is typically developed in Amniota as a sac blind posteriorly, but opening in front of or near the naso-palatine canal. It lies near the septum, supported by the paraseptal cartilage, and overlying the prevomer (Figs. 261, 264). It appears to be an organ of smell, in communication with the buccal cavity. (Broom, 1895–8; Seydel, 1885, 1891; Symington, 1891.) See page 367.

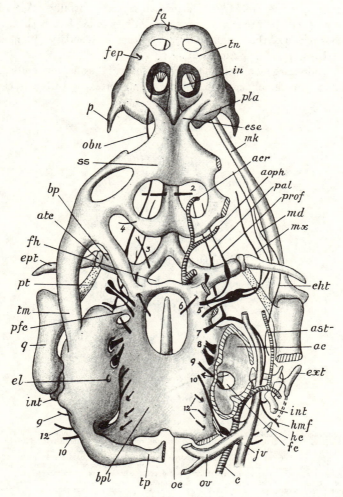

FIG. 263.

Diagram of cartilages, nerves, and blood-vessels of head of embryo *Lacerta*; dorsal view. Cartilages partly removed on right. Nerves black, arteries cross-lined, veins shaded. *ac*, Auditory capsule; *acr*, anterior cerebral; *aoph*, ophthalmic; *ast*, stapedial; *atc*, meniscus pterygoideus; *bp*, basitrabecular process; *bpl*, basal plate; *c*, internal carotid; *cht*, chorda tympani; *cse*, commissura spheno-ethmoidalis; *el*, endolymphatic foramen; *ept*, epipterygoid flattened out; *ext*, extrastapedial; *fa*, foramen apicale; *fc*, fenestra cochleae; *fep*, fenestra epiphaniale; *fh*, foramen hypophyseos; *hc*, top of hyoid cornu; *hmf*, hyoid branch of facial; *in*, fenestra for internal nostril; *int*, intercalary (dorsal process); *jv*, jugular vein; *md*, mandibular branch; *mk*, Meckel's cartilage on right side; *mx*, maxillary branch; *obn*, orbito-nasal branch of profundus; *oc*, occipital condyle; *ov*, occipito-vertebral; *p*, process (perhaps = part of pterygoquadrate); *pal*, palatine; *pfc*, prefacial commissure; *pla*, planum antorbitale; *prof*, profundus; *pt*, procartilage, vestigial part of palatoquadrate; *q*, quadrate; *ss*, septum supraseptale; *tm*, taenia marginalis posterior; *tn*, tectum nasi; *tp*, tectum posterius or synoticum. An arrow passes through foramen perilymphaticum internum (recessus scalae tympani). 2-12, cranial nerves.

run close to, but separate from, the median septum ventrally. According to Gaupp (343) the fusion which may sometimes take place between this posterior region of the capsules and the septum is secondary (Mammalia, and perhaps some other Amniota).

The Amniota generally tend to develop folds of the olfactory epithelium extending into the olfactory chamber and supported by turbinals, cartil-

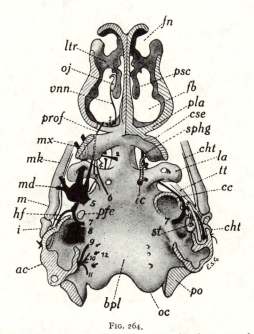

FIG. 264.

Reconstruction of chondrocranium of embryo *Trichosurus vulpecula*, 17·5 mm. long; dorsal view. On left side dorsal parts removed by horizontal cut; on right side cut at more ventral level (except root of facial nerve). *ac*, Auditory capsule; *bpl*, basal plate; *cc*, cochlear region; *cht*, chorda tympani; *cse*, commissura spheno-ethmoidalis; *fb*, fenestra basalis; *fn*, fenestra nasalis; *hf*, hyomandibular branch of facial; *i*, incus; *la*, lamina ascendens; *ltr*, lamina transversalis anterior; *m*, malleus; *md*, mandibular branch; *mk*, Meckel's cartilage; *mx*, maxillary branch; *oc*, occipital condyle; *oj*, organ of Jacobson; *pfc*, prefacial commissure; *pla*, planum antorbitale (lamina infracribrosa); *po*, pila occipitalis; *prof*, profundus; *psc*, paraseptal cartilage; *sphg*, sphenopalatine ganglion; *st*, stapes; *tt*, m. tensor tympani; *vnn*, vomero-nasal nerve. Nerves black and numbered.

aginous extensions from the wall of the capsule. Reptiles have only one main turbinal fold supported by the inturned edge of the paries nasi; while in Birds there is developed in addition another upper turbinal above it. In the Mammals, and especially in the Ditremata possessed of a well-developed sense of smell, these folds acquire much greater importance. Besides the lower maxillo-turbinal corresponding to the reptilian, there are naso-turbinals in front and ethmo-turbinals projecting forwards from the more posterior and median wall, Fig. 262.

The general history of the nasal capsule would seem to be as follows : Two lateral processes spread outwards from each trabecula ventrally : the cornu at its anterior extremity and the antorbital process farther back. The former is not much developed in Pisces ; but the latter encloses the olfactory foramen by growing round outside the olfactory nerve and fusing above with the septum nasi. The antorbital process expands vertically to form the planum antorbitale separating nasal from orbital cavity. In Tetrapods the cornu forms the anterior transverse lamina of the solum nasi, and fusing laterally with the outer edge of the tectum completes a zona annularis surrounding the nasal cavity. This anterior lamina also separates ventrally the external from the internal nasal openings. As the nasal cavity is enlarged in Amniotes the planum antorbitale tends to bulge more and more, and retreat backwards, the lengthening antorbital process extending alongside the septum as a paraseptal cartilage underlying Jacobson's organ. A narrow fissure remains between paraseptal and septal cartilages, which in Mammalia is bridged posteriorly by a broad fusion. The paraseptal cartilage may then be interrupted behind and even also in front of Jacobson's organ. In Mammalia, the ventral edge of the planum antorbitale forms the posterior transverse lamina now fused to the median septum, and the enlarged primitive olfactory foramen becomes subdivided into the many openings of the cribriform plate.

The auditory capsules of the Gnathostomata usually develop from independent cartilages which soon fuse with each other and with the parachordals and envelop the membranous labyrinth.

In the Selachii the ventral floor originates as a lamina basiotica chondrifying in continuity with the parachordal (Sewertzoff, 384 ; van Wijhe, 397 ; Goodrich, 349) ; but, according to van Wijhe, the rest of the capsule is derived from two independent cartilages—one antero-lateral appearing over the ampullae of the anterior and horizontal semi-circular canals, and the other, postero-lateral, over the ampulla of the posterior semicircular canal. These soon meet and complete the capsule, except for its inner wall which grows up chiefly from the parachordal, leaving foramina for the endolymphatic duct and the 8th nerve.

Stöhr and de Beer in *Salmo* (388, 422) and Pehrson in *Amia* (372) describe the origin of the auditory capsule from an independent ventro-lateral otic cartilage which becomes connected below with the parachordal, Figs. 250, 253. This primitive ventral connexion, the commissura basicapsularis anterior, corresponding to the lamina basiotica of the Selachian, appears very early, is present from the first in *Gasterosteus*, according to Swinnerton (389), and at least in the form of a procartilaginous commissure in other fish. The otic cartilage grows round the labyrinth; but the inner wall of the capsule remains membranous in Teleostomes. A long basicapsular or metotic fissure remains behind the commissure separating the capsule from the basal plate.

This fissure is usually more or less completely obliterated by fusion of the plate with the capsule in later stages, Fig. 255; but its fate differs in different forms. It is always closed above the vagus nerve by the junction of the pila occipitalis with the capsule. In Pisces the fissure is obliterated leaving a foramen for the 9th nerve and a jugular foramen for the 10th and posterior cerebral vein (see further, p. 261).

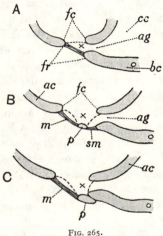

FIG. 265.

Diagrams showing relations of *fenestra cochleae* and *fenestra rotunda* in Reptilia, A, and Mammalia, B and C; transverse sections of vestibular region of pars cochlearis of left auditory capsule, *ac*, and parachordal basal late, *bc*. ×, Recessus scalae tympani occupied by saccus perilymphaticus; *ag*, aquaeductus perilymphaticus leading to cranial cavity, *cc*; *fc*, fenestra cochleae; *fr*, fenestra rotunda closed by secondary tympanic membrane, *m*; *p*, processus intraperilymphaticus (from front edge of f. cochleae), cutting off a portion of membrane, *sm*, as in B; or fused to edge of basal plate as in C.

It is characteristic of the Tetrapoda that the independent first rudiments of the cartilaginous auditory capsule (single on each side in the Amphibia, anterior and posterior in the Amniota) spread round the labyrinth, leaving in addition to the usual apertures for the 8th nerves and the ductus endolymphaticus two constant and important fenestrae, Figs. 263, 265-6-7-8. One, on the outer side of the capsule, is the fenestra ovalis or vestibulae; it is closed by a membrane in which fits the columella auris or stapes (p. 451); the other is the more ventral and posterior fenestra cochleae (perilymphatica) in the region of the opisthotic bone, or of the exoccipital in modern Amphibia (Parker, 358-9; Gaupp, 504).

This primary fenestra is situated opposite the free edge of the basal plate, which is here separated from the capsule by what remains of the front end of the basicapsular fissure, the recessus scalae tympani. A space is thus delimited opening above into that region of the general cavum capsularis known as the cavum vestibulare, inwards and above the basal plate by the secondary foramen perilymphaticum (aquaeductus perilymphaticus) into the cavum cranii, and outwards below the basal plate by the fenestra rotunda. Into this space passes the saccus perilymphaticus, sending a ductus (aquaeductus cochleae) to open through the perilymphatic foramen into the subarachnoid spaces below the brain, and abutting through the fenestra rotunda against a thick closing membrane, the membrana tympani secundaria. Pressure exerted at the fenestra ovalis by the base of the columella or stapes is compensated by the bulging outwards of the secondary tympanic membrane usually situated close to the wall of the tympanic cavity. The edge of the basal plate separates the two secondary foramina in the Amphibia, Figs. 260-61, 266-7, and the outer fenestra rotunda may be suppressed in the more aquatic forms.

Except in the Monotremata (*Echidna*, Gaupp, **511**), the primary fenestra cochleae becomes subdivided in the Mammalia into two by the growth of a bridge from the wall of the capsule (processus recessus), the upper and inner opening leading to the foramen perilymphaticum, the lower and outer to the fenestra rotunda, Fig. 265. The membrana tympanica secundaria now is attached to the rim only of the fenestra rotunda instead of partly to the edge of the basal plate as in the Reptile (Gaupp, **506**, **511**; Rice, **376**; Versluys, **769**; Fischer, **496**; Brock, **326**; de Beer, **422a**).[1]

While the Amphibian auditory capsule appears to develop from a single independent otic cartilage (Stöhr, **386-387**; Platt, **374**), in the Amniota, at all events in Birds (Sonies, **385**) and Mammals (Noordenbos, **506**; Terry, **390**), an anterior basiotic element

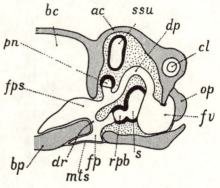

FIG. 266.

Diagrammatic transverse section of auditory capsule of larval *Rana* (partly from H. S. Harrison, 1902). *ac,* Auditory capsule; *bc,* brain-cavity; *bp,* basal plate; *cl,* lateral semicircular canal; *dp,* ductus perilymphaticus; *dr,* ductus reuniens; *fp,* fenestra rotunda; *fps,* ductus perilymphaticus; *fv,* spatium sacculare projecting through fenestra ovalis (vestibuli); *mts,* membrana tympanica secundaria; *op,* operculum in membrana tympanica; *pn,* pars neglecta (in adult a rod of cartilage lateral to it subdivides the fenestra cochleae between the two perilymphatic canals); *rpb,* pars basilaris; *s,* sacculus; *ssu,* sinus superior utriculi.

---

[1] This description does not quite agree with that generally adopted. Gaupp concluded that the membrana tympani secundaria is not homologous in Reptiles and Mammals. The name fenestra cochleae has here been retained for that opening in the floor of the pars cochlearis of the capsule which can be homologised throughout the Tetrapoda; it leads into the recessus scalae tympani. In Mammals where the cochlear region is greatly expanded this fenestra enlarges, the processus recessus grows across below it, and the recessus becomes continuous with the cavity of the pars cochlearis. The fenestra rotunda, often misnamed f. cochleae, is in all Tetrapods the lateral opening of the recessus closed by the same membrane. The recessus may become confluent with the jugular foramen, and the 9th nerve may pass through it in Reptiles.

forms the pars cochlearis, and a postero-lateral element contributes the
pars canalicularis. The fissura basicapsularis is more or less obliterated,
an anterior basicapsular commissure forming in front of the recessus
scalae tympani, and often a posterior commissure behind.

The Primitive Nerve Foramina.—Primitively the olfactory nerves
pass from the brain cavity into the nasal capsules through large olfactory

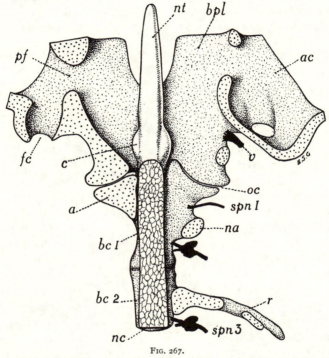

Fig. 267.

Salamandra maculosa, late larva. Partial reconstruction of posterior portion of skull, and first two
vertebrae; dorsal parts removed by horizontal cut, at lower level on left. *a*, Articular region of atlas;
ac, auditory capsule; *bc*, body of vertebra; *bpl*, parachordal basal plate; *c*, vestige of centrum (last
occipital?); *fc*, fenestra rotunda; *na*, neural arch; *nc*, cut end of notochord, and *nt*, its tip in basicranial fenestra; *oc*, occipital joint; *pf*, fenestra perilymphatica; *r*, rib; *spn*, spinal nerve; *v*, vagus.

foramina, one on either side of the median septum. Each foramen,
bounded below by the free edge of the planum antorbitale, is enclosed
by the growth upward and outside the nerve of the planum to join the
tectum and septum above, Figs. 246, 263. This opening persists as the
olfactory foramen in all Gnathostomes except the Mammalia (p. 258); but
becomes modified in those Teleostomes which develop an extensive
interorbital septum. In all forms with a tropibasic skull there is a
tendency for the anterior part of the cranial cavity to be narrowed and

reduced from below upwards by the enlarging orbits and the formation of the septum, and for the bulk of the brain to be pushed backwards. In the higher Teleostei this process is carried to an extreme; the orbital walls meeting in the middle line, the cranial cavity is obliterated in front and the brain retreats far back. Since the nasal sacs remain forward and the olfactory lobes retreat with the brain, the olfactory nerves become much lengthened and come to pass on either side of the thin septum

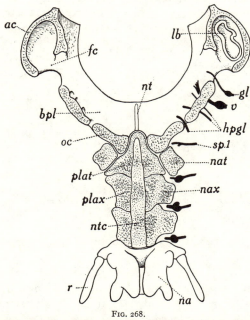

FIG. 268.

Lacerta. Diagrammatic reconstruction of portion of occipital region of skull and of first four vertebrae of a late embryo; dorsal parts removed by horizontal cut, except from 4th vertebra. Nerves in black on right. *ac*, Portion of auditory capsule; *bpl*, basal plate; *fc*, fenestra cochleae; *gl*, glossopharyngeal; *hpgl*, hypoglossal; *lb*, portion of labyrinth; *na*, neural arch; *nat*, neural arch of atlas, and *nax*, of axis; *nt*, notochord; *ntc*, notochordal 'cartilage'; *oc*, occipital arch; *plat*, pleurocentrum of atlas; *plax*, pleurocentrum of axis; *r*, rib; *sp*[1], first spinal nerve.

through the anterior corner of the orbit to reach the nasal sacs, Fig. 269 B. Each nerve then leaves the cranial cavity by a foramen olfactorium evehens, and, passing between the oblique muscles, pierces the planum antorbitale through a foramen olfactorium advehens. Intermediate conditions in the formation of the septum and subdivision of the olfactory foramen are found in the Cypriniformes, Fig. 287 (Sagemehl, 378; Gaupp, 343).[1] Another and divergent form of specialisation

[1] The explanation given above is that of Sagemehl. Gaupp regarded the passage of the olfactory nerves through the orbit as due rather to the fenestra-

has long been known in the Gadiformes, where the septum is formed below the brain-cavity, which persists as a narrow channel running forward to the nasal sacs. Here the olfactory lobes remain close to the sacs, and it is the olfactory tracts of the brain which are drawn out into slender strands while the nerves are short, Fig. 269 A (Goodrich, 35).

In Birds, where the orbits are also very large, the olfactory nerves may likewise pass through the anterior corner of the orbits.

In the Mammalia the original olfactory foramen becomes subdivided by extensions from its margin to form the lamina cribrosa (absent, however, in *Ornithorhynchus*), and the planum antorbitale fusing with the median

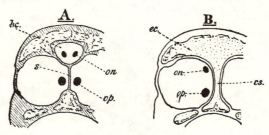

FIG. 269.

Diagrams of a transverse section through the front of the orbit—A, of *Gadus*, B, of *Perca*—to show the position of the olfactory nerves and tracts. *bc*, Brain-cavity of cranium; *cs*, cartilaginous interorbital septum; *ec*, ethmoid cartilage; *on*, olfactory nerve in B, olfactory tract in A; *op*, optic nerve; *s*, membranous interorbital septum. (From Goodrich, *Vert. Craniata*, 1909.)

septum forms the perforated vertical front wall of the brain-case in the Ditremata.

With regard to the orbito-temporal region, we find that whereas in the lower Craniata, such as the Chondrichthyes and most Amphibia, the cartilaginous side wall, chiefly derived from the orbital cartilage, Fig. 249, is so completely developed that the cranial nerves issue by separate foramina, there is a tendency in the higher groups for the wall to remain or become membranous. Increasingly large fenestrae are thus formed, and the cartilage may be reduced to a framework of slender bars (Lacertilia), though the individual foramina may be to a considerable extent replaced later by apertures in bones, with which, however, we are not concerned at present (cf. Figs. 249, 259, 274).

It will be understood that, since the side wall of the cranium occupies

tion of the lateral wall of the orbito-nasal canal at the side of the lengthening median septum. De Beer has recently suggested that the nasal septum lengthens backwards, carrying with it the olfactory foramen, while the planum antorbitale remains in front, leaving as it were the nerve exposed in the orbit (422).

the same general morphological position as the neural arches farther back, it bears the same topographical relation to the segmental cranial nerves as these do to the spinal nerves. In other words, the ventral and dorsal roots issue separately, and the ganglia on the latter lie primitively outside the wall (this applies to the otic and occipital region as well). In Pisces the ganglia of the trigeminal and facial nerves may retreat into a recess in the cranial wall through the foramina (p. 273).

It is a fundamental character of the Craniate skull that the auditory capsule lies morphologically between the facial and the glossopharyngeal nerves. The foramen for the facial (main branch: palatine + hyomandibular) is closed off in the Gnathostomes, by a prefacial commissure extending from the basal plate to the capsule, Figs. 270-71. This commissure, separating the 7th from the 5th nerve, may be lost in certain Chondrichthyes (Scyllioidei and some others), and disappears in the Teleostomi. But it is constant in Tetrapods with few exceptions (such as the adult frog). Primitively it is situated in front of the base of the auditory capsule; but

FIG. 270.

Diagram of portion of left orbito-temporal region of a Selachian skull showing relation of exit of cranial nerves to pila antotica, marked with a +; prefacial commissure, *pfc*, auditory capsule, *ac*, and basis cranii, *bcr*. 50, Ophthalmic, 5*mx*, maxillary, and 5*md*, mandibular branches of trigeminal; 70, ophthalmic, 7*b*, buccal, 7*p*, palatine, and 7*hm*, hyomandibular branches of facial; *prf*, profundus nerve; *ff*, facial foramen; *pof*, pro-otic foramen.

in the Amphibia becomes overgrown by the capsule, and in these the 7th nerve appears to pass through the cavity of the capsule, owing seemingly to the failure of its wall to chondrify in this region, Figs. 272-3 (see further, p. 278). In the Reptiles to a slight extent the capsule may extend below it (Rice, 376); in the Mammals the prefacial commissure is so much undermined by the cochlear region of the capsule that the nerve seems to pass through a foramen in its upper wall (see further, p. 272).

It has already been explained (p. 244) how, owing to the formation of a pila antotica and connexions with the capsules, the orbital cartilage separates a primitive fenestra pseudoptica in front from a fenestra pro-otica behind. Through the former issue the 2nd and 3rd nerves; through the latter the profundus, trigeminal, abducens, and lateral line branches of the facial (when present in Pisces and Amphibia). The 4th nerve is usually enclosed in a special foramen ; but it may, as in Lacertilia, issue through the anterior fenestra, or through the posterior fenestra as in

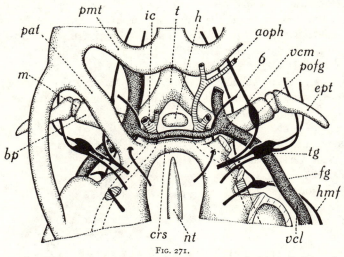

FIG. 271.

Enlarged view of orbito-temporal region of diagram, Fig. 263. *crs*, Crista sellaris (acrochordal) ; *fg*, facial ganglion ; *h*, hypophysis ; *ic*, internal carotid ; *nt*, notochord in fenestra basicranialis posterior ; *pat*, pila antotica ; *pmt*, pila metoptica: *pofg*, profundus ganglion ; *t*, fused trabeculae ; *tg*, trigeminus ganglion ; *vcl*, vena capitis lateralis ; *vcm*, vena capitis medialis. Other letters as in Fig. 263.

Mammalia. In some Urodeles the 4th nerve comes to pierce the parietal bone (Gaupp, 514). By the extension of the orbital cartilage to join the

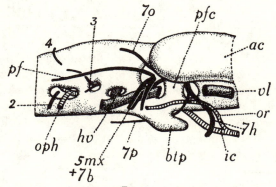

FIG. 272.

Diagram illustrating structure of orbito-temporal region in early stage of embryo Urodele ; left-side view. Cranial nerves numbered : *5mx*, maxillary branch of trigeminal ; *7b*, buccal, *7o*, ophthalmic branches of facial ; *pf*, profundus ; *vl*, vena capitis lateralis interrupted to show prefacial commissure, *pfc* ; *oph*, ophthalmic artery ; *or*, orbital or stapedial artery ; *hv*, hypophysial vein.

neighbouring cartilages the wall may be completed and all the nerves come to pass through separate foramina, as in the Elasmobranchii. But,

in the Teleostomi, the orbital wall is usually much less completely chondrified, the fenestrae remain, and even the pila antotica ceases to be formed
(absent in Amioidei, Teleostei, see Figs. 283, 284). The orbito-nasal
fissure remains as a small foramen, allowing the profundus nerve to pass
out of the orbit on to the surface of the snout in Pisces.

In the Reptilia the orbital wall is usually much less complete, especially
in the Lacertilia and Ophidia. The optic foramen is closed by a pila
preoptica in front and a pila postoptica behind, cutting off the fenestra
postoptica for the 3rd nerve. The trigeminal (with the profundus) may
alone pass through the diminished fenestra prootica (*Crocodilus*, Shiino,
597), the 4th nerve then issuing
through a small foramen above;
and the 6th bores its way, so to
speak, through the base of the pila
antotica.[1]

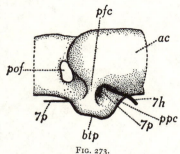

The profundus nerve in the
Tetrapods either passes by the
orbito-nasal foramen directly into
the nasal capsule, as in the Amphibian, or reaches the capsule
from the orbito-nasal fissure by the
olfactory foramen, as in Reptiles.[2]
In the capsule the nerve gives off
a branch which issues laterally
through the foramen epiphaniale,

FIG. 273.

Diagram illustrating relations in orbito-temporal
region of cartilages and facial nerve in late stage
of embryo Urodele; left-side view. *ac*, Auditory
capsule; *btp*, basitrabecular process; *pfc*, outer
portion of prefacial commissure; *pof*, pro-otic
foramen; *ppc*, postpalatine commissure; *7h*,
hyomandibular; *7p*, palatine branches of facial.

and a branch which passes forward to the foramen apicale (see p. 266
for Mammalia).

For the 8th nerve there are usually two foramina on the inner wall
of the auditory capsule; but, as mentioned above, a fenestration, already
seen in *Amia*, leads to the membranous condition of this wall in Teleosts.
The foramina for the 9th and 10th nerves through the occipital region
have already been described. They issue through the remains of the
fissura metotica; and although the 9th usually passes out separately, in

[1] The pila antotica may disappear in Ophidia.
[2] Primarily the orbito-nasal branch of the profundus enters the capsule
through the olfactory foramen, laterally to the olfactory nerve. In Pisces and
Amphibia, owing to the opening for the profundus becoming cut off from the
olfactory foramen by the extensive fusion of the vertical side wall (orbital
cartilage) with the planum antorbitale, the orbito-nasal foramen leads directly
from orbit to capsule. The less developed spheno-ethmoidal commissure in
Amniota arches over a large orbito-nasal fissure through which the profundus
runs to the olfactory foramen, passing across the corner of the brain-cavity
in Mammals.

Amphibia and Mammalia a jugular foramen serves for both. A varying number of segmental hypoglossal foramina remain for the roots of the 12th nerve.

The Chondrocranium in the Mammalia.—Certain special features distinguishing the Mammalia remain to be considered (Matthes, **544**). Con-

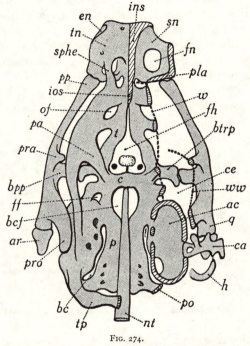

FIG. 274.

Diagram of the chondrocranium and first two visceral arches of a primitive *Tetrapod* ; dorsal view. On left a horizontal cut has removed dorsal parts.   *ac,* Auditory capsule ; *ar,* articular end of Meckel's cartilage ; *bc,* basicapsular fissure ; *bcf,* basicranial fenestra ; *bpp,* basal process ; *btrp,* basitrabecular process ; *c,* crista sellæ ; *ca,* columella auris ; *ce,* cavum epiptericum ; *en,* external nostril ; *ff,* facial foramen ; *fh,* foramen hypophyseos containing pituitary body and internal carotids ; *fn,* internal nostril ; *h,* top of hyoid cornu ; *ins,* median internasal septum ; *ios,* median interorbital septum ; *nt,* notochord ; *of,* optic foramen ; *p,* parachordal plate ; *pa,* pila antotica ; *pla,* planum antorbitale ; *po,* pila occipitalis ; *pp,* palatine process of palatopterygoid ; *pra,* processus ascendens ; *pro,* processus oticus ; *q,* quadrate ; *sn,* solum nasi ; *sphe,* commissura spheno-ethmoidalis ; *t,* trabecula ; *tn,* tectum nasi ; *tp,* tectum posterius ; *w,* limit of cranial cavity ; *ww,* limit of cavum epiptericum.

siderable doubt exists as to the development of trabeculae in mammals. Although typical paired trabeculae were described by Parker in the pig (**561**), this observation has not been confirmed in this or other forms. Certainly the base of the prechordal cranium appears as a rule in mammals in two regions : as paired islands of cartilage lateral or postero-lateral to the hypophysis (Gaupp, **511** ; Fawcett, **489-92** ; Noordenbos, **506**), and as a median rod extending from the hypophysis to between the nasal sacs.

Later the posterior cartilages join across and spread so as to floor the pituitary fossa, and unite with the anterior rod in front and the basal plate behind. The paired posterior elements have been considered as trabeculae by Levi (**364**), Fawcett, and others ; but Noordenbos compares them to

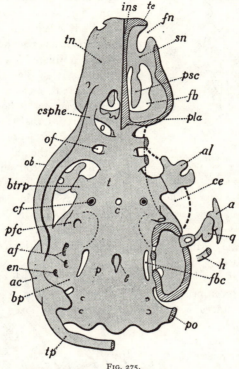

Fig. 275.

Diagram of the *chondrocranium* of a *placental mammal* ; dorsal view, with lateral parts somewhat flattened out. On left side dorsal parts removed by horizontal cut. *a*, Articular (malleus) ; *af*, foramina for auditory nerve ; *al*, ascending lamina of ala temporalis (epipterygoid) ; *b*, posterior basicranial fenestra ; *bp*, posterior basicapsular commissure ; *btrp*, basitrabecular process (prac. alaris) ; *c*, hypophysial foramen ; *ce*, cavum epiptericum (limited by thick broken line laterally, and dotted line medially) ; *cf*, carotid foramen (closed laterally by trabeculo-cochlear commissure) ; *en*, foramen endolymphaticum ; *fbc*, fissura basicapsularis ; *h*, hyoid ; *ins*, internasal septum ; *o*, orbito-nasal fissure ; *ob*, orbital (ala orbitalis), with pila preoptica and pila postoptica enclosing optic foramen, *of* ; *p*, parachordal basal plate ; *q*, quadrate (incus) ; *t*, trabecular plate ; *c*, cupula anterior ; *tn*, tectum nasi ; *tp*, tectum posterius. Other letters as in Fig. 274.

the polar cartilages of van Wijhe. They lie, however, at first between the internal carotids. Certainly this region of the skull in mammals is difficult to interpret. Whereas in all other Craniates the internal carotids primitively enter the cranial cavity from below on the median or inner side of the trabeculae, in the mammals the relative position of these parts seems at first sight to be reversed, since the carotids pass laterally to the central 'polar' or posterior 'trabecular' plate just described, and the

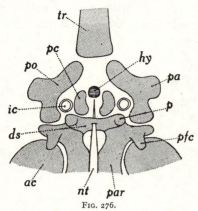

FIG. 276.

Diagram of cartilaginous elements of base of skull of *Mammal*; dorsal view. *ac*, Auditory capsule; *ds*, dorsum sellae (acrochordal cartilage); *hy*, hypophysis; *ic*, internal carotid; *nt*, notochord; *p*, point of attachment of pila antotica in Monotreme; *pa*, processus alaris (basitrabecularis); *pc*, cartilage on floor of sella turcica; *pfc*, prefacial commissure; *po*, posterior trabecular element (=polar cartilage?) sending comm. alicochlearis backwards; *tr*, anterior trabecular element.

carotid foramina are later completed by the growth from the trabecula of a commissura alicochlearis or trabeculo-cochlearis anterior, passing on the outer (lateral) side of the artery to join the cochlear region of the auditory capsule, Figs. 275-8. The unlikely view has been held by Voit that this region of the carotid in the mammal is not homologous with the internal carotid of other Craniates, a loop having been formed and the old vessel replaced by one on the outside. The commissura alicochlearis would then be secondary (Voit, 394; Toeplitz, 391). Gaupp, who describes paired trabeculae in the hypophysial region in *Echidna*, maintains that the internal carotid has cut its way

through the base of the trabecula to a position more lateral. But inspection of early stages in *Ornithorhynchus* [1] and Marsupials favours the view that the relations of the parts are essentially the same in Mammals as in other Craniates; and that in the former the basal pituitary region has been broadened, the carotids have been separated by the cartilaginous floor of the pituitary fossa now

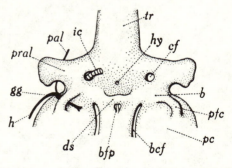

FIG. 277.

Diagrammatic reconstruction of central region of base of skull of *Placental Mammal*; dorsal view of later stage than that shown in Fig. 276. *b*, Anterior lateral region of parachordal which joins acrochordal and polar cartilages; *bcf*, basicapsular fissure; *bfp*, basicranial fontanelle; *cf*, carotid foramen; *gg*, geniculate ganglion on facial nerve; *h*, hyomandibular, and *pal*, palatine branches of facial; *hy*, vestige of foramen hypophyseos; *pral*, processus alaris (pr. basitrabecularis); *pc*, cochlear region of auditory capsule.

developed from separate elements (polar of Noordenbos, and trabeculae

---

[1] The trabeculae seem to be distinctly lateral to the carotids in the early procartilaginous stage in *Ornithorhynchus* (E.S.G.).

of other authors), and that the anterior trabeculo-cochlear commissure (commissura alicochlearis) really represents the original base of the trabecula (that region sometimes developed from a separate 'polar'

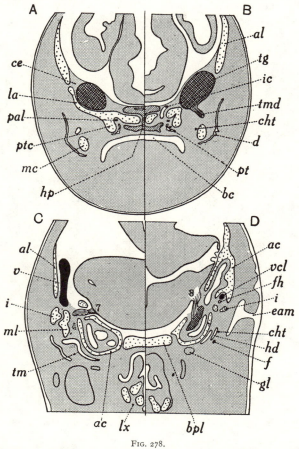

FIG. 278.

Perameles nasuta. Embryo of 7 mm. head-length. A-D, Four transverse half-sections of head; A, most anterior, D, most posterior. *ac,* Auditory capsule ; *al,* ala orbitalis ; *bc,* buccal cavity ; *bpl,* basal parachordal plate ; *ce,* cavum epiptericum ; *cht,* chorda tympani ; *d,* dentary ; *eam,* external auditory meatus ; *f,* hyoid branch, and *fh,* hyomandibular branch of facial nerve ; *gl,* glossopharyngeal ; *hd,* hyoid cornu ; *hp,* hypophysial stalk piercing basal plate ; *ic,* internal carotid ; *la,* lamina ascendens (pr. ascendens) ; *lx,* larynx ; *mc,* Meckel's cartilage ; *ml,* malleus ; *pal,* palatine nerve ; *pt,* pterygoid ; *ptc,* pterygoid cartilage ; *tg,* trigeminal ganglion ; *tm,* tympanic membrane ; *tmd,* mandibular nerve ; *v,* vein ; *vcl,* vena capitis lateralis.

element, see p. 237).   In Placentals especially it appears to have been pushed outwards, and so come to unite more with the capsule than with the basal plate.   According to this interpretation the trabeculae in the Mammalia preserve their usual position relative to the carotids, though it is possible that these have become to some extent surrounded by the trabecular cartilage; certainly cartilage develops between them, Fig. 276.

In the ethmoid region, the nasal capsule extends backwards along the lengthening internasal septum, while the interorbital septum becomes correspondingly shortened. As Gaupp has shown (343, 509), the planum

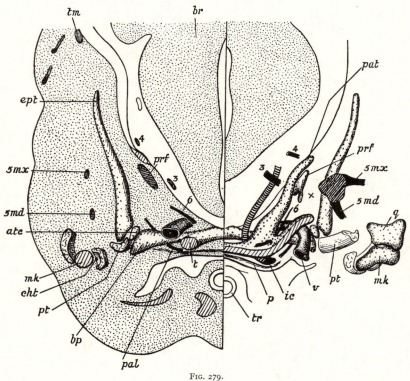

FIG. 279.

Partially reconstructed thick transverse section of head of late embryo *Lacerta*, posterior view cut through base of trabecula on left, and through parachordal plate on right. *atc*, Pterygoid cartilage (meniscus pterygoideus); *bp*, basitrabecular process; *br*, brain; *cht*, chorda tympani; *ept*, processus ascendens (epipterygoid cartilage); *ic*, internal carotid; *mk*, Meckel's cartilage; *p*, parasphenoid; *pal*, palatine nerve; *pat*, pila antotica; *prf*, profundus; *pt*, pterygoid; *q*, quadrate; *t*, cut end of trabecula; *tm*, taenia marginalis; *tr*, trachea; *v*, vena capitis lateralis; 3, 4, 5, 6, cranial nerves.

antorbitale forms now an oblique floor below the olfactory lobes (lamina infracribrosa). The cartilaginous cranial wall comes thus to include a small space, originally part of the orbit in the Reptile, which forms the extreme corner of the brain-cavity into which the orbito-nasal foramen now directly opens. The orbito-nasal branch of the profundus nerve thus passes through the cranial cavity to reach the cribriform plate and enter the nasal capsule, Figs. 264, 275, 281, and p. 258.

The optic foramen is closed by a pila preoptica and a pila postoptica (radix anterior and posterior of the 'ala orbitalis') in the Placentalia; but in the Monotremata and the Marsupialia the pila postoptica fails to develop. Consequently in these a large fenestra pseudoptica arises by the confluence of the foramen opticum with the foramen postopticum, Figs. 264, 275, 282.

Cavum epiptericum.—More important is the disappearance in Mammalia of the lateral cranial wall farther back which leads to the inclusion within the cranial cavity of a considerable space lying outside it in the skull of the lower Tetrapods (Gaupp, 343, 506, 508). Between the side wall of the chondrocranium on the inside and the palatoquadrate with its processus ascendens on the outside there is enclosed a space partially floored by the basitrabecular and basal processes and named by Gaupp the cavum epiptericum which it is important to recognise. It has one anterior, one lateral, and one posterior opening. Originally an extracranial space, it is derived from the posterior region of the orbit. Through it pass the internal jugular vein, the orbital and facial arteries, and the profundus, trigeminal and facial nerves, whose ganglia lie typically in it, Figs. 271, 274, 279, 280 A.

The anterior opening serves for the entrance of the internal jugular vein (v. capitis lateralis) or its branches, and the exit of the profundus nerve (also usually the abducens and trochlear nerves), which, therefore, passes out anteriorly to the ascending process. The more posterior lateral opening, between the ascending and the otic processes, serves for the exit of the maxillary and mandibular branches of the trigeminal (together with the superficial ophthalmic and buccal branches of the facial in Dipnoi and those Amphibia in which these lateral line nerves persist) and the branches of the facial artery, Fig. 261. The posterior opening, bounded below by the palatobasal articulation and above by the otic process, lets through the palatine nerve below, and the hyomandibular branch of the facial and internal jugular vein behind; it also lets in the facial artery. This opening, then, is the cranio-quadrate passage (p. 412). The disposition of these various structures entering and leaving the cavum epiptericum remains fundamentally unchanged throughout the Dipnoi and Tetrapoda, though considerably modified in many forms. The cavum epiptericum remains well defined in those Reptiles which have a well-developed upstanding epipterygoid, such as the Rhynchocephalia and Lacertilia; but tends to merge again with the orbit in those where the processus ascendens is reduced, such as the Aves, the Chelonia, and Crocodilia among Reptilia, and the Amphibia and lower forms (see further, Chapter VII.).

In the mammal, as the brain expands, the original side wall behind the optic region ceases to chondrify, and is reduced to a mere membrane through which pass the nerves into the cavum epiptericum. The latter

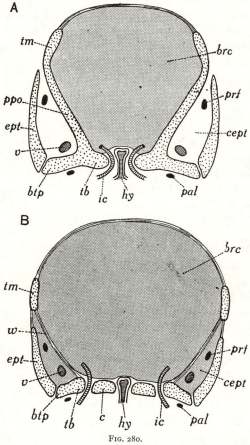

FIG. 280.

Diagrammatic transverse sections of orbito-temporal region of skull to show inclusion of extra-cranial cavum epiptericum of Reptile, A, in cranial cavity of Mammal, B. *brc*, Primitive brain-cavity ; *btp*, basitrabecular process ; *c*, cartilaginous floor of sella turcica ; *cept*, cavum epiptericum ; *ept*, processus ascendens (reptilian epipterygoid, mammalian lamina ascendens) ; *hy*, hypophysial stalk ; *ic*, internal carotid ; *pal*, palatine nerve ; *ppo*, pila antotica ; *prf*, profundus nerve ; *tb*, trabecula ; *tm*, taenia marginalis ; *v*, vena capitis lateralis ; *w*, membrane representing original side wall.

is now a part of the cranial cavity, since a new lateral wall becomes established at the level of the processus (Gaupp, 508 ; Broom, 327a, b). This new wall is represented by the lamina ascendens of the ala temporalis (p. 270), and a membrana spheno-obturatoria stretching from it, which may later become ossified. Thus the old nerve foramina, now mere holes in the inner membrane, are replaced by new fissures and foramina in the outer wall ; these become clearly defined when it is ossified. They are known as the sphenoidal fissure or foramen lacerum anterius, for the profundus (together with the 3rd, 4th, and 6th nerves) anterior to the lamina ascendens, the foramen rotundum for the maxillary branch of the trigeminal nerve

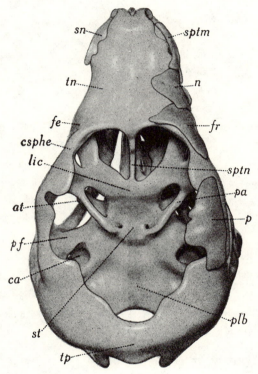

Fig. 281.

Dorsal view of skull of embryo *Echidna aculeata*, var. typica (from Gaupp-Ziegler model). Dermal bones on right. *at*, Ala temporalis ; *ca*, cavity of cochlear region of auditory capsule ; *csphe*, commissura spheno-ethmoidalis of orbital cartilage ; *fe*, foramen epiphaniale ; *fr*, frontal ; *lic*, lamina infracribrosa ; *n*, nasal ; *p*, parietal ; *pf*, prefacial commissure ; *pa*, pila antotica ; *plb*, planum basale ; *sn*, solum nasi ; *sptm*, septomaxillary ; *sptn*, septum nasi ; *st*, hind wall of sella turcica *tn*, tectum nasi ; *tp*, tectum posterius.

either through or in front of the lamina, and the foramen ovale for the mandibular branch farther back.

There is no better evidence of the fundamentally primitive character of the chondrocranium in Monotremes, in spite of its specialisations, than the fact described by Gaupp (**511**) and Wilson (**398**), that they alone among mammals preserve complete the pila antotica (taenia clino-orbitalis) of the original wall, Figs. 281-2. This pila has disappeared in the Ditremata, its last vestiges being perhaps represented by certain small cartilages found above the Gasserian ganglion in *Lepus, Felis,* and other mammals, Figs. 275, 282 (Voit, 394-5; Terry, 390).

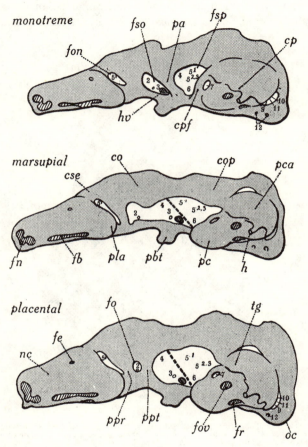

FIG. 282.

Diagrams of chondrocranium in *Mammalia*, left-side view. Exit of cranial nerves marked by Nos. 2-12. *co,* Orbital cartilage; *cop,* commissura orbito-parietalis; *cp,* crista parotica; *cpf,* comm. prefacialis; *cse,* comm. spheno-ethmoidalis; *fb,* fenestra basalis; *fe,* foramen epiphaniale; *fn,* fenestra narina; *fo,* fenestra optica; *fon,* fissura orbito-nasalis; *fov,* fenestra ovalis; *fr,* fenestra rotunda; *fso,* fenestra pseudo-orbitalis; *fsp,* fenestra pro-otica; *h,* laterohyal; *hv,* hypophysial vein; *nc,* nasal capsule; *oc,* condyle; *pa,* pila antotica, represented by broken line in Marsupial and Placental; *pbt,* processus alaris or basitrabecularis; *pc,* pars cochlearis, and *pca,* pars semicircularis of auditory capsule; *pla,* planum antorbitale; *ppr,* pila preoptica; *ppt,* pila metoptica; *tg,* tegmen tympani.

Ala temporalis.—The validity of this explanation of the fate of the cavum epiptericum in the mammalian skull depends to a great extent on our interpretation of the homology of the ala temporalis. The more horizontal region of this structure (processus alaris) is developed as an out-growth from the posterior trabecular rudiment, Figs. 275, 276-8. The 5th nerve and its ganglia and the vena capitis lateralis lie above it; the palatine arises from the facial behind it, and passes forward (joining a sympathetic nerve) as the vidian nerve below it. It thus has the same morphological relation to surrounding structures as the basitrabecular process of lower Gnathostomes, and is doubtless homologous with it (Gaupp, 508). This conclusion is generally accepted. But Broom sees in the mammalian lamina ascendens (later alisphenoid bone) the homologue of the Reptilian epipterygoid, Fig. 280, while Gaupp considered this lamina to be a new development scarcely begun in the Monotreme. The chief reason why Gaupp and others reject the homology with the processus ascendens is because this process in lower forms always separates the profundus from the maxillary and mandibular branches of the trigeminal nerve; whereas, in the Mammalia, the lamina of the ala temporalis usually grows up between the maxillary and mandibular branches, Figs. 449, 495. But, since in many mammals (both Marsupials, Fig. 264, and Placentals, Fig. 497) the lamina may pass also between the maxillary branch and the profundus, so as to enclose the latter in a foramen rotundum, the difference may be more apparent than real.[1] It may be supposed that the mammalian processus ascendens spread backward so as to pass on both sides of the maxillary nerve, and that then the anterior limb disappeared while the posterior persisted. Moreover, the lamina ascendens frequently develops separately from the basal processus alaris (Broom: in Marsupials. Levi, Fawcett: in Man. Wincza, Noordenbos, and others: in other Placentals). These facts, together with the evidence from the adult skull of the Theromorpha that the alisphenoid bone of the Mammal is derived from the epipterygoid of the Reptile, strongly support the view that the ala temporalis is formed of two elements: the basi-trabecular process from the cranium and the processus ascendens (the only remains in this region of the palatoquadrate bar; see Chapter VII.).

Other characteristics are found in the otic region of the mammalian chondrocranium. Speaking generally, the brain and brain-cavity are greatly enlarged; and as the brain bulges more and more at the sides, the auditory capsules, instead of standing upright, acquire a more hori-zontal position below the cranium (already seen in Theromorph Reptiles). Accompanying the great development of the cochlea this part of the capsule encroaches on the basal plate. Nevertheless, the pars cochlearis

---

[1] According to Fuchs (503), who supports Broom's view, the processus ascendens in young stages of *Didelphys* occupies the position of the reptilian epipterygoid; but in later stages, at all events, it surrounds the maxillary nerve as in many other mammals (Esdaile, 722; Toeplitz, 391).

chondrifies as a rule separately from the plate (Broom, 327b; Noordenbos, 370; de Beer, 442a), and doubtless belongs not to the basal plate, as Gaupp supposed, but to the capsule itself, Fig. 278. As already explained, it undermines the prefacial commissure, which comes to lie above it.[1] The facial nerve, issuing from the primary foramen below the commissure, bears a geniculate ganglion, gives off the palatine branch, which issues from the cavum epiptericum below, and continues as the hyomandibular branch round the capsule in a sulcus facialis. Posteriorly the sulcus is overhung by the crista parotica, anteriorly in the Ditremata by a new formation the tegmen tympani (Parker, Van Kampen, 741; see p. 465). The nerve leaves the sulcus by the primary stylo-mastoid foramen behind the region where the hyoid joins the crista, Fig. 277.

The side of the brain-case in the auditory region may be considerably strengthened by a cartilaginous extension from the orbito-parietal commissure to form a large 'parietal plate'. This may join the tectum synoticum, and between them the capsule is left a fissure (foramen jugulare spurium) for the passage of branch of the jugular vein.

Orbito-temporal Region.—There remain to be considered in various groups certain important specialisations of the orbito-temporal region. The fundamental constant relations of the cartilages to certain nerves and blood-vessels must first of all be understood, Figs. 245-7.

The jugular vein (p. 535) runs back close to the lateral wall of the skull, passing from the orbit, where it receives the hypophysial vein in front of the pila antotica. It passes dorsally to the subocular shelf (basitrabecular process), medially to the ascending process of the palato-quadrate, then through the cranio-quadrate passage (p. 412). The internal carotid passes forwards below the basal plate (partly enclosed in a para-basal canal between the plate and the parasphenoid in those forms where this bone is well developed); and, having given off an orbital artery which escapes outwards through the cranio-quadrate passage, the internal carotid passes inwards and upwards into the cranial cavity between the trabecula and the hypophysis, Figs. 261, 263.

The basitrabecular process, typically developed in Lacertilia, Figs. 263, 271, but present in a more or less modified form in all Tetrapoda, Osteichthyes, and probably also in Chondrichthyes, is anterior to the palatine nerve, which passes down behind it and then forwards below it, Fig. 284. As suggested by Veit (392-3), this process has probably been derived phylogenetically from that region of the subocular shelf, immediately in front

---

[1] A small space lodging the facial ganglion, known as the cavum supra-cochleare, originally outside the cranial cavity, may become included in it by the ossification of its outer wall.

of the palatine nerve, which occurs in most Selachii as an outward extension of the trabecula. In these fish the shelf may extend backwards and
enclose a foramen for the orbital artery and also more rarely a foramen
for the palatine nerve, Fig. 284. As a rule, however, this nerve passes
down outside the edge of the shelf. In most of the higher Teleostomes
the basitrabecular process is reduced and no longer meets the basal process of the palatoquadrate; but in *Lepidosteus,* among living forms, in
Crossopterygii (*Megalichthys,* Watson, 644; *Eusthenopteron,* Bryant, 465),
Coelacanthini (Stensiö, 605-6), and primitive Chondrostei and Amioidei

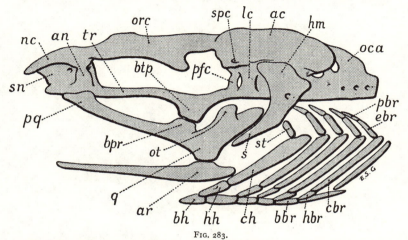

FIG. 283.

Diagram of chondrocranium and visceral arches of primitive Teleostome leading to Holostean type.
ac, Auditory capsule; *an*, antorbital process; *ar*, articular region; *bbr*, basibranchial; *bh*, basihyal;
bpr, basal process; *btp*, basitrabecular process; *cbr*, ceratobranchial; *ch*, ceratohyal; *ebr*, epibranchial; *hbr*, hypobranchial; *hh*, hypohyal; *hm*, hyomandibular; *lc*, lateral commissure; *nc*,
nasal capsule; *oca*, occipital arch; *orc*, orbital; *ot*, otic process; *pfc*, prefacial commissure; *pbr*,
pharyngobranchial; *pq*, palatine region of palatoquadrate; *q*, quadrate region; *s*, symplectic region;
sn, septum nasi; *spc*, spiracular canal through postorbital process; *st*, stylohyal; *tr*, trabecula.

it seems to have been well developed and to have articulated with the
pyterygoid region of the palatoquadrate, Fig. 283 (Chapter VII.).

Trigemino-facialis Chamber.—Having described these relations, we
may now deal with the trigemino-facialis chamber. Allis has given
this name to a space occurring in the side wall of the skull of Pisces
immediately in front of the auditory capsule (406-7). It is typically
developed in the higher Teleostomi, and may be described in the
Teleost. In Scomber, for instance (Allis, 404), is found a short horizontal canal in the pro-otic with anterior and posterior openings. The
jugular vein (v. capitis lateralis) passes through it from in front; the
orbital artery passes through it from behind. Out of its anterior
opening issue the trigeminal nerve with the lateral-line branches of

T

Heterodontus

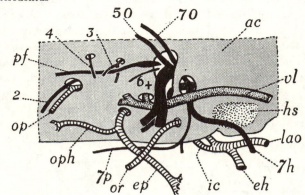

Squalus

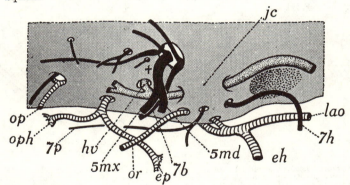

Polypterus

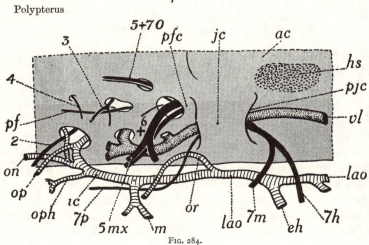

Fig. 284.

Lepidosteus

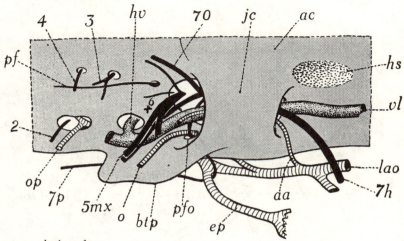

Amia embryo

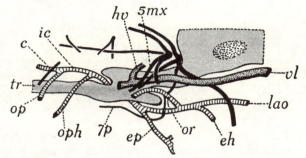

Amia, late stage

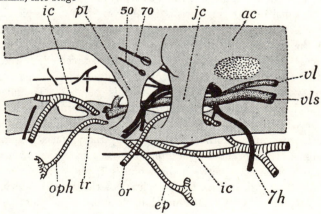

FIG. 284.

Salmo

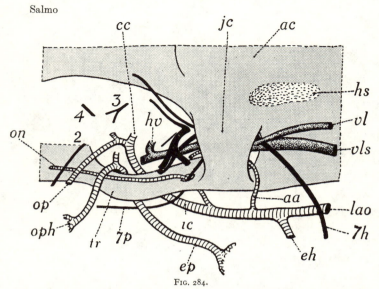

FIG. 284.

Diagrams of left side of orbito-temporal region of skull of fishes, showing relation of cartilages (stippled) to nerves (black), veins (dark), and arteries (cross-lined). Nerves : 2, Optic ; 3, oculomotor ; 4, trochlear ; 5mx, maxillary, and 5o, ophthalmic branches of trigeminal ; 6, abducens ; 7h, hyomandibular, 7m, mandibular, 7p, palatine ; 7o, superior ophthalmic branches of facial ; pf, profundus. Arteries : aa, o, and on (orbito-nasal) seem to represent the orbital and its branches in *Lepidosteus* and *Salmo* ; eh, efferent epibranchial of hyoid arch ; ep, efferent pseudobranchial ; c, cerebral ; cc, circulus cephalicus ; ic, internal carotid ; lao, lateral aorta ; m, mandibular ; op, optic ; oph, ophthalmic ; or, orbital. Veins: hv, Hypophysial ; vl, vena capitis lateralis and medialis = jugular or head vein ; vls, inner branch of same. Cartilages : ac, Auditory capsule ; hs, articular facet for hyomandibular ; jc, wall of jugular canal or lateral commissure ; pfc, prefacial commissure ; pfo, palatine foramen ; pjc, posterior opening of jugular canal ; pl, pila lateralis ; tr, trabecula. A + marks pila antotica. (Goodrich and de Beer.)

the facial. The hyomandibular branch of the facial issues through its posterior opening (with the sympathetic and ramus communicans from facial to trigeminal). Since the canal contains the jugular vein it is called a jugular canal or pars jugularis. The nerves enter it by piercing the bony wall separating the canal from the cranial cavity, and on the inner face of this wall is a recess lodging the trigeminal and facial ganglia. This recess, limited internally by the dura mater pierced by the nerve roots, is the pars ganglionaris. The pars jugularis and pars ganglionaris make up what Allis calls the trigemino-facialis chamber. To understand its morphology we must briefly describe it in other forms, Figs. 283-4.

In Selachii the main branches of the trigeminal and facial nerves emerge by foramina primitively separated by the prefacial commissure. The ganglia of these and the auditory nerve lie in a shallow acustico-trigemino-facialis recess between the commissure and dura mater, which latter may be considered as the true limit of the· cranial cavity. The ganglia are, then, strictly intramural in position. The palatine nerve branches off outside the facial foramen. The vena capitis lateralis, in early stages, runs freely above the subocular shelf and outside the auditory capsule, but later becomes enclosed in many Selachians (*Squalus*, Fig. 184) in a short jugular canal apparently formed by the upward growth of the hinder region of the subocular shelf which fuses above the vein with the auditory capsule (de Beer, 421). Meanwhile the shelf has also grown up so as to separate the original facial opening into an anterior palatine and a posterior hyomandibular foramen separated by a postpalatine commissure. Selachians show a condition in which the pars jugularis is still separated from the pars ganglionaris by the cranial wall (prefacial commissure).

Lepidosteus (Allis, 421; Veit, 392; de Beer, 421) shows an interesting structure in this region. It develops a prepalatine (basitrabecular process) and postpalatine subocular shelf. The two combine to close the palatine foramen, and grow up outside the vena capitis lateralis and nerves to join the capsule, thus forming a commissura lateralis or outer wall of the pars jugularis. This lateral commissure is in many Teleostomes strengthened by a lateral wing of the parasphenoid. The prefacial commissure is no longer developed in *Lepidosteus*, and the pars jugularis thus becomes confluent with the pars ganglionaris. The chamber then contains the vein which runs through it, the orbital artery which enters the palatine foramen, and the ganglia of the trigeminal and facial nerves. The branches of the trigeminal (and lateral-line branches of the facial) issue from its anterior opening, the palatine through its floor, the hyomandibular branch through its posterior opening. In *Amia* also the jugular canal and trigemino-facial recess are confluent; but in Teleostei they again become separated by the ossification of the intervening membrane, Fig. 733.

The recess and jugular canal in *Polypterus* (Allis, 410; Lehn, 542; de Beer, 421) and Palaeoniscoidei (Stensiö, 218) are in about the same condition as in Selachii; but in modern Chondrostei the prefacial commissure seems to have been lost (de Beer, 421a).

To sum up concerning the trigemino-facialis chamber of the Actinopterygii. It is formed by the confluence of an intramural recess (pars ganglionaris) with an extramural jugular canal (pars jugularis) due to the disappearance of the cartilaginous cranial wall which separates them in lower Pisces. Its outer wall is formed by the upgrowth of the subocular shelf. In Teleostei the two parts become secondarily separated by a bony septum.

The orbito-temporal region in the Dipnoi and Tetrapoda has evolved along different lines; for in these, as explained elsewhere (p. 267), there is formed a cavum epiptericum containing the jugular vein, orbital artery, trigeminal and facial ganglia. But although it happens to contain the same parts as the trigemino-facial chamber, and to be to some extent floored by the basitrabecular process, the cavum epitericum differs radically from the chamber in being a purely extracranial space limited externally by the palatoquadrate arch.

Nevertheless, there is developed in the modern Amphibia a structure somewhat resembling the piscine chamber. For in these Tetrapods there

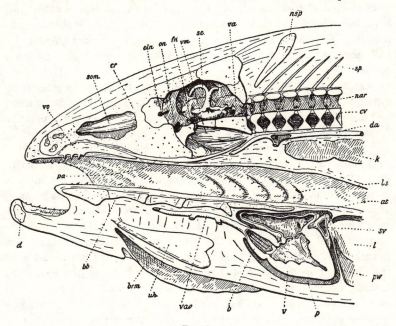

FIG. 285.

Median longitudinal section through the head of *Salmo salar*, L. (Modified, after Bruch.) *at*, Atrium; *b*, bulbus arteriosus; *bb*, basibranchial, *brm*, branchiostegal membrane; *bs*, branchial slit; *cr*, cranial cartilage; *cv*, vertebral centrum; *d*, dentary; *da*, dorsal aorta; *fn*, facial nerve; *k*, kidney; *l*, liver; *nar*, neural arch; *nsp*, enlarged radial; *oln*, olfactory nerve; *on*, optic nerve; *p*, pericardium; *pa*, parasphenoid; *pw*, septum between pericardial and abdominal coelom; *sc*, anterior vertical semicircular canal; *som*, superior oblique muscle of eye; *sp*, neural spine; *sv*, sinus venosus; *uh*, urohyal; *v*, ventricle—valves separate its cavity from that of atrium above and bulbus in front; *va*, vagus nerve; *vao*, ventral aorta; *vm*, rectus muscle of eye in eye-muscle canal; *vo*, prevomer. (From E. S. Goodrich, *Vert. Craniata*, 1909.)

is a subocular shelf forming a ' prepalatine' basitrabecular process and a postpalatine commissure in the floor of the cavum epitericum and enclosing the palatine nerve in a foramen. As the auditory capsule extends very far forwards it overhangs the prefacial commissure and fuses with the subocular shelf, so enclosing the hyomandibular branch of the facial nerve in a canal lateral to the true facial foramen (Urodela, Figs. 261, 272-3). The basitrabecular process disappears in Anura, leaving the palatine nerve free in front of the postpalatine commissure.

Myodome in Pisces.—Of considerable interest is the myodome, a space developed in the orbito-temporal and otic regions of the skull of Teleostomes for the accommodation of lengthened recti muscles of the eye. Strictly speaking, this space is the posterior myodome, since a similar anterior myodome is hollowed out for the oblique muscles in the ethmoid region by the enlargement of the orbito-nasal canal (p. 245). The structure and origin of the myodome has been studied by many modern anatomists

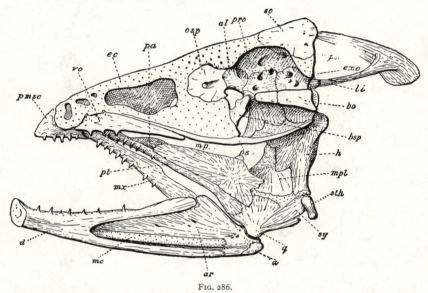

FIG. 286.

Skull of *Salmo salar*, L., cut longitudinally. (After Bruch.)   *a*, Angular; *al*, pterosphenoid; *ar*, articular; *bo*, basioccipital; *bsp*, basisphenoid; *d*, dentary; *ec*, ethmoid cartilage; *exo*, exoccipital; *h*, hyomandibular; *li*, ligament; *mc*, Meckel's cartilage; *mpl*, metapterygoid; *mp*, mesopterygoid; *mx*, maxilla; *osp*, orbitosphenoid; *p*, post-temporal; *pa*, palatine; *pmsc*, premaxilla; *pro*, pro-otic; *ps*, parasphenoid; *pt*, pterygoid; *q*, quadrate; *so*, supraoccipital; *sth*, stylohyal; *sy*, symplectic; *vo*, prevomer. (From E. S. Goodrich, *Vert. Craniata*, 1909.)

(Gegenbaur, 516; Sagemehl, 378; Gaupp, 343; de Beer, 421-2; and more especially Allis, 402, 404-5, 409, 413).

When fully developed, as in *Salmo*, the myodome in the dry skull is a large space between the floor of the brain-case (pro-otic and basioccipital) and the parasphenoid; it opens behind, and communicates in front with the orbits, Figs. 285-7. The myodome is supposed to have originated by the penetration into the enlarged opening for the pituitary vein of recti muscles originally inserted on the outer surface of the orbital wall. First the external (posterior) rectus, and later the internal (anterior) rectus, passed into the cranium at the side of the hypophysial fossa, dorsally to the trabecula and anteriorly to the pila antotica (so-called dorsal myodome).

The enlargement of this incipient myodome involves the disappearance of the pila antotica (p. 261) and confluence with the trigemino-facialis

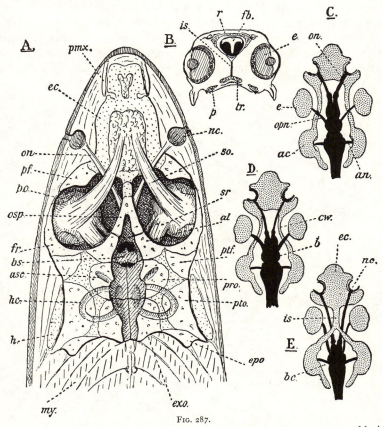

FIG. 287.

A, *Salmo salar*, L. ; longitudinal section through the head exposing the sense-organs, and brain-cavity viewed from above (after Bruch) ; B, transverse section of the head of a young *Salmo trutta*, L., in the region of the fore-brain ; C, D, and E, three diagrams showing the development of the inter-orbital septum. *ac*, Auditory capsule ; *al*, alisphenoid ; *an*, auditory nerve ; *asc*, cavity for anterior semicircular canal ; *b*, optic lobe ; *bc*, brain-cavity ; *bs*, basisphenoid ; *cw*, cranial wall ; *e*, eye ; *ec*, ethmoid cartilage ; *epo*, epiotic ; *exo*, exoccipital ; *fb*, fore-brain ; *fr*, frontal ; *h*, hyomandibular ; *hc*, cavity for horizontal semicircular canal ; *is*, interorbital septum ; *my*, myotome ; *nc*, nasal capsule ; *on*, olfactory nerve ; *opn*, optic nerve ; *osp*, orbitosphenoid ; *p*, palatine ; *pf*, prefrontal ; *pmx*, premaxilla ; *po*, ossification of optic capsule ; *pro*, pro-otic ; *ptf*, postfrontal ; *r*, cranial roof ; *so*, superior oblique muscle ; *sr*, superior rectus muscle ; *tr*, trabecula. (From E. S. Goodrich, *Vert. Craniata*, 1909.)

chamber (p. 277).[1] Extending their insertions further the muscles pass down through the hypophysial fenestra, and push their way backwards

---

[1] Owing to the confluence of myodome and chamber, the palatine nerve appears to pass through the former on its way down to its external foramen situated laterally to the original edge of the subocular shelf. The muscles enter the myodome ventrally to the jugular vein.

between the cranial floor and the parasphenoid (ventral myodome). In their course from the orbit the muscles remain at first outside the dura mater, then pass below the pro-otic bridge (p. 386), the right and left myodomal cavities being separated only by a median membrane, Figs.

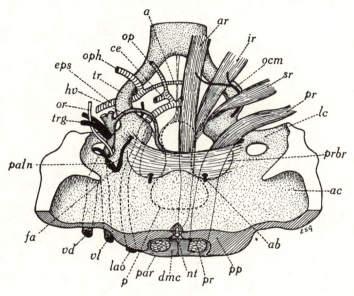

FIG. 288.

Salmo fario, embryo 22 mm. long. Reconstruction of orbito-temporal region of base of skull; blood-vessels on left, eye-muscles on right. Dorsal view. *a*, Transverse connexion between efferent pseudobranchial vessels ; *ab*, n. abducens ; *ac*, auditory capsule ; *ar*, anterior rectus muscle ; *ce*, cerebral artery ; *dmc*, dorsal chamber of posterior myodome ; *eps*, efferent pseudobranchial artery ; *fa*, n. facialis ; *hv*, hypophysial vein ; *ir*, inferior rectus muscle ; *lao*, lateral aorta ; *lc*, lateral commissure ; *nt*, tip of notochord in reduced basicranial fenestra ; *ocm*, n. oculomotorius ; *op*, optic artery ; *oph*, ophthalmic artery ; *or*, orbital artery ; *p*, dotted outline of pituitary body passing below pro-otic bridge ; *paln*, n. palatinus ; *par*, parasphenoid ; *pp*, parachordal ; *pr*, posterior rectus muscle ; *prbr*, membranous extension of pro-otic bridge ; *sr*, superior rectus muscle ; *tr*, trabecula ; *trg*, n. trigeminus ; *vd*, outer lateral vein ; *vl*, inner lateral vein.

288-9. Such in brief is supposed to have been the history of the posterior myodome.

Amia has a myodome less developed than that of typical Teleosts (Sagemehl, **378** ; Allis, **402** ; Pehrson, **372** ; de Beer, **421**). The cavity is confluent with the trigemino-facialis chamber, but only the external rectus muscle penetrates over the trabecula, and this but for a short way below the pro-otic bridge. In Teleosts (*Salmo*) the external rectus reaches farther back even to below the occipital region, and the internal rectus follows in a ventral compartment of the myodome separated from the dorsal by a horizontal septum usually membranous. The myodome is much less developed and even absent in some Teleosts (Siluridae, Anguilli-

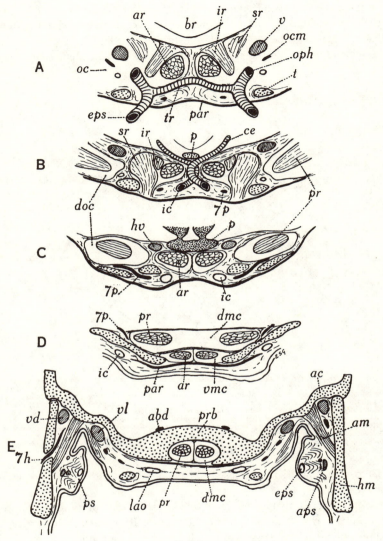

FIG. 289.

Salmo fario, embryo 28 mm. long. Series of transverse sections of basal region of head showing relations of myodome. A, Most anterior, E, most posterior section. *abd*, Abducens ; *ac*, auditory capsule ; *am*, adductor muscle ; *aps*, afferent pseudobranchial vessel ; *ar*, anterior rectus muscle ; *br*, brain ; *ce*, cerebral artery ; *dmc*, dorsal chamber of myodome—*doc*, its opening into orbit ; *eps*, efferent pseudobranchial vessel ; *hm*, hyomandibula ; *hv*, hypophysial vein ; *ic*, internal carotid ; *ir*, inferior rectus ; *lao*, lateral dorsal aorta ; *oc*, orbital cavity ; *ocm*, oculomotor ; *oph*, ophthalmic artery ; *p*, pituitary body ; *par*, parasphenoid ; *pr*, posterior rectus ; *prb*, pro-otic bridge ; *ps*, pseudobranch ; *sr*, superior rectus ; *t*, trabecula ; *tr*, transverse anastomosis ; *v*, head vein ; *vd*, vena capitis lateralis ; *vl*, secondary vein ; *vmc*, ventral chamber of myodome ; *7h*, hyomandibular, and *7p*, palatine branches of facial.

formes, some Gasterosteiformes, Gadiformes); but there can be little doubt that in these forms it has been lost. It is also absent or very little developed in *Polypterus, Lepidosteus,* and the modern Acipenseroidei; yet, since it has been described in Palaeoniscoidei (Stensiö, 218), Semionotidae (Woodward, 664; Frost, 500), and Saurichthyidae (Stensiö, 134), it very probably is an ancestral structure common to all the Actinopterygii (see p. 385).[1]

Conclusion.—From the description given above it will be seen that the chondrocranium of Gnathostomes consists essentially of a posterior basal plate to which the auditory capsules become attached, and an anterior trabecular region connected with the nasal capsules. Further, that the primitive side wall is derived from an orbital cartilage which joins the basal plate by means of a pila antotica. To these structures belonging to the neurocranium proper becomes connected the visceral palatoquadrate arch by an otic process to the auditory capsule above and a basal process to the trabecular region below (palatobasal or 'basipterygoid' articulation). On such a fundamental plan is built the skull of all Gnathostomes (with the possible exception of the palatobasal articulation in Chondrichthyes).

DERMAL BONES OF THE HEAD IN OSTEICHTHYES [2]

Covering the whole body, head, trunk, tail, and fins of Chondrichthyes are closely set dermal denticles; moreover, these penetrate by

---

[1] In spite of Stensiö's contention, it is not generally accepted that a true myodome existed in the Osteolepidoti and Coelacanthini.

[2] For the convenience of the reader a list of the names and synonyms of the chief bones of the head is given below :

Cranial dermal covering bones : Nasal, frontal, parietal, premaxillary (premaxilla), maxillary (maxilla), lacrimal (lachrymal), prefrontal, postfrontal, intertemporal, supratemporal (squamosal, suprasquamosal, supramastoid), squamosal (mastoid, supratemporal; the names squamosal and supratemporal have been interchanged, but it is now generally agreed to call the upper more dorsal bone supratemporal, and the lower outer bone squamosal); jugal (malar), quadratojugal (paraquadrate of Gaupp in Crocodile, quadrato-maxillary of Gaupp in Stegocephalia, etc.), postorbital, tabular (epiotic), post-parietal (dermo-supraoccipital, which with the tabulars may form a transverse row of bones sometimes called supratemporals), septomaxillary (may penetrate into nasal capsule). Certain of these dermal bones may have ingrowths invading the chondrocranium : the prefrontal may then be called lateral ethmoid, ectethmoid ; the postfrontal, sphenotic ; the supratemporal, pterotic.

Paired dermal bones of opercular fold and lower jaw: Opercular, preopercular, subopercular, interopercular, dentary (dentale), angular, supra-angular (surangular), splenial (opercular), postsplenial (preangular), prearticular (goniale), coronoid (complementary), lateral gulars (become branchiostegal

the mouth to the inner margin of the lips where they are modified into
teeth, and may spread further over that part of the lining of the buccal
cavity which is derived from the stomodaeum, and may also penetrate
through the gill-slits to the inner surface of the gill-bars (p. 441). Similarly,
denticles or their derivatives cover the head in Ostracodermi, and, in
those forms where underlying skeletal dermal plates are developed, these
are found especially on the head region where they may form large shields
arranged in definite patterns.   In the same way the bony covering of the
head, so characteristic of the Osteichthyes and their descendants the
Tetrapoda, is formed of plates originally of just the same histological
structure as the body scales and of the same complex origin.   They are
made up by the combination of superficial denticles with underlying
bony plates separately developed in the lower layers of the dermis.
In the early and more primitive Teleostomes (Osteolepids) and Dipnoi
(*Dipterus*) the head plates, like the body scales, are of cosmoid structure ;
but the outer cosmoid layer tends to disappear, leaving in later forms
and their modern survivors only bone (p. 304).   The early and more
primitive Actinopterygii likewise started with head plates similar in
structure to the ganoid scales ; and in this group also the plates soon
become simplified, losing the outer ganoine and retaining only the deeper
bony layers (p. 294).   The denticles may also be present (Goodrich, 35).

   Primitively these scales and plates in Osteichthyes formed a complete
covering to the head, leaving only openings for the mouth, nostrils, paired
eyes, pineal eye, spiracles, and gills.   They acquire a larger size than the
body scales either by growth or by fusion, and are usually more rigidly
connected with each other ; allowing, however, for the necessary motion
of parts, such as the lower jaw and opercular fold covering the openings
of the branchial slits.   The chief dermal bones are relatively few in number

---

rays).   Median dermal bones : Internasal or rostral (median ethmoid), interfrontal, interparietal (probably not homologous in fishes and Tetrapods),
intergular.
   Paired dermal bones of palate : Prevomer (vomer, anterior paired vomer),
palatine, ectopterygoid (transverse, transpalatine), endopterygoid (mesopterygoid, pterygoid of Tetrapod).   Median parasphenoid (parabasal basi-
temporal, vomer of mammal).
   Ossifications of chondrocranium : supraoccipital, exoccipital (lateral
occipital, pleuroccipital), basioccipital, opisthotic (paroccipital), pro-otic
(otosphenoid, together with opisthotic probably = mammalian petrosal or
periotic), epiotic of Teleostome, basisphenoid, orbitosphenoid, presphenoid of
doubtful occurrence, mesethmoid, pleurosphenoid (' alisphenoid ', laterosphenoid of Crocodile and bird), pterosphenoid of Teleostome.   Endochondral
bones of palatoquadrate are the quadrate, metapterygoid, autopalatine, and
epipterygoid = alisphenoid of mammal.

and disposed according to a regular and bilaterally symmetrical pattern. A fundamental plan can be made out common to all the Osteichthyes and even to the Tetrapoda, although variable in detail and subject to much modification owing to divergent specialisation in various groups. For the identification and comparison of individual bones and the successful tracing out of homologies in spite of such modifications, it is important to notice that the distribution of the lateral-line organs and

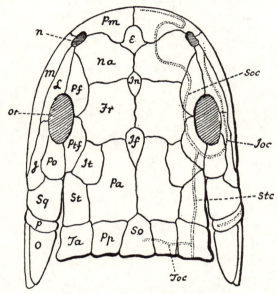

FIG. 290.

Diagram of fundamental plan of roofing-bones of skull in *Pisces*. Dorsal view of primitive representative of Osteichthyes (E. S. Goodrich, *Linn. Soc. J. Zool.*, 1925). *E*, Ethmoid or median rostral; *Er*, frontal; *If*, interfrontal; *In*, internasal; *Ioc*, infraorbital canal; *J*, jugal; *L*, lacrimal; *M*, maxillary; *N*, external nostril; *Na*, nasal; *O*, opercular; *Or*, orbit; *P*, preopercular; *Pa*, parietal; *Pf*, prefrontal; *Pm*, premaxillary; *Po*, postorbital; *Pp*, postparietal; *Ptf*, postfrontal; *So*, dermal supraoccipital; *Soc*, supraorbital canal; *Stc*, postorbital and temporal canal; *Sq*, squamosal; *Ta*, tabular; *Toc*, transverse occipital canal. Course of lateral-line canals shown on right side.

course of the lateral-line canals are on the whole remarkably constant in all the aquatic Gnathostomes. The organs also conform to a common plan with supratemporal, transverse occipital, supraorbital, infraorbital, postorbital, jugal, preopercular, oral and mandibular canals on the head (Chapter XIV.). Now, just as the lateral-line canal of the trunk comes to pierce a row of scales, so the canals of the head become embedded in dermal bones ; consequently a relation becomes established between a canal and certain particular bones lying along its course. Although subject to minor alterations among more specialised forms, and although the lateral-line

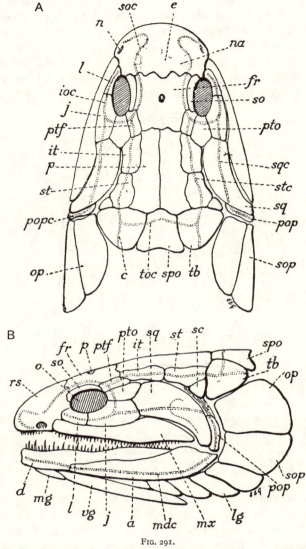

FIG. 291.

Restoration of head of *Osteolepis macrolepidotus*, dorsal view, A; left-side view, B. Course of lateral-line canals in bones shown in dotted lines (E. S. Goodrich, *Linn. Soc. J. Zool.*, 1919). *a*, Angular; *c*, main trunk canal; *d*, dentary; *e*, ethmoid included in rostral shield; *fr*, frontals fused and enclosing pineal foramen; *ioc*, infraorbital canal; *it*, intertemporal; *j*, jugal; *l*, lacrimal; *lg*, lateral gular; *mdc*, mandibular canal; *mg*, median gular; *mx*, maxillary; *n*, nostril; *na*, nasal included in rostral shield; *o*, orbit; *op*, opercular; *p*, pineal opening; *pop*, preopercular; *popc*, preopercular canal; *ptf*, postfrontal; *pto*, postorbital; *rs*, rostral shield; *sc*, plate overhanging hyomandibular; *so*, supraorbital; *soc*, supraorbital canal; *sop*, subopercular; *spo*, dermal supraoccipital; *sq*, squamosal; *sqc*, jugal canal; *st*, supratemporal or pterotic; *stc*, postorbital and temporal canal; *tb*, tabular; *toc*, transverse occipital canal; *vg*, ventral paired gular.

system may frequently become secondarily freed from the bones when these sink below the surface in higher forms, yet the disposition of the canals is of much value in determining the homologies of the cranial bones in the lower fish and amphibians. Many of the dermal bones also become closely associated with certain parts of the underlying chondrocranium and visceral arches.

The general plan of the dermal bones of a hypothetical primitive fish is shown in Fig. 290. A double series of paired nasals, frontals, and parietals form a roof to the chondrocranium. A prefrontal, postfrontal, lacrimal, postorbital, and jugal complete the orbit; while the margin of the upper jaw is strengthened by the premaxillary and maxillary bearing teeth. The external nostril is apparently on the ventral surface of the snout. Over the occipital region lies a row of bones containing a median occipital and paired postparietals and tabulars. An intertemporal and supratemporal overhang the space occupied by the jaw muscles, which is covered at the side by one or more 'cheek plates' (squamosal and quadratojugal?). The latter cover the quadrate region. The lower jaw has a marginal dentary bearing teeth, a large angular passing

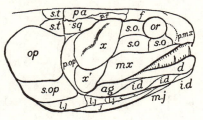

FIG. 292.

Restoration of head of *Rhizodopsis sauroides* (after Traquair, from A. S. Woodward, *Outlines of Vert. Palaeont.*, 1898). *ag*, Angular; *f*, frontal shield; *i.d*, infradentaries; *j*, paired inferior gular; *l.j*, lateral gulars; *m.j*, median inferior gular; *pa*, parietal; *p.f*, postfrontal; *s.o*, suborbital plates; *sq*, supratemporal; *s.t*, tabular and dermal supraoccipital; *x*, squamosal?; *x'*, quadratojugal?

back over the articulation, and probably a supra-angular as well. In front of the angular should perhaps be added infradentaries.

The hyoid region, behind the spiracle, has a preopercular over the hyomandibula, and a series of bones supporting the opercular fold which reaches forwards ventrally between the rami of the lower jaw; these are the opercular, subopercular, interopercular, and lateral gulars. Paired ventral gulars and a median anterior gular fill up the space between them.

Such are the chief covering bones, most of which can be identified in the majority of primitive fish. But there are other elements less constant but nevertheless to be considered as primitive. These are a number of plates on the snout between the premaxillaries and nasals, of which a median ethmoid or rostral is the most important, and an occasional small internasal and interfrontal which together with the occipital are probably the remains of the median series of trunk scales continued on to the head.

Teleostomi.—As an example of a primitive Teleostome we may take *Osteolepis* (Pander, 1860; Huxley, 533; Traquair, 619; A. S. Woodward, 231; Gregory, 523; Watson and Gill, 648; Goodrich, 518). In all essentials it agrees with our hypothetical form, Fig. 291.

The cheek, however, is covered by a single bone instead of by two as in some Osteolepids, and *Holoptychius, Glyptopomus,* and the Rhizodontids, *Rhizodopsis,* and *Eusthenopteron,* Fig. 292.

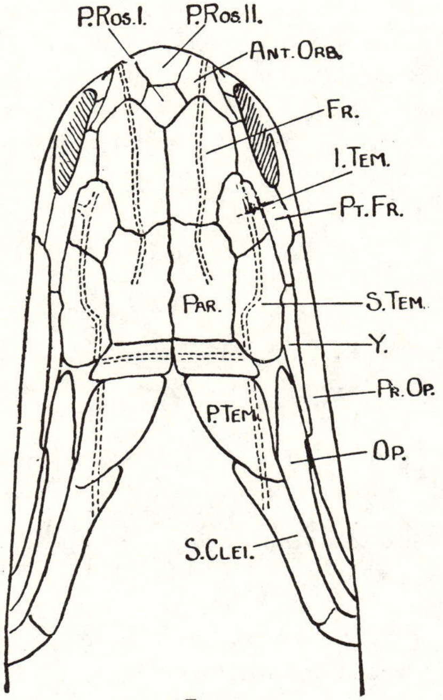

FIG. 293.

Dorsal view of skull of *Cheirolepis trailli,* Middle Old Red Sandstone, Scotland; restored (from D. M. S. Watson, *Proc. Zool. Soc.,* 1925). *Ant.Orb,* Antorbital (nasal); *Fr,* frontal; *I.Tem,* intertemporal; *Op,* opercular; *P.Ros.I, P.Ros.II,* postrostrals; *P.Tem,* posttemporal; *Par,* parietal; *Pr.Op,* preopercular; *Pt.Fr,* postfrontal; *S.Clei,* supracleithrum; *S.Tem,* supratemporal; *Y,* a dorso-lateral bone.

The Coelacanthini have departed considerably from the primitive plan (Huxley, 533; A. S. Woodward, 663; Wellburn, 649; Stensiö, 605, 218; Watson, 645). They are remarkable for the extension of the parietals and frontals apparently at the expense of some of the neighbouring bones, the large size of the operculum, and absence of lateral gulars and pineal foramen. According to Stensiö two nasal apertures are present on the nasal (?) in *Axelia.* Two conspicuous series of five or more parafrontals border the frontal and nasal regions. A transverse row of six bones represents the postparietal and tabular series. All the Crossopterygii seem to have been distinguished by the formation of a transverse joint between frontals and parietals allowing for some bending in this region (Watson, 644-5).

While the Actinopterygii conform in essentials to the fundamental plan, yet they have become specialised in a manner which diverges from that adopted by the Crossopterygii on the one hand and the Dipnoi on the other, perhaps in association with the development of hyostylic jaws, and the shifting of the nostrils on to the side of the snout. There is no intertemporal. Like the Coelacanthini and Dipnoi they have lost the pineal foramen.

The most primitive structure is, of course, displayed by the early fossil Chondrostei (Traquair, **614, 616, 620**; A. S. Woodward, **663**;

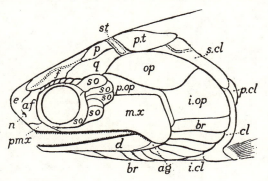

FIG. 294.

Restoration of head of *Palaeoniscus macropomus* (after Traquair, from A. S. Woodward, *Outlines of Vert. Palaeont.*, 1898). *af*, Prefrontal; *ag*, angular; *br*, lateral gulars (branchiostegal rays); *cl*, cleithrum; *d*, dentary; *e*, rostral; *f*, frontal; *i.cl*, clavicle; *i.op*, subopercular; *m.x*, maxillary; *n*, nostril; *op*, opercular; *p*, parietal; *p.cl*, postcleithrum; *pmx*, premaxillary; *p.op*, preopercular; *p.t*, post-temporal; *q*, supratemporal (pterotic); *s.cl*, supracleithrum; *so*, circum- and post-orbitals; *st*, postparietal (+tabular ?).

Watson, **646**; Stensiö, **218**). The Palaeoniscoidei differ from the Osteolepids in the absence of a median occipital, and of large paired ventral gulars, and in the characteristic modification of the bones on the cheek. Here the maxilla is greatly expanded, and above it is a peculiar bent plate which probably represents the squamosal and preopercular. In the Devonian *Cheirolepis* these appear to be separate, Figs. 293-4. The rostrals, of which several are still present in Palaeoniscids, become much reduced in other Actinopterygii. In these the double nostrils are generally between the nasal and antorbital, and the latter disappears in the Teleostei. All the dermal plates have a covering of ganoine in Palaeoniscoidei.

The dermal head plates of *Polypterus* have been studied by Traquair (**613**), Pollard (**575**), Allis (**410**), and others, and generally compared to those of the Osteolepids,

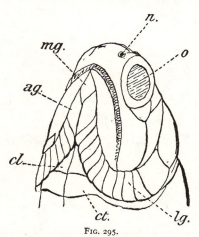

FIG. 295.

Oblique ventral view of the head of *Gonatodus punctatus*, Ag.; Calciferous Sandstone, Wardie. (After Traquair.) *ag*, Enlarged anterior lateral gular; *cl*, clavicle; *ct*, cleithrum; *lg*, lateral gular; *mg*, median gular; *n*, nostril; *o*, orbit. (From Goodrich, *Vert. Craniata*, 1909.)

U

with which extinct fishes Huxley believed the Polypterini to be allied (533). But *Polypterus* resembles the Palaeoniscids much more

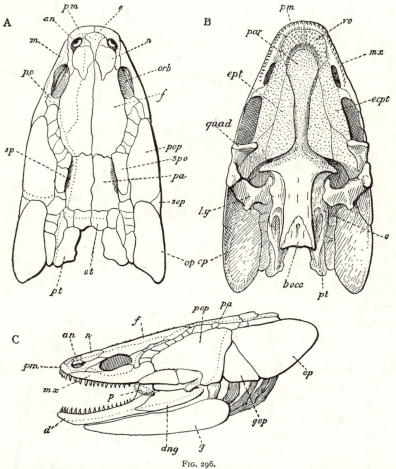

FIG. 296.

Skull of *Polypterus bichir*, Geoffr. A, Dorsal, and C, lateral view (modified from J. Müller and Allis) ; B, ventral view, without the lower jaw. *an*, Adnasal ; *ang*, angular ; *bocc*, basioccipital ; *d*, dentary ; *e*, mesethmoid ; *ecpt*, ectopterygoid ; *ept*, endopterygoid ; *f*, frontal ; *g*, paired gular ; *gop*, suboperculum ; *hy*, hyomandibular ; *m*, maxilla ; *n*, nasal ; *o*, opisthotic ; *op*, opercular ; *orb*, orbit ; *p*, labial cartilage (dotted) ; *pa*, parietal ; *par*, parasphenoid ; *pm*, premaxilla ; *po*, postorbital ; *pop*, preopercular ; *pt*, post-temporal ; *quad*, quadrate ; *sop*, suboperculum ; *sp*, spiracle ; *spo*, spiracular plate ; *st*, postparietal ; *vo*, vomer ? A dotted line indicates the course of the lateral-line canal. (From Goodrich, *Vert. Craniata*, 1909.)

closely in the number and disposition of the plates, as it does in so many other characters (Goodrich, 35, 520).   A ganoid layer covers their surfaces. The supratemporal appears to have fused with the parietal ;

there is the same bent plate extending over the hyoid and lateral temporal regions (certainly representing the preoperculum, and probably the squamosal as well). The rostral region is very similar, and a series of small plates extend from the orbit to the supratemporal, overlying the spiracular opening ('spiracular ossicles'). The median and lateral gulars have disappeared with the exception of two large ventral plates [1] (Goodrich, 35, 520).

Still more modified are the Acipenseroidei, Figs. 298, 434. Accompany-

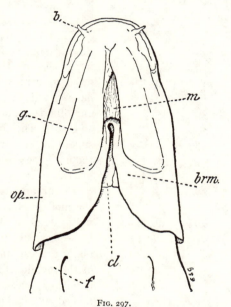

FIG. 297.

Polypterus lapradii, Stdr. Ventral view of head. *b*, Barbel; *brm*, branchiostegal membrane; *cl*, clavicle; *f*, pectoral fin; *g*, paired gular plate; *m*, intergular membrane; *op*, opercular region. (From Goodrich, *Vert. Craniata*, 1909.)

ing the development of a large rostrum and the reduction of the mouth, jaws, and teeth, the plates round the orbit are reduced, the rostral plates replaced by numerous ossicles, and the cheek plates, median gular, and angular lost. *Chondrosteus* of the Lias still has several lateral gulars, but in living genera these have disappeared and the opercular bones are more or less vestigial (Traquair, 620). Moreover, in Sturgeons the sucking mouth has very reduced jaws and no teeth in the adult. Although the

[1] Since the maxillary in *Polypterus* is traversed by the infraorbital lateral-line canal, and develops medially to the lip fold, Allis and Sewertzoff believe it to be a suborbital.

roofing bones are well developed, they, like all the exoskeleton, have lost all trace of ganoine.

The Saurichthyidae (Belonorhynchidae) are a specialised group which seem to be allied to the Palaeoniscidae and Acipenseroidei. They possess a very elongated rostrum and lower jaw, large frontals and supratemporals, but reduced parietals. and no median or lateral gulars (A. S. Woodward, 663 ; Stensiö, 134).

The Holostei show great diversity of detail in the covering bones of the head. The more primitive forms (Amioidei) have not departed much from our fundamental plan, Figs. 299-301. The bones are still superficial, except the prefrontal, though the original ganoine present in early fossils tends to disappear and is quite lost in the modern *Amia*. The transverse series of occipital plates is reduced to two, the cheek plates are lost, and the hind end of the maxilla is freed (a supramaxillary (cheek plate ?) is attached to its upper border). The opercular bones have become established as a set of four present in all Holostei except in very specialised forms. These are a large preopercular attached to the hyomandibular and quadrate, below it an interopercular

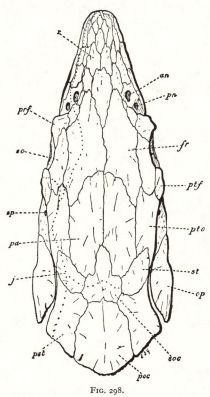

FIG. 298.

Skull of *Acipenser sturio*, L. ; dorsal view. A dotted line indicates the lateral line according to Collinge. *an,* Anterior nostril ; *fr,* frontal ; *j,* junction of postorbital, occipital, and trunk branches of lateral-line system ; *op,* opercular ; *pa,* parietal ; *pn,* posterior nostril ; *poc,* postoccipital ; *prf,* prefrontal ; *pst,* post-temporal ; *ptf,* postfrontal ; *pto,* pterotic ; *r,* rostral plates ; *so,* supraorbital ; *soc,* supraoccipital ; *sp.* spiracle ; *st,* tabular ? (From Goodrich, *Vert. Craniata,* 1909.)

of doubtful homology, a broad opercular, and a subopercular. The median gular is well developed in Amioidei ; but the lateral gulars are very narrow and have become converted into the so-called branchiostegal rays characteristic of the Holostei (Franque, 1847 ; Shufeldt, 85 ; Bridge, 429 ; Allis, 401-3).

The highly specialised Pycnodontidae have a remarkable and inconstant arrangement of the bones. Median plates separate the parietals, and numerous small plates cover the occipital region, the snout, and the space between the rami of the lower jaw (Woodward, 231, 663).

The modern *Lepidosteus*, which appears to be a specialised derivative of the Semionotidae (Traquair), with a long snout carrying the nostrils

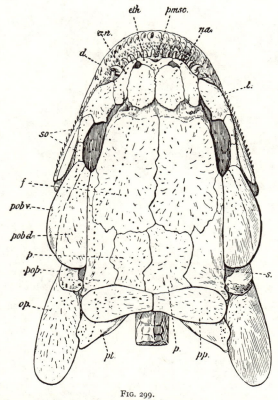

FIG. 299.

Dorsal view of the skull of *Amia calva*, L. (after Allis). The course of the lateral-line system is indicated by a dotted line on the left side. *an*, Adnasal; *d*, dentary; *eth*, mesethmoid; *f*, frontal; *l*, lachrymal or first suborbital; *na*, nasal; *op*, opercular; *p*, parietal; *pmsc*, premaxilla; *pobd* and *pobv*, dorsal and ventral postorbitals; *pop*, preopercular; *pp*, postparietal; *pt*, post-temporal; *s*, pterotic (supratemporal); *so*, suborbitals. (From Goodrich, *Vert. Craniata*, 1909.)

at its extremity, is also much modified (J. Müller, 1846; Balfour and Parker, 2; Collinge, 476; Regan, 577). The premaxillaries extend over a great part of the snout and the maxillaries are represented by series of small bones. Small plates cover the cheek. Owing to the forward position of the articulation of the jaws, Fig. 302, the opercular apparatus is strangely modified, the preopercular being reduced and situated in front of the enlarged interopercular. However, the homology of these bones is still doubtful, the posterior bone which contains the 'hyomandibular' lateral-line canal being considered by many to be the

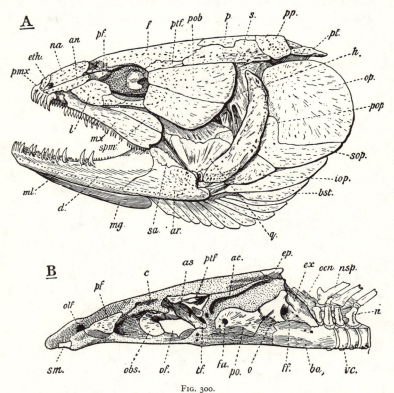

FIG. 300.

Amia calva, L. (after Allis, slightly altered). A, Left-side view of the skull; B, left-side view of the cranium, from which the dermal bones have been removed. Cartilage is dotted. *ac*, Auditory capsule; *an*, adnasal; *ar*, derm-articular; *as*, pterosphenoid; *bo*, basioccipital; *bst*, branchiostegal ray; *c*, cartilaginous cranium; *d*, dentary; *ep*, epiotic; *eth*, mesethmoid; *ex*, exoccipital; *f*, frontal; *fa*, foramen for facial nerve; *ff*, foramen for vagus; *h*, hyomandibular; *iop*, interopercular; *l*, lacrimal; *mg*, median gular; *ml*, lateral line in mandible; *mx*, maxilla; *n*, neural arch; *na*, nasal; *nsp*, neural spine; *o*, opisthotic; *obs*, orbitosphenoid; *ocn*, foramen for spino-occipital nerve; *of*, vacuity with optic foramen in front; *olf*, olfactory capsule; *op*, opercular; *p*, parietal; *pf*, prefrontal; *pmx*, premaxilla; *po*, pro-otic; *pob*, postorbital; *pop*, preopercular; *pp*, post-parietal; *pt*, post-temporal; *ptf*, postfrontal; *q*, quadrate; *s*, pterotic; *sa*, supra-angular; *sm*, 'septomaxillary'; *sop*, subopercular; *spm*, supramaxilla; *tf*, trigeminal foramen; *vc*, vertebral centrum. (From Goodrich, *Vert. Craniata*, 1909.)

preopercular, Fig. 302. The ganoine layer is well developed on all the superficial dermal bones.

Lastly, in the Teleostei the thin covering of ganoine, still present in the extinct Leptolepidae, soon disappears altogether, and the dermal bones tend to sink more and more below the soft tissues. This leads in higher forms to the subdivision of those which harbour lateral-line canals into deep-lying plates and independent, more superficial, narrow grooved 'canal bones' (Goodrich, 35). The latter remain near the surface forming more or less complete chains of ossicles protecting the canals, Fig. 303 (Bruch, 1862 ; Allis, 403-5 ; Cole, 475). The supratemporal

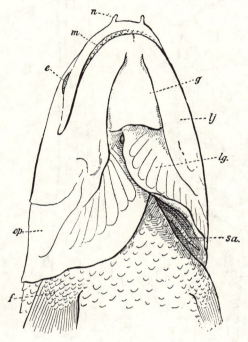

FIG. 301.

Head of *Amia calva*, L. ; oblique ventral view.   *e*, Eye ;   *f*, pectoral fin ;   *g*, median gular plate ; *lg*, lateral gulars or branchiostegal rays ;   *lj*, lower jaw ;   *m*, mouth ;   *n*, nostril ;   *op*, operculum ; *sa*, serrated appendage.   (From Goodrich, *Vert. Craniata*, 1909.)

(pterotic) invades the posterior region of the auditory capsule as the prefrontal (sphenotic) does in front (p. 376).

The tabulars and postparietals become reduced and finally disappear (except in so far as they may be represented by canal bones). The lower jaw retains only a small angular behind (the supra-angular has perhaps fused with the articular). A median gular persists very rarely (*Elops*, Fig. 304).

While in Cypriniformes the roofing bones may still be superficial (forming a posterior shield in combination with the post-temporal and supracleithral in some Siluridae), in most Teleosts the frontals and parietals sink deeply below the soft tissues and prolongations of the

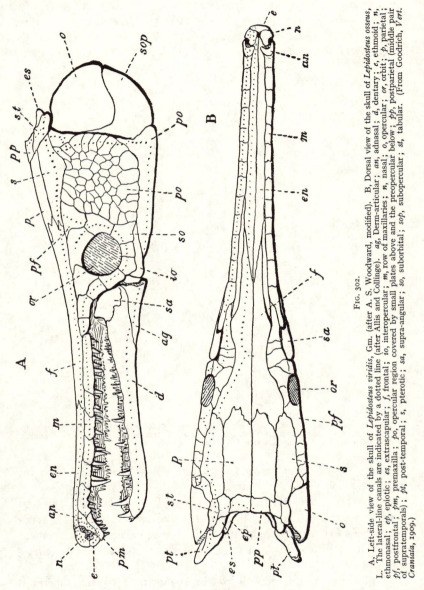

FIG. 302.

A, Left-side view of the skull of *Lepidosteus viridis*, Gm. (after A. S. Woodward, modified). B, Dorsal view of the skull of *Lepidosteus osseus*, L. The lateral-line canals are indicated by a dotted line (after Allis and Collinge). *ag*, Derm-articular; *an*, adnasal; *d*, dentary; *e*, ethmoid; *n*, ethmonasal; *ep*, epiotic; *es*, extrascapular; *f*, frontal; *io*, interopercular; *m*, row of maxillaries; *n*, nasal; *o*, opercular; *or*, orbit; *p*, parietal; *pf*, postfrontal; *pm*, premaxilla; *po*, opercular region covered by small plates above and the preopercular below; *pp*, postparietal (middle pair of supratemporals); *pt*, post-temporal; *s*, pterotic; *sa*, supra-angular; *so*, suborbital; *sop*, subopercular; *st*, tabular. (From Goodrich, *Vert. Craniata*, 1909.)

anterior myomeres, Figs. 303, 305. Ordinary scales may then secondarily extend over the greater part of the head in higher Acanthopterygii. In these also the parietals usually become separated by the supraoccipital (Sagemehl, 378; Bridge, 434; Boulenger, 426; Ridewood, 578-81; Gegenbaur, 516; Allis, 405; Goodrich, 35; etc.).

The marginal jaw bones often become much specialised in Teleosts. In Siluroids the maxilla may be reduced to a nodule supporting the cartilaginous axis of the barbel. Most of the higher groups have enlarged premaxillaries bordering the mouth, and the maxillaries modified into a toothless bone lying behind it separately articulated to the ethmoid

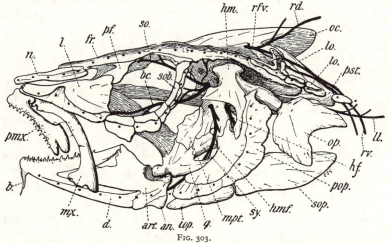

FIG. 303.

Left-side view of skull of *Gadus morrhua*, showing branches of facial nerve, and course of lateral-line system (by a series of dots), partly after Cole. *an*, Angular; *art*, articular; *b*, barbel; *bc*, buccal branch of facial supplying suborbital canal; *d*, dentary; *fr*, frontal; *hf*, hyoidean branch; *hm*, hyomandibula; *hmf*, hyomandibular branch of facial supplying its lateral-line canal; *iop*, inter-opercular; *l*, 'lachrymal'; *ll*, lateral-line branch of vagus; *lo*, lateral-line ossicles; *mpt*, meta-pterygoid; *mx*, maxilla; *n*, nasal; *oc*, crest of supraoccipital; *op*, opercular; *pf*, prefrontal; *pmx*, premaxilla; *pop*, preopercular; *pst*, post-temporal; *q*, quadrate; *rd*, branch supplying region of dorsal fin; *rfv*, dorsal recurrent branch of facial; *rv*, branch supplying region of pectoral fin; *so*, superior ophthalmic branch supplying supraorbital canal; *sob*, suborbital; *sop*, subopercular; *sy*, symplectic.

region of the skull, and no longer forming the margin of the mouth, Figs. 286, 303.

Palate in Teleostomi.—The dermal bones on the palate are more constant than the outer covering bones. Most of them can easily be traced from the earliest Osteichthyes to the Tetrapoda. Originally they were doubtless developed as basal plates to support patches of teeth, and similar bones occur regularly on the inner surface of the lower jaw and even in many Teleostomes spread over the inner surface of the gill bars. The teeth, however, may be lost in specialisation. These internal dermal bones, of course, never develop a cosmoid or a ganoid layer.

Below the chondrocranium is the median parasphenoid, found in

Crossopterygii below the orbito-temporal region only, Fig. 306 B (Bryant, 465 ; Watson, 644 ; Stensiö, 606) ; but extending farther back in Actino-pterygii, Figs. 296, 307, sometimes even beyond the occipital region as in Acipenseroidei. It tends to develop a strong transverse process support-ing the basitrabecular process of the basisphenoid region, and another more posterior lateral process or wing[1] supporting the outer wall of the trigemino-facialis chamber (p. 277). Primitively the parasphenoid closely adheres to the basis cranii, though allowing the internal carotids to reach the fenestra hypophyseos by the parabasal canals, Fig. 308 ;[2] but in Teleosts where the ventral chamber of the myodome becomes much developed, it may become widely separated from the floor of the brain-cavity (p. 279). Immedi-ately in front of the parasphenoid is found in all Osteichthyes (except the Acipenseroidei and Teleostei) a pair of prevomers, underlying the ethmoid region and generally strongly toothed (p. 284). *Acipenser* and the Teleostei are distinguished by the possession of a median pre-vomer, Fig. 286 ; since, however, it shows signs of paired origin in *Salmo* (Gaupp, 343) and is stated by Walther (630) to arise from paired rudiments in *Esox*, this bone prob-ably represents the two prevomers fused.

FIG. 304.

Head of *Elops saurus*, L.; oblique ventral view. *asc*, Axillary scale ; *br*, branchiostegal ray ; *ch*, preopercular region ; *e*, fold of skin over eye ; *g*, lower jaw ; *gp*, gular plate ; *iop*, interopercular region ; *mx*, maxilla ; *op*, oper-cular ; *pf*, pectoral fin ; *sop*, subopercular. (From Goodrich, *Vert. Craniata*, 1909.)

[1] This wing is often called the ' process ascendens ', inappropriately since this name is applied to a dorsal process of the palatoquadrate (p. 423).

[2] When the bony parasphenoid is developed below the basis cranii there become enclosed between them the internal carotid arteries running forwards to the fenestra hypophyseos and the accompanying sympathetic nerves in so-called parabasal canals (Gaupp, 508, Fig. 308). The palatine nerve may also enter the canal and continue forward between the basitrabecular process and the basipterygoid process of the parasphenoid in a prolongation of the para-basal canal, which may be distinguished as the Vidian or basipterygoid canal (enclosing Vidian nerve and palatine artery, see p. 272).

The remaining internal dermal bones are developed on the visceral arches, and many of them are closely associated with corresponding cartilage bones in these arches. On the anterior end of the palatoquadrate is a dermal palatine generally much toothed (accessory palatines may

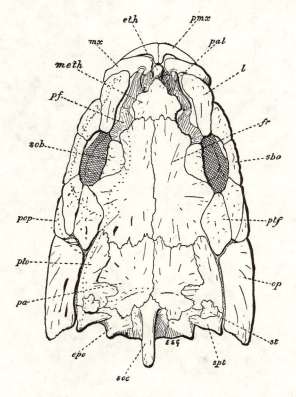

FIG. 305.

Dorsal view of the skull of *Cyprinus carpio*, L. A dotted line indicates the lateral-line canals on the left side. *epo*, Epiotic; *eth*, pre-ethmoid (rostral); *fr*, frontal; *l*, lacrimal; *meth*, mesethmoid; *mx*, maxilla; *op*, opercular; *pa*, parietal; *pal*, palatine; *pf*, prefrontal; *pmx*, premaxilla; *pop*, preopercular; *ptf*, postfrontal; *pto*, pterotic; *sob*, supraorbital; *soc*, supraoccipital; *spt*, tabular? *st*, anterior supratemporal. (From Goodrich, *Vert. Craniata*, 1909.)

occur as in *Amia*, Fig. 429). Further back the palatoquadrate bears several ' pterygoid ' bones, of which three are typically present throughout the Teleostomes. These are a large pterygoid (endopterygoid), an ecto-pterygoid joining it to the maxilla, and a metapterygoid. The first two are dermal bones, usually toothed ; the metapterygoid is developed as a cartilage bone (Parker, 560 ; Gaupp, 343 ; and others). This metaptery-goid is primitively articulated to the basitrabecular process (p. 421). The

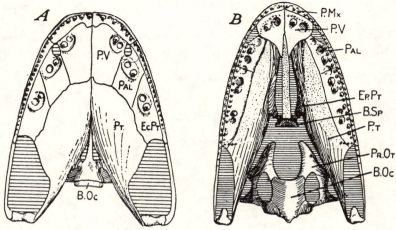

FIG. 306.

Ventral view of skull of A, *Baphetes Kirkbyi*, Middle Coal Measures, Fifeshire ; B, *Eusthenopteron*, Upper Old Red Sandstone, Canada (from D. M. S., Watson, *Tr. Roy. Soc.*, 1926). *B.Oc*, Basi-occipital ; *B.Sp*, basisphenoid ; *Ec.Pt*, ectopterygoid ; *Ep.Pt*, epipterygoid ; *Pal*, palatine ; *P.Mx*, premaxillary ; *Pr.Ot*, pro-otic ; *Pt*, pterygoid ; *P.V*, prevomer. Parasphenoid covers and projects forwàrds from basisphenoid. Internal nostril between *P.V* and *Pal*.

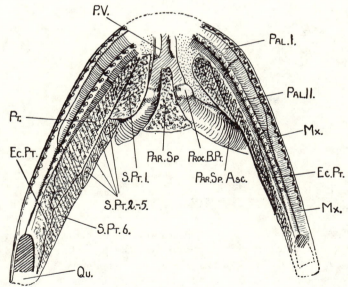

FIG. 307.

Elonichthys binneyi : ventral view of palate (from D. M. S. Watson, *Proc. Zool. Soc.*, 1925). *Ec.Pt*, Ectopterygoid ; *Mx*, maxillary ; *P.V*, prevomer ; *Pal.I., II.*, palatines ; *Par.Sp*, parasphenoid ; *Par.Sp.Asc*, ascending ramus of parasphenoid ; *Proc.B.Pt*, basipterygoid process ; *Pt*, pterygoid : *Qu*, quadrate ; *S.Pt.1-6*, suprapterygoids (1 = autopalatine, 6 = metapterygoid).

metapterygoid is toothed in *Polypterus* (van Wijhe, 654; Pollard, 575;

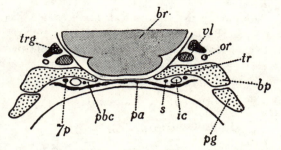

FIG. 308.

Lepidosteus osseus, embryo 18 mm. long. Transverse section of base of skull at level of basal processes, *bp*, of trabeculæ, *tr*. *7p*, Palatine nerve; *br*, brain; *ic*, internal carotid; *or*, orbital artery; *pa*, parasphenoid; *pbc*, posterior parabasal canal; *pg*, palatoquadrate; *trg*, trigeminal ganglion; *s*, sympathetic nerve; *vl*, vena capitis lateralis.

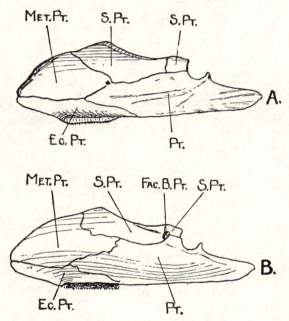

FIG. 309.

Nematoptychius greenocki, palatoquadrate apparatus. A, Outer view of right side; B, inner view of left side (from D. M. S. Watson, *Proc. Zool. Soc.*, 1928). *Ec.Pt*, Ectopterygoid; *Fac.B.Pt*, facet of basal process for articulation with basipterygoid (basitrabecular) process; *Met.Pt*, metapterygoid; *Pt*, Pterygoid; *S.Pt*, suprapterygoid.

Allis, 410), and it is probable that here also there are really two elements to which the same name metapterygoid has been given : a dermal bone (usually absent), and a corresponding cartilage bone (always present, p. 299).

Recently the palate of the Palaeoniscids has been described (Stensiö, 218; Watson, 646). Parasphenoid and prevomers are present as usual; and the palatoquadrate (in addition to the endochondral autopalatine, and quadrate) is provided with one or two dermal palatines, an ectopterygoid, a large pterygoid, and (in *Elonichthys*, Watson) along

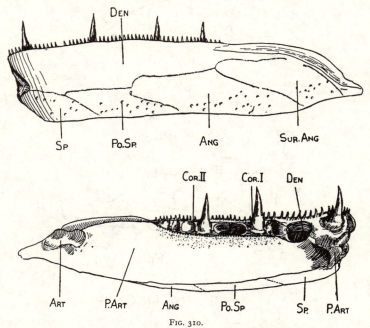

FIG. 310.

Megalichthys: above outer view, and below inner view, of left ramus of lower jaw (from D. M. S. Watson, *Tr. Roy. Soc.*, 1926). *Ang*, Angular; *Art*, articular; *Cor*, coronoid; *Den*, dentary; *P.Art*, prearticular; *Po.Sp*, postsplenial; *Sp*, splenial; *Sur.Ang*, supra-angular.

its medial border a varying series of dermal and endochondral suprapterygoids, of which the hindmost may represent the metapterygoid, Figs. 307, 310. The palate of *Polypterus* must be interpreted in the light of these discoveries.[1]

Lower Jaw.—Our knowledge of the structure of the lower jaw in fishes is by no means complete, and the nomenclature of its constituent

[1] Van Wijhe considers that the ectopterygoid of *Polypterus* contains a palatine element; Allis, 410, that the bone usually called vomer is a dermal palatine. It is possible that the prevomers, already reduced in some Palaeoniscids, are absent in *Polypterus*, the so-called 'vomer' belonging to covering bones of the palatoquadrate.

bones is in some confusion. In the early forms there are numerous bones, while in modern Teleostei they are reduced to three by the disappearance of some and the fusion of others. These three are the 'dentary', 'articular', and 'angular'. The lower jaw of the early primitive Teleostomes (Crossopterygii) closely resembled that of the more primitive Tetrapoda (Stegocephalia), and was covered externally by the dentary, supra-angular, and angular; below by the splenial, and internally by the prearticular; while the Meckelian fossa was roofed over by the coronoid. The angular and splenial seem to be the enlarged posterior and anterior elements of a series of external infradentaries (Traquair, 615; Bryant, 465; Watson and Day, 647) several of which were present in *Rhizodopsis*, *Eusthenopteron*, and some Osteolepids; and the coronoid the posterior element of a more dorsal internal series, Figs. 292, 309. Most of the chief elements are still present in Coelacanthini (Stensiö, 605; Watson, 645) and Palaeoniscioidei (Traquair, 614, 616, 618; Stensiö, 218; Watson, 646). They become much reduced in the Acipenseroidei, only the dentary remaining usually in *Acipenser*. The Amioidei (Sagemehl, 378; Allis, 402; Regan, 577; van Wijhe, 654; Bridge, 429) preserve in addition to the dentary, angular, and supra-angular, a series of toothed coronoids, Figs. 300, 431. The large posterior bone generally called 'splenial' seems to be the prearticular. *Lepidosteus* differs not in essentials from *Amia* (Collinge, 476; Regan, 577; van Wijhe, 654; Parker, 566). It has been mentioned above that in Teleosts the number of bones is much reduced; the coronoids have disappeared and even the 'angular' may vanish. The so-called articular is made up of an endochondral articular fused to an outer dermal element usually called 'derm-articular'. The latter corresponds in position to the large angular of Amia and is probably its homologue; in that case the 'angular', which develops from an endochondral and a dermal element (Ridewood, 578-9), has been wrongly named. The 'dentary' is also of compound origin, being formed of a true dermal dentary and a small anterior element probably representing the mento-Meckelian (Parker, 560; Schleip, 584; Gaupp, 343).

Dipnoi.—The Dipnoi form a well-defined group which can be followed from the Old Red Sandstone to the present day. (Agassiz, 1833-44; Pander, 1858; Hancock and Atthey, 1868 and 1871; Miall, 1878; Traquair, 617; A. S. Woodward, 663; Günther, 1871; Bridge, 433; Dollo, 119; Watson and Gill, 648.) In general build the early forms approach the Osteolepids; but even in the earliest known genus, *Dipterus*, the characteristic specialised dentition was already established with powerful grinding compound toothed plates, accompanied by a shortening of the jaws and reduction of the marginal teeth. The Dipnoan head is broad and depressed, the snout blunt with ventral external and internal nostrils, the orbit small and about midway, the cheek region narrow, and the operculum far forward. There is no pineal foramen, and no open spiracle. The head of *Dipterus* is well covered with dermal plates; and these are provided with an external layer of typical cosmine like that on the cosmoid body scales (Pander, 1860; Goodrich, 36, 519). This layer disappears in later forms as the bones sink below the skin, and, except in such Devonian genera as *Dipterus* and *Scaumenacia*, the lateral-

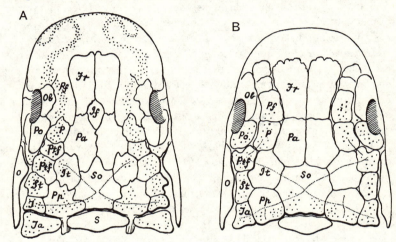

FIG. 311.

Diagrams of dorsal view of head of A, *Dipterus*, and B, *Scaumenacia*, showing dermal roofing-bones and pores of lateral-line system. *Fr*, Frontal; *If*, interfrontal; *It*, intertemporal; *O*, opercular; *Ob*, supraorbital; *P*, lateral plate; *Pa*, parietal; *Pf*, prefrontal; *Po*, postorbital; *Pp*, postparietal; *Ptf*, postfrontals; *S*, median postoccipital through which runs transverse lateral-line canal; *So*, median posterior plate or occipital; *St*, supratemporal; *T*, pretabular; *Ta*, tabular. Snout covered with plates more or less fused.

line canals are no longer enclosed. The much-reduced dermal bones of modern Dipnoi are deeply sunk, and, together with the lateral-line canals, are secondarily covered by large scales, which have spread over them from the trunk (K. Fürbringer, 25; Goodrich, 35). The disposition of the bones can be seen in the appended Figs. 311-12, 318-19. The cranial roof is remarkable for the large number of elements composing it in early forms, especially at the sides, and for the presence and increasing dominance of a series of median bones, perhaps remnants of a median series of scales extending along the trunk (Woodward, 231; Whiteaves, 229; Watson and Day, 647; Watson and Gill, 648; Goodrich, 519).

The roof of the skull in early Dipnoi is distinguished by the presence, in addition to the usual paired frontals and parietals,[1] of a large median occipital, separated by paired intertemporals (?) and postparietals from lateral rows of small elements extending from the prefrontal to the tabular region and harbouring the supraorbital, postorbital, and temporal lateral-line canals. In *Dipterus, Scaumenacia,* and probably other Devonian genera, a transverse posterior occipital bone containing the occipital canal appears to represent, together with the anterior median occipital, the median occipital of the Osteolepids ; this bone, however, seems to

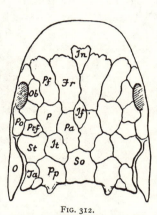

Fig. 312.

Ctenodus: restoration of dorsal view of skull (partly from Watson and Gill; E. S. Goodrich, *Linn. Soc. J. Zool.,* 1925). Lettering as in Fig. 311.

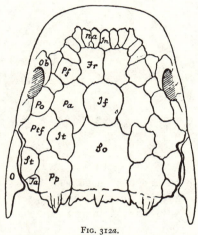

Fig. 312a.

Sagenodus: restoration of dorsal view of skull (partly from Watson and Gill; E. S. Goodrich, *Linn. Soc. J. Zool.,* 1925). Lettering as in Fig. 311.

have been only loosely connected with the cranial shield of the Dipnoi. The bones on the snout of large specimens of *Dipterus platycephalus* become fused into a shield, much as in many Osteolepids, which shield probably represents premaxillaries, maxillaries, nasals, and ethmoid. *Dipterus* is also provided with the usual bones round the orbit and an inner set of small circumorbitals. Two small bones appear to have been present on the cheek. A large angular, small post-splenial, splenial, and toothless dentary covered the lower jaw, Fig. 313. The opercular flaps were supported by opercular, subopercular, and lateral gular bones ; while a median and two ventral gulars filled the space between them (Watson

[1] These are considered by Watson, Day, and Gill to be the nasals and frontals ; while the parietals they suppose to have been included in the median occipital, which, however, shows no trace of compound origin, Fig. 319 c. *Scaumenacia* has large frontals and parietals meeting normally without the intervention of an interfrontal, Fig. 312.

and Gill, **648**). Dipterus is the only Dipnoan preserving the gulars. There is a remarkable resemblance to the Osteolepids in this region of the head.

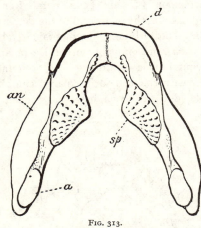

FIG. 313.

Dorsal view of the lower jaw of *Dipterus*. *a*, Articular; *an*, angular; *d*, dentary; *sp*, splenial tooth on the prearticular? (From Goodrich, *Vert. Craniata*, 1909.)

Reduced toothed premaxillary and maxillary bones may be made out in *Dipterus* (Watson and Day, **647**). Its palate seems to be nearly as specialised as in *Ceratodus*. Only the prevomers, with teeth, and the pterygoids, with large tooth plates, remain besides the parasphenoid. The inner face of the mandible is covered by the prearticular (generally called splenial), bearing the ventral tooth plate, Fig. 313.

Later forms tend to lose the bones on the snout, the marginal bones of the jaws, and the lateral ventral and median gulars. Meanwhile the median bones of the roof, so characteristic of the group, gradually enlarge. Already in the Upper Carboniferous *Sagenodus* the interfrontal meets the occipital; in *Ceratodus*, the most primitive of living Dipnoi, there remain above, in addition to the very large occipital and more anterior 'ethmoid' plate (probably interfrontal), only an elongated lateral plate (occupying the position of the frontal, parietal, and intertemporal, etc.), a postfrontal (postorbital?), and an outer plate covering the quadrate region (probably supratemporal, but often called squamosal). At the side is a small sub- and postorbital, an opercular, a much-reduced subopercular. The lower jaw retains only the angular and so-called dentary (probably splenial; Watson and Gill, **648**) on its outer surface. *Protopterus* and *Lepidosiren* are still further

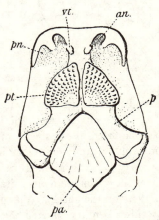

FIG. 314.

Ventral view of the palate of *Dipterus* restored. *an*, Anterior nostril; *p*, palatopterygoid bone; *pa*, parasphenoid; *pn*, posterior nostril; *pt*, palatine tooth; *vt*, vomerine tooth. (From Goodrich, *Vert. Craniata*, 1909.)

specialised in that the roofing has almost disappeared, the narrowed
lateral plates alone remaining superficial posteriorly, while the median
occipital ('fronto-parietal') has sunk below the muscles and spread

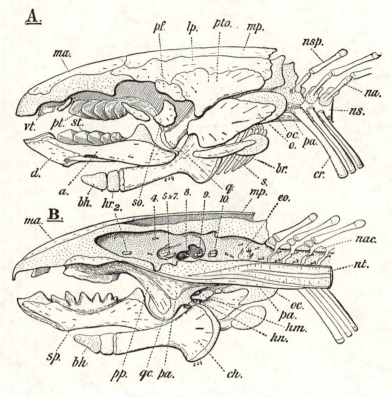

FIG. 315.

Ceradotus Forsteri, Krefft. A, Outer view of left half; B, inner view of right half. *a*, Angular;
bh, basihyal; *br*, fifth branchial arch; *ch*, ceratohyal; *cr*, 'cranial' rib; *d*, dentary; *eo*, 'exoccipital'; *hm*, hyomandibular; *hn*, hyomandibular nerve; *hr*, hypohyal; *lp*, lateral plate;
ma, median anterior, and *mp*, median posterior plate; *na*, neural arch; *nac*, cartilage of neural
arch; *ns*, notochordal sheath; *nsp*, neural spine; *nt*, notochord; *o*, opercular, and *oc*, its cartilage;
pa, parasphenoid; *pf*, postfrontal; *pp*, pterygo-palatine; *pt*, palatine tooth; *pto*, pterotic (?),
and *q*, its downward process covering the quadrate cartilage, *qc*; *s*, subopercular; *so*, suborbital;
sp, splenial (prearticular?); *st*, splenial tooth; *vt*, vomerine tooth. (From Goodrich, *Vert. Craniata*,
1909.)

over the brain-case (Wiedersheim, 653; Bridge, 433; Goodrich, 35),
Fig. 317.

The living Dipnoi, then, are highly specialised; but their extinct
predecessors, more especially the Devonian *Dipterus*, approach the Osteolepids in general structure and in the disposition of the bones on the head.
In the early Dipnoi, the fundamental plan of the covering bones can be

recognised, though the homology of many of the elements cannot be
determined with certainty. The two groups converge in this as in so

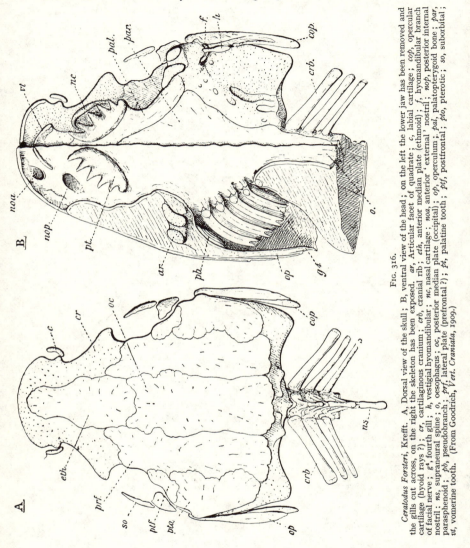

FIG. 316.

Ceratodus Forsteri, Krefft. A, Dorsal view of the skull; B, ventral view of the head; on the left the lower jaw has been removed and the gills cut across, on the right the skeleton has been exposed. *ar*, Articular facet of quadrate; *c*, labial cartilage; *cop*, opercular cartilage (hyoid rays?); *cr*, cartilaginous cranium; *crb*, cranial rib; *eth*, anterior median plate (ethmoid); *f*, hyomandibular branch of facial nerve; *g⁴*, fourth gill; *h*, vestigial hyomandibular; *nc*, nasal cartilage; *noa*, anterior ' external ' nostril; *nop*, posterior internal nostril; *nou*, posterior ' external ' nostril; *o*, oesophagus; *op*, operculum; *pal*, palatopterygoid bone; *par*, parasphenoid; *pb*, pseudobranch; *prf*, lateral plate (prefrontal?); *pt*, palatine tooth; *ptf*, postfrontal; *pto*, pterotic; *so*, suborbital; *vt*, vomerine tooth. (From Goodrich, *Vert. Craniata*, 1909.)

many other characters. Whether the multiplicity of the bones contributing to the roof of the skull in *Dipterus* is secondary, or a reminiscence
of a primitive condition when the head was covered with small scales
like those on the trunk, cannot at present be determined.

TETRAPODA : DERMAL BONES OF SKULL

Skull of Amphibia.—The dermal covering of the head in Tetrapoda is very similar to that of early Osteichthyes, from which it has doubtless been derived. Only in the hinder and ventral region are there any important differences ; for there is no evidence in the Tetrapod, even in the most primitive yet discovered, of any opercular or gular plates.

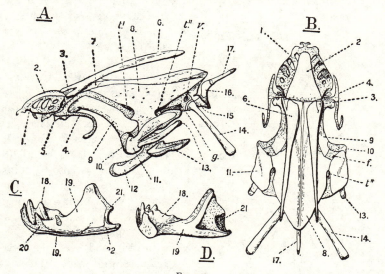

Fig. 317.

Lepidosiren paradoxa, Fitz. (after Bridge). A, Left-side view, and B, dorsal view of the skull ; C, outer view of the left, and D, inner view of the right ramus of the lower jaw. 1, Nasal capsule ; 2, ethmoid ; 3 and 7, process of pterygo-palatine ; 4, antorbital cartilage ; 5, palatine tooth ; 6, lateral (dermal lateral ethmoid) ; 8, occipital (fronto-parietal) ; 9, pterygo-palatine ; 10, quadrate ; 11, squamosal ; 12, ceratohyal ; 13, subopercular and opercular ; 14, cranial rib ; 15, parasphenoid ; 16, neural arch ; 17, supraneural spine ; 18, splenial tooth ; 19, splenial ? ; 20, Meckel's cartilage ; 21, articular cartilage ; 22, angular ; *f*, foramen for facial, *g*, for glosso-pharyngeal, *t*, for trigeminal, and *v*, for vagus nerve.

Either they were never developed in the ancestor, or were lost before the Amphibian stage was reached.

Stegocephalia.—The primitive condition is best seen in the Stegocephalia, which include the earliest known Tetrapods from the Carboniferous Epoch (v. Meyer, 1847 ; Fraas, 497 ; Miall, 1874 ; Cope, 1869 ; Williston, 657-61, 95 ; Bransom, 427 ; Broom, 448 ; Moodie, 548-9 ; Burmeister, 1840 ; Huxley, 1862–9 ; Ammon, 1889 ; Fritsch, 23 ; Jaekel, 51 ; Gaudry, 28 ; Thévenin, 610 ; Watson, 631, 644 ; A. S. Woodward, 231 ; and others). Here the superficial bones overlie the endoskeletal skull and jaws built on the autostylic plan (p. 409). Dorsally they form a roof firmly fixed to the chondrocranium, spreading laterally round the orbits and over the temporal space occupied by the jaw muscles, attached to the marginal bones of the upper jaw and outer end of the quadrates, and further buttressed by the paroccipital processes of the opisthotics. The roofing was thus complete, pierced only by external nostrils, orbits, and pineal foramen, Figs. 318-21. That the bones in many forms were still superficial is evidenced by their ornamented surface, and the frequent presence of well-marked grooves indicating the course of lateral-line canals disposed much as in primitive fish (Moodie, 548, 550; see p. 741).

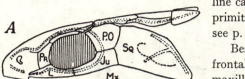

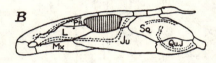

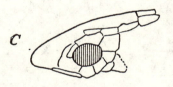

FIG. 318.

Left lateral view of skull of: A, Osteolepis; B, Palaeogyrinus; and C, Dipterus (from D. M. S. Watson, *Tr. Roy. Soc.*, 1926). Compare Figs. 291, 319, 311.

Besides the paired nasals, frontals, and parietals, are premaxillaries and maxillaries, prefrontals and postfrontals, postorbitals, jugals, and lacrimals, all easily derived from the similar bones found in fish (p. 287). There is also a transverse posterior row of four bones representing the postparietals and tabulars. The supratemporal can also be identified, and even the intertemporal in some; but the bones on the cheek are less easy to compare with those in primitive Teleostomes, possibly because the latter have a hyostylic suspension of the jaws in which the quadrate region is less developed. In the Tetrapod this part is covered by a squamosal and quadratojugal presumably derived from the cheek bones of the fish (cf. Figs. 291, 318). The lateral temporal roof covered a space for jaw muscles open to the orbit in front, and opening behind by the post-temporal foramen as in fish. The tabular usually projects at the hinder outer corner of the cranial shield, overhanging a notch in the posterior margin of the squamosal marking, no doubt the position of the tympanum (p. 483). This tympanic notch is of course a new feature in the Tetrapod skull. Such is the general plan of the dermal bones in Stegocephalia. It resembles somewhat that of early Dipnoi, but more closely that of primitive Teleostomes, not only in its completeness, but also in the elements which compose it. Indeed it affords the strongest evidence for the origin of land vertebrates from a fish-like ancestor in which this general plan had already been well established.

Certain special features remain to be considered. A constant and

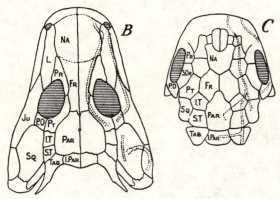

FIG. 319.

Dorsal view of B, skull of *Palaeogyrinus decorus*, Carboniferous, and C, *Dipterus valenciennessi*, Middle Old Red Sandstone, Caithness (from D. M. S. Watson, *Trans. Roy. Soc.*, 1926). In C, *Fr*, parietal (=frontal of Watson); *Par*, dermal occipital (=parietals fused of Watson); *Na*, frontals fused (=nasals of Watson). Cf. Fig. 311.

unexplained difference between Osteichthyes and Tetrapoda is that the pineal foramen is between the frontals in the former, and between the parietals in the latter.

In Osteichthyes the homologue of the lacrimal bone of Tetrapods

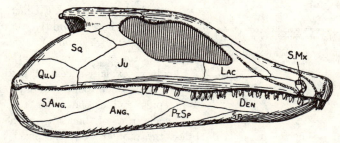

FIG. 320.

Orthosaurus pachycephalus, right lateral aspect of skull and lower jaw (from D. M. S. Watson, *Tr. Roy. Soc.*, 1926). *Ang*, Angular; *Den*, dentary; *Ju*, jugal; *Lac*, lacrimal; *Pt.Sp*, postsplenial; *Qu.J*, quadratojugal; *S.Ang*, supra-angular; *S.Mx*, septomaxillary; *Sp*, splenial; *Sq*, squamosal.

appears to be the dermal element at the anterior ventral edge of the orbit, lodged between the prefrontal and the jugal and belonging to the circumorbital series, Figs. 291, 318. When the ancestors of Tetrapods took to terrestrial life the necessity arose to keep the surface of the eye moist out of water, hence eyelids were developed and certain skin glands were specialised as lacrimal glands. A duct was formed to remove excess of fluid. In modern Tetrapods this lacrimal duct develops as a groove of the outer skin which closes and sinks in (Amniota), or as a solid ingrowth which is nipped off and hollowed out (Amphibia), so as to form a tube leading from the orbit to just inside the external nostril. The duct thus comes to pierce the later developed lacrimal bone, and open below it into the nasal cavity. In the more primitive Stegocephalia and Reptilia the lacrimal extends from orbit to nostril, and in *Micropholis*, according to Watson (633), this bone has a longitudinal groove indicating that the duct was still superficial. Many Stegocephalia and most of the higher Tetrapods have the lacrimal excluded from the nostril by the junction of the nasal and maxillary; and it may also be excluded from the orbit by the prefrontal and jugal (Gregory, 524).

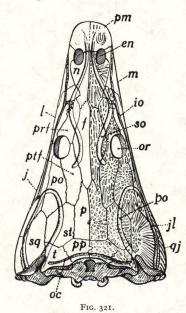

Fig. 321.

Dorsal view of skull of *Trematosaurus Brauni* (from H. Burmeister, 1849, modified). Dermal bones : *f*, Frontal ; *j*, jugal ; *l*, lacrimal ; *m*, maxillary ; *n*, nasal ; *p*, parietal ; *pm*, premaxillary ; *po*, postorbital ; *pp*, postparietal ; *prf*, prefrontal ; *ptf*, postfrontal ; *qj*, quadratojugal ; *sq*, squamosal ; *st*, supratemporal ; *t*, tabulare. Lateral-line grooves : *io*, Infraorbital ; *jl*, jugal ; *oc*, transverse occipital ; *po*, postorbital ; *so*, supraorbital. *en*, External nostril ; *or*, orbit.

The occurrence of an intertemporal in certain Rachitomi (*Trimerorhachis*, Williston, 660 ; *Palaeogyrinus, Micropholis*, Watson, 633, 644), Branchiosauria (*Melanerpeton*, Credner, 484), and again in the primitive reptile (?) *Conodectes (Seymouria)*, Fig. 333 (Williston, 95 ; Watson, 643), is difficult to explain except as a survival of the similar bone found in Osteolepids and Dipnoi. Of the transverse row of bones over the occipital region the outer is doubtless the tabular; but the exact phylogenetic connexion between the paired postparietals of the Tetrapod and the similarly placed median occipital of Crossopterygii is uncertain. In many Teleostomes (Coelacanthini, Palaoniscoidei, etc.) the postparietal elements are paired, while in Dipnoi (p. 305) there is a median occipital as well. Possibly there were here five bones of which the middle one has been lost in the Tetrapods. Both the postparietal and the tabular may spread on to the occipital surface, and the latter usually sends down a posterior plate which joins and may partially cover the paroccipital process of the opisthotic.

More anterior median bones sometimes occur recalling those of the Dipnoi (p. 304). Thus a median internasal (interfrontal) is described in

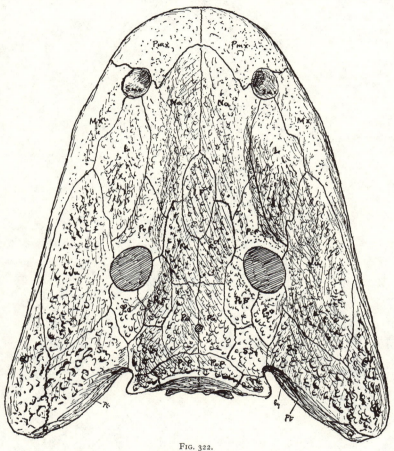

FIG. 322.
Dorsal view of skull of *Eryops megacephalus* (from R. Broom, *Am. Mus. Nat. Hist.*, 1913).

Ricnodon, Sclerocephalus, and *Eryops,* Fig. 322, among Stegocephalia, and a small rostral (internasal) in *Micropholis.*

Little change, except in shape and proportion, occurs in the roofing of the skull in Embolomeri, Rhachitomi, and Stereospondyli. It becomes narrow and elongated in *Cricotus* and *Archegosaurus,* or broad as in *Eryops.* Some show a foramen between the premaxillaries (*Trematops,*

Zatrachys), perhaps due to the presence of a gland ; others have the roof pierced (as in some Crocodilia) for the accommodation of two tusks of the lower jaw. In *Trematops* the roof is excavated backwards from the

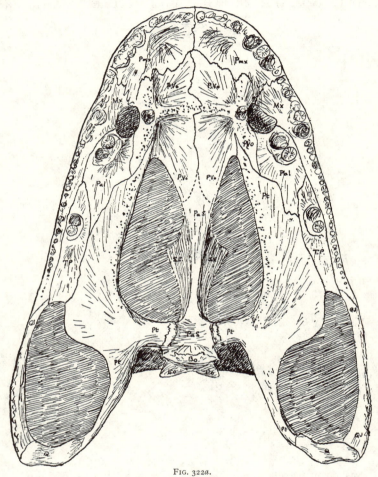

Fig. 322*a*.
Ventral view of skull of *Eryops megacephalus*, Permian, N. America (from R. Broom, *Bull. Am. Mus. Nat. Hist.*, 1913).

nostril, and forwards from the tympanic notch ; and the notch is then closed behind by the union of the tabular with the squamosal (Williston, 660-61).

The Branchiosauria have a broad flattened skull, approaching in general appearance that of the modern Amphibia, of which they may possibly be the ancestors (Fritsch, 23 ; Credner, 484 ; Case, 474 ; Moodie,

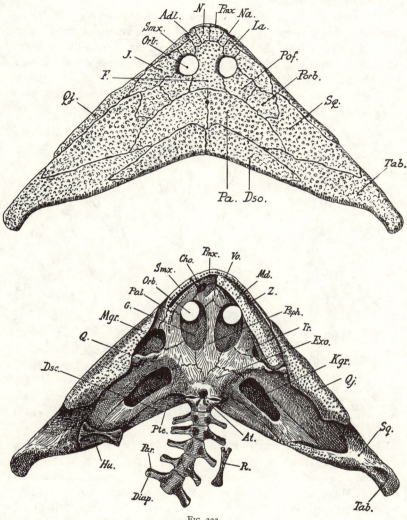

FIG. 323.

Diplocaulus magnicornis. Permian of Texas. Dorsal and ventral views of skull (from O. Abel, 1920). *Adl*, Lacrimal (adnasal) ; *At*, atlas ; *Diap* and *Par*, transverse process of vertebra ; *Dso*, postparietal ; *Exo*, exoccipital ; *F*, frontal ; *G*, palatal groove ; *Hu*, humerus ; *Kgr*, groove for m. retractor mandibulae ; *La*, prefrontal (lacrimal) ; *N*, nasal ; *Na*, external nostril ; *Md*, lower jaw ; *Mgr*, groove for m. adductor mandibulae ; *Orb*, orbit ; *Pa*, parietal ; *Pal*, palatine ; *Pmx*, premaxillary ; *Pof*, postfrontal ; *Porb*, postorbital ; *Pte*, pterygoid ; *Psph*, parasphenoid ; *Q*, quadrate ; *Qj*, quadratojugal ; *R*, rib ; *Smx*, maxillary ; *Sq*, squamosal ; *Tab*, tabular ; *Tr*, ectopterygoid ; *Vo*, prevomer ; *Z*, maxillary teeth.

549 ; Bulman and Whittard, 465a). The orbits are enlarged and also the

corresponding vacuities in the palate, Fig. 325. The endochondral bones seem to be much reduced or absent.

The skull of the Ceraterpetomorpha varies considerably. In Diplocaulidae it becomes enormously expanded at the sides posteriorly, Fig. 323; in Urocordylidae the tabular is produced into a horn-like process; while, in the snake-like Aistopodidae the bones become slender and loosely connected together, Fig. 323a A.

There is found in many Tetrapods a dermal bone named the septo-maxillary.[1] It is situated at the postero-ventral edge of the external

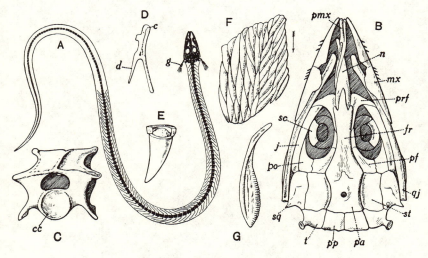

FIG. 323a.

Dolichosoma longissimum, L. Permian, Bohemia; restored (from A. Fritsch, 1885). A, Whole skeleton; B, skull, dorsal view; C, vertebra; D, rib; E, tooth; F, scales; G, cloacal plate. *mx,* Maxillary; *pa,* parietal; *pf,* postfrontal; *pmx,* premaxillary; *sc,* scleral plate. Other letters as in Fig. 321.

nostril, and tends to grow inwards so as to roof over and protect the organ of Jacobson in Amniota (Broom, **439, 451**; Gaupp, **343**; Fuchs, **502**; Lapage, **541**). First discovered by Parker in Frogs, Lizards, and Snakes, where it is developed internally, it has since been found in Stegocephalia, Urodela, Rhynchocephalia, and many extinct Reptiles, Fig. 320. In those forms, like the Chelonia and Crocodilia, where Jacobson's organ is ill-developed it does not occur. Often it is conspicuous on the outer surface of the skull in Theromorpha (Therocephalia,

---

[1] The little anterior paired bone found in certain Teleostomes, such as *Amia,* and often called the septomaxillary, is an endochondral ossification of the floor of the nasal capsule. It does not seem to be homologous with the true septomaxillary of Tetrapods, and is of quite different origin, Fig. 300.

Dromasauria, etc.). In Monotremes it fuses early with the premaxillary forming an important part of its outer surface (Gaupp, 511, in *Echidna*), and among Placentals it occurs as a separate internal bone in Armadillos (Broom, 439). There can be little doubt that it is an ancient bone which has persisted from the earliest Tetrapods, and has probably been derived from one of the dermal plates bordering the external nostril in their fish-like ancestors.

The study of the dermal bones of the palate confirms the view that the Stegocephalia approach the primitive Teleostomes, more especially the

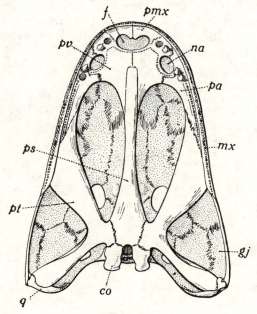

FIG. 324.

Ventral view of skull of *Cyclotosaurus robustus* (from E. Fraas, 1889). *co*, Condyle on ex-occipital ; *f*, foramen for teeth of lower jaw ; *gj*, quadratojugal ; *mx*, maxillary ; *pa*, palatine ; *pmx*, premaxillary ; *ps*, parasphenoid ; *pt*, pterygoid ; *pv*, prevomer ; *q*, quadrate.

Osteolepids. The Lower Carboniferous Embolomeri (Hancock and Atthey, 1869–71 ; Huxley, 1863–9 ; Watson, 631, 644) have, instead of the typical Amphibian palate with large vacuities, an extensive bony palate formed of prevomers separating the internal nostrils, palatines, and ectopterygoids disposed as in Crossopterygians, and large pterygoids which meet in front and are separated from the slender parasphenoid only by a narrow space. These pterygoids reach the well-developed basitrabecular processes of the basisphenoid and spread backwards over the inner side of the quadrates. They represent the pterygoids (so-called

endopterygoids) of fish (cf. *Eusthenopteron,* Fig. 306), and bear small
teeth ; while large tusk-like teeth, each with a replacement tooth by its
side, are present on the palatines and prevomers of these and many other
Stegocephalia, Figs. 306, 322a.

The parasphenoid in the Embolomeri was comparatively small, closely
connected with the basisphenoid posteriorly, and extending forward below
the sphenethmoid as a narrow grooved blade (Watson, **644**).

In the more advanced types (Rhachitomi, Stereospondyli) the interpterygoid palatal vacuities become progressively enlarged, the parasphenoid
expands behind and becomes immovably sutured to the pterygoid,
Figs. 323, 324, 325. The latter does not meet its fellow in front, and
becomes shortened until it no longer reaches the prevomer or even the

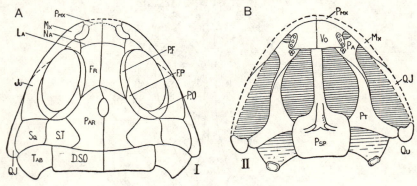

FIG. 325.

Skull of *Branchiosaurus amblystomus,* reconstructed. A, Dorsal ; B, ventral view (from Bulman
and Whittard, *Proc. Zool. Soc.,* 1926). *D.S.O,* Dermal supraoccipital=postparietal ; *Psp,* parasphenoid ; *Vo,* prevomer. Other letters as in previous figures.

palatine. The pterygoid is produced backwards as a wide flange which
tends to surround the inner side of the quadrate below and meet the
squamosal behind, a disposition which survives only in such primitive
Reptiles as *Seymouria, Pareiasaurus,* and *Varanosaurus* (Williston, **95** ;
Broili, **435, 437** ; Watson, **644** ; Sushkin, **768**).

The lower jaw in the more primitive Stegocephalia has, besides the
endochondral articular, a number of dermal covering bones comparable to
those of primitive Teleostomes (Williston, **659, 95** ; Bransom, **427** ; A. S.
Woodward ; Broom, **450** ; Watson, **632, 644**). The large dentary holds
powerful teeth, the outside of the jaw is completed by an angular and
supra-angular behind, and a splenial and postsplenial below (comparable
to the infradentaries of early fishes). The dentary and splenial enter
the symphysis. A prearticular covers the inner side, with a series of
more dorsal coronoids (coronoid, intercoronoid, precoronoid) roofing the

Meckelian cavity and limiting the supramandibular fossa. These coronoids may bear teeth.

The modern Amphibia, sometimes grouped together as Euamphibia, have several cranial characters in common. The pineal foramen has disappeared; the dermal covering bones are more or less reduced in extent, thickness, and number; the temporal region of the roofing is absent from the orbit backwards, leaving exposed the otic bones (in Apoda, however, the small orbit may be surrounded by bone). The postfrontal, postorbital, supratemporal, tabular, postparietal, and ectopterygoid have disappeared.

The Urodela are in some respects the least specialised, but, except in

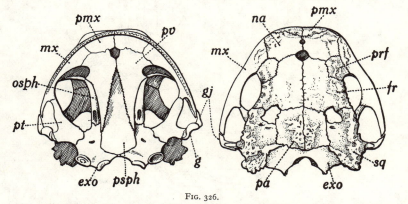

Fig. 326.

Ventral and dorsal views of skull of *Tylototriton verrucosus* (from H. Riese, 1891). *fr*, Frontal; *g*, quadrate; *osph*, orbitosphenoid; *pa*, parietal; *prf*, prefrontal. Other letters as in Figs. 290 and 324.

Tylototriton (Riese, 582), the margin of the roof is lost, there being neither jugal nor quadratojugal, the squamosal over the quadrate being the only dermal element left in this region, Figs. 326-8. The junction, in Salamandridae, of frontal and squamosal is probably secondary. In front of the orbit the lacrimal is lost (except in *Ranidens* and *Hynobius*) and the prefrontal also in *Necturus*. The nasal, frontal, parietal, premaxillary, and maxillary usually remain (Parker, 562, 573; Wiedersheim, 651). In the Perennibranchiata even the maxillaries and nasals are lost. The dermal bones of the lower jaw are also much reduced, there being, as a rule, only a dentary, a prearticular, and an inner tooth-bearing bone generally called splenial but more probably representing the coronoids, Figs. 327-8. An angular may also be present in Caducibranchiates. The palate in Urodela is remarkable for the width of the parasphenoid and the disappearance of the palatine, the short pterygoid being disconnected from the prevomer.

According to Winterbert (1910) the usual statement that the palatine shifts at metamorphosis and joins the vomer near the parasphenoid (Parker, Wiedersheim) is erroneous; the palatine disintegrates and the prevomer is altered by reduction at its outer border and growth at its median border.

The Anura have a still more specialised skull (Parker, 558, 559); but in them the margin remains complete with premaxillary, maxillary, and

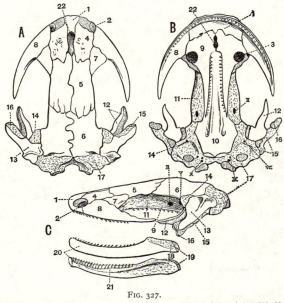

FIG. 327.

A, Dorsal; B, ventral; C, lateral views of skull of *Molge cristata* (after W. K. Parker, from S. H. Reynolds, *Vertebrate Skeleton*, 1913). Cartilage dotted, cartilage bones marked with dots and dashes, membrane bones left white. Upper view of mandible in C shows outer surface, lower shows inner surface. 1, Premaxillary; 2, anterior nares; 3, posterior nares; 4, nasal; 5, frontal; 6, parietal; 7, prefronto-lacrimal; 8, maxillary; 9, prevomero-palatine; 10, parasphenoid; 11, orbitosphenoid; 12, pterygoid; 13, squamosal; 14, pro-otic region of exoccipito-periotic; 15, calcified portion of quadrate region; 16, uncalcified portion of quadrate region; 17, exoccipital region of exoccipito-periotic; 18, calcified portion of articular region; 19, uncalcified portion of articular region; 20, dentary; 21, splenial; 22, space for glands; II, V, VII, IX, X, foramina for exit of cranial nerves.

quadratojugal bones. Nasals are present, also small septomaxillaries, but the superior fontanelle is covered by a single pair of bones called fronto-parietals. As a rule the sphenethmoid in front and the bones of the auditory capsules behind are exposed (the lateral temporal roofing having, of course, been lost), and the junction of the squamosal with the fronto-parietal and maxillary in such forms as *Calyptocephalus*, Fig. 329, is probably secondary.[1] The palate has enormous vacuities, and the small

[1] The bones which provide these extensions in Anura and also in *Tylototriton* have a rough outer surface resembling that of the secondary dermal ossifications on the pectoral girdle of some related forms, and are probably of the same origin.

palatine takes up a transverse position behind the prevomer. Only two dermal bones remain on each ramus of the lower jaw : a dentary, and a

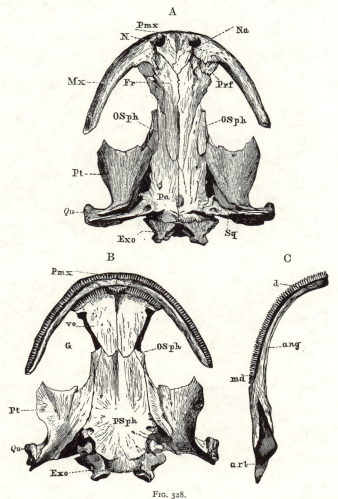

FIG. 328.

Skull of *Cryptobranchus japonicus*, v. d. Hoeven. A, Dorsal, and B, palatal aspects ; C, lower jaw. *Pmx*, Premaxilla ; *Mx*, maxilla ; *Na*, nasal ; *Prf*, prefrontal ; *Fr*, frontal ; *Pa*, parietal ; *OSph*, orbitosphenoid ; *Exo*, exoccipital ; *Qu*, quadrate ; *Sq*, squamosal ; *Pt*, pterygoid ; *PSph*, parasphenoid ; *Vo*, prevomer ; *G,* palatine vacuity ; *N*, external nares. (From K. Zittel, *Palaeontology*.)

larger bone which represents either the prearticular or a combination of prearticular and angular (?). It is often named angulosplenial.

The more complete ossification and more rigid structure of the skull in Apoda seems to be related to their burrowing habits. But in this

Y

group, as in other living Amphibia, the dermal bones have been much reduced in number and the lateral temporal roof lost, Fig. 330. The apparently complete roofing in *Ichthyophis* (Wiedersheim, 652; Sarasin, 763), where the gap between the squamosal and parietal is narrowed to a mere slit, is probably due to secondary consolidation, not to the survival of a stegocephalian roof.

The interesting Upper Carboniferous fossil *Lysorophus* (Cope, 1877;

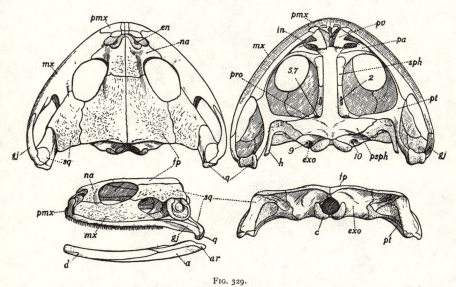

FIG. 329.

Dorsal, ventral, left-side, and posterior views of skull of *Calyptocephalus gayi* (from Parker, 1881). *a*, Angulosplenial; *ar*, articular cartilage; *c*, condyle; *d*, dentary; *en*, external nostril; *exo*, exoccipital; *fp*, frontoparietal; *gj*, quadratojugal; *h*, hyoid cornu; *in*, internal nostril; *mx*, maxillary; *na*, nasal; *pa*, palatine; *pmx*, premaxillary; *pro*, pro-otic; *psph*, parasphenoid; *pt*, pterygoid; *pv*, prevomer; *q*, quadrate cartilage; *sph*, sphenethmoid; *sq*, squamosal. Foramina: 2, of optic; 5 and 7, of trigeminal and facial; 9, of glossopharyngeal; 10, of vagus nerves.

Case, 472; Broili, 1904; Williston, 656; Moodie, 549; Huene, 531a; Sollas, 603) seems to be an elongated Amphibian with reduced limbs (Finney, 1912) allied to the Urodela or possibly to the Apoda. But it is far more primitive than any modern Amphibian, and retains the tabular and supra-temporal on the skull, and a supra-angular on the lower jaw. Moreover, the chondrocranium was more fully ossified, with supra- and basioccipital, and an orbitosphenoid (posterior pillar of Sollas).

Dermal Bones of the Skull in the Amniota.—The classification of the larger groups included in the Class Reptilia presents great difficulties, but ever since the pioneer work of Baur (414-17), Cope (479, 480), and

Osborn (554), the structure of the dermal roofing of the skull has been recognised as of the greatest value in tracing out the affinities of the Reptiles not only with each other, but also with the Birds on the one hand and the Mammals on the other (Broom, 459; Gaupp, 505; Versluys, 626; Williston, 655, 95; Woodward, 231; Boas, 424). It is now admitted that the early primitive Reptilia had a complete roofing like that of Stegocephalia and primitive Osteichthyes, pierced only for orbits, nostrils, and pineal eye; and further, that secondary openings have been formed

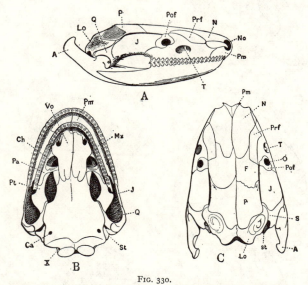

Fig. 330.

Skull of *Ichthyophis glutinosa*, × 3. A, Lateral; B, ventral; C, dorsal view. A, Posterior process of the os articulare; *Ca*, carotid foramen; *Ch*, choana or posterior nasal opening; *F*, frontal; *J*, jugal; *Lo*, exoccipital; *Mx*, maxilla; *N*, nasal; *No*, nostril; *O*, orbit; *P*, parietal; *Pa*, palatine; *Pm*, premaxilla; *Pof*, postfrontal; *Prf*, prefrontal; *Pt*, pterygoid; *Q*, quadrate; *S*, paraquadrate (squamosal); *St*, stapes; *T*, tentacular groove; *Vo*, vomer; *x*, exit of vagus nerve. (After Sarasin, from Parker and Haswell, *Zoology*.)

in this roofing in different ways in the various diverging phylogenetic lines. Thus have arisen openings variously called vacuities, foramina, or fossae, leaving where necessary buttresses or arches to strengthen the skull, support the jaws and quadrates, and delimit the orbits. It is agreed that there are at least three main types of dermal roofing: (1) with one lateral temporal fossa; (2) with two lateral temporal fossae, one superior and the other inferior; (3) with no fossa at all, Fig. 331. The latter 'stegocephalian' condition is found in those primitive Carboniferous Permian and Triassic Reptiles which retain the original complete roofing intact, usually with all or most of the original dermal bones (Microsauria,

Seymouriomorpha, Cotylosauria). Such primitive forms, in which the whole skull is built on essentially the same plan as in Stegocephalia, may be provisionally included in the Division Anapsida (Williston) ; probably

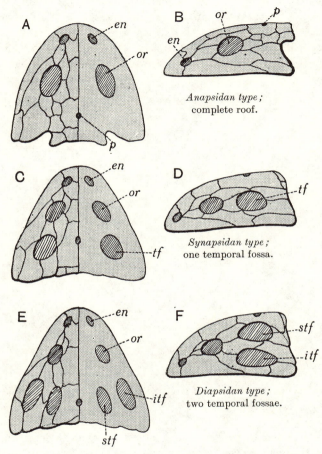

A

B   *Anapsidan type;*
complete roof.

C

D   *Synapsidan type;*
one temporal fossa.

E

F   *Diapsidan type;*
two temporal fossae.

Fig. 331.

Diagrams showing three principal types of cranial roof in *Tetrapoda*. Dorsal views on left, lateral views on right. *en,* External nostril ; *or,* orbit ; *p,* pineal foramen ; *itf, stf, tf,* inferior, superior, and single lateral temporal fossae.

a polyphyletic assemblage containing scarcely differentiated early representatives of several divergent groups. The Microsauria still have a frankly stegocephalian structure and are often included in the Amphibia (Baur, 420 ; Moodie, 549).

Of all the Permian Reptiles hitherto discovered, *Seymouria (Conodectes)* is perhaps the most primitive (Cope, **481-2**; Williston, **661**; Watson, **643**; Sushkin, 1925). Not only is the roofing complete, but it still preserves an intertemporal (as mentioned above), a bone which has disappeared in other forms, Figs. 332-3. Its palate is more reptilian, the internal nostrils being near the middle line and the parasphenoid relatively small. There are well-developed basipterygoid (basitrabecular) articulations.[1] The pterygoid still spreads under and behind the quadrate to meet the squamosal.

The occiput shows a tripartite condyle, and a post-temporal fossa reduced by downgrowths of the tabular and post-parietal (as in Rhachitomi). The lower jaw retains the postsplenial and three coronoids.

The Cotylosauria are more distinctly reptilian. (Seeley, **586**; Williston, **95, 661**; Huene v. Branson, **428, 530**; Case, **466, 470, 473**; Watson, **637, 639, 643**; Broom, **444, 449, 452, 460**.) The quadrate becomes more vertical, the supratemporal may disappear, and likewise the tympanic notch (Pareiasauria, Captorhinomorpha). The descending flange of the postparietal and tabular may almost obliterate the post-temporal opening (Captorhinomorpha, Diadectomorpha). All these Cotylosaurians, and especially the Procolophonida, have a more reptilian palate

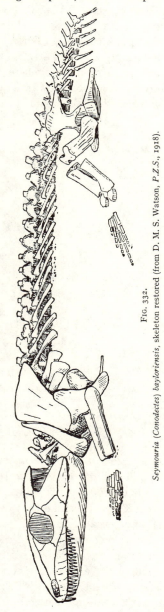

Fig. 332.

Seymouria (Conodectes) baylorinsis, skeleton restored (from D. M. S. Watson, *P.Z.S.*, 1918).

---

[1] A small bone intervenes between the basitrabecular process and the pterygoid (Watson, **643**). It may represent the meniscus pterygoideus of Rhynchocephalia (Howes and Swinnerton, **528**), Lacertilia (Gaupp, **506**), and Chelonia (Fuchs, **503**) in an ossified condition. Possibly this element is the remains of the endochondral metapterygoid of Teleostomes (Broom, **455**). According to Sushkin *Seymouria* is a Stegocephalian, 1925.

with pterygoids converging anteriorly, and with a pronounced ventral flange, Fig. 334.

From some 'Anapsidan' must have evolved the other two chief types—the 'Synapsidan' and the 'Diapsidan' (Osborn, 554). The Synapsida, typically represented by the Theromorpha, are distinguished by the possession of a single temporal fossa, Fig. 340. The cranial roof bones are at first complete; but the supratemporal remains only in primitive Pelycosauria, such as *Varanosaurus*, Fig. 336. The tabulars and postparietals become plastered on to the posterior occipital surface,

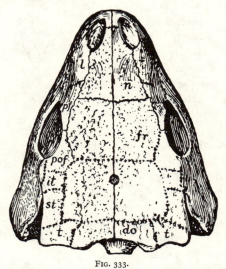

FIG. 333.

Dorsal view of skull of *Conodectes* (*Seymouria*) *bayloiensis*. Cotylosaur skull. *n*, Nasal; *l*, lacrimal; *pf*, prefrontal; *pof*, postfrontal; *fr*, frontal; *ti*, intertemporal; *do*, postparietal (dermo-occipital); *t*, tabulare. (From Williston, *Osteology of Reptiles*, 1925.)

the latter bones being often fused to form a median interparietal; and the post-temporal openings tend to become reduced. The single temporal fossa seems to have originated between the postorbital and jugal, and is still enclosed by these and the squamosal in Pelycosauria and Deino-cephalia, Figs. 336-8. But in more advanced forms it extends upwards as in Mammalia and is then bounded above by the parietal (Dicynodontia, Theriodontia), Figs. 340, 342. The transition to the usual Mammalian type is brought about by the enlargement of the fossa, and loss of the prefrontal, postfrontal, postorbital, and quadratojugal (already much reduced or vestigial in Theriodonts). The tabular is finally lost, or in-distinguishably fused with the parietal, while the two postparietals may remain as an interparietal often fused with the supraoccipital.

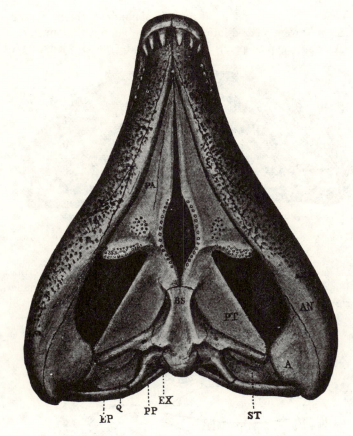

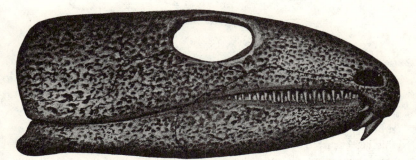

FIG. 334.

Ventral and right-side view of skull of *Labidosaurus hamatus* (from Williston, 1910). *A*, Articular; *AN*, angular; *BS*, basisphenoid (+ parasphenoid?); *EP*, tabular; *EX*, exoccipital; *PA*, palatine; *PP*, postparietal; *PT*, pterygoid; *Q*, quadrate; *ST*, stapes.

Meanwhile the post-temporal fossa is obliterated, the post-temporal bar being as it were merged into the hinder region of the brain-case, Fig. 343. As this region·expands to accommodate the growing brain, the

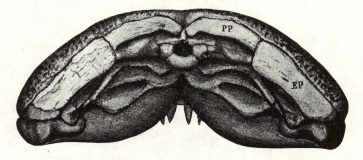

FIG. 335.

Posterior view of skull of *Labidosaurus hamatus* (from Williston, 1910). *EP*, Tabular; *PP*, postparietal.

parietals and squamosal spread over its surface below the powerful jaw muscles, and are withdrawn from the more superficial layers representing the original roofing. Lastly, in the Mammalia, the postorbital bar is inter-

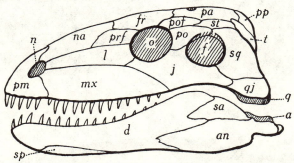

FIG. 336.

Pelycosauria. Diagram of skull and lower jaw. Openings in cranial roof: *f*, Lateral temporal fossa; *n*, external nostril; *o*, orbit; *if*, inferior temporal fossa; *pf*, pineal foramen; *pl*, preorbital fossa; *sf*, superior temporal fossa; *q*, quadrate; *a*, articular. Dermal bones: *an*, Angular; *co*, coronoid; *d*, dentary; *fr*, frontal; *ip*, interparietal (fused postparietals); *j*, jugal; *l*, lacrimal; *mx*, maxillary; *na*, nasal; *pa*, parietal; *pm*, premaxillary; *po*, postorbital; *pof*, postfrontal; *pp*, postparietal; *prf*, prefrontal; *qj*, quadratojugal; *s*, probably squamosal; *sa*, supra-angular; *sm*, septomaxillary; *sp*, splenial; *sq*, squamosal; *st*, supratemporal.

rupted, leaving the orbit confluent with the lateral temporal fossa (already this has occurred in the Cynodont *Bauria*, Fig. 499 [Broom, 443-5, 451, 458], but probably independently). Strangely enough, an analogous postorbital bar is re-formed in many mammals by the junction of the frontal with the jugal (later Ungulates, Primates, *Tupaja*).

The canal, arched over by the squamosal, at the side of the occipital region of the skull in Monotremes, is supposed to represent a vestige of the post-temporal opening occupying about the same position in Reptiles, Fig. 403.

Other orders of reptiles besides the Theromorpha have a single lateral

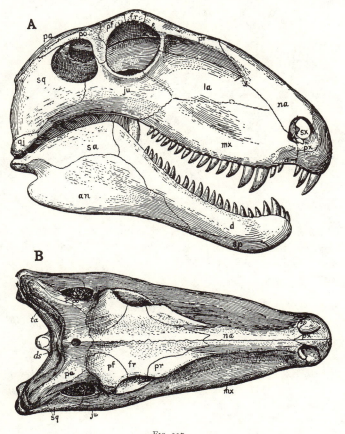

FIG. 337.

Skull of *Sphenacodon*; A, left-side, and B, dorsal view (from Williston, *Osteology of Reptiles*, 1925).

temporal fossa. Among these are the Sauropterygia (including Meso-sauria), Placodontia, Thalattosauria, and Ichthyosauria, and the important question arises whether in all or any of these the fossa is the same as that of the Theromorpha and Mammalia, or has been independently acquired, Fig. 344.

The temporal opening in the Sauropterygia (and Placodontia, which are probably allied to them) is large and bounded by the postfrontal, postorbital, squamosal, and parietal (the supratemporal, tabular, and post-

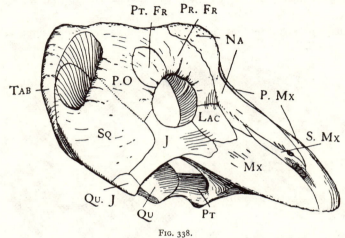

FIG. 338.

Mormosaurus seeleyi, right-side view of skull (D. M. S. Watson, *P.Z.S.*, 1914).

parietal have been lost). It, therefore, extends farther dorsally than in primitive Theromorpha, and is closed below by the postorbital meeting the squamosal, Figs. 344-6. Nevertheless, it appears to be essentially

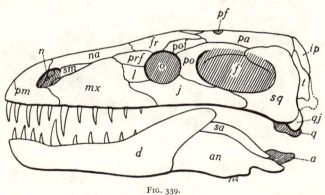

FIG. 339.

Therocephalia. Diagram of skull and lower jaw. Lettering as in Fig. 336, p. 328.

the same fossa, though some authors claim that it has had a separate origin (Versluys, **626** ; Broom, **456, 459**).

A single temporal fossa, usually of small size, occurs in the Ichthyosauria. Here the original reptilian roofing bones are complete excepting for the postparietal and tabular,[1] and the fossa lies between the parietal above and the postfrontal and supratemporal below, Figs. 348-9, 350.

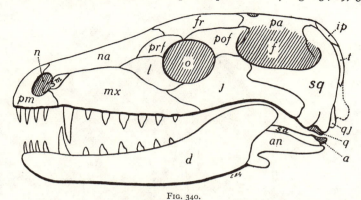

FIG. 340.

Cynodontia.   Diagram of skull and lower jaw.   Lettering as in Fig. 336, p. 328.

The affinities of this group are still very obscure. The great elongation of the snout, the enlargement of the orbit and consequent reduction of the posterior roofing and other characteristic features, appear to be

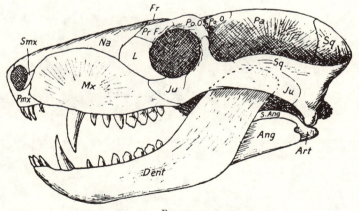

FIG. 341.

Thrinaxodon, left-side view of skull and lower jaw, restored (from R. Broom, *P.Z.S.,* 1911).

adaptations to an aquatic and predaceous mode of life, and are less pronounced in some of the early Triassic forms (Merriam, 547). The fossa

---

[1] According to Broom (447, 456, 459) it is the supratemporal which is lost and the tabular preserved, not only in Ichthyosauria, but also in *Youngina* and Lacertilia.

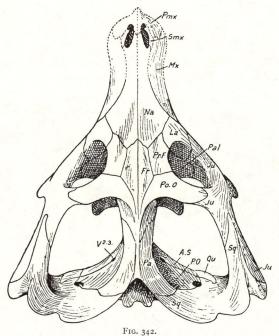

FIG. 342.

Gomphognathus minor, dorsal view of skull, restored (from R. Broom, *P.Z.S.*, 1911).

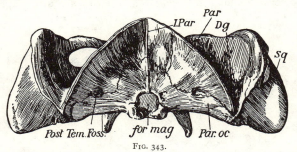

FIG. 343.

Diademodon browni, posterior view of skull (from D. M. S. Watson, *Ann. Mag. N.H.*, 1911)
Dg, Digastric groove.

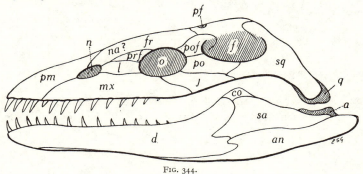

FIG. 344.

Plesiosauria. Diagram of skull and lower jaw. Lettering as in Fig. 336, p. 328.

in these is larger than in the later fossils, and appears not to differ essentially from that of other Synapsida.

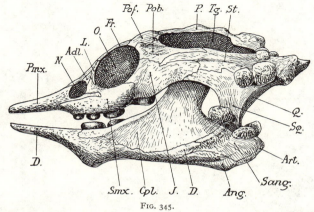

FIG. 345.

Placochelys placodonta, Trias of Hungary; left-side view of skull (after O. Jaekel, from O. Abel, 1920). *Ang*, Angular; *Art*, articular; *Cpl*, coronoid; *D*, dentary; *O*, orbit; *Sang*, supra-angular; *Tg*, temporal fossa. Other letters as in Fig. 336.

Quite distinct from the foregoing are the Diapsida with two temporal fossae and two lateral temporal arches on each side. This group

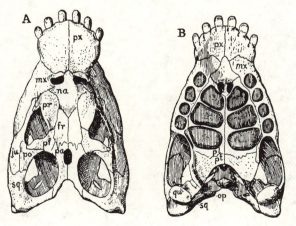

FIG. 346.

Skull of *Placodus*; A, dorsal, and B, ventral view (after Broili, from Williston, *Osteology of Reptiles*, 1925).

includes the Rhynchocephalia, Pseudosuchia, Phytosauria, Crocodilia, Pterosauria, and 'Dinosaurs' (Saurischia and Ornithischia), and leads towards the Aves. Broom discovered *Youngina*, the earliest and perhaps the most primitive known Diapsidan, in the Permian of S. Africa (Broom, 457). In general build the skull of this remarkable reptile resembles

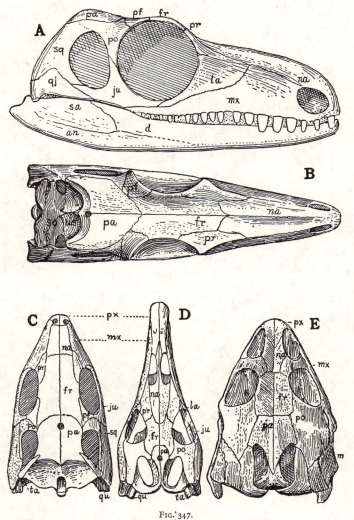

FIG. 347.

A, Left-side, and B, dorsal view of skull of *Mycterosaurus*; C, dorsal view of skull of *Araeoscelis*; D, of *Pleurosaurus*; E, of *Sauranodon* (from Williston, *Osteology of Reptiles*, 1925).

that of *Sphenodon*, but has preserved all the roofing bones except one, either tabular or supratemporal. The superior temporal fossa lies between the parietal, postorbital, and supratemporal (tabular ?) ; and the inferior

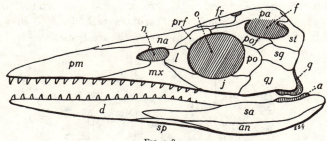

FIG. 348.

Ichthyosauria. Diagram of skull and lower jaw. Lettering as in Fig. 336, p. 328.

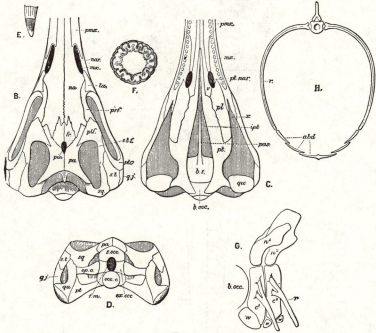

FIG. 349.

Diagrams illustrating structure of *Ichthyosaurus* (from A. S. Woodward, *Vertebrate Palaeontology,* 1898). B, C, D, dorsal, ventral, and posterior views of skull; E, tooth; F, its transverse section; G, atlas and axis; H, dorsal. *b.occ,* Basioccipital; *b.s,* basisphenoid; *d,* dentary; *ex.occ,* exoccipital; *f.m,* foramen magnum; *fr,* frontal; *ipt,* interpterygoid vacuity; *j,* jugal; *la,* lacrimal; *mx,* maxillary; *na,* nasal; *nar,* external narial opening; *occ.c,* occipital condyle; *op.o,* opisthotic; *pa,* parietal; *pas,* parasphenoid; *pin,* pineal foramen; *pl,* palatine; *pmx,* premaxillary; *prf,* prefrontal; *pt,* pterygoid; *pt.nar,* posterior nares; *ptf,* postfrontal; *pto,* postorbital; *q.j,* quadratojugal; *qu,* quadrate; *s.ag,* surangular; *s.occ,* supraoccipital; *s.t,* supratemporal (prosquamosal); *s.t.f,* supra-temporal vacuity; *scl,* sclerotic plates; *spl,* splenial; *sq,* squamosal; *x,* space for ectopterygoid (?); *v,* vomer. *abd,* Abdominal ribs; *c¹,* centrum of atlas; *c²,* centrum of axis; *n¹,* paired neural arch of atlas; *n²,* single neural arch of axis; *r,* ribs; *w,* subvertebral wedge-bones (intercentra or hypocentra).

fossa between the postorbital, jugal, quadratojugal, and squamosal, Figs. 351, 353 A. Once established, the Diapsidan type diverged in various directions. The Rhynchocephalia, one of the least specialised of these

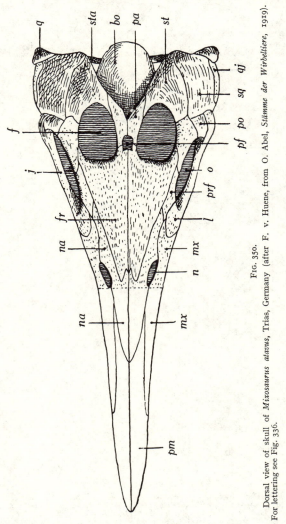

FIG. 350.

Dorsal view of skull of *Mixosaurus atavus*, Trias, Germany (after F. v. Huene, from O. Abel, *Stämme der Wirbeltiere*, 1919). For lettering see Fig. 336.

branches, survive at the present day in the single species *Sphenodon punctatum* of New Zealand (Günther, 1867; Siebenrock, 598; Ossawa, 552; Howes and Swinnerton, 528; Schauinsland, 583).

It has lost the lacrimal, postorbital, supratemporal, tabular, and post-

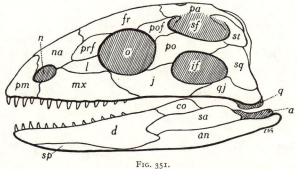

FIG. 351.

Eosuchia. Diagram of skull and lower jaw. Lettering as in Fig. 336, p. 328.

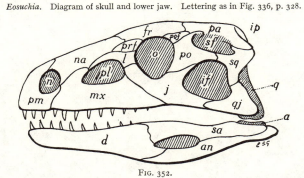

FIG. 352.

Pseudosuchia. Diagram of skull and lower jaw. *pl*, Preorbital fossa. Other lettering as in Fig. 336, p. 328.

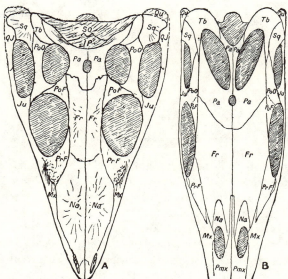

FIG. 353.

Dorsal view of restoration of skull of : A, *Youngina capensis* ; B, *Thalattosaurus alexandrae* (from R. Broom, *Proc. Zool. Soc.*, 1925).

Z

parietal. There are large post-temporal openings, the brain-case being attached to the roofing parietals only near the middle line, as in most Lacertilia; and a post-temporal arch formed by the squamosal and a process of the parietal extends on each side to the fixed quadrate. Large outstanding paroccipital processes lend further support to the quadrate. The teeth are firmly ankylosed to the premaxillary, maxillary, and dentary (acrodont), Fig. 354.

Under the name Thecodontia (with teeth set in sockets) or Archosauria

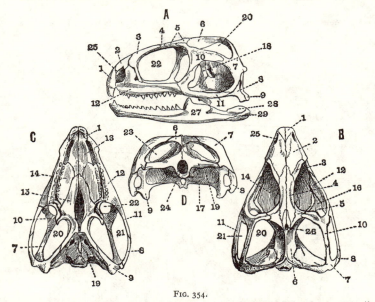

FIG. 354.

A, Lateral; B, dorsal; C, ventral; and D, posterior view of skull of *Sphenodon punctatus* (from S. H. Reynolds, *Vertebrate Skeleton*, 1913.) 1, Premaxillary; 2, nasal; 3, prefrontal; 4, frontal; 5, postfrontal; 6, parietal; 7, squamosal; 8, quadratojugal; 9, quadrate; 10, postorbital; 11, jugal; 12, maxillary; 13, prevomer; 14, palatine; 15, pterygoid; 16, transpalatine; 17, exoccipital; 18, epipterygoid; 19, basisphenoid; 20, supratemporal fossa; 21, infratemporal or lateral temporal fossa; 22, orbit; 23, post-temporal fossa; 24, foramen magnum; 25, anterior nares; 26, pineal foramen; 27, dentary; 28, supra-angular; 29, articular.

are sometimes grouped the Eosuchia, Parasuchia, 'Dinosaurs' (Saurischia and Ornithischia), and Pterosauria, all Diapsidan reptiles allied to the Crocodilia. Except the Crocodilia, they all possess a new opening in the dermal roofing between the lacrimal and maxillary, known as the pre-orbital or antorbital fossa. In most Crocodiles it appears to have been secondarily closed up, persisting, however, in some Teleosauridae. Even the superior temporal fossa may be much reduced in extent and rarely quite closed in modern Crocodiles, apparently by the deposition on the skull of new bony matter comparable to the bony dermal scutes developed over the rest of the body. This new growth adds a secondary rough superficial sculpturing to the skull (excepting the postorbital bar in all

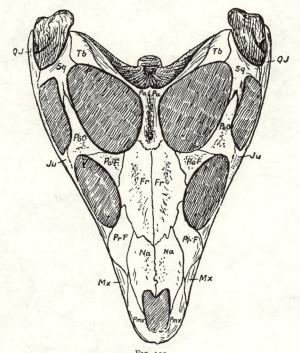

Fig. 355.

Dorsal view of skull of *Mesosuchus browni* (from R. Broom, *Proc. Zool. Soc.*, 1925).

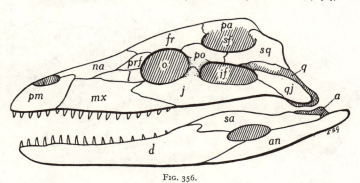

Fig. 356.

Crocodilia. Diagram of skull and lower jaw. Lettering as in Fig. 336, p. 328.

later forms) bearing a deceptive resemblance to the primitive sculpturing of Stegocephalia, Figs. 356-9, and p. 354.

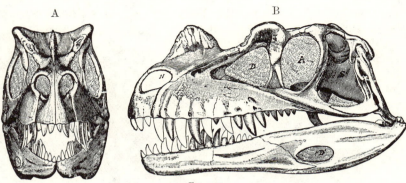

FIG. 357.

Ceratosaurus nasicornis, Marsh. Upper Jura : Colorado. Skull from anterior (A) and lateral (B) aspects, ⅕. *A*, Orbit ; *D*, antorbital vacuity ; *D'*, mandibular vacuity ; *N*, external nostril ; *S*, supratemporal vacuity. (After Marsh, from K. Zittel, *Palaeontology*.)

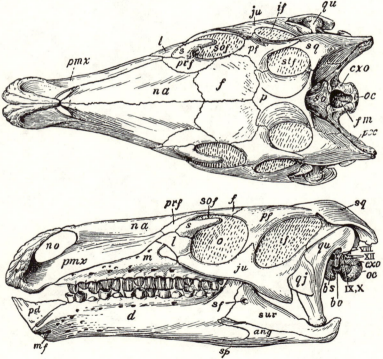

FIG. 358.

Dorsal and left-side view of skull of *Camptosaurus* (after C. W. Gilmore, '09, from O. Abel, *Stämme der Wirbeltiere*, 1919). *ang*, Angular ; *bo*, basioccipital ; *bs*, basisphenoid ; *d*, dentary ; *exo*, exoccipital ; *f*, frontal ; *fm*, foramen magnum ; *if*, inferior temporal fossa ; *ju*, jugal ; *l*, lacrimal ; *m*, maxillary ; *mf*, mental foramen ; *na*, nasal ; *no*, nasal opening ; *o*, orbit ; *oc*, occipital condyle ; *pd*, predentary ; *pf*, postfrontal ; *pmx*, premaxillary ; *poc*, opisthotic ; *prf*, prefrontal ; *qj*, quadrato-jugal ; *qu*, quadrate ; *s*, supraorbital ; *sf*, external vacuity ; *sp*, splenial ; *sq*, squamosal ; *stf*, superior temporal fossa ; *sur*, supra-angular.

The Archosauria, at all events the Crocodilia, Saurischia, Ornithischia, and Pterosauria, are further characterised by the fact that the inferior

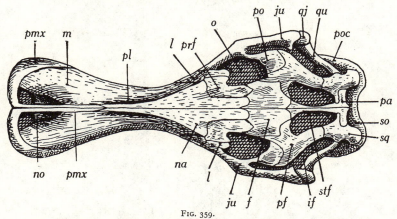

FIG. 359.

Reconstruction of skull of *Trachodon mirabilis*, Upper Cretaceous, N. America (from O. Abel, *Stämme der Wirbeltiere*, 1919). *pl*, Preorbital fossa; *po*, postorbital. Other letters as in Fig. 358.

temporal fossa is usually open behind, Figs. 356, 360-61. For the quadrato-jugal fails to join the squamosal (supratemporal ?),[1] thus exposing part of the quadrate (p. 433).

Besides the Reptilia dealt with above, which may confidently be

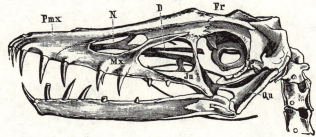

FIG. 360.

Scaphognathus crassirostris, Goldf. sp., Upper Jura, Eichstädt, Bavaria. *D*, Antorbital vacuity; *Fr*, frontal; *Ju*, jugal; *Mx*, maxilla; *N*, narial opening; *Pmx*, premaxilla; *Qu*, quadrate. (From K. Zittel, *Palaeontology*.)

classified into Synapsida and Diapsida, there remain certain important Orders provisionally included in the Parapsida whose position is still

[1] In most Archosauria and also in Rhynchocephalia only one bone persists between the parietal and the quadratojugal, and it is difficult, if not impossible, on existing evidence to decide whether it is the supratemporal or the squamosal. Both views have been held. On the whole, it seems best to call this bone the squamosal.

doubtful.   The Lacertilia have a skull with a single lateral temporal fossa,

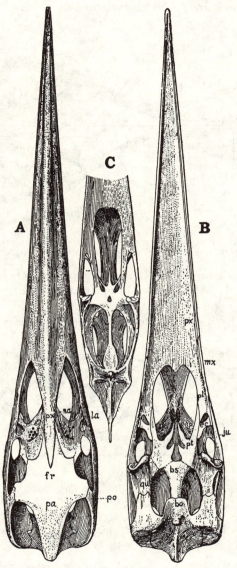

FIG. 361.

Skull of *Nyctosaurus:* A, from above, B, from below; and of *Pteranodon:* C, from below (after Eaton, from Williston, *Osteology of Reptiles,* 1925).

and the interesting question arises whether this opening corresponds to

the single one in Synapsida, or to the upper fossa of the Diapsida,[1] Figs.
362-6.   According to the first interpretation the lateral temporal bar of

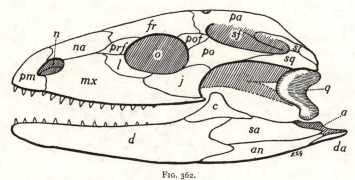

FIG. 362.

Lacertilia.   Diagram of skull and lower jaw.   Lettering as in Fig. 328.

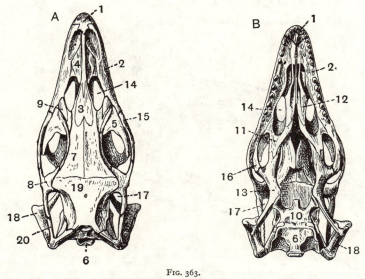

FIG. 363.

A, Dorsal, and B, ventral view of cranium of *Varanus* sp. (from S. H. Reynolds, *Vertebrate Skeleton*, 1913).   1, Premaxillary ; 2, maxillary ; 3, nasal ; 4, septomaxillary ; 5, supra-orbital ; 6, basioccipital ; 7, frontal ; 8, postfrontal ; 9, lacrimal ; 10, basisphenoid ; 11, palatine ; 12, prevomer ; 13, pterygoid ; 14, anterior narial opening ; 15, jugal ; 16, transpalatine ; 17, supratemporal fossa ; 18, quadrate ; 19, parietal ; 20, squamosal.

the Lacertilian would represent that of the Synapsidan, and would then be
formed presumably of a jugal and a quadratojugal ; while, according to

[1] The postorbital and temporal bars may be interrupted in degenerate
lizards, such as the Geckonida, and in the Amphisbaenidae they disappear.

the second, this bar would represent the upper one of the Diapsidan skull formed by the postorbital joining the squamosal. Huxley was of the opinion that the Lizards were derived from some form like *Sphenodon*, which they closely resemble in many characters, and he believed that the lower temporal bar has disappeared (the quadratojugal being now

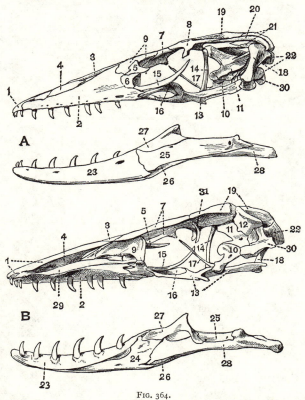

FIG. 364.

A, Lateral view, and B, longitudinal section of skull of *Varanus varius* (from S. H. Reynolds, *Vertebrate Skeleton*, 1913). 1, Premaxillary; 2, maxillary; 3, nasal; 4, septomaxillary; 5, supra-orbital; 6, lacrimal; 7, frontal; 8, postorbital; 9, prefrontal; 10, basisphenoid; 11, pro-otic; 12, supraoccipital; 13, pterygoid; 14, epipterygoid (columella cranii); 15, jugal; 16, transpalatine; 17, parasphenoid; 18, quadrate; 19, parietal; 20, squamosal; 21, supratemporal; 22, exoccipital; 23, dentary; 24, splenial; 25, supra-angular; 26, angular; 27, coronoid; 28, articular (+pre-articular); 29, vomer; 30, basioccipital; 31, orbitosphenoid.

represented only by a ligament), setting free the quadrate (streptostylic condition, p. 433). This view has been adopted by many (Broom, **456**; Versluys, **796**; and others). Williston (**95**) and Watson (**638**) derive the Lacertilia from some form with a single temporal fossa and a wide bar, like the Permo-Carboniferous *Araeoscelis* or the Jurassic *Pleurosaurus*, by the narrowing of the bar from below, Fig. 347. However, the affinities of these extinct genera are uncertain, and the whole ques-

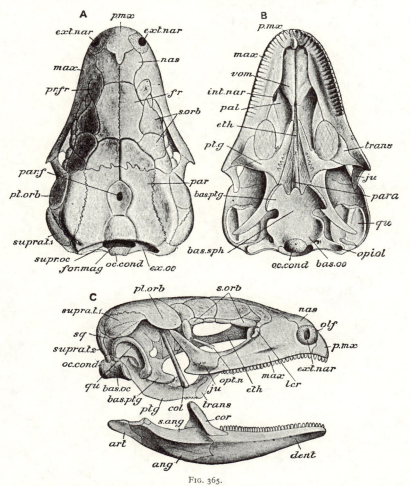

Fig. 365.

Skull of *Lacerta agilis* (from Parker and Haswell's *Zoology*, after W. K. Parker). A, From above ; B, from below ; C, from the side, showing secondary ossifications closing superior temporal fossa. *ang*, Angular ; *art*, articular ; *bas.oc*, basioccipital ; *bas.ptg*, basipterygoid processes ; *bas.sph*, basisphenoid ; *col*, epipterygoid ; *cor*, coronary ; *dent*, dentary ; *eth*, ethmoid ; *ex.oc*, exoccipital ; *ext.nar*, external nares ; *for.mag*, foramen magnum ; *fr*, frontal ; *int.nar*, internal nares ; *ju*, jugal ; *lcr*, lacrimal ; *max*, maxilla ; *nas*, nasal ; *oc.cond*, occipital condyle ; *olf*, olfactory capsule ; *opi.ot*, opisthotic ; *opt.n*, optic nerve ; *pal*, palatine ; *par*, parietal ; *para*, parasphenoid (anterior rostrum ; posterior plate fused to basisphenoid) ; *par.f*, parietal foramen ; *p.mx*, premaxillae ; *pr.fr*, prefrontal ; *ptg*, pterygoid ; *pt.orb*, postorbital ; *qu*, quadrate ; *s.ang*, supra-angular ; *s.orb*, supraorbitals ; *sq*, paraquadrate (squamosal) ; *supra.oc*, supra-occipital ; *supra.t.*[1], supratemporal ; *supra.t.*[2], supratemporal [2] ; *trans*, transpalatine ; *vom*, prevomer.

tion of the origin of the Lacertilia remains open for the present ; but their general resemblance to other living Diapsidan reptiles (Rhyncho-

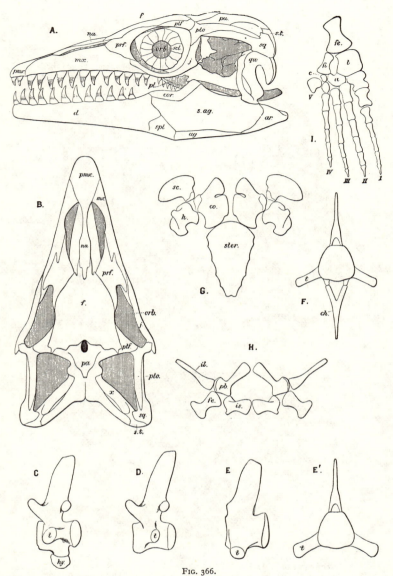

Fig. 366.

Diagram illustrating principal characters of Pythonomorpha.

A, B, *Platecarpus coryphaeus*. Skull from the lateral and superior aspects. U. Cretaceous, Kansas (slightly restored after Merriam). *ag*, Angular ; *ar*, articular ; *cor*, coronoid ; *d*, dentary ; *f*, frontal ; *j*, jugal ; *mx*, maxillary ; *na*, nasal ; *orb*, orbit ; *pa*, parietal ; *pmx*, premaxillary ; *prf*, prefrontal ; *pt*, pterygoid ; *ptf*, postfrontal ; *pto*, postorbital ; *qu*, quadrate ; *s.ag*, surangular ; *s.t*, supratemporal (prosquamosal) ; *scl*, sclerotic ; *spl*, splenial ; *sq*, squamosal ; *x*, exoccipital, etc.

C, *Mosasaurus camperi*. Cervical vertebra, left lateral aspect. U. Cretaceous, Maastricht, Holland. *hy*, Hypocentrum ; *t*, transverse process.

D, *Ditto*. Anterior dorsal vertebra, left lateral aspect. *Ibid. t*, Transverse process.

E, E¹, *Ditto*. Posterior dorsal vertebra, left lateral and hinder aspects. *Ibid. t*, Transverse process.

F, *Ditto*. Anterior caudal vertebra, hinder aspect. *Ibid. ch*, Chevron bone ; *t*, transverse process.

G, *Edestosaurus dispar*. Pectoral arch, ventral aspect. U. Cretaceous, Kansas (after Marsh). *co*, Coracoid ; *h*, humerus ; *sc*, scapula ; *ster*, calcified sternum.

H, *Lestosaurus simus*. Pelvic arch. U. Cretaceous, Kansas (after Marsh). *fe*, Femur ; *il*, ilium ; *is*, ischium ; *pb*, pubis.

I, *Mosasaurus lemonnieri*. Pelvic limb. U. Cretaceous, Belgium (after Dollo). *a*, Astragalo-central ; *c*, calcaneum ; *fe*, femur ; *fi*, fibula ; *t*, tibia ; *I.-V.*, the five digits, the fifth represented only by its metatarsal. (From A. S. Woodward, *Vertebrate Palaeontology*, 1898.)

cephalia, etc.), more particularly the structure of the heart (p. 572), other viscera, and hind foot (Goodrich, 517), affords strong corroborative evidence in favour of Huxley's view. Another controversy has taken place about the homology of the two lower bones limiting the single temporal fossa

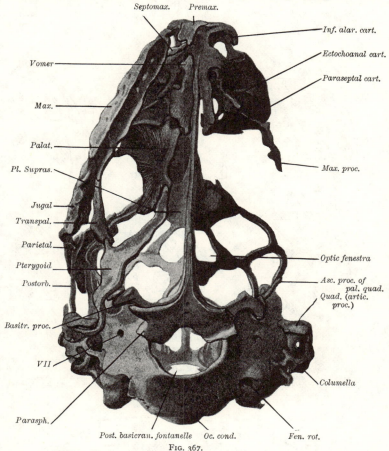

FIG. 367.

Ventral view of Gaupp-Ziegler model of skull of embryo *Lacerta agilis*, 4½ mm. in length. (From Wiedersheim, *Comp. Anatomy*.)

behind in lizards.[1] The outer bone, which joins the jugal, is considered by many to be the squamosal (Huxley, Boulenger, Gadow, Siebenrock, Broili, v. Huene, Broom, Jaekel, Williston, and others); it is called para-quadrate by Gaupp (505), prosquamosal by Baur (416-17) and Case, quadratojugal by Watson (638). The inner bone next to the parietal is

[1] The homology and nomenclature of these bones is discussed by Thyng (611).

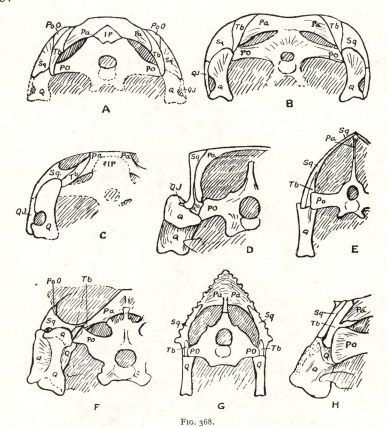

Fig. 368.

Posterior views of skull of : A, *Youngina capensis* ; B, *Mesosuchus browni* ; C, *Paliquana whitei* ; D, *Tiliqua scincoides* ; E, *Chameleo quilensis* ; F, *Chlamydosaurus kingi* ; G, *Lophosaura pumila* juv. ; H, *Varanus albigularis* (from R. Broom, *Proc. Zool. Soc.*, 1925). *I.P*, Interparietal (postparietal ?) ; *Pa*, parietal ; *PO*, paroccipital process of opisthotic ; *PoO*, postorbital ; *Q*, quadrate ; *Sq*, squamosal ; *Tb*, tabular, or supratemporal.

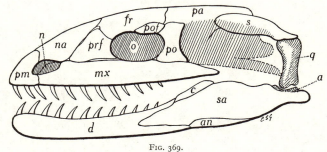

Fig. 369.

Ophidia. Diagram of skull and lower jaw. Lettering as in Fig. 336, p. 328.

called supratemporal by the majority, squamosal by Gaupp and Watson,

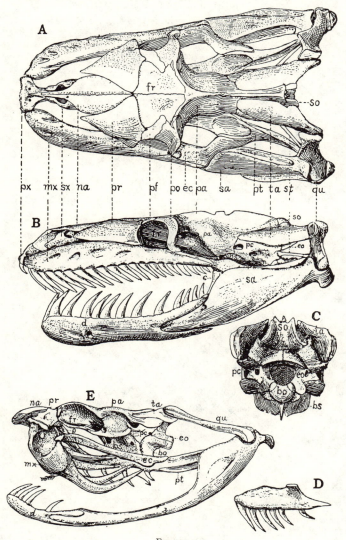

FIG. 370.

A, Dorsal, B, left-side, and C, posterior view of skull of *Python*; D, palatine of same; E, left-side view of skull of *Crotalus* (from Williston, *Osteol. of Reptiles*, 1925). *bo*, Basioccipital; *bs*, basisphenoid; *d*, dentary; *ec*, ectopterygoid; *eo*, exoccipital; *fr*, frontal; *mx*, maxillary; *na*, nasal; *p*, palatine; *pa*, parietal; *pc*, pro-otic; *pf*, postfrontal; *po*, postorbital; *pr*, prefrontal; *pt*, pterygoid; *px*, pre-maxillary; *qu*, quadrate; *sa*, supra-angular; *so*, supraoccipital; *st*, stapes; *sx*, septomaxillary; *ta*, tabular ?

tabular by Broom, Williston, and v. Huene.   The identity of these two bones remains uncertain, but if Huxley's view be correct they are probably the squamosal and supratemporal, Fig. 368.

If the generally accepted opinion that the Ophidia are closely related

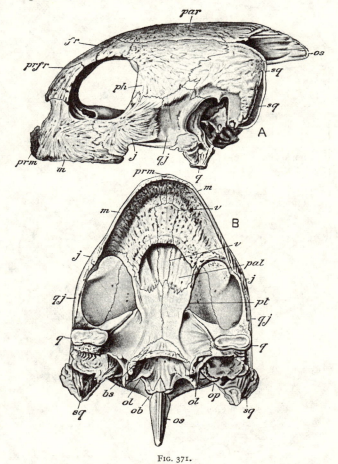

FIG. 371.

Ventral view of the skull of *Chelone mydas*. *bs*, Basisphenoid ; *fr*, frontal ; *j*, jugal ; *m*, maxilla ; *ob*, basioccipital ; *ol*, exoccipital ; *op*, opisthotic ; *os*, supraoccipital ; *pal*, palatine ; *par*, parietal ; *ph*, postfrontal ; *prfr*, prefrontal ; *pt*, pterygoid ; *prm*, premaxilla ; *q*, quadrate ; *qj*, quadrato-jugal ; *sq*, squamosal ; *v*, vomer. (After Hoffmann, from Parker and Haswell, *Zoology*.)

to the Lacertilia be correct, we must suppose that the temporal roofing has almost entirely disappeared, and that the fossae have been opened out completely. The single bone which remains connecting the quadrate to the cranium above may be either a squamosal, or more probably a supratemporal, or again possibly a tabular, Figs. 369, 370.

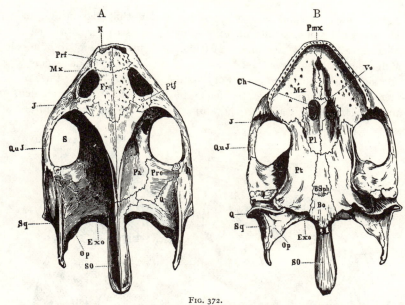

FIG. 372.

Trionyx gangeticus, Cuvier. Recent ; India. Superior (A) and palatal (B) aspects of skull, reduced. *Bo*, Basioccipital ; *BSph*, basisphenoid ; *Ch*, internal nares ; *Exo*, exoccipital ; *Fr*, frontal ; *J*, jugal ; *Mx*, maxilla ; *N*, external nostril ; *Op*, opisthotic ; *Pa*, parietal ; *Pl*, palatine ; *Pmx*, premaxilla ; *Prf*, prefrontal+nasal ; *Pro*, pro-otic : *Ptf*, postfrontal ; *Q*, quadrate ; *QuJ*, quadratojugal ; *S*, palatine vacuity ; *SO*, supraoccipital ; *Sq*, squamosal ; *Vo*, vomer. (From K. Zittel, *Palaeontology*.)

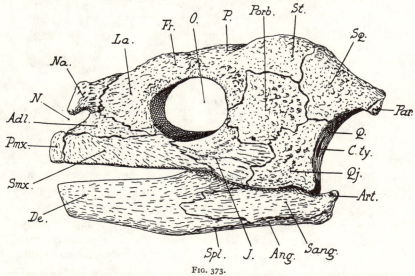

FIG. 373.

Triassochelys dux, Trias of Halberstadt ; left-side view of skull (after O. Jaekel, from O. Abel, 1920). *Adl*, Lacrimal ; *C.ty*, tympanic cavity ; *La*, prefrontal ; *Par*, paroccipital process ; *St*, supratemporal.

The Order Chelonia is another group of doubtful relationship, and whose cranial roofing is difficult to interpret. The structure of the heart (Chapter X.), viscera, and hind foot (Goodrich, 517, 826) point to affinity with other Diapsida; yet the temporal roofing of the Chelonidae seems at first sight to be complete as in Cotylosauria, Fig. 371. In the Chelonia there is never an enclosed fossa. The orbit is almost always surrounded, and usually the jugal joins a quadratojugal overlying the fixed quadrate.

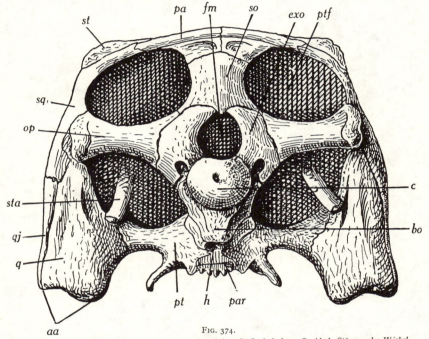

FIG. 374.

Posterior view of skull of *Triassochelys dux* (after O. Jaekel, from O. Abel, *Stämme der Wirbeltiere*, 1919). *bo*, Basioccipital; *c*, condyle; *exo*, exoccipital; *fm*, foramen magnum; *h*, hypophysial pit; *op*, opisthotic; *par*, parasphenoid; *pt*, pterygoid; *ptf*, post-temporal fossa; *so*, supraoccipital; *sta*, stapes.

The fact that in the aquatic Chelonidae and Dermochelydidae these bones join the postfrontal and parietal to form a practically complete temporal roofing with a wide post-temporal opening, Figs. 374-419, has led many zoologists to the conclusion that these Chelonia have retained the original Stegocephalian roofing, which has been more or less reduced (but not pierced) in most of the terrestrial forms (Baur, 414-16; Rabl, 1903; Boas, 424; Williston, 655, 95; and others). In the Chelydidae the roofing seems to have been, so to speak, eaten away from the lower margin, while in the majority it seems to have been eaten away from behind, leaving in many only narrow postorbital and inferior bars, Fig. 372. The roofing in the most primitive known form, *Triassochelys* (Jaekel, 536), is almost as complete as in *Chelone* and contains a supratemporal

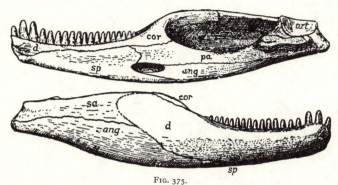

FIG. 375.

Right lower jaw of *Labidosaurus Namatus* Cope. Inner view above, outer view below. (From Williston, *Osteology of Reptiles*, 1925.)

in addition. Yet, the view that this covering is the primitive one is difficult to reconcile with the fact that the bones are much reduced in number and that the pineal foramen is closed. Not only are there no

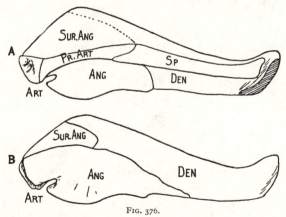

FIG. 376.

Lower jaw of *Dimetrodon*. A, Inner view of left ramus ; B, outer view of right ramus. (From D. M. S. Watson, *Ann. Mag. N.H.*, 1912.) Lettering as in Fig. 377.

supratemporal (except in *Triassochelys*), postparietal, and tabular, but there is only one bone to represent the postfrontal and postorbital and one to represent the nasal prefrontal and lacrimal. Taking into account the evidence from other parts as to the general affinities of the Chelonia (see above), it would seem more probable that the roofing

2 A

is secondary and, after many of the original dermal bones had disappeared, has been reacquired (by extension of the parietal, postfrontal, and squamosal in *Chelone*, and of the parietal, jugal, and quadratojugal in *Podocnemis* (Goodrich, 517)). The remaining roofing bones may have been extended by the addition of secondary bony matter of the same nature as the dermal bony scutes so conspicuously developed over

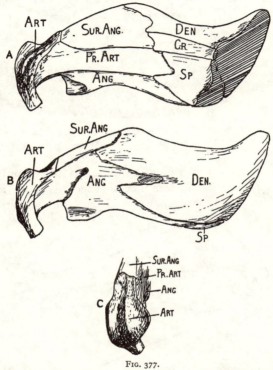

Fig. 377.

Lower jaw of *Dicynodon*. A, Inner view of left ramus ; B, outer view of right ramus ; C, posterior view of left articular region (from D. M. S. Watson, *Ann. Mag. N.H.*, 1912). *Ang*, Angular ; *Art*, articular ; *Cr*, coronoid ; *Den*, dentary ; *Pr.Art*, prearticular ; *Sp*, splenial ; *Sur.Ang*, supra-angular.

the rest of the body. Indeed a similar secondary covering of the temporal region by dermal scutes is known to occur among Lacertilia (*Lacerta*, Fig. 365), and even in Crocodilia and Anura (see above, pp. 320, 338).

The lower jaw in Reptilia preserves, besides the articular, a considerable number of the dermal bones present in Stegocephalia (Baur, 418 ; Broom, 450, etc.; Williston, 659; Watson, 632; Gaupp, 515). In most of the more primitive forms angular, supra-angular, dentary, coronoid, splenial, and prearticular bones can be distinguished, Figs. 362, 375-7, though the latter may often fuse with the articular (Lacertilia) or disappear

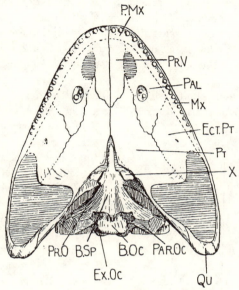

FIG. 378.

Seymouria (Conodectes) bayloriensis, ventral view of skull. (From D. M. S. Watson, *P.Z.S.*, 1918.)

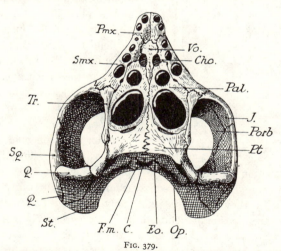

FIG. 379.

Cyamodus tarnowitzensis, Triassic ; ventral view of skull (after O. Jaekel, from O. Abel, 1920).
C, Median basioccipital condyle ; *Cho*, internal nostril ; *F.m*, foramen magnum ; *Op*, opisthotic ;
Porb, postorbital ; *Smx*, maxillary ; *St*, supratemporal ; *Tr*, ectopterygoid ; *Vo*, prevomers.

(Crocodilia). The bone generally considered as splenial no longer enters

the symphysis in Lacertilia, Ophidia, Crocodilia, and Chelonia.   An
opening between angular, supra-angular, and dentary is characteristic of
the Thecodontia.   Among Theromorpha, the Theriodontia are remarkable
for the progressive reduction of all the bones excepting the dentary, which
increases in size, acquires a large coronoid process, and comes more and
more to resemble the only remaining bone in the Mammalian lower jaw
(see p. 476).

The palate of primitive Reptiles differs little from that of primitive
Stegocephalia and even of early Teleostomes.   The same prevomers,
palatines, pterygoids, ectoptery-

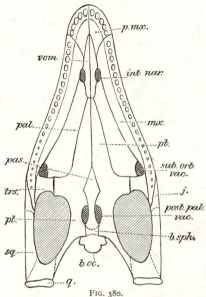

FIG. 380.

Ventral view of skull of *Plesiosaurus* (from C. W.
Andrews, *Q. J. Geol. Soc.*, 1896).   *int nar*, Internal
nostril; *sub.orb.vac*, suborbital vacuity; *pas*, parasphenoid; *post.pal.vac*, interpterygoid vacuity; *trs*,
transverse or ectopterygoid; *vom*, prevomer.

goids (transverse or transpalatine bones) occur, and the
median parasphenoid underlies
the brain-case.   All these bones
may bear teeth.   Considerable
divergence of structure, however, occurs in the various
specialised orders, Figs. 378-
382, 390; see authors quoted
above; Lakjer, 540a.

Usually the parasphenoid is
inconspicuous and fused to the
basisphenoid in the adult.   Its
anterior blade may form a
slender 'basisphenoid rostrum',
while its lateral wings fusing
with the basitrabecular processes enclose canals for the
vidian nerve (p. 298).   Parabasal
canals between the more posterior parasphenoid plate and
basisphenoid allow the internal

carotids to reach the pituitary fossa (p. 298), Figs. 279, 367.

The pterygoids generally converge and meet anteriorly, diverging
farther back; they articulate primitively with the basitrabecular processes, but their posterior quadrate limb is progressively less developed
than in Stegocephalia.   Nevertheless, the pterygoids usually reach and
help to support the quadrates.   A characteristic ventrally directed flange
is often formed by the pterygoid and ectopterygoid combined.   In
Lacertilia the palatines are interpolated between the pterygoids and the
prevomers, Fig. 363.

An interesting adaptation to aquatic life is seen in the higher Crocodilia, where an extensive false palate is developed carrying the choanae far back to behind the pterygoids. The nostrils are forward near the end of the snout, so that they are the first part to emerge above water. The true internal nares extend far back, as usual separated by the prevomers. The premaxillaries, maxillaries, palatines, and pterygoids have ventral palatal plates meeting in the middle line to form a bony false palate, the floor of two long nasal canals, separated from each other by vertical downgrowths from the prevomers and pterygoids, Figs. 383, 415-16. These canals run below the true palate from the original internal nostrils to a posterior median opening opposite the glottis. In the living crocodile two fleshy folds, a dorsal velum palati and a ventral basihyal valve, serve to keep the water out of the glottis when it opens its mouth below the surface,[1] Fig. 383.

The Phytosauria have a deep median groove. Intermediate stages in the formation of the false palate are known in fossil forms, Figs. 384, 385. In *Pelagosaurus*, for instance, a premaxillary and

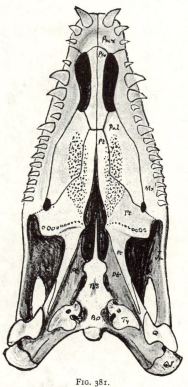

Fig. 381.

Ventral view of skull of *Dimetrodon*, restored (from R. Broom, *Bull. Am. Mus.*, 1910). *Bo*, Basioccipital; *Bs*, basisphenoid; *Ju*, jugal; *Pal*, palatine; *Pmx*, premaxillary; *Pt*, pterygoid; *Pv*, prevomer; *Q*, quadrate; *Q.J.*, quadratojugal; *Ty*, stapes.

[1] The Eustachian tubes, also, are modified in that they sink between the basisphenoid and basioccipital into bony canals opening behind the choanae on either side of the aperture leading into a median tympanic canal. This canal, running up through the basioccipital, bifurcates and opens into a complicated system of spaces communicating with the tympanic cavities, perforating the bones of the occipital region and penetrating into the lower jaw. Since these spaces contain air they lighten this part of the heavy skull, especially when immersed. A similar system of air spaces occurs in some Dinosaurs and in Birds.

The early fossil Crocodilia have a tympanic canal, but only grooves to hold the Eustachian tubes (Teleosauridae).

maxillary false palate is already developed, but the palatines scarcely meet

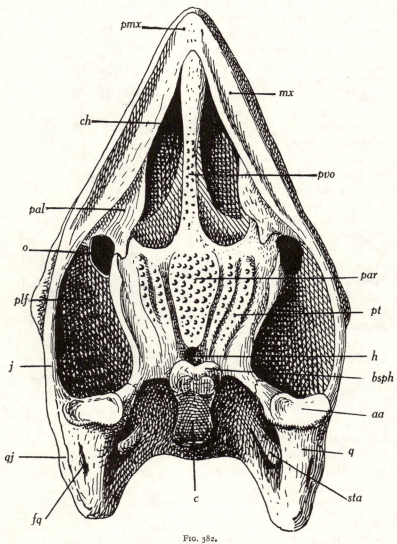

FIG. 382.

Ventral view of skull of *Triassochelys dux* (after O. Jaekel, from O. Abel, *Stämme der Wirbeltiere*, 1919). *aa*, Articular surface for lower jaw; *bsph*, basisphenoid; *fq*, quadrate foramen; *par*, parasphenoid with teeth; *plf*, lateral palatal space; *pvo*, prevomer; *sta*, stapes.

and the groove opens about midway, Fig. 385. *Steneosaurus*, with its complete palatine region, leads to the specialised modern type.

There can be little doubt that the Class Aves is related to the Diapsidan Reptiles, is indeed closely allied to the Crocodiles and Dinosaurs.
This conclusion is firmly based on a comparison of the general anatomy
of these groups. But the avian skull is highly specialised, and the
original roofing of the Diapsidan type has been much modified, chiefly
owing to the great enlargement of the brain-case. Unfortunately the
structure of the dermal roofing in Palaeornithes is but very imperfectly
known, but there is some reason to believe that it was more complete

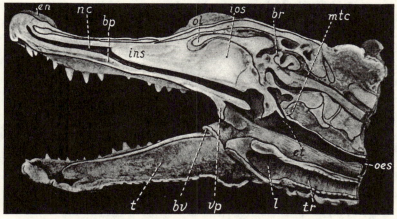

FIG. 383.

Nearly median longitudinal section of head of *Alligator mississipiensis*, showing right nasal canal,
nc, running above false palate, *bp*, and opening far back into pharynx opposite glottis. *br*, Brain ;
bv, basihyal valve ; *en*, position of external nostril ; *et*, opening of Eustachian canal ; *ins*, internasal
septum ; *ios*, interorbital septum ; *l*, larynx ; *mtc*, median tympanic canal, leading into spaces in
skull bones communicating with tympanic cavity ; *oes*, oesophagus ; *ol*, olfactory lobe ; *t*, tongue ;
tr, trachea ; *vp*, velum palati.

than in the Neornithes (Owen, 1863 ; Dames, 486). Renewed study of the
Berlin specimen of *Archaeopteryx* (*Archaeornis*) has revealed that the orbit
is completely surrounded by bone, that the inferior temporal fossa is clearly
delimited, and that there are signs of a superior temporal fossa (Heilmann,
527). Assuming, then, that it is derived from that type of Archosaurian
skull which possesses a preorbital and two lateral temporal fossae, the
roofing may be interpreted as follows in the Neornithes. The preorbital
fossa is typically present, but becomes confluent with the external nostril
in Ratitae where the nasal is small. A single bone (prefrontal or
lacrimal), pierced for the lacrimal duct, separates the fossa from the
orbit. The lower temporal bar remains complete, with jugal and quadratojugal loosely applied to the quadrate. The postorbital bar is incomplete, and represented only by a postfrontal process of the frontal. The

upper temporal bar has disappeared, leaving the temporal fossae confluent with each other and also with the orbit ; moreover, they are freely open behind owing to the wide gap between the squamosal (supratemporal ?) and quadratojugal, thus further exposing the quadrate which is loosely articulated to the brain-case (streptostylism, p. 432), Figs. 386-90.

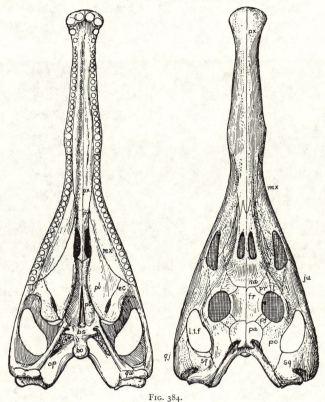

FIG. 384.

Skull of *Machaeroprosopus*, dorsal and ventral view.   (From Williston, *Osteology of Reptiles*, 1925.)

The bones of the skull tend to fuse in adult Carinates, but their boundaries can be made out in the young, and in Ratites (Parker, 556 ; T. J. Parker, 555). As in many Reptiles, but to an even greater extent, the parietals and frontals form not only the roof, but also much of the lateral walls of the large brain-case. The squamosals also come to contribute to it below them. The premaxillaries send backward processes between the nostrils. Although the lower jaw is slender it shows in the young most of the reptilian bones (articular, dentary, splenial, angular, supra-angular). A prearticular has recently been identified by Killian.

The two rami are fused at the symphysis and form together with the fused

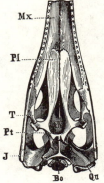

Fig. 385.

Pelagosaurus temporalis, Blv. sp.   Upper Lias: Curcy, Calvados.   Posterior half of skull, showing typical mesosuchian palate, ½.   *Bo*, Basioccipital; *Ch*, internal nares; *J*, jugal; *Mx*, maxilla; *Pl*, palatine; *Pt*, pterygoid; *Qu*, quadrate.   (From K. Zittel, *Palaeontology*.)

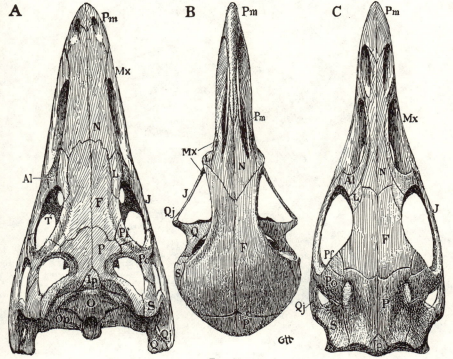

Fig. 386.

Dorsal view of skull of : A, *Euparkeria capensis* (after R. Broom) ; B, *Columba domestica* ; C, *Archaeornis siemensi*, restored.   (From G. Heilmann, *Origin of Birds*, 1926.)

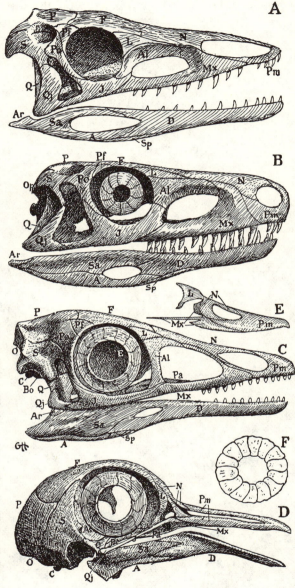

FIG. 387.

Right-side view of skull of : A, *Aëtosaurus ferratus*, Triassic (after F. v. Huene) ; B, *Euparkeria capensis*, Triassic (after R. Broom) ; C, *Archaeornis siemensi*, Jurassic, restored ; D, *Columba domestica* ; E, upper jaw of *Fuligula ferina* ; F, sclerotic ring of *Thalattasaurus alexandrae* (after Merriam).   *Al*, lacrimal ; *L*, prefrontal (lacrimal in D ?) ; *S*, squamosal.   (From G. Heilmann, *Origin of Birds*, 1926.)

premaxillaries the characteristic bony beak provided with a horny sheath above and below. There is a lateral fossa in the lower jaw, as in Crocodilia and certain related reptiles. Except in Palaeornithes and early fossils of the Cretaceous (*Ichthyornis, Hesperornis*, etc., Marsh, 1880), the teeth have been lost, Fig. 226.

The palate becomes remarkably specialised (Huxley, 533; Parker, 568, 572; Pycraft, 576). The parasphenoid (basitemporal of Parker) is very

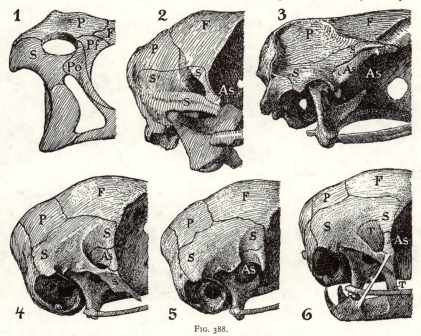

FIG. 388.

Right-side view of hind part of skull of: 1, *Aëtosaurus ferratus* (after F. v. Huene); 2, *Xenorhynchus*; 3, *Colymbus glacialis*; 4, *Phasianus*; 5, young *Phasianus*; 6, *Gallus*. *As*, Pleurosphenoid; *F*, frontal; *P*, parietal; *Pf*, postfrontal; *S*, squamosal; *T*, temporal muscle. (From G. Heilmann, *Origin of Birds*, 1926.)

large, but soon fuses indistinguishably with the basisphenoid, enclosing, as in Crocodiles, the Eustachian tubes in bony canals which open by a median aperture. The slender rostrum extends below the interorbital septum. In front the palate is, to a great extent, formed by processes of the premaxillaries and maxillaries, the prevomers being far back, usually fused, often very slender, and sometimes absent. The internal nostrils open in a deep groove. Primitively well-developed basitrabecular processes support the pterygoids, which articulate with the large palatines in front (Ratitae, and more primitive Carinatae), Fig. 390 A. In many modern forms, however, these processes are reduced or absent, and then both pterygoids and palatines come to rest on the parasphenoid (basisphenoid) rostrum, Fig. 391. There are no ectopterygoids.

Huxley first used the structure of the palate for the classification of Birds (533a). He distinguished four types: Dromaeognathous (Ratitae); Desmognathous (Hawks, Parrots, Ducks, etc.); Schizo-gnathous (Gulls, Game-birds, etc.); Ægithognathous (Passeres). Later Pycraft (576) defined the dromaeognathous type, with large 'vomer' (prevomers) reaching back (except in Struthio) to support the palatines and pterygoids, with the latter bones rigidly connected with the palatines

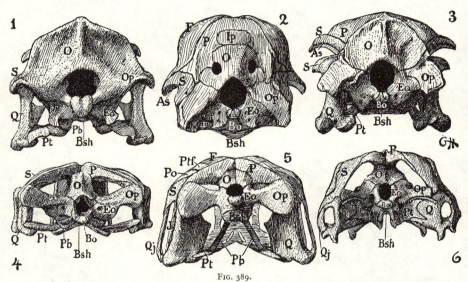

FIG. 389.

Posterior view of skull of: 1, *Hesperornis regalis*; 2, *Anser ferus*; 3, *Colymbus glacialis*; 4, *varanus*; 5, *Scaphognathus purdoni* (after E. T. Newton); 6, *Sphenodon* (after Zittel). *Pb*, Basipterygoid articulation. (From G. Heilmann, *Origin of Birds*, 1926.)

(as in most Reptilia), and with strong basipterygoid processes, as a primitive 'Palaeognathous' form of palate common to all the Ratites and the Tinamous, Fig. 390 A. The various types of palate found in other birds he defined as 'Neognathous'. In these the prevomers are small; the maxillopalatine processes contribute to the palate passing dorsally to the palatines; the latter and the pterygoids are movably connected and converge towards the parasphenoidal rostrum, Fig. 391.

The Mammalian Palate, False Palate, and Vomer.—Characteristic of the Mammalia is the bony false palate forming the ventral floor of two naso-pharyngeal passages separated by the vomer and opening behind by the choanae bounded in front and at the sides by the palatines. It is continued backwards in the living by a fleshy false palate which ends opposite the epiglottis, a fold supported by a cartilage in front of the glottis also characteristic of the mammal.   Yet this bony false palate was already present in the related Theromorph Reptiles (Seeley, 589-91 ; Broom, 441-5, 458 ; Watson, 634, 641) and undoubtedly evolved in that

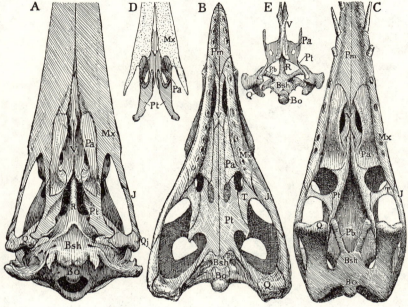

FIG. 390.

Ventral view of skull of : A, *Apteryx australis* (after Pycraft) ; B, *Ornithosuchus woodwardi* (after Broom and Newton) ; C, *Rhamphorhynchus gemmingi* (after A. S. Woodward and v. Huene) ; D, dorsal view of palate of *Apteryx* (after Pycraft) ; E, palate of *Rhea* (after Pycraft). *Bsh,* Basisphenoid +parasphenoid, *R,* its rostrum ; *Bo,* basioccipital ; *Mx,* maxillary ; *Pa,* palatine ; *Pb,* basipterygoid process ; *Pm,* premaxillary ; *Pt,* pterygoid ; *Q,* quadrate ; *T,* transverse (ectopterygoid) ; *V,* pre-vomers.   (From G. Heilmann, *Origin of Birds,* 1926.)

group, whose more primitive forms have the ordinary reptilian palate, Figs. 394-6.

The Therocephalia have in front a rudimentary false palate, the naso-pharyngeal passages being represented further back by a deep groove, and in the Dicynodontia (Seeley, 588 ; Lyddeker, 1888 ; Newton, 1893 ; Sollas, 602 ; Broom, 441 ; Huene, 532) the maxillae and palatines tend to form palatine processes enclosing the grooves ; but it is not until the Cynodontia that these processes are fully developed and join across the middle line.   In this group also are found a large median vomer separat-ing the nasal passages, pterygoids no longer reaching the quadrate, and alisphenoids passing back towards the tympanic cavities—all mammalian characters.

The secondary or false palate develops in the embryo mammal, behind the original internal nostrils on the true palate, by the growth from the sides of ridges enclosing palatal plates from the maxillae and palatines which meet ventrally and join a vertical plate of the vomer.   Into the

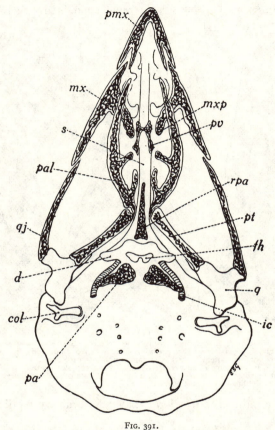

FIG. 391.

Ventral view of skull of late embryo of *Passer domesticus* showing ossification of palatal bones. *col*, Columella auris; *d*, infrapolar cartilage; *fh*, foramen hypophyseos; *ic*, internal carotid; *mx*, maxillary; *mxp*, maxillopalatine process; *pa*, paired posterior parasphenoidal wing; *pal*, palatine; *pmx*, premaxillary; *pt*, pterygoid; *pv*, prevomer; *q*, quadrate; *qj*, quadratojugal; *rpa*, parasphenoidal rostrum; *s*, interorbital septum.

air passages so enclosed the internal nostrils now open, only a pair of small anterior palatine foramina or naso-palatine canals remaining as parts of the original communications in front, Figs. 392, 397-9.   Into these canals open Jacobson's organs (Seydel, 595; Broom, 1895–98).

The interpretation of the palate of the mammal involves the important question of the homology of the mammalian vomer still under dispute.

For long it was thought that it represented the paired vomers (prevomers) of lower forms, which typically extend between the internal nostrils, and form their postero-medial boundary.[1] But Albrecht (1883), Sutton (609), and W. K. Parker (567) concluded that the prevomers of lower forms correspond to the palatal processes of the premaxillae of most mammals. Further, Parker claimed to show that these processes appear in many of the lower Mammalia (567) as separate bones which later fuse

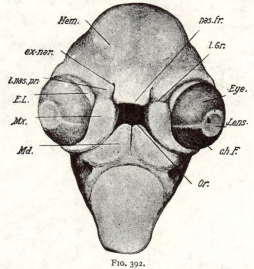

FIG. 392.

Head of chick embryo of about 5 days from the oral surface (N.L. 8 mm.) (from F. R. Lillie, *Devel. Chick*, 1919). *chF*, Choroid fissure; *E.L*, eyelid (nictitating membrane); *ex-nar*, external nares; *l.Gr*, lacrimal groove; *l.nas.pr*, lateral nasal process; *Md*, mandibular arch; *Mx*, maxillary process; *nas.fr*, naso-frontal process; *Or*, oral cavity.

with the premaxillae, Fig. 399 A. These observations have to some extent been confirmed by Broom on the bat *Miniopterus* and by Fawcett on *Tatusia*; but as a rule the premaxillary processes develop in continuity with the premaxillaries.

Jacobson's organ, which appears to serve for smelling the liquid contents of the buccal cavity, occurs in the Amniota as a small sac, blind behind and opening in front towards the palatal surface at or near the internal nostril, Figs. 264, 399. It is derived in development from the wall of the nasal cavity, and is supported below by the paraseptal cartilage and the prevomer which underlies it, Figs. 400-1. When

[1] These usually paired bones, now known as prevomers (Broom), occur from the earliest Teleostomes; but they fuse to a median bone in Teleostei, Chelonia, and most Birds. Even in the embryo chelonian there is evidence of paired origin (Kunkel, 540).

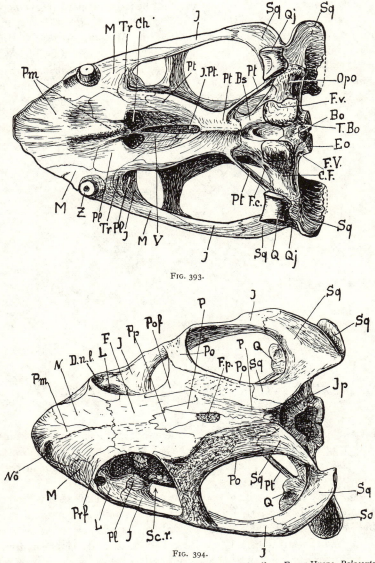

FIG. 393.

FIG. 394.

Fig. 393, ventral, and Fig. 394, dorsal aspect of skull of *Dicynodon* (from F. von Huene, *Palaeontol. Zeitschr.*, 1922). *Bo*, Basioccipital; *Bs*, basisphenoid; *C.F*, Canalis Fallopii, facial foramen; *Ch*, choana, internal nostril; *D.n.l*, ductus naso-lacrimalis; *Eo*, exoccipital; *Ep*, epipterygoid; *F*, frontal; *F.c*, carotid foramen; *F.p*, pineal foramen; *F.v*, fenestra ovalis; *F.V*, vagus foramen; *J*, jugal; *I.p*, interparietal; *I.Pt*, interpterygoidal vacuity; *L*, lacrimal; *M*, maxillary; *N*, nasal; *Nö*, external nostril; *Opo*, opisthotic; *P*, parietal; *Pa*, parasphenoid; *Pl*, palatine; *Pm*, premaxillary; *Po*, postorbital; *Pof*, postfrontal; *pp*, preparietal or interfrontal; *Pro*, pro-otic; *Ps*, 'presphenoid' = basisphenoid?; *Pt*, pterygoid; *P.T.F*, post-temporal fossa; *Q*, quadrate; *Qj*, quadratojugal; *Sc.r*, sclerotic ring; *Sm*, septomaxillary; *So*, supraoccipital; *Sq*, squamosal; *Tb*, tabular; *T.Bo*, tuber basioccipitale; *Tr*, ectopterygoid; *V*, prevomer; *Z*, broken tooth.

Wilson and Martin (1893–94) showed that the separate ' os paradoxum ' of *Ornithorhynchus* is of paired origin and occupies the same position with regard to Jacobson's organ and the paraseptal cartilage, Broom (**439, 440, 451**) homologised this bone in Monotremes and the premaxillary processes in other mammals with the prevomers, and the vomer of mammals with the parasphenoid of lower forms, Figs. 402, 404. Already in the more mammal-like Theriodontia the prevomers and

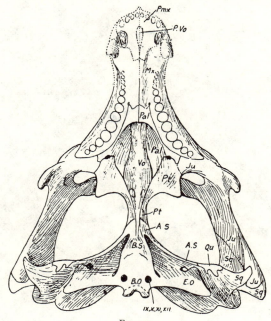

FIG. 395.

Gomphognathus minor, ventral view of skull, restored (from R. Broom, *P.Z.S.*, 1911). *As*, Alisphenoid; *BO*, basioccipital; *BS*, basisphenoid; *E.o*, exoccipital; *Pal*, palatine; *Pt*, pterygoid; *P.Vo*, prevomer; *Vo*, vomer. Other letters as in Fig. 396.

parasphenoid assume much the same form and position as the pre-maxillary processes and median vomer respectively of the mammal, Figs. 395, 413. Although Broom's interpretation has not yet been quite conclusively proved, yet there is much to be said in its favour, and it affords an explanation of the fate in the Mammalia of the parasphenoid, a bone which hitherto has generally been supposed to have unaccountably disappeared.[1]

[1] But the mammalian vomer, which does not extend backwards over the basisphenoid region, would appear to represent only the anterior rostral part of the parasphenoid situated below the median septum; the posterior plate may be separately represented, p. 370.

2 B

Another bone in the mammalian palate the homology of which has given rise to much controversy of late is the pterygoid. When Gaupp (511, 513) discovered a separate bone in *Echidna* ventral to that usually called pterygoid, Figs. 405, 402-3, he concluded that it is homologous with the reptilian pterygoid. The more dorsal bone he considered to represent on the one hand the mammalian pterygoid and on the other the lateral wing of the reptilian parasphenoid (see footnote, p. 369).

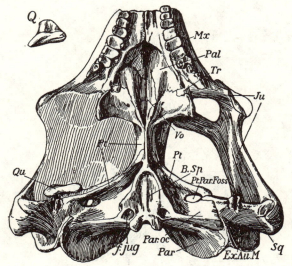

FIG. 396.

Diademodon browni, ventral view of skull (from D. M. S. Watson, *Ann. Mag. N.H.*, 1911). *B.Sp*, Basisphenoid; *Ex.Au.M*, external auditory meatus; *f.jug*, jugular foramen; *Pt*, pterygoid; *Pt.Par.Foss*, foramen between alisphenoid and paroccipital process of opisthotic; *Vo*, vomer; *Q*, figure of quadrate, anterior aspect.

Comparing the Monotreme with the higher mammal, the newly discovered ventral bone of the former ('Echidna pterygoid' of Gaupp) is found to correspond in position to the 'pterygoid cartilage' of the latter, Figs. 405-6. It has long been known that the 'pterygoid' bone in Mammalia may be formed of two elements, separate at all events in development (Fawcett, 488; Lubosch, 543); a more important dorsal element of dermal origin, and a more ventral related to the 'pterygoid cartilage' and forming the hamulus. Now this 'pterygoid cartilage', Figs. 406, 487, appears from its histological character to be of secondary nature (similar to that related to the dermal bones of the lower jaw) and to give rise to dermal bone. There are, therefore, probably two dermal elements related to the basitrabecular region in all mammals: the more dorsal 'mammalian pterygoid' of Gaupp, and the more ventral

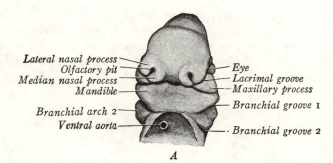

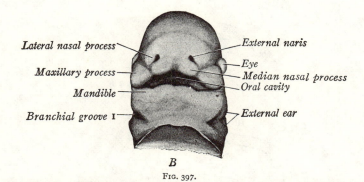

FIG. 397.

Two stages showing development of face in pig embryos. ×7. A, Ventral view of face of 12 mm. embryo; B, of 14 mm. embryo. (After Prentiss and Arey, *Text-Book of Embryology*, 1917.) Branchial grooves mark vestigial gill-slits.

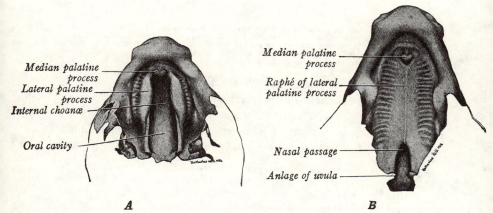

FIG. 398.

Dissections to show development of hard palate in pig embryos. ×5. A, Ventral view of palatine processes of 22 mm. pig embryo, mandible removed; B, same of 35 mm. embryo showing fusion of palatine processes. (From Prentiss and Arey, *Text-Book of Embryology*, 1917.)

'Echidna pterygoid' of Gaupp (de Beer, **422a**). The latter remains as a separate bone in the adult only in the Monotremata (Gaupp, van Bemmelen, **423**); in the Ditremata it is fused to the dorsal bone and gives rise to its hamular process. Gaupp showed that the dorsal element bears much the same relation to the basitrabecular process (processus

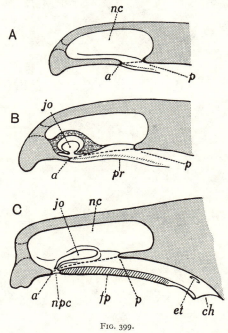

Fig. 399.

Diagrammatic longitudinal sections of anterior region of head to left of median line, showing relations of true and false palate. A, Amphibian; B, Reptile; C, Mammal. (Modified from O. Seydel, 1899.) *a* and *p*, Anterior and posterior limit of primitive internal nostril (marked by broken line); *ch*, choana; *et*, opening of Eustachian tube; *fp*, false palate; *jo*, organ of Jacobson; *nc*, nasal cavity; *npc*, nasopalatine canal; *pr*, palatine ridge.

alaris) and to the anterior parabasal or vidian canal as does the lateral wing of the parasphenoid in lower vertebrates. The ventral element, on the other hand, would seem to represent the degenerate vestige of the reptilian pterygoid (already diminished in the most mammal-like Cynodonts).[1] The facts that in *Ornithorhynchus* this bone is loose, and

[1] Broom (**451**) has described a separate ventral bone in a foetal *Tatusia*, and believes it and the 'Echidna pterygoid' to represent the ectopterygoid of Reptiles; more probably it is the pterygoid. Watson (**642**) argued that the ventral (Echidna) pterygoid is derived from the quadrate ramus of the epipterygoid of Cynodontia, and the mammalian pterygoid is homologous with the reptilian. Certainly the three bones (alisphenoid, dorsal, and ventral pterygoids) have not been found separate in Monotremes, where the ali-

that a slip of muscle is attached to it, may be taken as evidence in favour of this conclusion. The homology of the epipterygoid is discussed elsewhere (p. 456).

OSSIFICATIONS OF THE CHONDROCRANIUM

Osteichthyes.—It has already been mentioned above that the cartilaginous skull becomes more or less thoroughly replaced and com-

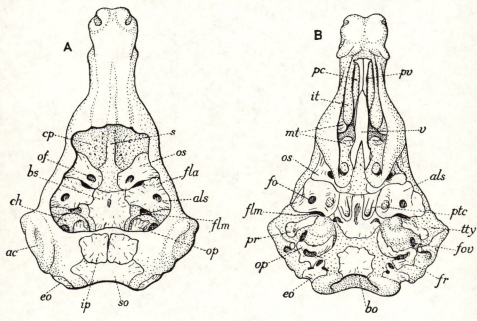

FIG. 399a.

Skull of embryo *Erinaceus europaeus*, 3 inches long (after W. K. Parker, 1885). A, Dorsal; B, ventral view. *ac*, Auditory capsule; *als*, alisphenoid; *bo*, basioccipital; *bs*, basisphenoid; *cp*, cribriform plate; *eo*, exoccipital; *fla*, foramen lacerum anterius; *flm*, f. lac. medium; *fo*, for. ovale; *fov*, fenestra ovalis; *fr*, fenestra rotunda; *ip*, 'interparietal'=postparietal?; *it*, inferior turbinal; *mt*, maxillo turbinals; *of*, optic foramen; *op*, opisthotic?; *os*, orbitosphenoid; *pc*, paraseptal cartilage; *pr*, pro-otic?; *pv*, prevomer; *ptc*, pterygoid; *s*, median septum; *so*, supraoccipital; *tty*, tegmen tympani; *v*, vomer. Cartilage dotted.

pleted by bone in the Osteichthyes. These endochondral bones spread from various 'centres' which do not correspond to the original cartilaginous elements of the skull; nor are they restricted to the

sphenoid (true epipterygoid) seems to have disappeared or fused with one of its neighbours (wing of periotic?). According to Van Kampen (1922) the mammalian pterygoid is homologous with the reptilian, but the vestige of the parasphenoid forms the processus tympanicus of the adult basisphenoid in Ditremata.

elements in which they first arise. The ossification of the chondrocranium
and visceral arches progresses not only in the ontogeny of the individual,
but also often in the phylogeny of the main groups ; and, speaking quite
generally, the most thoroughly bony are among the most highly specialised orders. But, although this holds good for the highest Teleostei,
the most bony of all fishes, yet it is by no means always the case.

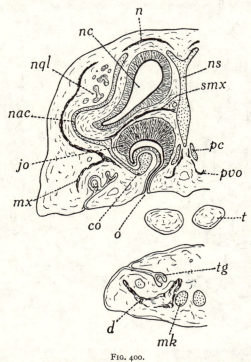

Fig. 400.

Lacerta, late embryo. Transverse section through snout, lower jaw, and Jacobson's organ, *jo*.
co, Lower edge of nasal capsule ; *d*, dentary ; *mx*, maxillary ; *n*, nasal ; *nac*, nasal capsule ; *nc*,
nasal cavity ; *ngl*, nasal gland ; *ns*, median nasal septum ; *mk*, Meckel's cartilage ; *o*, opening of
Jacobson's organ ; *pc*, paraseptal cartilage ; *pvo*, prevomer ; *smx*, septomaxillary ; *t*, tongue ;
tg, tooth-germ. Cartilage dotted.

Frequently in several important groups, such as the Dipnoi and Chondrostei, there may be on the contrary a progressive reduction of ossification,
and apparently a partial return to a cartilaginous condition due to the
retention or even to an increase in the adult of the cartilaginous parts
developed in the embryo (pp. 382, 389).

Taking first the Holostean Teleostomes (Amioidei, Lepidosteoidei,
Teleostei) as containing the best known and, in some ways, least specialised
forms, we find in the occipital region a basioccipital and paired exoccipital

bones, Figs. 429, 436. The former extends well forward in the basal para-chordal plate ; surrounds the notochord, here practically obliterated in the adult, and bears a posterior concave face, like that of a centrum, to which the vertebral column is attached by ligament without distinct articulation. In *Amia* two pairs (and in *Polypterus* one) of neural arches rest on the

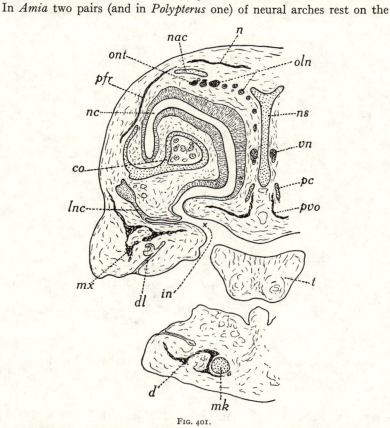

FIG. 401.

Lacerta, late embryo. Transverse section through head at level of internal nostril, *in. co*, Concha ; *dl*, dental lamina ; *lnc*, lacrimonasal duct ; *oln*, olfactory nerve ; *ont*, orbitonasal nerve ; *pfr*, prefrontal ; *vn*, vomeronasal nerve. Other letters as in Fig. 400.

hinder region of the basioccipital, which includes the corresponding centra. The exoccipitals more or less completely enclose the foramen magnum above and at the sides, let through hypoglossal nerves, and define the hinder limit of the vagal foramen. In the higher Teleostei they may have a facet for the first vertebra. A median supraoccipital appears only in the Teleostei, where it is generally provided with a characteristic keel projecting backwards between the right and left anterior myomeres. The history of the supraoccipital is still obscure ; there is no good evidence that it has been derived from the median dermal occipital of lower fish (Crossopterygii, Chondrostei). It develops as an endochondral bone, and may possibly correspond to the neural spines further back. Recently, however, Watson has described a supraoccipital

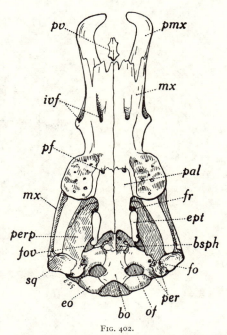

FIG. 402.

Ornithorhynchus paradoxus, ventral view of skull. *bsph*, Basisphenoid ; *fov*, foramen ovale ; *ivf*, infraorbital foramina ; *of*, fenestra occipitalis (for. jugulare and for. praecondyloideum) ; *pal*, palatine ; *perp*, periotic plate ; *pf*, palatine foramen ; *pv*, prevomers combined to ' dumb-bell shaped bone '. Other letters as in Fig. 403.

in an Osteolepid, a Coelacanth, and a Palaeoniscid, Fig. 408 ; it may after all be a primitive bone, perhaps homologous with that of Tetrapods.

It is well known since the work of Huxley and Parker that the auditory capsule in Teleostei is typically ossified from 5 points, Figs. 285, 429, 436. A pro-otic occupies its antero-ventral region, generally surrounding the facial foramen and lodging the anterior semicircular canal. It surrounds the jugular canal (p. 276). A postero-ventral opisthotic overlies the posterior corner reaching the vagal foramen in *Amia*. This bone varies greatly in development (Vrolick, 629 ; Sagemehl, 278 ; Allis, 401-2) ; fairly well developed in primitive forms it becomes very large in the Gadidae, but is reduced in many Teleostei, and sometimes even absent (Mormyridae, etc., also absent in *Lepidosteus* ?). Dorsally the capsule may become ossified by the prefrontal from in front, and the supratemporal ('squamosal') from behind, the former (sphenotic) overlying the anterior semicircular canal and the latter (pterotic) the horizontal canal.[1] The fifth bone is the epiotic which develops dorsally on the posterior medial corner of the

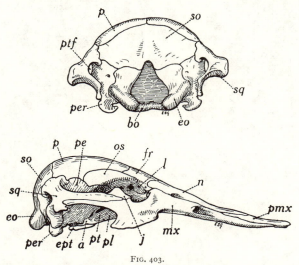

FIG. 403.

Ornithorhynchus paradoxus, right-side view and hind view of skull. *a*, Alisphenoid?; *bo*, basi-occipital; *eo*, exoccipital; *ept*, ventral pterygoid; *fr*, frontal; *j*, jugal; *l*, lacrimal fused to *fr*; *mx*, maxillary; *n*, nasal; *os*, orbitosphenoid; *p*, parietal; *pe*, wing of periotic; *per*, periotic; *pl*, palatine; *pt*, dorsal pterygoid; *ptf*, post-temporal foramen; *pmx*, premaxillary; *so*, supraoccipital; *sq*, squamosal.

capsule over the posterior semicircular canal. To the epiotic and opisthotic are usually attached the two limbs of the post-temporal bone by strong ligaments. In *Amia* and *Lepidosteus* the supratemporal has not invaded the capsule (Allis, **402**; Parker, **566**).

[1] There is still some uncertainty as to the phylogenetic history of the postfrontal and the supratemporal and their true relation to their deep-lying 'perichondral' and 'endochondral' parts (sphenotic and pterotic). Some authors consider that the fusion of the superficial plate (related to the lateral line) and the deeper ossification of the auditory capsule is secondary. But no primitive form is known possessing a separate supratemporal and pterotic in the adult, and their separate origin in early stages of development (dermo- and auto-squamosum of Schleip, **584**, and Gaupp, **343**) would appear to be secondary, and may be compared to the separation of 'canal bones' from dermal bones in other parts. The postfrontal also may develop from separate superficial and deep (autosphenotic) ossifications which fuse later, as in *Amia* and some Teleosts, or fail to fuse as in many of the higher Teleostei. But in *Polypterus* there is only one ossification (Allis, **410**; Lehn, **542**), and no separate autosphenotic is known in primitive fossil forms.

In Teleostomes the cavity of the auditory capsule is extensive and its inner wall is chiefly membranous, especially in Teleosts.

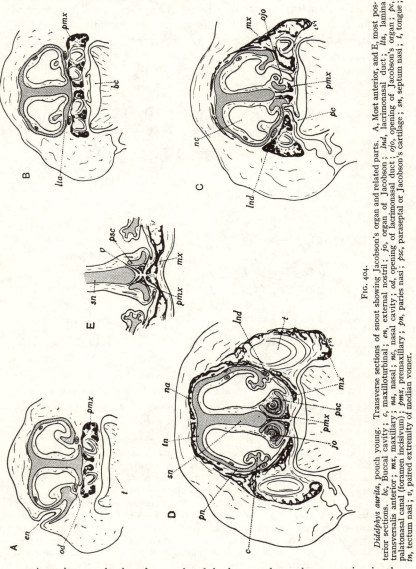

Fig. 404.

Didelphys aurita, pouch young. Transverse sections of snout showing Jacobson's organ and related parts. A, Most anterior, and E, most posterior sections. *bc*, Buccal cavity; *c*, maxilloturbinal; *en*, external nostril; *jo*, organ of Jacobson; *lnd*, lacrimonasal duct; *lta*, lamina transversalis anterior; *mx*, maxillary; *na*, nasal; *nc*, nasal cavity; *od*, opening of lacrimonasal duct; *ojo*, opening of Jacobson's organ; *pc*, palatonasal canal (foramen incisivum); *pmx*, premaxillary; *pn*, paries nasi; *psc*, paraseptal or Jacobson's cartilage; *sn*, septum nasi; *t*, tongue; *tn*, tectum nasi; *v*, paired extremity of median vomer.

Anterior to the basal parachordal plate and to the pro-otics is the basisphenoid.   Although arising from paired centres in the trabeculae in *Amia* (Allis, 402), it appears as a median bone with paired wings, Figs.

286-7. In the more platybasic skulls (Cypriniformes) the basisphenoid

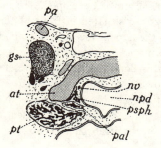

FIG. 405.

Echidna aculeata, var typica ; portion of transverse section of left side of head passing through base of ala temporalis, *at*, and bones of palate. *gs*, Gasserian ganglion ; *npd*, nasal canal ; *nv*, vidian nerve ; *pa*, pila antotica ; *pal*, palatine ; *psph*, dorsal pterygoid (=parasphenoid ?) ; *pt*, ventral pterygoid. (From E. Gaupp, 1911.)

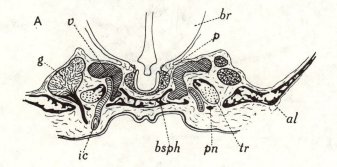

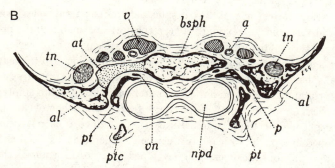

FIG. 406.

Didelphys aurita, pouch young.   Transverse sections through head in region of basisphenoid, *bsph*. A, Through sella turcica. *br*, Brain ; *ic*, internal carotid ; *p*, pituitary ; *pn*, palatine nerve ; *tr*, trabecula. B, Section farther back through ala temporalis, *at*. *a*, Artery ; *al*, alisphenoid ; *npd*, nasal passage ; *p*, palatine ; *pt*, pterygoid ; *ptc*, pterygoid cartilage ; *tn*, trigeminal ; *v*, vein ; *vn*, vidian nerve.

is flat or U-shaped in section, but becomes Y-shaped in the majority where the interorbital septum is developed. In fishes with a myodome and Teleosts generally it is situated in front of the pituitary fossa, and does not extend below the hypophysis as in Tetrapods, hence its homology has been doubted (Kindred adopts Hallmann's name suprasphenoid, 537). Nevertheless, this bone appears to represent the true basisphenoid pushed forward owing to the development of the myodome and growth of the pro-otics (see p. 279). In *Polypterus* and some more specialised Teleosts it may be absent or fused to the parasphenoid. Parker showed that the basitrabecular processes of *Lepidosteus* ossify

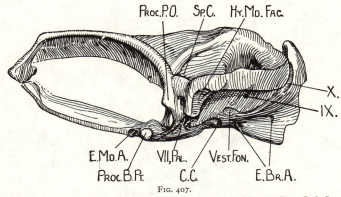

FIG. 407.

Left-side view of neuro-cranium of *Palaeoniscid* (from D. M. S. Watson, *Proc. Zool. Soc.*, 1925). *Hy.Md.Fac*, Facet for hyomandibular. Other letters as in Fig. 408.

separately at first and later fuse with the parasphenoidal wings (566); doubtless the former bones represent the basisphenoids.

Resting on the wings of the basisphenoid are the paired bones generally called alisphenoids, extending between the optic foramen and the pro-otic and up to the roof to form the side walls of the brain-cavity in this region, Figs. 286-7, 436. That this bone in the fish is not the homologue of the mammalian alisphenoid is obvious, since it is an ossification in the primitive cranial wall (orbital cartilage) and not in the outer wall of the cavum epiptericum as in the mammal (p. 437). Nor can it be the exact homologue of the pleurosphenoid of Reptiles and Birds (p. 392) since it is not primarily developed in a pila antotica. It must be considered as a special element, the pterosphenoid (de Beer, 421) developed as an ossification of the pila lateralis, a structure only found in the Holostei. This pterosphenoid is, then, a bone formed in what is apparently a secondary downgrowth of the orbital cartilage (pedicel of the alisphenoid of Allis, 402), which fuses with the subocular shelf laterally to the vena capitis lateralis, and behind the exit of the profundus and abducens nerves. Nevertheless, its more dorsal part may correspond to that of the pleurosphenoid.

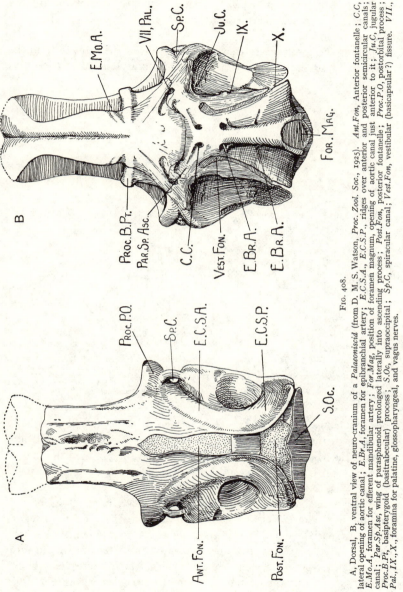

FIG. 408.

A, Dorsal, B, ventral view of neuro-cranium of a *Palaeoniscid* (from D. M. S. Watson, *Proc. Zool. Soc.*, 1925). *Ant.Fon*, Anterior fontanelle; *C.C*, lateral opening of aortic canal; *E.Br.A*, foramen for epibranchial artery; *E.C.S.A., E.C.S.P.*, ridges over anterior and posterior semicircular canals; *E.Mo.A*, foramen for efferent mandibular artery; *For.Mag*, position of foramen magnum, opening of aortic canal just anterior to it; *Ju.C*, jugular canal; *Par.Sp.Asc*, wing of parasphenoid prolonged laterally into ascending process; *Post.Fon*, posterior fontanelle; *Proc.P.O*, postorbital process; *Proc.B.Pt*, basipterygoid (basitrabecular) process; *S.Oc*, supraoccipital; *Sp.C*, spiracular canal; *Vest.Fon*, vestibular (basicapsular?) fissure. *VII., Pal., IX., X.*, foramina for palatine, glossopharyngeal, and vagus nerves.

In the wall of the brain-case in front of the optic foramen are developed the orbitosphenoids, of paired origin, Figs. 286-7, 436. They meet below the brain and above the parasphenoid fusing to form the interorbital septum, when such a septum is present and ossified ; and, thus, come to close the brain-cavity in front and surround the olfactory nerves on their way to the orbit in Teleosts (p. 257).

Just as the postfrontal extends inwards behind the orbit to form the ' sphenotic ', so the prefrontal invades the antorbital cartilage and forms

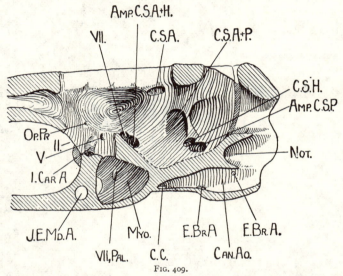

FIG. 409.

Brain-case of *Palaeoniscid* in sagittal section, inner view of right half (from D. M. S. Watson, *Proc. Zool. Soc.*, 1925). *Amp.C.S.A.+H*, Ampulla for anterior and horizontal semicircular canals ; *Amp.C.S.P*, for posterior s. canal ; *C.C*, lateral opening of aortic canal ; *C.S.A*, hole into cavity for ant. semic. canal ; *C.S.A.+P*, space for united vertical s. canals ; *C.S.H*, for posterior end of horizontal s. canal ; *Can.Ao*, aortic canal ; *E.Br.A*, foramen for epibranchial artery ; *I.Car.A*, groove for internal carotid ; *J.E.Md.A*, foramen for junction of efferent mandibular arteries.

the so-called ecto-ethmoid or 'lateral ethmoid' of Parker, which usually comes to surround the olfactory foramen, close the orbit in front, and support the palatoquadrate below, Fig. 436. An anterior median ethmoid or rostral spreads over the nasal cartilage and downwards into the nasal septum, Figs. 299, 305.

Such are the bones found replacing the chondrocranium of the Holostei. Turning to the lower Teleostomes we find that the living Chondrostei are all highly specialised and that in them the endochondral bones have been much reduced. In *Acipenser* only traces of them remain in the form of opisthotics, pro-otics, and orbitosphenoids, which appear late in life (Parker, 565). But the chondrocranium of Palaeoniscids was more fully ossified, Figs. 407-9 (Stensiö, 218 ; Watson, 646). The occipital and auditory regions were ossified much as in the higher forms, but the orbito-temporal ossifications further forward seem to have been fused to a single bone, the ' sphenoid ', apparently representing the basisphenoid, ptero-

Osteolepis

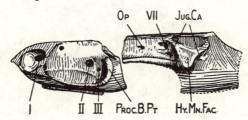

Palaeogyrinus

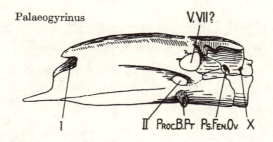

Eryops

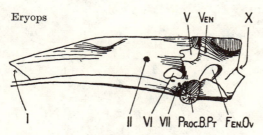

FIG. 410.

Left lateral aspects of ossified parts of brain-case of *Osteolepis macrolepidotus*, Middle Old Red Sandstone, Banffshire; *Palaeogyrinus decorus*, Middle Carboniferous, Fifeshire; *Eryops megacephalus*, Lower Permian, Texas (from Ḍ. M. S. Watson, *Tr. Roy. Soc.*, 1926). *Fen.Ov*, Fenestra ovalis ; *Hy.Mn.Fac*, facet for articulation of hyomandibula ; *Jug.Ca*, openings of jugular canal ; *Op*, superior ophthalmic foramen ; *Proc.B.Pt*, basipterygoid process ; *Ps.Fen.Ov*, pit corresponding to fenestra ovalis ; *Ven*, venous foramen. I-X, cranial nerve foramina.

sphenoids (?), and orbitosphenoids (and in some the ' lateral ethmoids ' as well). The basisphenoid region was more normal than in higher forms, and had a fossa for the pituitary body and well-developed paired basi-trabecular processes for the articulation of the basal processes of the palatoquadrates. This primitive joint persists in *Lepidosteus* (p. 421), but is lost in *Amia* and all modern Teleostei.[1] As in so many other respects, *Polypterus* resembles the Palaeoniscids in having a large

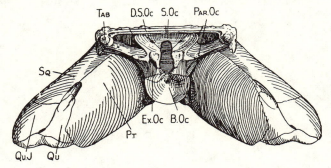

FIG. 411.

Orthosaurus pachycephalus; reconstruction of occipital aspect of skull (from D. M. S. Watson, *Tr. Roy. Soc.*, 1926). *B.Oc*, Basioccipital; *D.S.Oc*, dermal supraoccipital = postparietal; *Ex.Oc*, exoccipital; *Par.Oc*, paroccipital process of opisthotic; *Pt*, posterior plate of pterygoid; *Qu.J*, quadratojugal; *Qu*, quadrate; *S.Oc*, supraoccipital; *Sq*, squamosal; *Tab*, tabular.

'sphenoid' extending from the front end of the brain-cavity to the basisphenoid and even the pro-otic region (Traquair, 613; Allis, 410;

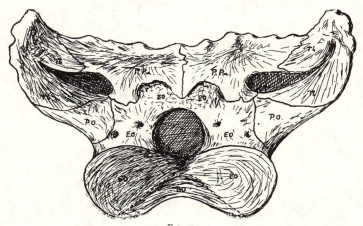

FIG. 412.

Occiput of *Eryops megacephalus*. (From R. Broom, *Am. Mus. Nat. Hist.*, 1913.)

Lehn, 542; Budgett, 10; Pollard, 575). It is pierced by the olfactory optic and trigeminal foramina, and passes behind the infundibulum over the hypophysial fossa to form a 'pro-otic bridge' (see below). A well-

---

[1] The connexion between parasphenoid and endopterygoid in Osteoglossidae, described by Bridge, 432, and Ridewood, 580, is almost certainly secondary.

developed independent opisthotic has not yet been identified in either

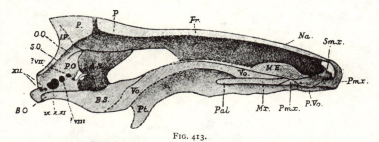

FIG. 413.

Median section of skull of *Diademodon* (after R. Broom, *Proc. Zool. Soc.*, from W. K. Gregory, *J. Morph.*, 1913). *AS*, Alisphenoid=epipterygoid ; *BS*, basisphenoid ; *ME*, mesethmoid ; *OO*, opisthotic ; *PO*, pro-otic ; *P.Vo*, prevomer ; *Smx*, septomaxillary ; *Vo*, vomer.

Palaeoniscids or *Polypterus* ; but small ossifications in the latter may perhaps represent it (Stensiö, **409**).

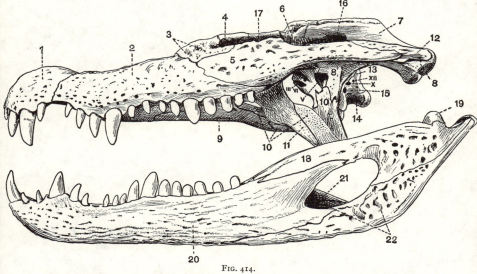

FIG. 414.

Lateral view of skull of *Caiman latirostris* (from S. H. Reynolds, *Vertebrate Skeleton*, 1913). 1, Premaxillary ; 2, maxillary ; 3, lacrimal ; 4, prefrontal ; 5, jugal ; 6, postfrontal ; 7, squamosal ; 8, quadrate ; 9, palatine ; 10, pterygoid ; 11, transpalatine (ectopterygoid) ; 12, quadratojugal ; 13, exoccipital ; 14, basioccipital ; 15, carotid foramen ; 16, external auditory meatus ; 17, frontal ; 18, supra-angular ; 19, articular ; 20, dentary ; 21, coronoid ; 22, angular ; *III, VI*, opening for exit of oculomotor and abducens nerves ; *V*, foramen ovale ; *X*, pneumogastric foramen ; *XII*, hypoglossal foramen.

It has recently been shown by Stensiö (**218**) and Watson (**646**) that a ' dorsal myodome ' is present in the pituitary region of Palaeoniscids, but has not yet in these early fish extended backwards to form a

2 C

' ventral myodome ' below the basis cranii, Fig. 409. It must be concluded, as previously mentioned, that the myodome was already developed
in the ancestral Actinopterygian and has been lost in those higher forms
in which it is absent (Chondrostei, Lepidosteoidei, certain Teleostei ; see
p. 283) ; for its development leads in this region of the skull to important
modifications found in all Actinopterygii whether a functional myodome
be present or not. These modifications involve the restriction of the

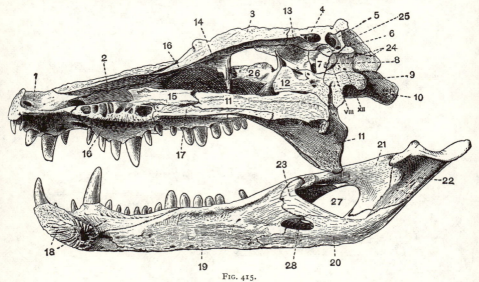

FIG. 415.

Longitudinal section through skull of *Caiman latirostris* (from S. H. Reynolds, *Vertebrate Skeleton*, 1913). 1, Premaxillary ; 2, nasal ; 3, frontal ; 4, parietal ; 5, supraoccipital ; 7, pro-otic ; immediately in front of the figure 7 is prominent foramen for trigeminal nerve ; 8, opisthotic ; 9, basioccipital ; 10, quadrate ; 11, pterygoid ; 12, basisphenoid ; 13, pleurosphenoid (' alisphenoid ') ; 14, prefrontal ; 15, prevomer ; 16, maxillary ; 17, palatine ; 18, dentary ; 19, splenial ; 20, angular ; 21, supra-angular ; 22, articular ; 23, coronoid ; 24, exoccipital ; 25, squamosal ; 26, jugal ; 27, external mandibular vacuity ; 28, internal mandibular foramen ; *VIII*, internal auditory meatus *XII*, hypoglossal foramen.

basisphenoid to a position in front of the pituitary fossa to close the
myodome anteriorly, and the roofing over of its cavity by a ' pro-otic
bridge ' continuing the cranial floor in front of the parachordal plate
and behind the infundibulum. This remarkable bridge, typically formed
by the junction of processes from the pro-otic bones, is partly preformed
in cartilage occupying the position of the acrochordal cartilage of
Tetrapods (p. 240), and is probably homologous with it. When a ventral
division of the myodome is developed in Teleostei between the basis
cranii and the parasphenoid, its side walls are strengthened by downgrowths from the pro-otics and basioccipital (p. 279 and Figs. 286, 289).

Some interorbital septum (more or less developed and often partly

membranous, p. 256) is present in Palaeoniscids and in all Holostean
Teleostomes, excepting Amioidei and a few Teleosts. It is associated with

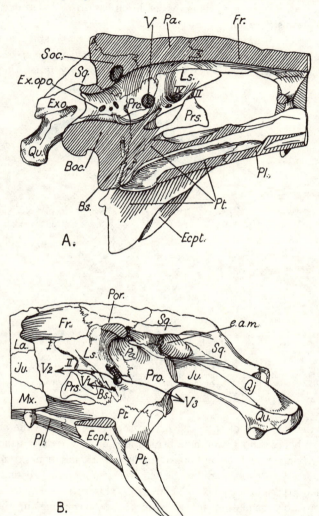

FIG. 416.

A, Inner view of left half of posterior region of skull of *Crocodilus*; B, outer view of same (from
Gregory and Noble, *J. Morph. and Physiol.*, 1924). *Bs* and *Prs*, Basisphenoid; *e.a.m*, external
auditory meatus; *Ls*, pleurosphenoid (laterosphenoid).

the growth of large eyes and orbits bringing the originally separate cranial
side walls together in the mid-line. How far the absence of the septum

and more platybasic build of the skull in *Polypterus, Amia,* and the Cypriniformes is primary or secondary is at present uncertain ; but there is some reason to suppose that this condition has arisen independently in several groups (p. 235).

The recent work of Bryant on *Eusthenopteron* (465), and more especially of Stensiö (218, 606) and Watson (644) on *Dictyonosteus, Osteolepis,* and other related fossils, has brought to light many interesting facts

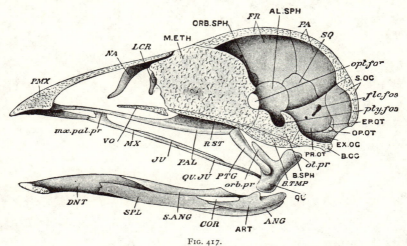

FIG. 417.

Sagittal section of a bird's skull (diagrammatic).   Replacing bones : AL.SPH, pleurosphenoid ; ART, articular ; B.OC, basioccipital ; B.SPH, basisphenoid ; EP.OT, epiotic ? ; EX.OC, exoccipital ; M.ETH, mesethmoid ; OP.OT, opisthotic ; ORB.SPH, orbitosphenoid ; PR.OT, pro-otic ; QU, quadrate ; S.OC, supraoccipital.   Investing bones : *ANG*, angular ; *B.TMP*, basi-temporal ; *COR*, coronary ; *DNT*, dentary ; *FR*, frontal ; *JU*, jugal ; *LCR*, lacrymal ; *MX*, maxilla ; *NA*, nasal ; *PA*, parietal ; *PAL*, palatine ; *PMX*, premaxilla ; *PTG*, pterygoid ; *QU.JU*, quadrato-jugal ; *RST*, rostrum ; *S.ANG*, supra-angular ; *SPL*, splenial ; *SQ*, squamosal ; *VO*, vomer. *flc.fos*, Floccular fossa ; *mx.pal.pr*, maxillo-palatine process ; *opt.for*, optic foramen ; *orb.pr*, orbital process ; *ot.pr*, otic process ; *pty.fos*, pituitary fossa.   (From Parker and Haswell, *Zoology*.)

about the structure of the skull in Crossopterygii. While the occipital and otic bones tend to combine behind and become rigidly attached to the overlying cranial roofing, the sphenoid and ethmoid bones combine to form a massive ' sphenoid ' connected with the anterior roofing above and the parasphenoid below, Fig. 410. This orbito-temporal region was probably somewhat movable on the posterior region, the joint being marked between the frontals and parietals at all events in Osteolepids. Jugular canals, basitrabecular processes, and paroccipital processes are present.

Evidence of close affinity between the Coelacanthini and Osteolepids is afforded by the skull (Stensiö, 605 ; Watson, 645) ; for both groups appear to have the same frontoparietal hinge. The anterior ethmoid region was mainly cartilaginous in Coelacanths, but there is a 'sphenoid' including the well-ossified basisphenoid.[1] A large pro-otic (which possibly represents the combined pro-otic and opisthotic), basioccipital, exoccipital, and supraoccipital bones have been found in *Macropoma* and others. The chondrocranium of the Devonian genus *Diplocercides* was solidly ossified in both its anterior and posterior regions (Stensiö, 606).

Little is known about the condition of the cranium in early Dipnoi beyond the fact that it was more fully ossified than in later forms (Traquair, 622 ; Watson and Day, 647). In existing genera it is almost completely cartilaginous, with only a pair of posterior bones considered to be exoccipitals (Huxley, 535 ; Bridge, 433 ; Wiedersheim, 653).

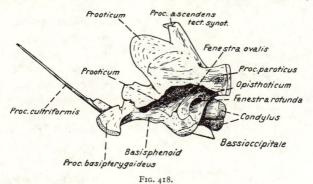

FIG. 418.

Occipital region, etc., of skull of *Tupinambis teguexin* (from Versluys, *Vergl. Anat.*, 1927). *Proc. cultriformis* = rostrum of parasphenoid.

Tetrapoda.—The chief ossifications in the chondrocranium of Tetrapods were doubtless derived from those of their fish-like ancestors. Indeed, most of the endochondral bones found even in the human skull can be traced back to the primitive Osteichthyes. But, although there is a fundamental similarity in these ossifications throughout the Tetrapods, specialisation has led to the fusion or loss of many of them in most modern forms. Among Amphibia it is to the early (Carboniferous and Permian) Stegocephalia that we must turn for the primitive structure, for in these the chondrocranium is much more completely ossified than in later forms and thus approaches in structure that of the early Reptiles. The early Embolomeri, such as *Palaeogyrinus* and *Orthosaurus* (*Loxomma*), have the foramen magnum closed above by a supraoccipital, and below by a basioccipital, which forms the greater part of a hollow occipital condyle

---

[1] The basitrabecular processes of the basisphenoid are brought high up at the sides in the Coelacanths ; but the depressions below them are probably not myodomes as described by Stensiö (Watson, 645).

completed at the sides by the exoccipitals, Figs. 410, 411-12. An opisthotic and a pro-otic are present in the auditory region ; the former provided with a strong paroccipital process abutting against the tabular region of the dermal roof, and probably also supporting the otic process of the quadrate. A basisphenoid lodging the pituitary fossa, and closely connected or fused to the underlying elongated parasphenoid, has pronounced basitrabecular (basipterygoid) processes with facets for the pterygoid and epipterygoid.

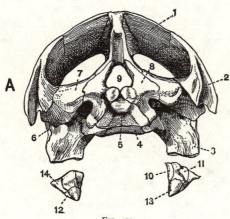

Farther forward extends a large ' sphenethmoid ', surrounding the anterior region of the brain-case, pierced in front for the olfactory nerves, and sheathed below by the narrowing parasphenoid. It meets the pro-otics above and the basisphenoid below, while the enclosed lateral openings are divided by a stout bar joining the basisphenoid and doubtless representing the ossified pila antotica. This ' s p h e n e t h m o i d ' bone, formed presumably by the fusion of two orbitosphenoids (ossified orbital cartilages), is of special interest since its

FIG. 419.

Posterior half of skull of *Chelone midas* (from S. H. Reynolds, *Vertebrate Skeleton*, 1913). 1, Parietal ; 2, squamosal ; 3, quadrate ; 4, basisphenoid ; 5, basi-occipital ; 6, quadratojugal ; 7, opisthotic ; 8, exoccipital ; 9, foramen magnum ; 10, prearticular ; 11, articular ; 12, dentary ; 13, angular ; 14, supra-angular.

ventral portion forms a thick low interorbital septum, indicating apparently that in these primitive Stegocephalia the skull was less platybasic than in modern Amphibia, and so approximated to the Amniote structure (Watson, 631, 644). Important also is the presence of a hypoglossal foramen in the exoccipital of such forms as *Orthosaurus, Trimerorhachis,* and *Eryops,* as evidence that some at least of the hypoglossal roots, as in Amniotes, passed through this region, and therefore that this nerve probably was composed of more segmental roots than in the higher Stegocephalia and modern Amphibia (p. 226), and that the occipital region of the skull perhaps contained more skeletal segments.

The later and more specialised Stegocephalia show traces of the reduction in ossification which is so characteristic of all modern Amphibia (Fraas, 497 ; Broili, 435 ; Williston, 657-8, 660 ; Broom, 448 ; Watson, 644). The supra- and basioccipital, still present in *Eryops,* usually are no longer developed, and the occipital condyles are then paired and formed exclusively by the exoccipitals which may meet in the middle line below. There are no hypoglossal foramina (p. 226).

The modern Amphibia have lost not only the supra- and basioccipital, but also the opisthotic (except perhaps *Necturus*, where a bone occurs called 'epiotic' by Huxley, 534). It has become either supplanted by or fused with the exoccipital. The very flat basis cranii has no basi-

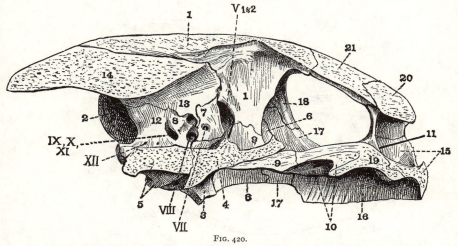

FIG. 420.

Longitudinal vertical section through the skull of *Chelone midas* (from S. H. Reynolds, *Vertebrate Skeleton*, 1913). 1, Parietal; 2, squamosal; 3, quadrate; 4, basisphenoid; 5, basioccipital; 6, quadratojugal; 7, pro-otic; 8, opisthotic; 9, pterygoid; 10, palatine; 11, rod passed into narial passage; 12, exoccipital; 13, 14, supraoccipital; 15, premaxillary; 16, maxillary; 17, jugal; 18, postfrontal; 19, prevomer; 20, prefrontal; 21, frontal; *V, VII, VIII, IX, X, XI, XII,* foramina for the exit of cranial nerves.

sphenoid. Anterior ossifications persist either in the form of a single sphenethmoid 'girdle-bone' (Anura) or of paired orbitosphenoids (Urodela and Apoda). There are few occipital segments, no hypoglossal foramina, and the paired exoccipital condyles are widely separated, Fig. 329.

As already mentioned, the most primitive Amniota (Cotylosauria) differ but little from the early Stegocephalia (Embolomeri) in the endochondral ossifications of the occipital and auditory regions, Fig. 410. A 'sphenethmoid' still surrounds the brain-cavity in front in such forms as *Diadectes* and *Pareiasaurus* (Case, 470; Williston, 661, 95). It persists anteriorly in some Theromorpha (Dicynodontia) as a median bone pierced for the olfactory nerves, lodging the olfactory lobes of the brain and spreading down into the median septum, Fig. 422 (Sollas and Sollas, 602; Broom, 1912–13). In the Theriodontia it seems to have been less developed. The occipital condyle in these Reptiles (Cotylosauria and Theromorpha) is

usually distinctly tripartite (formed by the basioccipital and two exoccipitals) and sometimes concave, as in *Pareiasaurus* (Seeley, 586).

The remainder of the Reptilia diverge considerably from this primitive
structure. Except in the Ophidia (p. 237) the skull is markedly tropibasic,
the interorbital septum high and thin ; consequently the sphenethmoid
ossification is completely lost as in Chelonia and Crocodilia, or reduced to
disconnected vestigial bones near the optic foramen and usually lost in

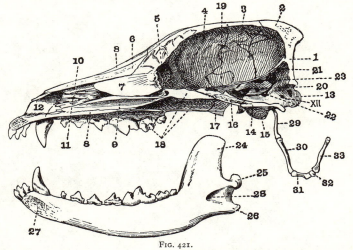

FIG. 421.

Vertical longitudinal section taken a little to left of middle line through skull of *Canis familiaris*
(from S. H. Reynolds, *Vertebrate Skeleton*, 1913). 1, Supraoccipital ; 2, interparietal ; 3, parietal ;
4, frontal ; 5, cribriform plate ; 6, nasal ; 7, mesethmoid ; 8, maxilla ; 9, vomer ; 10, ethmo-turbinal ;
11, maxillo-turbinal ; 12, premaxillary ; 13, occipital condyle ; 14, basioccipital ; 15, tympanic
bulla ; 16, basisphenoid ; 17, pterygoid ; 18, palatine ; 19, alisphenoid ; 20, internal auditory meatus ;
21, tentorium ; 22, foramen lacerum posterius ; 23, floccular fossa ; 24, coronoid process ; 25, condyle ;
26, angle ; 27, mandibular symphysis ; 28, inferior dental foramen ; 29, stylo-hyal ; 30, epi-hyal ;
31, cerato-hyal ; 32, basi-hyal ; 33, thyro-hyal ; *XII*, condylar foramen.

dry skulls. Such vestiges occur in *Sphenodon*, Lacertilia, and Ophidia
(orbitosphenoids, ' alisphenoids' of Parker, postoptics of Cope ;
Siebenrock, 599, 600).[1] Down-growing flanges of the parietals and
frontals to some extent replace the ' sphenethmoid ' in supporting the
sides of the brain-case.

An apparently new bone makes its appearance in the Crocodilia as an
ossification of the posterior lateral wall of the brain-case (orbital cartilage,
and pila antotica) extending downwards to the basisphenoid between the
optic foramen in front and the trigeminal foramen behind (Parker, 570 ;
Shiino, 597 ; de Beer, 421). It was called the alisphenoid by older authors,
but is certainly not homologous with the Mammalian alisphenoid, since it

---

[1] The absence of orbitosphenoids in so many fossil Reptilia may be due
to their having dropped out of the skull.

belongs to the true cranial wall and not the outer wall of the cavum epiptericum (Gregory and Noble, 524a; and see p. 267). This bone probably occurred in related fossil reptiles, such as the Phytosaurs (v. Huene), and is present in Birds. It may be called the laterosphenoid (Gregory and Noble) or better pleurosphenoid, Figs. 414-16; but it probably represents the hinder region of the extensive sphenethmoid described above in Stegocephalia which may be supposed to include an anterior (true orbitosphenoid) and this posterior element. Both elements are well developed in the avian skull, Figs. 388, 417; the pleurosphenoid occupy-

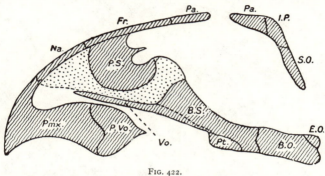

FIG. 422.

Median section of skull of *Dicynodon sollasi* (from R. Broom, *Proc. Zool. Soc.*, 1926). *Na,* Nasal; *Fr,* frontal; *Pa,* parietal; *I.P,* interparietal; *S.O,* supraoccipital; *P.S,* presphenoid or sphenethmoid; *B.S,* basisphenoid; *B.O,* basioccipital; *Pmx,* premaxillary; *pp,* its palatine process; *Vo,* vomer; *Mx,* maxillary; *Pal,* palatine; *P.Vo,* prevomer; *E.O,* exoccipital. (Cp. Fig. 423.) (From Broom, *Proc. Zool. Soc.,* 1926.)

ing a large part of the antero-lateral wall of the swollen brain-case, and the orbitosphenoids fusing to a median ethmoid usually forming an extensive bony interorbital septum (Parker, 557; Tonkoff, 612; Sushkin, 608; T. J. Parker, 555; de Beer, 421).

The opisthotics (paroccipitals) in Sauropsidan Reptiles usually bear strong paroccipital processes, particularly well developed in Rhyncho-cephalia and Lacertilia, Figs. 354, 368, 418, 419; but they are shorter in Crocodilia and still more reduced in Aves. The single occipital condyle characteristic of Reptiles and Birds is still tripartite in most Chelonia, Fig. 319, and to a less extent in Lacertilia and Ophidia; but in the other Orders it is typically a prominent articular knob on the basioccipital alone (Osborn, 553).

The ossifications of the Mammalian chondrocranium may be derived from those of the Theromorpha. The characteristic dicondylic condition arises from the primitive monocondylic[1] by the enlargement of the

[1] The condyle is more or less bilobed and paired in many of the mammal-like Theromorpha, Fig. 343.

exoccipital and gradual withdrawal of the basioccipital element. An intermediate ∪-shaped stage, in which the two lateral condyles are still joined by a narrow articular region, occurs in Monotremes, more especially in the embryo (Osborn, 553; Gaupp, 512). In the adult Mammal the opisthotic and pro-otic are always fused to one bone, the periotic or petrosal, enclosing the membranous labyrinth. This periotic is often but loosely connected with the surrounding bones in lower forms and may

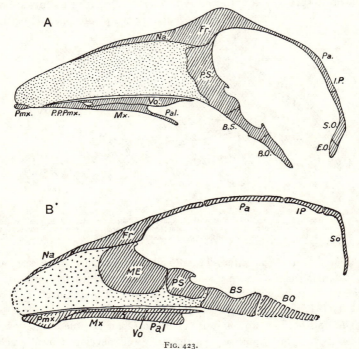

Fig. 423.

Median section of skull of A, young goat, *Capra*; B, young *Procavia capensis*. *ME*, Mesethmoid. Other letters as in Fig. 422. (From Broom, *Proc. Zool. Soc.*, 1926–7.)

fall out in macerated skulls; but in the higher mammals it may fuse with the squamosal and tympanic (= ' temporal bone ' of Man).

There is still some doubt about the derivation of the Mammalian periotic. The earlier authors described three centres of ossification in foetal man, and Huxley named them pro-otic, opisthotic, and epiotic, believing them to be homologous with these bones in the Reptile and Fish. They all three fuse later to the single bone of the adult. Parker (567), Strong, and Thomas (1926) also describe these three early centres in other mammals. But Vrolik (628) found four centres of ossification in the calf and pig, and from three to as many as six in the human foetus. While it may safely be held that the periotic as a whole is homologous with the reptilian pro-otic and opisthotic combined, it is doubtful whether the centres so named by Huxley alone represent them, or whether each bone may develop from several secondary centres.

The paroccipital process is little developed, being represented by the posterior portion of the crista parotica (p. 272). The basisphenoid harbours the pituitary body in a deep fossa or sella turcica which may in Rodents retain its hypophysial opening; this bone in Marsupials is pierced by the internal carotids (as in the lower Tetrapods), but in Placentals they pass to the pituitary fossa on either side of it.

In many Placentals two bones occur in the anterior region of the floor of the brain-case : one just in front of the basisphenoid has a median presphenoid region and two lateral orbitosphenoid wings ; the other and more anterior is the mesethmoid, an ossification of the anterior wall of the brain-case (including the lamina cribrosa) and median nasal septum, Figs. 421, 423 B. The latter bone arises from a median centre, while the former is formed apparently by the junction ventrally of two orbito-sphenoid ossifications. Some Placentals, however, and the Marsupials have only one bone occupying the place of presphenoid, mesethmoid, and orbitosphenoids,[1] Fig. 423 A. There appears to be no separate 'presphenoid' bone in any adult Mammal. Thus the sphenethmoid of Theromorphs would seem to be represented by a single bone in the adult skull of some Mammals ; but in many Placentals, including the Primates, there is an anterior mesethmoid in addition to the posterior ossification (fused orbitosphenoids). In these forms, then, either a new anterior ossification has appeared, or the old one has been divided into two (Broom, 462a and b).

---

[1] This bone develops from two orbitosphenoid centres as a rule (Parker) ; but in some Placentals there may appear an additional median anterior ossification of doubtful significance. According to Broom there is a mesethmoid and a 'presphenoid' in the Orders Tubulidentata, Pholidota ?, Hyracoidea, Rodentia, Insectivora, Carnivora, Menotyphla, Cheiroptera, Dermoptera, Primates ; but not in the Orders Xenarthra, Artiodactyla, Perissodactyla, Sirenia, Chrysochloridea, and Monotremata.

CHAPTER VII

SKELETAL VISCERAL ARCHES AND LABIAL CARTILAGES

THE SKELETAL VISCERAL ARCHES

THE skeletal visceral arches of Gnathostomes, already mentioned on
p. 129, are originally quite independent of the chondrocranium proper
(neurocranium). They belong to the visceral, not to the axial skeleton,
and are developed not from skeletogenous tissue surrounding the central
nervous system and notochord, but from that which envelops the alimentary canal. Hence they are derived from splanchnic mesoblast, and
primarily lie in the pharyngeal wall, internal to the main blood-vessels
and branchial nerves, and necessarily internal to the coelom, Fig. 142
(represented in this region by the pericardial coelom, and by canals in
the embryonic gill bars, see p. 490). The skeletal arches, then, primarily
are situated internally to the dorsal aorta, cardinal veins, and lateral
aortic arches, and meet ventrally above and medially to the pericardium,
heart, and ventral aorta. The ventral junction is perhaps secondary in

396

ontogeny, the earliest prochondral rudiments being paired, even probably the median copulae (Gibian, **684**).

Behind the mandibular and hyoid arches are the branchial arches

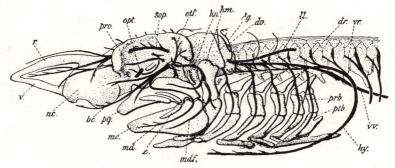

FIG. 424.

Skeleton and nerves of the head of *Mustelus laevis*, Risso (the nerves partly from Allis). *bc*, Buccal nerve ; *dg*, dorsal branch of glossopharyngeal ; *dr*, dorsal root of spinal nerve ; *dv*, dorsal branch of vagus ; *e*, labial cartilage ; *g*, glossopharyngeal ; *hm*, hyomandibular—the spiracle is indicated by a dotted line in front, and the prespiracular cartilage is shown in front of the spiracle ; *hn*, hyomandibular nerve ; *hy*, hypoglossal nerve ; *ll*, lateral-line branch of vagus ; *mc*, Meckel's cartilage ; *md*, mandibular nerve ; *mdf*, mandibular branch of facial nerve ; *nc*, nasal capsule ; *opt*, optic nerve ; *otf*, otic branch of facial ; *r*, dorsal rostral cartilage ; *sop*, superior ophthalmic branch of trigeminal and facial ; *v*, ventral rostral cartilage ; *vr*, ventral root ; *vv*, visceral branch of vagus. (From Goodrich, *Vert. Craniata*, 1909.)

(p. 440), loosely attached below the occipital region and vertebral column, Fig. 424. These branchial arches typically become subdivided in Pisces into dorsal pharyngeal, epibranchial, lateral ceratobranchial, and ventral hypo-

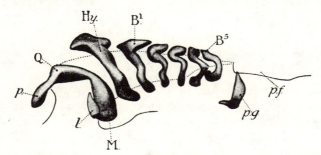

FIG. 425.

Skeleton of visceral arches and pectoral girdle of 20·5 mm. embryo of *Spinax*. (After Braus, 1906, from Kerr, *Embryology*.) B^1, B^5, Branchial arches ; *Hy*, hyoid ; *l*, labial cartilage ; *M*, mandibular arch ; *p*, palatopterygoid bar ; *p.f*, rudiment of pectoral fin ; *p.g*, pectoral girdle ; *Q*, knob for attachment to trabecular region of skull.

branchial segments joined together in the mid-ventral line by an unpaired basibranchial or copula. Even among fishes these elements undergo considerable modification in number and development, and it is doubtful whether the homology of the separate pieces can be traced to the branchial arches of Amphibia and still less of higher forms, where the arches are much altered in structure and function. The various elements appear to develop in a continuous procartilaginous crescentic rudiment as independent chondrifications (van Wijhe, 397, in *Squalus*; Sonies, 385, in birds), and not owing to the secondary subdivision of a cartilage rod as originally described by Dohrn in Selachians and Stöhr in Teleosts, Fig. 425.

Beginning with the Selachii, in which the gill-arches are in some respects in the most primitive condition, we find that they are typically composed of the several elements mentioned above, Figs. 424, 426-7, 439.

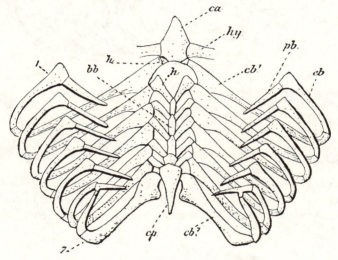

FIG. 426.

Branchial arches of *Heptanchus*. (After Gegenbaur.) *bb*, Basibranchial; *ca*, basihyal; *cb*¹⁻⁷, ceratobranchials; *cp*, fused 6th and 7th basibranchials; *eb*, epibranchial; *h*, hypobranchial; *hy*, ceratohyal; *pb*, pharyngobranchial. (From Goodrich, *Vert. Craniata*, 1909.)

Moreover, each arch, excepting the last, bears a single row of cartilaginous branchial rays attached to the epi- and ceratobranchial and radiating outwards in the gill-septum between successive gill-slits. These rays may fuse at their bases and extremities. Outside them, in many Selachians, are curved extrabranchials, whose homology has given rise to much controversy. By some they have been considered as outer branchial arches;[1] but Gegenbaur's view that they are merely the enlarged dorsal

[1] Rathke's view (1832) that the Cyclostomes have outer arches, and the Gnathostomes inner arches (the outer arches being represented by the extrabranchials of Selachii), was adopted by J. Müller, Duvernoy, Stannius, Balfour, and Gegenbaur at first. Later Dohrn established the modern view that the arches are homologous throughout the Craniata (333), and was followed by Goette, Gegenbaur, and others.

and ventral branchial rays is generally accepted and agrees with their structure and development (Gegenbaur, 166; Dohrn, 333).

Small outer septal cartilages (extraseptals) may also occur in the gill-septa of the Rajidae (K. Fürbringer, 681).

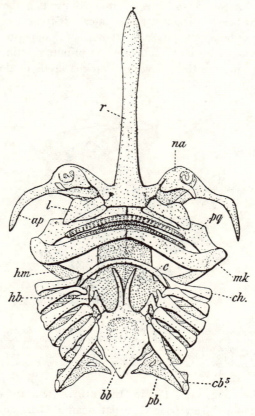

FIG. 427.

Skull and visceral arches of *Raja*, ventral view. (After Gegenbaur.) *ap*, Antorbital process; *bb*, compound basibranchial; *c*, basihyal; *cb⁵*, fifth ceratobranchial; *hb*, hypobranchial; *hm*, hyomandibular; *l*, labial; *mk*, Meckel's cartilage; *na*, nasal capsule; *pb*, pharyngobranchial; *pq*, palatoquadrate; *r*, rostrum. (From Goodrich, *Vert. Craniata*, 1909.)

The branchial arches in other Chondrichthyes (Cladoselachii, Dean, 154, 155; Pleuracanthodii, Fritsch, Jaekel, 186) seem to have been built on the same plan as in the Selachii. In Acanthodii the pharyngobranchial is in two pieces (Jaekel, 690).

In the Osteichthyes the branchial arches are also of essentially the same structure, Fig. 436. The chief modifications they undergo in these and other fishes will be dealt with farther on (p. 440).

The Mandibular and Hyoid Arches in Pisces.—The chief interest centres round the first two visceral skeletal arches, which become closely connected with the chondrocranium. The rudiment of the first or mandibular arch bends round the corner of the mouth and gives rise to two cartilages found in all Gnathostomes : the palatoquadrate bar above and the mandibular bar below. The latter is often called Meckel's cartilage. In Chondrichthyes these two bars meet those of the opposite side in the middle line below the snout, and form the primitive biting upper and lower jaws bearing the teeth developed on the inside of the margin of

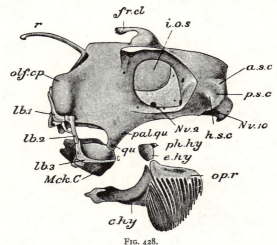

FIG. 428.

Skull of *Chimaera monstrosa*, lateral view. (From Parker and Haswell's *Zoology*, after Hubrecht.) *a.s.c*, Position of anterior semicircular canal ; *c.hy*, ceratohyal ; *e.hy*, epihyal ; *fr.cl*, frontal clasper ; *h.s.c*, position of horizontal semicircular canal ; *i.o.s*, interorbital septum ; *lb.* 1, *lb.* 2, *lb.* 3, labial cartilages ; *Mck.C*, mandible ; *Nv.* 2, optic foramen ; *Nv.* 10, vagus foramen ; *olf.cp*, olfactory capsule ; *op.r*, opercular rays ; *pal.qu*, palatoquadrate ; *ph.hy*, pharyngohyal ; *p.s.c*, position of posterior semicircular canal ; *qu*, quadrate region ; *r* rostrum.

the mouth, Figs. 424, 427. But in Osteichthyes (with the exception of the Chondrostei) and Tetrapoda the palatoquadrate bars do not meet in front, but usually and primitively become connected with the lateral ethmoid region (antorbital process), and although the quadrate region of the palatoquadrate behind always serves to support Meckel's cartilage, the more anterior pterygopalatine region becomes chiefly concerned in roofing the palate, Figs. 429, 437. Meanwhile the function of supporting the biting teeth is taken on by the marginal dermal bones (p. 287). Meckel's cartilage is always developed in young Osteichthyes ; but, except at its articular end, loses its importance in the adult, being always functionally replaced by dermal bones. Finally in Tetrapods the primitive upper and lower cartilages no longer take an important share in the support of the adult jaws except at their extreme posterior ends (quadrate and articular). They are more or less completely developed

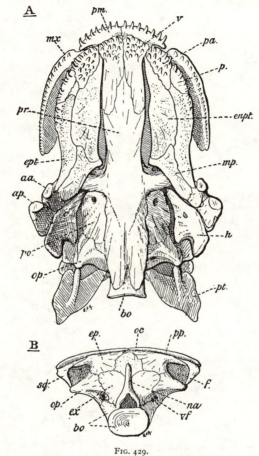

Fig. 429.

Amia calva, L. A, Ventral view of skull and upper jaw; B, posterior view of skull. *aa*, Articular surface of quadrate, *ap*, of symplectic; *bo*, basioccipital; *enpt*, entopterygoid; *ep*, epiotic; *ept*, ectopterygoid; *ex*, exoccipital; *f*, posterior temporal fossa; *h*, hyomandibular; *mp*, metapterygoid; *mx*, maxilla; *na*, neural arch, whose centrum is fused to basioccipital; *oc*, occipital cartilage; *op*, opisthotic; *p* and *pa*, palatine bones; *pm*, premaxilla; *po*, pro-otic; *pp*, postparietal?; *pr*, parasphenoid; *pt*, post-temporal; *sq*, pterotic; *v*, vomer; *vf*, vagus foramen. (From Goodrich, *Vert. Craniata*, 1909.)

in the embryo, but become greatly specialised and modified in the higher groups (p. 423).

Both the palatoquadrate and mandibular cartilages become ossified in varying degree in Osteichthyes, Figs. 430, 436. In the former there is generally an anterior autopalatine, a posterior quadrate, and a meta-

2 D

pterygoid between, at all events in Coelacanthini and Actinopterygii (recalling somewhat the three calcified pieces described by Jaekel in Acanthodians, **690, 443**).   But in the more primitive Crossopterygii

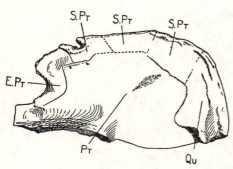

FIG. 429a.

Megalichthys; outer view of left palatopterygoid complex, with palatal part broken away (from D. M. S. Watson, *Tr. Roy. Soc.*, 1926).   *E.Pt*, Epipterygoid ; *Pt*, pterygoid ; *Qu*, quadrate ; *S.Pt*, suprapterygoids.

(Osteolepidoti) and Palaeoniscids the cartilage was usually more completely ossified.   Thus in Megalichthys (Watson, **644**) a continuous series of endochondral suprapterygoid bones extends along the pterygoid, the first representing the epipterygoid and the last reaching the quadrate, Figs. 309, 429. The epipterygoid bone extends into the basal process and articulates with the basitrabecular process of the basisphenoid.   From this condition that of the lower Tetrapoda may easily be derived (see pp. 428, 431).   Meckel's

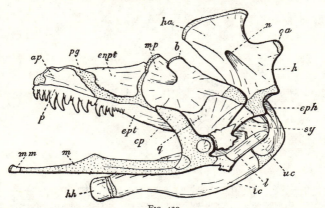

FIG. 430.

Amia calva, L.   Skeleton of the left jaws and hyoid arch, from which the dermal bones of the lower jaw have been removed.   (After Allis, slightly modified.)   The cartilage is dotted.   *ap*, Endochondral palatine ; *b*, otic process ; *cp*, coronoid process ; *enpt*, endopterygoid ; *eph*, epihyal ; *ept*, ectopterygoid ; *h*, hyomandibular ; *ha*, its articular head ; *hh*, hypohyal ; *ic*, ventral segment of ceratohyal ; *l*, ligament ; *m*, Meckel's cartilage ; *mm*, mento-Meckelian ; *mp*, metapterygoid ; *n*, foramen for hyomandibular nerve ; *oa*, articular head for opercular ; *p*, palatine (dermal) ; *pg*, palatopterygoid cartilage ; *q*, quadrate ; *sy*, symplectic ; *uc*, upper segment of ceratohyal.   (From Goodrich, *Vert. Craniata*, 1909.)

cartilage is generally ossified in the articular region ; in some, such as *Amia*, several small bones occur, Figs. 430-1, including an endochondral articular, ' coronoid ', and ' angular ' (van Wijhe,

654; Bridge, 429; Allis, 402), besides an anterior mento-mandibular (mento-Meckelian). Some of these persist in the higher forms; while

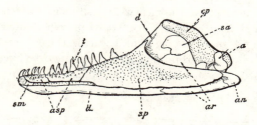

FIG. 431.

Inner view of the lower jaw of *Amia calva*, L. (After Allis.)   *a*, Articular; *an*, angular; *ar*, dermarticular (angular); *asp*, anterior coronoids; *cp*, coronoid cartilage; *d*, dentary; *sa*, supra-angular; *sm*, mento-Meckelian; *sp*, coronoid or prearticular with minute teeth; *t*, marginal tooth. (From Goodrich, *Vert. Craniata*, 1909.)

in *Saurichthys* the whole cartilage is replaced by a single bone (Stensiö, 134).

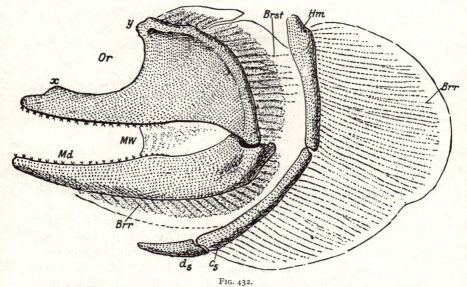

FIG. 432.

Pleuracanthus sessilis, L. Permian, Germany (from O. Jaekel, *Morph. Jahrb.*, 1925).   *x*, *y*, Anterior and posterior articulation of palatoquadrate with chondrocranium; *Brr*, branchial rays; *Brst*, outer dermal rays?; *c₅*, ceratohyal; *d₅*, hypohyal; *Hm*, hyomandibula; *Md*, mandibular cartilage; *MW*, skin at corner of mouth; *Or*, orbit.

Comparing the mandibular arch with the branchial in fishes it is clear that the mandibular cartilage somewhat resembles the ceratobranchial, and the palatoquadrate the epibranchial. Opinions differ as to their exact homology. Whether they deserve the names epimandibular and cerato-mandibular respectively is doubtful. There is no definite evidence from ontogeny or palaeontology that the mandibular arch was ever formed of the same elements as the branchial arches. Some evidence, however, has been found of the existence of vestigial pharyngo-, hypo-, and basi-mandibulars. A small median ventral cartilage has been described in *Chlamydoselachus* (K. Fürbringer, 681), in *Laemargus* (White, 702), and in Teleosts, and interpreted as a basimandibular. Small paired anterior

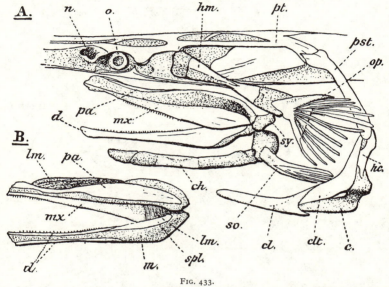

FIG. 433.

Polyodon folium, Lac. (After Traquair.) A, Left-side view of skull, jaws, and pectoral girdle ; B, inner view of right jaws. *c*, Coracoid ; *ch*, ceratohyal ; *cl*, clavicle ; *clt*, cleithrum ; *d*, dentary ; *hc*, postclavicle ; *hm*, hyomandibular ; *lm*, levator muscle ; *m*, Meckel's cartilage ; *mx*, maxilla ; *n*, olfactory capsule ; *o*, optic capsule ; *op*, opercular ; *pa*, palatine ; *pst*, post-temporal ; *pt*, pterotic ; *so*, subopercular ; *spl*, splenial ; *sy*, symplectic. (From Goodrich, *Vert. Craniata*, 1909.)

cartilages described by White in *Laemargus* (702) and by Gegenbaur (166), K. Fürbringer (681), and Sewertzoff (594, 701) in *Hexanchus* are considered by the last author to be vestigial hypomandibulars. The pharyngo-mandi-bular Sewertzoff believes to be represented by the orbital process of the palatoquadrate in most Selachians, and by a small dorsal cartilage dis-covered in the 'ethmoid ligament' of *Scaphirhynchus* by Ivanzoff (1887) and of *Laemargus* by Sewertzoff and Disler (594). These homologies cannot, however, be taken as well established.[1]

---

[1] Allis (670) sees the pharyngo-mandibular and pharyngo-hyal in the trabecular and polar cartilages ; but there are serious objections to this view, since these latter cartilages appear to belong to the axial skeleton (p. 238).

The same question arises with regard to the homology of some of the elements composing the hyoid arch. In Pisces in general there are two main paired elements : the dorsal hyomandibula and the more ventral ceratohyal, which seem to correspond to the epibranchial and cerato-branchial respectively. A ventral median copula or basihyal is usually well developed (p. 397) ; but is apparently fused with the first basi-branchial in Acipenseroidei, Polypterini, Amioidei, and *Lepidosteus*. Separate hypohyals are not usually found in Elasmobranchs ; though Braus has described them in an embryo *Heptanchus* (673), and Sewertzoff has found vestiges in the adult (701). Well-developed hypohyals have been

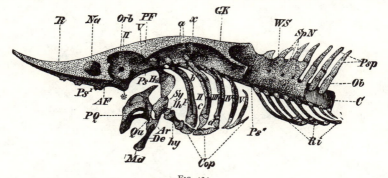

FIG. 434.

Cranial skeleton of sturgeon (*Acipenser*) after removal of exoskeletal parts. *Ar*, Articular ; *C*, notochord ; *Cop*, copulae ; *De*, dentary ; *GK*, auditory capsule ; *Hm*, hyomandibular ; *hy*, hyoid ; *I* to *V*, first to fifth branchial arches, with double pharyngobranchial (*a*), epibranchial (*b*), cerato-branchial (*c*), and hypobranchial (*d*) ; *Ih*, interhyal ; *II*, optic foramen ; *Md*, mandible ; *Na*, nasal cavity ; *Ob*, neural arches ; *Orb*, orbit ; *PF*, *AF*, postorbital and antorbital processes ; *PQ*, palato-quadrate ; *Ps*, *Ps'*, *Ps"*, parasphenoid ; *Psp*, supraneural spines ; *Qu*, quadrate ; *R*, rostrum ; *Ri*, ribs ; *SpN*, apertures for spinal nerves ; *Sy*, symplectic ; *WS*, vertebral column ; *x*, vagus foramen ; *, ridge on basis cranii. (From Wiedersheim, *Comp. Anatomy*.)

found in Pleuracanthini and Acanthodii (Jaekel, 690), Fig. 432, and are constantly present in Teleostomes living and extinct, and in Dipnoi, Fig. 436.

In the Holostei the ceratohyal has separate upper and lower ossifica-tions ; the more ventral is the main bone, and the smaller more dorsal has often been called the 'epihyal'. This name it certainly does not deserve, since it clearly forms part of the original cartilaginous ceratohyal. A rudiment of the dorsal bone has been found in Palaeoniscids (Stensiö, 218). All the Actinopterygii have, between the hyomandibular and ceratohyal elements, a small separate segment, the stylohyal or interhyal (its presence in Palaeoniscoidei is doubtful). It has sometimes been compared to the epibranchial, but probably is a new formation. The large hyomandibular cartilage of the Holostei ossifies in two pieces—a hyomandibular bone above articulating with the auditory capsule, and a symplectic below connected with the quadrate. It is probable that the small dorsal and large ventral ossifications of the hyomandibula in *Polypterus* do not correspond to the two in other forms, Fig. 435; but a symplectic has been described in some Palaeoniscids (Broom; Stensiö, **218**). This symplectic becomes an independent element in the Acipenseroidei, where it may be very large; but is lost in a few Teleosts, such as the Siluridae and Anguilliformes.

Whether the hyomandibular cartilage of fishes represents the epihyal element only, as generally held, or includes a pharyngohyal as well, can hardly be determined on the evidence available.[1] A very small cartilage

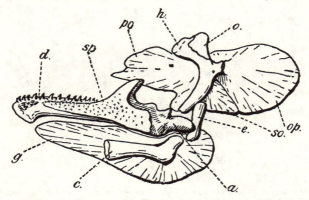

FIG. 435.

Polypterus bichir, Geoffr. Inner view of the lower jaw and hyoid arch. *a*, Articular; *c*, ceratohyal; *d*, dentary; *e*, epihyal; *g*, inferior gular; *h*, hyomandibular; *o*, small dermal ossicle; *op*, opercular; *po*, preopercular; *so*, subopercular; *sp*, coronoid or prearticular. (From Goodrich, *Vert. Craniata*, 1909.)

has been described as a vestigial pharyngohyal in *Stegostoma* by Luther (**695**), and a small pharyngohyal occurs in Holocephali (Dean, **17**; Schauinsland, **583**; Sewertzoff, **594**, **701**), Figs. 428, 444.

We must now consider the important question of the connexion of the first two visceral arches with the chondrocranium. In all the Gnathostomes (except the Mammalia, Chapter VIII.) the lower jaw is articulated to the quadrate region of the upper jaw by means of the posterior articular end of the mandibular cartilage, ossified as the articular bone in Osteichthyes and Tetrapoda. This articulation persists in Mammals, but serves another function (Chapter VIII.). Now the mode of support

---

[1] The term hyomandibula has been applied to structures which are not strictly homologous: to the cartilaginous dorsal element in Selachians, to the dorsal bone in the originally continuous hyomandibular cartilage in Actinopterygii, and to that cartilage before ossification; also to the separate dorsal cartilage (ossified in the adult) of Acipenser.

of this articulation and of the jaws, and the connexion of the first two visceral arches with the skull, varies in different fishes (and to a less extent in Tetrapods, p. 423).

When the palatoquadrate bar is complete it is connected in front to the chondrocranium, by fusion as in Dipnoi (and Amphibia) or by movable articulation as in Chondrichthyes and Teleostomes. But this articulation

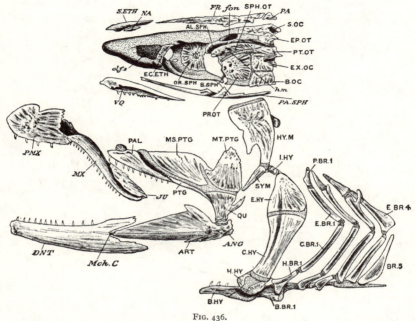

FIG. 436.

Salmo fario. Disarticulated skull with many of the investing bones removed. The cartilaginous parts are dotted. *fon*, Fontanelle ; *h.m*, articular facet for hyomandibular ; *Mck.C*, Meckel's cartilage ; *olf.s*, hollow for olfactory sac. Replacing bones : AL.SPH, pterosphenoid ; ART, articular ; B.BR.1, first basibranchial ; B.HY, basihyal ; B.OC, basioccipital ; BR.5, fifth branchial arch ; B.SPH, basi-sphenoid ; C.BR.1. first ceratobranchial ; C.HY, ceratohyal ; EC.ETH, ecto-ethmoid ; E.BR.1, first epibranchial ; E.HY, epihyal ; EP.OT, epiotic ; EX.OC, exoccipital ; H.BR.1, first hypobranchial ; H.HY, hypohyal ; HY.M, hyomandibular ; I.HY, interhyal ; MS.PTG, mesopterygoid ; MT.PTG, metapterygoid ; OR.SPH, orbitosphenoid ; PAL, palatine ; P.BR.1, first pharyngobranchial ; PTG, pterygoid ; PT.OT, pterotic ; QU, quadrate ; S.OC, supraoccipital ; SPH.OT, sphenotic ; SYM, symplectic. Investing bones : *ANG*, angular ; *DNT*, dentary ; *FR*, frontal ; *JU*, jugal ; *MX*, maxilla ; *NA*, nasal ; *PA*, palatine ; *PA.SPH*, parasphenoid ; *PMX*, premaxilla ; *VO*, vomer. (From Parker and Haswell, *Zoology*.)

is not necessarily the same in all fish. In the apparently more primitive Selachians the palatoquadrate acquires a well-marked articular process (orbital or ethmo-palatine process) which articulates with the trabecular basal part of the skull (palatobasal articulation (?), see p. 411), either far back in the postorbital region as in *Scymnus*, or about the middle of the orbit as in *Squalus* and *Notidonus*, or far forwards near the antorbital region as in *Mustelus* or *Heterodontus* (Gegenbaur, 166), Figs. 424, 439, 440.

In these the process usually fits in a well-marked groove ; but in others, such as *Scyllium*, the process is reduced and only loosely attached by the 'ethmo-palatine' ligament.   In Rajidae there is no longer any definite articulation.   It is characteristic of the Selachians that the bars are prolonged beyond the articulation as pre-palatine processes which, as already mentioned, meet in front.[1]

On the other hand, the palatoquadrate bars of other Gnathostomes do not typically so meet, but their anterior extremities articulate or fuse

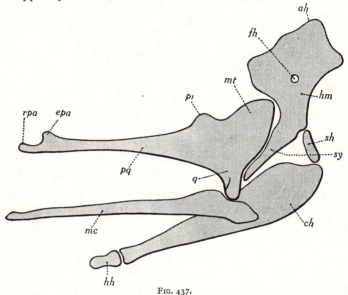

FIG. 437.

Salmo fario, late embryo.   Left-side view of paired cartilages of first two visceral arches.   *ah*, Articular head ; *ch*, ceratohyal ; *epa*, ethmo-palatine articular process ; *fh*, foramen for hyomandibular nerve ; *hh*, hypohyal ; *hm*, hyomandibular ; *mc*, mandibular ; *mt*, metapterygoid region ; *pq*, palatoquadrate ; *pr*, vestigial basal process ; *q*, quadrate region ; *rpa*, rostropalatine articular process ; *sh*, stylohyal ; *sy*, symplectic region.

with the lateral region of the antorbital processes.   Here in Teleostomes an articular surface is formed, simple primitively, but tending in many Teleostei to become subdivided into an anterior pre-palatine process meeting the pre-ethmoid cornu, and a posterior postpalatine process meeting the parethmoid cornu of this ethmoid region, Fig. 437.   In some of the most specialised Teleostei, such as the Gasterosteiformes and Plectognathi, the posterior articulation disappears (Swinnerton, 389).

[1] If the orbital process represents, as seems probable, the basal process of other forms, then this anterior prolongation may be compared to the palatine region of the palatoquadrate of Teleostomes and Tetrapods.   In Chlamydoselache it lies below and is connected with the nasal capsule (Sewertzoff, 701).

Before discussing the homology of these anterior connexions the posterior connexions of the palatoquadrate bar must be described. In Pisces the quadrate region may be directly articulated to or fused with the base and auditory region of the skull, or it may be supported away from the skull by the hyomandibula, or again it may be connected with the auditory capsule both by a dorsal articulation and by the hyomandibula. Huxley (535), who distinguished these three types of jaw suspension, named them autostylic, hyostylic, and amphistylic respectively. While the amphistylic type occurs only among the more primitive Chondrichthyes (Notidani and early Heterodonti among Selachii, Pleura-

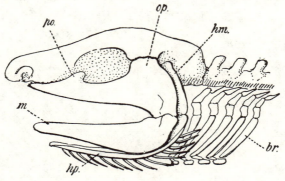

FIG. 438.

Diagram of the skull and visceral arches of an Acanthodian. *br*, Fifth branchial arch; *hm*, hyomandibula; *hp*, calcified plate bearing similar branchial rays (?); *m*, Meckel's cartilage; *op*, otic process of the palatoquadrate; *po*, ethmoid process. (From Goodrich, *Vert. Craniata*, 1909.)

canthodii, Acanthodii, Cladoselachii, Figs. 438-9), the hyostylic type is found in the majority of Selachii and in all Teleostomi, though not necessarily strictly homologous in these two groups.[1] Among Pisces, the true autostylic type occurs only in the Dipnoi, Figs. 441-2, and possibly in Holocephali, Fig. 444; but it is the type on which are built all the Tetrapoda, although it becomes much altered in the Amniota.

It should be noted that in the autostylic type the hyomandibula becomes much reduced, takes no share in the support of the jaws, and may even disappear altogether (*Protopterus, Lepidosiren*). In the hyostylic type, on the contrary, it becomes large and important, forms the chief support, and may give rise to the symplectic. The amphistylic

[1] In Gnathostomes generally the first or spiracular gill slit is early reduced from below upwards, and either closes altogether or remains only as a small dorsal spiracle (Chapter IX.): this enables the lower end of the hyomandibula to become strongly bound by ligaments to the quadrate region of the mandibular arch below the slit.

is, perhaps, the least specialised of the three types, and here the hyo-mandibular is only moderately developed.

Fully to appreciate the importance of Huxley's three types we must first of all study the autostylic more closely. In this type of suspension

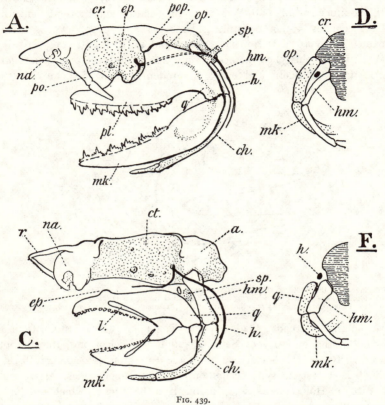

FIG. 439.

Diagrams of: A and D, an amphistylic skull (*Heptanchus*), and C and F, a hyostylic skull (*Scyllium*). A and C, left-side views; D and F, from behind (the mandibular arch being dotted, and the cranium shaded). *ch*, Ceratohyal; *cr*, cranium; *ep*, ethmoid process; *h*, hyomandibular branch of facial nerve; *hm*, hyomandibula; *l*, labial; *mk*, Meckel's cartilage; *na*, nasal capsule; *op*, otic process; *pl*, palatoquadrate cartilage; *po*, preorbital process; *pop*, postorbital process; *q*, quadrate region; *r*, rostral process; *sp*, spiracle. (From Goodrich, *Vert. Craniata*, 1909.)

the palatoquadrate bar is primitively connected with the chondro-cranium at three points (more clearly seen in the Amphibia than in the Dipnoi, Figs. 441-2, 449): by its anterior palatine process to the ant-orbital (parethmoid) process of the ethmoid region; by the basal process (or pedicle of Huxley and Parker) of its middle pterygoid part to the trabecula; and by a dorsal otic process of the posterior or quadrate part to the auditory capsule or its postorbital process.[1] A fourth possible connexion is through the processus ascendens of the palatoquadrate, which is dealt with later (pp. 413, 428). These three attachments are present in Dipnoi and Amphibia, where the palatine process usually fuses with the antorbital, the otic process with the auditory capsule, and the basal process with the trabecula (in Dipnoi and Urodela, but not in Anura or

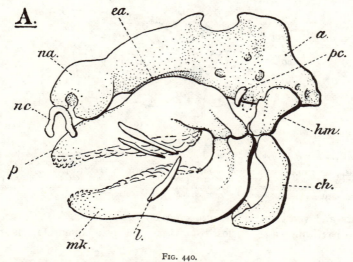

FIG. 440.

Skull, jaws, and hyoid arch of *Heterodontus (Cestracion) Philippi*, Lac. (after Parker, modified). *a*, Auditory capsule; *ch*, ceratohyal; *ea*, ethmoid articulation; *hm*, hyomandibula; *l*, labial; *mk*, Meckel's cartilage; *na*, nasal capsule; *nc*, nasal cartilage; *p*, palatoquadrate; *pc*, prespiracular. A dotted ring indicates the spiracle. (From Goodrich, *Vert. Craniata*, 1909.)

Apoda). In the Amniota and also in many Urodela the palatine process (pterygoid process of many authors) is little developed and fails to reach the ethmoid region (p. 425). From the trabecula[2] there grows out a basitrabecular process (basipterygoid process of many authors) to meet the basal process; it is little developed in modern Dipnoi and Amphibia, but usually prominent in Reptilia, Aves, and Mammalia (p. 270).

The fundamental relations of these connexions to the blood-vessels and nerves are as follows (Figs. 449, 495, 493): The palatine branch of

---

[1] The articulation of the otic process appears to shift along the crista parotica or lateral edge of the capsule. In Selachii it meets it anteriorly at the postorbital process, and in Reptilia posteriorly where the crista forms a paroccipital process.

[2] This basitrabecular process seems to grow out of that region of the trabecula which is developed from the polar cartilage, p. 237.

the facial nerve passes down posteriorly to the basitrabecular and basal processes, then forwards below them. The jugular vein or vena capitis lateralis (p. 532) runs back dorsally to the basitrabecular and basal pro-

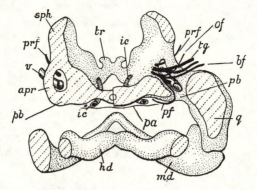

FIG. 441.

Ceratodus forsteri, embryo about 13 mm. long. Reconstruction from transverse sections of anterior part of head skeleton, seen from behind. On left, cut through ascending process and basal process; on right, cut farther back through quadrate. Nerves black, cartilage stippled. *apr*, Processus ascendens; *bf*, buccal; *hd*, hyoid; *ic*, internal carotid; *of*, ophthalmic; *pa*, parasphenoid; *pb*, processus basalis of palatoquadrate; *pf*, palatine; *prf*, profundus; *q*, quadrate; *sph*, sphenethmoid region of orbital; *tg*, trigeminal; *tr*, trabecular cornu.

cesses and ventrally to the otic process, passing through the cranio-quadrate passage enclosed between the palatoquadrate, its two

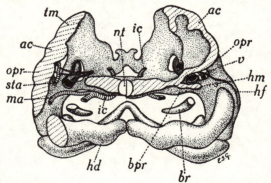

FIG. 442.

Ceratodus forsteri, same as in Fig. 441. On left, cut through otic process; on right, through auditory capsule. *ac*, Auditory capsule; *bpr*, basal process; *br*, branchial arch; *hf*, hyoid branch of facial; *hm*, remains of hyomandibula; *ma*, mandibular artery; *nt*, notochord; *opr*, otic process of palatoquadrate fused to capsule; *sta*, orbital or stapedial artery; *tm*, commissura orbito-parietalis *v*, vena capitis lateralis passing through cranio-quadrate passage with artery and nerve.

processes, and the chondrocranium. The vein is accompanied by the hyomandibular branch of the facial nerve, which passes over the spiracular slit to the hyoid arch. The internal carotid passes morphologically dorsally to the spiracular slit, ventrally to the hyomandibular attachment, but gives off an orbital artery (external carotid of many authors) which runs forwards to the orbit through the cranio-quadrate passage, Figs. 241-2.

Yet another important process is present in Dipnoi and Tetrapoda. It is the processus ascendens typically developed from the pterygoid region of the palatoquadrate bar, near the origin of the basal process, as a dorsal cartilage passing vertically upwards between the profundus nerve and the maxillary branch of the trigeminal, and laterally to the vena

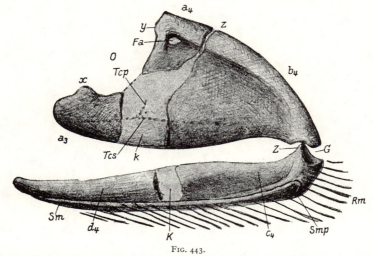

FIG. 443.

Acanthodes bronni, L. Permian, Germany. Outer view of left mandibular arch (from O. Jaekel, *Morph. Jahrb.*, 1925). *x*, *y*, Anterior and posterior articulation of palatoquadrate with chondrocranium ; *z*, fracture ? ; *a₃*, *a₄*, *b₄*, calcifications in palatoquadrate, and *c₄*, *d₄*, in mandibular cartilage, *k* ; *G*, articulation ; *O*, position of orbit ; *Rm*, rays ; *Smp*, submandibular bony plate ; *Tcp*, *Tcs*, scales of supra- and infra-orbital lateral lines.

capitis lateralis, Figs. 449, 495. Its upper end fuses with the orbital wall of the cranium in Dipnoi, Fig. 441 (Greil, 525). It is found neither in Chondrichthyes nor in modern Teleostomi except possibly as a vestige. Whether it occurred in early primitive Teleostomes is not yet certain, but Watson describes traces of it in an Osteolepid (644). In Dipnoi and Tetrapods the ascending process forms the outer wall of the cavum epiptericum which is described above (p. 267).

Comparing now the amphistylic with the autostylic type, it may safely be concluded that the dorsal process of the palatoquadrate in Notidani represents the otic process, since it bears the same relation to the nerves and blood-vessels, Figs. 438-9. But it is by no means so easy to determine the homology of the anterior 'orbito-palatine' (ethmo-

palatine) connexion of the Selachian. Is it formed by the basal process
which has moved forwards, or by an originally anterior palatine process
which has moved backwards along the base of the skull ? [1]   Huxley held
the former view, which on the whole seems the more probable ; [2] yet, the
anterior articulation of the palatoquadrate seems to have been very far
forward in such early forms as Acanthodii and Pleuracanthini (judging
from the reconstructions of Jaekel, 186), if it is really the palatobasal and
not an ethmoid connexion. The fact that it articulates really with the

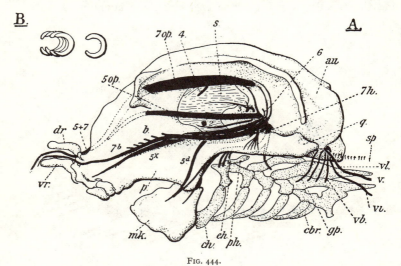

FIG. 444.

Callorhynchus antarcticus, Lac. A, Skeleton and nerves of the head of a young specimen (after
Schauinsland's figures). *au*, Auditory capsule ; *b*, region of fusion of palatoquadrate with nasal
capsule ; *cbr*, 5th ceratobranchial ; *ch*, ceratohyal ; *dr*, dorsal median rostral cartilage ; *eh*, epihyal ;
gp, glossopharyngeal nerve ; *mk*, Meckel's cartilage ; *p*, palatoquadrate region ; *ph*, pharyngohyal ;
q, quadrate region (otic process ?) ; *s*, interorbital septum ; *sp*, spinal nerve-roots ; *vb, vi, vl*,
branchial, intestinal, and lateral-line branches of vagus nerve ; *vr*, ventral paired rostral cartilage ;
2, optic, 3, oculomotor, 4, pathetic, 5, trigeminal, 6, abducent, and 7, facial nerves. B, Calcified
skeletal supports of the lateral line. (From Goodrich, *Vert. Craniata*, 1909.)

basal or trabecular and not antorbital region of the skull is strong
evidence that it is a true basal process. Further study of the fossils
may enable this point to be decided.

That the hyostylic type of the Selachian is derived from the amphistylic by the loss of the otic process, and accompanying shortening and
strengthening of the hyomandibula, now the chief support of the jaws,

[1] Since both the palatine and the orbital processes would occupy the same
position as the basal process with regard to the internal jugular vein and
palatine nerve if brought sufficiently far back, these relations do not help us
to decide on their homology.

[2] Allis (667, 412) believes the basal process to be represented in Selachians
by a medial ridge on the hinder region of the palatoquadrate.

it is easy to suppose,[1] Figs. 439, 440. While early Heterodonti have an otic articulation (Woodward, 703), the modern genus has lost it; similarly the articulation is lost and the otic process much reduced in *Chlamydoselachus* among Notidani.

A very specialised type of jaw suspension occurs in the Holocephali (Huxley). The whole structure has been profoundly modified in connexion with the development of permanent grinding tooth plates adapted for hard food, Figs. 428, 444. The jaws are shortened and strengthened; the rami of the lower jaws are fused in front, while the palatoquadrate is fused to the ethmoid and orbital region of the skull in front and to the auditory capsule behind (Dean, 17; Schauinsland, 583). The cranio-

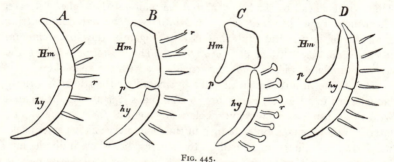

FIG. 445.

Diagram showing modification of hyoid arch in Selachians according to theory of Gegenbaur. (After C. Gegenbaur, from A. Sedgwick, *Zoology*, 1905.) A, Notidanus; B, pentanchal Selachian; C, Torpedo; D, Raja. *Hm*, Hyomandibula; *hy*, hyoideum or ceratohyal; *r*, branchial rays.

quadrate passage is almost obliterated. Since the hyoid arch remains free and takes no share in the suspension of the jaws, the result is a type of autostylism, which, however, has probably been developed independently of that found in Dipnoi and Tetrapods, and has been called 'holostylic' by Gregory (688).

A further important question arises as to which represents the original attachment of the mandibular arch to the chondrocranium—the otic or the basal process. Gegenbaur (1872) argued that the connexion of the otic process to the postorbital process was primitive and that the palato-

---

[1] In Rajidae the hyoid arch appears much modified, and, while the hyomandibula reaches down as usual to the jaw articulation, a rod at the dorsal end of the ceratohyal passes up behind to the skull. Gegenbaur believed this rod to have been extended from the ceratohyal which had become disconnected from the hyomandibula (see Fig. 445); others that it represents an additional post-hyomandibular arch (see footnote, p. 440). But Krivetski has recently given good evidence that, while the larger anterior supporting cartilage is the true hyomandibula, the posterior rod is derived from the fused bases of branchial rays (694); a conclusion supported by embryology (de Beer, 324).

quadrate secondarily developed a palatine region and orbital process. But Huxley's view (1876) is now generally accepted (Sewertzoff, **384**; Luther, **695**; Gaupp, **343**; and others) that the articulation of the orbital process with the trabecular region of the basis cranii represents in Selachians the original point of attachment. The Selachian hyomandibula is attached at a corresponding level, morphologically dorsal to the aorta and ventral to the jugular vein and exits of the nerves.

The hyomandibula of Teleostomes articulates dorsally with the lateral surface of the auditory capsule, the broad facet for its reception being usually shared by the postfrontal (sphenotic) and supratemporal (pterotic). In *Polypterus* it is formed by the opisthotic and supratemporal. The articular head of the hyomandibula in many of the higher Teleostei (Acanthopterygii) becomes subdivided into two, the anterior abutting against the postfrontal and the posterior against the supratemporal (Swinnerton, **389**; Allis, **667, 669, 671**). The posterior edge of the hyomandibula acquires a knob for articulation with the opercular.

It is by no means certain that the hyomandibular cartilage is strictly homologous in all fishes, Fig. 446. In Selachians it is always articulated with the capsule at a point morphologically ventral to the vena capitis lateralis and hyomandibular branch of the facial nerve as it passes backwards over the spiracle and cartilage to the hyoid bar where it divides into mandibular and hyoid branches. On the contrary, in all Teleostomes the hyomandibular cartilage is articulated at a point morphologically dorsal to the vena capitis lateralis or its derivatives, and the hyomandibular nerve to reach the bar passes medially and ventrally to the cartilage and then outwardly behind it, or through it as in the Holostei (Allis, **667, 669**).[1] It is clear that an important difference exists in the relations of these structures, and various theories have been set forth to explain it (Schmalhausen, **699, 700**; de Beer, **421**; Edgeworth, **678-9**; Allis, **671**). Either the hyomandibula is not strictly homologous in the two cases, or there has been some radical shifting of the parts concerned.

A further difficulty is presented by *Polypterus*, in which alone the hyomandibular nerve divides near its exit so that the mandibular branch

---

[1] In some Teleosts, such as the Siluroids (Pollard, **697**; Kindred, **537-8**), the hyomandibular cartilage at an early stage is continuous with the quadrate. Pollard concluded that the Teleostean hyomandibula corresponds to the Selachian otic process, and the Selachian hyomandibula to the Teleostean stylohyal. This conclusion is not consistent with the morphological position of these elements described above; that the temporary connexion of hyomandibula and palatoquadrate is secondary is shown by the structure of the more primitive Teleosts and lower Teleostomes (Norman, **551**; Berrill, **672**).

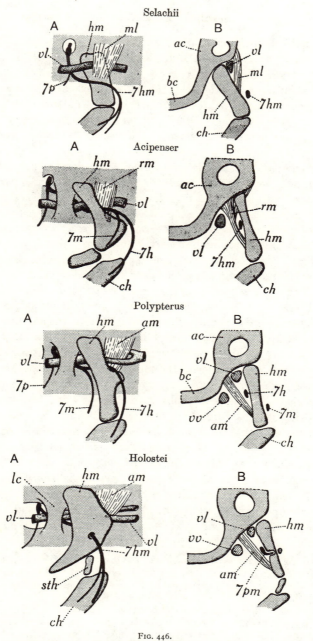

FIG. 446.

Diagrams illustrating relation of *hyomandibula* to skull, nerves, veins, and muscle in *Pisces*. A, Left-side view; B, transverse section. *ac*, Cut surface of auditory capsule, and *bc*, of basis cranii; *am*, m. adductor hyomandibulae; *ch*, ceratohyal; *hm*, hyomandibular cartilage; *ml*, m. levator hyomandibulae; *rm*, m. retractor hyomandibulae; *sth*, stylohyal; *vl*, vena capitis lateralis, and *vv*, its medio-ventral branch; *7h*, hyoid, *7hm*, hyomandibular, *7m*, mandibular, *7p*, palatine branches of facial nerve; *7pm*, hyomandibular nerve piercing cartilage. *am*, *ml*, and *rm* appear to be, at all events, partially homologous muscles.

passes outwards in front of, and the hyoidean branch behind, the hyo-mandibula (van Wijhe, 654 ; Allis, 410 ; Budgett, 10). This unusual position of the mandibular nerve (if it is the strict homologue of the mandibular branch of other forms) may perhaps be due to its having worked its way through from behind. The fact that in some Palaeoniscids the hyomandibula is perforated (Stensiö, 218 ; Watson, 646) suggests that the nerve may have reached half-way through in these fish.

Taking, first of all, the Holostei (Amioidei, Lepidosteoidei, Teleostei), we find that, as already mentioned, the main hyomandibular nerve pierces the broad hyomandibula before dividing. It has been suggested that the nerve has here, so to speak, eaten its way into the cartilage. But, according to Allis, while the Selachian hyomandibula (except in skates) is an epihyal having the usual relations to nerves and blood-vessels, the Holostean hyomandibula is made up of an anterior portion derived from branchial rays, and a posterior portion derived from more posterior branchial rays of the hyoid arch fused together so as to enclose the nerve. Schmalhausen's explanation is that the epihyal (hyomandibula) was originally connected with the auditory capsule by a ventral infrapharyngeal and a dorsal suprapharyngeal, the vein and nerve passing backward between them ; that in Selachians the infrapharyngeal has fused with the epihyal forming its ventral attachment, the suprapharyngeal disappearing ; and that in the Holostean the suprapharyngeal has fused to form the articular head of the epihyal, the infrapharyngeal disappearing. Moreover, he claims that both pharyngeals are present at a certain stage of develop-ment in *Acipenser* and *Ceratodus*.[1] De Beer, adopting Allis's suggestion that the otic process of the palatoquadrate is derived from branchial rays, concluded that the hyomandibula also primitively articulated by two processes, ' basal ' and ' otic ', with the vein and nerve between them. In the Selachii only the ' basal ' and in the Teleo-stomi only the ' otic ' process of the hyomandibula would then be preserved ; while traces of both would be present in *Ceratodus* at an early stage.[2]

---

[1] The relations of the two branches of the dorsal end of the hyoid arch in *Ceratodus* to the nerves and veins are, however, not the same as that of the inferior and superior pharyngeal of the branchial arch of *Acipenser* (see Fig. 447).

[2] It should also be mentioned that a muscle (C_2hd of Ruge, 1897; Allis, Luther, 695) passes in all Pisces from skull to hyomandibula dorsally to the vena cap. lateralis and ventrally to the hyomandibular branch of the facial. In Selachians it lies on the outer dorso-lateral face of the hyomandibula, while in all Teleostomes it is inserted more or less on its inner or ventro-median surface, Fig. 446.

Edgeworth, who has shown that the hyomandibula of Teleostomes is in early stages a cartilaginous rod situated entirely in front of the hyomandibular nerve (except in *Polypterus*) and that later it may surround the nerve by backward growth, maintains that this hyomandibula is homologous throughout the fishes, but offers no explanation of the different relation the articulation bears to vein and nerve in Selachii and Teleostomi.

On the whole, for the usually accepted view that the hyomandibula is homologous in all these fishes, there is good evidence not only from

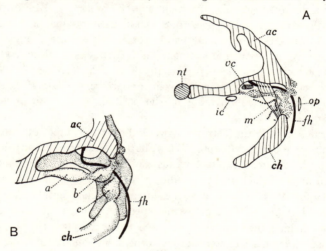

FIG. 447.

Ceratodus Forsteri, embryo 18·5 mm. long. A, Reconstructed thick transverse section seen from behind. *ch*, Ceratohyal; *fh*, hyomandibular branch of facial nerve; *m*, muscle; *op*, opercular. Spiracular pouch shown in dotted line at *m*. B, Diagrammatic reconstruction of cartilages *a*, *b*, *c*, derived from dorsal end of hyoid arch (cp. Fig. 446).

embryology but also from palaeontology; for it appears to have been well developed in Crossopterygii (Osteolepidoti) (Traquair, 619; Bryant, 465; Watson, 644, 647)[1] and early Chondrostei. That the articular head dorsal to vein and nerve in Teleostomes is an 'otic process' is doubtful; but it is not impossible that an articulation, originally ventral in Selachians, may have moved up to a new position by passing over the bridge forming the outer wall of the jugular canal (p. 376) into which the vein and nerve have sunk in Teleostomes (Stensiö, 218).[2] Thus we could account for a

---

[1] The hyomandibula is perforated in Rhizodopsis (Watson and Day, 647).

[2] The condition in *Polypterus* might perhaps be explained on the supposition that the mandibular nerve passed in front through the wall of the jugular canal before the migration of the hyomandibular.

change from a more primitive ventral articulation in the Selachian to the more specialised condition with dorsal articulation in the Teleostome, a change perhaps related to the reduction of the otic process of the quadrate region. The inclusion of the hyomandibular nerve might still be due either to backward growth of the hyomandibula, or to its fusion with branchial rays.

Nothing is yet known about the condition of the hyomandibula in early Dipnoi, in which it was probably small and unossified; but it appears to persist in *Ceratodus* partly as a small ventral cartilage having the relations of the Selachian hyomandibula (Krawetz, 362; K. Fürbringer, 682; Allis, 671; infrapharyngeal of Schmalhausen, otoquadrate of Edgeworth, 678). On the other hand, the small cartilage apparently described by Huxley as the hyomandibula (Huxley, 535; Ridewood, 698; Sewertzoff, 592; Goodrich, 35; interhyal of Edgeworth and Schmalhausen, 700) lies laterally to the vein, and there are lateral cartilages related to the operculum more in the position of an 'otic process',[1] Figs. 315, 447.

Turning now to the Tetrapoda, we find that the stapes of Amphibia (Anura and Apoda, see Chapter VIII.) abuts against the auditory capsule ventrally to the vena capitis lateralis and hyomandibular nerve, in the position of the Selachian hyomandibula, Fig. 449. Further, in modern Reptilia and Aves, not only does the stapes (columella auris) lie ventrally to these, but often its outer extra-stapedial portion acquires a dorsal process reaching the parotic process above them. The vein and nerve then pass through a gap comparable to the cranio-quadrate passage. In Mammalia also the extra-stapedial region abuts against the crista parotica enclosing the vein and nerve in a similar passage (Goodrich, Schmalhausen). These facts support the view that the hyomandibula of fishes is represented by the columella or stapes, and perhaps also by that dorsal region of the hyoid arch, just mentioned, which becomes attached to the crista parotica or quadrate in Tetrapods (Chapter VIII.). If further

---

[1] Edgeworth (678-9) has shown that a strand of tissue extends at an early stage from the auditory capsule to the ceratohyal in a position resembling that of the Selachian hyomandibula. Later its proximal end is chondrified as the 'otoquadrate', attached to the auditory capsule at its inner end and the otic process of the quadrate at its outer end. The middle part of the strand forms the 'interhyal' cartilage, while the outermost part of the strand forms the hyosuspensorial ligament binding the ceratohyal to the quadrate. It is possible that individuals vary, for in the specimen here figured, Fig. 447, there are three cartilages, and the middle piece has a dorsal connexion with the crista parotica behind the hyomandibular nerve.

research bears this out, it may yet be shown that a double connexion of the hyomandibular cartilage enclosing the nerve and vein is an ancient and primitive character.

That the palatobasal articulation with the trabecular region is a very primitive feature is shown not only by its occurrence in Dipnoi and Tetrapods, but also in many of the more primitive Teleostomes (p. 273). In these last, where the palatoquadrate becomes ossified, the articulation

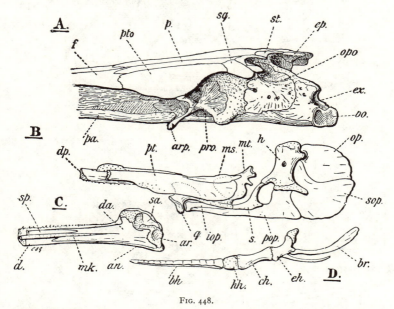

FIG. 448.

Lepidosteus osseus, L. A, Oblique view of the skull from behind ; B, inner view of the right opercular bones and upper jaw broken short in front ; C, inner view of the hind end of the lower jaw ; D, lower portion of hyoid arch, belonging to B. *an*, Angular ; *ar*, articular ; *arp*, articular process for metapterygoid ; *bh*, basihyal ; *bo*, basioccipital condyle ; *br*, branchiostegal ; *ch*, cerato-hyal ; *d*, dentary ; *da*, prearticular ; *dp*, palatine ; *eh*, epihyal ; *ep*, epiotic ; *ex*, lateral wing of basioccipital (fused neural arches) ; *f*, frontal ; *h*, hyomandibular ; *hh*, hypohyal ; *iop*, inter-opercular ; *mk*, Meckel's cartilage ; *ms*, endopterygoid ; *mt*, metapterygoid ; *op*, opercular ; *opo*, exoccipital (probably including opisthotic) ; *p*, parietal ; *pa*, parasphenoid ; *pop*, preopercular ; *pro*, pro-otic ; *pt*, ectopterygoid ; *pto*, postfrontal (supratemporal ?) ; *q*, quadrate ; *s*, symplectic ; *sa*, supra-angular ; *sop*, subopercular ; *sp*, splenial ? ; *sq*, tabular ? ; *st*, postparietal. (From Goodrich, *Vert. Craniata*, 1909.)

occurs between the metapterygoid and the basitrabecular process of the basisphenoid (sometimes combined with the transverse basipterygoid process of the parasphenoid (p. 423)). It is clearly developed in *Lepidosteus* (Parker, 2), but lost in the other modern Holostei and also in modern Chondrostei, where it is represented by a ligament. The basal process is probably represented by the ventral process of the metapterygoid in *Amia* and Teleostei (Swinnerton, 389 ; Allis, 402, 404), often well developed and directed towards the basis cranii.[1] That the more dorsal

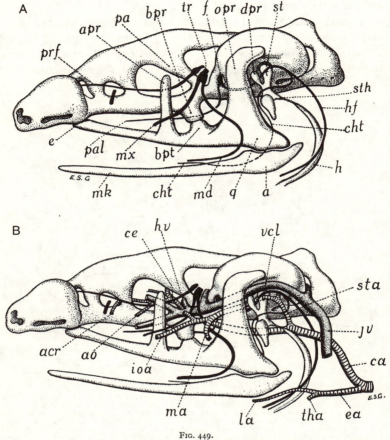

FIG. 449.

Chondrocranium, mandibular, and hyoid arches. Diagram of primitive Tetrapod. A, Cartilage and nerves; B, with blood-vessels in addition. Cartilage dotted, nerves black, arteries cross-lined, veins darkly shaded. *a*, Articular region; *acr*, ophthalmic; *ao*, supraorbital; *apr*, ascending process; *bpr*, basitrabecular process; *bpt*, basal process; *ca*, internal carotid; *ce*, cerebral; *cht*, chorda tympani; *dpr*, dorsal process; *e*, ethmoid articulation of palatoquadrate with antorbital process; *ea*, external carotid; *f*, main branch of facial; *h*, hyoid cornu; *hf*, hyomandibular branch; *hv*, hypophysial; *ioa*, infraorbital; *jv*, jugular; *la*, lingual; *ma*, mandibular; *md*, mandibular branch; *mk*, Meckel's cartilage; *mx*, maxillary branch; *opr*, otic process; *pa*, pila antotica; *pal*, palatine branch; *prf*, profundus; *q*, quadrate; *st*, stapes (columella auris); *sta*, stapedial; *sth*, stylohyal process; *tha*, thyroid artery; *tr*, trigeminal; *vcl*, vena capitis lateralis.

process of the metapterygoid of Teleostomes ('metapterygoid process') represents the reduced otic process is not so well established. Yet this

[1] The basitrabecular process is correspondingly reduced. It forms the floor of the anterior opening of the myodome.

is probably the correct interpretation of its homology, since there is some evidence that it met the postorbital process in Palaeoniscids (Stensiö, 218), and the space included between the two processes of the metapterygoid (ventral basal and dorsal ' metapterygoid ') bears the right relation to blood-vessels and nerves (Allis, 404, 678), and seems to represent the antero-ventral region of the cranio-quadrate passage, Fig. 436. If a functional otic connexion existed in primitive Teleostomes, as well as in primitive Chondrichthyes and Dipnoi, it must be a very ancient character derived from the common ancestor of all Gnathostomes.

But in later and more specialised Teleostomes the otic process is always reduced, and they evolve a special form of hyostyly (called methyostyly by Gregory, 688). It is typically developed in the Holostean series, in which the flattened hyomandibula and symplectic bones become firmly attached to the metapterygoid and quadrate, and together with the preoperculum and interoperculum form a rigid support for the jaws, Figs. 286, 303,

The hyostyly of the Chondrostei has evolved on rather different lines. For in them the hyomandibula is freer and retains a more cylindrical and elongated shape, Fig. 434. *Polypterus* resembles them in this, Fig. 435.

THE PALATOQUADRATE IN TETRAPODA

General Structure.—Of great importance for the proper understanding of the morphology of the skull of Tetrapods is the history of the palato-quadrate and the dermal bones associated with it (see further: palate, p. 317). We have seen above (p. 410) that the Osteichthyes have probably been derived from early forms in which the palatoquadrate was connected at three points with the chondrocranium: (1) In front its palatine process articulated with the antorbital region; (2) behind and dorsally the otic process joined the auditory capsule above the cranio-quadrate passage; and (3) ventrally the basal process formed with the basitrabecular process a palatobasal or ' basipterygoid ' articulation[1] below the cranioquadrate passage, Fig. 449. An additional connexion was probably formed with the side wall of the brain-case in

---

[1] The palatobasal articulation may be defined as that between the palato-quadrate and the trabecula. The term ' basipterygoid articulation ', often applied to it, may be restricted to the later underlying and supporting con-nexion between the lateral wing of the parasphenoid and the pterygoid bone. The former original articulation is more or less superseded by the later one in some Tetrapoda.

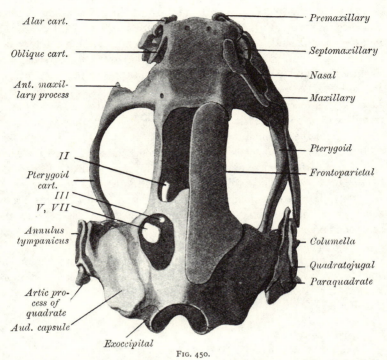

Alar cart.

Oblique cart.

Ant. maxillary process

II

Pterygoid cart.

III

V, VII

Annulus tympanicus

Artic process of quadrate

Aud. capsule

Exoccipital

Premaxillary

Septomaxillary

Nasal

Maxillary

Pterygoid

Frontoparietal

Columella

Quadratojugal

Paraquadrate

FIG. 450.

Skull of a young *Rana temporaria*, 2 cm. in length, just after metamorphosis, from the dorsal side. The investing bones are removed on the left side (×abt. 11). (After Gaupp, from a model by Fr. Ziegler.) Investing bones, *yellow*. (From Wiedersheim, *Comp. Anatomy*.)

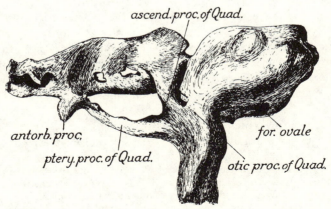

ascend. proc. of Quad.

antorb. proc.

ptery. proc. of Quad.

for. ovale

otic proc. of Quad.

FIG. 451.

Left-side view of model of skull of *Hynobius* larva 20 mm. long (from F. H. Edgeworth, *J. of Anat.*, 1923).

the orbital region by the ascending process. Moreover, dermal palatine, ectopterygoid, and pterygoid (endopterygoid) bones strengthen the palatal surface of the palatoquadrate. The primitive Tetrapoda have a palatoquadrate complex built on the same plan; but the processus ascendens is typically better developed and may become

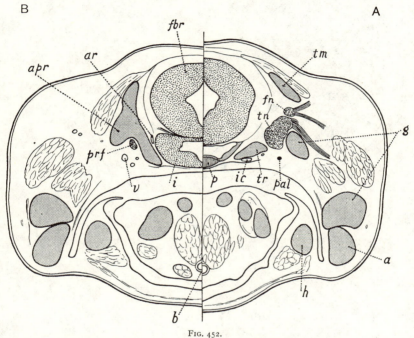

FIG. 452.

Salamandra maculosa, young larva. Transverse sections of head : A, More anterior, and D, more posterior ; A, posterior region of orbit ; B, through ascending process, *apr*. Cartilage grey. *a*, Articular region ; *a.c*, auditory capsule ; *an*, auditory nerve ; *ar*, artery ; *b*, basibranchial ; *bp*, basal plate ; *br*, ceratobranchial ; *brn*, brain ; *fbr*, fore-brain ; *fn*, facial ganglion and nerve ; *q*, quadrate ; *h*, hyoid arch ; *i*, infundibulum ; *ic*, internal carotid ; *m*, hind-brain ; *ms*, muscle ; *n*, dorsal lateral-line branch of facial ; *p*, pituitary body ; *pal*, palatine nerve ; *pc*, floor of facial canal ; *prf*, profundus nerve ; *s*, squamosal ; *sta*, stapedial (orbital) artery ; *tm*, taenia marginalis ; *tn*, trigeminal ganglion ; *tr*, trabecula ; *v*, vena capitis lateralis.

ossified. While the prevomer, palatine, ectopterygoid, and pterygoid firmly attached to the premaxillary and maxillary usually form a rigid bony palate, the anterior region of the palatoquadrate bar here tends to become reduced. It persists complete in some adult modern Amphibia, such as the Anura and *Ranodon* among Urodela, and, as Edgeworth has recently shown (676), is present in the young of *Crypto-branchus*, *Menopoma*, and *Hynobius*, and even in the larval Apodan *Ichthyophis*, Figs. 450-1, 506. But in the majority of adult Urodela

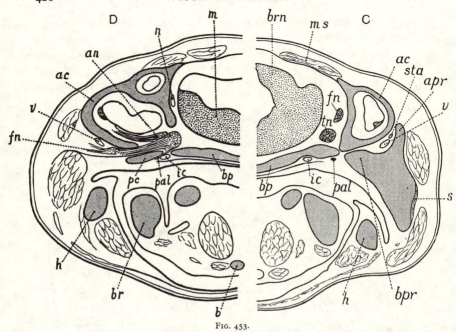

FIG. 453.

Salamandra maculosa, transverse sections of head of same larva as in Fig. 452; C, through basal process; D, more posterior through auditory capsule and facial foramen. Lettering as in Fig. 452.

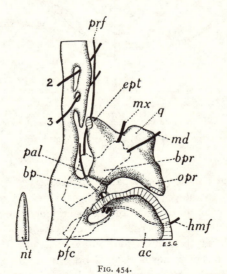

FIG. 454.

Necturus maculatus. Reconstruction of right orbito-temporal region of young larva, dorsal view. Broken lines indicate position of large Gasserian and geniculate ganglia which have been removed. 2, Optic, and 3, oculomotor nerve. *ac*, Auditory capsule; *bp*, basitrabecular process; *bpr*, basal process; *ept*, ascending process; *pal*, palatine nerve; *hmf*, hyomandibular branch of facial; *md*, mandibular, and *mx*, maxillary branches of trigeminal; *nt*, tip of notochord; *opr*, otic process; *pfc*, prefacial commissure; *prf*, profundus; *q*, quadrate cartilage.

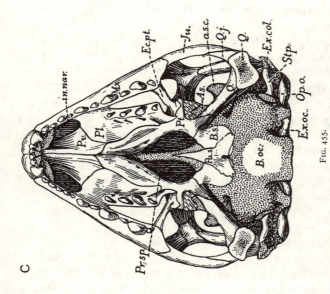

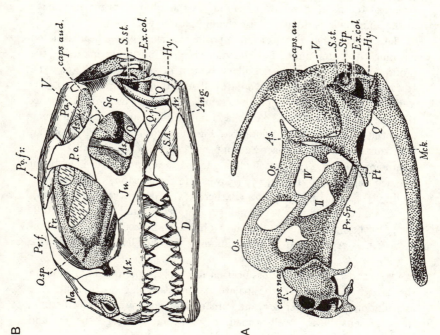

Fig. 455.

Sphenodon punctatus (after Howes and Swinnerton, *Tr. Zool. Soc.*, from W. K. Gregory, *J. Morph.*, 1913). A, Developing chondrocranium, etc.; B, developing skull; left-side view, and C, ventral view. *Ang,* Angular; *Av,* articular; *As,* processus ascendens of palatoquadrate cartilages in A, ossified as alisphenoid or epipterygoid in B; *B.oc,* basioccipital; *B.s,* basisphenoid; *caps.au,* auditory capsule; *caps. nas,* nasal capsule; *D,* dentary; *Ec.pt,* ectopterygoid; *Ex.col,* extra-stapedial; *Fr,* frontal; *Hy,* hyoid cornu; *Ju,* jugal; *Mck,* Meckel's cartilage; *Mx,* maxillary; *Na,* nasal; *Op.o,* opisthotic; *Os,* orbital cartilage; *O.sp,* orbito-sphenoid; *Pa,* parietal; *Pl,* palatine; *Pa.s,* parasphenoid; *Po.f,* postfrontal; *P.o,* postorbital; *Pr.f,* prefrontal; *Pr.sp,* pterygoid process in trabecular region of interorbital septum; *S.A,* supra-angular; *S.g,* squamosal; A; *Q,* quadrate; *Q.j,* quadratojugal; *Pt,* pterygoid, pterygoid process in *S.st,* extra-stapedial; *Stp,* stapedial. Between *IV* and ascending process is seen the pila antotica in A.

the palatopterygoid region fails to reach the antorbital cartilage and is represented merely by a short forwardly produced 'pterygoid process', more reduced still in Apoda. The processus ascendens is reduced in adult Urodela and disappears in Anura.

In the Amniota the palatoquadrate is always incomplete in front, being represented anteriorly to the basal process by at most a pterygoid process, some isolated fragments of cartilage, and perhaps by a backwardly projecting process of the antorbital cartilage in *Lacerta*, Figs. 445-6 (Gaupp, 506). Further, in Lacertilia and Ophidia the region between the basal process and body of the quadrate is reduced, being generally present only as a transient cartilaginous or procartilaginous rod (*Lacerta*) in the young (Gaupp, 506; Broom, 460). It is retained, however, in Rhynchocephalia, and at least in young stages in Chelonia (Parker, 574; Kunkel, 540; Fuchs, 503) and Crocodilia (Parker, 570; Shiino, 597).

A processus ascendens is well developed in the young of Rhynchocephalia and Lacertilia, is rather short in Chelonia, and much reduced or vestigial in Crocodilia and Ophidia. In Birds it appears to be absent (Filatoff's process is probably a processus basalis (1906)), only the quadrate region of the palatoquadrate remaining in the adult. Its representative in the Mammalia is discussed elsewhere (Chapter VIII.).

This processus ascendens is an important element of the head skeleton marking the outer limit of the cavum epiptericum, and separating the profundus nerve from the maxillary branch of the trigeminal (p. 271). It was doubtless present in the young of all primitive Amniotes (Gaupp, 343; Versluys, 625). In the majority of Reptiles it ossifies as a distinct bone, the epipterygoid (so-called columella cranii of older authors), and as such has been described in Cotylosauria, Fig. 460, *Seymouria*, Phytosauria, Dinosauria, Ichthyosauria, Sauropterygia, Pelycosauria, and other Theromorpha, Fig. 413. Among living forms the epipterygoid is found in Rhynchocephalia, Figs. 354, 455 (Howes and Swinnerton, 528; Schauinsland, 583), Lacertilia, Fig. 364 (Parker, 564; Gaupp, 506; Rice, 376), and Chelonia, where it is very small (Parker, 574; Kunkel, 540; Filatoff, 337; Fuchs, 503); but is lost in Ophidia and Crocodilia. The epipterygoid is represented in the Mammalia by the alisphenoid bone (see p. 436).

A second endochondral ossification appears as the quadrate bone in the posterior region of the palatoquadrate cartilage. It bears the articulation for the lower jaw, and extends upwards towards the cranium; it is present in the well-ossified, more primitive Stegocephalia and modern Amphibia (except Anura), and in all the Amniota.

The quadrate and epipterygoid, together with the dermal pterygoid, form a rigid palatoquadrate complex arch extending over the roof of the

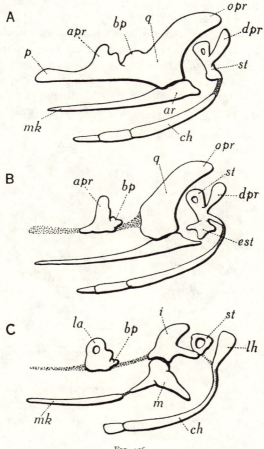

Fig. 456.

Diagrams illustrating fate of cartilaginous mandibular and hyoid arches in: A, primitive tetrapod; B, modern reptile; C, mammal.   *apr*, Ascending process (epipterygoid pr.); *ar*, articular region; *bp*, basal process; *ch*, ceratohyal; *dpr*, dorsal process (supra-stapedial pr.); *est*, extra-stapedial; *i*, incus (quadrate); *la*, lamina ascendens (ascending pr.); *lh*, laterohyal; *m*, malleus (articular); *mk*, mandibular (Meckel's cartilage); *opr*, otic process; *p*, palatine region; *q*, quadrate region; *st*, stapes.   Those regions which disappear in course of development are indicated in dots.

buccal cavity, Figs. 324, 455; connected in front with the prevomer and palatine, at the side with the ectopterygoid, and behind with the dermal temporal roofing (p. 317). It may be supposed that primitively this rigid arch was somewhat movably attached (by the otic process, and basal process) to the cranium. But, as already explained (p. 411), in the Amphibia it tends to become immovable, the basipterygoid articulation being obliterated and the otic, palatine process, and even the ascending process fusing at their points of contact with the chondrocranium (in Urodela the basal process also fuses).

It has been mentioned above (p. 401) that in primitive Osteichthyes, such as the Crossopterygii, the palatoquadrate has a series of endo-chondral ossifications, of which an anterior seems to represent the epipterygoid (and metapterygoid?) and a posterior the quadrate, Fig. 429a. The first (processus ascendens) tends to pass up between the profundus and trigeminal nerves ; the second bears the basal articulation;

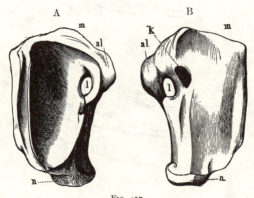

Fig. 457.

Mosasaurus camperi, v. Meyer. Quadrate. A, Outer, and B, inner aspect, ¼. *k*, Stapedial pit; *l*, meatus for columella?; *m*, superior margin; *n*, inferior margin. (After Owen, from K. Zittel, *Palaeontology*.)

the third supports the articular below and passes into the otic process above. These elements in Crossopterygii are somewhat inconstant in number, and in *Eusthenopteron* the first and second are represented by a single bone (Watson, 644).

Both the epipterygoid and quadrate bones of Tetrapods have, no doubt, been derived from those of their fish-like ancestors ; but whether the early Tetrapod also possessed a middle metapterygoid element is very doubtful. There appears to be no trace of such a separate bone in Amphibia. But in Reptilia, where this region of the palatoquadrate may persist as a little cartilage between the base of the epipterygoid and basitrabecular process (meniscus pterygoideus of Howes and Swinnerton in *Sphenodon*, and of Gaupp in *Lacerta*), it may possibly be represented by a little bone found in a similar position in certain fossil forms (*Seymouria* (*Conodectes*)), Watson, 643-4; Broom, 455; *Ichthyosaurus* (?), Sollas, 602), Figs. 263, 279, 378.

From the researches of Watson, 644, Sushkin, 678, and others on the Stegocephalia, it appears that in these primitive Amphibia the epipterygoid was often a large and important bone in the outer wall of the cavum epiptericum, Figs. 458, 459, 460. It may sometimes extend into the basal process itself. Moreover, in some forms (*Capitosaurus*, Sushkin) it extended backwards and upwards behind a notch for the trigeminal nerve to acquire an articulation with the pro-otic region of the auditory capsule (otic process ?).

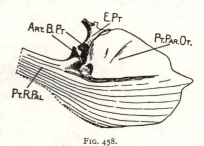

FIG. 458.

Palaeogyrinus decorus; inner aspect of right pterygoid and epipterygoid, *E.Pt.*, parotic wing, *Pt.Par.Ot*, and palatine ramus, *Pt.R.Pal*, of pterygoid; *Art.B*, basipterygoid articular surface; *Pt*, basal articular surface of epipterygoid. (From D. M. S. Watson, *Tr. Roy. Soc.*, 1926.)

On the other hand, in Amphibia the otic process of the quadrate, which bone develops at the distal end of the cartilage, does not reach the capsule as it does in the Amniota. Nevertheless, Sushkin's contention

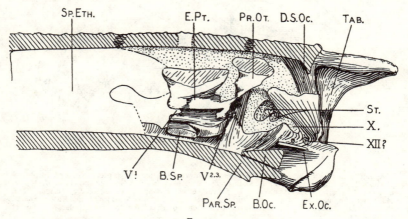

FIG. 459.

Inner view of right half of brain-case of *Capitosaurus* sp. (from D. M. S. Watson, *Phil. Trans.*, 1920). *B.Oc*, Basioccipital; *B.Sp*, basisphenoid; *D.S.Oc*, post-parietal (dermal supra-occipital); *E.Pt*, epipterygoid; *Ex.Oc*, exoccipital; *Par.Sp*, parasphenoid; *Pr.Ot*, pro-otic; *Sp.Eth*, sphenethmoid; *St*, stapes ?; *Tab*, tabular.

that the 'process' of the Reptilia is radically different from the more ancient 'pro-otic process' of the Amphibia seems scarcely justified, since they are both ossifications in an original otic process of the primitive cartilaginous palatoquadrate. They indicate, however, a divergence between the two groups.

Monimostylism, Streptostylism, and Kinetism.—Quite independently the palatoquadrate complex has become modified in the Amniota in divergent directions (Versluys, 625). On the one hand, the quadrate may become firmly fixed to the auditory region in the adult (monimostylic

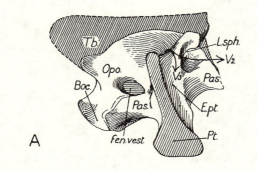

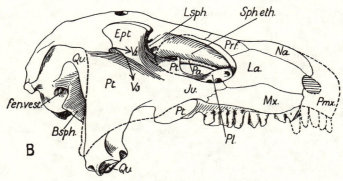

FIG. 460.

A, Left-side view of brain-case of *Eryops* sp., showing relations of epipterygoid, *Ept*, to pleurosphenoid (laterosphenoid), *Lsph* ; B, skull of *Diadectes molaris* (from Gregory and Noble, *J. Morph. and Physiol.*, 1924). *Bsph*, Basisphenoid ; *fen.vest*, fenestra ovalis ; *Ju*, jugal ; *La*, lacrimal ; *Mx*, maxillary ; *Na*, nasal ; *Pa*, parasphenoid ; *Pl*, palatine ; *Pt*, pterygoid ; *Qu*, quadrate ; *Spheth*, sphenethmoid. Arrows *V₂ V₃* indicate inferred course of maxillary and mandibular branches of trigeminal nerve.

skull, Stannius, 1856); on the other, it may become freely movable (streptostylic skull, Stannius, 1856). The fully monimostylic condition is established in the Chelonia and Crocodilia, among modern Reptilia, by the sutural connexion of the otic process of the bony quadrate with the pro-otic and opisthotic reinforced by the sutural connexion of the pterygoid below accompanied by the obliteration of the basipterygoid articulation and the fixation of the palatoquadrate complex to the base of the cranium, Figs. 416, 419. The typical streptostylic condition is seen in the Lacertilia, where the bony quadrate is isolated from the epipterygoid and only connected somewhat loosely to the pterygoid below and the parotic process above by ligament, Figs. 462-3. In the Ophidia the quadrate is still freer, and supported away from the skull by the squamosal. Another variety of streptostylism occurs in the Archosaurian branch, where the quadrate tends to become exposed (p. 341) and the epipterygoid to be reduced. This type culminates in the Aves, where the quadrate is again freely movable. The third type

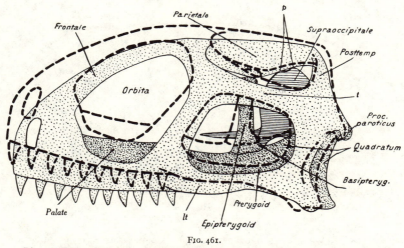

FIG. 461.

Diagram of Rhynchocephalian skull showing motion of 'maxillary segment' (dotted) on 'occipital segment' (hatched) as example of kinetism (from Versluys, *Vergl. Anat.*, 1927). *p*, Posterior hinge; *lt*, lower, and *t*, upper temporal arch.

of streptostylism develops in the Theromorpha and Mammalia, where the quadrate also becomes isolated from the epipterygoid region, and finally as the incus becomes quite freed from the skull, Figs. 456, 497 (Chapter VIII.).

Related to streptostylism is what Versluys has called the 'kinetism' of the skull (624-5, 627). This consists in a certain degree of looseness and consequent power of motion between the dermal roofing bones and portions of the underlying chondrocranium. For instance, although in *Sphenodon* the quadrate is firmly attached, the palatoquadrate complex is rigidly connected to the dermal roofing, and the basipterygoid articulation is less perfect in the adult than in the embryo (Schauinsland, 583; Howes and Swinnerton, 528); yet when the pterygoid muscles contract the

2 F

roofing and complex together form a rigid whole slightly movable on the posterior brain-case, joined to it only by cartilage and ligament,[1] Fig. 461.

It is probable that the primitive early Reptiles had slightly movable quadrates. In Lacertilia the kinetism may be very pronounced. Here, as in *Sphenodon*, the interorbital septum is weakly developed and partly membranous while the posterior brain-case is only loosely joined to the parietals (metakinetic). Further, in such forms as *Tupinambis* and *Uromastix*, a bending of the roof becomes possible in front of the orbit (mesokinetic), Figs. 462-3. In Ophidia, also, the preorbital region of the

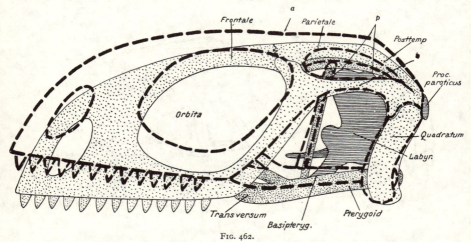

FIG. 462.

Diagram of Lacertilian skull showing motion of 'maxillary segment' (dotted) on 'occipital segment' (hatched), as example of kinetism (from Versluys, *Vergl. Anat.*, 1927). *a*, Anterior, and *p*, posterior movable hinges; *t*, upper temporal arch; *posttemp*, post-temporal arch.

skull can be lifted by the thrusting forward of the pterygoids. The Dinosaurian skull develops both metakinetic and mesokinetic specialisations, Fig. 464. Finally, in Aves, where the free quadrate is only loosely held by the delicate quadratojugal arch, metakinetism is pronounced. The pterygoid muscles are usually well developed as well as a m. orbito-quadratus (Gadow, 1891). Parrots have a markedly mesokinetic joint allowing the upper beak to be raised (Versluys, 625). The skull is rigid from the occipital region to the preorbital, the interruption of the

---

[1] These muscles are the m. protractor pterygoidei, m. pterygosphenoidalis posterior, and m. pterygoparietalis extending from pterygoid to pro-otic and basisphenoid in fully kinetic skulls (Lacertilia). They are reduced or absent in the adults with akinetic skulls (Versluys, 769; Edgeworth, 718; Cords, 1904; Bradley, 426a).

preorbital and postorbital bars frees the maxillo-quadratojugal bar, and the front region is lifted when the quadrate is brought forward.   In the

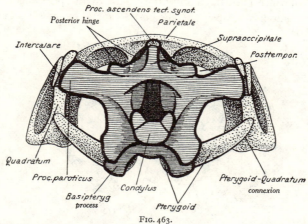

FIG. 463.

Posterior view of skull of Lacertilian showing 'occipital sigment' (hatched) movably attached to 'maxillary segment' (dotted) (from Versluys, *Vergl. Anat.*, 1927).

higher birds the basipterygoid articulation tends to disappear, the palato-quadrate complex sliding on the basisphenoid rostrum.

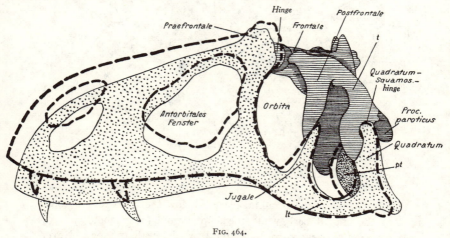

FIG. 464.

Diagram of kinetic skull of a carnivorous Dinosaur, *Creosaurus* (from Versluys, *Vergl. Anat.*, 1927). *pt*, Process of quadrate joining pterygoid.   'Occipital segment' (hatched) in this skull includes postfrontal and squamosal regions ; cf. Fig. 357.

A special akinetic type of skull is developed in Theromorpha, and leads to the Mammalian structure.   Here the dorsal roofing bones remain

rigidly connected to the chondrocranium, and the parietals as well as the
frontals take a large share in the covering of the brain; likewise the
squamosal, which is enlarged and firmly fixed by its spreading squamous
plate to the auditory capsule. The post-temporal opening is obliterated

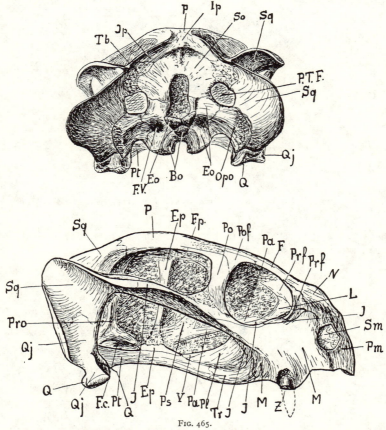

FIG. 465.

Posterior and right lateral aspect of skull of *Dicynodon* (from F. von Huene, *Palaeontol. Zeitschr.*, 1922). Lettering as in Fig. 394.

in the formation of the occipital plate. Below the pterygoid becomes
fixed to the cranial base, while the epipterygoid (alisphenoid, p. 437)
itself contributes largely to the lateral cranial wall (p. 262). Nevertheless, the mammalian skull is not monimostylic, for the palatoquadrate
complex is interrupted, and the quadrate becomes free (p. 433).

It has been pointed out above (p. 271) that the cartilaginous lamina
ascendens of the ala orbitalis of mammals, which often is separately

chondrified, in all probability corresponds to the processus ascendens of lower Tetrapods, and that both originally lay in the outer wall of the cavum epiptericum outside the brain-case (p. 271). In the mammalian skull this cartilage lamina becomes ossified as the true alisphenoid, which spreads and contributes to the side wall to the cranium a new element not homologous with the various bones in fishes, birds, and reptiles to which the name alisphenoid has been applied (p. 392). The processus ascendens becomes ossified as the epipterygoid in reptiles, Figs. 455, 460, 465, and in Theriodontia this bone may be considerably expanded and develop a posterior tympanic process (Watson, 642). In Cynodontia it has a vertical plate suturally connected with the parietal and forming the lateral wall of the brain-case, while its

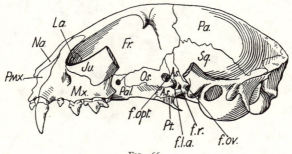

FIG. 466.

Left-side view of skull of cat, with jugal arch partly removed to show alisphenoid, *As*, and crowding together of exits for branches of fifth cranial nerve (from Gregory and Noble, *J. Morph. and Physiol.*, 1924). *f.l.a*, Foramen lacerum anterius (orbital fissure) for 3rd, 4th, 6th, and profundus nerves ; *f.ov*, foramen ovale for mandibular branch ; *f.r*, foramen rotundum for maxillary branch.

long posterior process reaches the tympanic region and quadrate, recall-ing the structure of the alisphenoid in Marsupials, Figs 394-6, 413, 467-9. At the same time the pterygoid loses its reptilian posterior limb, and takes up a position below the basitrabecular region much as the ventral pterygoid does in Mammalia (p. 370). There is little doubt that Broom's contention that the mammalian alisphenoid is derived from the epipterygoid of lower Tetrapods is correct (Broom, 327a and b, 451).

Summary.—Summarising some of the more important points with regard to the mandibular and hyoid skeletal arches in Gnathostomes, it may be pointed out that there is no convincing evidence that they were ever subdivided into the four paired pieces typical of the branchial arches, though traces of pharyngeal and hypo-arcual elements have been described in some fishes. The exact homology of the dorsal elements (palatoquad-rate and hyomandibula) with the epibranchials is uncertain. The hyoman-dibular cartilage can be traced throughout the whole piscine series ; but varies much in size, in function, and in its attachments (amphistylic, hyostylic, and autostylic types). In Selachians, Dipnoi, and Tetrapods it retains its primitive dorsal attachment ventral to the vena capitis lateralis and hyomandibular nerve, and it becomes converted in Tetra-

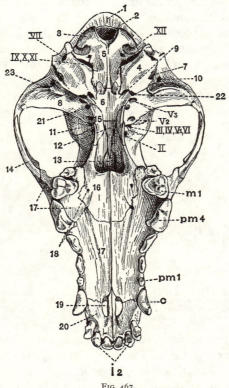

FIG. 467.

Ventral view of cranium of *Canis familiaris* (from S. H. Reynolds, *Vertebrate Skeleton*, 1913). 1, Supraoccipital ; 2, foramen magnum ; 3, occipital condyle ; 4, tympanic bulla ; 5, basioccipital ; 6, basisphenoid ; 7, external auditory meatus ; 8, glenoid fossa ; 9, foramen lacerum medium and anterior opening of carotid canal ; 10, postglenoid foramen ; 11, alisphenoid ; 12, presphenoid ; 13, vomer ; 14, jugal ; 15, pterygoid ; 16, palatal process of palatine ; 17, maxilla (palatal portion) ; 18, posterior palatine foramina ; 19, anterior palatine foramen ; 20, premaxilla ; 21, alisphenoid canal ; 22, Eustachian foramen ; 23, postglenoid process of squamosal ; *II*, optic foramen ; *III, IV, V₁, VI*, foramen lacerum anterius ; *V₂*, foramen rotundum ; *V₃*, foramen ovale ; *VII*, stylomastoid foramen ; *IX, X, XI*, foramen lacerum posterius ; *XII*, condylar foramen ; *i* 2, second incisor ; *c*, canine ; *pm* 1, *pm* 4, first and fourth premolars ; *m* 1, first molar.

pods into the columella auris (see Chapter VIII.). But in the Teleostomi its articulation with the auditory capsule shifts, possibly over the jugal canal, to a more dorsal position above the vein and nerve ; and, further, the nerve passes through the hyomandibula in Holostei.

The palatoquadrate bar extends primitively from the ethmoid to the otic region. It is usually articulated or fused in front to the antorbital process and connected further back with the chondrocranium at two points : a ventral basal process meets the trabecular region (basitrabecular process), and a dorsal otic process arching over a cranio-quadrate passage (for the vena capitis lateralis and hyomandibular nerve) meets the auditory capsule (post-orbital process, crista parotica, paroccipital process). The basal articulation is probably represented in Selachians by the orbital (ethmo-palatine) articulation, and the otic process is generally lost in these fishes. The Dipnoi alone among living fishes preserve both the basal and otic connexions. It is probable that both were present in the common ancestor of the Osteichthyes. The basal articulation occurs in Crossopterygii, early Chondrostei, and Actinopterygii ; it persists in *Lepidosteus*, but is lost in other modern Teleostomes. The otic connexion no longer remains in any of the Teleostomes (Crossopterygii ?), though traces of an otic process are found in some of the more primitive forms.

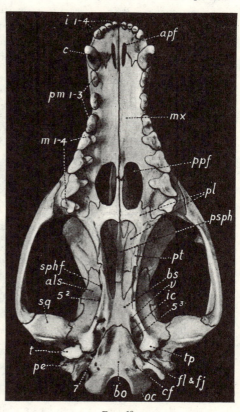

Fig. 468.

Ventral view of skull of *Thylacinus cynocephalus*. *als*, Alisphenoid ; *apf*, anterior palatine foramen ; *c*, canine ; *cf*, condylar foramen ; *fl* and *fj*, foramen lacerum posterius ; *i* 1-4, incisors ; *ic*, internal carotid foramen ; *m* 1-4, molars ; *mx*, maxillary ; *oc*, exoccipital condyle ; *pm* 1-3, premolars ; *ppf*, posterior palatine foramen ; *psph*, presphenoid region ; *tp*, tympanic process of alisphenoid. See Fig. 469 for other letters and details of posterior region.

In primitive Tetrapods the palatoquadrate is complete with antorbital, basal, and otic articulations. It is further provided with a dorsal processus ascendens in the outer wall of the cavum epiptericum (p. 267), between the profundus and trigeminal nerves. This ascending process is probably a primitive structure since it is well developed in Dipnoi, and traces of it occur in the earliest Teleostomes (Crossopterygii).

The palatoquadrate cartilage may become ossified in its anterior (autopalatine), middle (metapterygoid), and posterior (quadrate) regions in Teleostomes ; but in the early Crossopterygii it is more or less completely ossified throughout. In Tetrapods the cartilage is retained complete only in some Amphibia ; and in Amniota it always breaks down, the palatine region and generally also the pterygoid regions disappearing.

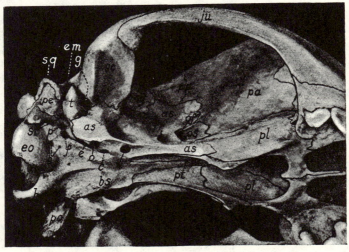

FIG. 469.

Oblique ventral view of skull of *Thylacinus cynocephalus*. *as*, Alisphenoid ; *bo*, basioccipital ; *bs*, basisphenoid ; *c*, internal carotid foramen ; *e*, Eustachian canal ; *em*, external auditory meatus ; *eo*, exoccipital condyle ; *f*, foramen caroticum posterius ; *fr*, foramen rotundum ; *g*, postglenoid foramen ; *gc*, glenoid cavity ; *h*, hypoglossal foramina ; *ju*, jugal ; *o*, for. caroticum anterius ; *of*, foramen ovale ; *os*, orbitosphenoid ; *p*, foramen lacerum posterius ; *pa*, parietal ; *pe*, periotic ; *pl*, palatine ; *ps*, presphenoid ; *pt*, pterygoid ; *sq*, squamosal ; *st*, stylomastoid foramen ; *t*, tympanic ; *v*, venous foramen ; *vo*, vomer.

Thus are left the processus ascendens and basal process articulating with the basitrabecular process, and the quadrate cartilage with the crista parotica. The former ossifies as the epipterygoid bone (alisphenoid of Mammal), and the latter as the quadrate bone (incus of Mammal).

THE HYOBRANCHIAL APPARATUS

Excepting for its dorsal hyomandibular element the hyoid arch is generally in Gnathostomes closely associated with the branchial arches.[1] Hence the whole system is called the hyobranchial skeleton.

[1] The structure of the hyoid arch in Batoidei (p. 415, footnote 3) and the presence of a slender rod of cartilage between the hyoid and first branchial arch in Dipnoi led Dohrn (333) and others to adopt van Wijhe's conclusion,

It has been mentioned above that the branchial arches of Selachii consist of paired pharyngobranchials, epibranchials, ceratobranchials, and hypobranchials, and of median basibranchials or copulae. But this typical structure, almost fully carried out in such primitive forms as the Notidani, is never quite complete in the posterior arches, generally much modified by fusion or reduction, Figs. 424, 426-7.

The segmentation of the arches is perhaps secondary; it is probably related to the development of special branchial muscles and allows the walls of the pharynx to be expanded and contracted for breathing and eating purposes.

The last two pharyngobranchials are generally fused, and usually only a ceratobranchial remains in addition in the last arch. The copulae are most variable. Usually situated between the successive pairs of hypobranchials, their true morphological relations and phylogenetic origin remain obscure. Gegenbaur finally concluded that each basibranchial belongs to and is derived from the arch in front of it. Only in *Heterodontus* has a copula been described between the 1st and 2nd branchial arches. The copulae tend to disappear or fuse up from before backwards, that between the 4th and 5th arch (or 6th and 7th arch in Heptanchus) remaining as a large plate for attachment of the ventral muscles and for the protection of the pericardium. This 'cardio-branchial' is possibly of compound origin, formed of two or more posterior basibranchials. The branchial arches of the Holocephali closely resemble those of sharks.

It is characteristic of the branchial arches of the Elasmobranchii that they are bent in a ⋝-shaped curve, the pharyngobranchial and the hypobranchial pointing backward. On the other hand, the arches of the Teleostomi, although ⋟-shaped, are in one plane, the pharyngo- and hypobranchials pointing forward. Those of *Ceratodus* are intermediate, only the hypobranchial pointing backward (Allis, 1925).

The usual four paired lateral and median ventral elements are typically present in the Teleostome branchial arch, and generally ossified; but are subject to considerable modification and reduction, especially in the posterior arches, Figs. 436, 599. The 5th arch, usually of one piece, frequently becomes conspicuously toothed forming the ' os pharyngeus inferior'; while the pharyngobranchials of the last three arches similarly form a toothed ' os pharyngeus superior '. Often these become

---

based on embryology, that the hyoid region includes two arches (p. 223). But the abnormal structure of the hyoid skeleton of skates is better explained otherwise (p. 415), and the rod in Dipnoi appears to be formed by the fusion of the base of branchial rays (K. Fürbringer, 682).

powerful masticating plates armed with large teeth. In addition to the usual pharyngobranchial (inferior pharyngobranchial) a small superior

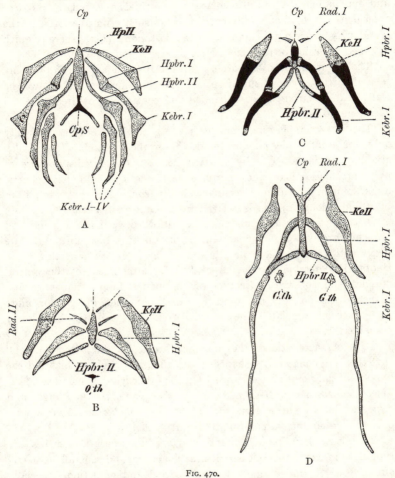

FIG. 470.

Hyobranchial apparatus of Urodeles. A, Axolotl (*Siredon* stage of *Amblystoma*); B, *Salamandra maculosa*; C, *Triton cristatus*; D, *Spelerpes fuscus*. *Cp, Cps, O.th,* basihyobranchial or copula; *G.th,* thyroid gland; *Hpbr. I* and *II,* first and second hypobranchial; *HpH, Rad. I,* hypohyal; *Kebr. I–IV,* first to fourth ceratobranchial; *KeH,* ceratohyal ('anterior cornu' of hyoid in Caducibranchs —the 'posterior cornu' being made up of *Hpbr. I* and *II* and *Kebr. I*). *Rad. II* arises in *Salamandra* secondarily during metamorphosis. (From Wiedersheim, *Comp. Anatomy.*)

pharyngobranchial may be present on some of the anterior arches, perhaps segmented off from the epibranchials, in *Acipenser*, Fig. 434, and some other Teleostomes (van Wijhe, 654; Allis, 667-71; de Beer, 421a). The copulae are well developed and usually along the base of all the arches. *Amia* has three separate median elements, apparently representing the five basi-branchials, while Teleosts have in addition a basihyal. A median urohyal, attached in front below the hypohyals and passing back between the sternohyoid muscles, is peculiar to the Teleostei. It is generally considered

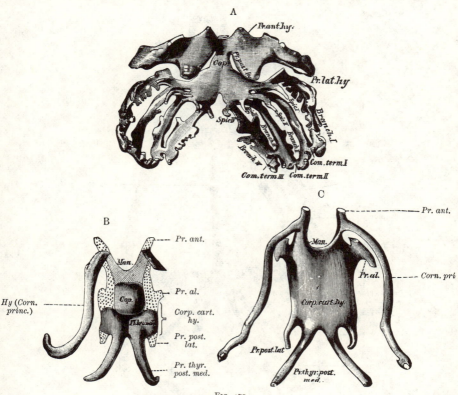

FIG. 471.

A, Hyobranchial skeleton of a larval *Rana temporaria*, 29 mm. in length, from the dorsal side ; B, the same of a larva, 15 mm. in length, at the end of metamorphosis, after disappearance of the tail ; C, hyoid cartilage of a young frog, 2 cm. in length, from the ventral side. (All these figures are from wax models after Gaupp.) A and B (in part), *Branch I-IV*, branchial arches ; *Com. term. I-III*, terminal commissures of same ; *Cop*, basal plate (copula) ; *Hy*, hyoid ; *Pr.ant.hy*, *Pr.lat.hy*, *Pr.post.hy*, anterior, lateral, and posterior processes of the hyoid ; *Spic. I-IV*, cartilaginous processes. B (in part) and C, *Corp.cart.hy*, body of hyoid cartilage ; *Corn. princ.*, anterior cornu ; *Man*, ' manubrium ' ; *Pr.al*, alary process ; *Pr.ant*, anterior process ; *Pr.post.lat*, postero-lateral process ; *Pr.thyr.post.med*, thyroid or postero-medial process (posterior cornu). (From Wiedersheim, *Comp. Anatomy*.)

to be an ossification of the median ligament, Fig. 285. Similar but paired bones occur in *Polypterus,* possibly enlarged branchial rays.

The hyobranchial skeleton of Amphibia is still concerned with respiration, bearing gills in the larva and also in the adult of those Urodela which have become readapted to aquatic life. In these the

apparatus preserves many fish-like characters (Dugés, 1834 ; Wiedersheim, 651 ; Parker, 562, 573 ; Drüner, 675 ; Gaupp, 683). The Urodelan
larva may have as many as four paired ceratobranchials, two hypo-

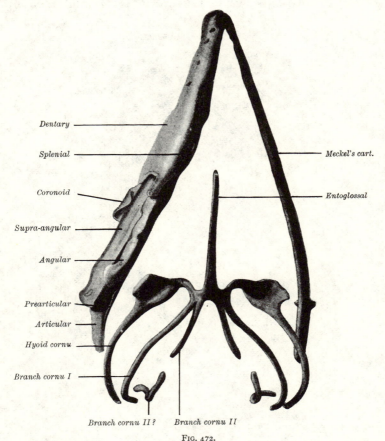

Dentary

Splenial

Coronoid

Supra-angular

Angular

Prearticular

Articular

Hyoid cornu

Branch cornu I

Meckel's cart.

Entoglossal

Branch cornu II ?   Branch cornu II

FIG. 472.

Hyobranchial skeleton and lower jaw of *Lacerta* (after Gaupp-Ziegler model, from Wiedersheim, *Comp. Anatomy*).

branchials, and behind the basihyal a copula with a posterior stilus
probably representing several basibranchials.

In adult terrestrial Amphibia the hyobranchial apparatus becomes
altered and adapted to support a projecting muscular tongue as well as to
serve for the attachment of muscles for lowering and raising the floor of the
buccal cavity in respiration (Chapter XI.). While the adult *Salamandra*
has a copula with two short ' radii ' in front, and two arches combining

to a single dorsal cornua behind, the perennibranchiate *Necturus* preserves three branchial arches, Fig. 470.

The Anuran larva possesses a cartilaginous basket-work consisting of
the ceratohyal and four branchial arches continuous with each other
dorsally and ventrally and with a median copula. At metamorphosis
the basibranchial and hypobranchial regions become converted into an
expanded ' corpus ' or ' body ' with four paired processes, most of the
arches having disappeared. The first process is prolonged into the slender
hyoid cornu attached dorsally to the skull. The last process is large and

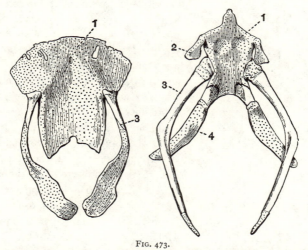

FIG. 473.

Hyoids of *Caiman latirostris* (to the left) and of *Chelone midas* (to the right) (from S. H. Reynolds,
Vertebrate Skeleton, 1913).   Cartilaginous portions dotted.   1, Basilingual plate or body of the hyoid ;
2, hyoid arch ; 3, first branchial arch (anterior cornu) ; 4, second branchial arch (posterior cornu).

usually ossified (processus thyreoideus), Figs. 471, 506.   The Apoda
(*Ichthyophis*) have slender remnants of four branchial arches in the adult.

The hyobranchial apparatus of the Amniota serves no longer for,
breathing purposes, but chiefly for the support of the tongue, into the
base of which projects in Reptiles and Birds a processus entoglossus or
lingualis from the copula or corpus.   All the median elements combine
to form this corpus, from which usually extend three tapering arches or
cornua—a hyoid and two branchial in the more primitive forms, Fig.
472.   The whole apparatus is reduced to slender remnants in Ophidia.

Chelonians have the hyoid cornua reduced to short ventral processes,
and they are further reduced in Crocodiles and Birds.   In these also the
second branchial cornua are lost, Figs. 473, 473*a*.

The Mammalian hyobranchial apparatus has been specialised along lines different from those adopted in the Reptiles and Birds. It consists typically of a broad body or corpus, two long anterior hyoid cornua, and two short posterior cornua. The latter represent the first branchial arches. The hyoid cornu fuses at its dorsal extremity with the crista par-otica (Chapter VIII.), and becomes usually subdivided into several pieces sometimes named hypohyal, ceratohyal, stylohyal, and tympanohyal (Reichert, 1837; Flower, 1870). The small tympanohyal or laterohyal joined to the crista helps to enclose the hyomandibular branch of the facial nerve in a long facial canal and complete the primary stylomastoid fora-men. The next or stylohyal element, often partly ligamentous, may fuse with the fixed laterohyal and so, in man, form a prominent styloid process.

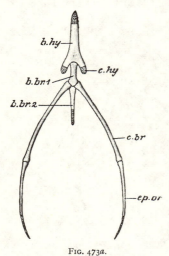

Fig. 473a.

Columba livia. Hyoid apparatus. The cartilaginous parts are dotted. *b.br.* 1, Basibranchials; *b.hy,* basihyal; *c.br,* ceratobranchial; *c.hy,* hyoid cornu; *ep.br,* epibranchial. (From Parker and Haswell, *Zoology.*)

But the hyobranchial apparatus we have so far described in the Tetrapods represents only part of the original branchial skeleton, for the posterior arches become more and more specialised to support the top of the trachea and are gradually converted into the skeleton of the larynx (Henle, 1839; Reichert, 1837; His, 1880; Göppert, 686-7; Gegen-baur, 170; Wilder, 947; Dubois, 1886; Kallius, 691). The mammalian glottis is protectea in front by an epiglottis strengthened by a cartilage provided in lower forms with lateral wings which are represented in man by separate cunei-form cartilages. The trachea is surrounded by cartilaginous rings, the main bronchial tubes being stiffened by similar, usually incomplete cartilages. The larynx itself in Placentals is provided with a large crescentic thyroid cartilage, small paired arytenoids, and a circular cricoid, Fig. 482. It has been held by Gegenbaur and others that the thyroid is derived from the 2nd and 3rd, the epiglottis cartilage from the 4th, the cricoid, arytenoid, and perhaps the tracheal rings from the 5th branchial arch. The evidence, however, from comparative anatomy and embryology with regard to the two last arches is still somewhat incomplete.

The Amphibian larynx is strengthened by paired arytenoids, which are sometimes in Anura greatly developed and ossified to form a ' reson-ance box '. These appear to represent the 5th branchial arch (7th visceral arch). They, together with laryngotracheal cartilages extending down the trachea, are developed in Urodela from a single pair of rudiments (Wilder, 947), but in Anura chondrify separately (Märtens, 1898). A distinct cricoid appears in Reptilia in addition to the arytenoids, but no true thyroid cartilage.

The Monotremata possess a much more primitive laryngeal skeleton than the Ditremata, for in them the thyroid is represented by two distinct arches, Fig. 474. The double origin of the Ditrematous thyroid is clear

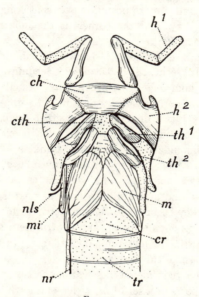

FIG. 474.

Echidna aculeata; ventral view of larynx, trachea, etc. (after E. Göppert, 1901). *ch*, Basihyal (+1st basibranchial ?) copula; *cr*, cricoid, mostly hidden by muscles; *cth*, thyroid copula (2nd and 3rd basibranchial); *h*[1], hyoid arch; *h*[2], 1st branchial arch (posterior cornu of hyoid); *m, mi*, muscles; *nls*, superior laryngeal nerve; *nr*, recurrent nerve; *th*[1], *th*[2], 4th and 5th branchial arches (=thyroid cartilage of higher mammals).

in development, and even in the adult Marsupial may be betrayed by a foramen on either side. That the cricoid is also a modified arch is still apparent in the embryo of Marsupials (Esdaile, 722); it seems to represent the 4th. But the origin of the epiglottis from an arch is more than doubtful (Kallius, 691).

Thus can be traced throughout the Gnathostomes the gradual modification of the branchial arches from their primitive branchial function to such diverse uses as the support of the tongue and the enclosure of the larynx; and the evidence is derived not only from the structure and development of the skeleton, but also from the vascular arches, the musculature, and the distribution of the originally branchial branches of the vagus nerve.

THE LABIAL CARTILAGES

Certain small cartilages of doubtful significance are found in the lips or marginal folds of the mouth in Elasmobranchs and some Teleostomes. They are known as labial cartilages, and in the former group consist usually of an anterior pair of simple rods or plates outside the palato-quadrate cartilage and a more posterior pair consisting of an upper and a lower segment bent round the corner of the mouth, Figs. 424, 427, 440 (Gegenbaur, 166 ; Parker, 569 ; Allis, 668).

Similar, but more elaborately fashioned, labials occur in the Holocephali, Fig. 428 (Hubrecht, 1877 ; Dean, 17 ; Schauinsland, 583). Labial cartilages, perhaps homologous with them, have been described in various Teleostomes, but are not universally present. An upper labial occurs in *Polypterus*, Fig. 296 ; two or more upper labial cartilages are found in various Teleosts (Stannius in Characinidae, Sagemehl, 378 ; Siluridae, Pollard, 696 ; Salmonidae, Parker, 560 ; Gadidae, Brooks ; Gasterosteidae, Swinnerton, 389 ; Serranidae, Kesteven, 692, etc.), and sometimes a pair of lower labials as well. In some cases the anterior upper labials seem to fuse and give rise to a median rostral cartilage (Berrill, 672), but usually they become closely connected with the premaxillaries and maxillaries lying outside them, and contribute a smooth surface to their articulations with the mesethmoid cartilages. In the Siluridae they may support the base of the barbels.

While Pollard (696) considered these various cartilages to be the remains of the skeletal supports of a set of primitive oral cirrhi such as are found still in *Amphioxus* and Myxinoids, Figs. 518, 520, others, like Sewertzoff (701), believe them to represent vestiges of the visceral arches of two segments in front of the mandibular. For the former view there is a good deal to be said, though the evidence in favour of it is by no means decisive. Against the theory maintained by Sewertzoff it may be urged that there is no good evidence of the existence at any time of gill-pouches, arches, etc., anterior to the mandibular, that the labials are too superficial to be of visceral nature, and that the supposed vestiges of gill-pouches corresponding to them apparently occur anteriorly to the pharynx (endodermal gut). Possibly the labial cartilages are merely secondary in Gnathostomes and of no great morphological importance.

THE MIDDLE EAR OF TETRAPODS, AND HISTORY OF THE MAMMALIAN EAR-BONES

FEW problems in morphology have aroused more interest than that of
the origin of the ear-bones. It is a familiar fact that the Mammalia
differ from all the lower Gnathostomes in having a lower jaw composed of
a single dermal bone, the dentary, articulating with another dermal bone,
the squamosal, fixed to the skull ; also in having a chain of three ossicles
(stapes, incus, and malleus) serving to convey vibrations from the
tympanic membrane to the labyrinth lodged in the auditory capsule.
These two characteristics are intimately related, and the explanation of
the second carries with it that of the first.

In all Gnathostomes, excepting the Mammalia, the articulation for
the jaws lies between the quadrate region of the palatoquadrate above
and the articular region of Meckel's cartilage below. In all above the
Chondrichthyes, the former is usually represented by a bone, the quadrate,
and the latter by another separate bone, the articular. The remainder
of the two cartilaginous bars may be more or less reduced, leaving
some cartilage at the joint. Moreover, in Tetrapods, a tympanic membrane (see, further, pp. 460, 477) is typically present behind the quadrate
region, closing the tympanic cavity which opens internally into the
pharynx by the Eustachian tube. The membrane is composed of an

outer layer of epidermis, continuous with the general surface, a thin sheet of connective tissue, and an inner layer of endodermal epithelium continuous with that lining the tympanic cavity, the tube, and the pharynx. This epithelium, closely applied to the auditory capsule, also covers the membrana tympanica secundaria closing the fenestra ovalis (see p. 254 and Figs. 475, 481).

Concerning the homology of the auditory ossicles many different views have been held into which it is not necessary now to go in detail (see Gaupp, 731). Suffice it to say that Peters, 1868, and Albrecht, 1883 (whose main contention has been supported by Cope, 711; Gadow, 730; Dollo, 713; Baur, 704; and more recently by Drüner, 716, and Fuchs, 728), held that these ossicles were derived from the Reptilian columella auris, and that the quadrate and articular are still represented in the articulation of the lower jaw by the cartilage covering the facet of the squamosal and the condyle of the dentary, and the meniscus between them. Among the many difficulties to be met by this theory, we may now mention the fact pointed out by Gaupp that there is no meniscus and practically no cartilage in the jaw articulation of the Monotreme.

Huxley (739) held that the stapes was derived from the reptilian columella, the incus from the supra-stapedial part of the hyoid, and the malleus from the quadrate, while the articular he supposed to be represented in the condyle of the dentary. Such a view could only be put forward when the development of these parts was little understood.

The theory now generally adopted is that based on the researches of Carus (1818), Meckel (1820), and Reichert (1837), and supported by the results obtained by a vast number of workers since, among whom may be mentioned Gaupp, who has recently given a comprehensive account of the evidence (732).[1] According to the modern version of Reichert's theory, the mammalian stapes derived from the reptilian columella auris, the incus from the quadrate, and the malleus from the articular, while the dentary has acquired a new articulation with the squamosal.

A complete and satisfactory theory should account not only for the fate of the quadrate and articular, and the change in the articulation of the lower jaw, but also : (1) for the position of the ossicles with regard to the tympanic cavity ; (2) for their developmental history ; (3) for their relation to the facial nerve and chorda tympani, (4) to the blood-vessels, and (5) to the muscles attached to them ; (6) for the nerve supply of these muscles ; and (7) for the origin of the tympanic bone. In fact, evidence should be sought from the structure and development of skeleton,

[1] Gaupp has reviewed the literature up to 1899 (731-2), and van der Klaauw from 1899 to 1923 (748a).

gill-pouches, muscles, nerves, and blood-vessels, if the account of the middle-ear region is to be completely elucidated.

It will be well to begin with a description of the middle ear of Reptiles, studied by Peters (1869), Parker (564), Gaupp (506), and others, but more particularly by Versluys in the Lacertilia (769, 770). The typical columella auris extends from the fenestra ovalis of the auditory capsule to the tympanic membrane, Figs. 475-9. It consists, in the Lacertilia, of a proximal or stapedial region (stapes), and a distal extra-stapedial region

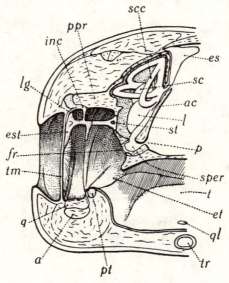

FIG. 475.

Left half of head of *Lacertilian* cut transversely through tympanic cavity and seen from behind (from figures of J. Versluys). *a*, Articular; *ac*, inner wall of auditory capsule; *es*, endolymphatic sac; *est*, extra-stapedial cartilage; *et*, Eustachian tube; *fr*, fenestra rotunda; *q*, quadrate; *inc*, interçalary, end of dorsal process; *l*, lagena; *p*, ductus perilymphaticus; *ppr*, paroccipital process of opisthotic; *pt*, pterygoid; *gl*, glottis; *sc*, sacculus; *scc*, semicircular canal; *sper*, saccus perilymphaticus; *st*, stapes; *t*, tongue; *tm*, tympanic membrane; *tr*, trachea.

(extra-stapes), generally called the extra-columella. The stapedial region is made up of a bony rod with a cartilaginous foot-plate embedded in the membrane closing the fenestra ovalis. The cartilaginous extra-stapedial region has its expanded outermost part embedded in the tympanic membrane; it also bears a processus internus (quadrate process) passing downwards and forwards in the roof of the tympanic cavity, and a more important processus dorsalis (supra-stapedial of Parker, and pr. paroticus of Gaupp). This latter dwindles to a ligament in the adult, except for its upper end, which remains as a nodule (intercalary of Versluys) lodged between the quadrate and paroccipital process (crista parotica) of the auditory capsule. A strong ligament passes from the dorsal process to the outer side of the extra-stapedial. Within the Order Lacertilia the columella varies but little ; but it should be noted that the internal process is lost in the Geckonidae, Uroplatidae, and Amphisbaenidae, that in these and

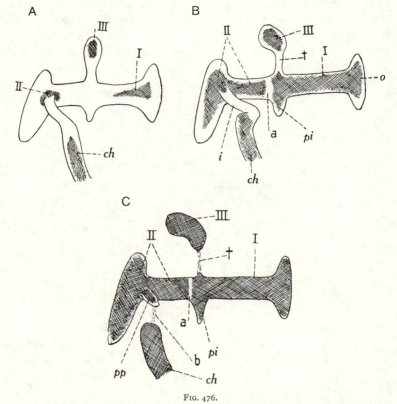

FIG. 476.

Diagrams of development of left columella auris in a *Lacertilian*; posterior view (from J. Versluys, *Zool. Jahrb.*, 1903). A, Earliest, and C, latest stage. Blastema outlined, cartilage shaded. *I*, Otostapedial ; *II*, hyostapedial ; *III*, pars dorsalis of dorsal process (intercalary) ; *a*, temporary limit between *I* and *II* ; *ch*, cornu hyale ; *i*, pars interhyalis ; *o*, foot of otostapes fitting in fenestra ovalis ; *pi*, processus internus ; *pp*, processus posterior ; †, ligamentous vestige of base of proc. dorsalis.

a few others the ligamentous base of the dorsal process disappears, and that the base of the stapedial may be pierced by the facial (stapedial) artery in some Geckonids (Versluys, 770).

The columella is developed from the dorsal end of the hyoid arch, which is bent so as to abut against the auditory capsule at the fenestra ovalis. Its exact origin has given rise to much discussion, since many have held that the capsule contributes to its formation. In the blastematous, and even in the procartilaginous stage, the columella forms a continuous rudiment from the basal plate to the extra-stapes, and the latter is itself continuous with the top of the hyoid arch. As shown by Versluys in *Lacerta*, cartilage then appears in the proximal region of the stapes, in the dorsal process, in the extra-stapedial, and in the more ventral cornu of the hyoid arch. The proximal cartilage spreads outwards, forms the internal process, and becomes connected with the dorsal process; this whole region is known as the otostapes (Hoffmann, 737). The outer extra-stapedial element forms the processus inferior inserted in the tympanum, a small interhyal process, and grows inwards to meet the otostapes, from which it is distinguished as the hyostapes (Hoffmann, 737). The procartilaginous connexion of the hyoid cornu with the interhyal process dwindles to a mere ligament. In later stages the hyostapes fuses with the otostapes, the proximal region of the latter ossifies as a slender rod (stapes), and a new joint may be

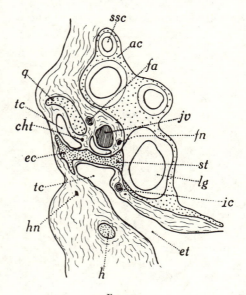

FIG. 477.

Transverse section through left auditory region of head of embryo *Lacerta*. *ac*, Auditory region; *cht*, chorda tympani; *ec*, extra-stapedial region; *et*, Eustachian tube; *fa*, stapedial (facial) artery; *fn*, hyomandibular nerve; *h*, hyoid; *jv*, jugular vein (v. capitis lateralis); *lg*, lagena; *q*, quadrate; *ssc*, semicircular canal; *st*, stapedial region of columella in fenestra ovalis; *tc*, tympanic cavity.

formed between its outer end and the cartilaginous adult extra-stapedial, Figs. 475-9.

This account of the development of the columella in *Lacerta* has been corroborated by observations on its origin in other Sauropsida since Reichert (1837) first described it in the bird, and Rathke (1839) in the snake. (Parker, Hoffmann, Gaupp, Versluys, Goodrich, 734, Rice, 376, in Lacertilia; Howes and Swinnerton, 528, Schauinsland, 583, Wyeth, 773, in *Sphenodon*; Parker, 570, Shiino, 597, in Crocodilia; Fuchs, 503, Kunkel, 540, Smith, 766, Bender, 706, Noack, 754, in Chelonia; Parker, 563, Möller, 753, Peyer, 759, Okajima, 755, Brock, 326, in

Ophidia ; Parker, 557, 568, 572, Sushkin, 608, G. W. Smith, 765, Sonies, 385, Goodrich, 734, in Aves.)

Although there is no doubt that the distal portion of the columella of Reptiles and Birds is derived from the hyoid arch ontogenetically as well as phylogenetically, it is not so certain that the whole of the proximal region is of the same origin. Several authors hold that the otostapes, or at all events its proximal base, is developed in continuity with the auditory capsule from which they believe it to be derived (Hoffmann,

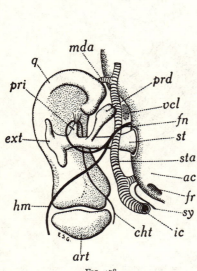

FIG. 478.

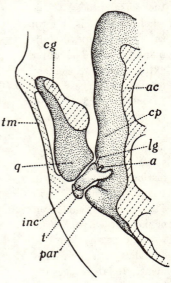

FIG. 479.

Lacerta embryo; diagrammatic reconstruction of posterior view of auditory region showing columella auris, etc. *ac*, Auditory capsule ; *cht*, chorda tympani ; *ext*, extra-stapedial ; *fn*, hyomandibular branch of facial nerve ; *fr*, fenestra rotunda ; *hm*, hyoid nerve ; *ic*, internal carotid ; *mda*, mandibular artery ; *prd*, dorsal process ; *pri*, internal process ; *q*, quadrate ; *st*, stapedial region of columella fitting into fenestra ovalis ; *sta*, stapedial artery ; *sy*, sympathetic cord ; *vcl*, vena capitis lateralis in cranio-quadrate passage.

Lacerta embryo ; ventral view of articulation of quadrate to auditory capsule, showing intercalary cartilage wedged in between them. *a*, Artery ; *ac*, auditory capsule ; *cg*, cut surface of quadrate ; *cp*, crista parotica ; *inc*, extremity of dorsal process of columella (intercalare of Versluys) ; *lg*, ligament ; *par*, paroccipital process ; *q*, otic process of quadrate ; *t*, ventral end of supratemporal bone ; *tm*, tympanic membrane. Cartilage stippled.

Fuchs, in Lacertilia ; Möller, Okajima, in Chelonia ; and others). This view, however, seems to be conclusively disproved by careful observations on the very earliest appearance of the blastema of the columella in several forms (Versluys, 770, in Lacertilia ; Wyeth, 773, in *Sphenodon* ; G. W. Smith, 765, in *Gallus*), where it has been shown to arise separately. At most the capsular wall contributes to the formation of the cartilaginous base in some cases.

The structure of the columella is essentially the same throughout the

Reptilia ; but usually the extra-stapedial (extra-columellar) region is less elaborately developed than in the Lacertilia, and the dorsal process is not separated off. Its various processes are not clearly formed in either Chelonians or Snakes ; and in the latter the extra-stapedial region is not developed, and the distal end of the stapes becomes connected to the quadrate by a ligament or cartilaginous process probably representing the dorsal process.

The condition in *Sphenodon* is interesting (Huxley, **739** ; Peters, 1874 ; Gadow, **730** ; Versluys, **769** ; Howes and Swinnerton, **528** ; Kingsley, **745** ; Osawa, **756** ; Wyeth, **773**). Here the top of the hyoid cornu is continuous in the adult with the extra-stapedial ; and this continuity is not secondary, as some supposed (Peters, Baur, Gadow, Osawa), but due to the retention of the original continuity visible in the procartil-aginous stage in *Sphenodon* as in other

Fig. 480.

Columba livia. The columella auris (magnified). The cartilaginous parts are dotted. *e.st*, Extra-stapedial; *i.st*, infra-stapedial; *s.st*, supra-stapedial; *st*, stapes. (From Parker's *Zootomy*.)

Reptilia (Schauinsland, **583** ; Wyeth, **773**). The dorsal process is joined to the extra-stapes by two limbs enclosing a foramen ; in development the outer gives rise to the intercalary knob, while the inner grows down from it (Wyeth). Which of the two represents the base of the dorsal process in *Lacerta* is uncertain ; probably the inner limb.

When the hyoid cornu becomes detached from the ' interhyal ' or infra-stapedial process of the columella, it may remain free (some Lacertilia, Ophidia, Chelonia), or it may become secondarily attached to the skull, generally by means of the intercalary (Geckonida).

In general structure and in development the columella auris of the Crocodilia resembles that of the Lacertilia (Parker, **579** ; Shiino, **597** ; Goldby, **733**). At first the extra-stapes is continuous by its infra-stapedial process with the ' ceratohyal ' or cornu of the hyoid ; but later this middle region of the arch separates from the ventral portion and acquires a temporary connexion with the hind end of Meckel's cartilage. This remarkable fusion of parts of the hyoid and mandibular arches is soon dissolved, the connexion with Meckel's being lost in the adult. It seems to be secondary and of no great significance, Fig. 483 A.

In the embryo Chelonian the blastema of the extra-stapes may attempt, so to speak, to join the articular, but continuity is not established (Smith, **766** ; Shaner, **596**).

The columella auris of Birds bears some resemblance to that of

Sphenodon on the one hand, and of *Crocodilus* on the other, Fig. 480. For the extra-stapedial region, though it may branch into separate processes (as in *Gallus*), is often pierced by a foramen, and carries in addition to the dorsal process (supra-stapedial) a long infra-stapedial process which may remain in continuity with the 'stylohyal' region of the hyoid

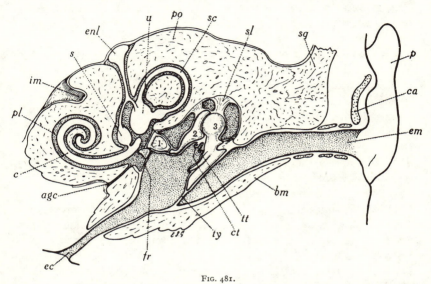

FIG. 481.

Diagram of *auditory organ* of *mammal*, seen in section. *agc*, Aquaeductus cochleae ; *bm*, bony meatus ; *c*, ductus cochlearis ; *ca*, cartilage ; *ct*, chorda tympani ; *em*, external meatus ; *enl*, saccus endolymphaticus ; *fr*, fenestra rotunda ; *im*, foramen for auditory nerve ; *p*, external pinna ; *pl*, perilymphatic cavity ; *po*, periotic ; *s*, sacculus ; *sc*, semicircular canal ; *sl*, suspensory ligament ; *sq*, squamosal ; *tt*, attachment of tensor tympani ; *ty*, tympanic membrane ; *u*, utriculus ; 1, stapes ; 2, incus ; 3, malleus.

arch. From the extra-stapes stretches forward Platner's ligament to the quadrate.

The early fossil Reptiles had a columella stretching from the fenestra ovalis to the quadrate, Figs. 334, 381. There is evidence of a dorsal process ; but the presence of an extra-stapes on the tympanum is doubtful since only the proximal region is ossified. The Cotylosauria have a stout bony stapes, perforated for the stapedial artery, and articulated to a distinct facet on the quadrate. In Ophiacodontidae and related forms a dorsal process reaches the paroccipital crest ; but in the more advanced Theromorphs the stapes approaches the mammalian type (p. 474). On the other hand, the stapes of the Seymouriamorpha is widely separated from the bony quadrate and resembles that of the Stegocephalia (Sushkin, 768).

Turning now to the Mammalia, we find that three ear ossicles occur throughout the class (p. 449). The malleus has a processus anterior (processus Folii, or gracilis ; see p. 476) projecting forwards, and a manubrium fixed to the tympanic membrane ; by its head it is attached to the next ossicle, the incus, whose processus longus articulates with the outer end of the stapes (Frey, **724**). The latter, a pierced stirrup-shaped bone, has a basal plate fitting into the fenestra ovalis,[1] Figs. 481, 497-8.

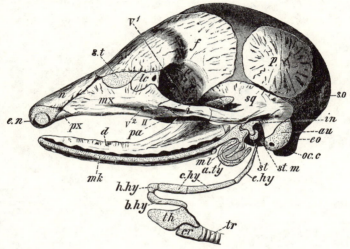

FIG. 482.

Skull of embryo of armadillo (*Tatusia hybrida*). (Modified from a drawing by W. K. Parker.) *a.ty*, Tympanic annulus ; *au*, auditory capsule ; *b.hy*, basihyal ; *c.hy*, ceratohyal ; *cr*, cricoid ; *d*, dentary ; *e.hy*, epihyal ; *e.n*, external nostril ; *eo*, exoccipital ; *f*, frontal ; *h.hy*, hypohyal ; *i*, jugal ; *in*, incus ; *lc*, lacrimal ; *mk*, Meckel's cartilage ; *ml*, malleus ; *mx*, maxilla ; *n*, nasal ; *oc.c*, occipital condyle ; *p*, parietal ; *pa*, palatine ; *px*, premaxilla ; *so*, supraoccipital ; *st*, stapes ; *s.t*, ethmoturbinal ; *st.m*, stapedius muscle ; *sq*, squamosal ; *th*, thyroid ; *tr*, trachea ; *II*, optic foramen ; *V¹*, *V²*, foramina through which the first and second divisions of the trigeminal pass out from the orbit. (From Wiedersheim, *Comp. Anatomy*.)

A very large number of observers have studied the development of the ear ossicles in various mammals since Reichert, 1837 (see Gaupp for literature up to 1913 (**731-2**), and van der Klaauw from 1899 to 1924 (**748a**)). Though the earlier workers often reached very discordant conclusions, the results of the best modern research have on the whole strongly supported the theory of Reichert. It may be considered as firmly established that the incus, malleus, and cartilage of the mandible

[1] The ossicles differ considerably in bulk and shape in the various groups (Hyrtl, Doran, **714**). In many Marsupials (Dasyuridae, Phalangistidae, Peramelidae) and in the Monotremes the stapes is imperforate and columelliform, but this condition appears to be secondary since, in *Ornithorhynchus* and *Trichosurus* at all events, it surrounds the stapedial artery in the blastematous stage (Goodrich, **734**).

are developed from an originally continuous blastematous rudiment (Salensky, 762; Gradenigo, Baumgarten, Dreyfus; Kingsley and Ruddick, 746; Broman, 708; Jenkinson, 740; Goodrich, 734), that the incus chondrifies separately and later comes close to and even articulates with the crista parotica of the auditory capsule dorsally and the malleus ventrally, and that the cartilage of the malleus remains in continuity with the rest of Meckel's cartilage until a quite late stage, a striking fact known since the time of Meckel (1820). These and other facts to be mentioned later receive their only natural explanation

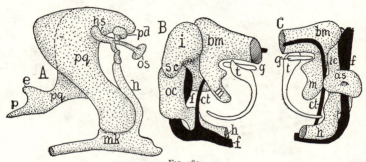

FIG. 483.

Development of auditory ossicles and related parts in A, *Crocodilus*, and B and C, *Man*. A, Outer view of palatoquadrate, *pq*; Meckel's cartilage, *mk*; and columella of late embryo. *e*, Processus ascendens; *h*, hyoid; *hs*, hyostapes; *p*, pterygoid process; *pd*, dorsal process; *os*, otostapes. B, Lateral, and C, medial view of human embryo, 16 mm. long. *as*, Stapes; *bm*, body of malleus; *ct*, chorda tympani; *f*, facial nerve; *g*, prearticular (gonial); *i*, incus; *lc*, long crus of incus; *m*, manubrium mallei; *mk*, Meckelian cartilage; *oc*, auditory capsule; *sc*, short crus of incus; tympanic bone. (A after Parker, '83; B and C after Bromann, '99; from J. S. Kingsley, *Vertebrate Skeleton*, 1925.)

on the supposition that the incus and malleus represent the quadrate and articular, Figs. 482, 487-9, 494, 496-7.

On the other hand, it has been conclusively shown that the stapes develops from the hyoid arch, the earliest rudiment of which forms a continuous blastema separate from that of the mandibular arch (Kingsley, 745; Broman, 708; Jenkinson, 740; Goodrich, 734). The top end of the rudiment is bent at an angle so that its innermost extremity is continuous with the tissue closing the fenestra ovalis, Figs. 483-4. Here develops the stapes, as a ring round the stapedial artery, and it chondrifies separately from the ear capsule. The more distal (interhyal of some authors) 'hyostapedial ligament' remains for a time uniting the stapes to the dorsal end of the 'laterohyal' ('tympanohyal') region of the hyoid cornu, but disappears later. Meanwhile, the laterohyal and more ventral region of the cornu form a continuous cartilage which fuses with the paroccipital process of the auditory capsule. Later the dorsal region of the cornu below the laterohyal degenerates into a ligament, leaving, however, in man a considerable portion to form the styloid process (p. 446).

Comparing the mammal with the reptile, the laterohyal region in the former is seen to correspond to the dorsal process of the extra-stapedial region of the Reptilian columella, Figs. 495-6. The mammalian stapes would, then, represent the proximal region of the columella without its extra-stapedial bar and insertion in the tympanum. However, two little

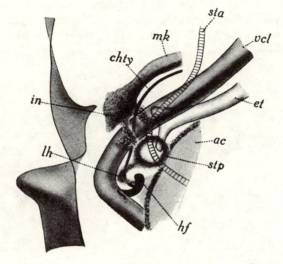

FIG. 484.

Dorsal view of reconstructed thick transverse section of left auditory region of embryo mouse, *Mus musculus*, showing ear ossicles in blastematous stage : *stp*, stapes continuous with blastema of hyoid arch ; *in*, incus and malleus continuous with each other and remainder of mandibular arch. *ac*, Auditory capsule ; *chty*, chorda tympani ; *et*, Eustachian tube ; *lh*, laterohyal ; *mk*, Meckel's cartilage ; *sta*, stapedial artery ; *vcl*, vena capitis lateralis. (E. S. Goodrich, *Quart. Jour. Micr. Sci.*, 1915.)

cartilages have been found which may represent remnants of parts of the extra-stapedial region, as suggested by van Kampen (1915) and van der Klaauw (748). The first is minute and situated in the tendon of the stapedial muscle. Discovered by the Dutch anatomist Paauw in 1645, it has since been described in man and a large number of Placentals and Marsupials (Parker, Hyrtl, Rauber), and originates from the outer end of the stapes (v. d. Klaauw, 748). The second cartilage is of doubtful origin ; it was first described by Spence in the cat (1890), and lies above the chorda tympani, between the stapes and the hyoid cornu. It has since been found in various other Mammalia (v. Kampen, 741 ; Bondy, 707 ; v. d. Klaauw, 748).

In order to establish the theory of Reichert we must now examine the corroborative evidence from other parts in the middle ear. That the Eustachian tube and tympanic cavity correspond in a general way to the first or spiracular gill-slit has long been held (Huschke, 1831); but it is no longer supposed that the tympanic membrane represents the wall between the outgrowing endodermal pouch and the ectodermal ingrowth which becomes pierced in the development of an ordinary gill-slit. In the Reptilia and Aves the spiracular slit is pierced as an elongated cleft which soon closes up from below upwards, that is, ventro-dorsally (Versluys, 770; Kastschenko, 788; Cords, 712; Goodrich, 734). The point of final closure at the dorsal end is marked by the ectodermal epibranchial placode, a proliferation contributing to the geniculate ganglion, Figs.

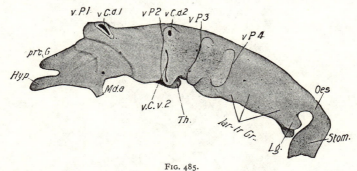

FIG. 485.

Reconstruction of fore-gut of a chick of 72 hours (after Kastschenko, from F. R. Lillie, *Develt. Chick*, 1919). *Hyp*, Hypophysis; *lar-tr.Gr*, laryngotracheal groove; *Lg*, lung; *Md.a*, mandibular arch; *Oes*, oesophagus; *pro.G*, preoral gut; *Stom*, stomach; *Th*, thyroid; *v.C.d,*1, 2, dorsal division of first and second visceral clefts; *v.P.*1, 2, 3, 4, first, second, third, and fourth visceral pouches.

485-6, 495. The tympanic cavity arises from the more ventral and inner region of the endodermal gill-pouch as a tympanic diverticulum which expands and, becoming applied to the ectoderm, forms the tympanic membrane.[1] In *Lacerta*, as the diverticulum enlarges it tends to surround the columella which comes to project into the tympanic cavity from above and behind, suspended by a fold of the lining membrane except in the extra-columellar region, where the fold is obliterated by the junction of an anterior with a posterior pocket. The closure of the spiracular slit, development of the tympanic cavity from the more ventral region of the gill-pouch, and formation of the membrane are very similar in other Reptiles and in Birds, Figs. 485-6.

Thus the tympanic cavity and tympanum lie between the mandibular and hyoid arches, and the columella is morphologically posterior to them.

---

[1] The view of some authors (for instance, Wyeth in *Sphenodon*, 773; Bender in *Testudo*, 706) that the second gill-slit contributes to the tympanic cavity does not seem to be well founded.

Just as in the Reptilia, so in the Mammalia the tympanic cavity and Eustachian tube are developed from the first or spiracular gill-pouch (Kastschenko, 787; Hammar, 736; Goodrich, 734). Although the cleft does

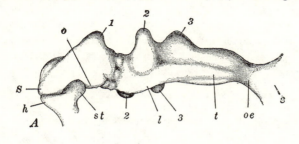

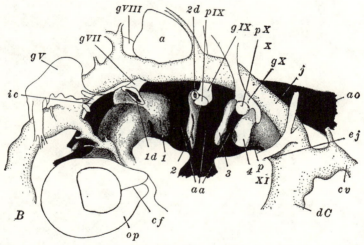

FIG. 486.

Models of pharynx and associated structures in chick (after Kastschenko, from W. E. Kellicott, *Chordate Develt.*, 1913). A, Ventro-lateral view of pharynx at beginning of third day. B, Later view of pharynx and associated nervous and vascular structures at end of third day. Nervous structures are left unshaded; arteries in solid black; veins lightly stippled; pharyngeal structures darkly stippled. *a*, Auditory sac; *aa*, aortic arches; *ao*, dorsal arches; *cf*, choroid fissure; *cv*, posterior cardinal vein; *dC*, ductus Cuvieri; *ej*, external jugular vein; *gV*, Gasserian ganglion of V. cranial nerve; *gVII*, geniculate ganglion of VII cranial nerve; *gVIII*, acustic ganglion of VIII. cranial nerve; *gIX*, ganglion petrosum of IX. cranial nerve; *gX*, ganglion nodosum of X. cranial nerve; *h*, hypophysis; *ic*, internal carotid artery; *j*, internal jugular vein; *l*, rudiment of larynx; *o*, oral evagination of fore-gut; *oe*, oesophagus; *op*, optic vesicle enclosing lens; *p*, pulmonary artery; *pIX*, placode of IX. cranial nerve; *pX*, placode of X. cranial nerve; *s*, stomach; *S*, Seessel's pocket (preoral gut); *st*, stomodaeum; *t*, rudiment of trachea; 1-4, first to fourth visceral pouches (or their ventral position, in B); *1d*, *2d*, dorsal portions of first and second visceral pouches; *IX*, glossopharyngeal nerve; *X*, vagus nerve; *XI*, hypoglossal nerve.

not appear ever to be pierced, yet the endodermal pouch comes in contact with the ectoderm in early stages between the blastematous rudiments of the mandibular and hyoid arches. The pouch soon peels off, the last contact being at that dorsal point where the ganglionic proliferation takes place. A tympanic postero-ventral diverticulum grows out, and becoming applied to a deep ectodermal invagination (external auditory meatus) forms the tympanum (see, further, p. 469). As the tympanic cavity expands it tends to surround the developing ossicles, so that the incus and malleus come to project into it from above and in front, and the stapes from above and behind. The relative position of these parts is essentially the same as that borne by the tympanic cavity to the quadrate, articular, and columella in the Reptilia (see, however, p. 477).

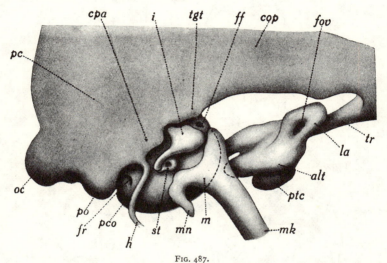

FIG. 487.

Trichosurus vulpecula, embryo 17 mm. long. Reconstruction of posterior region of skull; right-side view showing developing auditory ossicles and surrounding parts in cartilage. *alt*, Ala temporalis ; *cop*, commissura orbito-parietalis ; *cpa*, crista parotica ; *ff*, facial foramen ; *fov*, foramen ovale ; *fr*, fenestra rotunda ; *h*, hyoid cornu ; *i*, incus ; *la*, lamina ascendens ; *m*, malleus continuous with Meckel's cartilage, *mk* ; *mn*, manubrium ; *oc*, occipital condyle ; *pc*, pars canalicularis, and *pco*, pars cochlearis of auditory capsule ; *po*, parotic process ; *ptc*, pterygoid cartilage ; *st*, stapes ; *tgt*, tegmen tympani ; *tr*, trabecula.

.Of particular importance in this connexion is the position of the facial nerve and its branch the chorda tympani (Versluys, **769**; Gaupp, **732**; Kingsley, **745**; Broman, **708**; Goodrich, **734**). The facial nerve in Reptilia issues from the cranial cavity by the facial foramen, and having given off the palatine branch, passes back outside the auditory capsule, runs above the columella, below the crista parotica, medial to the dorsal process and hyoid cornu, then continues down the hyoid, Figs. 477, 478, 495. On its downward course this main hyomandibular branch of the facial gives off the chorda tympani, which passes forwards lateral to the hyoid, over the tympanic cavity and columella, then downwards along the posterior surface of the quadrate to the inner side of the articular. Thence it runs forwards along Meckel's cartilage,

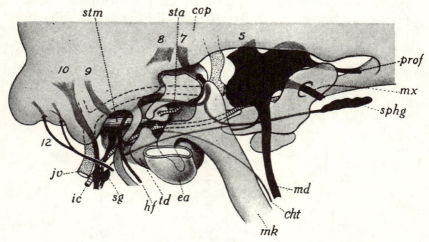

FIG. 488.

Trichosurus vulpecula. Same as Fig. 487, but with nerves, blood-vessels, etc., added. Nerves black; arteries cross-lined; veins stippled. Cartilage represented as transparent. *cop*, Commissura orbito-parietalis; *ea*, inner part of external auditory meatus; *md*, *mx*, mandibular and maxillary branches; *prof*, profundus branch of fifth nerve; *sg*, superior cervical sympathetic ganglion; *td*, outer wall of tympanic diverticulum. Other letters as in Figs. 487 and 489.

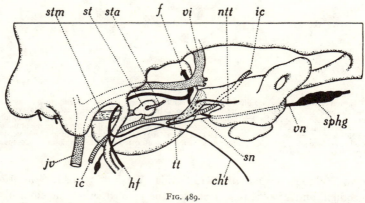

FIG. 489.

Trichosurus vulpecula. Same view as in Fig. 487, but with incus and malleus removed, and muscles, nerves, and blood-vessels inserted. *cht*, Chorda tympani; *f*, facial nerve; *hf*, hyoid branch; *ic*, internal carotid; *jv*, jugular vein (v. capitis lateralis); *ntt*, branch of trigeminal nerve; *sphg*, sphenopalatine ganglion; *sn*, sympathetic nerve from superior cervical ganglion; *sta*, stapedial artery; *stm*, stapedial muscle innervated from facial; *tt*, m. tensor tympani; *vi*, middle cerebral vein; *vn*, vidian nerve formed by union of sympathetic with palatine.

anastomosing with the mandibular branch of the trigeminal nerve and supplying the lining of the mouth in the region of the lower jaw and tongue (see p. 752).[1] Further, it should be noticed that the chorda tympani when crossing over the extra-stapedial passes outside the dorsal process (and usually outside the extra-stapedial ligament as well). It may, however, slip forward, so to speak, and pass more directly downwards in front of the columella in those forms in which the dorsal

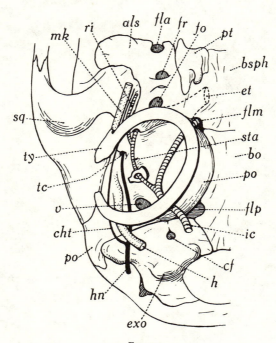

FIG. 490.

Diagram of ventral view of auditory region of skull of *Mammal* (after v. Kampen, '05, modified). *als*, Alisphenoid ; *bo*, basioccipital ; *bsph*, basisphenoid ; *cf*, condylar foramen ; *cht*, chorda tympani ; *et*, position of Eustachian tube drawn in broken line ; *fla*, foramen lacerum anterius ; *flm*, foramen lacerum medium ; *flp*, foramen lacerum posterius ; *fo*, foramen ovale ; *fr*, foramen rotundum ; *h*, hyoid cornu ; *hn*, hyomandibular branch of facial nerve ; *ic*, internal carotid ; *mk*, vestigial Meckel's cartilage ; *po*, periotic ; *pt*, pterygoid ; *ri*, infraorbital artery ; *sq*, squamosal ; *sta*, stapedial artery ; *tc*, outer facial foramen.

process is absent (*Amphisbaena*, Geckonida : Versluys, 769). In *Sphenodon*, Crocodilia, and Aves it passes outside the pierced supra-stapedial plate (dorsal process).

Such is the course of the facial nerve and chorda tympani in the middle ear of the embryo and adult of Reptiles and Birds. Now the chorda tympani is a postspiracular nerve, and its peculiar position in the

---

[1] In the adult the chorda runs inside the lower jaw, a position it reaches usually by piercing the prearticular bone (goniale), as Gaupp has shown (732).

adult, close behind the quadrate and in front of the tympanum, needs elucidation. The explanation is that, although in early stages the chorda is found developing as a distinctly postspiracular branch of the hyomandibular nerve, when the slit closes the chorda gets pushed forward and upward as the tympanic diverticulum expands behind it, Fig. 494*a* (Versluys, **770**; Goodrich, **734**).

In the adult mammal the chorda tympani runs forwards over the roof of the tympanic cavity in a fold of its lining membrane, passing between

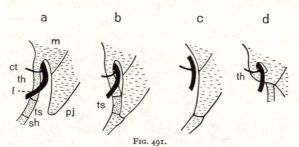

FIG. 491.

Diagrams showing fixation of hyoid to periotic, and its relation to facial nerve and chorda tympani in various Mammalia (after P. N. van Kampen, '05, from M. Weber, *Saügetiere*, 1927). *ct*, Chorda tympani; *f*, facial; *m*, periotic; *pj*, paroccipital process; *sh*, stylohyal; *th*, tympanohyal; *ts*, tympano-styloid cartilage.

the crus longus incudis and the manubrium mallei, Figs. 481, 490; it then turns·downwards behind and below these ossicles to the lower jaw. The general relations of the facial nerve are the same in the mammalian embryo as in the reptilian, Figs. 484, 488-9, 496. Issuing by the facial foramen, it gives off the palatine branch, then runs round the outer side of the auditory capsule below the overhanging tegmen tympani and below the crista parotica in a groove, which may be more or less completely converted into a canal. From this it escapes behind the stylohyal by the primitive stylomastoid foramen. On its way down the hyoid cornu the hyomandibular nerve gives off the chorda tympani, which passes forwards outside the cornu; except in some Marsupials, such as *Didelphys* (Gaupp; Toeplitz, **391**), *Dasyurus* (Cords, Esdaile, **722**), *Phascolarctus* (v. d. Klaauw, **748**), and in *Manis* (v. Kampen, **741**), where it passes medially.[1] Similarly the hyomandibular nerve and the chorda may in some mammals pass freely outside the hyoid cornu, Fig. 491. Running over the tympanic cavity, the chorda tympani does not come into close relation with the stapes, since there is no extra-stapedial; but it passes 'morphologically' over it and then posteriorly to the incus to reach the inner side of the malleus, whence it continues

---

[1] This unusual position of the nerve seems to be secondary.

along Meckel's cartilage. In spite of many statements to the contrary, it has been conclusively shown that the chorda tympani in the mammal develops early as a postspiracular branch of the facial nerve (Froriep, **725**; Broman, **708**; Emmel, **719**; Goodrich, **734**); but in later stages it runs anteriorly to the tympanum and for the same reason throughout the Amniota. The chorda tympani, then, is postspiracular in the embryo and pretympanic in the adult.[1] Its relation to the ossicles can only satisfactorily be accounted for on the supposition that the incus and malleus are homologous with the quadrate and articular.

Evidence from the muscles gives further support for the theory of Reichert, Figs. 488-9, 492. The malleus is provided with a muscle (m.

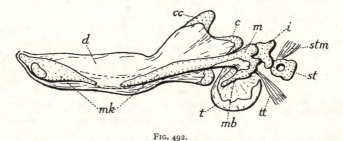

FIG. 492.

Right ramus of lower jaw of embryo *Erinaceus europaeus*, 3 inches long (after W. K. Parker, 1885). *d*, Dentary; *c*, condylar cartilage; *cc*, coronoid cartilage; *i*, incus; *m*, malleus with dermal bone (prearticular); *mb*, manubrium; *mk*, Meckel's cartilage; *st*, stapes; *stm*, stapedial muscle; *t*, tympanic; *tt*, tensor tympani muscle.

tensor tympani) attached to its medial surface and passing forwards towards the base of the skull; and to the stapes is attached a stapedial muscle passing backwards medially to the paroccipital region (Hagenbach, 1833; Parker, **561**, **567**; Eschweiler, **740-41**; Bondy, **707**). Now the former is innervated by a twig of the trigeminal nerve, and the latter by a twig from the facial nerve, which is in agreement with the view that the malleus is derived from the prespiracular mandibular arch, and the stapes from the postspiracular hyoid arch. The musculus tensor tympani is generally considered to have originated in phylogeny from the m. pterygoideus, attached to the articular region of the lower jaw in the Sauropsida, a conclusion borne out by its development in the mammal from the same rudiment as the m. pterygoideus internus (Killian,

---

[1] It is a remarkable fact that the chorda tympani in the chick develops as a prespiracular nerve (Goodrich, **734**). In adult Gallinaceous birds it passes downwards to the lower jaw in front of the tympanic cavity (Magnien, **751**; G. W. Smith, **765**). An explanation of its anomalous course and exceptional development may be that it is in these birds composed of sympathetic fibres only, the postspiracular facial fibres having disappeared.

742 ; Eschweiler, 720-21 ; Edgeworth, 718). The relation of its ligament to the chorda tympani varies. Usually it reaches the malleus below the nerve ; but in Sciurus and Equus it is pierced by the nerve, and in some (including Man and the Apes) it passes above (Bondy, 707).

The mammalian stapedial muscle is probably homologous with the similar extra-stapedial muscle of crocodiles and birds (Killian, 742 ; Goodrich, 734), and possibly with the Lacertilian 'extra-columellar muscle' of the adult Geckonida and embryonic Lacerta (Versluys, 769-70), which, however, is lateral to the hyomandibular nerve. The musculus stapedius is derived from the same source as the Sauropsidan depressor mandibulae ; but appears to be really a levator hyoidei, since it is first attached to the stylohyal and then shifts to the stapes (Edgeworth, 718).

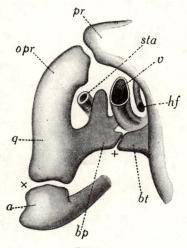

For the determination of the morphological relations of the parts of the middle ear the blood-vessels are scarcely less important than the nerves. In the reptilian and avian embryos the vena capitis lateralis (jugular vein, see p. 532) runs back from the orbit alongside the hyo-

FIG. 493.

Diagram of auditory region of primitive Reptile (embryonic stage), looking from behind through cranio-quadrate passage. Lettering as in Fig. 494.

mandibular branch of the facial nerve, below the crista parotica to the neck. On its way it is medial to the quadrate, ventral to the attachment of the otic process, dorsal to the tympanic cavity and columella, and medial to the latter's dorsal process. The vena capitis lateralis of the mammalian embryo follows the same course, medial to the incus below its connexion with the auditory capsule, dorsal to the stapes and tympanic cavity, and medial to the laterohyal and cornu, Figs. 477, 489, 490, 493-6, 498.

The internal carotid, on its way to the hypophysial space, runs in all Tetrapods between the base of the skull above and the pharynx below. It gives off in the embryonic mammal a branch which passes outwards over the tympanic cavity, pierces the stapes, runs on the medial side of the incus below its connexion with the skull to the orbit, where it divides into supraorbital, infraorbital, and mandibular branches (see p. 529). This stapedial artery is developed at or near the upper end of the second or hyoidean arterial arch, and is always present in the embryo, though the region piercing the stapes may be suppressed in the adult (as in man, see p. 531). An arteria stapedialis or facialis, similar in its general distribution and origin to the a. stapedialis of mammals, occurs also in the lower Tetrapods (see p. 530). Typically it runs over the tympanic cavity, medially to the quadrate and below the otic process forwards, giving off mandibular infraorbital and supraorbital branches. It corresponds in the

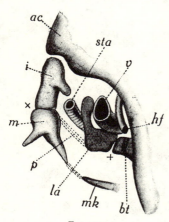

FIG. 494.

Diagram of auditory region of Mammal (embryonic stage) looking from behind through cranio-quadrate passage. *a*, Articular region of mandibular; *ac*, auditory capsule; *bp*, basal process; *bt*, basitrabecular process (pr. alaris); *i*, incus; *hf*, hyomandibular branch of facial nerve; *la*, lamina ascendens; *m*, malleus; *mk*, anterior part of Meckel's cartilage; *opr*, otic process; *p*, part of palato-quadrate which has disappeared; *pr*, parotic process; *q*, quadrate region of palato-quadrate; *sta*, stapedial artery; *v*, vena capitis lateralis; +, basal (basipterygoid) articulation; ×, original articulation of lower jaw.

middle-ear region exactly to the mammalian a. stapedialis, but varies in its relation to the base of the columella. In certain Geckonids (*Pachydactylus, Hemidactylus, Tarentola*; Versluys, 769) and certain birds (*Procellaria, Aquila*; Breschet, 1844; Doran, 714) it· pierces the columella, but in certain other Geckonids (Versluys, 770) and in *Sphenodon* (Versluys, 770; Wyeth, 773) it runs ventrally to it, while in the majority of reptiles it passes dorsally to it. The first condition is probably primitive and the other two secondary. There can be little doubt that a stapedial artery passed through the foramen often found in the stapes of early fossil Reptilia and Amphibia (pp. 474 and 483).

Summary.—The chief points in the evidence with regard to the origin of the mammalian ear-bones may be summarised as follows: A comparison of the structure and development of the parts of the middle ear in Reptilia and Mammalia shows a remarkable similarity in the two groups.

From the continuous blastema of the mandibular arch, anterior to the first gill-pouch, are developed a dorsal element which becomes connected with the auditory capsule by the otic process, and a ventral Meckelian cartilage. The former becomes the quadrate in the Reptilia and the incus in the Mammalia. The posterior end of Meckel's cartilage forms the articular bone in the Reptilia and the malleus in the Mammalia, which becomes disconnected from the rest of the degenerating Meckelian cartilage. The primitive quadrato-articular joint is represented between the incus and malleus.

From the continuous blastema of the hyoid arch posterior to the first gill-pouch are developed a dorsal columella auris and a more ventral cornu hyale in the Reptilia. The columellar blastema forms later a proximal stapedial region fitting into the fenestra ovalis and a distal extra-stapedial region, which usually becomes disconnected from the hyoid cornu. From this extra-stapedial region extend a dorsal process, which connects with the paroccipital process of the skull, a quadrate

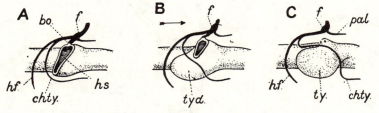

FIG. 494a.

Diagrams illustrating relation in *Amniota* of chorda tympani, *chty*, to hyoidean or spiracular slit, *hs*, and tympanic diverticulum, *tyd*, as shown in development. A, earliest, C, latest stage; right-side view. *bo*, Branchial placode; *f*, facial ganglion; *hf*, hyomandibular branch; *pal*, palatine branch. (From Goodrich, *Quart. Jour. Micr. Sci.*, 1915.)

process, and an outer part inserted in the tympanum. In the mammal the hyoid gives rise to a proximal stapes fitting into the fenestra ovalis. There is no insertion in the tympanum ; but a dorsal process for a time continuous with the hyoid cornu fuses with the crista parotica. This process (laterohyal or stylohyal) may be compared to the dorsal process of the Reptilia ; but the region between it and the stapes (' hyostapedial ligament ') does not persist. The extra-stapedial region is thus less developed in the mammal ; the stapes is disconnected, and articulates now only with the incus (quadrate).[1]

[1] The acceptance of Reichert's theory does not necessarily commit us to the view that the Mammalia have been derived from forms, like the modern Reptilia, with a columella provided with a well-developed extra-stapedial region inserted in the tympanum. The position and fate of the laterohyal, the chief surviving element of the extra-stapes in the mammal, suggests that it was always posterior to the membrane. The ancestral stapes was probably stout, perforated, and attached distally to the quadrate (cp. early fossil Reptilia and Apoda). We have no clear evidence from the fossils whether or not its cartilaginous extremity spread over the tympanum, but an extra-stapedial cartilage may well have been present (Gregory, **735**). At all events the change from the reptilian to the mammalian condition was probably due to the reduced and loosened quadrate and articular being pushed in, so to speak, between the stapes and the tympanic membrane so as to form a chain of three firmly united elements connecting it to the fenestra ovalis. The retro-articular process of the reptilian articular may be compared to the manubrium mallei of the mammal.

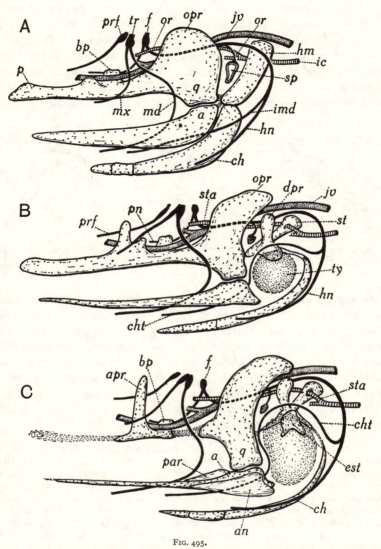

FIG. 495.

Diagrams showing relation of nerves (black), arteries (cross-lined), and veins (shaded) to *Mandibular and Hyoid* skeletal *visceral arches* (dotted) in : A, primitive autostylic Fish, B, primitive Tetrapod, and C, primitive Lacertilian. *a*, Articular region, *an*, angular bone ; *apr*, ascending process ; *bp*, basal process ; *ch*, ceratohyal ; *dpr*, dorsal process ; *f*, facial ; *hm*, hyomandibula ; *hn*, hyomandibular nerve ; *ic*, internal carotid ; *imd*, internal mandibular ; *jv*, jugular (v. capitis lateralis) ; *md*, mandibular ; *mx*, maxillary ; *opr*, otic process ; *or*, orbital = stapedial ; *p*, palatine region of palato-quadrate; *par*, prearticular bone ; *pn*, palatine ; *q*, quadrate region ; *sp*, spiracle (closed in B and C) ; *st*, stapes (base of columella) ; *sta*, stapedial ; *tr*, trigeminal ; *ty*, tympanic membrane.

The dorsal end of the hyoid cornu in Reptilia, which almost invariably becomes detached from the extra-stapedial, may become secondarily connected to the skull.[1]

The view that the stapes represents the proximal region of the columella, and that the incus and malleus represent the quadrate and articular, is further borne out by a consideration of the relation these ossicles bear to surrounding structures. In both Mammalia and Reptilia, as the spiracular or first endodermal gill-pouch separates from the ectoderm from

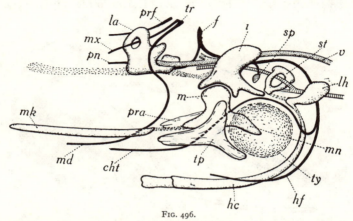

Fig. 496.

Diagram of the fate of the first two visceral arches and of their relation to nerves and blood-vessels in a *Mammal*. Left-side view. *cht*, Chorda tympani ; *f*, facial nerve ; *hc*, hyoid cornu ; *hf*, hyoid branch of facial ; *i*, incus ; *la*, lamina ascendens (epipterygoid cartilage) ; *lh*, latero-hyal, fuses with auditory capsule ; *m*, malleus ; *md*, mandibular branch ; *mk*, Meckel's cartilage ; *mn*, manubrium mallei ; *mx*, maxillary branch ; *pn*, palatine nerve ; *pra*, prearticular (processus anterior) ; *prf*, profundus nerve ; *sp*, position of closed spiracular slit and its placode ; *st*, stapes ; *tp*, tympanic (angular) ; *tr*, trigeminal nerve ; *ty*, tympanic membrane ; *v*, vena capitis lateralis. Stapedial artery, branch from internal carotid, passes through stapes ; these arteries are cross-lined.

below upwards, an ectodermal proliferation contributing to the facial ganglion marks the last point of contact (see p. 764). Whereas the proximal region of the pouch forms the Eustachian tube, a postero-ventral diverticulum of the pouch enlarges distally to form the tympanic cavity, and, becoming applied to the ectoderm, gives rise to the tympanic membrane between the mandibular and hyoid arches. This is not the original closing membrane of the spiracular slit ; but a new formation, developed in essentially the same manner and position in all Amniotes. The expanding tympanic cavity tends to surround the skeletal structures. As the cavity enlarges behind the quadrate in Reptiles the columella appears to sink into it from above and behind ; while in mammals, not

[1] The homology of the columella with the hyomandibula of fishes, of the chorda tympani, and of the stapedial artery is discussed elsewhere, see pp. 420, 464, 529, 752.

only does the stapes sink in from above and behind, but the incus and malleus from above and in front.

The relation of the facial nerve in its backward course over the columella in the Reptile is the same as that borne to the stapes in the Mammal. Of particular importance is the chorda tympani, a post-spiracular but pretympanic branch of the facial which bears the same relation to the quadrate and articular as it does to the incus and malleus, and the same relation to the columella as it does to the stapes.

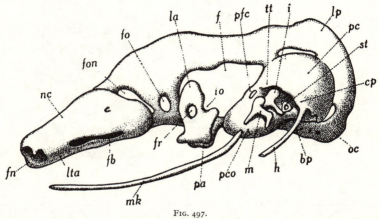

FIG. 497.

Diagram of cartilaginous skull of primitive Placental Mammal. *bp*, Fenestra rotunda ; *cp*, crista parotica ; *f*, fenestra pro-otica ; *fb*, fenestra basalis closed behind by lamina transversa posterior ; *fn*, fenestra narina ; *fr*, foramen rotundum ; *fo*, optic foramen ; *fon*, fenestra orbitonasalis ; *h*, hyoid cornu ; *i*, incus ; *io*, incomplete foramen ovale ; *la*, lamina ascendens ; *lp*, lateral plate ; *lta*, lamina transversa anterior ; *m*, malleus ; *mk*, Meckel's cartilage ; *nc*, nasal capsule ; *oc*, occipital region ; *pa*, basitrabecular process (pr. alaris) ; *pc*, canalicular region of auditory capsule, and *pco*, its cochlear region ; *pfc*, prefacial commissure ; *st*, stapes ; *tt*, tegmen tympani.

The development and innervation (from the trigeminal) of the tensor tympani muscle, attached to the malleus, is in agreement with the view that it represents part of the pterygoid muscle attached to the articular region in the Sauropsida, and belongs to the prespiracular musculature of the mandibular arch ; while the development and innervation of the stapedial muscle, attached to the mammalian stapes, is in agreement with the view that it belongs to the postspiracular musculature of the hyoid arch, and is probably homologous with a similar muscle belonging to the columella in certain Sauropsida. Probably it is a derivative of the depressor mandibulae, of which it appears to be the only remnant in the Mammal (Gaupp, 732 ; Edgeworth, 718 ; Futamura, 1907).

The course of the blood-vessels is the same in the two groups. The vena capitis lateralis, as it runs backward along the hyomandibular nerve, passes between the skull and the quadrate (that is to say, through the cranio-quadrate passage, see p. 412) ventrally to the articulation of the otic process and over the columella in the Reptile. In the Mammal it passes between the incus and the skull below the connexion of the incus to the crista parotica, and above the stapes. More striking is the fact that in the mammalian embryo the stapedial artery, derived from the root of the second aortic arch, runs forwards over the tympanic cavity,

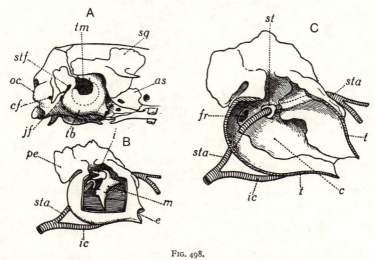

FIG. 498.

Right auditory region of skull of *Mus decumanus*, showing relation of *stapedial artery* to tympanic and auditory ossicles. A, Outer view; B, with window cut in tympanic bulla; C, cavity of bulla further exposed and incus and malleus removed. B and C more enlarged than A. *as*, Alisphenoid; *c*, periotic wall; *cf*, condylar foramen; *fr*, fenestra rotunda; *i*, incus; *ic*, internal carotid; *jf*, jugular foramen; *m*, malleus; *oc*, occipital condyle; *pe*, periotic; *sq*, squamosal; *st*, stapes in fenestra ovalis; *sta*, stapedial artery; *stf*, stylomastoid foramen; *t*, tympanic; *tb*, tympanic bulla; *tm*, tympanic membrane, limited by dotted line and partly concealed.

piercing the stapes, and passes forwards between incus and cranium to supply the jaws and other parts. A similar facial or stapedial artery in the Reptile pierces the stapedial region of the columella, or passes above or below it, and runs forward between quadrate and cranium to the jaws and other parts. The structure, development, and relative position of the parts are essentially the same in Birds as in Reptiles.

New Articulation of the Lower Jaw.—It is clear that the evidence derived from the study of the embryology and anatomy of the middle-ear region is strongly in favour of the 'Reichert theory'. But if the original articulation between the quadrate and articular is indeed represented by the joint between the incus and malleus, we are driven to the conclusion that the mammalian articulation for the lower jaw between the derma squamosal and dentary bones is a new one. How can such a radical change have taken place without interrupting the proper function of the jaws ? Embryology has thrown little light on this point ; but it is one

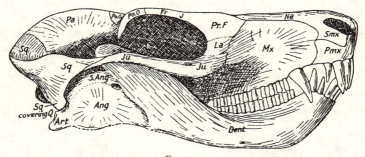

FIG. 499.

Bauria cynops, right-side view of skull and lower jaw restored (from R. Broom, *P.Z.S.*, 1911). *Art*, Articular ; *Dent*, dentary ; *Fr*, frontal ; *Ju*, jugal ; *L*, lacrimal ; *Mx*, maxillary ; *Na*, nasal ; *Pa*, parietal ; *Pmx*, premaxillary ; *Po.O*, postorbital ; *Pr.F*, prefrontal ; *Qu*, quadrate ; *S.Ang*, supra-angular ; *Smx*, septomaxillary ; *Sq*, squamosal.

of the triumphs of the long series of researches on the extinct Theromorph reptiles, begun by Owen (1845), and continued by Seeley, Broom, and Watson, to have revealed the intermediate steps by which the change

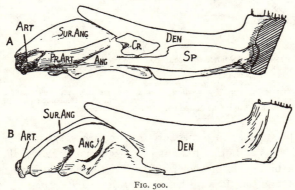

FIG. 500.

Lower jaw of *Scymnosuchus whaitsi*. A, Inner view of left ramus ; B, outer view of right ramus (from D. M. S. Watson, *Ann. Mag. N.H.*, 1912). Lettering as in Fig. 377.

may have occurred from an inner quadrate to an outer squamosal articulation (Gregory, **735**).

In the more reptilian Theromorphs the stapes is found as a stout or slender rod extending from the fenestra ovalis to the quadrate to which its distal end is articulated (Dicynodontia, Broom, **709**, **710** ; Sollas, **602**. Pelycosauria, Case, **471**. Theriodontia, Watson, **640**), but in the higher forms the stapes may be perforated as in mammals (Thrinaxodon, Cynognathus).

In the more primitive and reptilian Theromorphs the quadrate is

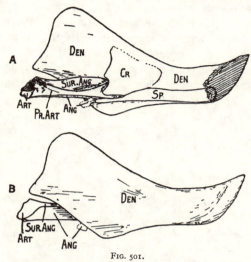

FIG. 501.

Lower jaw of *Cynognathus crateronotus*. A, Inner view of left ramus; B, outer view of right ramus. (From D. M. S. Watson, *Ann. Mag. N.H.*, 1912.) Lettering as in Fig. 377.

fixed laterally to the squamosal, by its otic process to the auditory capsule, and by an anterior process to the pterygoid; the articular is

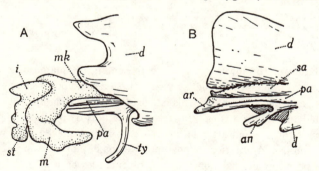

FIG. 502.

A, Outer view of posterior region of lower jaw of late embryo *Perameles*, and B, inner view of same region of *Cynodont* reptile (from R. W. Palmer, 1913). *an,* Angular; *ar,* articular; *d,* dentary; *i,* incus; *m,* malleus; *mk,* Meckel's cartilage; *pa,* prearticular in B, processus anterior mallei in A; *sa,* supra-angular; *st,* stapes.

large and of the usual type. The squamosal and quadratojugal bear much the same relation to the quadrate as in Sphenodon. Passing from the less to the more mammal-like Theriodontia, the squamosal is seen to enlarge, the quadratojugal to dwindle. The quadrate diminishes in relative size, retreats inwards together with the reduced quadratojugal below the overhanging squamosal, loses its pterygoid process, becoming loosened from the pterygoid complex. Meanwhile the dentary becomes progressively larger, and the hinder region of the lower jaw reduced, the articular following the quadrate. The small angular, supra-angular, and inner prearticular remain close to the articular, while the coronoid and splenial may disappear. Thus, in the most mammalian Cynodontia, the quadrate and articular dwindle in importance as supports for the

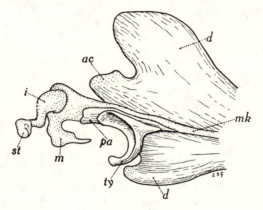

FIG. 503.

Reconstruction of inner view of posterior end of lower jaw and auditory ossicles of a pouch young of *Didelphys aurita*, left ramus. Cartilage dotted. *ac*, Secondary cartilage on condyle ; *d*, dentary ; *i*, incus ; *m*, malleus ; *mk*, Meckel's cartilage ; *pa*, prearticular (anterior process of malleus) ; *st*, stapes ; *ty*, tympanic.

articulation of the lower jaw ; while the enlarged squamosal is met by a backward process of the dentary, and these dermal bones come to share in the articulation. From this stage the transition to the typical mammalian condition is easy, Figs. 339, 340, 394-6, 499-504.

But if the original articulation has thus been replaced by a new one more lateral and anterior in position, we might hope to find remnants of the posterior dermal bones of the reptilian lower jaw in the middle-ear region of the mammal. The tympanic and processus anterior of the malleus appear to be such bones. This anterior process has long been known to develop as a separate dermal bone situated ventrally and close to Meckel's cartilage (Parker ; Broman). Later it becomes ankylosed to the ossified malleus, and has been considered to represent either the angular (Parker, 567), supra-angular (v. Kampen, 741 ; Palmer,

757), or prearticular (Gaupp, **732**). Since, like the prearticular of the lower Tetrapods, the anterior process is sometimes pierced by the chorda tympani, the last seems the more probable homology.

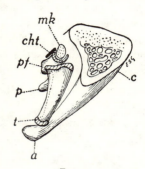

The tympanic, that dermal bone peculiar to and so characteristic of the Mammalia, which encircles and stretches the tympanum, may be derived from the reptilian angular. It arises as a crescentic bone near the malleus, with which it remains closely associated in lower Placentals and in Marsupials, and may become fused in Monotremes (Doran, **714**; Peters, 1868). In the foetal *Orycteropus* (Gadow, **730**) and Marsupial (Palmer, **757**; Toeplitz, **391**) it has a horizontal limb along Meckel's cartilage, and a ventral crescentic process, Figs. 482, 496, 502-5. The very similar angular of *Cynognathus* (Watson, **640**; Palmer) suggests that already in the Theromorph the angular was notched behind to stretch the tympanum below the level of the articular, thus leading to the mammalian condition. Already in Pelycosauria the angular is somewhat notched and extended behind in a manner suggesting that it shared in

Fig. 504.

Didelphys aurita, late embryo. Portion of thick transverse section reconstructed to show relations of skeletal parts near hind end of lower jaw; posterior view of right side. *a*, Angular, and *c*, condylar processes of dentary bone; *cht*, chorda tympani; *mk*, Meckel's cartilage; *p*, anterior process of tympanic; *pf*, anterior process (prearticular ?) of malleus; *t*, tympanic ring of which posterior part has been cut away; a dotted line shows attachment of tympanum.

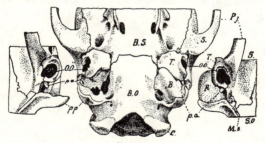

Fig. 505.

Paradoxurus musanga, hinder portion of young skull, ventral and lateral views (from M. Weber, 1927). *B*, Entotympanic part of bulla (removed in left figure); *B.O*, basioccipital; *B.S*, basisphenoid; *C*, occipital condyle; *M.s*, periotic; *OO*, external bony meatus; *pa*, processus postanditivus; *Pj*, processus jugalis; *pp*, processus paroccipitalis; *S*, squamosal; *S.o*, supraoccipital; *T*, tympanic.

the support of the tympanum (Watson, **640**). As Gregory has pointed out, a groove on the hinder surface of the squamosal in many Theromorphs seems to indicate the position of an external auditory meatus, and suggests that the tympanum was deeply sunk and obliquely set below this region of the skull as in mammals (Gregory, 735; Watson, 631a). For Gaupp's view that the tympanic membrane, situated dorsally to the retro-articular process and angular in modern Reptilia, and partly below this level in Mammalia, has been independently evolved in the two groups, there seems to be no justification. How the mammalian condition could have been evolved from the reptilian is clearly shown in the Theromorpha.[1]

THE MIDDLE EAR IN AMPHIBIA

From what has been said above it will be gathered that the various parts of the middle-ear region in Reptilia and Mammalia can be shown to be homologous, and to occupy the same fundamental relative positions in the two groups. Difficulties, however, arise when the comparison is extended to the Amphibia. To begin with, the existing forms are all highly specialised, and the terrestrial Anura alone preserve a tympanum. In the Urodela and Apoda the auditory apparatus is more or less degenerate, and the Eustachian tube, tympanic cavity, and tympanum are lost, no doubt as the result of the burrowing habits of the latter and re-adaptation to aquatic life of the former. Nevertheless, the structures of the middle ear can be shown to be arranged on the same general plan, to develop in much the same way, and to occupy the same morphological position in Amphibia as in Amniota, with one notable exception, the chorda tympani.

The large tympanic membrane of the Anura is supported by a cartilaginous annulus tympani situated behind the quadrate, and attached to the crista parotica and squamosal, Figs. 507, 511-12. Developed apparently as an outgrowth from the quadrate, this annulus is peculiar to the Anura; its phylogenetic origin is unknown (Parker, 558-9; Gaupp, 504).

Extending from membrane to fenestra ovalis is a columella auris of

---

[1] In modern Reptilia the primitive position of the tympanum seems to be vertical and nearly flush with the surface; but in the Crocodilia it sinks somewhat, and is protected by a movable flap. Among the Lacertilia a shallow external meatus is often developed; its opening may be much narrowed. In many families the tympanum is frequently more or less completely obliterated by secondary modifications, especially among burrowing or creeping forms. This may take place by thickening, by overgrowth, or by closure of the external meatus (Versluys, 769). The Chamaeleontids and Amphisbaenids have no tympanic membrane, and the latter have lost the cavity as well. In Sphenodon there is no membrane; and all the Ophidia have lost it, together with the cavity and Eustachian tube. Yet the faculty of hearing is preserved, vibrations being communicated probably from the ground through the lower jaw. In the Class Aves a considerable external auditory meatus is present.

complex structure (Parker, 558, 559 ; Gaupp). It consists of a proximal
element known as the 'operculum' (stapes of Parker), usually cartilaginous, and closing the posterior region of the fenestra but overlapping
its rim. Articulating with the operculum and plugging the anterior
region of the fenestra is the 'plectrum' (Gaupp); from the latter's
swollen base projects a rod whose distal end is continued into an extra-
stapedial cartilage inserted into the tympanum. The rod is usually

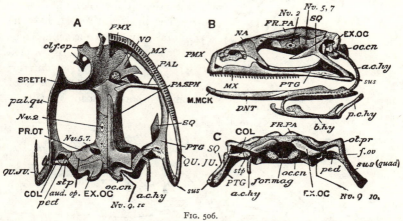

FIG. 506.

Rana temporaria. The skull. A, From beneath, with the investing bones removed on the right
side (left of figure) ; B, from the left side, with mandible and hyoid ; C, from behind, the investing
bones removed at *sus. a.c.hy,* Anterior cornu of hyoid ; *aud.cp,* auditory capsule ; *b.hy,* body of hyoid ;
COL, columella ; *DNT,* dentary ; EX.OC, exoccipital ; *for.mag,* foramen magnum ; *FR.PA,* fronto-
parietal ; M.MCK, mento-Meckelian ; *MX,* maxilla ; *NA,* nasal ; *Nv. 2,* optic foramen ; *Nv. 5, 7,*
foramen for fifth and seventh nerves ; *Nv. 9, 10,* foramina for ninth and tenth nerves ; *oc.cn,* occipital
condyle ; *olf.cp,* olfactory capsule ; *ot.pr,* otic process ; *PAL,* palatine ; *pal.qu,* palatoquadrate ;
PA.SPH, parasphenoid ; *p.c.hy,* posterior cornu of hyoid ; *ped,* pedicle ; *PMX,* premaxilla ; PR.OT,
pro-otic ; *PTG,* pterygoid ; *QU.JU,* quadratojugal ; SP.ETH, sphenethmoid ; *SQ,* paraquadrate ;
stp, stapes ; *sus (quad),* suspensorium (quadrate) ; *VO,* prevomer. (After Howes, slightly altered.) A
minute investing bone, the *septo-maxillary,* which is present above the maxilla, close to the nostril, is
not here represented. (From Parker and Haswell, *Zoology.*)

ossified (meso-stapedial of Parker), and from the extra-stapedial cartilage
comes off a supra-stapedial process (dorsal process ?) which unites with
the paroccipital process and quadrate, Figs. 506-7, 512. The operculum
is almost always present, but the plectrum may be reduced or absent, as
for instance in the adult *Bombinator* and *Phryniscus*, where the annulus
and tympanic membrane are also lost. Many authors have studied the
development of the columella and with varying results (Reichert, 1837 ;
Parker ; Huxley, 738 ; Cope, 711 ; Villy, 771 ; Fuchs, 727 ; Gaupp,
504). The latest and most careful researches seem to have established
that the operculum arises as an independent rudiment in the membrane
closing the fenestra ovalis (Gaupp, 343, 732 ; van Seters, 382 ; Peter, 758).[1]

[1] Nevertheless, many observers believe the operculum to be 'cut
out' of the wall of the auditory capsule. Fusion with the edge of the

Later (during metamorphosis) the plectrum appears in the same way, as a rudiment which joins the edge of the fenestra and grows outwards in a pre-existing strand of tissue to the quadrate and beyond to the tympanum. The outermost region of the extra-stapedial chondrifies separately (Cope, Gaupp).[1]

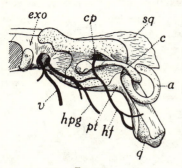

FIG. 507.

Bufo aqua; posterior view of right half of skull (from Parker, 1881, modified with cranial nerves added). *a*, Annulus; *c*, columella; *cp*, cranio-quadrate passage between auditory capsule and quadrate, *q*; *exo*, exoccipital; *hf*, hyomandibular branch of facial; *hpg*, glossopharyngeal; *pt*, ptery-goid; *sq*, squamosal; *v*, vagus.

The modern Urodela, like the Anura, have two elements composing the columella auris : a posterior operculum and an anterior plectrum (columella and stapes of some authors) bearing usually a

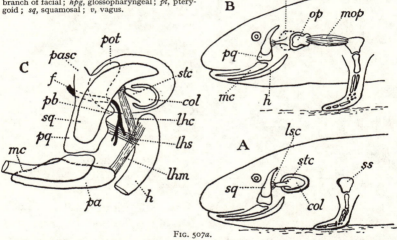

FIG. 507a.

Diagrams illustrating mechanism of communication between exterior and inner ear in Urodela. A, Aquatic larvae and aquatic adults; B, terrestrial adults; C, relation of skeleton, ligaments, and nerve in majority of Urodela (from figures of Kingsbury and Reed, *J. of Morph.*, 1909). Columella auris represented by its stylus, *stc*, its base, *col*, and separate opercular element, *op*. *f*, Hyomandibular branch of facial nerve; *h*, ceratohyal; *lhc*, ligamentum hyo-columellare; *lhm*, l. hyo-mandibulare; *lhs*, l. squamoso-columellare; *lsc*, l. squamoso-columellare; *mc*, Meckel's cartilage; *mop*, musculus opercularis; *pa*, prearticular; *pasc*, processus ascendens; *pb*, pr. basalis; *pot*, pr. oticus; *pq*, palato-quadrate; *sq*, squamosal; *ss*, suprascapula.

fenestra certainly takes place very early ; but the first rudiment of the oper-culum is probably independent. A separate operculum does not appear in Aglossa (van Seters, **382**).

[1] The development of the Eustachian tube and tympanic cavity of the frog is much modified owing to the very specialised structure of the head of the tadpole larva. They arise from the rudiment of the first gill-pouch as a slender

projecting 'stylus' which reaches or is connected by ligament with
the squamosal in early stages but generally with the quadrate later.
Another more ventral ligament unites the base of the columella
with the ceratohyal, Fig. 570 A. A muscle usually stretches from
the operculum to the suprascapula (m. opercularis): it may possibly
be homologous with the stapedial muscle of the Amniota (p. 466).
Both operculum and plectrum may be ossified. The structure and
development of this auditory apparatus has recently been very
thoroughly studied by Kingsbury and Reed (**744**). In the ' Amblystomid
type' (*Amblystoma, Salamandra*), while the operculum remains free, the
plectrum tends to fuse in the
adult with the anterior edge of
the fenestra ovalis, and in Triton
it is vestigial without stylus. In
the ' Plethodont type' the basal
plate of the plectrum fuses with
the operculum (*Necturus, Spelerpes*) and again by means of
a narrow ' isthmus ' with the
edge of the fenestra (*Desmognathus, Amphiuma*). The
muscle is often lost (*Necturus,
Cryptobranchus, Siren*, etc.),
and vibrations probably pass by
way of the lower jaw and squamosal in these aquatic forms

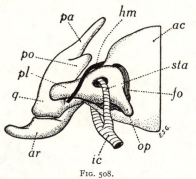

FIG. 508.

Ichthyophis glutinosa, embryo 3·5 cm. long. Reconstruction of auditory region, posterior view. Cartilage
shaded with dots. *ac*, Auditory capsule; *ar*, articular
region; *fo*, fenestra ovalis; *hm*, hyomandibular branch
of facial nerve; *ic*, internal carotid; *op*, operculum;
pa, ascending process; *pl*, plectrum (columella); *po*,
otic process; *q*, quadrate; *sta*, stapedial artery.

when the head rests on the ground. Where present the muscle possibly
communicates vibrations from the suprascapula and fore-limb in terrestrial forms, Fig. 570 A.

Clearly the auditory apparatus is specialised and degenerate in the
Urodela. That the operculum in Urodeles develops from the wall of the
auditory capsule seems to be fully established, though it appears often
to be not so much ' cut out ' of the cartilaginous wall (Huschke, 1824;
Reichert, Parker, Huxley) as to grow forwards from it in the membrane
closing the fenestra ovalis (Stöhr, **386**; Kingsbury, **743**; Reed, **760-61**).
Nevertheless, since the opercular muscle is inserted on the operculum, it
is difficult to believe that this element did not originally belong to a
movable visceral arch. The plectrum, on the other hand, arises separ-

tube which only takes up its definitive position and expands to form the cavity
and membrane at metamorphosis (Villy, **771**; Spemann, **767**; Fox, **723**).

ately in a strand stretching from the membrane of the fenestra to the squamosal and quadrate (Kingsbury and Reed, 744).[1]

The Apoda are provided with a columella consisting of a small operculum and a larger ossified plectrum, pierced by the stapedial artery in *Ichthyophis*, but not in *Hypogeophis* or *Siphonops*, Figs. 508-9, 510. Its distal end is articulated to the quadrate (Sarasin, 763; Marcus, 752; E.S.G.). According to Marcus, the rudiment of the columella of *Hypogeophis* is in the early blastematous stage continuous with the remainder of

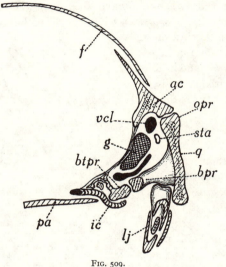

FIG. 509.

Reconstructed thick transverse section of head of *Siphonops braziliensis*, embryo 7 cm. long. *bpr*, Basal process of palatoquadrate; *btpr*, basitrabecular process; *f*, frontal; *g*, geniculate ganglion in cranio-quadrate passage; *lj*, lower jaw; *opr*, otic process of quadrate region, *q*.

the hyoid arch. This is an important observation, since it is the only case of such continuity described in an Amphibian, but confirms the opinion founded on comparative anatomy that the columella of the Amphibian represents the dorsal end of the hyoid arch and is homologous with that of the Amniota. Probably this primitive connexion has been obscured or lost in Urodela owing to the degenerate condition of the columella, and in Anura owing to the great disturbance of this region during larval development.

In attempting to work out the general homology of the Amphibian columella auris it should be remembered that, however degenerate and

[1] Parker (562, 573) believed the plectrum, and Witebsky (772) the whole columella, to be derived from the hyoid arch; but no proof of this origin has yet been brought forward in any Urodele or Anuran.

modified it may be in modern forms, it has probably been derived from a columella similar to that of the primitive Reptilia. In the more primitive Stegocephalia (Embolomeri, Rhachitomi) the stapes has an ossified expanded proximal end fitting into the fenestra ovalis, and pierced apparently for the stapedial artery, and a more slender distal end passing upwards towards the tympanic notch. A process, comparable to the dorsal supra-stapedial process of Reptiles, is attached to the parotic process of the opisthotic and tabular bone of the cranial roof near the tympanic notch, and there is a short extra-stapedial process, which may have been connected by cartilage to the tympanum. But the bony stapes is widely separated from the quadrate bone. The more specialised Stegocephalia have a more slender and rod-like stapes, preserving the same general disposition (Watson, 644; Sushkin, 768).

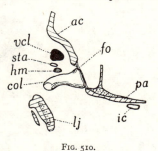

FIG. 510.

Siphonops braziliensis, embryo 7 cm. long. Transverse section through left auditory region. *ac*, Auditory capsule; *col*, columella; *fo*, fenestra ovalis; *hm*, hyomandibular branch of facial nerve; *ic*, internal carotid; *lj*, lower jaw; *pa*, parachordal plate; *sta*, stapedial artery; *vcl*, vena capitis lateralis.

In all the living Amphibia the vena capitis lateralis passes above and the internal carotid passes below the columella and fenestra ovalis, as in Amniota, Figs. 511-12. Moreover, the stapedial artery (facial artery, p. 529) passes through or above the columella on its way forwards through the cranio-quadrate passage to the cavum epitericum and orbit (p. 412). Also, as in Amniotes, the course of the main hyomandibular branch of the facial nerve is dorsal to the columella in Anura and Apoda. Only in Urodela does a serious difficulty arise in the comparison, for here this nerve and its branches pass either ventrally to the ' stylus ' and its ligament, Fig. 570 A, or, as in *Necturus* and *Proteus*, it divides into its jugular and alveolar (internal mandibular, chorda tympani) branches anterior to the ' stylus ', the jugular nerve passing backwards dorsally and the alveolar anteriorly and ventrally. The ligament, then, cannot correspond to the true columella of the other Amphibia and Amniota, Fig. 595, but may after all represent in Urodela its dorsal process and attachment.

The fact already referred to (p. 478), that in Anura not only the

¹ What significance is to be attached to a small cartilaginous rod recently described by Litzelmann (750) behind the quadrate in the larval *Molge*, and claimed by him to represent the hyomandibula, remains to be seen. It passes anteriorly to both the jugular and the internal mandibular nerves, and may possibly be a remnant of the distal region of the columella.

hyomandibular branch of the facial nerve (jugular branch) but also the
internal mandibular branch (chorda tympani, p. 752) pass down behind
the tympanum, while in all Amniotes the chorda tympani passes down in
front of the tympanum, is a serious difficulty in comparing the two groups.
It has already been explained that the chorda tympani in the Amniota

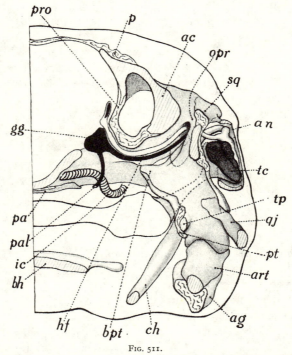

Fig. 511.

Rana temporaria, young ; same reconstruction as in Fig. 512, but cut transversely farther forward
(anterior view), and showing cranio-quadrate passage into which run hyomandibular branch of facial
and accompanying v. cap. lateralis. *bpt*, Basal process ; *gg*, Gasserian ganglion ; *opr*, otic process ;
ptp, pterygoid process ; *pa*, parasphenoid ; *pal*, palatine nerve ; *pro*, pro-otic ; *pt*, pterygoid ; *qj*,
quadratojugal ; *tc*, tympanic cavity exposed (Eustachian tube behind quadrate indicated by dotted
line).

is post-trematic but pretympanic ; in the Anura it is both post-trematic
and post-tympanic, Figs. 507, 512. Gaupp (**732**) concluded that the
tympanum is not homologous throughout the Tetrapoda and has arisen
independently in the two different positions. It is unlikely, however, that
two structures so similar as the tympanum of a frog and of a lizard, each
situated behind the quadrate and having imbedded in it the distal end
of the columella, should have been independently developed. Moreover,
the presence of a tympanic notch and a columella in the Stegocephalia
is clear evidence that these primitive Tetrapods, not far removed from

the common ancestor of both modern Amphibia and Amniota, already possessed an auditory apparatus provided with a tympanum. Rather may we suppose that there has been in the ancestral Amniota a shifting downwards of the tympanum and forwards of the chorda tympani across the tympanum perhaps at a time when this membrane was still thick.[1]

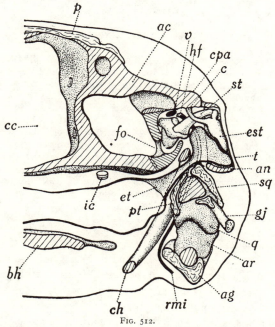

FIG. 512.

Rana temporaria, young. Reconstruction of thick transverse section of auditory region of left side; anterior view. *ac*, Auditory capsule; *ag*, angulosplenial; *an*, annulus; *ar*, articular cartilage; *bh*, basihyal; *c*, connexion between extra-stapedial and cranial cartilages; *cc*, cranial cavity; *ch*, ceratohyal; *cpa*, crista parotica; *est*, extra-stapedial cartilage; *et*, Eustachian tube; *fo*, fenestra ovalis; *gj*, quadratojugal; *hf*, hyomandibular branch of facial nerve; *ic*, internal carotid; *p*, parietofrontal; *pt*, pterygoid; *q*, quadrate cartilage; *sq*, squamosal; *st*, stapes; *rmi*, ramus mandibularis internus; *t*, tympanic membrane; *v*, position of vena capitis lateralis passing through cranio-quadrate passage.

Fully to account for the derivation of the Amniote from the Anuran condition, assuming the latter to be primitive, we should also have to suppose that the chorda tympani shifted dorsally over the end of the extra-stapedial cartilage, Fig. 495, or that the extra-stapedial region has been developed independently in the two groups. On the other hand, it is perhaps more probable that the Anuran condition is secondary.

[1] There can be little doubt that the Eustachian tube and tympanic cavity are homologous throughout the Tetrapoda, as evidenced by their development and innervation (Bender, **705**).

CHAPTER IX

VISCERAL CLEFTS AND GILLS

THE VISCERAL CLEFTS AND GILLS

OF all the characteristic features of the vertebrate phylum few are more fundamental than the gill-slits. Serving primarily for the passage of a respiratory and nutritive current of water from the pharynx to the exterior, they occur not only in the Cephalochorda, Cyclostomata, and Pisces, but also in Amphibia, and in a more or less reduced condition in the embryonic stages of all Amniota. Moreover, their presence in Tunicata, Enteropneusta, and Pterobranchia is evidence that gill-slits were possessed by the common ancestor of all these groups.

The structure of the gills of *Amphioxus* is so well known that it need not here be described in detail ; suffice it to say that each primary slit is subdivided by a tongue bar into two secondary slits and these again by delicate synapticula, and that the whole sieve-like structure is supported by an internal chitinoid cuticular skeleton [1] (Willey, **94** ; Benham, 1893 ; Boveri, **814**). The atrium into which the gill-slits open arises as a ventral invagination of the ectoderm (Lankester and Willey, **792**), and has probably been derived phylogenetically from paired

---

[1] The gills of the Enteropneusta so closely resemble those of Amphioxus that, in spite of Spengel's objections (1893), we seem justified in supposing that in many respects they are strictly homologous (tongue-bars, synapticula, etc.). In a general way the atrium of Amphioxus, of Tunicata, and the outer gill-chambers of Enteropneusta correspond.

longitudinal grooves joining the external gill-apertures. Although the gill-slits of *Amphioxus* are no doubt respiratory, yet their great development and large number are probably specialisations related to another function, namely, the creating by means of their cilia and sifting of a current of water carrying food particles which are caught in mucus secreted by the endostyle. The food material so entangled is driven upwards along the peripharyngeal ciliated grooves and obliquely over the inner surface of the gill-bars to the dorsal groove, whence it passes backwards to the intestine (Orton, 796). In aquatic Craniata the respiratory function is dominant and the gill-bars become provided with thin-walled lamellae ; but even in them gill-rakers are generally present along their inner edges which serve to prevent food from passing out of the pharynx, and in some fishes the bars may be armed with teeth which help in mastication, Fig. 599.

The gill-slits develop by the meeting of paired outgrowths of the endodermal wall of the pharynx with the outer ectoderm, and the piercing of the thin membrane so formed, Figs. 514, 517. Usually corresponding shallower ectodermal ingrowths meet the pharyngeal pouches. When the membrane is pierced the limit between ectoderm and endoderm soon becomes obscured and lost. The gill-slits develop from before backwards, and their number is increased or decreased at the end of the series (though the first pair may be closed, see below, p. 519), and behind the open slits there are usually to be found some incompletely developed owing to the failure of the pouch to reach the ectoderm, or of the closing membrane to become pierced. Speaking quite generally, the number of gill-slits is larger in the lower than in the higher vertebrates. But if we ask how many were present in the ancestral Vertebrate, no definite answer can yet be given. The very large number of slits seen in *Amphioxus*, some 180 pairs in the adult, can hardly be truly primitive. The whole pharynx and its slits have probably been greatly extended in adaptation to the mode of feeding by ciliary action ; indeed, new slits seem to be added throughout life, or at all events so long as the animal continues to grow. Nevertheless, it is probable that the slits were originally more numerous than in modern Pisces, in which there are never more than eight and usually only six pairs of visceral clefts.[1] For in the Cyclostomes, where the first cleft is obliterated in ontogeny (Dohrn, 333, 780 ;

---

[1] In Amphioxus the first gill-slit is closed in development (Willey, 94). The first slit in Craniata is always modified, and usually closed, but may remain as an open spiracle in some Pisces ; it corresponds to the facial nerve, is situated between the mandibular and the hyoid arches, and usually goes by the name of spiracular slit or pouch (p. 519). The more posterior truly respiratory slits are called branchial.

Dean, 778), there may remain as many as 14 pairs of open branchial slits: 7 in Petromyzontia, 6 (and an oesophageal duct) in *Myxine,* and

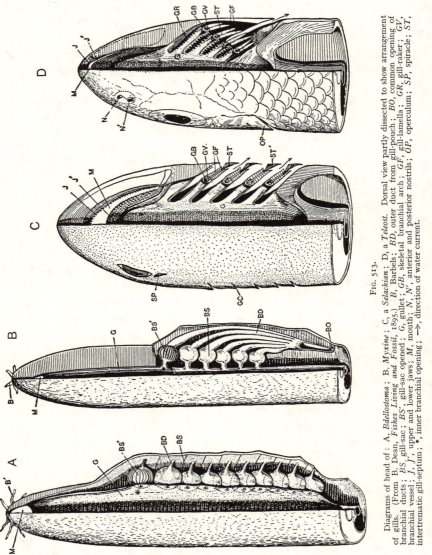

FIG. 513.

Diagrams of head of : A, *Bdellostoma*; B, *Myxine*; C, a *Selachian*; D, a *Teleost*. Dorsal view partly dissected to show arrangement of gills. (From B. Dean, *Fishes Living and Fossil*, 1895.) *B*, Barbels; *BD*, outer duct from gill-pouch ; *BO*, common opening of branchial ducts ; *BS*, gill-sac ; *BS'*, gill-sac opened ; *G*, gullet ; *GB*, skeletal branchial arch ; *GF*, gill-lamella ; *GR*, gill-raker ; *GV*, branchial vessel ; *J, J'*, upper and lower jaws ; *M*, mouth ; *N, N'*, anterior and posterior nostrils ; *OP*, operculum ; *SP*, spiracle ; *ST*, intertrematic gill-septum ; *, inner branchial opening ; →, direction of water current.

6 to 14 (and a duct) in *Bdellostoma*, Figs. 518-19. In the majority of Pisces there are 5 pairs of branchial slits ; but among Selachians there may be more (6 in *Hexanchus* and *Chlamydoselache*, 7 in *Heptanchus*, 6 in

Pliotrema, an aberrant Pristiophorid (Regan, **798**)). The earliest known fossil fish, the Ostracodermi, appear to have usually possessed not more than ten pairs of slits. Traces of a posterior vestigial slit or visceral arch have been described in Selachii by various authors (Hawkes, **689,** Daniel, **674,** in *Heterodontus;* Goodey, **685,** Braus, **673,** K. Fürbringer, **681,** in Notidanids, etc.). Moreover, the suprapericardial body (ultimo-branchial body) is considered to have been derived from a vestigial pouch behind the last branchial slit. In the Tetrapoda, also, four or five visceral pouches are usually developed behind the spiracular (5 in Amphibia, 4 in Aves, 5 in Mammalia). A posterior vestigial sixth pair has been described

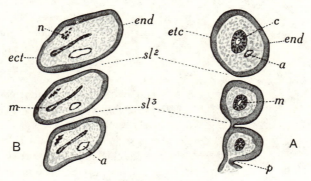

FIG. 514.

Horizontal sections through wall of pharynx of embryo *Scyllium canicula,* cutting gill-bars transversely. A, Stage I; B, stage K. *a,* Vascular arch; *c,* coelomic canal; *ect* and *etc,* ectoderm on outer surface; *end,* endoderm on inner surface; *m,* tube of lateral plate mesoderm in A, giving rise to flattened muscular plate in B; *n,* post-trematic nerve; *p,* endodermal gill-pouch; *sl,* gill-slit.

in Urodela, and even a seventh in Apoda (Marcus, **752**). We may conclude that the original number of slits in the ancestral Gnathostome was not very large, possibly only seven or eight including the spiracular.

Whether the gill-slits were originally segmental is an interesting question. In no adult Vertebrate does branchiomerism correspond strictly with the metamerism of the body as indicated by the myotomes. When in the embryo an endodermal gill-pouch grows outwards it meets and pierces the lateral plate mesoderm to reach the ectoderm (p. 487), and since the primary segmentation has already disappeared in this plate, the exact relation the pouch may have borne to the mesoblastic segments is obscured. Nevertheless, it is probable that they were originally intersegmental in position. At their first appearance the first few pouches alternate with the segmental somites above them; but, owing to differential growth, the great enlargement of the gill-pouches and arches, and their tendency to spread backwards, any such original correspondence is soon lost, especially in Gnathostomes. Strong evidence that the visceral clefts were originally intersegmental is afforded by the nerves ;

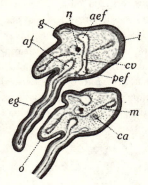

FIG. 515.

Scyllium canicula, embryo 32 mm. long ; portion of horizontal section through wall of pharynx showing two branchial arches cut transversely. *af*, Afferent vessel ; *aef, pef*, anterior and posterior efferent vessels ; *ca*, mesenchymatous rudiment of skeletal arch ; *cv*, cross commissural vessel ; *eg*, external gill-filament ; *g*, anterior gill-lamella ; *i*, inner endodermal surface ; *n*, posttrematic nerve ; *o*, outer ectodermal surface.

for behind each passes the main branch of one segmental (really intersegmental) dorsal-root nerve in regular sequence (see p. 219). This is particularly striking in the larval *Amphioxus* (Goodrich, 1909) and in *Petromyzon*.

Leaving aside for the present the specialised spiracular slit, we find that the branchial clefts are separated from each other by branchial bars (gill-arches), covered by ectoderm on their outer and endoderm on their inner surface, and containing lateral plate mesoderm with a coelomic cavity. This cavity, the remains of the coelom of the lateral plate in this region, persists in *Amphioxus*, Fig. 707, but is soon obliterated in the embryonic stages of Craniata, Figs. 514, 517. From its walls, however, develop branchial muscles (p. 219). In the aquatic gill-bearing Gnathostomes the large branchial pouches are compressed and open into the pharynx by dorso-ventrally elongated slits. The intervening bars are differentiated

into an inner region containing the skeletal arch, and an outer region or septum passing obliquely backwards towards the surface, containing nerves and blood-vessels and bearing the respiratory lamellae. Primitively, no doubt, the visceral clefts all opened separately to the exterior as they still do in modern Selachii, Fig. 513.

The respiratory lamellae of Pisces are typically thin folds extending from

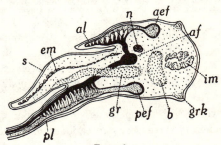

FIG. 516.

Section across gill-bar of *Scyllium canicula*, late embryo 32 mm. long, showing blood supply to lamellae. *aef* and *pef*, Anterior and posterior efferent vessels ; *af*, afferent vessel ; *al*, anterior lamella ; *b*, branchial bar ; *em*, external constrictor muscle ; *gr*, gill-ray ; *grk*, gill-raker ; *im*, adductor branchialis muscle ; *n*, nerve ; *pl*, posterior lamella continued into external filament, cut short ; *s*, gill-septum.

the gill-septum, and set transversely to the axis of the gill-bar, Figs 515-17. They bear on their upper and under surfaces numerous small transverse secondary folds (Riess, **799** ; Plehn, **797** ; Moroff, **793-4**). It is in these secondary folds that the respiratory exchange chiefly takes place. They are thin plates the walls of which consist of a delicate flattened outer ectodermal epithelium and a subepithelial layer of very thin underlying connective tissue ; covered by these epithelia is a middle vascular layer with a network of capillaries through which blood flows from the

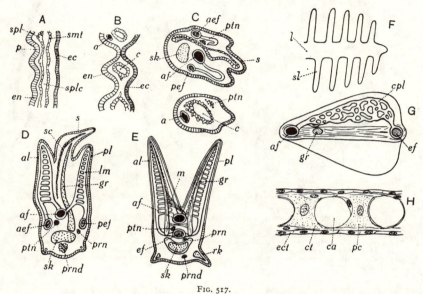

FIG. 517.

Diagrams illustrating structure and development of *gills* of a *Selachian*, A, B, C, D, and a *Teleostean*, E, F, G, H.   A, Horizontal section of wall of pharynx at early stage when endodermal gill-pouches, *p*, have not yet met ectoderm, *ec*, and splanchnocoele in lateral plate is still continuous, *splc* ; B, later stage with lateral plate interrupted (cp. Fig. 514) ; C, still later stage (cp. Fig. 515) ; D and E, diagrammatic transverse sections of gill-bars ; F, section along gill-lamella, *l*, to show disposition of secondary lamellae, *sl* ; G, section across gill-lamella showing circulation in secondary lamella ; H, section of secondary lamella at right angles to surface.   *a*, Embryonic arterial arch ; *aef*, anterior efferent vessel of arch ; *af*, afferent vessel of lamella ; *al*, anterior lamella ; *c*, coelom ; *ct*, cutis ; *ect*, ectoderm ; *ef*, efferent vessel ; *en*, endoderm ; *gr*, gill-ray ; *lm*, capillary network ; *m*, lamellar muscle ; *pc*, pilaster cells ; *pef*, posterior efferent vessel ; *pl*, posterior lamella ; *prn*, pre- and post-trematic nerves ; *prnd*, branchial muscle ; *rk*, gill-raker ; *s*, outer region of septum ; *sk*, skeletal arch ; *smt*, somatoblast ; *spl*, splanchnopleure.

afferent to the efferent limb of the vascular loop.   Joining the opposite walls and in the interstices of the network are peculiar pilaster cells first described by Biétrix (**775**).   This histological structure and the pilaster cells occur not only in Pisces, but also in Cyclostomata, and even in Amphibia (Faussek, **781**), and may be taken as evidence that the gills are homologous throughout the Craniata (p. 504).

The lamellae are disposed in close-set rows on the anterior and posterior faces of a bar ; each row forms a hemibranch, while the front and back row on a bar together make up a holobranch.

We may now briefly review the chief modifications of the gill-clefts,

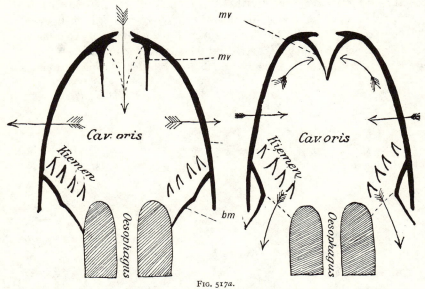

FIG. 517a.

Diagram illustrating the mechanism of respiration in Teleosts (after Dahlgren). A, Phase of inspiration; B, phase of expiration. In both figures the anterior oral part (*cav. oris*) represents a vertical section, and the posterior pharyngeal part enclosing the gills (*Kiemen*) a horizontal section. The arrows indicate the direction of the water-current and pressure, and those passing through the walls of the oral cavity the expansion and contraction of the opercular apparatus. In A, the maxillary and mandibular valves (*mv*) are open, and the branchiostegal membrane (*bm*) closed; in B, this condition is reversed.

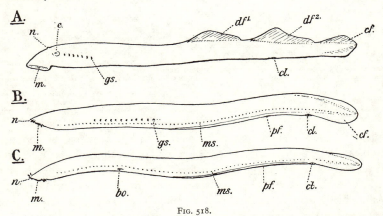

FIG. 518.

A, *Petromyzon fluviatilis*, L., the river Lamprey or Lampern. B, *Bdellostoma Dombeyi*, Lac. C, *Myxine glutinosa*, L., the Hag-fish. *bo*, Branchial opening; *cf*, caudal fin; *cl*, cloacal aperture; df^1 and df^2, first and second dorsal fin; *e*, eye; *gs*, gill-slit; *m*, mouth; *ms*, mucous sac; *n*, nostril; *pf*, preanal fin. (From Goodrich, *Vert. Craniata*, 1909.)

gill-bars, and gills in the various groups of Craniata (Oppel, **795**; Dohrn,

333). The gills of Cyclostomes are highly specialised structures, the branchial lamellae being enclosed in the adult in rounded sacs with narrow ducts leading into them from the alimentary canal, and out of them to the exterior; hence the name Marsipobranchii often applied to this group, Fig. 513. Their external openings are in the form of small rounded pores. Similar pores have been found in some Ostracodermi (Anaspida).

Taking the Myxinoidea first (J. Müller, 1841; Dean, 155, 779),

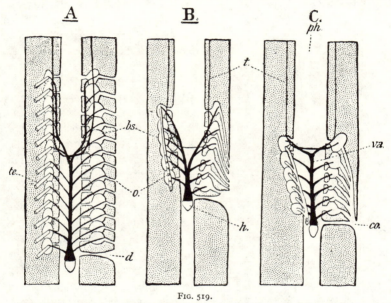

FIG. 519.

Diagram of the gills and their afferent blood system in A, *Bdellostoma* (*Homea stouti*); B, *Paramyxine*; and C, *Myxine* (after Dean). Ventral view. *co*, Common opening of six gill-sacs and oesophageal duct; *bs*, gill-sac; *d*, oesophageal duct; *h*, heart; *o*, external opening of gill-sac; *ph*, pharynx; *t*, outline of rasping 'tongue'; *te*, tube leading to exterior; *va*, ventral aorta. (From Goodrich, *Vert. Craniata*, 1909.)

we find from 6 to 14 pairs of gill-sacs in the genus *Bdellostoma*, and 6 (sometimes 7) in *Myxine*. But while in the more primitive *Bdellostoma* the external ducts open separately, in *Myxine* they combine to a single opening posteriorly. An interesting intermediate condition has been described by Dean in *Paramyxine*, where the pores are closely approximated. The condition in *Myxine* is obviously secondary, Figs. 518-19.

In addition to these gill-sacs there is in all Myxinoids a 'pharyngocutaneous duct' passing on the left side from the pharynx to the exterior behind the last sac; its opening is adjacent or confluent with the last

left gill-pore, Figs. 519-20. It is conjectured that this gill-less duct has
been derived from a visceral cleft. Breathing takes place by the expansion and contraction of the muscular walls of the sacs; water being drawn in through the median ' nostril ', Fig. 518, which leads into a hypophysial canal opening into the pharynx behind, and expelled to the exterior along the expiratory tubes. When the Myxinoid is feeding and its head is buried in its prey, water may still pass in and out by these tubes. It is no doubt to facilitate this respiration that the gill-sacs in Myxinoids are situated very far back and the expiratory tubes lengthened so as to carry the external openings still farther from the head. This quite exceptional disposition of the gills is secondary. The visceral clefts develop in the usual position ; but, excepting for the first pair which remain stationary and later disappear, they shift backwards (together with the heart, etc.) at a later stage (Dean, 778).

The Petromyzontia preserve the primitive separate external pores, but in the adult the sacs open internally into a median branchial tube situated below the oesophagus. This branchial tube is blind behind and opens in front into the pharynx just behind the valve derived from the larval velum. In respiration water may enter the mouth ; but when this is not possible, owing to the oral sucker being fixed or closed, water is pumped in and out of the

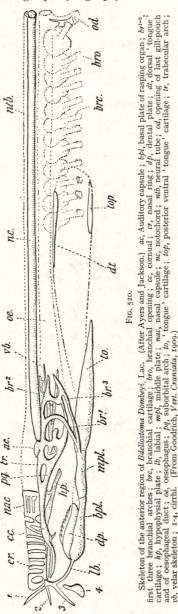

FIG. 520.

Skeleton of the anterior region of *Bdellostoma Dombeyi*, Lac. (After Ayers and Jackson.) *ac*, Auditory capsule ; *bpl*, basal plate of rasping organ ; *br*[1-3], first three branchial arches ; *brc*, branchial cartilage ; *bro*, branchial opening ; *cc*, cornual ; *cr*, nasal ring ; *dp*, dental plate ; *dt*, dorsal ' tongue ' cartilage ; *hp*, hypophysial plate ; *lb*, labial ; *mpl*, middle plate ; *nac*, nasal capsule ; *nc*, notochord ; *ntb*, neural tube ; *od*, opening of last gill-pouch and of oesophageal duct ; *oe*, oesophagus ; *pq*, suborbital arch ; *to*, ' tongue ' cartilage ; *top*, posterior ventral ' tongue ' cartilage ; *tr*, trabecular arch ; *vb*, velar skeleton ; 1-4, cirrhi. (From Goodrich, *Vert. Craniata*, 1909.)

contractile sacs through the expiratory tubes. The Ammocoete larval

stage possesses gill-pouches which open in the usual manner by wide
slits into the pharynx, and it is at metamorphosis that the blind

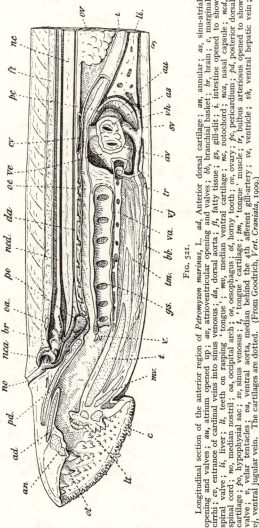

FIG. 521.

Longitudinal section of the anterior region of *Petromyzon marinus*, L. *ad*, Anterior dorsal cartilage; *an*, annular; *as*, sinu-atrial opening and valves; *au*, atrium opened up; *av*, atrioventricular opening and valves; *bb*, branchial basket; *br*, brain; *c*, marginal cirrhi; *cv*, entrance of cardinal veins into sinus venosus; *da*, dorsal aorta; *ft*, fatty tissue; *gs*, gill-slit; *i*, intestine opened to show spiral valve; *li*, liver; *lt*, teeth on rasping 'tongue'; *mv*, median ventral cartilage; *nc*, notochord; *nca*, nasal capsule; *ncd*, spinal cord; *no*, median nostril; *oa*, occipital arch; *oe*, oesophagus; *ot*, horny tooth; *ov*, ovary; *pc*, pericardium; *pd*, posterior dorsal cartilage; *po*, hypophysial sac; *t*, 'tongue' cartilage; *tm*, 'tongue' muscle; *tr*, bulbus arteriosus opened to show valve; *v*, velar tentacles; *va*, ventral aorta, median behind the 4th afferent gill-artery; *ve*, ventricle; *vh*, ventral hepatic vein; *vj*, ventral jugular vein. The cartilages are dotted. (From Goodrich, *Vert. Craniata*, 1909.)

respiratory tube becomes separated off below from the oesophagus which
extends forwards above, Figs. 521-3.

The gill-lamellae of Cyclostomes resemble those of Pisces in having
transverse secondary vascular folds across them on both sides, and are

set in an almost complete ring inside the sacs of the adult, interrupted, however, dorsally and ventrally by a fold. In the Ammocoete they are normally disposed as hemibranchs, one on each side of each gill-

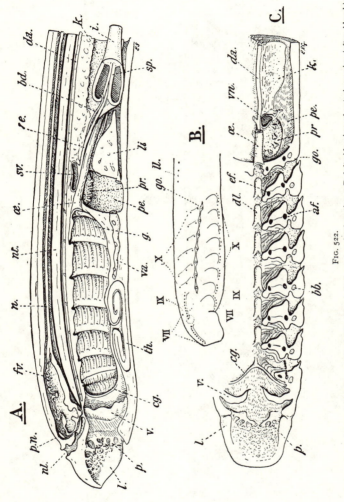

Fig. 522.

Ammocoete larva of *Petromyzon fluviatilis*, L. A, Enlarged sagittal section; B, left-side view of anterior region (after Alcock); C, horizontal longitudinal section passing through the gills, etc., represented on the right side only, seen from below. *af*, Afferent branchial vessel; *bb*, branchial basket; *bd*, bile-duct; *cg*, ciliated groove; *da*, dorsal aorta; *dl*, lobe of dorsal lamina; *ef*, efferent branchial vessel; *fv*, cavity of brain; *g*, gill; *go*, gill-opening; *i*, intestine; *l*, lip; *li*, liver; *ll*, trunk lateral-line organs; *k*, kidney (mesonephros); *n*, nerve-chord; *nl*, median nostril; *nt*, notochord; *oe*, oesophagus; *p*, papilla; *pe*, pericardium from which the heart has been removed; *pn*, pineal eye; *pr*, pronephros; *sp*, spiral valve exposed in intestine; *sv*, entrance to sinus venosus; *th*, thyroid gland; *v*, velum; *va*, ventral aorta; *ve*, intestinal vein; *vn*, sinus venosus; VII, IX, and X denote lateral-line organs supplied from the facial, glossopharyngeal, and vagus nerves. (From Goodrich, *Vert. Craniata*, 1909.)

pouch. The gill-bars, however, differ from those of Pisces: they project deeply into the cavity of the pharynx, and become differentiated into outer and inner regions; the skeletal arch being in the outer region close to the skin, and the respiratory lamellae and blood-vessels being borne by the inner region, Figs. 522-3-4 (see p. 504). This exceptional and characteristic relative position of arch and gill is preserved in the adult, where the arches combine in Petromyzontia to form a delicate 'branchial basket', and in Myxinoids remain as vestigial cartilages near the external gill-openings (separate in *Bdellostoma,* and combined to a complex cartilage in *Myxine*), Fig. 520.

In Pisces both the internal and external gill-openings are dorso-

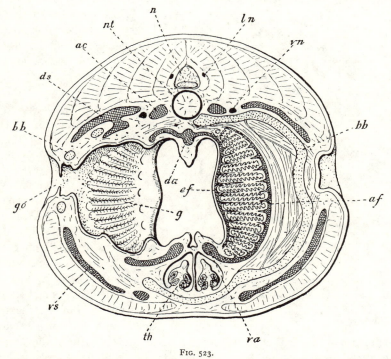

FIG. 523.

Transverse section of the gill region of an Ammocoete larva, somewhat diagrammatic. (Partly after Alcock.) *ac,* Anterior cardinal vein ; *af,* afferent artery ; *bb,* branchial basket ; *da,* dorsal aorta ; *ds,* dorsal blood-sinus ; *ef,* efferent artery ; *g,* gill-lamella ; *go,* gill-opening ; *ln,* lateral-line nerve ; *n,* nerve-cord ; *nt,* notochord ; *th,* thyroid gland ; *va,* ventral aorta ; *vn,* vagus nerve ; *vs,* ventral blood-sinus. (From Goodrich, *Vert. Craniata,* 1909.)

ventrally elongated and narrow, the pouches being compressed. In Selachii a hemibranch of gill-lamellae occurs on the anterior and posterior side of each branchial arch except the last, which is gill-less, Figs. 513, 532. A posterior hemibranch is also present on the hyoid arch (and on the mandibular arch, see below, p. 519). Water is taken in at the mouth and expelled through the slits ; but in bottom-living forms like the Rajiformes, where the branchial slits are ventral, it may enter through the enlarged spiracle provided with a movable valve (p. 519). The gill-bar is differentiated into an inner region containing the skeletal visceral arch, and a thin outer region containing the blood-vessels, etc., and bearing the gill-lamellae ; this gill-septum being complete passes beyond the lamellae into the outer denticle-bearing skin surrounding the opening, and is supported by a single posterior row of cartilaginous gill-rays (p. 398). In the young stages the external openings are elongated dorso-ventrally, but in the adult they often become relatively shortened.

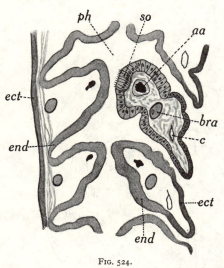

FIG. 524.

Petromyzon, young ammocoete larva ; slightly oblique horizontal section through pharynx and gill-slits, passing through openings on right and above them on left. *aa,* Arterial arch ; *bra,* skeletal arch ; *c,* coelomic canal ; *ect,* ectoderm ; *end,* endoderm ; *ph,* lumen of pharynx ; *so,* sense organ.

The more primitive forms, however, preserve wider openings over which the edge of the septum extends as a backwardly directed valvular flap particularly well-developed in *Chlamydoselachus,* Fig. 27.

The enlargement of this flap on the hyoid bar leads to the condition seen in Holocephali, where it forms a large operculum covering all the branchial slits. The gills of these specialised Elasmobranchs are built on the same plan as those of Selachii (there are a posterior hyoidean hemibranch, holobranchs on the first three branchial bars, and an anterior hemibranch on the fourth), but consequent on the development of the operculum the outer portions of the gill-septa are reduced. Cartilaginous gill-rays remain in the opercular flap, but are lost on the branchial bars, Figs. 428, 444. This operculum has, no doubt, been developed independently of that characteristic of the Osteichthyes. The provisional embryonic and larval external gills of Elasmobranchs and other forms will be dealt with later (p. 501).

All the Osteichthyes are primitively provided from the hyoid bar with a large opercular fold strengthened by opercular bones and covering all succeeding branchial slits, which open into a subopercular branchial chamber. Water is taken in at the mouth, passed through the slits, and expelled from the chamber into which project the gill-lamellae, through an opening now bounded behind by the dermal pectoral girdle. Appropriate movements of the branchial bars and operculum, helped by breathing valves or folds developed in many of the higher Teleostomes on the inner margin of the upper and lower jaws, propel the respiratory current backwards, Fig. 517a. This current may be strong and used in locomotion.

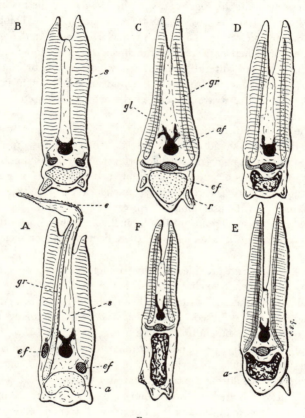

FIG. 525.

Sections across the gill-arch of A, *Mustelus*; B, *Ceratodus*; C, *Acipenser*; D, *Lepidosteus*; E, *Salmo*; F, *Polypterus*. *a*, Skeletal arch; *af*, afferent artery (black); *e*, septum reaching external surface; *ef*, efferent artery (cross hatched); *gl*, gill-lamellae; *gr*, supporting gill-ray; *r*, gill-raker; *s*, septum, largest in A, and smallest in E. Anterior lamellae to the right. (From Goodrich, *Vert. Craniata*, 1909.)

The septa on the branchial bars are reduced and never project beyond the lamellae, Fig. 525. Probably also on account of the great development of the operculum the hyoidean posterior hemibranch is retained only in *Ceratodus,* Chondrostei, and *Lepidosteus.* A mandibular pseudobranch is usually present (p. 519).

Among living Dipnoi the respiratory function of the gills seems to be on the decline and is supplemented by that of the lung-sacs (so-called air-bladder, p. 594), an adaptation to life in rivers apt to become dried up or foul in certain seasons of the year. *Ceratodus* retains the most primitive gills (four holobranchs on the first four branchial bars and a hyoidean pseudobranch), Figs. 534, 535 ; in *Lepidosiren* the hyobranchial or first branchial slit is closed and there are only three holobranchs. *Protopterus*, in which the gills are much reduced, has none at all on the first and second branchial bars, a holobranch on the third and fourth, and on the fifth bar an anterior hemibranch which is possibly merely an extension of the one in front. The hyoidean hemibranch is here a true gill. No gill-rays are present, and the lamellae extend beyond the septum.

In the Teleostomi the branchial bars become slender, and the slits between them narrow and much elongated ; the septum is progressively reduced, while the pointed lamellae project freely into the branchial chamber. Two rows of gill-rays occur on each arch, extending not into the septum but into the respiratory lamellae, Figs. 517, 525. Similar rays may occur in the pseudobranch (p. 520).

Polypterus has no hyoidean hemibranch, and the fourth branchial arch has an anterior hemibranch only, there being no slit or skeletal arch behind it.

The Chondrostei are provided with five branchial slits, four holobranchs, and a hyoidean hemibranch. Five slits and four holobranchs on the branchial bars are found in Amioidei, Lepidosteoidei, and Teleostei (with few exceptions in specialised forms). *Lepidosteus* alone among Holostei preserves the hyoidean hemibranch (see below, p. 517).

The primitive Tetrapoda no doubt inherited true gills from their fish-like ancestors, which gills served for respiration at all events in the aquatic larval stage during the transition to terrestrial life. It is probable that they also inherited an operculum, though no trace of opercular bones has yet been found in any fossil or living Tetrapod. Traces of four branchial arches have been described in the young of several Stegocephalia. That they were similar to the gill-bearing arches of fish, and were even retained as such in some of the adult Branchiosauridae, is evidenced by the disposition of the fossilised gill-rakers (Credner, 484 ; Fritsch, 23 ; Bulman and Whittard, 465a). Among modern Tetrapods internal gills occur only in the tadpole larvae of Anura in the form of two rows of branching filaments on the bars separating four open gill-slits, Fig. 541 ; hyoidean membranous opercular folds cover over the gill region and even the developing pectoral limbs. Paired branchial openings are formed in Aglossa, but in most other Anura where the two chambers communicate below only one opening remains on the left side until the fore-limbs emerge. In the young of other modern Amphibia there are no internal gills and no opercular folds of any size, and the extensive operculum of Anura is probably to a great extent a new formation.

The four branchial slits pierced in Anura and Apoda are closed in the adult where respiration is pulmonary and cutaneous (p. 547); but among the Urodela which are readapted to an aquatic adult life some of them may remain open (third in *Menopoma* and *Amphiuma*; second and third in Proteidae; first, second, and third in Sirenidae).

The Amniota have lost all trace of gills as respiratory organs, and the slits themselves open for but a short time in Reptiles and Birds, and not

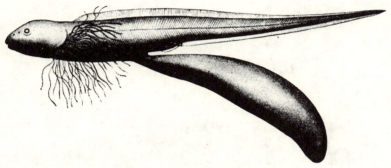

FIG. 526.
Larva of *Gymnarchus niloticus*; tenth day (from Kerr, *Embryology*, 1919).

at all in most Mammals. The embryonic vestigial slits may be marked externally by grooves, lodged in a deep lateral depression called the cervical sinus, which later is overgrown and closed by the hyoid arch and a fold from above and behind.

External Gills.—In addition to the true or internal gills there may be developed provisional superficial gills. These external gills are of two kinds, which may be distinguished as true external gills and external gill-filaments, Figs. 146, 515-16, 526. The latter occur in the early stages of all Elasmobranchs, and are long filamentous prolongations of the outer tips of the young internal gill-lamellae of all the posterior hemibranchs. Each filament is supplied with a vascular loop passing from the afferent to the efferent branchial vessel, and floats in the albuminous fluid within the egg-case. These filaments probably serve not only for respiration, but also for the absorption of food-material. They disappear when the fish hatches. Similar external gill-filaments have been found in certain larval Teleostomes (Chondrostei, some Teleostei, and especially Mormyridae: Budgett, **776**; Assheton, **774**).

True external gills occur in the embryonic or free larval stages of Dipnoi (except *Ceratodus*), of Polypterini, and of Amphibia, Figs. 527-8, 541-2. They usually develop early, even before the slits are pierced and before the opercular fold is formed, as outgrowths from the dorsal outer surface of one or more of the gill-bars. They contain vascular loops derived from the aortic arches, and are covered by ectodermal epithelium sometimes ciliated. Usually they are provided with muscles, are movable and retractile. Four pairs of pinnate external gills are developed on the four branchial bars of *Protopterus* and *Lepidosiren* (Budgett, 776 ; Kerr, 790). When the operculum develops they become crowded together above

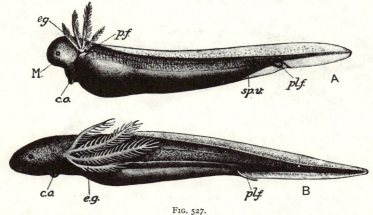

Fig. 527.

Stages in the development of *Lepidosiren paradoxa*. A, Stage 33 ; B, 35. *c.o*, Cement organ ; *e.g*, external gill ; *M*, mouth ; *p.f*, pectoral fin ; *pl.f*, pelvic fin ; *sp.v*, spiral valve of intestine. (From Kerr, *Embryology*, 1919.)

its posterior border. *Protopterus* is remarkable in that it usually retains the three posterior external gills in a reduced condition even in the adult, Fig. 535. The larva of *Polypterus* has only one pair of external gills (Steindachner, 1869 ; Budgett, 10, 776 ; Kerr, 791). They are large, pinnate, with a strong axis supported by a short cartilaginous ray, and are developed from the hyoid bar (Kerr). The external gills of Amphibia are usually purely larval organs (Anura and Apoda), but the Urodela being more or less readapted to an aquatic life tend to retain them in the adult as important organs of respiration. In the Perennibranchiata, then, they develop into permanent large arborescent gills on the first three branchial bars (Clemens, 777 ; Boas, 813 ; Maurer, 849).[1]

[1] A pair of slender processes, known as 'balancers', occur on the head of the larva of many Urodela (Amblystomidae, Salamandridae, and Hynobiidae), and were first described by Rusconi, 1821. They have been thought

The origin and morphological significance of the true external gills are still somewhat obscure. For Kerr's suggestion that they are ancient organs which preceded and perhaps even gave rise to the internal gills

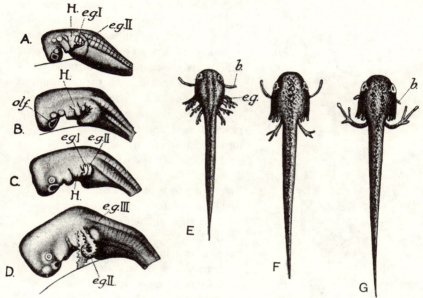

FIG. 528.

A-D, *Hypogeophis* embryos showing development of the external gills. (After Brauer, 1899.) *e.g*, External gill; *H*, hyoid arch; *olf*, olfactory organ. The rounded knobs seen projecting in B from the hyoid arch, and also from the mandibular arch in front of it, are possibly external gill rudiments which do not go on with their development. E-G, Three stages in larval development of a newt (*Triton taeniatus*) as seen from above. (After Egert, 1913.) *b*, Balancer; *e.g*, external gill of first branchial arch. In Fig. A what looks like a posterior external gill is the pectoral limb. In Figs. B and C the external gills have been cut away leaving only their basal stumps. (From Kerr, *Embryology*, 1919.)

there seems to be little evidence, since they are found neither in Cyclostomes nor in Elasmobranchs. Nevertheless, they possibly were present as larval organs in the common ancestor of the Osteichthyes and

by some to represent hyoidean external gills, but for this view there is no good evidence. The balancers have recently been studied in detail by Harrison (**785**). They are supporting and adhesive processes, are covered at their tips by glandular epithelium and strengthened by a thick basement membrane; a nerve runs up them and they contain no muscles. The blood-supply is derived from a branch of the hyoidean artery, but is returned not to the aorta but to the jugular vein (Maurer, **849**). The balancers are probably homologous with the adhesive organs of larval Anura. They develop early on the mandibular arch, are indeed borne on a knob of the palatoquadrate cartilage, and drop off when the pectoral limbs are sufficiently developed to support the head, Fig. 528.

Tetrapoda, and Bulman and Whittard have recently described three pairs of external gills in *Branchiosaurus* from the Permian (**465a**), a Stegocephalian apparently readapted to an aquatic life like the modern Urodele.

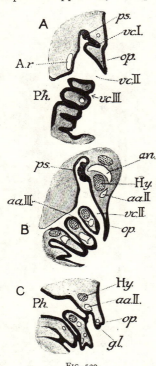

FIG. 529.

A, B, Horizontal sections through salmon embryos explaining position of pseudobranch on inner surface of operculum. (After Goette, 1901.) *A.r*, aortic root; *aa*, aortic arch; *an*, anastomotic vessel connecting aortic arches I and II; *Hy*, hyoid arch; *op*, operculum; *Ph*, cavity of pharynx; *ps*, pseudobranch; *vc*, visceral cleft. C, Horizontal section through branchial region of young *Acipenser* showing the ectodermal origin of the gill-lamellae. (After Goette, 1901.) *aa*, Aortic arch; *g.l*, rudiment of gill-lamella; *Hy*, hyoid arch; *op*, operculum; *Ph*, cavity of pharynx (From Kerr, *Embryology*, 1919.)

Following Boas (**812-13**) and Gegenbaur (**170**) we may consider the external gills as specialised dorsal gill-lamellae, belonging originally to the so-called internal gills, which have grown outwards together probably with a portion of the gill-septum forming the axis. These external gills would, then, be specialised regions of the gills enlarged and precociously developed for larval respiration.

Two further questions remain to be discussed concerning the gill-lamellae. Are they products of the ectoderm or of the endoderm, and are they homologous throughout the Craniata?

Goette (**782, 783**), having studied their development in various groups, concluded that they are of endodermal origin in Cyclostomes, but that they are of ectodermal origin in all other Pisces and in Amphibia, with the exception of the spiracular gill, which he believed to be endodermal. These conclusions have not been generally accepted and have given rise to much controversy. That Goette was mistaken in attributing a different origin to branchial and spiracular lamellae there can be little doubt; but whether in the Elasmobranchii they are covered by ectoderm or endoderm it is difficult to determine, since before they appear the slits have been pierced and the limit between the two germ-layers has already vanished.

Dohrn, while maintaining that the most important constituent of the gill is the blood-vessel, held that the lamellae are chiefly covered with ectoderm. Certainly in Teleostomes, where the lamellae sprout outwards before the membranes closing the slits have been broken through, as Goette showed in *Acipenser*, there can be no doubt that they are covered by ectoderm, Figs. 528-9. Kellicott (789) describes the ectoderm as growing inwards and covering the lamellae in *Ceratodus*. Even more obviously does ectoderm cover the external gills of Pisces and Amphibia. Greil, however, attempted to prove that both layers covered the gill-lamellae of Dipnoi and Amphibia, endodermal cells extending outwards beneath the ectodermal. This view has not been supported by later observers (Kerr, 840 ; Jacobshagen, 786).

But, while it may be considered as established that the gill-lamellae of all Gnathostomes are generally and probably always covered by ectoderm, and are homologous organs, there remains considerable doubt with regard to those of the Cyclostomes. In this group, as already explained, they are situated more deeply on the inner region of the gill-bars and would appear to develop within the endodermal area.[1]

On this account a sharp distinction is sometimes drawn between the ' endobranchiate ' Cyclostomata and the ' ectobranchiate ' Gnathostomata (Goette, 783 ; Sewertzoff, 701).

---

[1] Nevertheless, the evidence on this point is still uncertain; further investigation may well show that even here there is an early ingrowth of ectoderm, and that the gill-lamellae of Cyclostomata are wholly homologous with those of Pisces (see p. 496).

THE VASCULAR SYSTEM OF THE HEAD AND BRANCHIAL REGION

General Plan of the Vascular System.—No description of the gills and gill-bars would be complete without an account of their blood-vascular supply. The general plan of the vascular system of a primitive Craniate is as follows. The venous blood of the alimentary canal (stomach and intestine) is collected into a median longitudinal subintestinal vein passing forwards to the liver where it breaks up

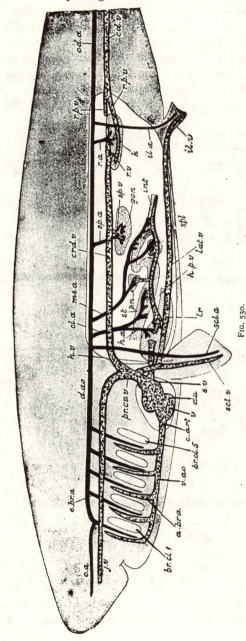

FIG. 530.

Diagram of the vascular system of a fish. Vessels containing aerated blood red,[*] those containing non-aerated blood blue.[**] *a.br.a*, Afferent branchial artery; *au*, atrium; *br.cl*, 1-5, branchial clefts ; *c.a*, carotid artery ; *c.art*, conus arteriosus ; *cd.a*, caudal artery ; *cd.v*, caudal vein ; *cl.a*, coeliac artery ; *crd.v*, cardinal vein ; *d.ao*, dorsal aorta ; *e.br.a*, efferent branchial artery ; *gon*, gonad ; *h.a*, hepatic artery ; *h.p.v*, hepatic portal vein ; *h.v*, hepatic vein ; *il.a*, iliac artery ; *il.v*, iliac vein ; *int*, intestine ; *j.v*, jugular vein ; *k*, kidney ; *lat.v*, lateral vein ; *lv*, liver ; *ms.a*, mesenteric artery ; *pm*, pancreas ; *pr.cv.v*, pre-caval vein ; *r.a*, renal artery ; *r.p.v*, renal portal vein ; *r.v*, renal vein ; *scl.v*, subclavian vein ; *sp.a*, spermatic artery ; *spl*, spleen ; *sp.v*, spermatic vein ; *st*, stomach ; *s.v*, sinus venosus ; *v*, ventricle ; *v.ao*, ventral aorta. (From Parker's *Elementary Biology*.)

*Instead of red, black.

**Instead of blue, coarsely stippled.

into a network of small vessels. These join again to form the hepatic veins carrying the blood to the posterior chamber of the heart or sinus venosus. The venous blood from in front of the heart is collected by cerebral and segmental veins from the head and body-wall into paired longitudinal anterior cardinal veins, and that from behind the heart by segmental veins into paired longitudinal posterior cardinal veins. The latter receive venous blood from the tail, body-wall, kidneys, limbs, etc. On either side the anterior and posterior cardinals, running dorsally to

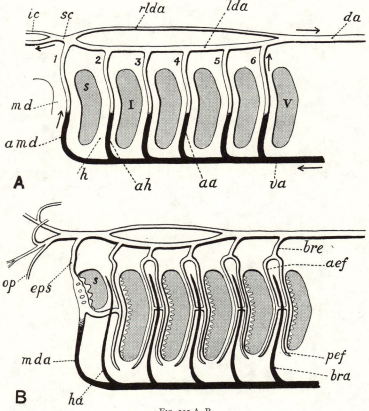

FIG. 531 A, B.

Diagrams illustrating development and fate of primitive embryonic arterial arches in a *Selachian*; left-side view completed. Afferent branchial vessels black, efferent vessels and other arteries white. (Partly from works of Dohrn, Allis, de Beer, and Sewertzoff.) A, Earliest stage; C, latest stage. 1-6, Primitive arches; *I-V*, branchial gill-slits; *S*, spiracular slit   *aef*, Anterior efferent; *ah*, afferent of hyoid bar; *amd* and *mda*, afferent of mandibular bar; *aps*, afferent pseudobranchial cross anastomosis; *bra*, afferent branchial vessel; *bre*, epibranchial artery; *ce*, cerebral; *co*, cross

the coelom below the notochord, meet to form a ductus Cuvieri which passes across by the septum transversum to the sinus venosus. Moreover,

in the Dipnoi and Tetrapoda, the venous blood from the posterior region
of the body-wall and from the kidneys is diverted into a new vein, the

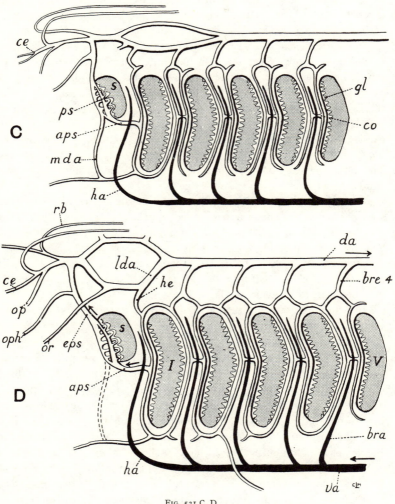

FIG. 531 C, D.

commissural vessel; *da*, median dorsal aorta; *eps*, efferent pseudobranchial (dorsal part of 1);
gl, gill-lamellae; *h*, hyoid bar; *ha*, afferent, and *he*, efferent vessel of hyoid bar; *ic*, internal carotid;
lda, left lateral dorsal aorta; *md*, mandibular bar; *mda*, its vascular arch; *op*, optic; *oph*,
ophthalmic; *or*, orbital; *pef*, posterior efferent; *ps*, mandibular pseudobranch; *rb*, posterior
cerebral; *rlda*, right lateral aorta; *sc*, sinus cephalicus; *va*, ventral aorta.

vena cava inferior, passing downwards and forwards through the liver
to join the hepatic veins and open into the sinus venosus. Thus in all

Craniata the venous blood from both the splanchnic and somatic systems reaches the heart and is pumped forward along a median longitudinal ventral aorta which bifurcates behind the thyroid gland. The ventral aorta gives off at intervals paired aortic arches which run up the visceral arches to join the longitudinal dorsal aorta, Fig. 530.

The aorta develops from paired rudiments which fuse below the notochord to a median dorsal aorta behind, but remain to a varying extent separate in the branchial region and head. The aorta distributes oxygenated 'arterial' blood to the body generally, including the head region, limbs, alimentary canal, kidneys and gonads, and gives off paired segmental arteries to the body-wall all along its course. All the vessels carrying blood to the heart are called veins, and all the vessels carrying blood away from the heart to all parts of the body are called arteries. The blood passes from the arterial system to the venous by minute capillary vessels.

The Aortic Arches in Cyclostomes and Fishes.—There are developed six pairs of primary aortic arches, one in front of each visceral cleft, in the embryo of all Craniata (with the exception of the Cyclostomes and Selachians with a larger number of visceral clefts, in which an additional arch occurs for each slit). But of these six, the second in the hyoid bar, and especially the first in the mandibular bar, become greatly modified, as will be explained later (p. 514).

It is important to notice that in Dipnoi, *Polypterus*, and *Amia* the so-called air-bladder is supplied by a 'pulmonary artery' from the last epibranchial artery (sixth aortic arch) as in all Tetrapods (see p. 583).

In all the gill-bearing visceral arches of Pisces the primary aortic arch becomes interrupted in such a way that the blood has to filter through the gill-lamellae where it is oxygenated, Fig. 533. Afferent and efferent branchial vessels are thus formed, the former bringing venous blood from the ventral aorta, the latter taking arterial blood to the dorsal aorta, and the two communicate only by a system of delicate loops in the lamellae of the gills. Goette (782-3) showed that the development of the afferent and efferent vessels takes place differently in Elasmobranchs and in Teleostei ; the greater part of the original arch forming the afferent vessel in the first group and the efferent vessel in the second.

In all Pisces the afferent vessel passes up the bar outside the efferent vessel, and the skeletal arch lies inside all these vessels. In each bar the afferent vessel is always single in Teleostei ; but in Elasmobranchs and Dipnoi there are two efferent vessels, one to each row of lamellae, Fig. 525. Moreover, the dorsal efferent arteries (epibranchial arteries) of adult Selachii owing to secondary shifting correspond no longer to their gill-bars but to the gill-slits (p. 514), whereas in other fishes they retain their original position.

In the embryonic branchial bar of a Selachian (Goette, 783; Dohrn, 333; Sewertzoff, 869) the anterior and posterior efferent branchial vessels develop separately parallel to the primary continuous aortic arch with which they become connected by capillary loops; the primary arch then becomes interrupted dorsally and its longer ventral portion is converted into the afferent vessel. The newly formed efferent vessels having joined the shorter dorsal portion, this now functions as the epi-branchial artery carrying oxygenated blood to the dorsal aorta (about

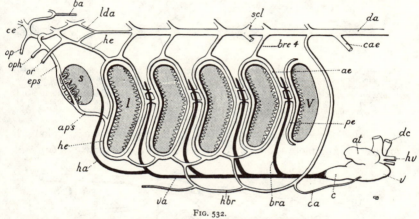

FIG. 532.

Diagram of branchial circulation of *Selachian*, left-side view completed.   *at*, Atrium; *ba*, basal artery; *c*, conus; *ca*, cardiac; *cae*, coeliac; *dc*, ductus Cuvieri; *hbr*, hypobranchial; *he*, posterior efferent of hyoid bar; *hv*, hepatic veins; *scl*, subclavian; *v*, ventricle. Other letters as in Fig. 531.

the middle of the bar transverse connexions early arise between the two efferent vessels). A transverse section of the branchial bar of an Elasmo-branch, then, shows two inner efferent vessels, each joined to the single outer afferent vessel by a series of loops passing into the lamellae, Figs. 517, 533. The transverse section of a Teleostean gill-bar, however, shows only two branchial vessels, an outer afferent and an inner efferent, connected by a double series of loops passing into the anterior and posterior gill-lamellae. In development it is found that here the primitive aortic arch is interrupted near its ventral base, and that while the greater part of the arch is converted into the efferent branchial vessel, the afferent vessel is a new formation joined to the base of the aortic arch. There is thus an important difference between the gill-vessels of these two groups of fishes; but intermediate conditions occur in the lower Teleostomes. In *Acipenser* the upper region of the gill-bar conforms to the Teleostean

and the lower to the Elasmobranch type.   The aortic arch is here interrupted about half-way ; its dorsal portion is continued ventrally into

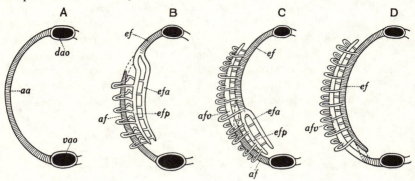

FIG. 533.

Diagrams illustrating development of adult branchial vessels in various fishes. A, Original
continuous embryonic arch cross-lined ; B, Selachian ; C, intermediate form such as *Acipenser* ; D,
Teleost. Newly developed vessels, white.   In B, C, D original arch interrupted.   *aa*, Embryonic
vascular arch ; *af*, afferent ; *afv*, newly formed afferent ; *dao*, dorsal aorta ; *ef*, efferent epibranchial ;
efa, anterior efferent ; *efp*, posterior efferent ; *vao*, ventral aorta ; small loops pass in branchial
lamellae from afferent to efferent vessels.

two newly formed efferent vessels, and its ventral portion is prolonged
upwards to complete the afferent vessel, Fig. 533 (Sewertzoff, 869).   A

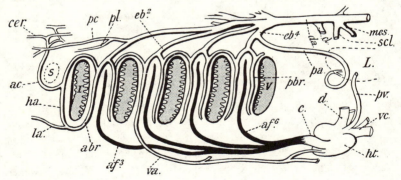

FIG. 534.

Diagram of branchial circulation of *Ceratodus forsteri* (chiefly from W. B. Spencer, 1892).   *I-V*,
Five branchial slits with gill-lamellae.   *abr*, Anterior efferent vessel of first branchial bar ; *ac*, efferent
'pseudobranchial', portion of mandibular ; *af³⁻⁶*, afferent vessels of four branchial bars, corresponding to original aortic arches 4-6 ; *c*, conus ; *cer*, cerebral artery ; *cl*, coeliac artery ; *d*, left ductus
Cuvieri ; *eb²* and *eb⁴*, second and fourth epibranchial arteries ; *ha*, hyoid artery, derived from second
aortic arch ; *L*, lung (air-bladder) ; *la*, lingual artery ; *mes*, mesenteric artery ; *pa*, left pulmonary
artery ; *pbr*, posterior efferent artery of fourth branchial bar ; *pc*, orbital artery ; *pl*, palatine artery ;
pv, pulmonary vein ; *s*, position of closed spiracle ; *scl*, subclavian artery ; *va*, hypobranchial artery.

similar disposition is found in *Lepidosteus* and in *Amia*, where a second
(posterior) efferent artery also occurs dorsally (Allis, 806).

Since the gill-bearing bars of Dipnoi are also provided with an outer afferent and two inner efferent vessels (Spencer, **872** ; Kellicott, **789** ; Parker, **854**), this would appear to be an ancient disposition preserved in both the Chondrichthyes and the primitive Osteichthyes, from which the Teleostomes have diverged owing to a progressive tendency for the aortic arch to become converted into the efferent vessel, Fig. 534.

It may here be mentioned that in *Protopterus* the three pairs of external gills are supplied from the vessels of the last three pairs of gill-bars (fourth, fifth, and sixth aortic arches). The third and fourth aortic

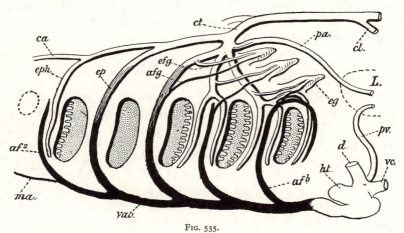

FIG. 535.

Branchial circulation of *Protopterus* (from Parker's figures). $af^{2\text{-}6}$, Afferent branchial arteries ; *afg*, afferent vessel to external gill ; *ca*, carotid ; *cl*, coeliac artery ; *ct*, left branch of dorsal aorta formed by junction of branchial efferent vessels ; *d*, ductus Cuvieri ; *efg*, efferent vessel of external gill ; *eg*, external gill ; *ep*, epibranchial region of arterial arch of first gill-less branchial arch ; *eph*, epibranchial vessel of hyoid arch ; *ht*, heart ; *L*, lung ; *ma*, lingual artery ; *pa*, pulmonary artery ; *pv*, pulmonary vein ; *vao*, ventral aorta ; *vc*, vena cava posterior. The five branchial slits are shaded ; the first two branchial arches are without gills. A dotted line indicates the position of the obliterated spiracle.

arches of *Protopterus* remain as continuous vessels running from the ventral to the dorsal aorta in the first and second branchial bars which have lost their gills, Fig. 535. This resemblance to the Amphibian structure is probably due to convergence, since it is not shown by *Ceratodus*. Other resemblances to the Amphibia in the arterial system of the Dipnoi, such as the gathering of the epibranchial arteries on each side before they unite to form the aorta and the shortening of the ventral aorta to form a truncus with subdivided lumen, seem to indicate true affinity with the lower Tetrapods (see also p. 552).

In all fishes the efferent branchial vessels tend to anastomose ventrally below the gill-slits, giving rise to more or less continuous vessels from which arise hypobranchial arteries supplying the ventral branchial region

2 L

and the heart. Similar dorsal anastomoses occur above the gill-slits from posterior to anterior efferent vessels in *Ceratodus* and Selachii (Parker, 853 ; Spencer, 872).

The secondary shifting in Selachiáns of the epibranchial artery, mentioned above, takes place owing to its thus acquiring a connexion with the posterior efferent vessel of the bar in front and then losing the earlier connexion with the anterior efferent of its own bar ; the epibranchial artery now receives blood from the anterior and posterior hemibranchs of one slit (Dohrn, 333 ; Sewertzoff, 869). It is interesting to note that in *Chlamydoselachus* the original connexions are retained in the adult (Allis, 804), Figs. 531-2, 536.

THE ARTERIAL SUPPLY OF THE FIRST TWO VISCERAL ARCHES AND HEAD IN PISCES

As was mentioned above, the originally continuous aortic arches of the embryonic mandibular and hyoid bars become greatly modified in the course of development. The divergence in fate of these and the more posterior aortic arches is partly due to the fact that the stream of arterial blood tends to become subdivided into two : the bulk of oxygenated blood needed to supply the trunk, limbs, and tail flows backwards in the median dorsal aorta ; but there is also a no less important stream flowing forward to supply the head, and this in fishes is derived to a considerable extent from the first two aortic arches.

The fundamental relations of the internal carotids in Craniates are as follows : the lateral dorsal aortae run forwards below the basis cranii to enter the cranial cavity on either side of the pituitary body through the median foramen hypophyseos between the posterior ends of the trabeculae cranii, Figs. 247, 271.[1] In fishes the name ' internal carotid ' is generally applied to that region of the vessel beyond the origin of the orbital artery (external carotid of some authors). The ' internal carotid ' of Tetrapods reaches down to its origin with the true external carotid from the ventral ' common carotid ' (ventral aorta).

Whether the true forward continuations of the lateral dorsal aortae are represented by the internal carotids, or, as suggested by Allis, by orbito-nasal arteries running forwards through the orbits in some Teleostomes, is doubtful. The true external carotids are considered in Tetrapods

[1] A few exceptions occur among fishes (*Polypterus, Amiurus*, Allis, 410, 803) and Amphibia (Anura), in which the internal carotid appears to enter the side wall of the brain-case farther forwards ; but they are probably always due to secondary modifications—the establishment of a lateral vessel and obliteration of a portion of the original internal carotid, as Gaupp showed in *Rana* (504). See also Mammalia, p. 263.

to be the forward continuations of the ventral aorta,[1] and are represented in fishes by small branches of the hypobranchial arteries and vestigial

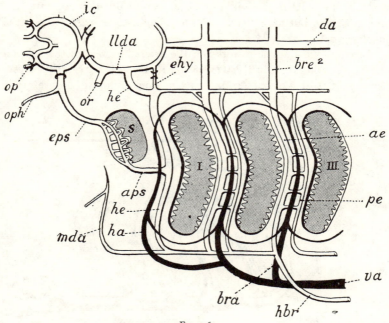

FIG. 536.

Chlamydoselachus anguineus: diagram of cerebral and anterior branchial circulation (modified from Allis, 1911). *bre*², Epibranchial of 2nd branchial bar; *ehy*, epibranchial of hyoid bar. Other letters as in Fig. 532.

ventral part of the mandibular arch to the thyroid gland and region of the lower jaw.

In Pisces the internal carotid gives off intracranially ophthalmic, optic, and cerebral arteries, the two former passing out of the cranial cavity

---

[1] There is considerable confusion in the nomenclature of the arteries of fishes. For the convenience of the reader the following synonyms may be given : Orbital artery (Carazii) =carotis facialis (Rathke), external carotid (Allis, and others), posterior carotid (T. J. Parker), art temporalis (Greil). Ophthalmic artery =art. ophthalmica magna, art. choroidalis (Dohrn), art. orbitalis (Greil). Optic artery (Allis) =art. centralis retinae (Dohrn), art. ophthalmica minor (Müller), art. ophthalmica (Greil). Afferent mandibular vessel =art. hyoidea (Wright), art. pseudobranchialis (Parker), art. thyreospiracularis and part of afferent spiracular (Dohrn). Efferent pseudobranchial vessel (Allis) =anterior carotid (Parker), part of afferent spiracular and carotis interna anterior (Dohrn). The common carotid of T. J. Parker is that part of the lateral dorsal aorta between its bifurcation into posterior and internal carotids and its junction with the efferent hyoidean vessel.

dorsally to the trabecula. The important orbital artery comes off the lateral dorsal aorta close to the second aortic arch (or from this arch) ventrally to the basis cranii, Figs. 532, 537-8. It does not enter the cranial cavity, but may pierce the subocular shelf or pass through the jugular canal (p. 276).

Selachii.—Of the many works on the arterial system of the Selachii one may mention those of Hyrtl (1858 and 1871), T. J. Parker (853), Ayers (807), Allis (804), Carazzi (818), on its anatomy ; and of Dohrn (333),

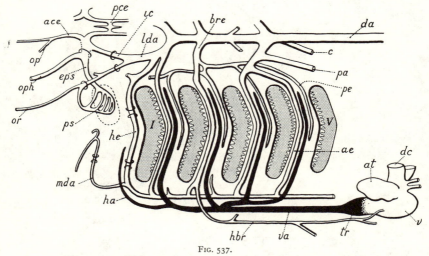

Fig. 537.

Amia : branchial circulation, left-side view completed (modified from Allis, 1912). *ace*, Anterior cerebral ; *pce*, posterior cerebral ; *pa*, pulmonary artery. Other letters as in Figs. 531 and 532.

Rückert (863), Platt (855), Raffaele (856), Scammon (380), de Beer (421), on its development.

The second or hyoidean aortic arch is soon interrupted, giving rise, as in the branchial bar, to an afferent vessel from the ventral aorta and a dorsal epibranchial vessel. Since only a posterior hemibranch is present only one corresponding efferent branchial vessel is developed in the hyoid bar. Very early the first or mandibular aortic arch becomes connected with the hyoidean efferent vessel by a commissural vessel below the spiracular slit. The mandibular aortic arch then becomes interrupted ventrally to this connexion, and its ventral portion secondarily united to the lower end of the efferent hyoidean vessel or anterior end of the hypobranchial artery ; it supplies in the adult the thyroid gland and lower jaws. As soon as the mandibular pseudobranch develops, the mandibular aortic arch divides into capillaries to supply its lamellae above the point where it receives the commissural vessel. The latter now functions as the afferent pseudobranchial vessel receiving arterial blood from the hyoidean efferent vessel. The more dorsal remainder of the mandibular aortic arch becomes the efferent pseudobranchial vessel which joins the lateral dorsal aorta (internal carotid), Figs. 531-2, 536. It is a remarkable fact, discussed elsewhere, that in Selachians alone this efferent vessel passes in to join the internal carotids dorsally instead of ventrally to the trabecula cranii (p. 529).

The lateral dorsal aortae unite in a median sinus cephalicus ventral to the tip of the notochord in the very early embryo, Fig. 234, and this union persists in the adult forming the anterior limit of a circulus cephalicus passing behind the pituitary body.[1] From the sinus spring the internal carotids, which receive the efferent pseudobranchials and give off optic and cerebral arteries. Intracranial branches of the latter meet below the brain to form the median basilar artery. The orbital artery, supplying the orbit, side of the head, and jaws, arises from the lateral dorsal aorta close to its junction with the efferent hyoidean vessel. The anatomical relations of the chief arteries are explained elsewhere (p. 528).

Teleostomi.—Where the hyoidean hemibranch persists, as in *Lepidosteus*, an afferent vessel supplies it from the ventral aorta ; but this vessel disappears in *Acipenser* where the hemibranch is a ' pseudobranch ' receiving only arterial blood, and also in other forms without a hyoidean hemibranch except *Amia* (Allis). As in Selachians the afferent mandibular vessel becomes secondarily connected to the hypobranchial arterial system ; it is retained in Chondrostei, *Lepidosteus*, and some Teleosts (*Gadus*), Fig. 539, but is usually interrupted before reaching it. The cross commissural vessel from the hyoidean to the mandibular arch seems to be always developed in the embryo, though it may disappear later (*Amia* and most Teleosts). The pseudobranch then receives its arterial blood from the mandibular arch (if complete) and a branch of the orbital artery which anastomoses with it (Allis). The efferent hyoidean and orbital arteries may be much developed to supply the opercular region. Optic, ophthalmic, and orbital arteries occur ; but the ophthalmic disappears when the chorioidal gland is absent (*Lepidosteus, Polypterus, Amiurus*). In connexion with this chorioidal gland, the name given to a rete mirabile on the eye-ball, it should be mentioned that Dohrn and Allis consider that it represents the vestige of a premandibular gill comparable to the pseudobranch of the mandibular arch. The existence, however, of a premandibular gill-slit has not yet been proved in any Craniate (p. 448).

---

[1] A circulus cephalicus is also completed behind the hypophysis in Dipnoi, and persists in the adult (Spencer, 872).

A characteristic feature of the Teleostei is the formation of a circulus cephalicus due to the reunion inside the skull of the divergent lateral dorsal aortae, Figs. 288-9, 538. Already in *Amia* the efferent pseudo-branchial tends to separate from the internal carotid and pass directly into the ophthalmic artery. The dorsal connexion with the carotid is obliterated, at all events in some individuals (de Beer, **421**). In Teleosts this is the usual definitive condition ; the two internal carotids come close together and fuse to complete the circulus cephalicus, while the separated

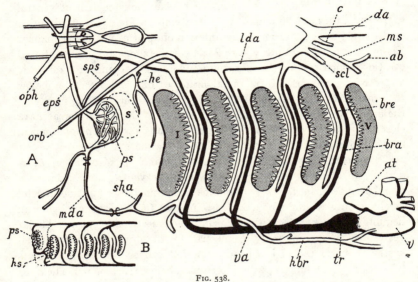

FIG. 538.

A, *Gadus* : branchial circulation, left-side view completed (modified from Allis, 1912). B, Branchial circulation of *Lepidosteus*. *ab*, Branch to air-bladder ; *ms*, mesenterial ; *s*, position of closed spiracle ; *sha*, secondary afferent hyoidean ; *sps*, secondary afferent pseudobranchial artery ; *tr*, bulbus. Other letters as in Figs. 531-2.

efferent pseudobranchial arteries continue forwards as the ophthalmics. A slender cross vessel also unites them in front of the circulus. Both these unions across the middle line are secondary and anterior to the pituitary body, and are therefore not homologous with the sinus and circulus of the Selachian (Allis, **807**).

The extent of the circulus cephalicus varies greatly in Teleosts, owing to the lateral dorsal aortae being more or less completely fused : in *Gadus* it is extensive and receives all four pairs of epibranchial arteries ; in *Clupea* it is much reduced and receives only the first pair (Ridewood, **680**).

THE PSEUDOBRANCH AND ITS VASCULAR SUPPLY

The Spiracle and Pseudobranch.—We may now return to the consideration of the first gill-slit in the Gnathostomata.   Even when it is widely
open in the embryo, as in Elasmobranchs, this spiracular slit is always
more or less completely closed in the adult.   The closure takes place from
its ventral end, and its dorsal region only may remain open.   It occurs
in Selachii as a small dorsal pouch passing outwards from the pharynx
between the mandibular and hyoid arches, and usually opening behind
the eye by a small pore or spiracle.   While it is minute or even closed in

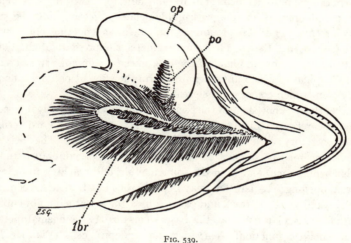

FIG. 539.

Ventro-lateral view of head of *Salmo salar* with right operculum, *op*, raised to show first branchial
arch, *ibr*, and pseudobranch, *po*.

some Lamnidae and Carchariidae, the spiracle is larger in Spinacidae, and
largest in Rajiformes, where it is provided on its anterior wall with a
movable valvular flap strengthened by the prespiracular cartilage, Fig. 440.
In these bottom living fish it may serve, when the mouth is applied to
the ground, for the passage in and out of a current of water for respiration.
On the front wall of the spiracular slit is the pseudobranch (p. 516).

The adult Holocephali and Dipnoi have lost both pseudobranch and
spiracular slit ; but in Teleostomes an open spiracle persists in Polypterini, *Acipenser*, and *Polyodon*.   In all other adult Teleostomes the
spiracle is closed, although a distinct vestige may remain of the pharyngeal
pouch in *Amia*, *Lepidosteus*, and some Teleosts (Wright, **882** ; Sagemehl,
378).   A pseudobranch is here developed in *Acipenser*, *Polyodon*,
Lepidosteus, *Amia*, and the majority of Teleosts. Figs. 537-9.

These pseudobranchs have attracted much attention, and have been studied by many observers since they were first discovered as a pair of gill-like organs on the inner surface of the opercular fold in Teleosts by Broussonet in 1782. The two pairs of subopercular gills (pseudobranchial and opercular hemibranchs) of *Acipenser* were later found by v. Baer (1827), and those of *Lepidosteus* by Valentin (1831). It was Hyrtl (1838) who first pointed out that the pseudobranch is supplied with arterial blood and therefore is not respiratory in function. In 1839 J. Müller gave a masterly account of these organs in Pisces, established the homology of the pseudobranch of Teleostomes with that of Selachians (spiracular or posterior mandibular hemibranch), and described two kinds of Teleostean pseudobranch, (1) free, with gill-lamellae projecting into the subopercular cavity, and (2) glandular, sunk below a covering epithelium. He suggested that the pseudobranch regulates the pressure of the blood supplied to the chorioid gland of the eye, since the afferent artery breaks up into capillaries which unite again to form the efferent pseudobranchial artery from which comes off the great ophthalmic artery.

Recently our knowledge of the pseudobranch has been greatly increased by the researches of Granel (827-30) and Vialli (880-81), who have described special secretory ' acidophil cells ' in the higher forms. The Selachii (except *Scymnus*) possess a well-developed gill-like pseudobranch (mandibular hemibranch) lodged in the spiracular slit and reaching to near its external opening. Its lamellae may be numerous but bear few secondary lamellae, and these are thick and covered by a thick epithelium of cubical cells. The vascular tissue at the base seems to have a haematopoietic and haemolytic function. There are no acidophil cells (Vialli). The pseudobranchs of *Acipenser* and *Polyodon* are of similar structure, but more deeply set in the spiracular slit. In *Lepidosteus*, where this slit is closed externally but widely open internally, the pseudobranch migrates in development into the branchial cavity to a position on the under surface of the operculum. In *Amia* the pseudobranch grows and bulges outwards on the wall of a saccular diverticulum of the spiracular slit Fig. 733. Although modified, it preserves the essential gross structure of a gill with lamellae supported by branchial rays as in the previously described forms, but, it is important to notice, resembles that of Teleosts in the appearance of a layer of acidophil cells. In the Teleostei the spiracle closes early, the mandibular pseudobranch arises near the interna opening of the spiracular slit, Fig. 529, but migrates and spreads into the subopercular cavity (Dohrn, 780).[1] It is at first a ' free ' gill-like

[1] The view that the pseudobranch of Teleosti is a hyoidean gill (Cole, 1901) lacks good evidence.

organ with projecting vascular lamellae provided with secondary lamellae, but tends to sink below the covering epithelium and lose its primitive structure in later stages in more specialised cases (Maurer, 848).

Four adult types may be distinguished (Granel, 829) : (1) a ' free ' pseudobranch with lamellae set along an axis and bearing secondary lamellae all covered by ectodermal epithelium (*Trachinus*, and many other genera) ; (2) the lamellae are still free, but the epithelium covers them without sinking between the secondary lamellae (*Chrysophrys*) ; (3) the lamellae no longer project, and have sunk below the general covering of ectodermal epithelium (*Phoxinus*) ; (4) the whole organ is not only buried, but has become more or less separated from the superficial epithelium by an overgrowth of connective tissue (*Gadus, Cyprinus*, and others with a ' glandular ' pseudobranch). Excepting for the separation from the thick ectodermal covering, the histological structure remains much the same in all four types, and may be interpreted as a modification of that of the ordinary gill with its middle vascular layer strengthened by pilaster cells, its subepithelial layer, and its now loosened ectoderm. The pseudobranch of *Amia* and the Teleostei is distinguished by the conversion of the subepithelial cells covering the middle layer into large granular secretory acidophil cells. Thus, far from being a mere useless vestigial organ, the pseudobranch seems in the higher Teleostomes to have acquired a new function, that of an endocrine gland, in addition to its original function of regulating the blood-supply to the eye. What this new endocrine action may be is not yet known.

The blood-supply to the pseudobranch is dealt with elsewhere (p. 516), but it may here be pointed out that, owing to the formation of an anastomosis (afferent pseudobranchial artery) with the efferent vessel of the hyoid arch and interruption of the ventral region of the mandibular arch, it receives only arterial blood, Figs. 532, 537-8. This is the condition in Selachians, and usually found in the lower Teleostomes, and many Teleosts (*Salmo*) ; but in *Polyodon* the ventral mandibular supply persists, and in *Gadus* is added to it a secondary afferent vessel coming from the circulus cephalicus. In *Amia* and such Teleosts as *Esox* this secondary vessel (which may be derived from the orbital artery) alone supplies blood to the pseudobranch (Hyrtl, 1858–72 ; F. W. Müller, 851 ; Maurer, 848 ; and especially Allis, 802-3, 806).

THE AORTIC ARCHES IN TETRAPODA

The fate of the aortic arches in Tetrapods affords material for one of the most instructive studies of the evolution of vertebrate structure. Among the various factors which contribute to their gradual modification

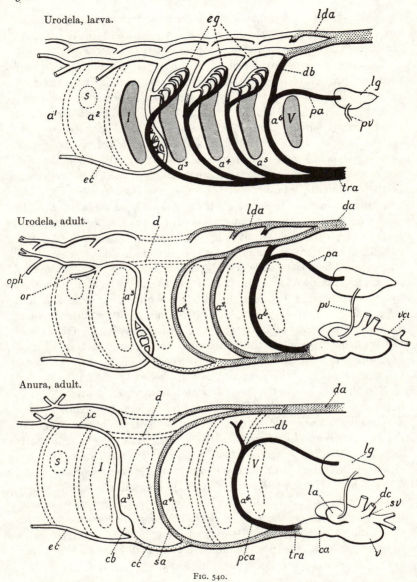

Urodela, larva.

Urodela, adult.

Anura, adult.

FIG. 540.

Diagrams illustrating development and fate of aortic arches in *Amphibia*, left-side view completed. Vessels carrying most arterial blood white, most venous blood black, and mixed blood stippled. a^{1-6}, Primary arterial arches; *ca*, conus arteriosus; *cb*, carotid gland; *cc*, common carotid; *da*, median dorsal aorta; *db*, ductus Botalli; *dc*, left ductus Cuvieri; *ec*, external carotid; *eg*, blood-supply to external gill; *ic*, internal carotid; *la*, left auricle; *lda*, lateral dorsal aorta (*d*, obliterated part=ductus caroticus); *lg*, lung; *oph*, ophthalmic; *or*, orbital; *pa*, pulmonary artery; *pca*, pulmo-cutaneous arch; *pv*, pulmonary vein; *s*, closed spiracular slit; *sa*, systemic arch; *sv*, sinus venosus; *tra*, truncus arteriosus (ventral aorta); *v*, ventricle; *vci*, vena cava inferior.

may be mentioned the establishment of pulmonary and loss of branchial respiration, the increasing separation in the heart of the venous from the arterial blood-stream, and the differentiation of the neck and consequent retreat of the heart to a more posterior position in the thoracic

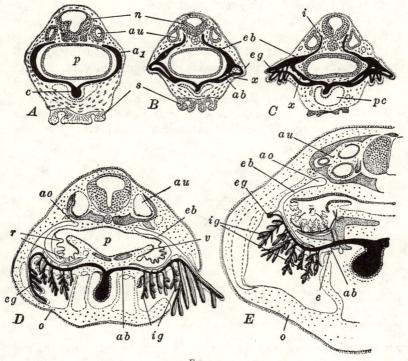

FIG. 541.

Sections through branchial region of tadpoles of *Rana esculenta*, showing development of gills and history of aortic arches (after Maurer, from W. E. Kellicott, *Chordate Develt.*, 1913). A, 4 mm. larva showing continuous first branchial aortic arch. B, 5 mm. larva showing anastomosis between the afferent and efferent portions of aortic arch. C, 6 mm. larva with vascular loops in external gills. D, 13 mm. larva. On the left opercular cavity is closed and external gill is beginning to atrophy, while on the right this cavity is still open and external gill well developed and projecting through opercular opening. E, 17 mm. larva, vessels of second branchial arch ; external gill represented only by a minute pigmented vestige. *a* 1, First branchial aortic arch ; *ab*, afferent branchial artery ; *ao*, root of lateral dorsal aorta ; *au*, auditory organ ; *c*, conus anteriosus ; *e*, epitheloid body ; *eb*, efferent branchial artery ; *eg*, external gill ; *i*, internal (anterior) carotid artery ; *ig*, internal gills ; *n*, nerve cord ; *o*, operculum ; *p*, pharynx ; *pc*, pericardial cavity ; *r*, gill-rakers ; *s*, oral sucker ; *v*, velar plate ; *x*, anastomosis between afferent and efferent branchial arteries.

region. Already in the Amphibia these changes have begun, but it is not till the Amniote grade of structure is reached that they are fully carried out.

Concerning the general development it may be mentioned that the paired rudiments of the dorsal aorta fuse to a median vessel posteriorly, but remain separate in the branchial region and lead to the internal carotid arteries.[1] The ventral aorta divides anteriorly into paired branches, giving rise to the external carotids, represented in Anura by the lingual arteries. The usual six pairs of lateral aortic arches are laid down in early stages ; and it should be noticed that the so-called aortic arches of adult Tetrapods do not strictly correspond to them alone, but may include a portion of the original ventral aorta below and of the lateral dorsal aorta above.

The early development of the blood-vessels in the Anuran tadpole is much specialised and differs considerably in different forms (Goette, 1875 ; Marshall and Bles, 846 ; Boas, 813 ; Maurer, 849) ; but the general results obtained chiefly by Marshall and Bles may be summarised as follows. The first or mandibular aortic arch persists but little as the ' pharyngeal ' artery. The second or hyoidean disappears. The third

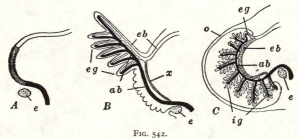

FIG. 542.

Diagrams of aortic arch of adult frog and tadpole (after Maurer, from W. E. Kellicott, *Chordate Dev't.*, 1913). A, Continuous aortic arch of adult, showing parts corresponding with larval vessels. B, First external gill and associated vessels in young tadpole.　C, Internal gill and associated vessels in tadpole after disappearance of external gills.　*ab*, Afferent branchial artery ; *e*, epithelioid body ; *eb*, efferent branchial artery ; *eg*, external gill ; *ig*, internal gill ; *o*, operculum ; *x*, direct anastomosis between afferent and efferent branchial arteries.

aortic arch remains as the so-called carotid arch, carrying arterial blood from the truncus arteriosus to the external and internal carotid arteries. The fourth aortic arch remains as the chief or systemic arch which joins its fellow to form the dorsal aorta. The fifth arch disappears, and the sixth gives rise to the pulmo-cutaneous arch taking chiefly venous blood to the skin by its cutaneous branch and the lung by its pulmonary artery, Fig. 540.

Important specialisations characteristic of Tetrapods generally have appeared. The task of carrying the forward stream of blood to the head is now assumed by the third aortic arch, and that part of the lateral dorsal aorta between the third and fourth arches, known as the ductus caroticus, is atrophied. Since, however, the ductus caroticus still persists

---

[1] The bifurcation of the aorta tends to be carried farther backwards in later stages owing to the secondary splitting of the median aorta in Tetrapods generally.

in Apoda, in some adult Urodela, such as *Triton* (Boas, **813**), and some adult Reptilia (*Sphenodon, Alligator* (?)), and many Lacertilia (Rathke, 1843, 1857; van Bemmelen, **809**; Beddard, 1904–6; O'Donoghue, **852**), its obliteration has probably taken place independently in Amphibia and Amniota. The same may be said of the ductus Botalli (ductus arteriosus), that dorsal part of the sixth aortic arch which joins the pulmonary artery to the dorsal aorta. Closed and reduced to a fibrous strand in adult Anura and the majority of Amniotes, it survives as an open vessel in Apoda, Urodela, *Sphenodon, Alligator,* and some Chelonia. *Sphenodon,* indeed, is the only Amniote normally preserving both these ducts in the fully developed adult.

The fifth aortic arch disappears entirely in all adult Tetrapods except the Urodela. It is always small even in the embryo, and vanishes so quickly that for a long time it was unidentified in Amniotes until van Bemmelen (1886) described it in the embryo of Reptiles and Birds, and Zimmermann (1888) in that of Man and other Mammals, Figs. 543-4, 548.

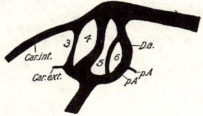

FIG. 543.

Aortic arches of left side of chick embryo, 4½ days old. From an injected specimen. (After Locy, from F. R. Lillie, *Develt. Chick,* 1919.) *Car.ext,* External carotid; *Car.int,* internal carotid; *D.a,* ductus arteriosus; *p.A,* pulmonary artery; 3, 4, 5, 6, third, fourth, fifth, and sixth aortic arches.

The ventral aorta always becomes spirally split to its base, being thus divided into a pulmonary trunk leading venous blood to the pulmonary arteries, and an aortic trunk leading to the systemic and carotid arches; for these two pairs of arches are always associated.

It is a familiar fact that in the Mammalia a complete aortic arch persists on the left side only. At an earlier stage in ontogeny the fourth aortic arches were complete on both sides, each giving off a subclavian artery; but later the arch on the right atrophies posteriorly to the origin of the subclavian. What remains of the arch as far as its junction with the common carotid is now called the subclavian artery, and the region of the lateral ventral aorta leading to them is called the innominate artery.

The specialisation of the aortic system in the modern Reptilia and Birds has proceeded on different lines from the Mammalian and must apparently have started before the obliteration of the ductus caroticus and ductus Botalli. In these groups not only has the pulmonary trunk been separated off from the original ventral aorta, but the remaining aortic trunk has also been spirally split to its base, so that two aortic trunks are formed.

The smaller left arch now comes from the right side of the heart, and the larger right from the left side of the heart. In Birds only the latter persists (see, further, p. 562). An important point to notice is that in all living Reptiles and in Birds the base of the carotid trunk, formed by the union of the right and left carotid arches (third aortic arches), opens into the

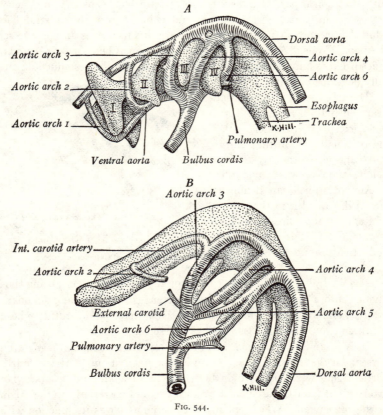

FIG. 544.

Aortic arches of human embryos : A, 5 mm. ; B, 7 mm. (after Tandler). *I-IV*, Pharyngeal pouches. (From Prentiss and Arey, *Text-book of Embryology*, 1917.)

larger right aortic trunk, and the whole aortic arch system is characteristically asymmetrical in a manner differing radically from the asymmetry of the mammalian system. The significance of this divergence will be discussed later (p. 572).

As the neck becomes lengthened the vessels of the carotid system become correspondingly elongated, Figs. 545-6. In Mammals all the arches retreat backwards with the heart and the internal and external carotids are lengthened ; and the same thing happens in short-necked Lacertilia, Chelonia, Crocodilia, and Aves. In the two last groups the external and internal carotids anastomose anteriorly, and if the latter fuse to a median vessel the right carotid arch may then atrophy (Crocodilia). In Birds the external carotids usually atrophy ; the internal carotids may fuse, and then the base of either the left or the right atrophy. Another

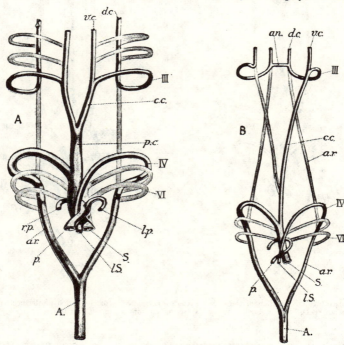

FIG. 545.

Illustrating modification of the carotid arteries, correlated with elongation of the neck region. A, Varanid lizard ; B, grass-snake (*Tropidonotus*). *p.c*, Primary carotid ; *A*, dorsal aorta ; *a.r*, aortic root ; *an*, anastomotic vessel ; *c.c*, common carotid ; *d.c*, dorsal (internal) carotid ; *l.p*, left pulmonary artery ; *l.s*, left systemic ; *p*, pulmonary ; *r.p*, right pulmonary ; *S*, systemic aorta ; *v.a*, ventral aorta ; *v.c*, ventral (external) carotid ; III, IV, VI, aortic arches.

modification takes place in certain Lacertilia (Varanidae) and in the Ophidia ; here the carotid arches remain near the head and so become widely separated from the systemic arches, Figs. 545-6.

Before entering the cranial cavity the internal carotid gives off a palatine branch which runs forwards below the basis cranii. This artery, absent in Elasmobranchs and rarely represented in Teleostomes, occurs in the Dipnoi. After having entered the cranial cavity the internal carotids provide as usual cerebral arteries and a basilar artery to the central nervous system, small arteries to the pituitary body, and an 'ophthalmic' artery on each side, which issues with the optic nerve and no doubt corresponds to the optic and ophthalmic artery of the fish.

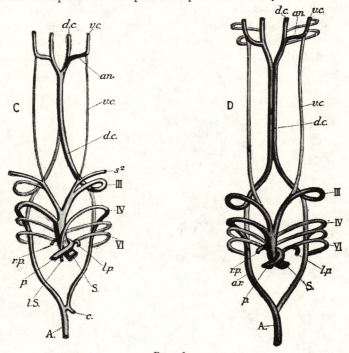

Fig. 546.

Illustrating modification of the carotid arteries, correlated with elongation of the neck region. C, Crocodile; D, bird. *c*, Coeliac artery; s^2, secondary subclavian. (Other letters as in Fig. 545.)

BRANCHES OF THE INTERNAL CAROTIDS : OPHTHALMIC AND STAPEDIAL ARTERIES

From the description given above of the arterial system of the head region, it appears that the specialisation of the cephalic arteries has had a considerable influence on the fate of the aortic arches and on the differentiation of the heart in the Vertebrate series. There remains to be discussed the homology of certain arteries and their relation to the skeletal and surrounding parts (Allis, 802-6; de Beer, 421).

The fundamental relations of the internal carotids have already been described (p. 272). Each internal carotid is a prolongation of the lateral dorsal aorta, and gives off after entering the cranial cavity not only cerebral arteries but also an optic artery (so-called ophthalmic of Tetrapods), which passes out to the retina of the eye with the optic nerve. In

Pisces it also gives off a more posterior ophthalmic artery to the chorioid plexus of the eye, Figs. 247, 288. Both these arteries issue from the skull in front of the pila antotica, and dorsally to the trabecula cranii. On the other hand the aortic arches join the lateral dorsal aorta ventrally to the basis cranii. The Elasmobranchs are the only exception to this rule (p. 517), for in them the mandibular aortic arch (efferent pseudobranchial) joining the ophthalmic outside passes into the cranial cavity through a foramen dorsally to the trabecula, Figs. 246, 247. This condition is probably due either to a difference in the assemblage of the cartilages of the posterior trabecular region (Allis, 670), or more probably to a loss of the original connexion of the mandibular arch with the carotid after its junction with the dorsally placed ophthalmic artery (de Beer, 421). No evidence of such a ventral connexion has, however, been found in the embryo of Selachians. In Teleostei these vessels are typically related in early stages, but, owing to the interruption of the original connexion with the lateral dorsal aorta of the efferent pseudobranchial, the latter passes intracranially directly into the ophthalmic artery in the adult, and this artery issues of course dorsally to the trabecula, Figs. 284, 288-9, 537-8 (see p. 518). The apparently abnormal relation of the internal carotid to the trabecula in Mammalia is discussed elsewhere (p. 263).

Of great significance in determining the homologies of the stapes and parts of the 'middle ear' in Tetrapods is the stapedial artery (p. 467). This artery can be traced with considerable certainty from Man down to the lowest fishes, where it appears to be represented by the orbital artery, the so-called external carotid of many authors (Allis, 802-6; Schmalhausen, 699; Versluys, 769-70; Goodrich, 734; de Beer, 421). In Gnathostomes generally an artery is found branching off from the dorsal end of the second or hyoidean aortic arch or from the lateral dorsal aorta in front of that arch, Fig. 547. Its primary morphological relations are as follows : it is dorsal and posterior to the first (spiracular) gill-pouch, ventral to the hyomandibula in fishes ; it passes forwards through the cranio-quadrate passage to the orbit, below the otic process of the quadrate (p. 412) ; its branches pass outwards above the palatoquadrate bar and behind the processus ascendens to supply the side of the head and jaws. There are three main branches : the supraorbital, infra-orbital to upper jaw, and mandibular to lower jaw (Tandler, 875-7 ; Grosser and Brezina, 832 ; Shiino, 870 ; O'Donoghue, 852 ; Twining, 878 ; Hafferl, 833-4). In Selachians the orbital artery may, owing to the outward growth of the subocular shelf, come to pierce the cartilage on its way to the orbit, and become partially buried in the wall of the auditory capsule. In *Amia* it is enclosed in the jugular canal, while in

Lepidosteus and most Teleosts it appears to be represented by two vessels,

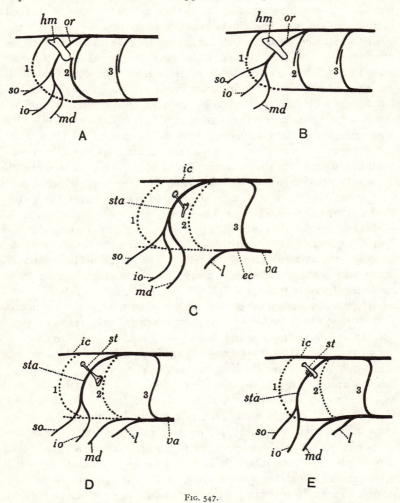

FIG. 547.

Diagrams illustrating fate of first three Aortic Arches, 1, 2, 3, and relation of Orbital or Stapedial Artery to Hyomandibula in: A, Selachian; B, Teleostome; C, Lacertilian; D, Crocodilian; E, Monotreme. Left-side views. *ec*, External carotid; *hm*, hyomandibula; *ic*, internal carotid; *io*, infraorbital; *l*, lingual; *md*, mandibular; *or*, orbital; *so*, supraorbital; *st*, stapes or columella auris; *sta*, stapedial artery; *va*, ventral aorta.

one passing through the jugular canal and the other through the subocular shelf farther forward, Fig. 284.

The orbital artery is typically developed in Dipnoi (*Ceratodus*) and in Tetrapods, where it is known as the stapedial artery and often pierces the stapes (p. 467). It has essentially the same relations and distribution as in Selachians, but complications arise owing to anastomosis with the ventral true external carotid, which tends to rob it of some or all of its branches. This external carotid is an extension of the ventral aorta supplying thyroid, lingual, and sometimes mandibular branches. Even

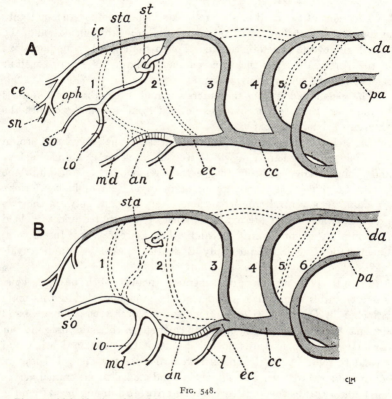

FIG. 548.

Diagram of left-side view of arterial arches and cephalic arteries in Rat, A, and Man, B (partly from J. Tandler). *an*, Secondary anastomosis; *cc*, base of carotid arch; *ce*, cerebral artery; *da*, dorsal aorta; *ec*, external carotid; *io*, infraorbital; *l*, lingual; *md*, mandibular; *oph*, ophthalmic; *pa*, pulmonary; *so*, supraorbital; *st*, stapes; *sta*, stapedial artery. 1-6, Six embryonic aortic arches.

in Reptilia it tends to join the mandibular branch of the stapedial (Crocodile); and in Mammalia, although the stapedial artery is developed in the embryo with its three main branches, in later stages an anastomosis is established with one or more of these, and their original connexion with the dorsal aorta through the stapedial artery is lost (Tandler, 876). Thus in the rat the stapedial artery persists in the adult, Fig. 498, but gives off only the supra- and infraorbital branches, while in Man and most Ditrematous Mammals all three branches join the external carotid and the stem of the stapedial is lost, Fig. 548. The morphological significance of the stapedial artery is further discussed above (p. 472).

THE CHIEF VEINS OF THE HEAD

The veins of the head region are of considerable importance not only as forming part of the vascular system, but also for the elucidation of the homologies of the skeletal elements. For in spite of many modifications in the Vertebrate series the veins on the whole preserve very constant relations to the surrounding structures.

The venous blood from the anterior region in Gnathostomes returns to the heart by a system of veins derived from or connected with the anterior cardinals (p. 508).[1]

On each side cerebral veins issuing from the cranial cavity and orbital veins join to a jugular vein, which, receiving a ventral external jugular and a subclavian vein from the pectoral region, passes back into the ductus Cuvieri (vena cava anterior or precaval vein of higher forms). In Mammals a ventral cross connexion often develops between the two precavals, and the left precaval may then be interrupted and the blood flow into the right as in Man and other Placentals (in Insectivora, Cheiroptera, Ungulata, and many Rodentia, however, both precavals persist even when a commissure is formed).

From the point of view of the general morphology of the head, the chief interest centres round the development of the jugular vein and its main branches in Craniates (Hochstetter, 835; Salzer, 866; Grosser and Brezina, 832; van Gelderen, 824). In early stages the embryonic anterior cardinal veins run forwards on either side of the notochord below and medial to the roots of the spinal nerves and the myomeres. They are continued forwards over the gill-pouches and medial to the cranial nerves and auditory sac as the venae capitis mediales draining blood from the brain. Just behind the developing hypophysis they become united by a transverse hypophysial vein passing ventrally to the original tip of the notochord, Figs. 242, 245, 256.

The tributaries of the longitudinal v. capitis medialis become differentiated into an anterior cerebral vein (from the eye and fore-brain), a middle cerebral vein (from the middle region of the brain including the cerebellum), and a posterior cerebral vein (from the remainder of the

[1] In Cyclostomata these veins, although built on the same general plan, undergo some peculiar modifications mentioned in connexion with the heart, p. 543.

hind-brain). The v. cap. medialis receives the last vein behind the auditory sac near the vagus nerve, the middle vein near the root of the facial nerve in front of the auditory sac, and reaches forwards to the anterior vein. This system of embryonic head veins is situated within the cranial cavity; but the v. capitis medialis soon becomes more or less completely replaced by another more lateral longitudinal vein, the vena capitis lateralis, which develops outside the cranial

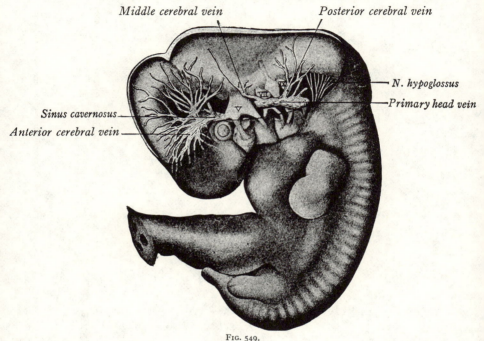

Middle cerebral vein   *Posterior cerebral vein*

N. hypoglossus

Primary head vein

Sinus cavernosus

Anterior cerebral vein

FIG. 549.

Veins of head of 9 mm. human embryo (after Mall.) × 9.   (From Prentiss and Avey, *Text-book of Embryology*, 1917.)

wall and flows into the anterior cardinal behind the vagus nerve. The v. cap. lateralis arises from a number of loops which grow round the auditory sac and cranial nerves from the v. cap. medialis, and fuse to a longitudinal vessel outside them. As this v. cap. lateralis becomes completed the inner vein tends to disappear, remaining only in so far as is necessary to enable the tributaries to communicate with the outer one. Thus the v. cap. medialis always disappears in the region of the auditory sac, but remains on the inner side of the trigeminal nerve to receive the anterior cerebral, ophthalmic, and hypophysial veins, and passes out through the wall of the skull with the facial nerve to join the v. cap. lateralis. The posterior cerebral vein flows into the v. cap. lateralis farther back, passing out of the skull with the vagus nerve through the jugular foramen, Figs. 245-6, 550.

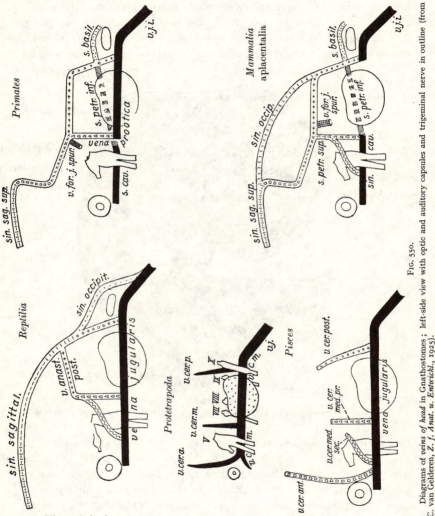

Fig. 550.

Diagrams of *veins of head* in Gnathostomes; left-side view with optic and auditory capsules and trigeminal nerve in outline (from C. van Gelderen, *Z. f. Anat. u. Entwickl.*, 1925).

The adult jugular vein (head vein, Stammvene) is, then, of mixed origin, and varies considerably in composition in different groups. Throughout all the Gnathostomes, however, its development is similar, and the transverse hypophysial vein passes out of the skull just anteriorly to the pila antotica (p. 244).

That part of the jugular vein derived from the v. cap. lateralis persists in Cyclostomata, Pisces, Amphibia, Reptilia,[1] Monotremata, and Marsupialia. It runs outside the auditory capsule backwards alongside the hyomandibular branch of the facial nerve through the cranio-quadrate passage (p. 412) and dorsally to the spiracular gill-pouch. In Selachians it runs dorsally to the hyomandibular cartilage (p. 416), and in the Tetrapods dorsally to the columella auris or stapes (p. 467). In Teleostomes it runs through a short jugular canal in the thickness of the wall of the auditory capsule (p. 276), and passes medially and ventrally to the hyomandibula, being apparently represented in this region by two veins, Fig. 446. A third venous loop occurs in some Teleostei passing outside the hyomandibular (de Beer, 421). Occasionally (as in Cyclostomata, most Selachii, Polypterini, and Anura) the v. cap. lateralis continues forward to the orbit outside the trigeminal nerve. But usually this nerve passes laterally to the head vein, which is then derived here from the v. c. medialis.

Among the Tetrapods there is a progressive tendency for the blood from the brain to drain into a newly formed median dorsal sinus passing backwards to the posterior cerebral veins. The connexion of the anterior cerebral vein with the v. cap. medialis is lost ; but the middle cerebral vein persists, except in the higher Mammalia, issuing into the orbit with the facial nerve. In the Lacertilia (except Amphisbaenidae) the original connexion of the posterior cerebral vein with the jugular is lost, and the blood is carried out behind the occipital arch ; in Chelonia, however, both outflows are present (Grosser and Brezina, 832 ; Versluys, 769 ; Bruner, 817).

While the v. cap. lateralis still persists in Monotremes and Marsupials, it disappears for the most part in adult Placental Mammalia. Anastomoses occur and it is replaced by intracranial sinuses which carry the blood back to the posterior cerebral vein and so out by the jugular foramen to the ' internal jugular vein '.

From the brief account given above it appears that the vena capitis medialis and the vena capitis lateralis are constantly developed in the embryo of all Craniata, and that they and their branches bear important relations to the skeletal and other parts. They always contribute to the formation of the adult jugular vein. The external jugular vein, another branch from the ductus Cuvieri, is more variable in extent, seems to develop independently, and drains usually the more ventral and lateral regions of the head, including sometimes the lower jaw.

---

[1] In Crocodilia and Aves it is replaced by a secondary outer vein (van Gelderen).

THE HEART

The heart of the Craniate Vertebrates is a specialised part of the primary longitudinal ventral vessel, and is a muscular pumping organ adapted in the first place to drive the venous blood forward and upward through the gills. Consequently it develops just behind the gills and in front of that point where the ventral vessel receives the venous blood from the body-wall and the alimentary canal by the ductus Cuvieri and hepatic veins (see p. 506).

The Acrania (Cephalochorda) represent an earlier stage in the phylogeny of Vertebrates before the differentiation of a heart, since in them the subintestinal vein (somewhat broken up on the liver) is continued forwards into the pharyngeal region as a ventral aorta without special enlargement. The venous blood in *Amphioxus* is propelled along this ventral aorta, which has contractile muscular walls, and up the branchial vessels of the primary gill-bars with the help of bulbous enlargements at their base. The ventral aorta is suspended in the coelom below the endostyle, and the main branchial vessel passes up the bar on the inner side of its coelomic canal,[1] Fig. 707.

The relations of the vessels and cavities in Craniates are fundamentally similar (p. 490), but the coelomic cavities of the bars are early suppressed, and the ventral coelomic cavity is concentrated as it were round the heart to form the pericardial coelom. The latter cavity is of course at first paired, the right and left cavities being separated by a primary longitudinal mesentery, and it is between the two folds of this mesentery (mesocardium) that the heart is developed. Very soon the ventral mesocardium disappears leaving at most some connecting strands, and later the dorsal mesocardium also breaks down except at the anterior and posterior ends where the cardiac tube pierces the wall of the pericardial coelom. Except at its two ends the heart now lies freely in the protective pericardial cavity, and is thus enabled to become coiled, to enlarge, and to pulsate. The early development of the heart is so well described in current text-books that only a few points need here be mentioned, Fig. 551.

In ontogeny as in phylogeny the heart first appears as a longitudinal

---

[1] From the afferent vessels three branches pass up each primary bar and join again to open into the dorsal aorta. A loop of two vessels extends down to near the ventral end of the secondary bar, receiving blood from the primary bar through the cross-bars or synapticula, and also opens into the dorsal aorta. Dorsally the efferent vessels on each side are joined by a longitudinal commissural vessel which breaks up at intervals into a capillary network supplying the nephridia (Spengel, Benham, Boveri, **814**).

median vessel : Cyclostomata, Elasmobranchii, Teleostomi (except Teleostei), and Amphibia. But in Teleostei and Amniota, where the embryo develops as a flattened blastoderm overlying a yolk-sac, and where the closure and separation of the alimentary canal is delayed, the heart

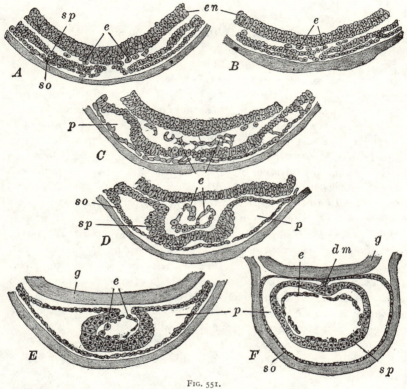

FIG. 551.

Sections showing formation of heart in frog.   A-D, Series of transverse sections through corresponding regions of a series of embryos of *R. temporaria* ; E, F, sections through same region in older embryos of *R. sylvatica*. (After Brachet, from W. E. Kellicott, *Chordate Develt.*, 1913.) A, 2·6 mm. embryo, mesoderm approaching mid-line ; endothelium appearing. B, Older embryo of same length as A. C, 3 mm. embryo showing enlargement of pericardial cavity and beginning of folding of somatic mesoderm. D, 3·2 mm. embryo, endothelial cells becoming arranged in form of a tube. E, Embryo of about 3 mm. F, Embryo of 5-6 mm. Heart tube established ; dorsal mesocardium still present. *dm*, Dorsal mesocardium ; *e*, cardiac endothelial cells ; *en*, endoderm ; *g*, wall of gut (pharynx) ; *p*, pericardial cavity ; *so*, somatic layer of mesoderm ; *sp*, splanchnic layer of mesoderm.

develops from paired rudiments which fuse in the middle line below the fore-gut at a later stage. The outer wall of the tubular heart is formed by the enveloping splanchnic mesoblast or visceral layer of the pericardium, and becomes the thin adult ' epicardium '. The thin inner lining is formed of a special endothelium, yielding the ' endocardium ' of the adult. It is derived from cells which gather between the splanchnopleure and the yoik-cells, and are in all probability of mesoblastic origin. The thick muscular layer between the endocardium and epicardium, and known as the 'myocardium', is developed from mesenchymatous cells proliferated from the medial surface of the splanchnopleure, Fig. 551.

The originally straight cylindrical cardiac tube becomes differentiated into four primary regions or chambers separated by constrictions—a posterior sinus venosus, atrium, ventricle, and anterior bulbus cordis,[1] Fig. 552. The sinus venosus bulges into the pericardium from its posterior wall, and receives all the venous blood from the hepatic veins and right and left ductus Cuvieri, which open freely into it ; also from the heart itself by the cardiac vein. The sinus opens into the atrium in front by an aperture guarded by two sagittal sinu-auricular valves. The atrium communicates with the ventricle by an atrio-ventricular aperture also provided with two valves, the region leading to this opening generally forming a narrow (atrial or auricular) canal. The tapering anterior end of the ventricle passes into the bulbus cordis, which is continued into the truncus or base of the ventral aorta outside the pericardial cavity. This bulbus in lower fishes is converted into the conus provided with longitudinal rows of pocket valves—a primitive condition (p. 540). The valves develop from thickenings or folds of the endocardium, are disposed with their concavity in front, and ensure the passage forwards of the blood from chamber to chamber. A backward flow of the blood tends to make them meet and close the lumen, while blood flowing forwards tends to separate them. Whereas the walls of the blood-vessels generally are provided with smooth muscle fibres, the walls of the four chambers of the heart in all Craniates have striated muscle-cells of a peculiar histological structure quite characteristic of this organ. The blood is propelled forwards from chamber to chamber by successive rhythmic waves of contraction starting in the sinus and ending in the bulbus. In the fully developed heart the wall of the sinus is thin, has little muscle, and is only slightly contractile, that of the atrium more muscular and very distensible, while the conus is primitively stiff, muscular, and contractile ;

[1] There has been considerable confusion in the nomenclature of the anterior region of the heart. Bulbus cordis is the name now generally applied by embryologists to the anterior chamber. But the name conus arteriosus, introduced by Gegenbaur to designate the anterior muscular region of the Selachian heart, is often given to it. Moreover, the Selachian conus does not correspond to that part of the heart so called in human anatomy. It is best, then, to apply the name bulbus cordis, introduced by A. Langer, to the embryonic structure throughout the Craniata, and keep the name conus arteriosus for the adult muscular contractile chamber derived from it in Pisces and Amphibia.

but the most effective chamber is the ventricle, whose powerful musculature forms a very thick wall. The lumen of the ventricle is partly invaded

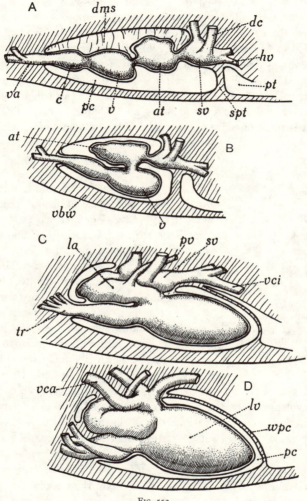

FIG. 552.

Diagrams illustrating the disposition in the pericardial cavity and phylogenetic modification of the chambers of the *heart* in Gnathostomes. A, Hypothetical primitive condition; B, Selachian stage; C, Amphibian stage; D, Amniote stage. *at*, Atrium; *c*, conus arteriosus; *dc*, ductus Cuvieri; *dms*, dorsal mesocardial fold; *hv*, hepatic vein; *la*, left auricle; *lv*, left ventricle; *pc*, pericardial coelom; *pt*, peritoneal coelom; *spt*, septum transversum; *sv*, sinus venosus; *tr*, truncus arteriosus; *v*, ventricle; *va*, ventral aorta; *vbw*, ventral body-wall; *vca*, vena cava anterior; *vci*, vena cava inferior; *wpc*, wall of pericardial coelom = *spt*.

and subdivided by muscle strands and trabeculae (columnae carneae) extending inwards, sometimes attached by tendons, and preventing undue dilatation. Similar but much less pronounced strands may be present in the atrium (musculi pectinati). Tendinous restraining strands (chordae tendineae) may be attached to the edges of the valves, often provided with muscles (musculi papillares) in the higher forms.

The cardiac tube of the embryo never retains its primitive longitudinal disposition in Craniates. Confined within the pericardium and attached at both ends, as it lengthens and thickens it necessarily bends and becomes somewhat spirally coiled. A ventral V-shaped loop to the right involving the ventricular and bulbar regions and directed backwards is constantly formed, and the whole tube acquires an S-shaped curve. The atrium thus comes to lie dorsally to the ventricle whose apex points backwards.

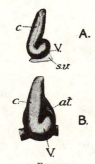

FIG. 553.

Two stages in the development of the heart of *Squalus* seen from the ventral side. (After Hochstetter, 1906.) *at*, Atrium; *c*, conus arteriosus; *s.v*, sinus venosus; *V*, ventricle. (From Kerr, *Embryology*, 1919.)

This curvature of the heart is more pronounced in higher than in lower forms, Figs. 553, 560, 572.

Coming now to the modifications of the heart in the various groups of fishes (Gegenbaur, 822-3 ; Röse, 862 ; Lankester, 843 ; Parker, 853 ; Goodrich, 35 ; Daniel, 487 ; and others), we find the most primitive form in the Elasmobranchs, Figs. 553-5. Here the heart is outwardly almost bilaterally symmetrical and is enclosed in a roomy pericardial cavity. The large sinus, which receives the hepatic veins and ductus Cuvieri, has its opening into the still larger atrium guarded by two sagittal valves. Two obliquely set valves are also present at the atrio-ventricular aperture situated somewhat on the left side. The well-developed contractile conus developed from the bulbus cordis is usually provided with three main longitudinal rows (of which one is dorsal and two ventro-lateral) of many pocket valves set in transverse rows or tiers. The number of valves varies considerably, and there may be vestigial valves between the main ones. So many as six tiers of valves may be present ; those of the first tier are often enlarged, and those of the middle and posterior tiers may be reduced. A gap may occur between anterior and posterior tiers, as in *Heptanchus* ; in *Chimaera* and some sharks only two tiers of well-developed valves remain. These valves develop in the bulbus cordis from four longitudinal endocardial ridges capable of closing the lumen on contraction of the wall, and which later become subdivided and hollowed out into pockets directed forwards. Owing to the reduction of the ventral ridge only three well-developed valves usually occur in each tier.[1] The conus leads into the elongated ventral aorta, which shows no differentiated truncus chamber.

The structure of the heart of the Teleostomi resembles that of the Elasmobranch, and has diverged from it chiefly in the region of the conus where important modifications occur, Figs. 555-6. The presence of a muscular contractile conus may be considered as a primitive feature (Gegenbaur, 823), and it persists in such lower Teleostomes as the Chondrostei, Polypterini, and Lepidosteoidei. That of *Acipenser*, with its three tiers of four or five valves, closely resembles the conus of Sela-chians; but in *Polypterus* and *Lepi-dosteus* it has become elongated and the number of valves greatly increased. *Lepidosteus* may have so many as seven longitudinal and eight transverse rows. On the contrary, in *Amia* and in the Teleostean series, as Gegenbaur showed, the conus tends to become reduced and replaced from in front by a non-contract-ile region with fibrous wall called the bulbus arteriosus. *Amia* has a large bulbus arteriosus, but still possesses a considerable conus with three tiers of four valves. In the typical Teleost, however, the conus has been practi-cally abolished, being represented by at most a narrow muscular zone bearing a single row of valves rarely more than two in number. Intermediate con-

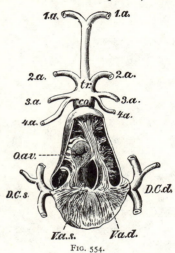

Fig. 554.

Heart of *Squalus acanthias*, from the dorsal side, with the atrium cut open. (After Röse.) *co, tr*, Truncus anteriosus; *D.C.d* and *D.C.s*, right and left precavals; *O.a.v*, atrio-ventricular aperture; *V.a.d* and *V.a.s*, right and left valve of the sinus venosus; *1a-4a*, afferent branchial arteries. (From Wiedersheim, *Comp. Anatomy*.)

ditions are found in some of the lower Teleosts (Clupeiformes); a distinct remnant of the conus with two transverse rows of valves occurs in *Albula* (Boas, 811), and also in *Tarpon* and *Megalops* (Senior, 868). Whether the Teleostean bulbus, which may acquire very thick fibrous walls, should be considered as a new formation, as a back-ward growth from the ventral aorta (truncus), or as a converted part of the conus itself, is doubtful. But in favour of the first view it should be said that there is little doubt that the surviving row of valves repre-

[1] The fact that four endocardial longitudinal ridges (right, dorsal, left, and ventral) occur in the embryonic heart not only of Elasmobranchii, Dipnoi, and Tetrapoda, but also of Acipenser, Amia, and Lepidosteus (E.S.G.), is evidence that the valves of the conus were originally disposed in four corre-sponding rows.

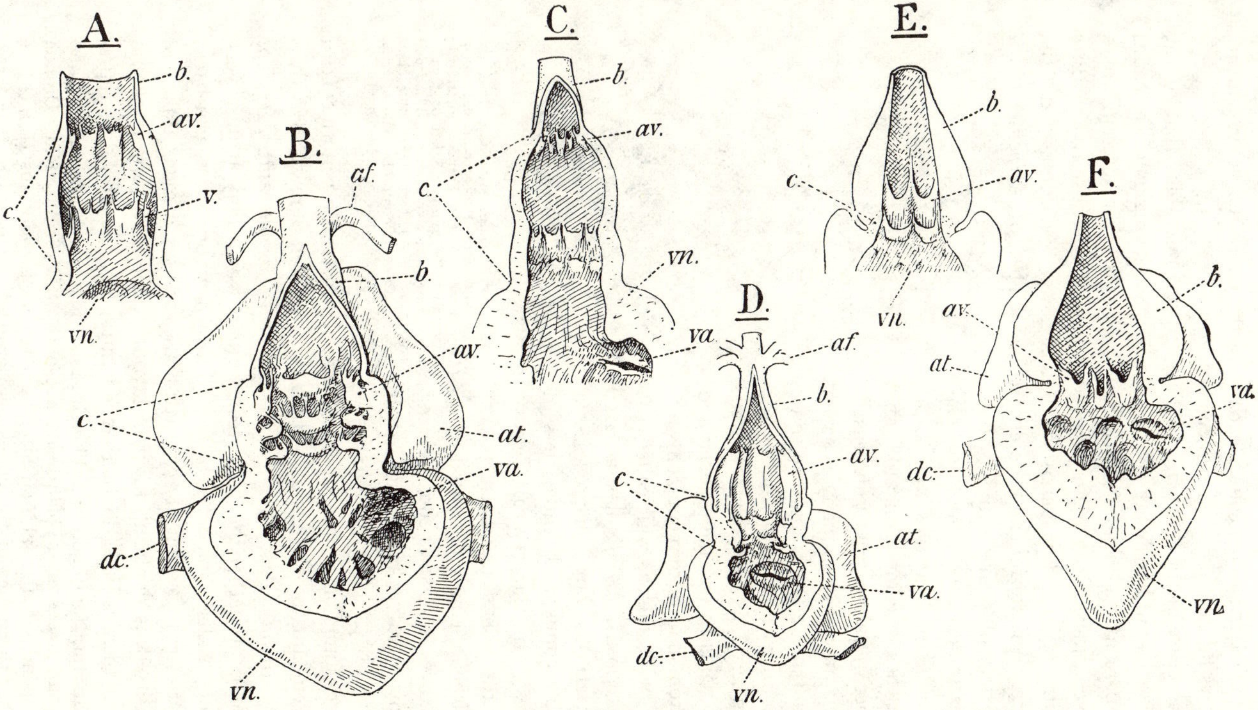

Fig. 555.

A, Conus arteriosus of *Chimaera monstrosa*, L.; B, heart of *Alopias vulpes*, Gm.; C, conus of *Acipenser sturio*, L.; D, heart of *Amia calva*, L.; E, bulbus of *Albula conorhynchus*, B, and E. (after Boas); F, heart of *Salmo salar*, L. Ventral view. *af*, Posterior afferent branchial vessel; *at*, atrium; *av*, anterior valve; *b*, base of ventral aorta which becomes the bulbus arteriosus in E and F; *c*, muscular contractile conus arteriosus which disappears in F; *dc*, ductus Cuvieri; *v*, posterior valve; *va*, atrio-ventricular aperture guarded by two valves; *vn*, ventricle. The bulbus conus and ventricle have been slit along the ventral mid-line, and stretched open to expose the valves. (From Goodrich, *Vert. Craniata*, 1909.)

sents the anterior and generally enlarged tier of more primitive forms, and Gegenbaur has pointed out that its valves are already attached in front to the wall of the truncus in Selachians. According to Senior and Hoyer (838) the conus is represented by the narrow muscle zone at the base of the valves which becomes joined to the Teleostean ventricle, and according to Hoyer and Smith the conus has been by a process of intussusception, so to speak, telescoped into the ventricle.

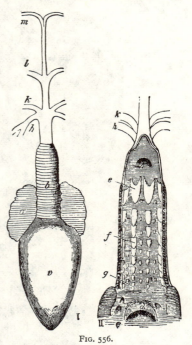

The Teleostean heart shows an extreme stage of specialisation peculiar to the group, and not leading to the structure of any of the higher vertebrates. This conclusion is amply supported by evidence derived from other organs such as the brain, alimentary canal, and gonads, all of which point to the Teleostei being a side branch of the phylogenetic tree.

The heart of Cyclostomes is in some respects highly specialised; chiefly perhaps owing to the strange fact, not yet explained, that only one ductus Cuvieri is preserved in the adult (J. Müller, 1833–43; Vialleton, 879; Röse, 862). The sinus venosus consequently takes up an almost vertical, dorso-ventral position, passing between atrium on the left and ventricle on the right. As Goette showed (825), the venous system develops normally in *Petromyzon* with paired ductus Cuvieri leading

Fig. 556.

Heart of *Lepidosteus osseus*, L. I, Ventral view; II, conus arteriosus opened. *a*, Atrium; *b*, conus; *e, f, g*, transverse rows of valves in conus; *h, k, l, m*, four afferent branchial vessels; *v*, ventricle. (After Günther, from Goodrich, *Vert. Craniata*, 1909.)

from paired cardinals to the sinus, which also receives blood from hepatic veins and a median jugular vein. Later an anastomosis develops dorsally between the right and left cardinals so that all their blood passes to the right ductus Cuvieri and the left ductus is obliterated. The sinus thus becomes a narrow vessel passing in the remaining portion of the meso-cardium to the right of the gut into the ductus above and into the jugular and hepatic veins below. The usual paired valves guard the sinu-atrial and the atrio-ventricular apertures. There is no well-developed conus; but this chamber is represented by a short region beyond the ventricle, containing in *Petromyzon* two longitudinal ridges comparable to those found in the embryonic conus of other forms (Vialleton). The passage

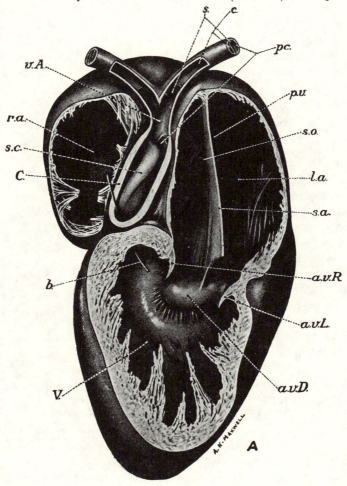

FIG. 557.

Heart of the frog. A, General dissection from the ventral side. *a.v.d*, Dorsal atrio-ventricular valve; *a.v.L*, left ditto; *a.v.R*, right ditto; *b*, bristle passed from ventricle into conus; *C*, conus; *c*, carotid; *l.a*, left auricle; *p.v*, opening of pulmonary vein; *pc*, pulmo-cutaneous; *r.a*, right auricle; *s*, systemic cavity; *s.a*, atrial septum; *s.c*, septum of conus; *s.o*, opening from sinus venosus; *V*, ventricle; *v.A*, ventral aorta. (From J. G. Kerr, *Zoology*, 1921.)

from the ventricle to the conus is provided with a right and left pocket valve. A remarkable fact is that, while it is the right ductus Cuvieri which alone persists in the adult Lampreys, it is the left ductus which is preserved in Myxinoids, Figs. 521, 680*a*.

Leaving for the present the consideration of the Dipnoan heart and passing to the Tetrapods, we find a general advance in specialisation, and a profound modification correlated with the appearance of pulmonary respiration and the disappearance of the gills. Already in the Amphibia, more especially in Anura, the heart acquires the shape characteristic of this organ in Tetrapods (Boas, 813 ; Gaupp, 821 ; Röse, 862 ; Rao and Ramana, 857 ; Rau, 858 ; Oliver, 1910). Moreover, the pericardial wall becomes thinned out and freed from the lateral body-wall, so that the heart is covered by a thin membrane extending above, behind, and at the sides, Figs. 552, 557-8. In the Anuran heart the S-shaped curvature is so pronounced that the atrium (auricles) is carried forwards in front of the large and very muscular ventricle and now opens backwards into it. The sinus venosus is also carried forwards dorsally and opens downwards into the atrium. An important innovation is the subdivision of the atrial cavity into a large right and smaller left auricle by an interauricular septum, a sagittal fold first arising from the anterior wall and extending posteriorly and to the auricular ventricular aperture. The oxygenated blood is brought directly from the lungs to the left auricle by a special pulmonary vein passing over the sinus. The sinus itself, still a nearly bilaterally symmetrical cavity formed by the junction of the two ductus Cuvieri and vena cava inferior, receives the venous blood from the rest of the body (and from the heart by the coronary vein) and pours it into the right auricle through an opening protected by right and left semilunar valves. The openings of the pulmonary vein and of the sinus are on opposite sides of the septum. Two large dorsal and ventral and two small right and left valves guard the single opening from the auricles to the ventricle. The spirally twisted conus is much specialised. The region of the embryonic bulbus cordis becomes in adult Anura differentiated into a posterior conus (pylangium) and short anterior ' truncus impar ' (synangium). The latter passes outside the pericardium into right and left branches, each of which soon splits into the three anterior arches. That the conus with its contractile wall provided with striated muscle belongs to the heart there can be no doubt ; but the anterior part or perhaps the whole of the truncus impar should be reckoned as representing the shortened ventral aorta. As usual the bulbus cordis is provided in the embryo with four longitudinal (spiral) internal endocardial ridges which give rise to valves. The dorsal, left, and ventral ridges are interrupted in the middle and give rise as a rule only to three pocket valves at the anterior end and three or fewer at the posterior end of the conus with a wide gap between. The better-developed right ridge becomes a continuous longitudinal fold (septum bulbi, spiral valve) projecting into the lumen so as to subdivide it incompletely into two channels—the cavum systemo-caroticum and cavum pulmo-cutaneum. Attached anteriorly to the dorsal surface of the conus wall, this fold turns to the left and ends posteriorly on the ventral surface near the ventricular opening. At its anterior end it expands to a hollow pocket-like valve, into which penetrates the extreme posterior end of the horizontal septum of the truncus (septum principale). This fold completely separates the lumen of the truncus into a dorsal passage continuing the cavum pulmo-cutaneum and a ventral passage continuing the cavum systemo-caroticum. Shortly in front of the anterior conus valves the

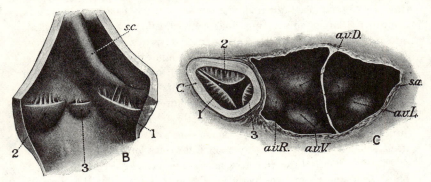

FIG. 558.

Heart of a frog. B, Ventricular end of conus slit open to show the pocket valves; C, atrio-ventricular valves (closed), etc., as seen in a heart cut transversely through the auricles and conus and viewed from the headward side. *a.v.D*, Dorsal atrio-ventricular valve; *a.v.L*, left ditto; *a.v.R*, right ditto; *a.v.V*, ventral ditto; *C*, conus; *s.a*, atrial septum; *s.c*, septum of conus; 1, 2, 3, pocket valves at ventricular end of conus. (From J. G. Kerr, *Zoology*, 1921.)

latter passage is again subdivided into paired systemic and carotid cavities leading to their respective arches. These various septa, subdividing the cavity of the truncus, including the septum principale, appear to be formed by the extension backwards of the walls separating the approximated bases of the arches. Concerning the exact mode of action of this complicated system of folds, valves, and septa, there is still some difference of opinion (Sabatier, 865; Gaupp, 821; Rau, 858); but the main result seems to be as follows. The venous blood received by the right auricle passes into the ventricle whence it is forced up the cavum pulmo-cutaneum, and the aerated blood, brought to the left auricle from the lungs, is passed through the ventricle and driven up the cavum systemo-caroticum. The arterial and venous streams are but little mixed in the ventricle owing to the sponginess of its wall, and to the bulk of the venous blood being expelled in the first phase of ventricular contrac-

tion and of the purest arterial blood in the last phase of contraction. The
anterior pocket valves help to distribute the blood to the several passages
in the truncus; and thus, while the carotid arch receives the purest
arterial blood and the pulmonary the most venous, the systemic supply
is somewhat mixed. But the blood may be oxygenated not only in the
lungs but also in the buccal cavity and in the skin supplied by the
cutaneous branch of the pulmonary arch (p. 524), so that the circulation

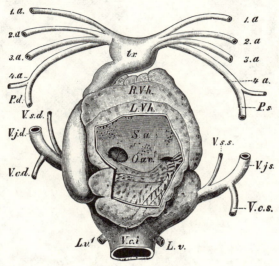

FIG. 559.

Heart of *Cryptobranchus japonicus*. From the ventral side. (After Röse.) The left auricle is cut
open. *L.v*, *L.v*[1], The two pulmonary veins, opening by a single aperture into the left auricle; *L.Vh*,
R.Vh, left and right auricles; *O.av*, atrio-ventricular aperture; *P.d* and *P.s*, left and right pulmonary
arteries; *S.a*, septum atriorum, perforated by numerous small apertures; *tr*, truncus arteriosus;
V.c.d, *V.c.s*, posterior cardinal veins; *V.c.i*, postcaval vein; *V.j.d* and *V.j.s*, jugular veins; *V.s.d*
and *V.s.s*, subclavian veins; 1a, 4a, the four arterial arches. (From Wiedersheim, *Comp. Anat.*)

fulfils its purpose in spite of the incomplete separation of the two streams
in the heart.

In the structure of the heart, as in so many other respects, the U r o d e l a
are somewhat degenerate or retain certain larval characters, Fig. 559.
While the sinus venosus, auricles, and ventricle resemble in general those
of the Anura, the heart is less compact and the conus is usually much
simpler (Boas, **813**). The terrestrial *Salamandra* preserves the pocket
valves and longitudinal spiral valve much as in the frog, but the horizontal septum of the truncus is not so much developed, does not grow
backwards so far into the conus. The lumen of the truncus is subdivided
into four paired channels corresponding to the four pairs of persistent
arches. In the more aquatic Urodela the tendency is for the conus to

remain undeveloped, and for the spiral valve to disappear more or less completely. Thus *Siren* has four posterior pocket valves and three anterior, with a fourth elongated into a short spiral fold ; but in *Meno-*

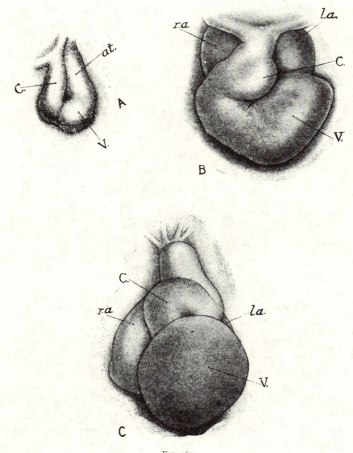

FIG. 560.

Views of the heart of *Lepidosiren* as seen from the morphologically ventral side. (B and C after J. Robertson, 1913.) A, Stage 32 ; B, stage 31 ; C, stage 35. *at*, Atrium ; *C*, conus arteriosus ; *l.a*, left auricle ; *r.a*, right auricle ; *V*, ventricle. (From Kerr, *Embryology*, 1919.)

branchus the spiral fold has disappeared leaving only four main pocket valves at each end of the conus, and in *Proteus* only two pairs of such valves are present. Moreover, the interauricular septum is thin and often perforated. Thus the venous and arterial streams become less completely separated, and it is clear that the heart tends to lose its specialisations as the branchiae remain more and more functional in the adult, and pulmonary respiration loses its importance. The same sort of degeneration occurs in the Gymnophiona (Apoda) where cutaneous respiration is so much developed ; for here also the spiral fold disappears, and even the posterior valves of the conus (Wiedersheim, 652 ; Boas, 813). In the aberrant lungless Salamanders, indeed, not only may the conus be simplified, but the interauricular septum may not be developed, a return to an almost fish-like condition (Bruner, 816 ; Hopkins, 387).

The study of the heart of the Dipnoi is of great importance for the understanding of the structure and phylogeny of the heart in the Tetrapoda

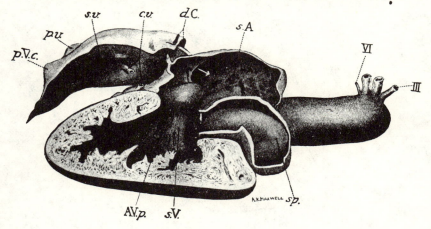

FIG. 561.

Heart of an adult *Lepidosiren* with the right side removed. (After J. Robertson, 1913.) *AV.p*, Atrioventricular plug ; *c.v*, coronary vein (cut) ; *d.C*, ducts of Cuvier ; *p.v*, pulmonary vein ; *p.V.c*, posterior vena cava at its opening into the sinus venosus ; *s.A*, atrial septum ; *s.V*, ventricular septum ; *s.v*, sinus venosus (its opening into the right auricle indicated by an arrow) ; *sp*, spiral valve ; III, VI, aortic arches cut near their ventral ends. (From Kerr, *Embryology*.)

(Boas, 812 ; Röse, 862 ; Robertson, 861 ; Kerr, 840 ; Goodrich, 35). The heart lies in a pericardial cavity behind the gills and still bounded by stiff walls. The curvature is more pronounced than in other fishes, and although the sinus venosus still starts near the posterior pericardial wall it lies to a great extent dorsally to the ventricle whose apex projects far back below it. The atrium is brought far forwards above the ventricle and base of the conus and opens downwards into the ventricle, Figs. 560-64. Thus the heart approaches in shape that of the Urodele Amphibian. The sinus venosus receives the venous blood from two ductus Cuvieri and a median vein formed by the union of a vena cava inferior with the hepatic veins ; but an important new departure is the separation of the return stream from the lung-like air-bladder. For the two pulmonary veins join, as in Amphibia, to a single vessel which, passing along the dorsal wall of the sinus, empties aerated blood directly into the left side of the atrium. Into the right side of the atrium the sinus empties the venous blood, and there extends from behind and above a muscular septum which subdivides the atrial cavity almost completely into larger right and smaller left auricular spaces. Immediately to the right of this interauricular septum is the sinu-auricular opening provided with a right valve, and on the left side of the septum is the opening of the pulmonary vein protected by a small fold on the left. Into the large atrio-ventricular opening fits a fibrous or partly cartilaginous plug arising from the posterior

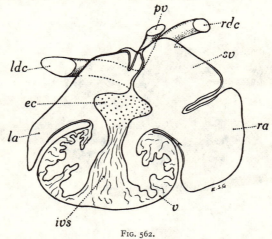

Fig. 562.

Protopterus annectens, transverse section of heart of larva through auriculo-ventricular openings. *ec*, Endocardial cushion ; *ivs*, interventricular septum ; *la*, left auricle ; *ldc*, left ductus Cuvieri ; *pv*, pulmonary vein ; *ra*, right auricle ; *rdc*, right ductus Cuvieri ; *sv*, sinus venosus ; *v*, ventricle.

margin of the opening. This highly characteristic structure is peculiar to the Dipnoi; it is attached above to the interauricular septum and below to a corresponding median ventral muscular interventricular septum which incompletely subdivides the ventricular cavity from behind. By means of these muscles the plug can be raised to open or lowered to close the aperture into the ventricle. No other valves are developed at the atrio-ventricular opening, and the plug probably represents an enlargement of the ventral of the two endocardial cushions here developed in other forms, Fig. 562.

The large conus, developed from the bulbus cordis, has but feebly developed musculature anteriorly and a marked spiral twist. Robertson describes the development in the bulbus cordis of *Lepidosiren* of four longitudinal ridges interrupted in the middle region and sharing in the

spiral twist.[1] From these arise the valves and ridges of the adult conus.
The dorsal and ventral ridges
are vestigial; posteriorly they
give rise to vestigial pocket
valves round the base of
the conus. The right main
ridge becomes continuous and
develops into the 'spiral
valve', which meets and even
fuses anteriorly with the opposing shorter left ridge. These
two ridges meeting subdivide
the lumen of the conus in front
into dorsal and ventral channels (as in Amphibia). Since
the hinder end of the right
ridge thus passes to the left
and becomes ventral, owing to
the spiral twist, it continues
the line of the interventricular and interauricular septa.
Thus the cavity of the whole
heart is incompletely divided
longitudinally into two channels. The venous stream on
the right is driven mainly into
the dorsal channel, and the
arterial stream on the left into
the ventral channel of the
anterior region of the conus.
In this way the first two adult
arterial arches, with which
are connected the carotid
arteries, are supplied mainly
with arterial blood; while the
more venous blood passes to
the remaining arterial arches,
including the last which gives

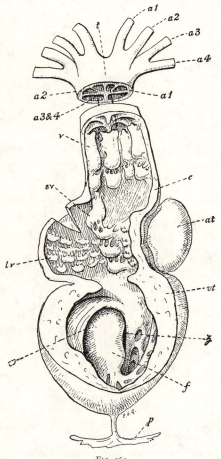

FIG. 563.

Ceratodus Forsteri, Krefft. Ventral view of the heart
dissected so as to expose the inside of the ventricle and
conus, and the disposition of the aortic arches. a^{1-4}, Four
aortic arches (a dotted line passes up the base of the 1st
and combined 3rd and 4th); *at*, atrium; *c*, cut wall of
conus; *f*, plug filling the atrio-ventricular opening; *lv*,
small posterior valves; *p*, portion of wall of pericardium;
sv, specialised row of enlarged valves; *t*, truncus; *v*,
anterior valve; *vt*, cut wall of ventricle; *w* and *z*, dotted
lines passing into the sinus venosus. (Compare Fig. 564,
p. 552.) (From Goodrich, *Vert. Craniata*, 1909.)

off the pulmonary artery (p. 583). For the ventral aorta of Dipnoi is

[1] The apparent spiral twist of this region in Dipnoi and Tetrapoda is
associated with the kinking of the elongating tube (Kerr, **840**; Bremer, **815**).

shortened up into a truncus remarkably like that of Amphibia, and as
in the latter the walls between the arches are carried back so as to sub-

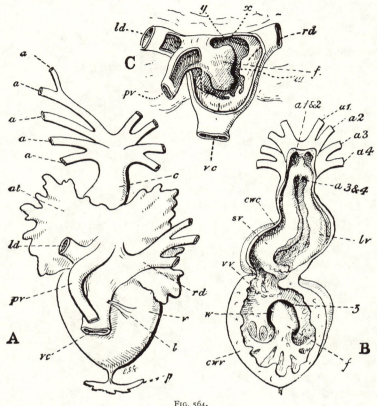

FIG. 564.

Heart of *Protopterus annectens*, Owen. A, Dorsal view; B, opened, ventral view; C, the sinus
venosus opened. *a*, Cut arterial arches; *a* 1 and 2, entrance from conus to first two arches; *a* 3 and
4, entrance to last two arches; *at*, atrium; *c*, conus arteriosus; *cwc*, cut wall of conus; *cwv*, cut
wall of ventricle; *f*, fibrous plug closing the passage from ventricle to atrium and passing into the
sinus; *l*, dorsal attachment to pericardial wall; *ld*, left ductus Cuvieri; *lv*, longitudinal ridge;
p, small portion of pericardial wall; *pv*, pulmonary vein; *rd*, right ductus Cuvieri; *sv*, longitudinal
compound valve; *v*, ventricle; *vc*, vena cava inferior; *vv*, row of small valves; *w*, dotted line
indicating course of venous blood from the shallow sinus venosus, through the atrium (C) into
the ventricle on the right of the plug (B). In C the dotted lines *ld*, *rd*, and *vc* pass into the sinus
venosus; the lines *x* and *y* into the cavity of the atrium opening widely into the sinus; the line *pv*
passes down the pulmonary vein to enter the ventricle on the left of the plug (*z* in B). (From
Goodrich, *Vert. Craniata*, 1909.)

divide its lumen into three paired channels : two ventral leading to adult
arches 1 and 2, and one dorsal leading to arches 3 and 4 on each side.

There is, then, a very striking resemblance in the structure of the
heart and truncus of the Dipnoi and Amphibia, which cannot be put
down entirely to convergence, and clearly points to the development of

a similar structure in their common ancestor when still at an aquatic and fish-like stage. Nevertheless, to some extent the resemblance may be due to convergence; for *Ceratodus*, the most primitive of living Dipnoi, has a less specialised heart than *Lepidosiren* or *Protopterus*. For instance, in *Ceratodus* the valves of the conus are more numerous, regular, and less fused into ridges; they are more like the valves of the Selachian conus. There are anteriorly four longitudinal rows of valves, of which the first tier is enlarged, and a gap separates the second tier from the four posterior tiers of numerous small valves in three of the rows.

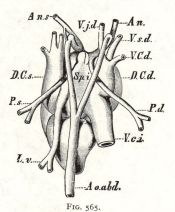

Fig. 565.

Heart of *Cyclodus boddaertei*. From the dorsal side. (After Röse.) *An, An.s*, Innominate arteries; *Ao.abd*, dorsal aorta; *D.C.d, D.C.s*, precaval veins; *L.v*, pulmonary vein; *P.s, P.d*, pulmonary arteries; *Sp.i*, spatium intersepto valvulare; *V.C.d*, posterior cardinal; *V.j.d*, jugular, and *V.s.d*, subclavian vein of the right side; *V.c.i*, postcaval vein. (From Wiedersheim, *Comp. Anatomy*.)

The main right longitudinal row is continuous and spiral, but its constituent six valves are not fused into a ridge. The tendency to form longitudinal ridges is more pronounced and the interventricular and interauricular septa are more developed in the specialised Dipneumones than in *Ceratodus*. In these respects, then, some of the Dipnoi have probably advanced beyond the level of specialisation reached by the ancestor common to them and the Tetrapoda. Indeed, although an interventricular septum reappears in all Amniota, it is very doubtful whether it is a common ancestral structure, since it is not found in any modern Amphibian.

The Amniote heart reaches an altogether higher grade of specialisation than the Amphibian. Even in the lower forms it is more compact, with the auricles carried forwards on to the anterior surface of the large conical ventricle, and the reduced sinus venosus on to the dorsal surface of the auricles, so that the bases of the arterial arches and great veins come nearer together, Figs. 552, 571-2. The sinus tends to disappear during development, being flattened and more or less merged, as it were, into the wall of the right auricle. An important advance, usually not sufficiently noticed, is the disappearance of the conus as such in all Amniotes. For the cavity of the bulbus cordis becomes subdivided by a longitudinal spiral septum down to its base, so that completely separated pulmonary and aortic channels are formed with separate openings, provided with semilunar pocket valves, into the ventricle (or ventricles). The former channel opens

backwards towards the right and the latter towards the left. The wall
of this region ceases to be contractile, loses its striated musculature,
and becomes fibrous like that of the large arteries with which it is continuous. Furthermore, except in some of the smaller and lower Reptiles,
the whole conus region be-
comes so completely sub-
divided that it is split into
two separate tubes or
trunks, of which the pul-
monary trunk passes for-
wards ventrally to the left
round the aortic trunk, and
then dorsally to the lungs.

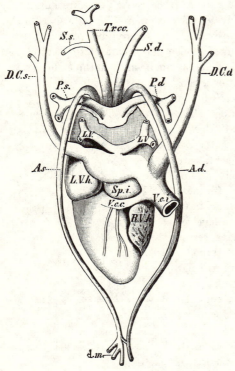

FIG. 566.

Heart of a young *Crocodilus niloticus*. From the dorsal
side (after Röse). *A.d* and *A.s*, Right and left aortic arches;
A.m, mesenteric artery; *L.V.h*, *R.V.h*, left and right atria;
S.d, *S.s*, subclavian arteries; *Tr.cc*, common carotid; *V.c.c*,
coronary vein; *D.C.d*, *D.C.s*, precaval veins; *L.V*, pulmonary vein; *P.s*, *P.d*, pulmonary arteries; *Sp.i*, spatium
intersepto-valvulare; *V.c.i*, postcaval vein. (From Wiiders-
heim, *Comp. Anatomy*.)

The development of the
Amniote heart has been
much studied; among re-
cent writers one may con-
sult Langer (841-2), Hochstetter (835-6), Greil (831)
for Reptiles and other
Amniotes; Masius (847),
Greil (831), Fuchs (820)
for Birds; Lockwood
(1888), Röse (862), Born
(1888–9), Langer (842),
Tandler, and others for
Mammals. It may here be
mentioned that the bulbus
in Mammals, and perhaps
also in other Amniotes,
seems to contribute to the
wall of the definitive right
ventricle, and that longitudinal endocardial ridges
(typically four in number) develop in it as in other Craniates. The
semilunar valves at the base of the large vessels are derived from their
posterior ends, while the dividing septum mentioned above is first formed
by the meeting across and fusion of the right and left ridges. This
septum pulmo-aorticum, then, may be compared to the spiral folds of
Amphibia and Dipnoi described above.

Another very important new feature is the formation of a septum

more or less completely subdividing the cavity of the ventricle into
right and left chambers. Already in Amphibia the venous and arterial

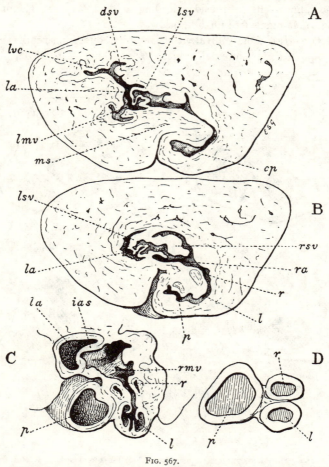

FIG. 567.

Chelone midas. Successive transverse sections through ventricle and base of aortic arches seen
from behind (posterior view) with dorsal edge above and ventral edge below. A, Most posterior,
and *D*, most anterior section. *cp*, Cavum pulmonale; *dsv*, muscles representing dorsal region of
interventricular septum; *ias*, interauricular septum; *l*, left aortic arch; *la*, left auriculo-ventricular
opening; *lmv*, left auriculo-ventricular marginal valve; *lsv*, left septal (medial) auriculo-ventricular
valve; *lvc*, left ventricular chamber; *ms*, incomplete muscular interventricular septum; *p*, pulmonary trunk; *r*, right aortic arch; *ra*, right auriculo-ventricular opening; *rmv*, right auriculo-
ventricular marginal valve; *rsv*, right septal (medial) auriculo-ventricular valve; *v*, ventricle. (Figs.
567-70 and 580, from Goodrich, *J. of Anat.*, 1919.)

blood-streams are separated so far as the ventricle by an interauricular
septum, but it is not until the higher Amniota that the separation already
foreshadowed in Dipnoi is at last accomplished by a similar partition in
the ventricle.

An interesting point of great phylogenetic significance and not usually appreciated is that this has been completed along two independent divergent lines—the Sauropsidan leading to Birds and the Theropsidan leading to Mammals (see further, below, p. 572).

In Reptiles, though already considerably reduced in size, the sinus is usually quite distinct internally, Figs. 565-6. The right ductus Cuvieri

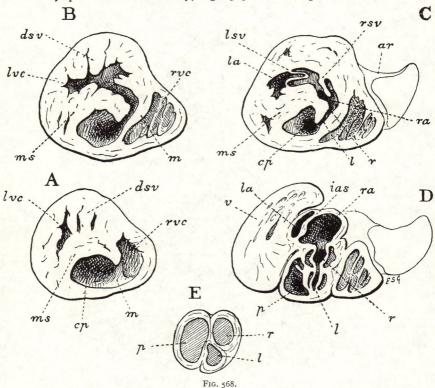

FIG. 568.

Varanus sp. Successive transverse sections through heart, seen from behind. A, Most posterior, E, most anterior section. Lettering as in Fig. 567.

(anterior vena cava dextra) tends to open anteriorly to the vena cava inferior, and the left ductus (ant. vena cava sinistra) passes across to open near the latter. A partial septum sinu-venosi may arise from the wall between these two openings and that of the right ductus. The well-developed valves guarding the sinu-auricular opening into the right auricle are set almost transversely, one anterior and one posterior. In the Reptilian heart, then, the sinus opens into the larger right auricle and the pulmonary vein into the smaller left auricle, and these two cavities are separated

by an interauricular septum always complete in the adult, but pierced
by secondary temporary perforations in the embryo which are closed
when pulmonary respiration is established (such perforation also occurs
in Birds and Mammals).

Along the free edge of this interauricular septum crossing the atrio-
ventricular opening are attached two membranous valves projecting into

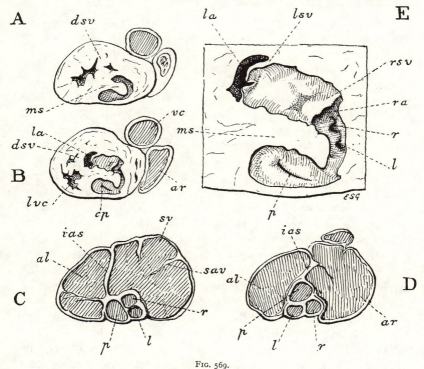

FIG. 569.

Python molurus. Successive transverse sections through heart, seen from behind. E, Portion of
Fig. B on a larger scale. *al*, Left auricle ; *ar*, right auricle ; *vc*, vena cava inferior. Other letters
as in Fig. 567.

the ventricular cavity ; the opening is thus subdivided into separate right
and left auriculo-ventricular apertures, Fig. 570. In ontogeny the valves
are derived from a posterior (dorsal) and an anterior (ventral) endocardial
cushion (doubtless representing the valves of lower forms), which meet
and fuse along the free edge of the septum. Into the cavity of the ventricle
projects in all Reptiles an interventricular septum, complete only in Crocodilia, Fig. 587. Leaving for the present the Crocodilia to be dealt with
later (p. 563), the second important new feature we meet in the reptilian
heart is this incomplete interventricular muscular septum, essen-

tially median and posterior in origin, and tending to divide the cavity into right and left chambers (cavum venosum and cavum arteriosum respectively). It is formed from ingrowing muscular trabeculae. Posteriorly, towards the apex of the ventricle, the septum is complete and dorsoventral ; farther forwards, where it is incomplete, its ventral base of attachment (following the external sulcus interventricularis) shifts to the left and becomes continuous with the inpushing due to the bulbo-auricular groove. Thus anteriorly, where the septum reaches near the base of the arterial trunks, it extends almost horizontally with its free edge to the right and joins the ventral endocardial cushion. In this region the cavum arteriosum opens over the edge of the septum freely into the cavum venosum, which is prolonged forwards and to the left into the ventral cavum pulmonale leading to the opening of the pulmonary trunk (see below). The cavum arteriosum (left ventricular cavity) leads antero-dorsally to the septum towards the opening of the right carotico-systemic trunk, while the opening of the left systemic trunk is situated almost opposite the free edge of the septum. Both auriculo-ventricular apertures are dorsal to the septum. From the left auricular opening a stream of blood passes back into the cavum arteriosum, and from the right auricular opening into the cavum venosum. On contraction of the ventricle the venous blood passes into the pulmonary trunk from the cavum pulmonale, mixed blood into the left systemic trunk, and the purest arterial blood into the right carotico-systemic trunk whose opening is a little farther forward. The position of the opening into the left systemic trunk varies a little in different forms, being nearer the opening of the right trunk in Ophidia, and of the pulmonary trunk in Chelonia ; but the general disposition of the three openings is remarkably constant throughout the Reptilia, Figs. 547-70.

For it is one of the most characteristic and important features of the heart of all living Reptilia that the cavity of the bulbus cordis becomes subdivided into three separate channels (p. 525), Fig. 571 B. Not only does the main septum pulmo-aorticum grow back to close off the pulmonary trunk (as in all Amniotes), but the remaining trunk is also subdivided by a longitudinal spiral aortic septum into two aortic trunks : one, the carotico-systemic trunk, coming from the left side of the ventricle passes to the right, gives off the carotid arteries, and continues as the right systemic arch ; while the other, coming from the right side of the ventricle, passes over to the left, and is continued as the left systemic arch. A pair of semilunar pocket valves protects the openings of each of the three trunks.

An important point to notice in the arterial system of all Reptilia (and also of Birds), related to the structure of the heart just described, is that the right and left carotid arches come off from the same trunk as the right systemic arch, that is, from the trunk which receives the purest blood from the left side of the ventricle.[1]

The way in which the lumen of the truncus and bulbus cordis becomes subdivided during ontogeny in Reptiles and Birds must now be explained (Langer, 841-2 ; Hochstetter, 836 ; Greil, 831 ; Kerr, 840). As in Dipnoi

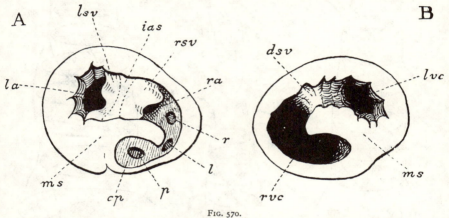

FIG. 570.

Diagrams of ventricle of *Reptilian heart* (not Crocodilian). A, Cut transversely and seen from behind ; position of interauricular septum is indicated by dotted lines. B, Cut transversely and seen from in front. Lettering as in Fig. 567.

and Amphibia, the lumen of the truncus becomes divided into dorsal pulmonary and ventral carotico-systemic channels by the backward growth of the wall between the bases of the fourth and sixth primary arches, Fig. 571. This horizontal septum joins the anterior ends of the right and left anterior endocardial ridges of the bulbus. Four such ridges develop in the anterior region of the spirally twisted bulbus cordis, and two approximately dorsal and ventral ridges in the posterior region.[2] The right anterior ridge is the largest and fuses across with the left, thus

---

[1] The roots or base of the subclavian arteries to the fore-limbs arise from a pair of segmental arteries, and primitively branch off from the dorsal aorta. Owing to the secondary extension backwards of the bifurcation of the two aortic stems, the subclavians may, as in Lizards, arise from the right aortic arch. A significant fact is that in Chelonia, Crocodilia, and Aves these primary subclavians are replaced in ontogeny by secondary subclavian arteries coming from the carotid arches, and whose roots have arisen from more anterior segmental arteries.

[2] It will be understood (see pp. 538, 550) that these longitudinal ridges represent originally continuous ridges or rows of valves, extending along the whole bulbar region, and perhaps even as ridges into the ventricle.

continuing backwards the subdivision into pulmonary and caroticosystemic channels ; and this aortico-pulmonary septum is completed spirally behind by the junction of the right anterior bulbar ridge with

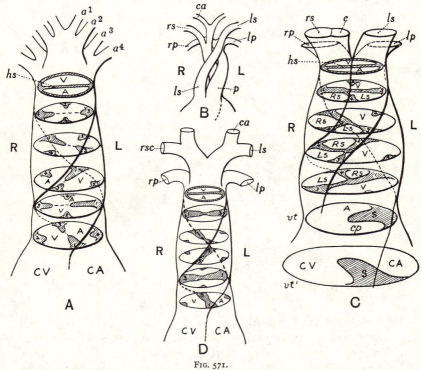

FIG. 571.

Diagrams illustrating the spiral subdivision of *Bulbus cordis* in Dipnoi, Amphibia, Reptilia, Aves, and Mammalia : ventral view ; transverse sections drawn at successive levels showing disposition of endocardial ridges and septa derived from them ; spiral lines indicate their attachment to wall along which tube becomes subdivided in higher forms (solid lines ventrally become broken lines when passing dorsally to bulbus). A, Primitive disposition of ridges in embryo Amniote and retained in adult Dipnoan and Amphibian ; B, oblique view of three trunks formed in Reptile ; C, ridges and septa in Reptile (in Bird ridges 3 and 4 come together and obliterate Ls) ; D, Mammal. Most anterior section shows A, arterial, and V, venous channels separated by horizontal septum, hs, in truncus ; A leads to left side of ventricle (CA, cavum arteriosum), V leads to right side (CV, cavum venosum). 1, 2, 3, 4, endocardial ridges of spiral bulbus (conus) ; a^{1-4}, adult arterial arches, venous channel leads to $a^{3, 4}$, arterial channel to $a^{1, 2}$; c, carotid trunk ; ca, carotid arch (a^1) ; cp, cavum pulmonale ; p, rp, left and right pulmonary arch (a^4) ; ls, Ls, left, rs, Rs, right systemic arch (a^2) ; p, pulmonary trunk ; rsc, right subclavian (base of rs) ; S, interventricular septum ; vt and vt^1, sections across anterior end of ventricle.

the ventral posterior and the left anterior ridge with the dorsal posterior, these two posterior ridges also fusing across.[1] The pulmonary channel now opens into the ventricle (cavum pulmonale) ventrally towards the

---

[1] The two posterior ridges are doubtless the hinder portions of the right and left anterior ridges, the originally continuous ridges having been interrupted (see p. 550).

right, and the carotico-systemic opens into the ventricle more dorsally. The further subdivision of the carotico-systemic channel in Reptiles into left systemic (opening to the right) and right carotico-systemic (opening to the left) is brought about by the formation and backward extension of a septum aorticum derived from the right and ventral endocardial

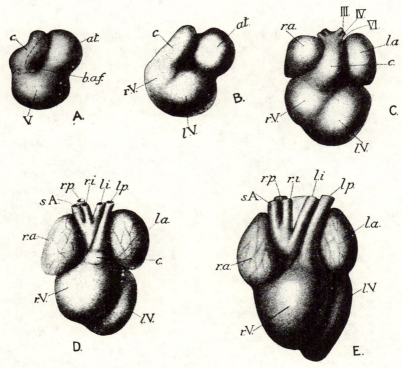

Fig. 572.

Illustrating the development of the heart in the fowl. (After original drawings by Greil.) *at*, Atrium; *b.a.f*, bulbo-auricular fold; *c*, conus; *l.a*, left auricle; *l.i*, left innominate artery; *l.p*, left pulmonary; *l.V*, left ventricle; *r.a*, right auricle; *r.i*, right innominate artery; *r.p*, right pulmonary; *r.V*, right ventricle; *s.A*, systemic aorta. (From Kerr, *Embryology*, 1919.)

ridges of the bulbus. The pairs of pocket valves at the base of each of these three channels are derived from the posterior ridges.

In Birds (Langer, **842** ; Takahashi, **874**) the anterior left and ventral ridges appear to have combined, and no subdivision of the carotico-systemic channel at the base of the trunk occurs, since the left systemic arch is obliterated before these septa are completed. Thus although the two systemic arches are present in early stages, the Bird, so far as known, does not normally pass through a Crocodilian stage with the left opening independently into the ventricle (this may occur occasionally as an abnormality [Bremer]).

The formation of the longitudinal septa from the bulbar ridges in Dipnoi and Tetrapoda is explained in the diagrams here given, Figs. 563, 571. Whether their spiral course is due to the actual twisting of the heart or merely to their kinking and spiral growth is doubtful; but if the bulbus cordis is twisted to the right the more posterior region must be twisted to the left, since the two extremities are fixed (Kerr, 840).

In the majority of modern Reptilia, the attempt, so to speak, to separate the arterial from the venous stream is only partially carried out in the heart itself, and in the peripheral arteries mixed blood is again brought into the circulation by the opening of the left systemic arch into the dorsal aorta. Along the Sauropsidan phyletic line only the Birds have succeeded in completely separating the two streams. For in these warm-blooded Amniotes not only is the arterial blood of the left ventricle completely shut off from the venous blood of the right ventricle by the completion of the interventricular septum, but the now useless, or actually harmful, left systemic arch is entirely suppressed, Figs. 572-3.

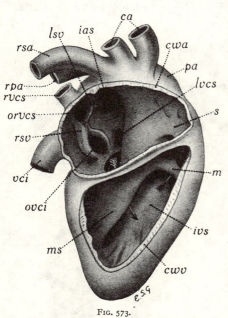

FIG. 573.

Heart of Swan, *Cygnus olor*; seen from right side, and opened so as to expose cavity of right auricle and of right ventricle (an arrow passes through auriculo-ventricular opening). *ca*, Carotid arches; *cwa*, cut wall of right auricle; *cwv*, cut wall of right ventricle; *ias*, interauricular septum; *ivs*, interventricular septum; *lsv*, left sinu-auricular valve; *lvcs*, opening of left vena cava superior; *m*, muscular band; *ms*, muscular portion of auriculo-ventricular valve; *orvcs*, opening of right vena cava superior; *ovci*, opening of vena cava inferior; *pa*, base of pulmonary arch; *rpa*, right pulmonary arch; *rsa*, right systemic arch; *rsv*, right sinu-auricular valve; *rvcs*, right vena cava superior; *s*, septal portion of auriculo-ventricular valve; *vci*, vena cava inferior. An arrow passes under *m* into base of pulmonary trunk.

In the avian heart the sinus venosus is so much reduced and sub-divided that the great veins appear to open separately into the right auricle (Röse, 862; Kern, 839). The two sinu-auricular valves may be much modified. The pulmonary veins may also open separately into the left auricle. The ventricles become very unequally developed, the wall of the left being thicker and more muscular than that of the right which partially surrounds it. The left auriculo-ventricular aperture is provided with membranous valves surrounding it (usually an outer and an inner or septal valve). The valve guarding the right auriculo-ventricular aperture is very characteristic. Instead of the membranous valve attached to the interauricular septum found in Reptiles, there is in Birds a large, almost entirely muscular valve attached round the ventral edge of the aperture and to the outer wall of the right ventricle, and stretching far into the ventricular cavity; it appears to be almost entirely developed from the ventricular muscle, and is held ventrally by a stout muscular bridge to the outer wall; the reptilian 'septal' valve seems to be represented by at most a membranous vestige passing round to the inner or medial wall of the ventricle. The only trunk issuing from the right ventricle is the pulmonary, whose opening is provided with three semilunar valves. The only trunk issuing from the left ventricle is the carotico-systemic, also having three semilunar valves at its base; this trunk passes dorsally to the right over the pulmonary, and splits into the carotid arteries and right systemic arch. Thus in the adult Bird the pulmonary and systemic circulations are completely separated, Figs. 573-9. The resemblances of the 'four-chambered' avian heart to that of the mammal are superficial and misleading, and the clue to its structure and origin must be sought in the crocodilian heart (Beddard and Mitchell, 808; Greil, 831; Goodrich, 517, 826). For in the Crocodilia also the interventricular septum has been completed to separate the cavity of a left ventricle with powerful muscular walls from that of a weaker right ventricle, in such a way that the pulmonary and left systemic trunks receive all the venous blood and the right carotico-systemic trunk all the arterial blood. As in other Reptiles the aortic trunks cross at their base, and the interventricular septum is finally completed by fusing with the wall formed between them; so that the right trunk opens into the left cavity, and the left trunk together with the pulmonary into the right cavity. A comparison of the avian and crocodilian hearts shows that they are built on the same plan, agree in almost every detail, except that the crocodilian is more primitive and preserves the left systemic arch and trunk. Near the base of the systemic trunks, just anterior to the semilunar valves, there is in the Crocodilia a small foramen (foramen of Panizza) allowing blood to pass from one arch to the other, and perhaps serving to equalise the pressure within them.[1] The muscular valve, at the right auriculo-

---

[1] According to Greil this is a secondary perforation formed late in development. It would seem more probable, however, that it is a remnant of the original communication.

ventricular aperture, so characteristic of Birds, is already developed in Crocodilia, which also preserve a considerable portion of the membranous septal valve, Figs. 580-82, 587.

Exactly how the interventricular septum becomes completed in Birds has been studied by Lindes (1865), Masius (847), Hochstetter (835), Lillie (845), and more recently by Takahashi (874). In the embryonic heart, by the time the endocardial cushions have fused to divide the right from the left auriculo-ventricular apertures,

FIG. 574.

Transverse section through the ventricles of *Grus cinerea*. *S*, septum ventriculorum ; *Vd*, right, and *Vg*, left ventricle.

the interventricular septum has reached the dorsal cushion above and the ventral cushion below ; but the cavum arteriosum still freely com-

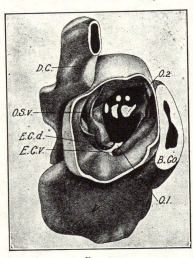

FIG. 575.

Reconstruction of heart of chick embryo of 5·7 mm. head-length, seen from right side. Part of wall of right auricle is cut away. (After Masius, from F. R. Lillie, *Develt. Chick*, 1919.) *B.Co*, Bulbus cordis ; *D.C*, duct of Cuvier ; *E.C.d*, *v*, dorsal and ventral endothelial cushions ; *O.S.v*, opening of sinus venosus into right auricle ; *O.1*, passage below interauricular septum ; *O.2*, secondary ostia in interauricular septum.

municates in front of the free edge of the septum with the cavum venosum on the right. This portion of the ventricular cavity now appears as a narrow passage running in a groove below the united cushions, Fig. 577. The interventricular septum next grows on to the endocardial rudiment of the right septal valve, and joining the septum aorticum of the bulbus finally closes the communication completing the wall between the arterial and venous streams, Fig. 578. The morphology of this wall is better understood by studying its development in the Crocodilia. Here the right septal valve is well developed, and its free membranous edge remains in the adult projecting into the right ventricle attached to the interventricular septum, Fig. 582 ; its base, however, is involved in the completion of the

septum. The closing of the venous from the arterial channel takes place by the centripetal growth inwards of the edge of the interventricular septum from behind and above, the septum aorticum (between the right and left systemic trunks) from in front, and the endocardial rudiment of the right septal valve, which all combine to complete the spiral partition, Figs. 580-81. The more or less complete disappearance of the right septal valve in Birds is probably related to its sharing in the formation of this partition.

The history of these atrio-ventricular valves may be briefly given as follows. In the Amphibia the dorsal and ventral endocardial cushions spread round the single atrio-ventricular aperture and give rise to two main dorsal and ventral valves, and two smaller right and left valves,

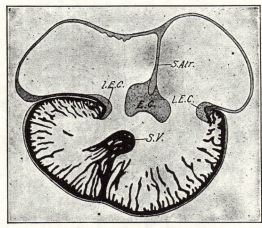

Fig. 576.

Frontal section of heart of chick embryo of 9 mm. head-length. After Hochstetter (from F. R. Lillie, *Develt. Chick*, 1919). *E.C*, Median endothelial cushion ; *l.E.C*, lateral endothelial cushion ; *S.Atr*, septum altriorum ; *S.V*, septum ventriculorum.

Fig. 558. In the Amniota the cushions meet and fuse across with each other and the free edge of the interauricular septum, thus dividing the aperture into right (venous) and left (arterial) openings. The dorso-ventral endocardial ridge thus formed gives rise to a 'septal' valve on either side in all Reptilia, Fig. 570. But in the Crocodilia, although similar septal valves are developed, the endocardial cushion on the left also spreads round the auriculo-ventricular aperture and forms an opposing valve attached to the outer wall of the left ventricle ; while at the right auriculo-ventricular opening the right septal valve is supplemented by a muscular valve from the outer wall of the ventricle. The atrio-ventricular valves of Birds resemble those of Crocodiles, but are more specialised.

The Mammalia also have succeeded in separating the venous from the arterial circulation, but in a different way. In the mammalian heart the sinus is more completely suppressed than in other Amniotes,[1] its remains being incorporated in development into the wall of the right

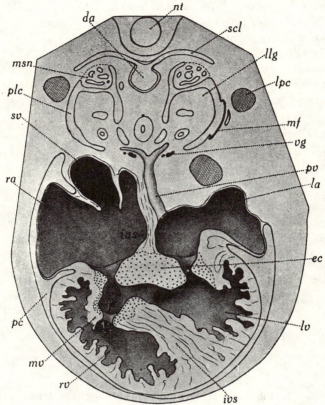

FIG. 576a.

Transverse section of trunk of embryo duck of 7 days, *Anas platyrhyncha*, showing heart cut through auriculo-ventricular openings and seen from in front. *da*, Dorsal aorta ; *ec*, endocardial cushion, and *ivs*, incomplete interventricular septum, with interventricular passage still widely open between ; *la*, *ra*, left and right auricles ; *llg*, left lung ; *lpc*, left cardinal vein ; *lv*, *rv*, left and right ventricles ; *mf*, lip of Müllerian funnel ; *msn*, mesonephros ; *mv*, muscular valve attached by strand to septum (arrow passes behind) ; *nt*, notochord ; *pc*, pericardial cavity ; *plc*, pleural cavity ; *pv*, pulmonary vein ; *scl*, primitive subclavian ; *sv*, sinus venosus ; *vg*, branch of vagus.

auricle, and the three great veins coming to open directly into the auricular cavity. The sinu-auricular valves, still fairly well shown in the Monotremata, are much reduced and modified in Ditremata. On the dorsal

[1] This suppression of the no longer necessary sinus venosus seems to have occurred to some extent independently in various groups, and the relative position of the venous apertures, as well as the modification of the valves, varies much in different forms (Röse). That portion of the auricular wall derived from the sinus is smoother and less muscular than the rest.

wall of the auricle opens the right vena cava superior, near the middle
the vena cava inferior, and more posteriorly and to the left the left

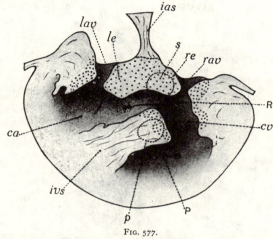

FIG. 577.

Embryo duck, *Anas* : Posterior portion of ventricles of heart shown in Fig. 576 seen from behind.
Upper arrow passes into base of right systemic trunk, *R* ; lower arrow into base of pulmonary trunk, *P*.
ca, cv, Cavum arteriosum and cavum venosum ; *lav, rav*, left and right auriculo-ventricular openings ; *le, re*, left and right endocardial rudiments of septal valves ; *p*, dotted line shows position farther
forward of pulmonary opening, and *s* position of systemic opening. Interventricular passage still
widely open.

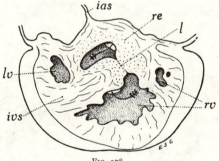

FIG. 578.

Ventricles of heart of *Passer* shown in Fig. 579, cut farther forward. Vestige of closure of
interventricular passage at dotted line, *l*. Upper arrow into base of right systemic trunk, lower arrow
into pulmonary trunk. Other letters as in Fig. 579. (Partly from wax model made by G. R. de Beer.)

v. c. superior.[1] The right sinu-auricular valve extends along the right
edge of the openings, and remains as a rule in the adult as a Eustachian valve protecting the opening of the v. c. inferior, and Thebesian

[1] In many mammals (Edentate, Carnivora, Primates) the left vena cava
superior disappears after the formation of anastomosis carrying the venous
blood from the left side into the right v. c. superior.

valve protecting the openings of the left v. c. sup. and coronary sinus.

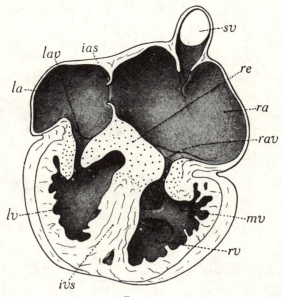

FIG. 579.

Heart of embryo *Passer domesticus*, cut transversely through auriculo-ventricular openings, *lav*, *rav*, at stage immediately after completion of interventricular septum, *ivs*; seen from behind (cp. Fig. 577). *ias*, Interauricular septum (pierced); *mv*, right muscular valve. Other letters as in previous figures. (Partly from wax model by G. R. de Beer.)

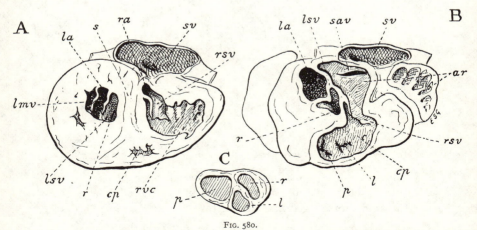

FIG. 580.

Caiman sclerops. Successive transverse sections through the heart, seen from behind : A, Most posterior ; C, most anterior. *s*, Interventricular septum ; *sv*, sinus venosus ; *sav*, sinu-auricular valve. Other letters as in Figs. 568-9.

The pulmonary veins come to open separately into the left auricle, owing to the absorption of their common stem into its wall, Fig. 583.

In development the first formed interauricular septum fuses with the dorsal and ventral endocardial cushions which join across, dividing the

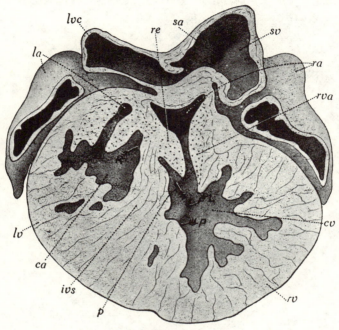

FIG. 581.

Heart of embryo *Crocodilus* sp. cut transversely through auriculo-ventricular openings at stage before completion of interventricular septum ; seen from behind. *ca*, Cavum arteriosum, anterior region of left ventricular cavity ; *cv*, cavum venosum, anterior region of right ventricular cavity ; *ivs*, muscular interventricular septum ; *L*, arrow into base of left systemic trunk ; *la*, *ra*, left and right auricles (wall partially cut away) ; *lv*, *rv*, cut wall of left and right ventricles ; *lvc*, left vena cava anterior ; *P*, arrow into base of pulmonary trunk ; *p*, arrow into interventricular passage over free edge of septum ; *R*, arrow into base of right systemic trunk ; *re*, right septal valve ; *rva*, right valve guarding auriculo-ventricular opening; *sa*, sinu-auricular opening; *sv*, sinus venosus. Endocardial cushions dotted.

ventricular aperture into right and left atrio-ventricular openings. This septum becomes perforated (foramen ovale) in the embryo (as in Reptiles and Birds) and is later completed by a second septal fold in combination with the left sinu-auricular valve. The cushions grow round the two openings to form the auricular ventricular valves (three tricuspid valves on the right, and two mitral valves on the left).[1] Meanwhile a longitudinal

[1] The conversion of the thickenings into membranous valves takes place by their being hollowed out into thin folds from the ventricular side leaving muscular cords stretching from the edge to the wall of the ventricle. Later the muscle of the valve and the cords is replaced by connective tissue. In

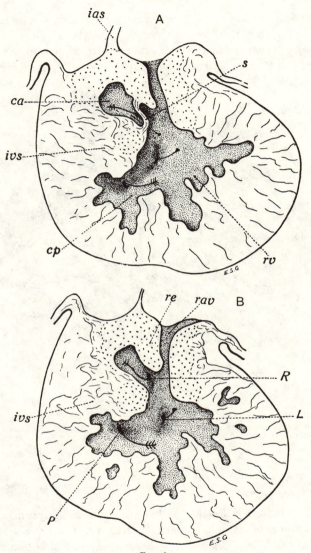

Fig. 582.

Ventricles of heart of *Crocodilus* shown in Fig. 581, cut farther forward in A and still farther forward in B.   A pin is shown in A, through interventricular passage which is fully opened up in B.   *s*, Thin fold which will by growing forward to right finally separate cavum arteriosum from cavum venosum ; *rav*, right auriculo-ventricular opening.   Other letters as in Fig. 581.

the Monotreme Ornithorhynchus, however, a considerable amount of muscle remains in the outer flap of the tricuspid valve, while the inner flaps are small, giving it a certain resemblance to the right valves of a bird (Lankester, 1882–3).

muscular interventricular septum grows forwards from the postero-
dorsal wall of the ventricle to meet and fuse with the endocardial cushions
between the two auriculo-ventricular apertures. Right and left ventricular
cavities are thus formed still communicating with each other in front
near the base of the arterial trunks, Figs. 584-5.

As usual in Amniotes, the primitive ventral aorta becomes subdivided into dorsal pulmonary and ventral aortic (carotico-systemic)

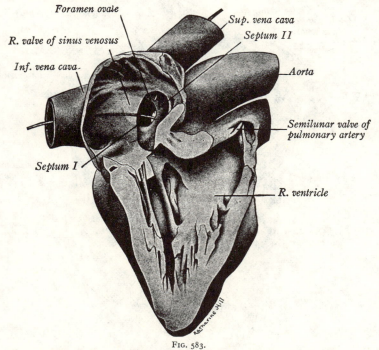

Foramen ovale

Sup. vena cava

Septum II

R. valve of sinus venosus

Inf. vena cava

Aorta

Semilunar valve of
pulmonary artery

Septum I

R. ventricle

Fig. 583.

Lateral dissection of heart of 65 mm. human foetus viewed from right side. × 12. (From Prentiss
and Arey, *Text-book of Embryology*, 1917.)

channels, and this subdivision is carried backwards throughout the bulbus
cordis by the formation of a spiral septum due to the fusing across of the
right and left endocardial ridges, Figs. 571, 584, 586. At their junction
with the ventricle the septum pulmo-aorticum, between the pulmonary
trunk on the right and aortic on the left, becomes vertical, and fusing
with the edge of the interventricular septum closes the interventricular
canal. The right ventricle now leads only to the pulmonary trunk, and
the left only to the carotico-systemic trunk. The division of the whole
heart into right and left halves is now completed, and the venous

pulmonary circulation separated from the arterial systemic circulation (except for the mixture in the embryo by means of the temporary foramen ovale). It should be noticed that the carotico-systemic vessels are at first symmetrical; from the median trunk come off right and left systemic arches (each giving off a subclavian artery), and right and left carotid arches. Later, while the left systemic or aortic arch remains

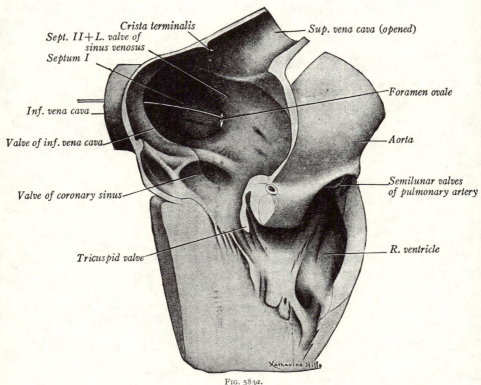

FIG. 583a.

Lateral dissection of heart of 105 mm. human foetus viewed from right side. × 7. (From Prentiss and Arey, *Text-book of Embryology*, 1917.)

complete and passes into the dorsal aorta, the right is interrupted dorsally and posteriorly, ceases to join the aorta, and forms in the adult mammal merely the base of the right subclavian. As usual, the pocket valves protecting the entrance to the pulmonary and aortic trunks are formed from the posterior ends of the endocardial ridges.

The seldom fully appreciated but nevertheless great significance in phylogeny of the structure of the heart may now be pointed out. Since all living Reptiles possess the reptilian type of heart, with its quite characteristic specialisations, they have in all probability been derived from

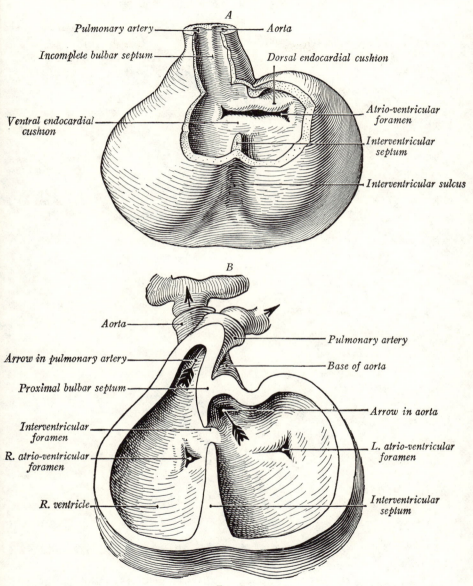

FIG. 584.

Ventral view of stages in development of heart to show differentiation of bulbus cordis into aorta and pulmonary trunk (Kollman) : A, Heart of 5 mm. human embryo ; B, of 7·5 mm. human embryo. (From Prentiss and Arey, *Text-book of Embryology*, 1917.)

a common ancestor in which the reptilian type had been developed. The
Birds also no doubt came from that same stock, and branched off not

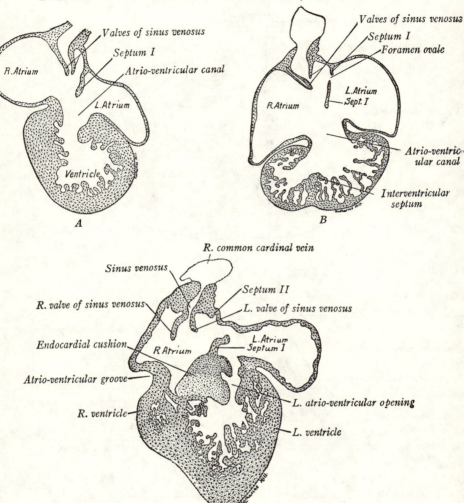

Fig. 585.

Horizontal sections through chambers of human heart : A, 6 mm. ; B, 9 mm. ; C, 12 mm. (A and B are based on figures of Tandler.) ×About 50. (From Prentiss and Arey, *Text-book of Embryology*, 1917.)

far from the Crocodilia. But the Mammalia must have branched off
and diverged from a common Amniote ancestor before the Reptilian
type of specialisation had begun, since, once committed to this line of

specialisation, the heart must inevitably evolve in the reptilian and crocodilian direction, Fig. 587. Now there is good reason to believe that certain extinct forms (Theromorpha) usually included in the Class Reptilia are, if not the ancestors of the Mammalia, at least closely allied to them. These so-called Reptiles, then, probably possessed a heart of the Mammalian type, or at all events capable of giving rise to this type. It follows that the Class Reptilia, as commonly understood, is an artificial polyphyletic group containing, besides certain primitive Amniotes (Cotylosauria), two distinct diverging branches : one leading to modern Reptilia and eventually also to Birds, and the other leading to Mammalia. These forms may be called the Sauropsidan and Theropsidan Reptiles respectively (Goodrich, 517). While the Reptilia Theropsida are all

Aorta

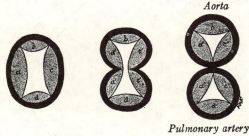

Pulmonary artery

FIG. 586.

Scheme showing division of bulbus cordis and its thickenings into aorta and pulmonary artery with their valves (from Prentiss and Arey, *Text-book of Embryology*, 1917). *a, b, c, d,* Anterior endothelial ridges.

extinct, the Reptilia Sauropsida[1] include all the modern Reptiles and many extinct ones. This conclusion is amply confirmed by evidence derived not only from the skeleton, but also from other parts, such as the lungs, coelomic septa, and brain.

Summary.—We may now briefly recapitulate the probable history of the heart. It developed from the median ventral vessel, immediately behind the gills and in front of the liver, as a muscular enlargement to pump the venous blood through the gills. It became subdivided into chambers (sinus venosus, atrium, ventricle, conus) which contract consecutively from behind forwards, and are separated by valves ensuring the flow of blood forwards. Lodged in the pericardial coelom, the heart acquired an S-shaped bend. From such a primitive condition the heart of Cyclostomes diverged with marked specialisations. In Pisces longitudinal rows of valves developed in the anterior chamber or bulbus cordis, from which is formed the conus.

[1] The term Sauropsida, originally used by Huxley to include all Reptiles as well as Birds, is here used in a restricted sense.

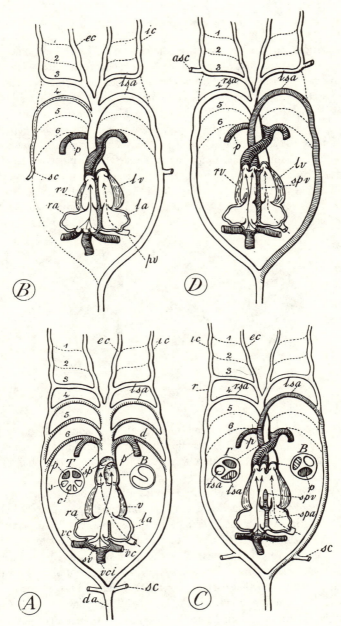

FIG. 587.

Diagrams of *heart* and *aortic arches* in an Amphibian, A; a Mammal, B; a Reptile (Chelonia, Lacertilia, Ophidia, Rhynchocephalia), C; and a Crocodile, D. Ventral view of heart represented as untwisted so as to bring chambers into a single plane, with sinus venosus behind and ventricle in front. B, Transverse section of region of bulbus cordis; T, transverse section of region of truncus arteriosus. *asc*, Anterior subclavian; *d*, ductus Botalli; *ec*, external carotid; *ic*, internal carotid; *la*, left auricle; *lsa*, left systemic arch; *lv*, left ventricle; *p*, pulmonary artery; *r*, portion of lateral aorta remaining open only in *Sphenodon* and some Lacertilia as ductus arteriosus; *ra*, right auricle; *spa*, interauricular septum; *spv*, interventricular septum; *sv*, sinus venosus; *v*, ventricle; *vc*, vena cava superior; *vci*, vena cava inferior. Arrows from sinus venosus indicate main stream of venous blood; arrows with dotted line from left auricles indicate stream of arterial blood. 1-6, Original series of six embryonic aortic arches. (Goodrich, 1916, modified.)

This conus became replaced in the higher Teleostomes by a non-contractile bulbus arteriosus. In Dipnoi and Amphibia the lumen of the heart began to be subdivided longitudinally into right venous and left arterial channels by means of longitudinal ridges and septa ; and the bulbus cordis became spirally twisted. The subdivision was further carried out in the Amniota where the atrio-ventricular opening is divided into two by the junction across of the endocardial cushions and interauricular septum ; the interventricular septum (absent in modern Amphibia) becomes definite, and the lumen of the bulbus cordis is completely divided into pulmonary and carotico-systemic channels. But the further completion of the interventricular septum and separation of the venous and arterial streams was carried out independently along two diverging phyletic lines, the one leading through the Theropsidan Reptiles to Mammalia, and the other through the Sauropsidan Reptiles to Birds.

CHAPTER XI

AIR-BLADDER AND LUNGS

THE AIR-BLADDER OF PISCES

ALL the Tetrapoda being air-breathing vertebrates in the adult condition are provided with lungs : essentially a bilobed outgrowth of the pharynx communicating with it by a median ventral glottis. The only exception to this rule is found in the Urodela, where some species have lost the lungs. Looking for the origin of the lungs in the lower forms, anatomists have long considered them to be represented by the air-bladder of fishes. Owen, having studied the African ' lung-fish ' *Protopterus*, maintained their homology, and it has since been generally adopted. But as it is by no means certain that the various kinds of air-bladder found in fishes are homologous, and since the origin of the bladder itself is not yet fully explained, the whole question remains even at the present time undecided. It is not so simple as appeared at first sight.

Except in certain Teleosts in which it has obviously been lost, some sort of air-bladder exists in all the Osteichthyes, and distinguishes this group from the other and lower fishes in which it is absent. For the suggestion by Miklucho-Maclay that it is represented in Selachians by a small oesophageal diverticulum has not been borne out by later work (Mayer, 850). Perhaps the most primitive air-bladder is seen in *Poly-*

pterus (Geoffroy Saint-Hilaire, 1825 ; Kerr, 840), where a bilobed sac, with a short left and a long right lobe, opens on the floor of the pharynx just behind the gill-slits, through a muscular vestibule leading to a glottis slightly to the right of the median line, Figs. 588 B, 592 A. The Dipnoi are provided with a very similar muscular vestibule opening into the oesophagus slightly to the right of the mid-ventral line, and protected by an epiglottis-like fold strengthened by connective tissue (Parker, 854 ; Göppert, 902). While in *Polypterus* the air-bladder lies ventrally to the oesophagus for the most part, and only the elongated right lobe takes up a position behind in the mesentery dorsal to the gut, in the Dipnoi it lies entirely dorsal to the alimentary canal in the mesentery, between it and the dorsal aorta. It is single in *Ceratodus* but bilobed in *Protopterus* and *Lepidosiren*, and communicates with the vestibule by means of a narrow pneumatic duct passing round the right side of the oeso-phagus,[1] Figs. 588 C, 591, 592 B.

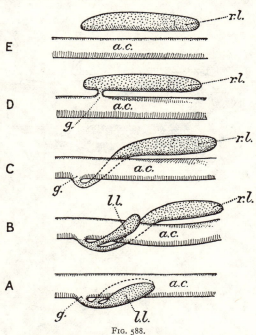

FIG. 588.

Diagram illustrating the lung in fishes ; as seen from the left side. A, Primitive symmetrical arrangement ; B, *Polypterus* ; C, *Ceratodus* ; D, physostomous Teleost ; E, physoclistic Teleost. *a.c*, Alimentary canal ; *g*, glottis ; *l.l*, left lung ; *r.l*, right lung. (From J. G. Kerr, *Zoology*, 1921.)

In the remainder of the Osteichthyes, that is to say in all the Actinopterygii, the air-bladder is essentially a median dorsal diverticulum of the alimentary canal lying between it and the dorsal aorta, and outside the coelom (retroperitoneal). It usually opens into the oesophagus by a ductus pneumaticus, originally described by Needham in 1667. This

---

[1] Wiedersheim (946) has suggested that there were originally two larynges : one dorsal and median (*Lepidosteus*), and one ventral and median (Dipnoi and *Polypterus*). But there is little to support this view, partly based on erroneous observations, and it has not been accepted.

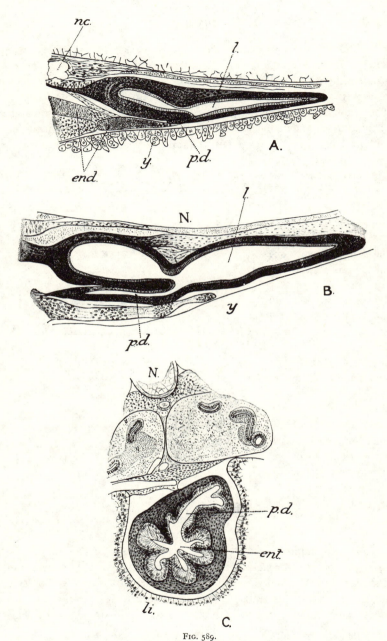

Fig. 589.

Development of the air-bladder of a Teleost. (After Moser, 1904.) A, *Rhodeus*, 5 mm., longitudinal section ; B, *Rhodeus*, 6 mm., longitudinal section ; C, *Rhodeus*, 7 mm., transverse section, showing small pouch-like outgrowth of pneumatic duct. *end*, Endoderm ; *ent*, enteric cavity ; *l*, air-bladder ; *li*, liver ; *N*, notochord ; *nc*, pronephric chamber ; *p.d*, pneumatic duct ; *y*, yolk. (From Kerr, *Embryology*, 1919.)

ductus passes down the mesentery, is short and wide in the lower forms, such as *Amia, Lepidosteus,* and the Chondrostei, but longer and narrower in the Teleostei. In *Lepidosteus* it is provided with a median dorsal muscular vestibule (Wiedersheim, 946). While it usually occurs as an open duct in the lower Teleostei, such as the Clupeiformes, Esociformes, Anguilliformes, and Cypriniformes, the pneumatic duct may become very narrow, reduced to a solid thread, or finally disappear in the more specialised groups. Adult Gasterosteiformes, Mugiliformes, Notacanthi-formes, and Acanthopterygii, with few exceptions, have no open duct. The Teleosts were therefore formerly subdivided into the Physostomi with an open duct (J. Müller, 1842) and Physoclisti with a closed bladder (Bonaparte), Figs. 588-9, 593.

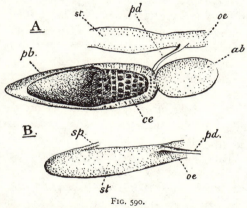

FIG. 590.

A, Air-bladder of *Lebiasina limaculata,* opened to show internal cells ; B, stomach of *Ichthyoborus niloticus,* showing entrance of pneumatic duct. *ab,* Anterior division ; *ce,* cellular wall ; *oe,* oesophagus ; *pb,* posterior division ; *pd,* pneumatic duct ; *sp,* pyloric end of stomach ; *st,* stomach. (After Rowntree from Goodrich, *Vert. Craniata,* 1909.)

However, since all these air-bladders develop as diverticula in open communication with the alimentary canal (Rathke, 1827 ; von Baer, 1834), the duct in later stages becoming more or less completely obliterated, and since physoclistous forms occur exceptionally among physostomous groups, and open ducts are retained sometimes in physoclistous groups (Berycidae among the Acanthopterygii), this classification has been abandoned as of little value. As a rule the pneumatic duct opens in the mid-dorsal line in the Actinopterygii ; but in many Teleosts, such as the Salmonidae, Siluridae, Cyprinodontidae, Percopsidae, Galaxiidae, it is somewhat to the right ; while in the Mormyridae, Notopteridae, Gym-notidae, Cyprinidae, Characinidae, it is rather to the left, Fig. 590. Indeed, in *Erythrinus* and Macrodon, belonging to the last family, the duct opens well on the left side of the oesophagus (Sagemehl, 378 ; Rowntree, 938).

It is clear that important differences exist between the various kinds of air-bladder found in the Osteichthyes ; but attempts have been made to derive them all from some primitive ancestral form. To Boas we owe the suggestion that an original single dorsal bladder, opening by a median dorsal duct, split into right and left halves, which separated, passed down each side of the gut, and reunited ventrally ; finally opening by a median ventral glottis as in *Polypterus*. There is little, however, to support this view. More plausible is the theory of Sagemehl, according to which, on the contrary, the bladder was originally ventral and bilobed, and shifted round to the right side in the Dipnoi and to the left side in the Actinopterygii. *Erythrinus*, and other fishes in which the duct is on the left, would represent intermediate stages in the shifting of the opening. We shall see that neither of these theories is satisfactory, and that the best solution of this morphological puzzle seems to be afforded by the theory of Spengel (941) based on Goette's original suggestion that the lungs are derived from a posterior pair of branchial pouches. The union of the right and left pouch ventrally would give rise to the condition in *Polypterus* (Dipnoi and Tetrapoda), while their more complete fusion dorsally would yield the single bladder of *Amia* and the other Actinopterygii, Figs. 588, 592.

Concerning the original position of the dorsal opening the evidence of embryology is ambiguous, as the bladder appears to arise sometimes on the right (*Amia*, Piper, 930 ; *Acipenser*, Ballantyne, 886 ; Cyprinoids, Salmonids, Moser, 922), sometimes on the left (Lophobranchs, Weber, '86). These appearances are, however, probably deceptive, and due to differential growth and secondary torsions which are known to occur in the later stages of development. Recently Makuschok has contended, probably rightly, that the dorsal air-bladder always first arises as a diverticulum in the middle line, and that its apparent development on the right side is due to the bending over to the left of the mesentery and gut as it rises off the yolk-sac (919). Of a double origin the embryo shows no distinct trace in Actinopterygians, though Ballantyne has brought forward some evidence of a bifurcation of the rudiment in the embryo. According to her it is the original right lobe which has moved round to a dorsal position, the left disappearing (886).[1]   It must

---

[1] It has been suggested that the preponderance of the right lobe and reduction of the left is due in large measure to the position of the stomach. Supposed vestiges of the left lobe have been described in *Amia* and *Acipenser* by Ballantyne (886).

be confessed that no convincing intermediate steps between the single dorsal bladder and the paired pouches which are supposed to have given rise to it have yet been discovered (see, however, *Polypterus* below). The problem remains unsolved; nevertheless, Spengel's theory seems to be the most promising.

The blood-supply affords important evidence. It is a remarkable fact that in *Polypterus* and the Dipnoi the blood is supplied to the air-bladder by paired afferent pulmonary arteries coming from the last

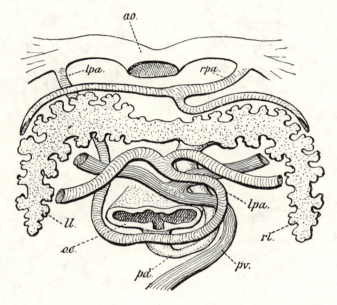

FIG. 591.

Diagram showing the relations of the oesophagus, *oe*, the pneumatic duct, *pd*, the bilobed air-bladder and its blood-supply in *Protopterus*, seen from behind. *ao*, Junction of aortic arches to dorsal aorta; *ll*, left lobe of air-bladder; *lpa*, left pulmonary artery; *pv*, pulmonary vein; *rl*, right lobe of air-bladder; *rpa*, right pulmonary artery. (From Goodrich, *Vert. Craniata*, 1909.)

(fourth) arterial arch, the sixth of the embryonic series. Paired efferent veins return the blood to the hepatic veins near the sinus venosus in *Polypterus*, and directly to the heart in the Dipnoi, where they join to a single pulmonary vein entering on the left side (p. 550). Thus the structure of *Polypterus* is consistent with the theory of Boas, but not with that of Sagemehl. On the other hand, that the bladder of the Dipnoi was originally ventral is proved not only by the course of the left pulmonary vessels, which pass round the oesophagus ventrally together with the duct to reach the now dorsal bladder, but also by the development of the bladder itself, Figs. 591-2. It arises ventrally and grows round in later stages, so that the original right edge comes to lie on the left (Neumayer, **926**; Kerr, **914**).[1] The fact that the large right lobe of the air-bladder of *Polypterus* has already come to occupy a dorsal position posteriorly (Kerr, **840**) would seem further to support Sagemehl's inter-

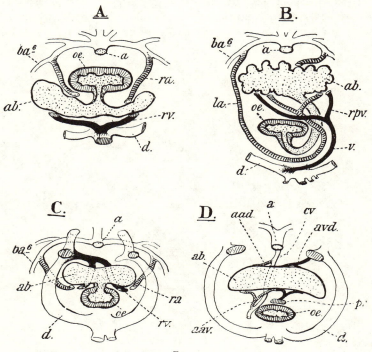

Fig. 592.

Diagrams illustrating the blood-supply of the air-bladder in A, *Polypterus*, B, *Ceratodus*, C, *Amia*, and D, a Teleost. The blood-vessels are seen from behind, and cut short in transverse section. *a*, Dorsal aorta; *aad*, anterior dorsal artery from the coeliac; *aav*, ant. ventral artery; *ab*, air-bladder; *avd*, anterior dorsal vein to the cardinal; *ba⁶*, 4th aortic arch (6th of the series); *cv*, coeliac artery; *d*, ductus Cuvieri; *la*, left pulmonary artery; *oe*, oesophagus; *pr*, portal vein receiving posterior vein from air-bladder; *ra*, right 'pulmonary' artery; *rpv*, right (branch of) 'pulmonary' vein; *rv*, right vein from air-bladder; *v*, left 'pulmonary' vein. (From Goodrich, *Vert. Craniata*, 1909.)

pretation. For it would only be necessary for the left lobe to disappear and for the glottis to shift on to the dorsal side with the shortening duct

[1] The monopneumonous condition of *Ceratodus* is probably due to the suppression of the original left lobe. A small vestige of this lobe has been described in the embryo; it soon merges ventrally with the ductus (Gregg Wilson, 1901; Neumayer, **926**; Ballantyne, **886**).

But if this vestige has been rightly identified, it is not clear why the left pulmonary artery should pass round to the dorsal lung.

to complete the Actinopterygian structure ; the shifting would be to the right, however, instead of the left side. But we now meet with a serious and perhaps fatal difficulty in *Amia*. For here, while the bladder and opening of the duct are median and dorsal, the blood-supply is essentially bilaterally symmetrical. It is true that the efferent veins join and enter the left ductus Cuvieri ; but *Amia* alone among the Actinopterygii has its bladder supplied by a right and left pulmonary artery from the last pair of arterial arches. There can, therefore, apparently have been no shifting of the bladder either to the right or the left, Fig. 592 c.

In all other Actinopterygii the air-bladder receives its blood from the dorsal aorta or its branches. Concerning the possibility of a change of arterial supply from branchial aortic arch to dorsal aorta or coeliac artery, interesting evidence may be derived from the development of *Gymnarchus*. Whereas in early stages the air-bladder receives blood from the posterior aortic arches on the left and the combined arches and coeliac on the right side, in later stages the coeliac and pulmonary arise from the aorta (Assheton, 885 ; Ballantyne, 886).

Important also is the nerve-supply, first studied by Czermack, 1850. The dorsal bladder of Actinopterygians appears to be always innervated from both the right and left vagus. Must it not, therefore, have preserved its original position ? Kerr has shown that in Dipnoi, where the left vagus does not pass round ventrally (like the pulmonary artery) but dorsally to the oesophagus, and crossing over the right vagus passes to the right side of the bladder, this strange condition must be due to secondary anastomosis. For in *Polypterus* also the branch of the left vagus crosses over the oesophagus to the large right lobe of the air-bladder ; although, of course, in this fish there has certainly been no shifting of the bladder from one side to the other. It follows that nerve-supply is no certain guide to the phylogenetic history of the air-bladder, but may be re-adjusted to suit varying adaptations (Kerr, 914).

About the functions of the Actinopterygian air-bladder there has been much controversy ever since Rondelet, in 1554, maintained that it helps the fish to swim. More than a century later Needham (1667) discussed the possibility of its serving as a float, a respiratory reservoir of air, or an organ secreting gas to help in digestion. Boyle, Mayow, Ray, Borelli, and many others in the seventeenth century studied the question experimentally, and the discussion was carried on with much zeal throughout the eighteenth and nineteenth centuries. Perrault (1680) and Monro (1785) showed that gas is secreted by the red gland in closed bladders ; and a few years later Priestley, Fourclay, Brodbelt, Lacépède, and Biot analysed the gas and found that the proportion of oxygen varies greatly, and is often greater than in air. Biot found that almost pure oxygen fills the bladder of deep-sea fish. Hitherto the bladder had been generally held to act as a hydrostatic organ enabling the fish to rise and sink ; but in 1809 Provencal and Humboldt pointed out that a fish can swim well after its bladder has been punctured or even removed. Delaroche (1809), and

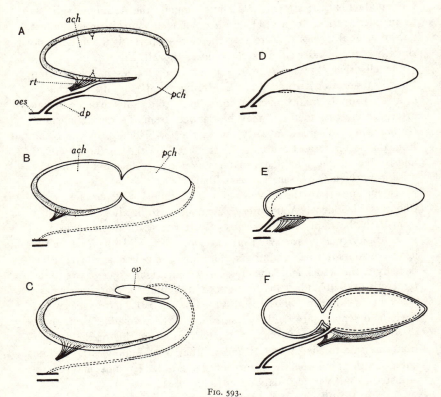

FIG. 593.

Diagram of various types of *air-bladder* found in *Teleostei* (from M. Rauther, 1923). A, Hypothetical ; B, physoclistous with posterior chamber, and C, with oval ; D, Salmonid ; E, Esocid ; F, Cyprinoid. *ach*, Anterior gas-secreting chamber ; *dp*, ductus pneumaticus ; *oes*, oesophagus ; *ov*, gas-absorbing oval ; *pch*, posterior chamber ; *rt*, rete mirabile related to gas-gland.

later Moreau (1876), did much to establish the modern view, according to which the air-bladder in the Teleostei does not actively cause the fish to rise and sink, this rapid vertical motion being brought about by means of the fins, but enables it so to alter its specific gravity as to keep it approximately equal to that of the water at any desired level. By varying the amount of gas inside the bladder and adjusting it to the pressure from outside, the fish is kept in a state of equilibrium in which it can maintain itself with minimum exertion in a plane of least effort (Hall, 907; Tower, 1902; Baglioni, 1908; Popta, 1910–12; Guyénot, 1909–12).

Further evidence that the bladder acts thus as an adjustable float is afforded by the fact that most of the Teleosts which have lost it are bottom-living forms, as for instance the Pleuronectidae. Cunningham has made the interesting observation that in *Rhombus* the bladder is present in the free-swimming larva, and it has since been found to occur in the young of other Pleuronectids and of *Uranoscopus* (Ehrenbaum, 1896; Thilo, 1899–1914).

Some fish with an open pneumatic duct can fill the bladder by swallowing air from the surface; but as a rule the gas is secreted from the blood, and the duct serves rather as a safety valve to let

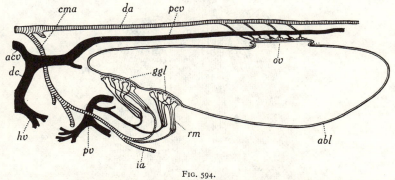

FIG. 594.

Diagram of blood-supply to physoclistous *Teleostean air-bladder*; left-side view. Veins black, arteries cross-lined. *abl*, Air bladder; *acv*, anterior cardinal; *cma*, coeliaco-mesenteric; *da*, dorsal aorta; *dc*, ductus Cuvieri; *ggl*, gas-gland; *hv*, hepatic; *ia*, intestinal; *ov*, oval; *pcv*, posterior cardinal; *pv*, portal; *rm*, rete mirabile.

out excess. The fact that oxygen can be much more easily secreted and absorbed than nitrogen accounts for the presence of oxygen in large and very variable proportion. The secretion of gas may take place in less specialised forms from the general inner surface of the bladder (ventral surface in Cypriniformes, anterior surface in most Clupeiformes), or from a special area of the wall (de Beaufort, 887; Rauther, 932). There is a tendency for the bladder to become differentiated into an anterior oxygen-producing and a posterior oxygen-absorbing region. Further, a special area in the former becomes differentiated for secreting gas and is known as the red body (Monro, 1785). This consists of the internal oxygen-secreting epithelium and a capillary network on the wall of the bladder together forming the gas gland proper, in more or less close connexion with a rete mirabile, a wonderfully complex structure of venous and arterial capillaries which do not communicate until they reach the gland (J. Müller, 1840–42; Quekett, 1842–4; Corning, **896**; Woodland, **949-50**).   The more primitive structure, found in physostomous fishes, in which the gland is covered with simple flat epithelium, has been named (together with its rete mirabile) the red body, while the higher type found in physoclistous fishes, in which the gland is covered with thick glandular

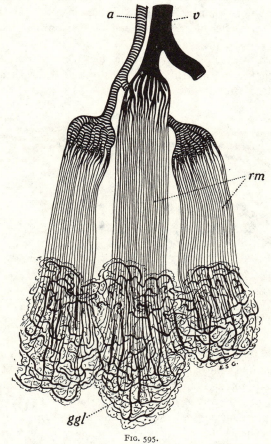

FIG. 595.

Labrus bergylta, portion of gas-gland, *ggl*, with vascular supply. *a*, Artery; *rm*, rete mirabile; *v*, vein.

epithelium thrown into folds or sunk in crypts, is named red gland (Coggi, **895**; Vincent and Barnes, **844**; Nusbaum, **928**; Reis and Nusbaum, **933**; Haldane, **906**; Woodland, **849**).   The remainder of the anterior chamber is lined with thick impermeable tissue, covered internally with simple epithelium, Figs. 593-7.

The posterior oxygen-absorbing region, derived apparently from the embryonic pneumatic duct itself, in physoclistous fishes is lined by a thin epithelium through which oxygen can easily pass to the rich network of vessels overlying it (Tracy, 942). In still more specialised forms, such as *Mugil, Balistes,* and the Gadidae, this posterior region becomes converted into a flattened 'oval' which can be closed off by a circular fold provided with sphincter and dilator muscles (Tracy, Woodland, Reis and Nusbaum).

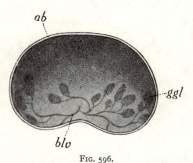

FIG. 596.

Inner view of anterior end of air-bladder of *Labrus bergylta,* seen from behind. *ab,* Cut wall; *blv,* blood-vessel ; *ggl,* gas-gland.

While the red glands are supplied from the coeliac artery and return their venous blood to the portal vein, the absorptive region and 'oval' receive branches from the dorsal aorta and return their venous blood to the posterior cardinals (Corning, 896). Moreover, secretion of oxygen would seem to be controlled by the vagus and absorption by the sympathetic (Bohr, 889). Thus can the pressure of oxygen inside even the closed air-bladder be delicately regulated, and if necessary be made vastly superior to its pressure outside or in the blood by the active participation of the glandular cells.

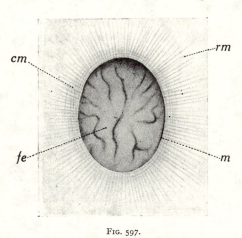

FIG. 597.

Inner view of roof of air-bladder of *Mugil chelo,* showing *oval. cm,* Circular closing muscles ; *fe,* folded epithelium of oxygen-absorbing area ; *m,* rim of oval pocket ; *rm,* radiating opening muscles.

The air-bladder undergoes many strange modifications in the Teleostei. It acquires a second opening to the exterior near the anus in many Clupeids such as the herring, *Clupea harengus, Pellona, Sardinella* (de Beaufort, 887), and in the horse-mackerel, *Caranx trachurus,* a small opening into the right branchial cavity. Often the bladder is provided with lobes and branches, sometimes of great complexity and of doubtful function (Gunther, 1880). They sometimes actually push their way into the perilymph cavity of the auditory capsule, Fig. 598 (Weber, 1820; Parker, 929; Ridewood, 934; de Beaufort, 887; Tracy, 943). In *Megalops* these diverticula are lodged in bony bullae of the pro-otic, in *Notopterus* they are separated from the perilymph by membrane alone, while in Hyodon they plunge into the perilymph itself. The diverticula in Clupeidae may pass ventrally beyond the labyrinth, and in Mormyridae may be applied to the wall

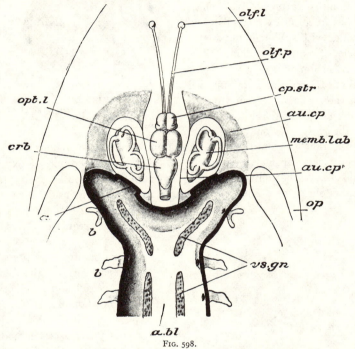

Fig. 598.

Horizontal section of posterior portion of head and anterior end of air-bladder in *Pseudophycis bachus*, one of the Gadidae or Cods (semi-diagrammatic). *a*, Thickened portion of air-bladder fitting into fenestra in posterior wall of auditory capsule; *a.bl*, air-bladder; *au.cp*, outer wall of auditory capsule; *au.cp'*, inner (membranous) wall; *b*, hollow offshoots of air-bladder; *cp.str*, corpora striata; *crb*, cerebellum; *memb.lab*, membranous labyrinth; *olf.l*, olfactory bulbs; *olf.p*, olfactory peduncles (olfactory tracts); *op*, operculum; *opt.l*, optic lobes; *vs.gn*, vaso-ganglia. (From Parker and Haswell, *Zoology*.)

of the sacculus. But by far the most interesting form of connexion between the bladder and the ear is that described by Weber in 1820, and which occurs in all the Cypriniformes (Ostariophysi), Figs. 599, 600. Here the bladder is enabled to communicate pressure to the perilymph by means of a series of ossicles compared by Weber to the ear ossicles of the mammal, but now known to be of quite different origin and to be formed by the specialisation of portions of the anterior vertebral segments, as first suggested by Geoffroy Saint-Hilaire, 1824. Accompanying the chain of ossicles is a singular modification of the ear itself. The right and

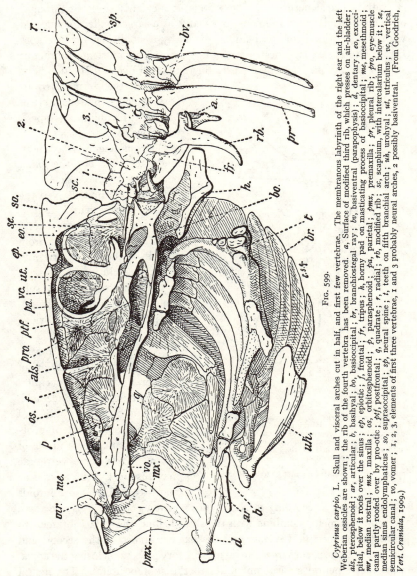

FIG. 599.

Cyprinus carpio, L. Skull and visceral arches cut in half, and first few vertebrae. The membranous labyrinth of the right ear and the left Weberian ossicles are shown; the rib of the fourth vertebra has been removed. *a*, Surface of modified third rib, which presses on air-bladder; *als*, pterosphenoid; *ar*, articular; *b*, basihyal; *bo*, basioccipital; *br*, branchiostegal ray; *bv*, basiventral (parapophysis); *d*, dentary; *eo*, exoccipital, below it roofs over the sinus; *ep*, epiotic; *f*, frontal; *fr*, tripus; *h*, horny pad on masticating process of basioccipital; *me*, mesethmoid; *mr*, median rostral; *mx*, maxilla; *os*, orbitosphenoid; *p*, parasphenoid; *p4*, parietal; *pmx*, premaxilla; *pr*, pleural rib; *pro*, eye-muscle canal partly roofed over by pro-otic; *ptf*, postfrontal; *q*, quadrate; *r*, radial; *rb*, modified rib; *sc*, scaphium, with intercalarium below it; *se*, median sinus endolymphaticus; *so*, supraoccipital; *sp*, neural spine; *t*, teeth on fifth branchial arch; *uh*, urohyal; *ut*, utriculus; *vc*, vertical semicircular canal; *vo*, vomer; 1, 2, 3, elements of first three vertebrae, 1 and 3 probably neural arches, 2 possibly basiventral. (From Goodrich, *Vert. Craniata*, 1909.)

left perilymph cavities become continuous below the brain, and the two membranous labyrinths join to a median canal from which arise and pass

backwards two sacculi and a median sinus endolymphaticus. The latter
is lodged in an extension of the perilymph cavity excavated in the basioccipital. The structure of Weber's apparatus is fairly constant throughout the Cypriniformes and consists usually of a large malleus connected
with the front wall of the bladder, joined by a small incus to a stapes
fitting into the atrial fenestra, and a claustrum lying on the membranous
wall of the atrium ; for these misleading names Bridge and Haddon have
proposed the terms: tripus (malleus), intercalarium (incus), scaphium (stapes), which we shall adopt.

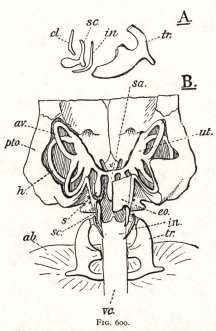

Fig. 600.

Macrones nemurus. A, The Weberian ossicles; B, portion of the skull, the labyrinth, and Weberian apparatus
diagrammatically represented from above (from the
figures of Bridge and Haddon). *ab*, Air-bladder ; *av*,
anterior vertical canal of the ear ; *cl*, claustrum ; *eo*,
exoccipital ; *h*, horizontal canal ; *in*, intercalarium ; *pto*,
pterotic ; *s*, sacculus ; *sc*, scaphium ; *tr*, tripus ; *ut*, utriculus ; *vc*, vertebral column. (From Goodrich, *Vert.
Craniata*, 1909.)

Various attempts have been
made to trace out the exact
homology of the ossicles. A.
Müller (1853) first derived the
scaphium from neural arch 1,
the intercalarium from neural
arch 2, and the tripus from
the rib of the third vertebra ;
and his conclusions have been
in the main confirmed by subsequent observers (Nusbaum,
927 ; R. Wright, **951** ; Bridge
and Haddon, **891** ; Bloch, **888**).
But the question is complicated by the assimilation of
vertebrae to the occipital
region of the skull (Sagemehl,
378), and by the formation of
rigid compound vertebrae behind the skull by the fusion of
centra. Thus three vertebrae
may fuse in Cyprinids, and as
many as five in Siluroids
(Wright, **951**). According to Nusbaum's recent account in *Cyprinus* three
vertebral segments combine with the skull behind the vagus foramen to
form the basioccipital region, and the haemal arches of the second and
third fuse to form the large ventral masticatory process enclosing the aorta,
while their neural spines contribute to the supraoccipital. The neural
arch of the third of these segments forms the scaphium, that of the
first post-occipital segment the intercalarium, and the haemal arch

(probably including the rib) the tripus. The claustrum, when present, would be derived from the intercalary of that segment (Wright, 951). The Cypriniformes, especially the Siluroids, often have a highly modified air-bladder, reduced in size, and sometimes partially enclosed in bony expansions of the vertebrae or ossified (see Bridge and Haddon, 891; Sörensen, 939; Bloch, 888). In many Cyprinidae, Characinidae, and Siluridae the bladder has paired extensions passing outwards to below the skin, forming a sort of tympanum behind the pectoral girdle.

Sörensen supports Weber's original suggestion that the apparatus serves to intensify sound vibrations and carry them to the ear, while also holding that it assists in the production of sounds by the fish. Hasse (909), on the other hand, considers that it acquaints the fish with the state of tension of the air in its bladder at various depths. When the air-bladder expands the stapes or scaphium is pushed in; when it contracts the scaphium is drawn out. Although it cannot yet be held that the function of Weber's apparatus has been thoroughly determined, yet it seems highly probable that, as Sagemehl (378) suggested, it transmits changes of pressure to the perilymph, and sets up reflex actions which allow gas to escape by the duct or be secreted into the bladder (Evans, 897; Evans and Damant, 898; Guyénot, 905).

It may further be noticed that the wall of the air-bladder is generally provided with a layer of smooth splanchnic muscle fibres, and in addition with extrinsic striated muscles supplied by anterior spinal nerves. These muscles seem to have little to do with the altering of the capacity of the bladder for hydrostatic purposes, but are specially developed in those fishes which emit sounds, such as *Zeus, Dactylopterus, Trigla* (Delaroche, 1809). An elaborate sound-producing organ known as the elastic spring apparatus is developed from the modified transverse processes of the fourth vertebra in connexion with the bladder in certain Siluroids (*Auchenipterus, Doras,* etc.; see Sörensen, 939; Bridge and Haddon, 891).

Summary.—We may conclude, then, that the functions of the air-bladder in fishes are many and various, but that it acts chiefly as an adjustable float to enable the fish to swim at any level with the least effort. But even among the Teleosts it sometimes acts also as a reserve of oxygen to be drawn upon in case of special need (Jacobs, 1898; Moreau, 1876–7). On the whole, it seems probable that the original function of the bladder was respiratory. A cellular lung-like bladder occurs in *Amia, Lepidosteus,* and less developed in certain Teleosts, such as *Megalops, Chirocentrus, Gymnarchus, Arapaima,* and Cyprinoids, Fig. 590 (Wiedersheim, 310a; de Beaufort, 887). Hyrtl (1856) showed that the bladder acts as a lung in *Gymnarchus,* and Jobert (1878) did

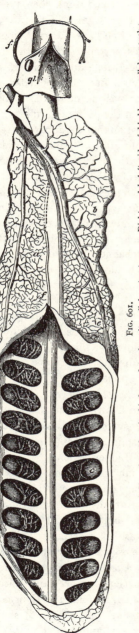

FIG. 601.

Air-bladder of *Ceratodus Forsteri*, Kr. Opened at its hinder end to show its cellular structure. *a*, Right, and *b*, left side of bladder; *c*, cellular pouch; *e*, pulmonary vein; *f*, pulmonary artery; *gl*, glottis exposed by opening the oesophagus, *oe*. (After Günther.)

the same for *Erythrinus*; nevertheless, it is possible that the lung-like function and structure has been reacquired in these Teleosts as a special adaptation for living in foul water. Such an explanation, however, does not appear to apply to the more primitive fish, like *Polypterus*, *Amia*, and *Lepidosteus*, where the respiratory function seems to be primary (Mark, 1890; Budgett, 776; Potter, 931). That the bladder of the Dipnoi resembles a lung both in structure and in function has long been known, Fig. 601, and in *Protopterus* the air-cells are more elaborately developed than in many Amphibia (Parker, 929; Spencer, 940).

As for the origin of the air-bladder, we may conclude that it was probably derived from a posterior pair of gill-pouches, although it must be confessed that so far no definite embryological evidence of this has been found. This theory alone accounts for the blood-supply from the arterial arch in Dipnoi, *Polypterus*, and *Amia*, and overcomes the difficulty of explaining the initial stages in phylogeny. Presumably when, in the Actinopterygii, either one or both of the pouches became dorsal and converted into a float into which gas was secreted, the arterial blood-supply was drawn more directly from the aorta. Moreover, the theory of Spengel, already referred to (941), is supported by the latest views on the first origin of lungs.

THE LUNGS OF TETRAPODA

That the lungs of the land vertebrates were originally derived from gill-pouches was suggested many years ago by Goette (1875), who showed that in *Pelobates* they arise from paired rudiments. Nevertheless, it has generally been held that the lungs of the Tetrapods develop as a median ventral diverticulum from which grow out a right and left lobe, and this is still stated to be the case in man and in the pig (Flint, **900**). Recently, however, much evidence has been brought forward to prove that in all Tetrapods the earliest trace of the lungs is in the form of paired pouches of the endoderm close behind the last-formed gill-slits, Fig. 604. Weber and Buvignier (**945**), Greil (**903**), and Makuschok (**918-19**) have shown that in Urodela and Anura there is developed behind the vestigial sixth pair of gill-pouches yet another pair of outgrowths, which soon join a median ventral depression, developed immediately in front, either at the same time or very soon after, and representing the rudiment of the larynx and trachea, Figs. 602-3. The tracheo-laryngeal groove, carrying the lungs with it, closes off from behind forwards, leaving the open glottis in front. This mode of development has now been followed not only in Amphibia, but also in Reptilia (Hochstetter, 1906), in Aves (Katschenko, **788** ; Rösler, **937** ; Locy and Larsell, **916**), and in Mammalia (Fol ; Weber and Buvignier, **945**). Nevertheless, some differences of opinion still persist as to the homology of the lungs. Greil con-

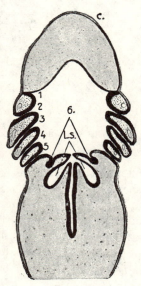

Fig. 602.

Diagram illustrating modification of gill-pouches in phylogeny of *Amphibia* (from M. Makuschok, *Anat. Anz.*, 1914). 1-6, Gill-pouches and slits ; *Ls*, lungs = 7th pair of pouches.

siders that the rudiments are not truly comparable to gill-pouches, being according to him more ventral and sometimes appearing before the last pair ; while Weber and Buvignier believe that they represent rather a reappearance of ancestral pouches than a persistent seventh pair. But although the sixth pair may be delayed in development in Amphibia, there can be little doubt that Makuschok is right in insisting that the lung rudiments are true gill-pouches. Just as in the case of the air-bladder, no other view harmonises so well with the fact that the afferent pulmonary vessel comes from the ventral aorta by way of the sixth embryonic aortic arch, or gets over so easily the difficulty of explaining the initial stages in phylogeny. For it is easy to imagine that an endodermal gill-pouch might, like the vestigial and evanescent sixth pair, fail to fuse with the ectoderm and remain as blind sacs in which air could be lodged. Real

Fig. 603.

Horizontal section through posterior region of pharynx of embryo *Triton* (from M. Makuschok, *Anat. Anz.*, 1911). *d*, Cavity of pharynx; *Lh*, splanchnocoele; 3, 4, 5, developing gill-slits; 6, vestigial sixth slit; +, rudiment of lung diverticula.

intermediate stages we could only expect to find in the long extinct ancestors of the Amphibia.

The organs of respiration undergo most interesting changes in the different classes of the Tetrapoda, and we may now briefly consider their structure (Oppel, **795**). We believe the lung of terrestrial Vertebrates to have started from a simple bilobed sac, not unlike that of *Polypterus*, with a vestibule in front opening by the glottis on the floor of the pharynx; the wall was thin, highly vascular, and covered outside by coelomic epithelium, contained smooth muscle fibres supplied by twigs of the vagus; the lining epithelium was ciliated, except in special regions where it was thin and flattened over the capillaries to allow easier respiratory exchange.

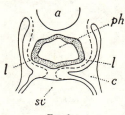

Fig. 604.

Transverse section of embryo *Gallus domesticus*, 24 somite stage (from Rösler, 1911), showing earliest paired rudiments of lungs, *l*. *c*, Splanchnic coelom; *ph*, pharynx.

Amphibia.—The lungs of the Amphibia have departed but little from this condition; but the vestibule becomes differentiated into a larynx (Göppert, **687, 902**; Wilder, **947**), and a stiff trachea becomes more or less distinctly marked off from the thin-walled distensible lung, the inner wall of which may be thrown into folds. In some Urodeles, such as *Necturus*,

Proteus, and *Triton*, where the lung is rather hydrostatic than respiratory in function (Camerano, 893), the inner surface is smooth. This may, however, be due to secondary simplification. But in others, and especially the Anura, the folds may be increased in depth, and the cavity of the lung be subdivided peripherally by primary and secondary trabeculae into large chambers and smaller irregular air-cells amply supplied with capillaries and affording a large respiratory surface. As a rule the trachea is little differentiated; but in the Gymnophiona and some Urodela (*Siren, Amphiuma*) it is definitely formed, and its wall strengthened by a series of paired cartilages, amounting to semi-rings in the former group.

Between the Amniota and the Amphibia there is an important divergence in the mechanism of respiration; for while in the former the lungs are filled according to the principle of a suction pump, in the latter the action is like that of a force pump. Townson (1794–5) was the first to study the mechanism in the Amphibia, which has since been worked out in detail by various authors (Cuvier, 1835; Haro, 1842; Panizza, 1845; Milne-Edwards, 1857; P. Bert, 1869; Gaupp, 821). But oxygenation of the blood takes place in the Amphibia not only in the lungs, but also in the buccal cavity, the lining of which is usually very vascular, and to an even greater extent in the skin, likewise well supplied with blood-vessels (Williams, 1859). For buccal respiration the glottis is closed, and water or air passed in and out of the buccal cavity through the nostrils by depressing and raising the floor of the buccal cavity.

The process of pulmonary respiration in the frog is as follows (Dakin, 1927). The mouth is kept shut, and at intervals the nostrils are closed and the buccal cavity enlarged by the action of muscles which depress the hyoid plate in its floor. Air is thus drawn out of the lungs and mixed with fresh air present in the buccal cavity. The elevation of the plate and floor of the buccal cavity now forces mixed air into the lungs through the open glottis. The nostrils are closed by pressure of the lower jaw in the Anura or by special valves in the Urodela (Wilder, 947; Anton, 884; Bruner, 892). The lungs behave as passive distensible sacs, though the contractility of their walls may help in expiration. Accompanying this peculiar mode of respiration in which the ribs take no part is their great reduction; in no living Amphibia do they meet the sternum (p. 78).

Before leaving the subject of the lungs in Amphibia, something must be said about the remarkable lungless Urodelous Amphibians recently described by Wilder (948) and Camerano (893). In various species and genera of the family Salamandridae (*Salamandrina, Plethodon, Spelerpes, Batrachoseps, Manculus, Aeneides, Desmognathus*), both in

Europe and in America, the lungs, trachea, and larynx have entirely vanished, being represented by at most a depression on the floor of the oesophagus in the embryo, and even this may disappear in the adult. Intermediate stages in the degeneration occur in *Salamandrina* (Camerano, 893 ; Lühe, 917 ; Lönnberg, 1899). Respiration in these lungless Urodeles is carried out by the vascularised surface of the skin, the buccopharynx, and even the oesophagus (Bethge, 1898 ; Barrows, 1900). Considerable modifications are entailed in the vascular system, both pulmonary vein and left auricle being reduced (Bruner, 892).

In the Amniota not only is the trachea strengthened by cartilaginous

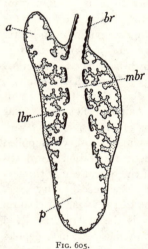

FIG. 605.

Diagram of longitudinal section through lung of Lacertilian. *br*, Lateral external bronchus ; *lbr*, lateral secondary bronchus leading to air-cells ; *mbr*, main internal bronchus ; *a*, anterior, and *p*, posterior saccular extensions of bronchi.

rings, incomplete dorsally in the Reptiles, but from its hinder end are differentiated two extra-pulmonary bronchi of similar structure leading to the lungs. The glottis can be firmly shut in Reptiles and Birds, and is closed in Mammals by a protective flap supported by cartilage, the epiglottis. As the neck becomes differentiated and lengthened, so does the trachea elongate and the lungs retire to the thoracic region with the heart. Their inflation is brought about according to the principle of the suction pump, by the expansion of the thoracic chamber whose walls are (except in Chelonians, see p. 600) provided with well-developed jointed movable ribs articulating for the most part with the ventral sternum. Contraction of the intercostal muscles drawing forwards and straightening the ribs expands the thoracic cavity, and air rushes into the lungs through the open glottis. Relaxation of the muscles and the collapse of the ribs accompanies expiration. These respiratory movements are further helped by the development of septa and diaphragms (see below).

In the course of adaptational evolution, the lungs of the Amniota become more and more specialised in divergent directions. Starting from such a mere saccular enlargement as we find in Amphibia, with thin slightly folded walls enclosing a spacious central cavity, the distinction between the air-passages and the truly respiratory region becomes more pronounced, the latter being ever increased to afford a larger surface for respiratory exchange. Numerous and small alveoli, lined with the thinnest epithelium, become set round internal chambers delimited by ingrowing septa. These chambers become regularly disposed so as to lead out from a central cavity which grows more defined and tubular by the regular arrangement of the edges of the septa, until they finally surround intrapulmonary passages or bronchi. The chambers themselves may be differentiated into mere air-passages leading from a bronchus to more numerous atria beset with alveoli. Thus gradually develops the spongy parenchyma of the higher types of lung, Figs. 605-606, 638.

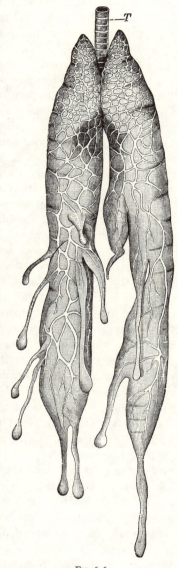

Reptilia.—The minute structure of the lung in Reptiles has been studied of late by Miller and Milani (**920**). As a rule in Lacertilia and Ophidia the more parenchymatous part is situated in the anterior region near the entrance of the bronchus, and gradually dwindles towards the apex, which may be thinwalled and saccular. In the Varanidae, Crocodilia, and Chelonia, where the parenchyma is more developed, distinct secondary bronchi expanding into chambers radiate from a tubular mesobronchus. Moreover, in the Chelonia and certain Lacertilia (Ascalobatae, Iguanidae, Varanidae) the terminal thin-walled sac may be prolonged into blind processes. These are especially well developed and numerous in Chamaeleons, where they extend among the abdominal viscera and help the animal to distend itself, Fig. 606. Probably as a consequence of the elongation of the body, snakes and snake-like lizards usually have a vestigial left and lengthened right lung.

FIG. 606.

Lungs of *Chamaeleo monachus*. *T*, Trachea. (From Wiedersheim, *Comp. Anatomy.*)

Owing to the presence of a hard carapace and plastron, often rigidly connected together, respiration cannot be carried out in Chelonians as it is in other reptiles by means of the movable ribs. It is therefore brought about by a special mechanism peculiar to the Chelonian (Townson, 1799 ; Weir-Mitchell, 1870 ; P. Bert, 1870). Inflation and deflation of the lungs, highly differentiated organs closely adpressed to the carapace, is due to some extent to the drawing in and out of the neck and limbs, but chiefly to the action of the pectoral and pelvic limb girdles, whose remarkable position and attachments within the ' shell ' allow a certain rotary movement (Sabatier, 1881 ; Charbonnel-Salle, 1883 ; François-Franck, 1906). The posterior oblique muscle and the post-hepatic septum help in the respiration of crocodiles (p. 641).

Aves.—Most interesting, however, is the respiratory apparatus of birds, whose body has to be kept at a constant high temperature, and whose great metabolic activity during violent muscular exertion has to be provided for. In the bird's lung, indeed, the respiratory exchange is probably more intense and more effectively carried out than in any other respiratory organ known, and the lungs become differentiated into what may be called the lungs proper, situated in the thoracic region close up against the ribs and vertebral column, and blind thin-walled distensible air-sacs extending among the viscera. Ever since Harvey, in 1651, proved that the apertures on the ventral surface of the lung lead into air-sacs, and Camper (1773) and Hunter (1774) further showed that these sacs lead into air-cavities in the bones, the breathing apparatus of birds has attracted the attention of a multitude of observers. But it is only quite recently that the peculiar structure of the avian lung has been fully appreciated. For our knowledge of the anatomy of the sacs we are chiefly indebted to the work of Sappey (1874), Campana (894), and Huxley (911) ; but much detail has been made known by Schulze (1871), Beddard (953), Guillot (904), Roché (936), Müller (923), Weldon (1883), Juillet (912), and others ; while their development has been revealed by Selenka (1866), Bertelli (955), Poole (974), Juillet (912), Locy and Larsell (916). Briefly the air-sacs may be described as, so to speak, the blown-out extremities of certain bronchial tubes, Figs. 607-9. There are usually five pairs : a cervical extending up the neck ; an interclavicular pair, usually fused to a median sac ventral to the oesophagus in the pectoral region (remaining separate in Vultures, *Ciconia*, and *Ardea*) ; an anterior and a posterior pair of thoracic or intermediate sacs, below the ribs and in front of the post-hepatic septum (p. 633) ; and lastly, two abdominal sacs, which push far into the abdominal cavity, except in *Apteryx* (Huxley, 911). Rarely there are three pairs of intermediate sacs (*Podargus*, Beddard, 1898). These thin-walled non-vascular sacs become pressed against each other, the ventral surface of the lungs and the oblique septum (pp. 633, 639), and are permanently distended with air to a greater or less extent. Much speculation has arisen about their

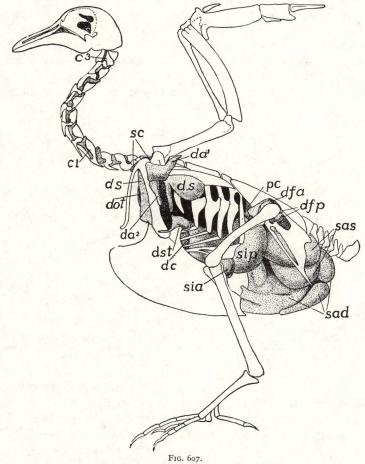

FIG. 607.

Air-sacs and canals leading into bones of pigeon (after B. Müller, from J. S. Kingsley, *Comp. Anat. of Vertebrates*, 1926). c^{1-5}, Intertransverse canals; da^{1-2}, axillary sac and its ventral diverticulum; *dc*, canal for ribs; *dot*, infraclavicular canal; *ds*, subscapular sac; *dst*, sternal canal; *pc*, pre-acetabular canal; *sad*, *sas*, right and left abdominal sacs; *sc*, cervical sac; *sia*, *sip*, anterior and posterior intermediate or thoracic sacs.

function. That their chief use is to help in respiration there can be no doubt (Harvey, 1651; Perrault, 1666; Sappey, 1847; Campana, 894). Placed between the skeleton above and an almost rigid sheet of connective tissue below (pulmonary diaphragm of Sappey, pulmonary aponeurosis of Huxley, ornithic diaphragm of Bertelli), the lungs proper hardly alter at all in volume in breathing; they are relatively small and enclosed in virtual pleural cavities (p. 639). This arrangement is to some extent foreshadowed in the Crocodilia, where the lungs are also enclosed in pleural cavities distinct from the abdominal coelom (p. 641). The original lung has in fact become differentiated into two regions: the closely packed parenchymatous vascular lung in which respiratory exchange takes place, and the saccular diverticula whose function it is to serve as reservoirs for pumping air through it. Campana adopted the then prevalent theory that the intermediate sacs received most of the

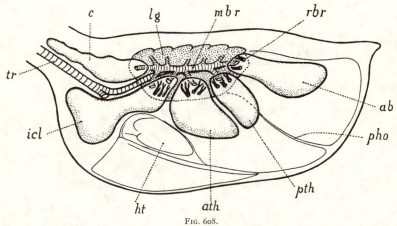

FIG. 608.

Diagram of respiratory organs of a Bird, left-side view. *ab*, Abdominal air-sac; *ath*, anterior thoracic air-sac; *c*, cervical air-sac; *ht*, heart; *icl*, interclavicular air-sac; *lg*, left lung; *mbr*, meso-bronchus; *pho*, post-hepatic septum; *pth*, posterior thoracic air-sac; *rbr*, recurrent bronchi; *tr*, trachea.

fresh air from the bronchi at inspiration, when the ribs are straightened and the sternum lowered; the more anterior and posterior sacs being filled from the intermediate sacs when the thorax contracts. The outer sacs would thus be expanded when the intermediate sacs are compressed, the small muscles of the pulmonary diaphragm serving to keep constant the volume of the lung at expiration. Thus by the alternate contraction of antagonistic sacs (Perrault, 1666) a constant flow of air would be kept without intermission through the lung at average tension and composition. Moreover, the large surface of the sacs serves to keep the air both warm and damp, and mixes the new air with the old. But this theory of antagonistic sacs has been considerably modified by later observers (P. Bert, 1870; Soum, 1896; Bär, 1896). To understand avian respira-

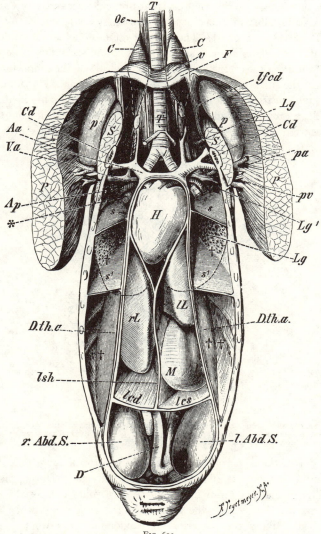

FIG. 609.

Abdominal viscera and air-sacs of a duck after the removal of the ventral body-wall. (From a drawing by H. Strasser.) *Aa*, *Va*, Innominate artery and vein with their branches ; *Ap*, pulmonary artery ; *C*, *C*, cervical sacs ; *Cd*, coracoid ; *D*, intestine ; *D.th.a*, oblique septum ; *F*, furcula ; *H*, heart, enclosed within the pericardium ; *lfcd*, coraco-furcular ligament ; *Lg*, *Lg¹*, lung ; *lsh*, suspensory (falciform) ligament ; *lcd*, *lcs*, right and left parts of post-hepatic septum ; *P*, pectoralis major ; *p*, axillary sac lying between the coracoid, scapula, and the anterior ribs, and communicating with the sub-bronchial air-sac ; *pa*, *pv*, pectoral artery and vein ; *r.Abd.S*, *l.Abd.S*, right and left abdominal (posterior) air-sac ; *rL*, *lL*, right and left lobes of liver : *S*, subclavius muscle ; *s*, *s*, partition walls between the anterior thoracic air-sacs and the unpaired sub-bronchial sac, lying in the anterior part of the body-cavity ; *s¹*, *s¹*, partition walls between the anterior and posterior thoracic air-sacs ; *T*, trachea ; *v*, portion of anterior wall of the body-cavity ; *, point of entrance of the bronchi into the lung ; †, anterior thoracic air-sac ; ††, posterior thoracic air-sac.

tion we must now describe the recurrent bronchi discovered by Campana (894), but the full significance of which was first made known by Juillet (912), Figs. 608, 612, 636-7. The main or meso-bronchus passes down to open into the abdominal sac, and gives off lateral secondary branches to each of the other sacs. In the case of the cervical sac, the simple bronchial orifice on the surface of the lung proper, the large primary ostium, is the only communication ; but each of the remaining four sacs has in addition, as a rule, a group of small secondary openings leading back from the sac into recurrent bronchi which penetrate and branch in the substance of the lung ; and it is by means of these that relatively pure air received

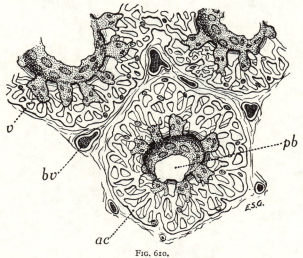

FIG. 610.

Diagram of a thick section of small portion of lung of a bird, much enlarged. *ac*, Air-capillaries with blood spaces between them ; *bv*, large blood-vessels ; *pb*, parabronchial cut across; *v*, vestibule, diverticulum into which open air-capillaries.

directly from the primary bronchus is returned to the lung from the sac, ensuring a thorough ventilation of the parenchyma.[1] During flight the interclavicular and anterior intermediate sacs would be specially active, being much influenced by the muscles of the wing ; but when at rest and when walking the posterior and abdominal sacs would be more effective. The mechanism regulating the flow of air from the sac back to the lung through the recurrent openings and not through the ostia into the large bronchi has not yet been satisfactorily elucidated. Probably

[1] No doubt when the avian lung was evolved from the reptilian, the constriction between the true lung and the air-sacs took place in such a way as to leave a few of the air-pockets on the side of the sac, and it is these distal diverticula which became converted into the recurrent bronchi.

muscular contraction of the ostia and valvular folds contributes to this end ; but the evidence of observers on this point is contradictory.

The lung itself has become marvellously adapted to secure the greatest possible surface for respiratory exchange and the most perfect ventilation of its parenchyma. It is chiefly to the later work of Fischer (**899**), Juillet (**912**), Locy and Larsell (**916**), and Larsell (**915**), that we are indebted for the elucidation of the unique structure of the bird's lung, which differs from that of all other vertebrates in that it contains no culs-de-sac, no blind air-cells, but only freely communicating anastomosing passages forming complete air-circuits. As already explained, the extra-pulmonary bronchus enters the lung and passes down to its hinder end as the main bronchus, giving off as a rule four large secondary ventral bronchi, and then eight secondary dorsal bronchi; to these may be applied Huxley's terms, mesobronchus, entobronchus, and ectobronchus respectively. There are, in addition to these, six laterobronchi and some dorsobronchi. These various secondary bronchi soon branch into tertiary parabronchial tubes of uniform bore which join end to end with those of neighbouring bronchi and of the recurrent bronchi. Further, as shown by Rainey and Williams

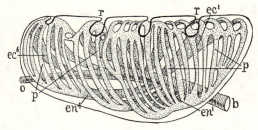

Fig. 611.

Diagrammatic side view of right lung of bird showing parabronchi connecting ecto- and endobronchi (after Locy and Larsell, from J. S. Kingsley, *Comp. Anat. of Vertebrates*, 1926). *b*, Bronchus ; *ec*, ectobronchus ; *en*, endobronchus ; *o*, opening into abdominal sac ; *p*, parabronchus ; *r*, impression of ribs.

(1859), the parenchyma without air-cells consists of hexagonal areas surrounding the parabronchi, supplied by blood-vessels, and pierced by a system of air-capillaries (Fischer, **899**), offshoots from the parabronchi forming an anastomosing network of minute air-circuits leading from one parabronchus to another, and from recurrent to excurrent parabronchi. The spongy mass of the lung thus consists of a complex network of interlacing blood-vascular and air-capillaries of great efficiency, Fig. 610.

The fact that the bones of birds often contain air instead of marrow is said to have been first mentioned by the Emperor Frederick II. in a treatise on the chase in the thirteenth century. This pneumaticity, due to the penetration of diverticula from the sacs, varies greatly in extent. Although generally well developed in good fliers, this is by no means always the case, since it is absent in gulls and little developed in small birds. Every gradation is known, from that of the Penguins in which no bones are pneumatic to that of the Frigate bird in which they all are down to the metatarsals. The early fossil birds *Archaeopteryx* and *Hesperornis* seem not to have possessed pneumatic bones (Marsh); but the vertebrae of some Dinosaurs were pneumatic. It is the cervical sac which supplies air to the cervical and thoracic vertebrae and ribs, the interclavicular to the pectoral girdle wing sternum and sternal ribs, and the abdominal to the hinder parts of the skeleton (Campana, **894**). More or less extensive diverticula may also spread underneath the skin and among the muscles, especially in good fliers and aquatic birds.

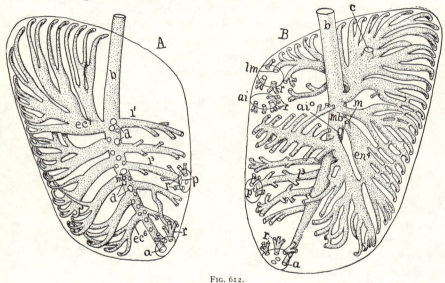

Fig. 612.

Diagrams of lung of hen, *Gallus domesticus* : A, Dorsal, B, ventral aspect (after Locy and Larsell, from J. S. Kingsley, *Comp. Anat. of Vertebrates*, 1926). *a*, Opening of mesobronchus, *mb*, into abdominal air-sac ; *ai*, region of recurrent bronchi from anterior intermediate sac ; *ai⁰*, opening of *b*, bronchus, into intermediate sac ; *c*, opening into cervical sac ; *d*, roots of dorsibronchi ; *ec*, ecto-bronchi ; *en*, entobronchi ; *l*, laterobronchi ; *lm*, *m*, openings into lateral and medial moieties of inter-clavicular sac ; *p*, opening into posterior intermediate sac ; *r*, recurrent bronchi from abdominal sac.

The development of the lungs and air-sacs takes place as follows. Starting from the paired evaginations already mentioned, the lung rudiments join a median tracheo-laryngeal groove which is constricted off from the oesophagus. The paired endodermal diverticula grow backwards, surrounded by a thick layer of mesenchyme covered with coelomic epithelium ; soon they bud off side branches which, penetrating the mesenchyme, give rise to the secondary bronchi we have described in the adult, while the main stem becomes the mesobronchus. Numerous parabronchi now budding off from the secondary bronchi fuse end to end with and open into branches from other bronchi. Still more numerous and finer outgrowths anastomose and complete the system of air-capillaries, thus establishing the characteristic air-circuits. Meanwhile the air-sacs grow out from the surface of the young lung as paired rudiments or buds. The cervical sac arises from the first entobronchus, the anterior intermediate sac from the third entobronchus, the posterior intermediate from the third laterobronchus, the abdominal sac from the extremity of the mesobronchus. The interclavicular sac has a double rudiment on each side, one from the first entobronchus and the other and larger from the third entobronchus ; these two fuse not only with each other, but also in most birds with the corresponding sac of the opposite side. In the meantime the recurrent bronchi have budded off from the necks of the developing sacs, and grow inwards to branch and finally anastomose with the other parabronchi, Figs. 633-7.

What can be the function of these extensive air-cavities, besides that already discussed of acting as accessory respiratory reservoirs to ensure a continuous instead of an intermittent flow of air, has long puzzled zoologists. That, being filled with warm air, they may serve as aerostatic organs was long ago suggested ; but Campana has conclusively proved that their lifting power is quite negligible. They may lower the specific gravity of a bird's body by 4 per cent at most. The replacement of marrow by air in the skeleton lessens the absolute weight ; but here again the difference is probably insignificant, and far surpassed by the variation in the relative weight of the skeleton in different species, and by the difference in weight due, for instance, to the taking in of food even at a single meal. In floating and diving birds, however, the air-cavities act as hydrostatic organs, also in singing birds as reservoirs of air, and generally they may serve to distribute the weight and so help to shift the centre of gravity to an advantageous point in flight. On the whole, the respiratory function seems to be by far the most important.[1]

Mammalia.—Although the differentiation of the lung has taken place in Mammals on the same principle, so to speak, as in Reptiles and Birds of increasing to the greatest possible extent the respiratory surface, yet the mammalian lung has been doubtless evolved along independent lines, and to trace its development we should probably turn back to the un-

---

[1] Another possible function (Campana, Vesiovi, Madarasz) is that of lowering the temperature by offering a large surface for evaporation. Since birds do not sweat and are covered with a non-conducting layer of feathers, they must have some way of keeping down their temperature during violent exertion. Indeed, the whole question of the mechanism for heat regulation in birds seems not yet to have been adequately worked out. It must differ radically from that of mammals, has been independently acquired, and would doubtless well repay investigation.

differentiated type of the Amphibian or lower Reptile such as *Sphenodon*,
Intermediate steps are lacking, since even the Monotremes are thoroughly
mammalian in their respiratory apparatus.

The lung of the mammal is characterised by the great development
of a branching tree-like system of intra-pulmonary bronchi. These
ramifying tubes, whose walls are strengthened by cartilages, are

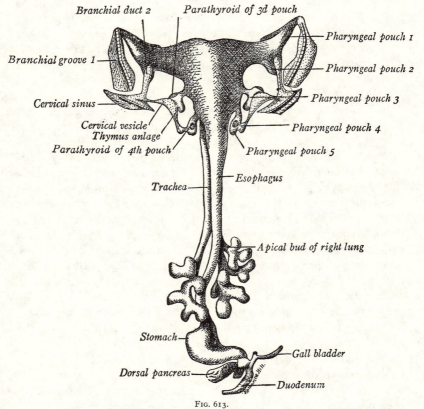

FIG. 613.

A reconstruction of pharynx and fore-gut of 11·7 mm. human embryo seen in dorsal view (after
Hammar). Ectodermal structures are stippled. (From Prentiss and Arey, *Text-book of Embryology*, 1917.)

provided with a layer of smooth muscle fibres, and a lining of columnar
ciliated epithelium reaching as far as the slender bronchioli entering the
lobules. The latter are separated from each other by connective tissue,
and in each of them the bronchiolus branches into respiratory bronchioli
leading to alveolar ducts which expand into atria. Finally the atrium,
and the infundibula or alveolar saccules coming from it, are beset with

minute alveoli or air-cells. Alveolar duct, atrium, saccule, and alveolus are all lined with a very thin flat epithelium, covered externally by a close network of capillaries. It is in this region that the respiratory exchange takes place (Oppel, **795**).

Aeby (**883**) was among the first to make a detailed systematic study of the bronchial tree, and to point out that in the vast majority of mammals it is built on a remarkably asymmetrical plan. From each stem bronchus arise secondary bronchi : a main outer lateral series (called ventral by Aeby and many authors since, but more correctly named lateral by Robinson, **935**) ; a dorsal series and a less complete ventral series (called accessory by Aeby, and supposed by him to be secondarily derived from the lateral). There may also be present a less developed and less regular series of internal lateral bronchi (d'Hardiviller, **908** ; Flint, **900**).

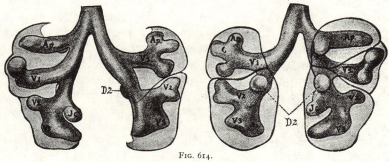

FIG. 614.

Ventral and dorsal views of lungs from human embryo of about 9 mm. (after Merkel, '02) (from Prentiss and Avey, *Text-book of Embryology*, 1917). *Ap*, Apical bronchus ; *D1, D2*, etc., dorsal, *V1*, *V2*, etc., ventral bronchi ; *Jc*, infracardial bronchus.

These secondary bronchi are given off at fairly regular intervals along the stem bronchus, and diminish in size from before backwards to the tip of the tree. The dorsal and ventral series correspond in position, but alternate with the lateral. The pulmonary vein enters the lung ventrally and branches, spreading over the ventral surface of the bronchi. On the contrary, the pulmonary artery, although likewise entering from below, passes round the outer side of the stem bronchus, runs back along its latero-dorsal surface, and spreads over the more dorsal aspect of the bronchial tree. Now Aeby, who held that the artery exerts a controlling influence over the distribution of the bronchi, pointed out that, whereas in the left lung not only the ventral regions, but also the anterior apex, are supplied by the lateral series coming off below the artery, and therefore termed hyparterial ; in the right lung the apical region is supplied by a distinct bronchus coming off more dorsally, passing above the artery, and termed eparterial, Figs. 613-14, 648. The asymmetry is further marked

by the great extension of the first of the ventral bronchi on the right side
to supply the infra-cardiac lobe, large in most mammals, but reduced in
some, as in man. This lobe is developed in a special diverticulum of the
right pleural cavity situated between the diaphragm and the heart, Figs.
644-5, 654.

Much controversy has taken place about the significance of the
eparterial bronchus. With few exceptions it occurs on the right side in
all species not only of the Placentalia but also of the Marsupialia and
Monotremata (Aeby, 883; Huntington, 910; Narath, 924-5). The
known exceptions are certain genera in which it occurs on both sides,
such as *Bradypus*, *Phoca*, *Equus*, *Auchenia*, *Elephas*, *Phocaena*, *Delphinus*, and *Cebus*; and *Hystrix* and *Taxidea*, in which it does not occur
at all. In the Artiodactyle Ungulates and the Cetacea it may arise from
the trachea; sometimes its origin is bronchial on the left and tracheal on
the right (*Auchenia*, *Delphinus*). Aeby concluded that the eparterial is
a special bronchus not usually represented on the left side, and therefore
that the apical lobes of the two lungs are not truly homologous.

Narath, whose results have been generally accepted, strongly opposes
Aeby's conclusions, denies, like Zumstein before him, the importance of
the position of the pulmonary artery, and maintains that the eparterial
bronchus is merely the specialised first bronchus of a dorsal series, and
that it is still, where apparently absent, attached to the first ventral
bronchus, of which it is probably the dorsal branch as suggested by
Willach.

Moreover, Narath holds that the dorsal series of bronchi are phylogenetically to be derived from the lateral bronchi, of which they are to
be considered as branches which have migrated on to the stem bronchi;
and the same suggestion is made about the ventral series. Development,
however, does not support this view, as it seems clearly established that
the series of secondary bronchi appear regularly on the stem bronchi
from before backwards as independent buds (Flint, in the pig; E.S.G.,
in *Trichosurus* and the mouse). Moreover, it seems far more probable
that the arrangement of the secondary bronchi in three or four series is
derived from the reptilian condition, where the outgrowths are more
numerous and less regular, by their reduction and specialisation. Huntington, extending Aeby's observations on the adult tree, concluded that
the symmetrical bilateral hyparterial type (*Hystrix*, *Taxidea*) is the
most primitive, leading through the type with a right eparterial to the
symmetrical eparterial type (*Cebus*, *Phoca*, etc.). But this theory cannot
seriously be maintained. The exceptional cases in which there appear
to be two eparterial bronchi are sporadically distributed among the higher

mammals, and are possibly due to the functional replacement of the original eparterial by the next pair of dorsal bronchi. Observations on the development of these mammals are greatly needed. The two forms without eparterial bronchus (*Hystrix*, *Taxidea*) possess lungs of very aberrant structure. On the other hand, the asymmetrical type is widely prevalent, fully developed in the most primitive groups, and appears in the very earliest stages in the ontogeny of the bronchi, not only of the Ditrematous but also of the Monotrematous mammals (*Echidna*, Narath, 924 ; *Ornithorhynchus*, E.S.G.). There can be little doubt that this type was established in the primitive ancestral mammal.

It is generally supposed that the asymmetry of the bronchi is an adaptation to allow for the backward motion of the heart and the development of the left aortic arch, leading either to the suppression of the eparterial bronchus on the left (Aeby), or to its non-separation from the first lateral (Willach, Narath). However, no trace of a left eparterial bronchus has been discovered in the embryo by most authors ; although d'Hardiviller described in an early stage of the rabbit a left eparterial, which vanishes later, and Bremer (890) observed the same in the embryo of the opossum, *Didelphys*. These observations need confirmation, and it seems not improbable that these authors described either abnormalities, or a first dorsal for the true eparterial bronchus. If we are unable to accept the view of Willach and Narath that the eparterial is homologous with the apical branch of the first left lateral, since there is no evidence that a bronchus can migrate from a hyparterial to an eparterial position, we may still hold that it is essentially merely the first of the dorsal series.

General Development and Phylogeny.—Turning now to the more general significance of the mode of development of the lung structure, we find that new light has been thrown on this problem by the researches of F. Moser (921), confirmed on Reptiles by Hesser (909). She holds that the subdivision of the lumen of the lung is brought about in ontogeny, not by the formation of ingrowing septa, but by the outgrowth of successive bud-like branches from a primitive mesobronchus in the lower as well as in the higher forms. The buds penetrate, sprout, and expand in the surrounding mesoblastic tissue, and thus the thicker this layer of mesenchyme the deeper will be the cavities and the more developed will be the walls separating them. The Amphibian lung-rudiment having but little mesenchyme, the central lumen is large and the peripheral chambers shallow ; moreover, the buds are irregularly distributed. On the other hand, in the Amniota the mesenchyme is progressively increased, and the primary buds fewer and more regular in their origin, but more subdivided distally, leading to the mode of development so admirably described by numerous observers in Mammals (His, Narath, Flint, etc.). We have already sketched this development in the case of the bird (p. 606). There remains only to mention the long controversy held as to whether, especially in the Mammalia, the ramification of the embryonic bronchus takes place according to a monopodial or a dichotomous plan (in the latter case by subdivision of the growing tip, in the former by lateral sprouting from the main stem). Most of the older authors (Remak, Koelliker), and more recently Ewart (1889), Minot (1892), Justesen (913), interpreted the growth as dichotomous; but v. Baer, Cadiat, Küttner, Aeby (883), on the contrary, held it to be monopodial. Some have maintained that both processes take place (Robinson, 935; d'Hardiviller, 1897; Flint, 900). The work of Moser, Narath, and Flint may be considered to have established that the monopodial is the chief, if not the only mode of branching in early stages, that the buds grow out regularly in succession from the base of the bronchus to near its blind end, and that the main buds appear just short of the growing tip, except perhaps in the case of accessory bronchi. In later stages and towards the extremity of the finer twigs it is probable that monopody merges into dichotomy. Obviously intermediate stages between monopody and dichotomy may occur.

Although this may well be a correct account of the embryonic development of the lung, there can be little doubt that it cannot represent its phylogenetic history. Indeed, the lung seems to present a remarkably clear case of an ontogeny which is not recapitulative. For the accumulation of a thick layer of mesenchyme round the narrow endodermal rudiment of thick columnar epithelium can surely in no way repeat a primitive stage, but is rather due to the precocious gathering together of the building material for future differentiation. On the contrary, the lung, to be an efficient organ of respiration, must from the first have had a thin wall abundantly supplied with a superficial network of blood-vessels, and have become progressively folded and pocketed to form the parenchyma of air-cells in higher forms. Thus the respiratory surface, which is the last to develop in the embryo, must have been present from the first and throughout phylogeny, although doubtless less perfected in the lower than in the higher terrestrial vertebrates.

CHAPTER XII

SUBDIVISIONS OF THE COELOM, AND DIAPHRAGM

THE SUBDIVISIONS OF THE COELOM. THE PERICARDIAL COELOM AND SEPTUM TRANSVERSUM

IN the Craniata the coelom always becomes transversely subdivided more or less completely into separate chambers fulfilling special functions. The most constant of these, and the first to appear in both phylogeny and ontogeny, is the pericardial chamber, Fig. 615. Surrounding the heart below the fore-gut, it becomes completely shut off from the more posterior trunk coelom in all adult Craniates, excepting the Selachii, Chondrostei, and certain Cyclostomes (see below, p. 619). The pericardium in fishes is situated far forward, below and behind the posterior gill-arches. Its sides are formed by the body-wall ; behind it is closed off by a firm, nearly vertical, septum transversum (see p. 616). It is protected above by the basi-branchials, and below by the pectoral girdle. Passing to the

Tetrapods we find that the pericardium moves farther and farther back-wards away from the gill-arches; its side walls become thinner, being freed from the body-wall, and the thin septum transversum takes up an oblique position from before backwards and above downwards. This change, only begun in the Amphibia, accompanies the differentiation of a well-defined neck in the Amniota, and the retreat of the heart into the thoracic region above the protecting sternum. The backward migration of the heart and pericardium is always repeated in ontogeny. Mean-while the pericardium tends to become more and more overlapped by the pleural divisions of the coelom, into which grow the lungs. Moreover, in the Sauropsida it becomes separated from the body-wall at the sides by the peritoneal coelom and liver, and in the Mammalia by the ventral

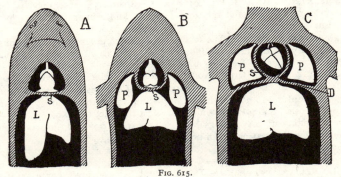

Fig. 615.

Diagram showing relations of coelomic cavities (black) in: A, Pisces; B, Amphibia and some Reptilia; C, Mammalia. (From J. S. Kingsley, *Comp. Anat. of Vertebrates*, 1926.) *L*, Liver; *P*, lung; *S*, septum transversum; *D*, diaphragm.

extensions of the pleural cavities, which in man, for instance, leave the pericardium attached below only in the middle line above the sternum.

The function of the pericardial chamber is no doubt manifold and has changed somewhat from its first appearance in the lowest Craniata. Primarily it affords protection to the heart from surrounding pressures, and a free space filled with fluid in which the heart can easily undergo contraction. The stiffness of its wall in Pisces must also serve to maintain a negative pressure helping the venous flow back to the sinus venosus (see further, p. 536).

The closing off of the pericardial coelom takes place in very much the same way in the embryo of all Gnathostomes. In early stages the coelom of the lateral mesoblastic plate extends on either side of the median mesentery from the trunk into the pericardial region. The rudiment of the heart or ventral splanchnic vein below the oesophagus receives blood from the yolk-laden mid-gut by paired omphalomesenteric veins; but as yet has no direct connexion with the somatic veins. Soon, however, the splanchnic wall, bulged outward by the omphalomesenteric veins, meets

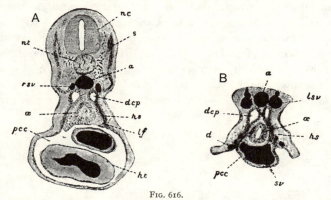

Fig. 616.

Transverse section of an embryo *Scyllium canicula*, Stage I, in A, posterior pericardial region, and B, region of sinus venosus; showing origin of horizontal septum or mesocardia lateralia, *hs*, by fusion of somatic and splanchnic walls leaving two dorsal coelomic passages above, *dcp*, and pericardial coelom below, *pcc*. *a*, Dorsal aorta; *d*, somatic rudiment of ductus Cuvieri; *ht*, heart; *lf*, lateral mesenteric fold; *lsv*, *rsv*, left and right somatic veins (future cardinals); *nc*, nerve cord; *nt*, notochord; *oe*, oesophagus; *s*, somite; *sv*, sinus venosus. (Figs. 616-19 and 621 from E. S. Goodrich, *Jour. of Anat.*, 1918.)

the somatic wall, and a bridge becomes established across the coelom on either side in which develop the ductus Cuvieri carrying blood from the

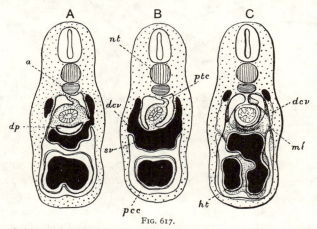

Fig. 617.

Embryo *Scyllium canicula*, Stage L. Transverse sections viewed from in front. A, Section through the median pericardial depression, *dp*, between the two lateral mesenterial folds; B, section farther back through the mesocardia lateralia and ductus Cuvieri; C, reconstructed thick section of same region, beginning in front of that drawn in A. *dcv*, Ductus Cuvieri; *ml*, mesocardium laterale; *ptc*, pericardio-peritoneal canal. Other letters as in Fig. 616.

somatic cardinal veins to that anterior part of the splanchnic vein destined to form the sinus venosus, Figs. 616-17. Thus arise the 'mesocardia lateralia' of Koelliker enclosing the ductus Cuvieri, and separating two dorsal and two ventral openings. On either side of the median mesentery is thus established a dorsal pericardio-peritoneal or pleuro-pericardial passage (recessus parietalis dorsalis of His), and a ventral pericardiaco-peritoneal passage (recessus parietalis ventralis). Into the anterior region of the ventral mesentery below the sinus venosus now grow the hepatic diverticula (see pp. 620-22), which soon by repeated branching subdivide the veins and develop into a bulky liver. Over the front face of this organ spreads the connective tissue of the mesentery to

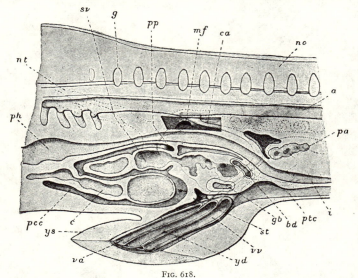

FIG. 618.

Inner view of right half of pericardial region of an embryo *Scyllium canicula*, 26 mm. long, reconstructed. Section cut to left of median mesentery through which a window has been cut to expose Müllerian funnel, *mf*, and the root of vitelline artery, *ca*. Behind pericardio-peritoneal opening, *pp*, is seen beginning of left mesenterial fold. An arrow passes behind mesohepatic ligament from pericardial to peritoneal cavity. *a*, Dorsal aorta; *bd*, bile-duct; *c*, conus; *g*, spinal ganglion; *gb*, gall-bladder; *i*, intestine; *pa*, pancreas; *ph*, pharynx; *st*, septum transversum; *va, vv*, vitelline artery and vein; *yd*, yolk-duct; *ys*, yolk-stalk. Other letters as in Figs. 616 and 619.

the sides, where it meets oblique lateral somatic ridges running downwards along the body-wall from the mesocardia lateralia to meet the ventral mesentery below, Figs. 618-20. Thus the narrow passages between the body-wall and the liver are closed and a complete septum transversum (of His) is formed, shutting off the pericardial cavity ventral to the ductus Cuvieri. Later on the liver is to a great extent separated off from the septum transversum, remaining attached to it by coronary ligaments connected with the hepato-enteric mesentery above and the falciform ligament below. The septum transversum is then left as a peritoneo-pericardial membrane.

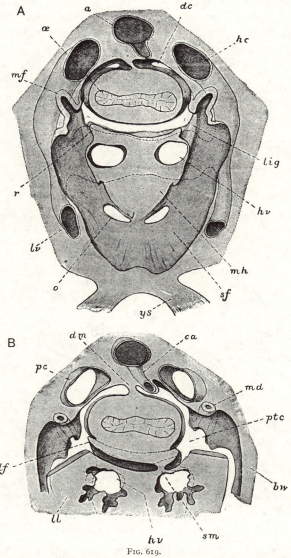

FIG. 619.

Posterior views of reconstructed thick transverse sections of an embryo *Scyllium canicula*, Stage O, about 28 mm. long, showing obliteration of dorsal coelom and formation of septum transversum. Posterior face of the section in Fig. A passes just in front of liver, and dotted line in Fig. B fits on to it; top corners of liver cut away. *bw*, Body-wall; *dc*, dorsal coelom; *dm*, dorsal mesentery; *hc*, and *pc*, posterior cardinal; *hv*, hepatic vein; *lf*, left mesenteric fold; *lig*, suspensory ligament of liver; *ll*, left lobe of liver; *lv*, lateral vein; *md*, Müllerian duct; *mf*, Müllerian funnel; *mh*, mesohepaticum anterius or falciform ligament; *o*, opening in incomplete septum transversum; *r*, ridge of mesenterial fold; *sf*, somatic fold of transverse septum; *sm*, lesser ventral or hepato-enteric mesentery.

The dorsal pericardio-peritoneal passages are later obliterated (incompletely in Cyclostomes and Selachians) by the approximation and final fusion of the splanchnic wall of the oesophagus with the somatic wall and mesocardia lateralia.

Although the septum transversum develops in the Cyclostomes much as in other Vertebrates (Goette, **964**), their pericardium differs in some

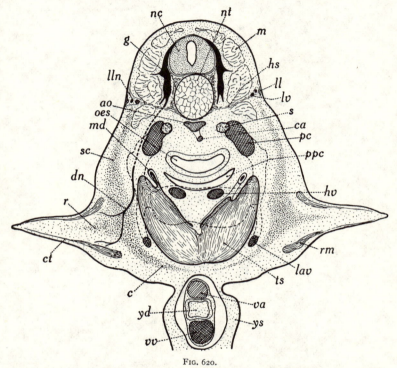

FIG. 620.

Scyllium canicula, embryo, 33 mm. long, cut transversely just in front of liver and seen from behind. *ao*, Dorsal aorta; *c*, coracoid region of procartilaginous pectoral girdle; *ca*, coeliac artery; *ct*, ceratotrich; *dn*, diazonal nerve; *g*, spinal ganglion; *hs*, horizontal septum; *hv*, hepatic vein; *lav*, lateral vein; *ll*, lateral line; *lln*, lateral-line nerve; *lv*, lateral-line vein; *nc*, nerve-cord; *nt*, notochord; *m*, myomere; *md*, Müllerian duct and funnel; *oes*, oesophagus; *pc*, posterior cardinal vein; *ppc*, passage to pericardial coelom; *r*, rudiment of radial; *rm*, radial muscle; *s*, mass of sympathetic and chromaffine cells; *sc*, scapular region; *ts*, transverse septum separating pericardial from peritoneal coelom; *va*, vitelline artery; *vv*, vitelline vein; *yd*, yolk-duct; *ys*, stalk of yolk-sac. Dotted lines show union of large veins to sinus venosus in front.

important respects. In Myxinoids the dorsal region persists in the adult at the side of the oesophagus lodging the pronephros, which extends in front of the ductus Cuvieri and is therefore supplied, in part at least, by the anterior cardinal vein (p. 508); and the pericardial cavity remains on the right in open communication with the trunk peritoneal coelom by a wide aperture, the persistent dorsal passage, Fig. 679. A similar condition obtains in the Ammocoete larva of the lamprey, Fig. 522 (and is said to persist in the adult *Petromyzon planeri* by Broman). The closed pericardial cavity of *Petromyzon* has its wall strengthened by a cartilaginous lamella forming the posterior part of the branchial basket within which the pericardium lies.

Every student of zoology doubtless knows that the pericardial cavity communicates with the more posterior peritoneal coelom in Elasmobranchs by a canal below the oesophagus, Fig. 621, which canal bifurcates and opens behind by right and left apertures above the liver (Monro, 1783). Hochstetter maintained that these pericardio-peritoneal openings are formed late in development by a secondary piercing of the completed posterior pericardial wall (968). But it has been shown

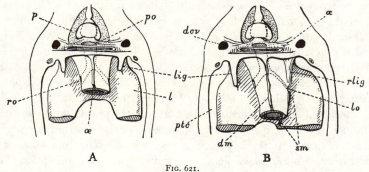

FIG. 621.

Diagrams showing relation of the suspensory ligaments of liver to mesenteries and pericardio-peritoneal canals in an adult Squalus, A, and *Scyllium*, B. Dorsal parts removed by horizontal cut ; dorsal view in which canals below the oesophagus are represented by dotted lines. *dcv*, Ductus Cuvieri ; *dm*, dorsal mesentery ; *lig*, lateral suspensory ligament of liver, *l* ; *lo, ro*, left and right openings of pericardio-peritoneal canals ; *p*, pericardial coelom, *po*, median opening into it of pericardio-peritoneal canal ; *rlig*, right lateral suspensory fold.

(Goodrich, 965) that, as Balfour suggested, they are narrow remnants of the original wider passages found in earlier stages dorsally to the mesocardia lateralia, and that in these fishes the closing off of the pericardial from the peritoneal cavities is never completed. The dorsal passages become constricted by the obliteration of the dorsal region of the pericardial coelom due to concrescence of the oesophagus with the roof and sides of the pericardial cavity, but remain as small openings, situated at first opposite and below the Müllerian funnels, and later shifting on to the ventral side of the oesophagus. A median depression of the hind wall of the pericardial coelom forms the anterior median part of the communication, and the original paired openings get carried back to their position in the adult.

Whether the pericardio-peritoneal opening in Chondrostei is likewise a remnant of the original communication does not seem to have been determined.

THE MESENTERIES AND BURSA OMENTALIS

The splanchnic coelom being of paired origin, the primitive mesentery first appears as a complete partition between the right and left coelomic cavities (Balfour, 317). It is formed of the two layers of coelomic epithelium and the intervening splanchnic mesoblast, mesenchymatous tissue surrounding the gut in the middle, the longitudinal dorsal blood-vessel or aorta above, and the longitudinal ventral vessel below, Figs. 1 and 5. Suspending the alimentary canal, the mesentery also affords access to it for blood-vessels, nerves, and lymph-channels, yet it never remains quite complete in the adult, being pierced more or less extensively by secondary perforations allowing communication from side to side. The primitive mesentery may be distinguished into a dorsal region above the gut and a ventral region below it. The dorsal mesentery remains complete in Mammals and Reptiles, but may be pierced in Birds and Amphibians, and is generally very much reduced in Selachians and Actinopterygians. With rare exceptions, such as the Anguilliformes, Dipnoi, and *Lepidosteus*, the ventral mesentery is so extensively perforated that it disappears almost entirely below the mid- and hind-gut, thus facilitating the coiling of the intestine and allowing free play for its peristaltic movements. The dorsal mesentery is often again subdivided for descriptive purposes in higher forms into the mesogastrium or great omentum, mesoduodenum, mesentery proper, mesocolon, and mesorectum, according as these various regions support the stomach, duodenum, small intestine, colon, and rectum. That portion of the ventral mesentery, lying in front of the septum transversum (see p. 536) and supporting the ventral blood-vessel which here becomes transformed into the heart, is known as the meso-cardium. It is a very transitory membrane, and soon both the dorsal mesocardium and the ventral disappear above and below the heart, Figs. 551-2.

Behind the septum transversum the ventral mesentery usually persists as the omentum minus or hepato-enteric ligament, between the gut and the liver ; and below the liver as the falciform or median ventral hepatic ligament, extending as far as the umbilicus. A small portion of the ventral mesentery may also remain below the rectum supporting the allantoic bladder. The digestive glands, being diverticula of the fore-gut immediately behind the septum transversum, grow out into the mesentery. A longitudinal ventral hepatic groove appears in the embryo and becomes subdivided into an upper anterior hepatic and a lower posterior cystic diverticulum, Figs. 618, 622, 652. The former, branching in the septum transversum and ventral mesentery, soon gives rise to a bulky

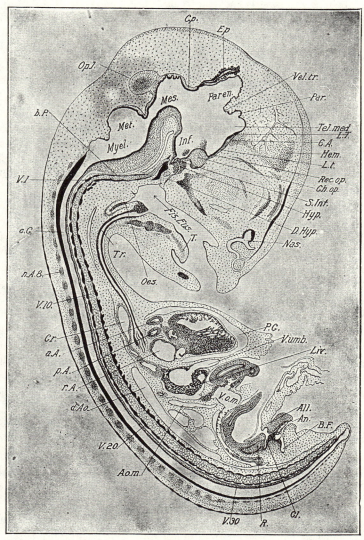

FIG. 622.

Median sagittal section of chick embryo of eight days (from F. R. Lillie, *Devel. Chick*, 1919). *a.A*, Aortic arch; *All*, allantois; *An*, anus; *A.o.m*, omphalomesenteric artery; *B.F*, bursa Fabricii; *b.P*, basilar plate; *C.A*, anterior commissure; *c.C*, central canal; *Ch.op*, optic chiasma; *C.p*, posterior commissure; *Cl*, cloaca; *Cr*, crop; *d.Ao*, dorsal aorta; *D.Hyp*, duct of hypophysis; *Ep*, epiphysis; *Fis.Eus*, fissura Eustachii; *Hem*, surface of hemisphere barely touched by section; *Hyp*, hypophysis; *L.t*, lamina terminalis; *n.A.8*, neural arch of eighth vertebra; *Nas*, nasal cavity; *Oes*, oesophagus; *p.A*, pulmonary arch; *par*, paraphysis; *P.C*, pericardial cavity; *Rec.op*, recessus opticus; *R*, rectum; *S.Inf*, saccus infundibuli; *T*, tongue; *Tel.Med*, Telencephalon medium; *Tr*, trachea; *V.*1, 10, 20, 30, first, tenth, twentieth, and thirtieth vertebral centra; *r.A*, right auricle; *Vel.tr*, velum transversum; *V.o.m*, omphalomesenteric vein; *V.umb*, umbilical vein.

liver bulging backwards into the trunk coelom. While the posterior diverticulum yields the gall-bladder, the base of the hepatic outgrowth becomes narrowed and lengthened to form the bile-duct or ductus chole-dochus, running along the ventral free edge of the lesser omentum to open into the duodenum. Two paired outgrowths from the base of the primitive hepatic diverticulum and one dorsal outgrowth from the fore-gut above them give rise to the pancreas.

The ventral pancreas, whose duct (ductus Wirsungnianus) opens at or near the base of the bile-duct, grows upwards next the stomach, and usually fuses with the dorsal pancreas ; the bulk of the ventral pancreas is derived almost exclusively from the right rudiment. The dorsal pancreas extending in the dorsal mesentery may retain its own duct (ductus Santorini).[1]

The endodermal (hypoblastic) gut of Vertebrates can be divided into fore-, mid-, and hind-gut ; the mid-gut being that portion in the ventral wall of which is stored the bulk of the yolk in the embryo. It is here that the yolk-stalk or vitelline duct is formed by the constriction of the embryo from the yolk-sac in those forms where the yolk is very abundant, and it is below this region of the gut that the ventral mesentery is invariably interrupted. From the fore-gut arise the pharynx with its gill-pouches and lungs, the oesophagus, and, immediately behind the septum trans-versum, the stomach ; lastly, the hepatic and pancreatic diverticula grow out from it just in front of the anterior portal or front end of the mid-gut. From this region backwards extends the mid-gut as far as the caecum ; it is a region variously differentiated in the different classes, and comprising the main part of the intestine. The hind-gut gives rise to the caecum, colon, and rectum, with the allantoic bladder in the higher forms.

Having thus outlined the main morphological relation of these parts, we may proceed to describe those developments which lead up to the formation of various recesses, and especially of the lesser peritoneal cavity or bursa omentalis opening into the main peritoneal cavity by the foramen of Winslow in the Mammalia.

All the Gnathostomata have a markedly asymmetrical fore-gut, due to the development of a large well-differentiated stomach from the region lying just behind the septum transversum and in front of the hepatic

---

[1] No ventral pancreas has been found in Cyclostomes and Selachians. The suggestion made by Goette and Laguesse that some of the hepatic diverticula may represent it seems not to be justified (Brachet, **993**). It is possible that the function of the ventral pancreas differs from that of the dorsal.

diverticulum. Since the anterior end of the duodenum remains, so to speak, firmly held in place by the lesser omentum and contained bile-duct, the gut is bent to the left to accommodate the growing stomach carrying the mesenteries with it. A depression is thus formed on the right side of the mesentery and extending over the dorsal side of the stomach and backwards between the stomach, spleen, and pancreas; it is the first beginning of a bursa omentalis, or rather that region of it distinguished later as the bursa omenti majoris.

This is the condition of the bursa omentalis in the Selachian embryo (Phisalix, 1885; Broman, 959), where it is a shallow depression widely

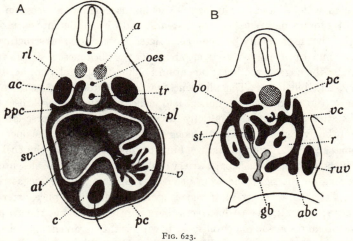

FIG. 623.

Mus musculus, embryo cut transversely and reconstructed. A, Through pericardial region, anterior view; B, through stomach, posterior view. *a*, Aorta; *abc*, abdominal (peritoneal) coelom; *ac*, right anterior cardinal; *at*, atrium; *bo*, bursa omentalis; *c*, bulbus cordis; *gb*, gall-bladder; *oes*, oesophagus; *pc*, right posterior cardinal; *pl*, pleural coelom; *ppc*, pleuro-peritoneal passage; *r*, right lobe of liver; *rl*, right lung bud; *ruv*, right umbilical vein; *st*, stomach; *sv*, opening to sinus venosus; *tr*, trachea; *v*, ventricle; *vc*, vena cava inferior.

open to the general trunk coelom on the right, and marked off above by an overhanging ridge of the dorsal mesentery, the plica meso-gastrica, but leading back to a deep blind pocket between the stomach and pancreas, the recessus pancreatico-entericus. This latter recess opens behind the coeliac artery, which becomes freed from the mesentery at a very early stage. In later development the mesentery becomes so defective that scarcely any trace of the bursa remains in the adult; and the same may be said of most of the Pisces.

Already in the Dipnoi, however, begin to appear the important changes and complications characteristic of the Tetrapoda, and correlated for the most part with the development of the lungs and the vena cava. In these higher Vertebrates, the stomach becomes more differentiated and more bent away from the middle line, and the bursa omentalis correspondingly enlarged and deepened, Figs. 628, 650. Pushing its way between the mesenteries and viscera, and tending to wrap round the alimentary canal, it forms special recesses which all communicate with the general trunk coelom on the right by an opening, the hiatus communis recessum (primitive foramen of Winslow), behind the liver. Dorsally the recessus pancreaticus-entericus extends back into a blind pocket spreading over the right surface of the stomach ; ventrally a recessus hepato-entericus, bounded below by the lesser omentum, extends forwards above the liver where it passes into the right pneumato-enteric recess, the origin of which will be described below. The hiatus communis recessum or primitive foramen of Winslow is at first large and without definite posterior margin ; but it gradually becomes constricted by surrounding structures. It is bounded in front by the liver, above by the plica mesogastrica, and below by the edge of the lesser omentum holding the portal vein, hepatic artery, and bile-duct. The plica mesogastrica contains the vena cava inferior (thus forming the dorsal fold called Höhlenvenengekröse by Hochstetter, and Vena-cava-falte by Ravn), and, as the right dorso-lateral lobe of the liver grows backwards along the vein into the fold, the opening becomes restricted from in front. Further, in the Amniota, the coeliac artery and its branch the hepatic artery draw out a fold from the mesentery projecting on the right and from behind, so that the plica mesogastrica now splits into an outer plica venae cavae and an inner plica arteriae coeliacae ; a small cavo-coeliac recess between them becomes the atrium or vestibulum bursae omentalis, just within the foramen epiploicum Winslowi in the mammal. The coeliac fold tends to separate the bursa omenti majoris on the left from the bursa omenti minoris on the right.

THE PULMONARY FOLDS AND RECESSES

An important new feature in the anatomy of the air-breathing vertebrates is the outgrowth from the wall of the oesophagus of an accessory mesentery, lateral mesenterial or pulmonary fold (Nebengekröse, pulmo-hepatic ligament of Butler, mésolatéral of Brachet), to support the lung. When the mesoblastic pulmonary thickening grows outwards and backwards from the pharynx a pocket is formed behind it into whose outer wall pushes the hypoblastic lung-bud. This outer wall becomes later developed into a membrane, subdivided by the lung into a dorsal ' pulmonary ligament ' attached to the median mesentery above, and a ventral ' pulmonary ligament ' attached to the edge of the lateral lobe of the liver below, Figs. 624, 630. We shall call these membranes, ligaments or accessory mesenteries, simply the right and left pulmonary folds. In front this pulmonary fold is attached to the oesophagus, the root of the lung, and the septum transversum above the sinus venosus; it passes, therefore, on the inner side of the original pericardiaco-pleuro-peritoneal communications, while the nephric fold passes to the outer side of these openings (see p. 631). Thus a deep recess may be cut off from the general peritoneal cavity on either side of the median mesentery between it and the pulmonary folds, or in other words between the oesophagus, lung, and liver, Figs. 628-9. These pulmonary or pneumato-enteric recesses (pulmo-hepatic recesses of Butler, 962-3) are blind in front, but open behind the free edge of the

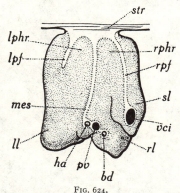

Fig. 624.

Diagram of dorsal surface of liver of a Tetrapod (modified from Broman); white areas bounded by dotted line indicate attachments of median mesentery, *mes*, left pulmonary fold, *lpf*, and right pulmonary fold, *rpf*. *bd*, Bile duct; *ha*, hepatic artery; *ll*, left lobe of liver; *lphr*, left pulmohepatic recess; *pv*, portal vein; *rl*, right lobe; *rphr*, right pulmohepatic recess; *sl*, dorsal lobe; *str*, hind surface of septum transversum; *vci*, vena cava inferior.

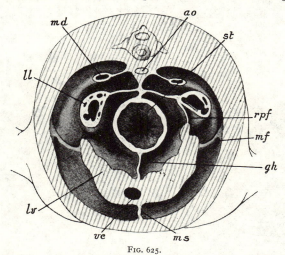

Fig. 625.

Salamandra maculosa, ♀; cut transversely through anterior trunk region and viewed from behind. *ao*, Dorsal aorta; *gh*, gastro-hepatic region of mesentery; *ll*, left lung; *lv*, liver; *md*, Müllerian duct suspended by nephric fold; *mf*, Müllerian (oviducal) funnel; *ms*, subhepatic mesentery; *rpf*, right pulmonary fold closing right pulmonary recess; *st*, stomach; *vc*, vena cava inferior.

pulmonary folds. Generally the lungs bulge on the outer surface

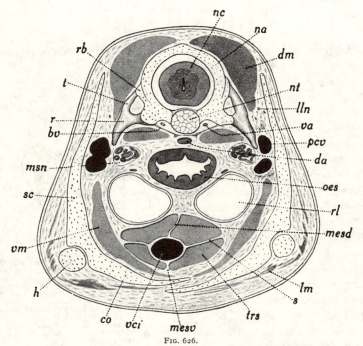

FIG. 626.

Salamandra, larva. Reconstructed thick transverse section, posterior view. *bv*, Ligament from rib to vestigial basiventral ; *co*, coracoid region ; *da*, dorsal aorta ; *dm*, dorsal muscles ; *h*, humerus ; *lln*, lateral-line nerve ; *lm*, limb muscles ; *mesd*, dorsal mesentery ; *mesv*, ventral mesentery ; *msn*, mesonephros ; *na*, neural arch ; *nc*, nerve cord ; *nt*, notochord ; *oes*, oesophagus ; *pcv*, posterior cardinal vein ; *r*, rib ; *rb*, rib-bearing cartilage ; *s*, lateral septum (part of nephric fold) ; *sc*, scapular region ; *t*, tuberculum ; *trs*, transverse septum closing pericardial cavity ; *va*, vertebral artery ; *vci*, vena cava inferior ; *vm*, ventral muscles.

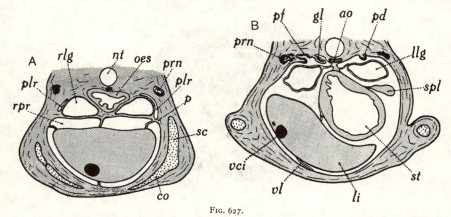

FIG. 627.

Amblystoma tigrinum, transverse sections of larva 28 mm. long. A, Through pectoral girdle ; B, farther back. *ao*, Dorsal aorta ; *co*, coracoid plate ; *gl*, glomus ; *li*, liver ; *llg*, left lung ; *nt*, notochord ; *oes*, oesophagus ; *p*, point of fusion between pulmonary fold and coelomic wall ; *pd*, pronephric duct ; *pf*, pronephric nephrocoelostome ; *rlg*, right lung ; *rpr*, right pulmonary recess ; *sc*, scapular region ; *spl*, spleen ; *st*, stomach ; *vci*, vena cava inferior ; *vl*, ventral mesentery.

of the pulmonary folds into the pleural region of the coelom, and their posterior tips may grow out freely. As a rule, the left pulmonary fold (ligamentum-hepato-pulmonale of Mathes, 973) is much less developed than the right (ligamentum-hepato-cavo-pulmonale of Mathes, 973); and although its dorsal region may stretch far back, its ventral pulmonary ligament bounding the wide aperture into the general peritoneal cavity is usually much shorter, so that the left recess becomes much reduced. Indeed in all the Mammalia the left pulmonary fold and recess are quite

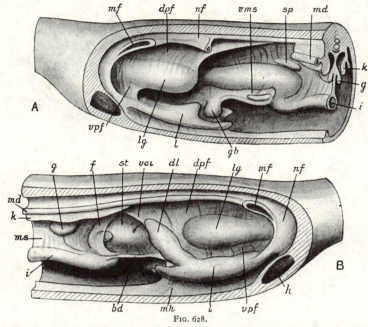

FIG. 628.

Diagram of anterior trunk region of a Tetrapod (Lizard), cut longitudinally so as to expose inner view of left side in A, and right side in B. *bd*, Bile duct; *dl*, dorsal lobe of liver in dorsal pulmonary fold, *dpf*; *f*, fold; *g*, gonad; *gb*, gall-bladder; *i*, intestine; *k*, mesonephros; *l*, liver; *lg*, lung; *md*, Müllerian duct (part of it cut away in A); *mf*, its funnel; *mh*, subhepatic mesentery; *nf*, nephric fold; *sp*, spleen; *st*, stomach; *vci*, vena cava inferior; *vms*, ventral mesentery; *vpf*, ventral pulmonary fold. An arrow passes into pulmonary recess.

vestigial even in the embryo, and unrecognisable in the adult. On the other hand, the right pulmonary fold is usually fully developed. Its dorsal fold extends backwards as the plica mesogastrica, and becomes converted into the plica venae cavae, being invaded by the vena cava and the dorso-lateral lobe of the liver growing up into its posterior margin. Therefore the right pneumato-enteric recess is continuous above the liver with the recessus hepato-entericus, and opens into the general peritoneal cavity, not independently like the left recess, but in common with the other mesenterial recesses by the primary foramen of Winslow (hiatus communis). The bursa omentalis extends, then, not only into the bursa omenti majoris and bursa omenti minoris, but also forwards into the recessus pneumato-entericus. That dorsal region of the liver which is included between the mesentery and the pulmonary fold, and forms the floor of the bursa, develops into the Spigelian lobe ; a corresponding lobe

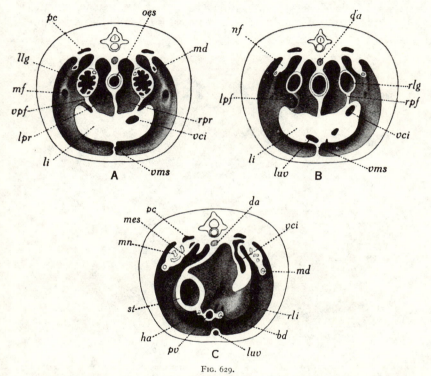

Fig. 629.

Diagrammatic transverse sections through a Tetrapod (Lacertilian) showing subdivision of splanchnocoele. A, Far forward ; B, through hind end of lungs ; C, behind lungs. All seen from behind. *bd*, Bile duct ; *da*, dorsal aorta ; *ha*, hepatic artery ; *li*, liver ; *llg*, left lung ; *lpf*, left pulmonary fold ; *lpr*, left pulmonary recess ; *luv*, left umbilical vein ; *md*, Müllerian duct, and *mf*, its funnel ; *mes*, dorsal mesentery ; *mn*, mesonephros ; *nf*, nephric fold ; *oes*, oesophagus ; *pc*, posterior cardinal ; *pv*, portal vein ; *rlg*, right lung ; *rli*, right lobe of liver ; *rpf*, right pulmonary fold ; *rpr*, right pulmonary recess ; *st*, stomach ; *vci*, vena cava inferior ; *vms*, ventral mesentery ; *vpf*, ventral part of pulmonary fold.

may be delimited on the left side in those forms where the left pulmonary fold is well developed.

Certain modifications of the mesenteries and recesses in the various classes of the Tetrapoda may now be noticed (Broman, **959, 960**).

Amphibia.—Although normally developed in the embryo, the recesses usually become much modified in the adult, especially in the Anura. In these amphibians the hiatus communis recessum persists ; but the re-

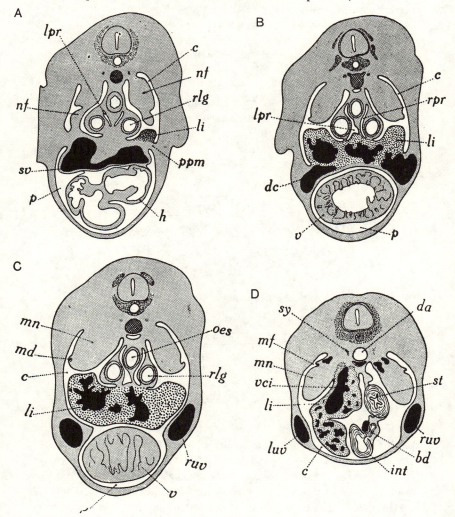

Fig. 630.

Transverse sections of embryo *Lacerta*, before closure of pericardial coelom.   A, Most anterior, through septum transversum ;   D, most posterior.   *bd*, Bile duct ;   *c*, general splanchnic coelom ; *da*, dorsal aorta ;   *dc*, ductus Cuvieri ;   *h*, heart ;   *int*, intestine ;   *li*, liver ;   *lpr*, left pulmonary recess ; *luv*, left umbilical vein ;   *md*, Müllerian duct ;   *mf*, Müllerian funnel ;   *mn*, mesonephros ;   *nf*, nephric fold ;   *oes*, oesophagus in mesentery between pulmonary recesses ;   *p*, pericardial coelom ;   *ppm*, peri-cardio-peritoneal membrane ;   *rlg*, right lung in pulmonary fold ;   *rpr*, right pulmonary recess ; *ruv*, right umbilical vein ;   *st*, stomach ;   *sv*, sinus venosus ;   *sy*, sympathetic ;   *v*, wall of ventricle ; *vci*, vena cava inferior.

cesses almost entirely disappear owing to the reduction of the pulmonary folds and the perforation of the omentum minus. The left pneumato-enteric recess generally is very small, as the accessory mesentery is much reduced ventrally, in the Urodela, Figs. 625-7, and Gymnophiona; but the bursa omentalis is well developed, formed as usual by the combination of the right pneumato-enteric recess with the hepato-enteric and the pancreatico-enteric recesses. However, except in *Cryptobranchus* and *Menopoma*, it seems always to be completely shut off on the right by the closure of the hiatus due to the fusion of the posterior edge of the pulmonary fold with the median mesentery. Communication with the general coelom is then brought about by the perforation of the dorsal mesentery (except in *Amphiuma*).

Reptilia.—As a rule the recesses are well marked in the adult, Figs. 628-

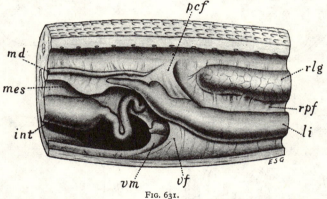

FIG. 631.

Egernia Cunninghami, ♀; right-side view of mid-trunk opened by removal of body-wall. *int*, Intestine; *li*, liver; *md*, Müllerian duct; *mes*, dorsal mesentery; *pcf*, posterior closing fold; *rlg*, hind end of right lung; *rpf*, right pulmonary fold; *vf*, ventral lateral fold; *vm*, ventral mesentery (subhepatic ligament).

631, the primitive foramen of Winslow remains large and open (except in Amphisboenids and some Chelonia, *Emys*), and the plica arteriae coeliacae is much developed, especially in Lacertilia, partially cutting off an extensive bursa omenti majoris. The lizard *Agama* has a dorsal right recessus pneumato-entericus communicating as usual with the recessus hepato-entericus, as well as a ventral right recessus pneumato-entericus opening independently to the general trunk coelom. In the Tejidae, Anguidae, *Gongylus*, and Amphisbaenidae, the right pneumato-enteric recess also opens independently (Broman).

The condition of the recesses and mesenteries in the Aves is complicated by the development of air-sacs and a post-hepatic septum, a structure already found in the Tejidae and Crocodilia, whose origin will be discussed later (p. 641), Figs. 633, 640.

Mammalia.—The left pneumato-enteric recess is here vestigial, disappearing in the adult, and often scarcely visible even in the embryo. On the right, all the recesses join to open by the clearly circumscribed foramen of Winslow ; but the anterior tip of the pneumato-enteric recess becomes nipped off by the diaphragm as a separate small closed cavity in the thorax (see p. 654). Accompanying the rotation of the stomach to the left there is a great development of the mesogastrium, forming a large ventral sac, the great omentum, enclosing an extension of the bursa omenti majoris. The walls of this sac usually come together in the adult so as to obliterate the cavity.

THE NEPHRIC FOLDS

Yet another peritoneal fold remains to be described which plays an important part in the subdivision of the trunk coelom in the higher Vertebrates (Bertelli, 954 ; Rabl, 1049 ; Hochstetter, 967). In the Selachian, Figs. 619, 669, 670, a longitudinal fold runs forwards on each side along the dorsal wall of the trunk coelom from the anterior end of the mesonephros to the body-wall just behind the ductus Cuvieri and septum transversum, along which it passes to the dorsal edge of the lateral lobe of the liver. Primitively this nephric fold (a name we prefer to that of pronephric or mesonephric fold applied to it by most authors) is derived from the nephric ridge in which develop the pronephros in front and the mesonephros farther back, Figs. 628-30. In the female sex it remains as the mesorchium, a mesenterial fold supporting the Müllerian funnel and duct. Now these nephric folds are constant and important features in the anatomy of Craniates, being especially well developed in the higher Gnathostomes, where they undergo many modifications, tending to separate off a pleural division of the coelom in Reptiles, and contributing to the diaphragm in Mammals. The early stages in their development are very similar in all Gnathostomata, see Figs. 620, 629, 633, where they are seen bearing the oviducal funnel on their free edge and cutting off an anterior coelomic recess on each side. But they are situated on the outer side of the embryonic dorsal pericardiaco-peritoneal passage, and must not be confused with the true accessory mesenteries or pulmonary folds of the air-breathing vertebrates which are situated on the inner side of these passages next to the oesophagus.[1]

---

[1] The paired hepatic mesenterial folds described by Goette in the larva of *Petromyzon* are therefore nephric folds and not accessory mesenteries as he and Maurer state.

THE COELOMIC SEPTA IN AVES

We may now pass to the consideration of the various septa which subdivide the coelom in Birds, a subject which has long attracted the interest of anatomists since the days of Aristotle, yet is still but incompletely understood.   The avian heart is situated far back, the pericardium resting immediately on the sternum or, as in Passeres, separated from it by a backward extension of the interclavicular air-sac ; the compact lungs are pressed close up against ribs and vertebral column, the intermediate air-sacs passing down from them along the inner side of the body-wall ; the large liver reaches forward, its lateral lobes extending on

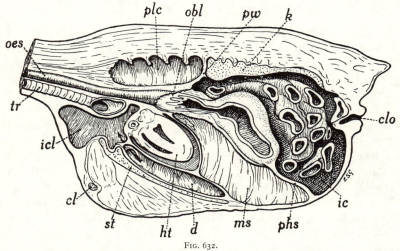

FIG. 632.

Corvus corone, inner view of right half cut longitudinally to left of middle line.   *cl*, Clavicle ; *clo*, cloacal aperture ; *d*, diverticulum of interclavicular sac ; *ht*, heart in pericardial coelom ; *k*, kidney; *ic*, intestinal chamber of coelom ; *icl*, interclavicular air-sac ; *ms*, ventral mesentery ; *obl*, top of oblique septum ; *oes*, oesophagus ; *phs*, post-hepatic septum; *plc*, left pleural coelom (lung removed) ; *pw*, posterior wall of pleural cavity ; *st*, sternum ; *tr*, trachea.

either side of the pericardium and backwards beyond it, Figs. 608-9. The disposition of the viscera is thus very different from that seen in the Mammalia, where the liver is pushed back and the pericardium flanked by the lungs.   In addition to the median mesentery and the pericardial wall or pericardio-peritoneal membrane, f o u r m e m b r a n o u s p a r t i t i o n s may be distinguished in birds subdividing the body-cavities—three longitudinal and one obliquely transverse.   The first longitudinal membrane stretches almost horizontally across the thoracic region, really in two halves running from its attachment to the median mesentery to the body-wall below the lungs, which adhere to its dorsal surface.   This more dorsal membrane, the pulmonary aponeurosis of Huxley, 911 (diaphragme pulmonaire of Sappey, diaphragmite antérieur of Milne-Edwards, diaframmo ornitico of Bertelli), in front joins the cervical aponeurosis, roots of the lungs and septum transversum, and the body-wall at the sides on the inner face of the ribs ; behind it passes upwards to below the vertebral column between the lungs and the kidneys, thus completely cutting off the lungs and remains of the pleural cavities above from the air-sacs below, being pierced only by ostia leading from the former to the latter. As a rule striated muscles—costo-pulmonary supplied from corresponding spinal nerves, and sometimes anterior sterno-costal muscles as well—extend from the ribs into the membrane, Figs. 609, 637.

The second longitudinal membrane is the oblique septum of Huxley (diaphragme thoraco-abdominal of Sappey, diaphragmite thoraco-abdominal of Milne-Edwards), attached dorsally to the median mesentery where it meets the pulmonary aponeurosis, and ventrally to the body-wall at or near the edge of the sternum. It may contain unstriated muscle fibres passing forward from the pubis, and extends forwards to the cervical aponeurosis forming the ventral inner wall of the intermediate air-sacs, and backwards to the body-wall as a very thin membrane covering the abdominal sacs. Behind the intermediate sacs the oblique septum meets the post-hepatic septum (see below, p. 639), beyond which in all birds except the *Apteryx* (Owen, 1863 ; Huxley, 911) the abdominal air-sacs project into the intestinal division of the peritoneal cavity. The space between the pulmonary aponeurosis and oblique septum is entirely occupied by the air-sacs, and is not of coelomic origin.

The next partition to be mentioned is the post-hepatic septum, so named by Butler (962-3), who first clearly described its relations and homologies, Figs. 608, 632. It is a transverse membrane stretching obliquely backwards from the oblique septum above to the lateral and ventral body-wall below (Milne-Edwards, 1867 ; Campana, 894 ; Weldon, 1883 ; Bignon, 1887-9 ; Butler, 962 ; Beddard, 953 ; Broman, 959 ; Hochstetter, 969 ; Poole, 974), and is formed of two halves passing outwards from the dorsal mesentery and the ventral mesentery. Dorsally the post-hepatic septum is continuous with the oblique septum and pulmonary aponeurosis, where these combine above the liver and behind the lungs ; this region is distinguished by Butler as the oblique abdominal septum (passing across between the abdominal sacs, and lined on both sides by coelomic epithelium) from the true oblique septum (underlying the thoracic sacs and lined with coelomic epithelium on the ventral side only). Ventrally the post-hepatic septum encloses the

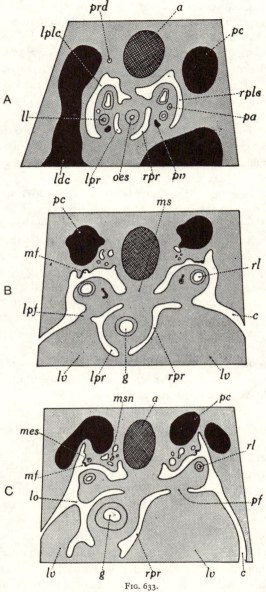

FIG. 633.

Transverse sections of 5-day chick, *Gallus* : A, Most anterior ; C, most posterior. *a*, Dorsal aorta ; *c*, splanchnocoele ; *g*, gut ; *ldc*, left ductus Cuvieri ; *ll*, left lung ; *lo*, opening of left pulmonary recess ; *lpf*, left pulmonary fold ; *lplc* and *rplc*, left and right pleural coelom still open behind ; *lpr* and *rpr*, left and right pulmonary recess ; *lv*, liver ; *mf*, Müllerian funnel ; *mes*, mesonephric duct ; *msn*, mesonephros ; *oes*, oesophagus ; *pa*, pulmonary artery ; *pc*, posterior cardinal ; *pf*, pulmonary fold ; *prd*, pronephric duct ; *pv*, pulmonary vein ; *rl*, right lung.

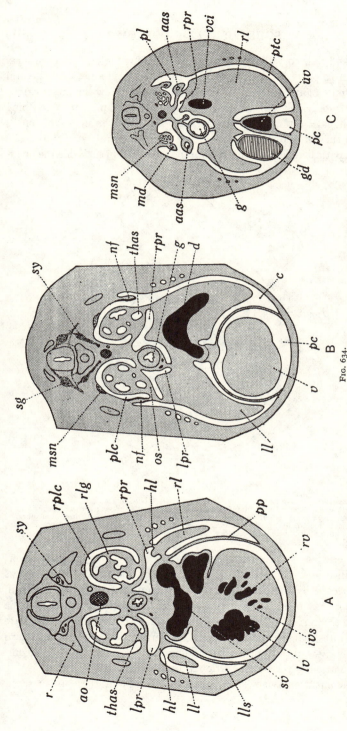

FIG. 634.

Passer domesticus, transverse sections of embryo showing development of air-sacs and septa. A, Through anterior thoracic; B, through posterior thoracic; and C, through abdominal air-sacs. *aas*, Abdominal sac; *ao*, dorsal aorta; *c*, coelom; *d*, junction of great veins; *g*, gut; *gd*, gizzard; *hl*, hepatic ligament (ventral pulmonary fold); *ll*, left lobe of liver; *lls*, left liver-sac; *lpr*, left pulmonary recess (open in C); *md*, Müllerian duct; *msn*, mesonephros; *nf*, nephric fold; *os*, top of oblique septum; *pc*, pericardial cavity; *pl*, intestinal coelom; *plc*, pleural coelom; *pp*, liver; *r*, rib; *rl*, right lobe of liver; *rlg*, right lung; *rplc*, right pleural coelom; *rpr*, right pulmonary recess; *sg*, spinal ganglion; *sv*, sinus venosus; *sy*, sympathetic ganglion and cord; *uv*, umbilical vein; *vci*, vena cava inferior.

gizzard, and thus completes a partition subdividing the original peritoneal
coelom into a posterior cavity enclosing the intestine, kidneys, and genital
organs, and an anterior cavity enclosing the liver.   Now the latter cavity
is further subdivided into four by the complete ventral mesentery reaching
the post-hepatic septum, and the third longitudinal membrane which may
be called the horizontal hepatic ligament (ventral part of Butler's
pulmo-hepatic ligament).   This membrane separates on each side the pneu-

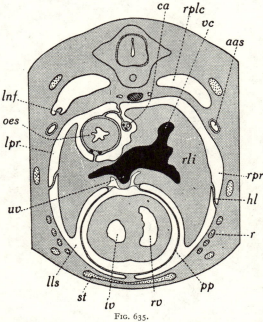

FIG. 635.

Passer domesticus, transverse section of late embryo.   *aas,* Abdominal air-sac ;   *ca,* coeliac artery ;
hl, hepatic ligament (ventral part of pulmonary fold) ;   *lls,* left liver-sac ;   *lnf,* left nephric fold ;
lpr, left pulmonary recess ;   *lv,* left ventricle ;   *oes,* oesophagus ;   *pp,* pericardio-peritoneal membrane ;
r, rib ;   *rli,* right lobe of liver ;   *rplc,* posterior region of right pleural coelom ;   *rv,* right ventricle ;
st, sternum ;   *uv,* umbilical vein ;   *vc,* vena cava inferior.

mato-enteric recess above from a ventral cavity into which bulges the liver,
the so-called liver-sac of Butler.   While the right pneumato-enteric recess
is quite closed by the obliteration of the foramen of Winslow, except in
so far as it may open into the left recess by secondary perforation of the
median mesentery, this left pneumato-enteric recess may still communicate
with the intestinal coelomic chamber by a narrow slit between the gizzard
and the body-wall (wall of the left abdominal air-sac), due to the incomplete formation of the post-hepatic septum at this point.

Development of Avian Septa.—The morphology of these avian septa can only be understood from a knowledge of their development, now fairly complete, thanks to the work of Butler (962), Ravn (977), Hochstetter (835), Bertelli (955), and Poole (974). The development of the septum transversum, studied in detail by Ravn and Brouha (961), differs in no important respect from that described above for the Tetrapods in general. Owing to the backward movement of the heart it becomes very oblique. The median mass, overspreading the anterior face of the liver and passing downwards and backwards into the median hepatic ligament or ventral mesentery, joins the lateral closing folds running down the body-wall from the mesocardia lateralia and

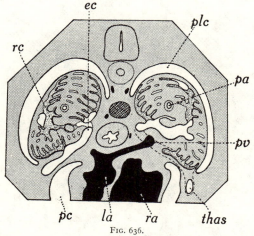

Fig. 636.

Passer domesticus, embryo late stage ; transverse section showing development of anterior thoracic air-sac in pulmonary fold. *ec*, Excurrent bronchus opening into sac ; *la*, left auricle ; *pa*, pulmonary artery ; *pc*, pericardial coelom ; *plc*, pleural coelom ; *pv*, pulmonary vein ; *ra*, right auricle ; *rc*, recurrent bronchus ; *thas*, thoracic air-sac.

enclosing the umbilical veins. Thus, on the eighth day in the chick, the lower coelomic passages are closed. As usual the dorsal passages are closed by the fusion of the ductus Cuvieri with the pulmonary thickenings of the oesophagus, and the pericardium is now shut off by the septum transversum extending over it as a thin pericardio-peritoneal membrane, from which the liver becomes separated off later. The membrane is further increased by the extension forward and downward of the peritoneal cavity (and liver) at the sides of the pericardium. The well-developed and nearly symmetrical pulmonary folds (accessory mesenteries) play an important part in the formation of the avian septa, Figs. 633-7. Forming, of course, the outer walls of the pneumato-enteric recesses, they spread outwards (chick, sixth day) attached to the edge of the lateral lobes of the liver, and give rise to the floor of the future pleural cavities, into which the lungs project on their dorso-lateral surface.   At this stage the pleural cavities still open widely behind, and also by narrow slits at the sides, between liver and body-wall, into the peritoneal cavity.   With this cavity the left pneumato-enteric recess communicates by a long aperture in front of the stomach, the ventral region of the pulmonary fold being less developed than the dorsal.   On the right side the usual bursa omentalis is developed ; but

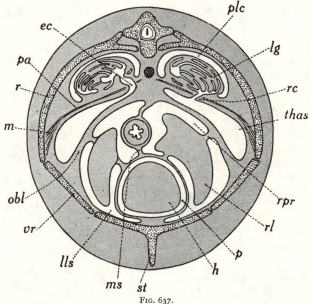

FIG. 637.

Diagram of transverse section through thorax of a Bird.   *ec*, Excurrent passage from lung to air-sac through pulmonary aponeurosis ; *h*, heart ; *lls*, left liver-sac ; *lg*, lung ; *m*, muscle ; *ms*, mesentery below oesophagus ; *obl*, oblique septum ; *p*, pericardial coelom ; *pa*, pulmonary aponeurosis ; *plc*, reduced pleural coelom ; *r*, dorsal rib ; *rc*, recurrent bronchus from sac to lung ; *rl*, right lobe of liver ; *rpr*, right pulmonary recess ; *st*, sternum ; *thas*, posterior thoracic air-sac ; *vr*, sternal rib.

the primitive foramen of Winslow is closed by the backward growth and fusion of the pulmonary fold, and the bursa omenti majoris may become cut off by fusion of the plica arteriae coeliacae, and subsequently more or less completely obliterated (Butler (962), Broman (959), in *Gallus*).

Meanwhile the pleural cavities become shut off as follows.   The pulmonary folds attached ventrally to the liver lobes are stretched out-wards, and fuse progressively from before backwards with the somatic wall.   The fusion, starting from the septum transversum, is helped by an ingrowing shelf of the body-wall (E.S.G.), and occurs just below the line of attachment of the nephric fold, which, however, takes little or no share in the process. In this way the pulmonary fold becomes bent and subdivided into a more dorsal pleuro-peritoneal membrane passing below the lungs to the body-wall, and a more ventral part, the future horizontal hepatic ligament, from the body-wall to the liver. The former is the roof and the latter the ventro-lateral wall of the pneumato-hepatic recess. Thus the pleural coelom above is cut off on each side from the coelomic cavity below destined to become the liver-sac. The pleural cavities become later completely closed behind by the extension backwards of the pleuro-peritoneal membranes which meet posterior closing folds, transverse growths of connective tissue from the median mesentery in front of the kidney and dorsal to the attachment of the pulmonary folds. At this stage, then, the median ventral peri-cardial and two dorsal pleural cavities have been closed off. Later on the lungs press against the ribs, and obliterate the pleural cavities.

In the meantime three paired hollow outgrowths from the bronchi, rudiments of the intermediate and abdominal air-sacs (see p. 600), penetrate into the dorsal region of the pulmonary fold. Here the intermediate sacs expand, growing outwards and downwards, thus splitting the fold into the upper pulmonary aponeurosis and the lower oblique septum (Butler, 963). Between the sacs the tissue is reduced to thin vertical septa. Into the post-pulmonary mass of tissue forming the thickened edge of the pulmonary folds extend the rudiments of the abdominal air-sacs. Each sac becomes applied to and fuses with the lateral abdominal wall, along which it grows. The unsplit portion of the pulmonary folds, stretching between the sacs, the median mesentery and gastric loop of the gut, form that dorsal region of the post-hepatic septum called oblique abdominal septum by Butler (see p. 633).

The development of the ventral region of the post-hepatic septum is difficult to make out. Butler (963), Poole (974), and Hochstetter (969) all derive it wholly or in part from the ventral mesentery. The gastric portion of this mesentery is greatly extended on the left side and fusing with the body-wall binds the stomach to it. The ventral gastric fold thus formed grows up so as to meet the oblique abdominal septum above and the wall of the abdominal air-sac at the side. As mentioned above, a communication remains at this point between the left recess and the intestinal coelom. The right portion of the ventral post-hepatic septum seems to be formed by the lateral and dorsal growth of the free posterior edge of the ventral mesentery along the body-wall, until it meets the oblique abdominal septum.

THE SUBDIVISION OF THE COELOM IN REPTILIA

We naturally turn to the Reptilia for an explanation of the origin of the avian septa. As a rule, in this class the pleural region of the coelom is not shut off from the remainder of the body-cavity ; but in the Lacertilia it becomes partially closed owing to the development of a very deep nephric fold, forming a pleuro-peritoneal membrane sometimes attached ventrally to the body-wall as well as the liver, Figs. 628, 631. Hochstetter has also described a vertical posterior closing fold extending from the mesentery behind the tip of the right lung (a vestige of a

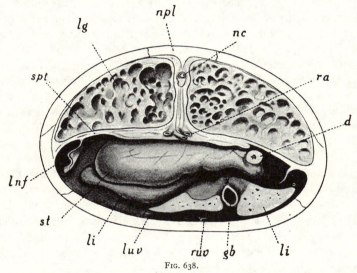

FIG. 638.

Testudo graeca, cut across and seen from behind. *d*, Duodenum ; *gb*, gall-bladder ; *li*, liver ; *lg*, lung ; *lnf*, left nephric fold ; *luv*, left umbilical vein ; *nc*, nerve-cord ; *npl*, neural plate ; *ra*, right aortic arch ; *ruv*, right umbilical vein ; *spt*, septum.

similar fold occurs in some species on the left), which he considers to be homologous with the similar fold found in the embryo of mammals (and birds ?), Fig. 631. When the nephric fold joins the posterior fold, as in *Stellio*, only a narrow aperture of communication remains from the pleural to the peritoneal cavity on the right ; in *Agama* even this opening is closed (Hochstetter, **967**). Although bearing a certain superficial resemblance to the dorsal diaphragm of the mammal, there can be little doubt that this pleural wall is a special development in the Lacertilian. With the oblique septum of the bird it, of course, has no connexion. On the other hand, in the Varanidae, where the pleural cavity is obliterated and the lungs are pressed up against the vertebral column and ribs, there is a membrane extending from the body-wall and covering the lungs below somewhat as in birds; but there is no evidence that it is really derived from the pulmonary folds. In the Chelonia, however, the lungs adhere to the body-wall above, and are partially (*Emys*) or entirely (*Testudo*) shut off from the peritoneal cavity by a septum which, according to Bertelli, is developed by the spreading outwards and fusion with the body-wall of the pulmonary folds, just as in birds, Fig. 638. These conditions deserve further study.

It is in the Crocodilia, which as Huxley maintained are of all living reptiles the most closely allied to birds, that we might expect to find the avian septa developed; and in the Crocodile, indeed, there is a complete post-hepatic septum shutting off pleural cavities, pulmo-hepatic cavities, and liver-sacs from an in-

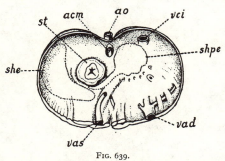

Fig. 639.

testical chamber behind (Huxley, **911**; Butler, **962**; Poole, **974**; Hochstetter, **969**). Nevertheless, the pleural cavity is held by Butler and Hochstetter not to correspond strictly to that of birds, since it appears to be cut off from the ventral liver-sac by the adhesion of the liver lobes to the body-wall (ventral oblique hepatic ligament), and not by the lateral fusion of the pulmo-hepatic liga-

View from behind of post-hepatic septum of *Crocodilus niloticus* (from Hochstetter, 1906). *acm*, Coeliaco-mesenteric artery; *ao*, dorsal aorta; *she*, limit of saccus hepato-entericus (left pulmonary recess), and *shpe*, of right saccus hepatopulmoentericus (right pulmonary recess) seen through septum; *st*, section of stomach; *vad*, right anterior abdominal vein; *vas*, left anterior abdominal vein; *vci*, vena cava inferior.

ment. A short fold passing inwards from the ventral body-wall between the liver and the lung is supposed to represent a rudiment of the 'avian diaphragm' (Butler, **962**). Yet it seems not impossible that the ventral oblique hepatic ligament really represents the outer region of the pulmo-hepatic ligament which has shifted downwards, in which case the septa and cavities of the Crocodile would correspond almost exactly with those of a bird, excepting, of course, for the absence of air-sacs, Figs. 639, 640.

As for the origin of the post-hepatic septum itself, Butler has shown that it exists in an incomplete state in the family Tejidae alone among Lacertilia. In *Tupinambis*, for instance, two folds diverge right and left from the hinder margin of the ventral mesentery; they are attached to the ventral and lateral body-wall, and have a free dorsal edge bounding a small opening on the left and a larger on the right, communications

2 T

between the pleural and the intestinal regions of the coelom. Were these folds completed dorsally, the crocodilian condition would be realised ; while, on the contrary, were they less developed, the condition found in most Lacertilia would result.

Butler has described a complete post-hepatic septum in the Ophidia, whereby two liver-sacs are closed off (the pleural cavities, and apparently also the pulmo-hepatic recesses, disappear in development).

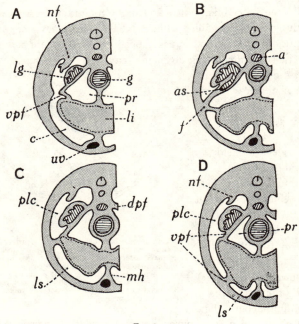

Fig. 640.

Diagrams illustrating development of coelomic septa in a bird, A and B, and a crocodile, C and D. A, early, and B, later stage ; C, stage corresponding to B : D, nearly adult. *a*, Dorsal aorta ; *as*, air-sac ; *c*, splanchnocoele ; *dpf*, dorsal pulmonary fold ; *f*, fusion of ventral pulmonary fold with body-wall ; *g*, gut ; *lg*, lung ; *li*, liver ; *ls*, liver-sac ; *mh*, hepatic ligament or ventral mesentery ; *nf*, nephric fold ; *plc*, pleural coelom ; *pr*, pulmonary recess ; *uv*, umbilical vein ; *vpf*, ventral pulmonary fold.

While the post-hepatic septum of the Crocodilia may be considered as homologous with that of Aves (although Hochstetter does not admit its full homology), it is difficult to account for the appearance of this septum in the Tejidae and Ophidia. Since it is absent in other Lacertilia and in *Sphenodon* it can hardly be considered as a primitive structure, and we are left with the unsatisfactory conclusion that it has been independently developed in two or more groups.

THE MAMMALIAN DIAPHRAGM

The development of the mammalian diaphragm is brought about by a very complicated process difficult to describe, partly owing to its being built up from several separate rudiments, partly from its connexion with the vascular system, and liver, and partly from the constant relative shifting of its parts and the associated organs during ontogeny, and the difficulty of representing, in three dimensions, the various curved surfaces involved except by means of models. Our modern knowledge of its development may be said to date from the work of Uskow on the rabbit

(979). Lockwood (971), Ravn (975-7), Bertelli (954), Mall (972), Broman (960), and especially Brachet (957-8), have confirmed and completed his account.

Very early, when the embryo is scarcely folded off from the extra embryonic layers, the mesocardium laterale appears, owing to the splanchnic mesoblast covering the omphalomesenteric veins meeting the somatic wall, and the ventral pericardioperitoneal passage is almost at once obliterated, Figs. 641-642, 645. A complete ventral septum transversum is thus early formed just in front of the anterior intestinal portal, from which it soon becomes separated owing to the narrowing of

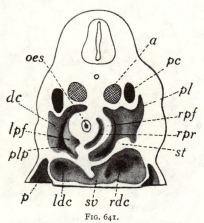

FIG. 641.

Reconstruction of embryo *Sus*, domestic pig, 5 mm. long, cut transversely through sinus venosus, *sv*, and seen from behind. *a*, Lateral aorta ; *dc*, ridge enclosing ductus Cuvieri ; *ldc*, entrance of left ductus ; *lpf*, left pulmonary fold ; *oes*, oesophagus ; *p*, peritoneal coelom ; *pc*, posterior cardinal ; *pl*, wall of future pleural cavity ; *plp*, pleuro-pericardial passage still widely open ; *rdc*, entrance of right ductus Cuvieri ; *rpf*, right pulmonary fold ; *rpr*, right pulmonary recess ; *st*, septum transversum.

the yolk-stalk and growth of the liver. As the latter organ enlarges spreading over the posterior surface of the septum towards the body-wall laterally and ventrally, the septum shifts not only backwards but also into a more upright position. At first, while its dorsal free edge, delimiting the dorsal pleuro-pericardial passages (ductus pleuro-pericardiacus and recessus parietalis dorsalis of His) on either side of the mesentery, is occupied by the ductus Cuvieri, the paired and symmetrical umbilical and omphalomesenteric veins reach the sinus venosus through the septum transversum (that dorsal region of it which represents the mesocardia lateralia). Later, the umbilical veins lose their anterior connexions, and the vena cava inferior alone pierces the septum from behind, combined in the embryonic stages with the ductus venosus Arantii.

The development of the pleuro-pericardial membrane, destined to close off the pericardium below from the pleural cavities above and at the sides, takes place owing to the backward shifting of the heart and forward growth of the pleural cavities by the extension of that dorsal region of the septum transversum which is not invaded by the liver. The ductus Cuvieri play an important part in the development of the membrane and the closure of the dorsal coelomic passages or pleuro-pericardial communi-

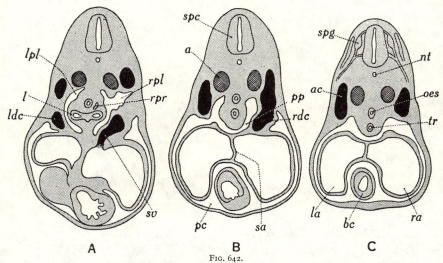

A   **B**   **C**

FIG. 642.

Transverse sections through pericardial region of embryo *Sus*, domestic pig, 5 mm. long. C, Most anterior; A, most posterior. *a*, Lateral aorta; *ac*, anterior cardinal; *bc*, bulbus cordis; *l*, lung bud; *la*, left auricle; *ldc*, left ductus Cuvieri; *lpl*, left pleural coelom; *nt*, notochord; *oes*, oesophagus; *pc*, pericardial coelom; *pp*, passage from pericardial to pleural cavities; *ra*, right auricle; *rdc*, right ductus Cuvieri; *rpl*, right pleural coelom; *rpr*, right pulmonary recess; *sa*, septum atriorum; *sv*, sinus venosus; *tr*, trachea.

cations. At first these veins run in the dorsal free edge of the septum transversum outwards and upwards in the body-wall; then, as the heart passes backwards, they lengthen and bend forwards, taking up a position more and more parallel and relatively nearer to the oesophagus and trachea as the pleural cavity expands. Thus the upper part of the septum becomes stretched out, so to speak, as a thin pleuro-pericardial membrane between the inner ductus Cuvieri, the outer body-wall, and the ventral region of the septum transversum attached to the liver. In this way is formed a membranous floor to the pleural cavity passing transversely and obliquely forwards and upwards. In early stages a pleural groove

extends on either side of the mesentery forwards on to the roof of the
pericardium ; the pleuro-pericardial passage over the transverse septum
becomes converted into a narrow slit between the ductus Cuvieri and the
pulmonary ridges or masses of splanchnic mesoblast into which grow the
lung-buds. These slits close from before backwards, leaving for a considerable time two minute pleuro-pericardial canals, Fig. 650, comparable

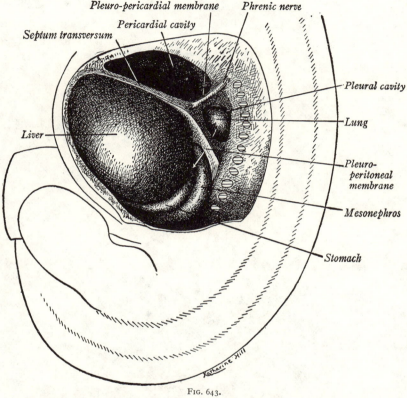

FIG. 643.

Reconstruction of 11 mm. human embryo to show development of diaphragm (from Prentiss
and Arey, *Text-book of Embryology*, 1917).

to the permanent pericardio-peritoneal canals of the Elasmobranch, p. 619.
Although these canals appear to be finally obliterated, they may occasionally persist open as abnormalities. The ductus Cuvieri have now sunk
below the membrane to reach the sinus venosus, and to the fusion of them
and the oesophagus with the sinus is due the final closure of the canals.
Thus is completed the floor of the anterior region of what we may call
the primitive pleural cavity. There remains to be described the develop-

ment of the very important phrenic nerve. It appears early, when the septum is situated about the level of the fifth cervical ganglion, as a

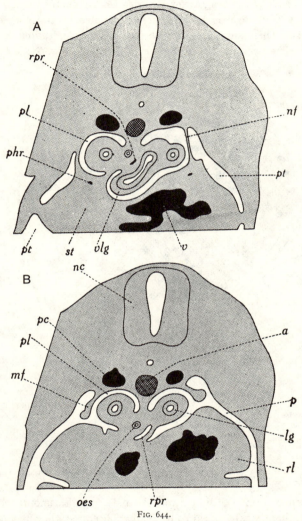

FIG. 644.

Mus musculus, transverse sections of embryo. A more anterior than B. *a*, Dorsal aorta ; *mf*, Müllerian funnel ; *nf*, nephric fold (pleuro-peritoneal membrane) ; *oes*, oesophagus ; *p*, peritoneal coelom opening into pleural coelom ; *pc*, post-cardinal ; *pt*, peritoneal coelom ; *phr*, phrenic nerve ; *pl*, left pleural coelom ; *rl*, right lobe of liver ; *rpr*, right pulmonary recess ; *st*, septum transversum on liver ; *v*, vein ; *vlg*, ventral lobe of right lung.

compound nerve formed from branches of the third, fourth, and fifth cervical nerves passing down outside and behind the ductus Cuvieri to the dorsal region of the septum. When the ductus Cuvieri lengthens as the heart recedes, the nerve stretches back at first obliquely, then almost longitudinally, in the pleuro-pericardial membrane to its insertion in the septum transversum. The latter connexion remaining constant forms an important topographical point when comparing the embryonic septum with the adult diaphragm, Figs. 643, 646, 649, 650.

The shutting off of the hinder region of the pleural cavities is brought about by the combination of three structures. The floor is formed from a thick layer of mesoblast (dorsal diaphragm of Uskow) overlying the liver and apparently derived partly from the mesentery between it and

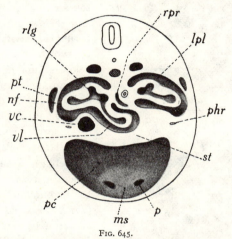

FIG. 645.

Diagrammatic reconstruction of embryo mouse, *Mus musculus*, before complete closure of peri-cardial cavity. Anterior view of embryo cut transversely through pericardial coelom, *pc. lpl*, Left pleural coelom ; *ms*, ventral mesentery ; *nf*, nephric fold ; *p*, passage from pericardial to peri-toneal cavities ; *phr*, phrenic nerve ; *pt*, abdominal or peritoneal coelom ; *rlg*, right lung ; *rpr*, right pulmonary recess ; *st*, septum transversum ; *vc*, vena cava inferior ; *vl*, ventral diverticulum of pleural cavity.

the gut (lesser omentum, gastro-hepatic mesentery), but mostly from a backward growth of the dorsal edge of the septum transversum itself, as is evidenced by the course of the dorsal branch of the phrenic nerve. This floor, passing obliquely backwards and upwards, forms an angle with the pleuro-pericardial membrane as it meets it on the top of the septum transversum, Fig. 643.

The second element contributing to the pleural wall is the pleuro-peritoneal membrane, derived from the nephric fold, Figs. 644, 646-7. It first appears as a dorsal ridge continuous with the meso-nephric fold behind, and running down the posterior surface of the ductus Cuvieri to the dorsal edge of the transverse septum, and so on to the lateral lobe of the liver, where it merges with the outer edge of the mesoblastic layer on the floor of the pleural cavity described above. From the vertical part of the ridge the pleuro-peritoneal membrane develops, and its dorsal and ventral extensions are known as its dorsal and ventral pillars (Uskow, 979). As the ductus Cuvieri bends forwards, the nephric fold remains behind attached to the body-wall, growing backwards as an almost vertical (sagittal) membrane, with a free posterior margin;

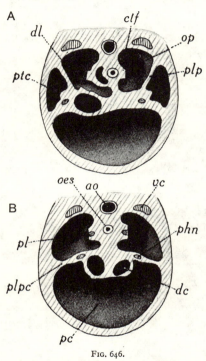

it thus tends to separate an inner pleural chamber above the liver from an outer peri-toneal recess, opening behind and below into the general peritoneal cavity. The extent to which the recess is de-veloped, extending over the pericardium, seems to vary considerably in different mam-mals; and appears to be of little morphological importance, being later on flattened out, so to speak, and merged with the general peritoneal space behind the diaphragm. In Marsupials (*Trichosurus*, E.S.G.) the recess is small; but in those forms like the Rabbit (Uskow and Brachet) or Guinea-pig (Hochstetter, 835), where it is large, the base of the pleuro-peritoneal membrane marks off an inner pleuro-pericardial from an outer peritoneo-pericardial membrane. Both the dorsal and the ventral pillars of the nephric fold tend to approach

FIG. 646.

Diagrams of *embryo Mammal* cut transversely through pericardial and pleural cavities, A, and more posteriorly, B; anterior view, lungs and heart removed. *ao*, Dorsal aorta; *clf*, left posterior closing fold; *dc*, ductus Cuvieri; *dl*, tissue overlying dorsal surface of liver; *oes*, oesophagus; *op*, opening from left pleural to peri-toneal coelom; *pc*, pericardial coelom; *phn*, phrenic nerve; *pl*, cut surface of fold supporting lung; *plp*, pleuro-peritoneal membrane; *plpc*, pleuro-pericardial membrane; *ptc*, peritoneal coelom; *vc*, cardinal vein.

the middle line behind, and the growing pleuro-peritoneal membrane narrows the originally wide communication of the pleural coelom with the peritoneal coelom to a small posterior slit. But the final closure of the pleural cavity is brought about on the right side by the fusion with the pleuro-peritoneal membrane of a transverse outgrowth of the pulmonary fold, bearing a cup-like depression to receive the tip of the lung. On the left side the membrane meets a similar outgrowth (posterior closing fold) from the dorsal mesentery, Figs. 646-7, 649, 650.

This closure of the pleural cavity takes place at a comparatively late stage; meanwhile the pleuro-peritoneal membrane, at first almost sagittal in position, has become flattened out and extended laterally, and moulded on the concave surface of the liver to acquire its definitive shape as the dorsal region of the diaphragm. At

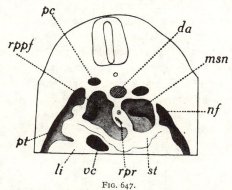

Fig. 647.

Mus musculus, rather late embryo, before complete closure of pleural cavities. Reconstruction cut transversely and showing from in front posterior ends of nephric folds, etc. *da*, Dorsal aorta; *li*, liver; *msn*, mesonephros; *nf*, nephric fold; *pc*, posterior cardinal; *pt*, peritoneal coelom; *rppf*, right posterior closing fold; *rpr*, right pulmonary recess; *st*, septum transversum; *vc*, vena cava inferior.

the same time still greater changes take place in front, where the small

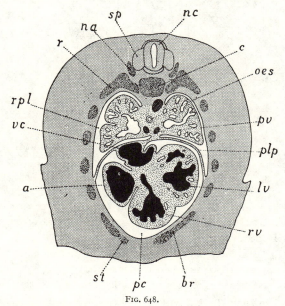

Fig. 648.

Trichosurus vulpecula, embryo 10 mm. long; transverse section of thorax. *a*, Auricle; *br*, sternal rib; *c*, centrum; *lv*, left ventricle; *na*, neural arch; *nc*, nerve-cord; *oes*, oesophagus; *pc*, pericardial coelom; *plp*, pleuro-pericardial membrane; *pv*, pulmonary vein; *r*, rib; *rpl*, right pleural coelom; *rv*, right ventricle; *sp*, spinal ganglion; *st*, sternal band; *vc*, vena cava inferior.

primitive pleural cavity expands rapidly outwards and downwards, carrying the outer margin of the pleuro-pericardial membrane ventrally, until in some cases it almost meets that of the opposite side below the heart. By the ingrowth of the coelomic epithelium at the sides the liver also becomes peeled off from the diaphragm, remaining attached to it by the gastro-hepatic mesentery above, the coronary ligaments at the sides, and the falciform ligament below, Figs. 651-3.

During this process a great expansion of the diaphragm takes place

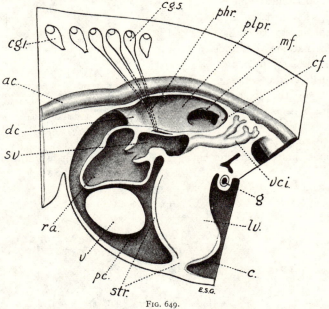

Fig. 649.

Mus musculus, 12-day embryo ; pericardial region cut longitudinally to right of middle line ; inner view of right side, reconstructed. *ac*, Anterior cardinal ; *c*, abdominal (peritoneal) coelom ; *cf*, posterior closing fold ; *cg*$^{1-5}$, cervical ganglia ; *dc*, ductus Cuvieri in transverse septum ; *g*, gall-bladder ; *lv*, liver ; *mf*, Müllerian funnel on nephric fold ; *pc*, pericardial coelom ; *phr*, phrenic nerve ; *plpr*, pleuro-peritoneal membrane (nephric fold) ; *ra*, right auricle ; *str*, septum transversum ; *sv* sinus venosus ; *v*, wall of ventricle ; *vci*, vena cava inferior.

at the sides. How much of this new membrane between the peritoneal and pleural cavities is derived from the body-wall and pleuro-peritoneal membrane respectively, it is difficult if not impossible to say. According to Brachet, the greater part of it would be of pleuro-peritoneal origin ; but there appears to be no good evidence that this membrane spreads down into the ventral region of the diaphragm. Rather would the ventral half of the diaphragm appear to be derived almost entirely from the body-wall and transverse septum, as indicated in the diagram, Fig. 654, showing the final constitution of the mammalian diaphragm.

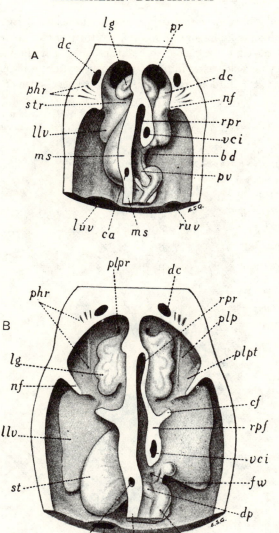

FIG. 650.

Diagrams of anterior trunk region of embryo mammal showing development of diaphragm : Fig. A early stage, Fig. B later stage. A horizontal cut has removed dorsal half and exposed splanchnocoele, etc. ; dorsal view. *bd*, Bile duct ; *ca*, coeliac artery ; *cf*, posterior closing fold ; *dc*, ductus Cuvieri passing down into transverse septum ; *dp*, dorsal pancreas ; *fw*, arrow passing through primitive foramen of Winslow ; *i*, intestine ; *lg*, left lung ; *llv*, left lobe of liver ; *luv*, left umbilical vein ; *ms*, median dorsal mesentery ; *nf*, nephric fold (*plpt*, pleuro-peritoneal membrane) ; *phr*, phrenic nerve ; *plp*, pleuro-pericardial membrane ; *plpr*, pleuro-pericardial canal, derived from large opening into pericardial coelom, *pr* ; *pv*, portal vein ; *rpf*, right pulmonary fold ; *rpr*, right pulmonary recess ; *ruv*, right umbilical vein ; *st*, stomach ; *str*, septum transversum ; *vci*, vena cava inferior.

It would be very interesting to know the exact derivation of the muscles of the diaphragm ; unfortunately their origin has not yet been definitely traced.   But there can be little doubt that the material from which they develop passes into the transverse septum from the myotomes belonging to the third, fourth, and fifth cervical nerves (which provide the

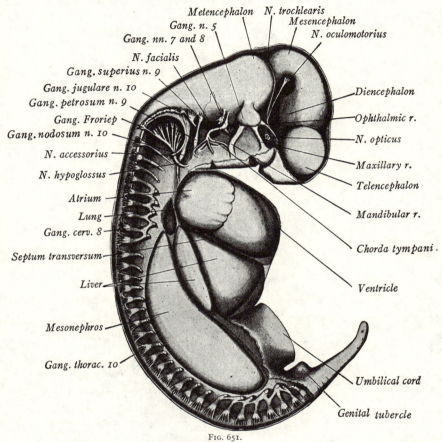

Metencephalon   N. trochlearis
Gang. n. 5   Mesencephalon
Gang. nn. 7 and 8   N. oculomotorius
N. facialis
Gang. superius n. 9
Gang. jugulare n. 10
Gang. petrosum n. 9
Gang. Froriep
Gang. nodosum n. 10
N. accessorius
N. hypoglossus
Atrium
Lung
Gang. cerv. 8
Septum transversum
Liver
Mesonephros
Gang. thorac. 10

Diencephalon
Ophthalmic r.
N. opticus
Maxillary r.
Telencephalon
Mandibular r.
Chorda tympani .
Ventricle
Umbilical cord
Genital tubercle

FIG. 651.

Lateral dissection of a 10 mm. pig embryo, showing viscera and nervous system from right side. The eye has been removed and otic vesicle is represented by a broken line.   Ventral roots of spinal nerves are not indicated.   × 10·5.   (From Prentiss and Arey, *Text-book of Embryology*, 1917.)   *n*, Nerve; *r*, ramus.

branches constituting the phrenic nerve of the adult) at an early stage when the septum is still far forward in the future neck region.   Bardeen, indeed, has found the rudiment of the diaphragm muscles in the region of the fifth cervical nerve in early human embryos (3).

To sum up the ontogenetic history of the diaphragm.   It is of compound origin ; formed in part ventrally from the transverse septum, dorsally from the pleuro-peritoneal membrane and transverse outgrowths or closing folds from the mesentery on the left and the pulmonary fold on the right. The body-wall itself, when the thorax expands and the

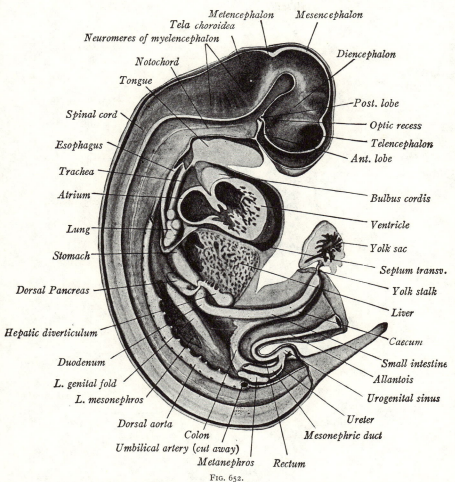

FIG. 652.

Median sagittal dissection of 10 mm. pig embryo, showing brain, spinal cord, and viscera from right side. ×10·5. (From Prentiss and Arey, *Text-book of Embryology*, 1917.)

pleural cavities grow downwards on either side of the pericardium, probably contributes largely to its formation, extending the pleuro-peritoneal membrane and the septum transversum at the sides. In addition to these elements of the diaphragm should also be mentioned the dorsal and ventral mesenteries, which join the right and left halves, tissue lying on the liver dorsally to the transverse septum, and lastly

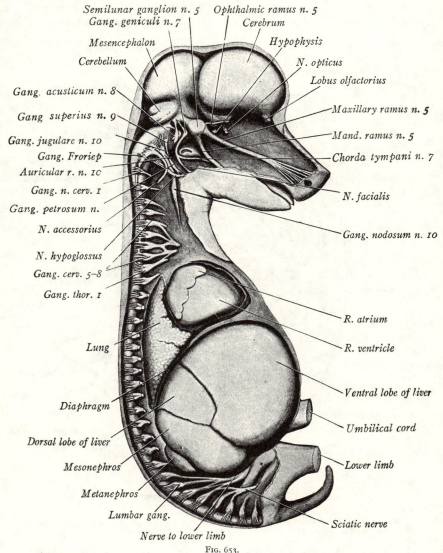

Semilunar ganglion n. 5
Gang. geniculi n. 7
Ophthalmic ramus n. 5
Cerebrum
Mesencephalon
Hypophysis
Cerebellum
N. opticus
Lobus olfactorius
Gang. acusticum n. 8
Gang superius n. 9
Maxillary ramus n. 5
Gang. jugulare n. 10
Mand. ramus n. 5
Gang. Froriep
Chorda tympani n. 7
Auricular r. n. 10
Gang. n. cerv. 1
Gang. petrosum n.
N. facialis
N. accessorius
Gang. nodosum n. 10
N. hypoglossus
Gang. cerv. 5–8
Gang. thor. 1
R. atrium
Lung
R. ventricle
Ventral lobe of liver
Diaphragm
Umbilical cord
Dorsal lobe of liver
Mesonephros
Lower limb
Metanephros
Lumbar gang.
Sciatic nerve
Nerve to lower limb

FIG. 653.
Lateral dissection of 35 mm. pig embryo to show nervous system and viscera from right side. × 4. (From Prentiss and Arey, *Text-book of Embryology*, 1917.)

a small contribution from the pulmonary fold (accessory mesentery) on the right of the oesophagus. For it must be remembered that the right pulmo-hepatic recess in early stages extends forwards into the pleural region in mammals as in other Tetrapods (see p. 631). As the heart moves backwards, the transverse septum shifts from its primitive position far forwards towards the hinder region of the thorax, carrying with it the phrenic nerve and the material from which develop the muscles of the diaphragm.

The history of the complex mammalian diaphragm is doubtless lost with the long line of extinct reptile-like ancestors. Although the elements out of which it has been built, septum transversum, nephric folds, and

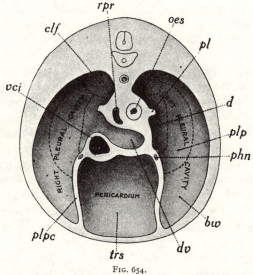

FIG. 654.

Diagram of *diaphragm* of *Mammal*; showing supposed approximate limits (indicated by dotted lines) of various regions derived from different sources. Thorax cut transversely and viewed from in front; heart and lungs removed. Part derived from body wall, *bw*; from posterior closing fold, *clf*; from median mesentery and connective tissue dorsal to liver, *d*; from pleuro-peritoneal membrane, *plp*; from ventral region of transverse septum, *trs*. *dv*, Diverticulum of right pleural cavity; *oes*, oesophagus; *phn*, phrenic nerve; *pl*, cut fold supporting lung; *plpc*, pleuro-pericardial membrane; *rpr*, right pulmonary recess; *vci*, vena cava inferior. (Modified from Broman.)

even posterior closing folds, may all occur in the modern Reptiles, it is not among these that we should expect to find its homologue. Huxley long ago rejected the loose comparisons made by the older authors between the diaphragm of the mammal and the various septa of birds. Indications of the origin of the diaphragm may, however, be sought among Amphibia, though the modern forms are too specialised and too far removed from the mammalian stem to afford much evidence. One of its most characteristic features is its derivation in great part from the septum transversum and the growth into it of somatic muscles supplied from in front by the phrenic nerve. It seems clear, therefore, that the mammalian diaphragm must have begun to develop as part of the respiratory mechanism when the heart was still far forward, and before the neck region had become differentiated. Now, ever since Breyer (1811) described muscles attached to the base of the lungs in the Surinam toad, *Pipa americana*, anatomists have compared certain muscles found in this region in the Anura with the muscular diaphragm of the mammal (Meckel, 1821 ; Dugés, 1834 ; Goette, 1875 ; Giglio-Tos, 1894, 1906 ; Beddard, 952 ; Nussbaum, 1896), and lately the question has been treated in detail by Keith (970). The muscular body-wall of the Amphibia, formed of the rectus abdominis ventrally, splits laterally into external and internal oblique muscles, and again into a superficial external oblique and an internal transversus muscle. While the anterior segment of the rectus abdominis profundus (supplied by the second spinal nerve) becomes attached in Anura Aglossa to the pericardial wall or septum transversum, below and at the sides, the anterior region of the transversus (supplied by the third spinal nerve) is attached to it more dorsally. These two muscles may well represent the material from which the ventral sterno-costal and spinal musculature (similarly supplied from the third, fourth, and fifth spinal nerves, Gössnitz, 966) have been phylogenetically derived. Much the same conditions obtain in the Urodela. Yet the conditions in modern Amphibia differ radically from the mammalian, in that the anterior muscles mentioned above (chiefly the transversus attached to the oesophagus, roots of the lungs, and pericardial wall) serve in them for expiration, and in that inspiration is not brought about by expanding the thorax (see p. 597).

Bertelli (955) denies the possibility of these Amphibian structures having given rise to the muscular diaphragm of the Mammalia. Nevertheless, it seems not impossible that some such extension of muscles into the transverse septum may have taken place in Theromorphous ancestors, and that by the motion backwards of this muscular transverse septum to a position behind the lungs and its combination with the nephric folds, into which the muscles may have penetrated, the mammalian diaphragm may have developed.

CHAPTER XIII

EXCRETORY ORGANS AND GENITAL DUCTS

THE EXCRETORY ORGANS AND GENITAL DUCTS OF CRANIATA

THE kidneys of adult Amniotes are compact organs situated on the roof of the abdominal cavity in the pelvic region. They are covered ventrally by the coelomic epithelium, and consist of a mass of urinary tubules opening into a duct or ureter which runs backwards to open into the cloaca, or its derivative, the urinogenital sinus. The Amphibia have a more extended kidney reaching further forward, and in Pisces and Cyclostomata it may extend almost up to the pericardium. Moreover, while in Amniotes the definitive kidney proper is devoted to the elimination of waste products, and there are separate genital ducts, in lower Gnathostomes the kidney and its duct serve also for the passage to the exterior of the spermatozoa. In fact there is reason to believe, as will be explained later (p. 714), that the whole excretory system of the Gnathostomata consisted originally of paired segmental tubules leading by means of a longitudinal duct from the coelom to the exterior, and that, while their original function was to serve as genital ducts in both sexes, their excretory function is secondary and was acquired as the original excretory organs or true nephridia disappeared (p. 718).

To understand the morphology and physiology of the genital and excretory organs of the Craniata it is necessary to study their development

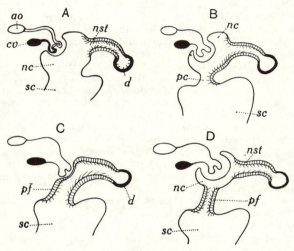

Fig. 655.

Diagrams showing relations of segmental *excretory tubule* of Craniate, and formation of *peritoneal funnel, pf,* and *nephrocoelostome, nst. ao,* Median dorsal aorta supplying artery to glomerulus ; *cv,* cardinal vein receiving vein from glomerulus ; *d,* transverse section of longitudinal duct ; *nc,* nephrocoele ; *pc,* communication between latter and general coelom or splanchnocoele.

and modification from Cyclostome to Mammal. But a thorough exposition of the subject would take a volume in itself, and it has been so well dealt with recently in several text-books (Felix, 1005-7 ; Kerr, 840 ; Brachet, 993) that it will be sufficient here to give only a general account of some of the more important facts and theories.

The whole excretory system, as it may be called, consists primarily of paired tubules opening on the one hand into the coelom and on the other into a longitudinal duct. The coelomic opening consists of a ciliated 'peritoneal funnel' leading into a narrow 'funnel tubule' which widens out into a thin-walled chamber (Bowman's capsule), one side of which is pushed in by blood-vessels (glomerulus). Capsule and glomerulus together make up a 'Malpighian body'. From the chamber another ciliated funnel (nephrocoelostome)[1] leads to the coiled 'glandular tubule' which opens into the duct, Fig. 665.

The chief function of the peritoneal funnel seems to be to drive fluid from the coelom towards the tubule, that of the glomerulus to filter fluids from the blood into the cavity of the capsule, and that of the walls of the glandular tubule to excrete nitrogenous waste into its lumen, and to reabsorb some of the fluid. Although the structure just described may be considered as the fundamental plan of the segmental units of the Craniate system, yet such complete tubules are rarely found in the full-grown kidney, owing to the specialisation of some parts and the disappearance of others.

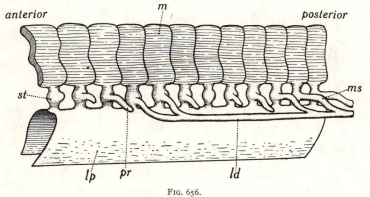

FIG. 656.

Diagram showing development of pronephric, *pr*, and mesonephric, *ms*, tubules from stalk or nephrotome, *st*, joining segmental somites, *m*, to unsegmented lateral plate, *lp*; *ld*, longitudinal duct.

Development.—In accordance with the general differentiation of the segments the tubules develop from before backwards. Usually present at some time in all the trunk segments, they fail to appear in the anterior (head) and posterior segments (tail). There is a tendency for those of the more anterior segments to develop and become functional in early life, and for those of the more posterior segments to develop and function in later life, accompanied by the more or less complete degeneration of the first-formed tubules. While the ancestral Craniate may be supposed to have been provided with a complete set of seg-

---

[1] The term 'nephrostome' has been applied to both the peritoneal funnel and the funnel leading from the chamber. The latter is here called nephrocoelostome to emphasise the fact that it is derived from and opens into the nephrocoele (p. 660), and to avoid confusion with the nephrostome of the Annelid nephridium with which it has nothing to do (p. 718).

mental tubules throughout the whole trunk region opening into a longitudinal duct leading to a posterior cloaca, in all existing Craniates a few
anterior tubules and their duct develop first, and are later succeeded
by a larger number of more posterior tubules opening into the same
duct (Cyclostomes, Pisces, Amphibia). Further, in the Amniota, this
second set is succeeded by yet more posterior tubules which make up the
adult organ. Thus an original complete **archinephros** with its archinephric duct became differentiated into pro-, meso-, and meta-nephros
(Lankester's terminology, Fig. 673).

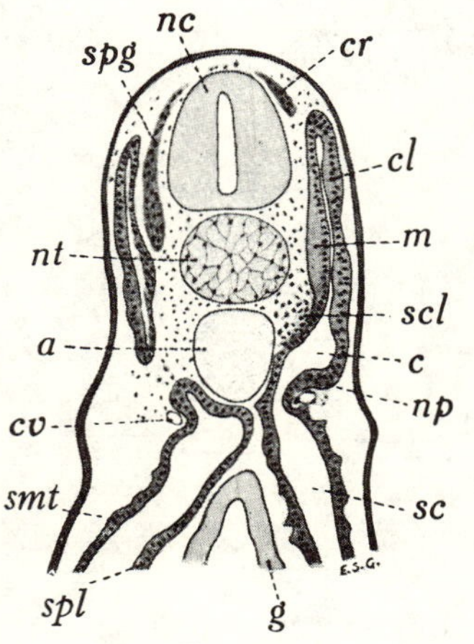

FIG. 657.

Transverse section of *Scyllium canicula*,
embryo stage J, taken in front of pronephros.
a, Dorsal aorta; *c*, nephrocoele; *cl*, outer
'cutis layer' of myotome; *cr*, neural crest;
cv, cardinal vein; *m*, myomere in inner wall
of myotome; *nc*, nerve cord; *np*, rudiment
of pronephric nephrocoelostome; *nt*, notochord; *sc*, splanchnocoele; *scl*, sclerotome;
spl, splanchnopleure; *smt*, somatopleure;
spg, spinal ganglion; *g*, wall of gut.

It has now been satisfactorily established that all the tubules are developed directly or indirectly from the
segmental stalk of mesoblastic tissue
which in early stages unites the dorsal
segmented somite to the more ventral
unsegmented lateral plate (pp. 3, 722).
This intermediate region, known as
the **nephrotome**, is often hollow from
the first or becomes so later, its cavity
communicating with the sclero-myo-
coele above and the splanchnocoele
or general coelom below. Soon the
nephrotome becomes separated off from
the somite and the communication of
its cavity (**nephrocoele**) with the sclero-
myocoele is cut off, while the opening
into the general coelom persists as the
peritoneal funnel. Meanwhile from
the lateral wall of the nephrotome
there grows out a diverticulum towards the ectoderm to form the

glandular tubule. The nephrocoele persists as the cavity of Bowman's
capsule. In the anterior or pronephric region, at all events, the
diverticula (blind at first) lengthen, bend backwards, and fuse at their
distal ends to a longitudinal duct, which in the majority of Gnathostomes
(Elasmobranchii, Apoda, Amniota) grows back freely between the
ectoderm and somatic wall of the coelom to the cloaca. The mesonephric
tubules which develop later find the duct ready made, join and open
into it. Thus the **primary longitudinal duct** functions at first as a
pronephric duct and later as a mesonephric duct behind the pronephric
region, Figs. 656, 660, 665.

The development described above becomes variously modified in the different regions and various groups owing to specialisations which often render it difficult to interpret, especially as regards the ducts (see below : Müllerian and Wolffian ducts, etc.).[1]

The pronephric tubules are usually degenerate, incompletely developed, and apparently functionless in Gnathostomes whose embryos are supplied with a large amount of yolk and have no free larval stage (Elasmobranchii, Reptilia, Aves, and in Mammalia), and are fully developed and functional only in those forms with comparatively little yolk and a larval stage (Sedgwick, 1055 ; Rabl, 1048).

The Pronephros.—Usually far more pronephric rudiments appear than ever reach full development and functional activity, and the best developed remain towards the middle of the pronephric region, the other rudiments at either end of the series soon disappearing. Nor are the rudiments constant in number or in position, although they usually begin only a few segments behind the auditory sac.

Among Gnathostomes a functional pronephros occurs only in Actinopterygii, Dipnoi, and Amphibia. But even in these complete tubules with peritoneal funnel, capsule, and glandular tube are but rarely found (Apoda among Amphibia ; Chondrostei, *Amia*, *Lepidosteus* among Actinopterygii). It is, indeed, character- istic of the pronephros that the glandular tube and its nephrocoelostome are usually the only parts typically developed, Figs. 658-68. The nephrocoeles usually expand into thin-walled sacs, and these generally combine on each

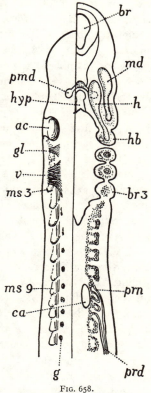

Fig. 658.

Scyllium canicula, embryo stage I. Partial reconstructions of longitudinal horizontal sections, the half on left more dorsal than that on right. *ac*, Auditory sac ; *br*, brain ; *br* 3, third branchial arch ; *ca*, ductus Cuvieri ; *g*, spinal ganglion ; *gl*, glossopharyngeal nerve ; *h*, hyoid somite ; *hb*, hyoid bar ; *hyp*, hypophysis ; *md*, mandibular somite ; *ms* 3-9, third to ninth metaotic somites ; *pmd*, premandibular somite ; *prd*, pronephric duct ; *prn*, first of three pronephric funnels (nephro-coelo-stomes) ; in front of them are four nephrocoeles.

---

[1] Further complications are brought about owing to different modes of development of the nephrotome. In many forms (Amphibia and Amniota) it is at first solid, it and the tubule rudiment becoming secondarily hollowed

side to a common pronephric chamber into which open the nephrocoelostomes, and project the glomerular vessels joined to a longitudinal vascular
ridge or ' glomus '. The corresponding peritoneal funnels disappear as
such, the openings becoming widened out and confluent, so that the
chamber is more or less completely merged with the general coelom as
in Anura. But, as a rule, by the meeting of the two coelomic walls
below the chamber it becomes cut off from the more ventral coelom by a
floor leaving a wide communication at the posterior or at both ends.
In Teleosts the chamber may be completely closed.

In *Polypterus* (Kerr, 1033) about nine nephrocoeles become enlarged,
from the first five of which tubules develop, and of these two survive and become functional. Their nephrocoeles fuse to a chamber
widely open to the general coelom. The pronephros of the Dipnoi
(Semon, 1062 ; Kerr, 840) is similarly developed. In *Lepidosiren* tubule
rudiments arise in metaotic segments 5-7 ; two become functional
and open into a chamber with a glomus. *Acipenser* (Salensky, 1880–81 ;
Jungersen, 1029 ; Maschkowzeff, 1040; Fraser, 1011) has some eight rudiments, of which six or seven survive ; in *Lepidosteus* (Balfour, 2 ; Beard,
989 ; Felix, 1005) three, and in *Amia* (Jungersen, 1029 ; Dean, 1002 ; Felix,
1005) three or four survive and become functional out of some eight
rudiments, of which the first corresponds to the third myomere. The
number of rudiments varies much in Teleosts (Swaen and Brachet, 1066 ;
Felix, 1005), beginning at the level of the second myomere in *Leuciscus*,
and fourth in *Salmo*. Of the six rudiments in *Salmo* three survive, but
in others only one tubule may develop. It is a remarkable fact that
the pronephros remains functional in some adult Teleosts (*Fierasfer*
and *Zoarces* : Emery, 121 ; *Lepadogaster* : Guitel, 1018).

In the Amphibia there is a well-developed pronephros with generally
three coiled tubules in Anura and two in Urodela, which with the coiled
anterior end of the duct form a conspicuous organ. The nephrocoelostomes open into a chamber with a glomus. But it is especially in the
Apoda that the pronephros is large, being in fact better differentiated
in this than in any other group of land vertebrates. Rudiments appear

---

out. The solid rudiment is known as the ' intermediate cell-mass ' (Balfour,
985). It may, as in Teleostei, become separated off very early from the
somite, thus seeming to belong to the lateral plate; or it may become early
separated from the lateral plate to reacquire its connexion later.
Sometimes, in Amniotes, the pronephric rudiments seem to arise from the
unsegmented somatopleure as a continuous ridge which only later becomes
subdivided into segmental outgrowths (Burlend in *Chrysemys*, 1000). But
this appearance seems to be secondary and related to the late differentiation
of the nephrotomes (de Walsche, 1068).

in segments 4-16, and of these the first eight may become fully developed

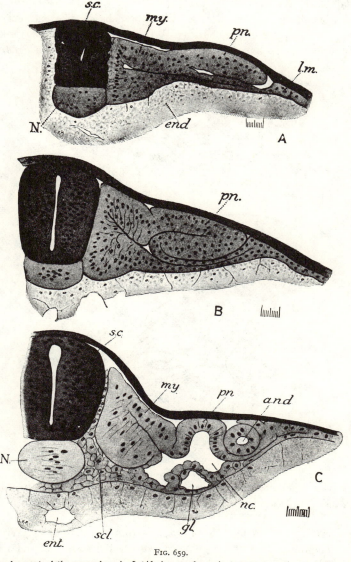

FIG. 659.

Development of the pronephros in *Lepidosiren* as shown in transverse sections. A, Stage 21 ; B, stage 21 ; C, stage 24+. *a.n.d*, Archinephric duct ; *end*, endoderm ; *ent*, enteric cavity ; *gl*, glomerulus ; *l.m*, lateral mesoderm ; *my*, myotome ; *N*, notochord ; *nc*, nephrocoele ; *pn*, pronephric tubule ; *s.c*, spinal cord ; *scl*, sclerotome. (From Kerr, *Embryology*, 1919.)

and functional (*Hypogeophis* : Brauer, **995**). All the parts of a typical tubule are represented, and the nephrocoeles (Bowman's capsules) remain usually separate, Figs. 665-7.

Although the nephrocoeles of Elasmobranchs may become enlarged in the first 14 trunk segments (v. Wijhe, 1073), only a few produce nephrocoelostomes and tubules (trunk segments 3, 4, and 5 in *Pristiurus*

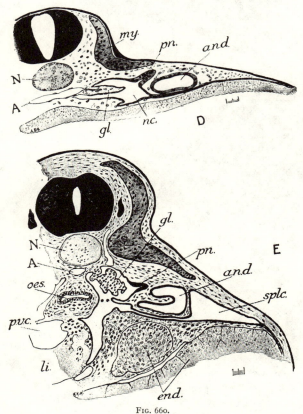

FIG. 660.

Development of the pronephros in *Lepidosiren* as shown in transverse sections. D, Stage 30 ; E, stage 31+. *A*, Dorsal aorta ; *a.n.d*, archinephric duct ; *end*, endoderm ; *gl*, glomerulus ; *li*, liver ; *my*, myotome ; *N*, notochord ; *nc*, nephrocoele ; *oes*, oesophagus ; *pn*, pronephric tubule ; *p.v.c*, posterior vena cava ; *splc*, splanchnocoele.   (From Kerr, *Embryology*, 1919.)

and *Scyllium* : v. Wijhe ; Rabl, 1048—5 tubules in *Raja* : v. Wijhe— 7 in *Torpedo* : Rückert, 1050-51).   Neither Bowman's capsules nor glomeruli are formed, although blood-vessels pass down between the successive tubules from the aorta to the wall of the gut where they join the subintestinal vein, Figs. 658, 661 (Mayer, 850 ; Rabl, 1048 ; Rückert, 1050).

In Reptiles the pronephros is usually vestigial, but in Chelonians and Crocodilians it is functional for a time (Wiedersheim, 1890; de Walsche, 1068); large nephrocoeles appear in many anterior segments (from about the fourth to thirteenth metaotic segment in *Lacerta*: Hoffmann, 1022; Kerens, 1031). Only the last five to seven of these nephrocoeles in Lizards and Snakes produce tubules, and neither glomeruli nor capsules are formed (Milhalkovics, 1041; Strahl, 1882; Schreiner, 1053; Kerens, 1031).

The very similar pronephros of Birds stretches over some twelve segments beginning about the third metaotic, Fig. 668 (Sedgwick, 1055; Renson, 1883; Schreiner, 1053; Felix, 1005; Keibel and Abraham, 1900; Soulié, 1902; Kerens, 1031).

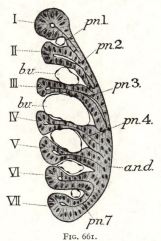

The mammalian pronephros, seen by earlier observers, but first clearly identified as such by Rabl, 1896, extends over about eight segments from probably the sixth metaotic; but is still more vestigial than in lower Amniotes, the tubules soon fusing to a solid rod or 'pronephric ridge' (Martin, 1039; Bonnet, 1887–8; Kerens, 1031; Fraser, 1010).

Fig. 661.

Horizontal section through rudiment of right pronephros of *Torpedo*. (After Rückert, 1888.) *a.n.d*, Archinephric duct; *b.v*, blood-vessel; *pn.1*, etc., pronephric tubules. The nephrotomes are numbered with Roman numerals. (From Kerr, *Embryology*, 1919.)

The Primary or Pronephric Duct.—As already mentioned, the anterior end of the primary or pronephric duct (segmental duct, primärer Harnleiter) is derived from the fusion of the outer ends of the pronephric tubules. The solid rod of cells so formed grows backwards in most Gnathostomes between the ectoderm and the somatic wall of the coelom, becomes hollowed out to a tube, fuses with the wall of the cloaca, and finally opens into its cavity, Figs. 656, 658-9, 664. The prolonged controversy as to whether the growing tip of the duct receives cells from the ectoderm to which it is often very closely applied (v. Wijhe, 1073; Beard, 988; Rückert, 1050; Laguesse, 1891; and others) has been set at rest, and it is now generally admitted that the duct is of purely mesodermal origin (Rabl, 1048; Field, 1008; Felix, 1005; Gregory, 1897; and others). Nevertheless, its exact mode of origin in many forms is still a matter of dispute; for, although its free growth backwards has been shown to occur in

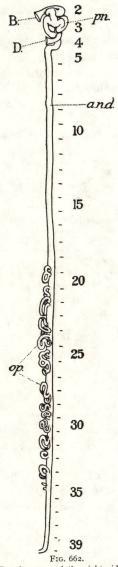

FIG. 662.

Renal organs of the right side of a *Protopterus* larva of stage 34. (From a reconstruction by M. Robertson.) *a.n.d* Archinephric duct; *op*, opisthonephric tubules; *pn*, pronephros. The capital letters indicate funnels and the figures metotic * mesoderm segments. (From Kerr, *Embryology*, 1919.)

* 'Metotic' = posterior to the otocyst.

Elasmobranchs (Balfour, 985; Rabl, 1048; Bates, 987), Apoda (*Hypogeophis*, Brauer, 995), and Amniotes (Weldon, 1071, Hoffmann, 1022, in Reptiles; Sedgwick, 1055-6, Schreiner, 1053, Felix, 1005, in Birds; Martin, 1039, in Mammals; Kerens, 1031, in Amniota), yet in other groups it appears to develop differently.

In Urodela and Anura, according to Mollier (1043) and Field (1008), it develops as a longitudinal thickening and folding off of the mesoderm of the nephrotomal region and only grows out freely at the extreme posterior end to join the cloaca. Much the same mode of development is described in Teleosts (Swaen and Brachet, 1066; Felix, 1005) and other Actinopterygii (*Acipenser*, Maschkowzeff, 1040; Fraser, 1011; *Polypterus*, Kerr, 1033). The morphology of the pronephric duct is further discussed below (p. 684).

The Mesonephros and its Duct.—The mesonephric tubules of Gnathostomes appear later than the pronephric, arising from more posterior nephrotomes. At first they are strictly segmental. In Pisces, Amphibia, and Reptilia peritoneal funnel, Bowman's capsule, glomerulus, and glandular tube are usually typically developed throughout the series, excepting for the first and last few rudiments which may be vestigial, Figs. 656, 673-4. (Transitory vestigial closed peritoneal funnels appear as a rule in Birds and Mammals, but open tubes are only fully developed in connexion with the testis, p. 686.) The growing outer tip of the tubule fuses with and opens into the ready-formed pronephric duct, which therefore becomes in this region the

mesonephric (Wolffian) duct, and is known by this name. The position
of the first-formed tubule varies in different forms, but appears usually

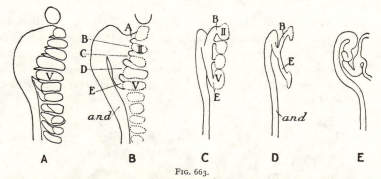

FIG. 663.

Dorsal view of pronephros of *Polypterus* at stages 20, 23, 24+, 25, and *about* 28. *a.n.d*, Archinephric
duct. The tubule rudiments are indicated by letters, the nephrotomes by Roman numerals. (From
Kerr, *Embryology*, 1919.)

close behind the last pronephric rudiment. It may belong to the next
segment ; or there may be a gap between pronephros and mesonephros,
which tends to widen with age
owing to the disappearance of
evanescent tubule rudiments
(generally of about 8 segments
in Selachii, 3 in *Acipenser*, 16
in *Amia*, 2-17 in Amphibia).

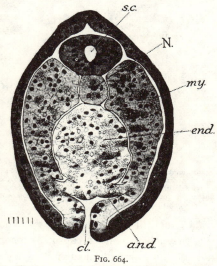

FIG. 664.

Transverse section through *Polypterus* of stage 23
at level of cloacal opening. *a.n.d*, Opening of archinephric duct into cloaca ; *cl*, opening of cloaca to
exterior ; *end*, alimentary canal rudiment ; *my*, myotome ; *N*, notochord ; *s.c*, spinal cord. (From Kerr,
Embryology, 1919.)

Many observers have helped
to work out the history of the
mesonephros, among whom
may be mentioned: Balfour
(985), Sedgwick (1056), v. Wijhe
(1073), Rückert (1050-51), Rabl
(1048), Borcéa (991), in Elasmobranchs ; Balfour and Parker
(2), Beard (1889), in *Lepidosteus* ; Salensky (1880–81),
Jungersen (1029), Maschkowzeff
(1040), in *Acipenser*; Jungersen
(1029), Dean (1002), Felix
(1005), in *Amia* ; Semon (1062),
Kerr (840), in Dipnoi ; Budgett
(10), Kerr (1033), in *Polypterus* ; Felix (1005), Audigé (982), Guitel (1018),
in Teleosts ; Fürbringer (1012), Field (1008), Brauer (995), in Amphibia ;

Hoffmann (**1022**), Sedgwick (**1056**), v. Milhalkovics (**1041**), Schreiner (**1053**), Gasser (1872–84), Gregory (1900), Abraham (1901), Fraser (**1010**), in Amniotes.

It is characteristic of the mesonephros of Gnathostomes that its primitive early segmentation is always lost owing to the development of secondary tubules, many of which may appear in each segment. These tubules are derived from embryonic tissue of the nephrotome, remaining after the production of the first series, and usually proliferating from near Bowman's capsule. Several successive generations of tubules may be

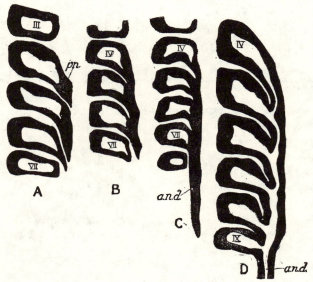

FIG. 665.

Early stages in the development of the pronephros of *Hypogeophis*. Each figure represents a longitudinal section, so arranged as to pass outwards through the nephrotomes, cutting them across, and viewed from the dorsal side. (After Brauer, 1902, slightly simplified.) A, From an embryo with 15 mesoderm segments; B, 12 segments; C, 16 segments; D, 27 segments. *a.n.d*, Archinephric duct; *pn*, pronephric tubule. The Roman figures are placed in the nephrocoeles. (From Kerr, *Embryology*, 1919.)

added from this tissue, each one more dorsal than the last formed. As a rule they develop only a capsule, glomerulus, and glandular tube, but do not acquire a peritoneal funnel. They come to open at the base of the primary tubule leading to the longitudinal duct. According to Borcéa (**991**) a portion of the collecting tubule or region next to the duct is derived from an outgrowth of the wall of the latter in Selachii, Figs. 669-75.

Except when they convey the spermatozoa from the testis in the male (p. 686), the peritoneal funnels are usually closed in the adult. In many

Selachii, however, in *Amia*, and in Amphibia, they remain open to the

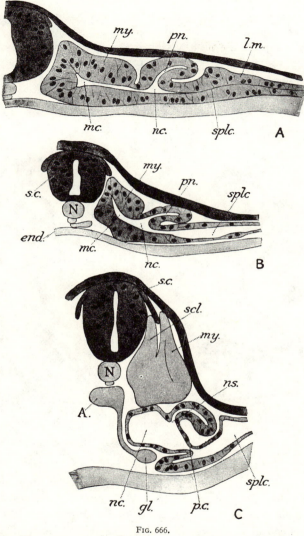

FIG. 666.

Development of pronephros of *Hypogeophis* as seen in transverse sections. (After Brauer, 1902.) A, Embryo with 22 segments; B, with 29 segments; C, with 44 segments. *A*, Dorsal aorta; *end*, endoderm; *gl*, glomerulus; *l.m*, lateral mesoderm; *mc*, myocoele; *my*, myotome; *N*, notochord; *nc*, nephrocoele; *ns*, nephrocoelostome; *p.c*, peritoneal canal; *pn*, pronephric tubule; *s.c*, spinal cord; *scl*, sclerotome; *splc*, splanchnocoele. (From Kerr, *Embryology*, 1919.)

coelom; but their connexion with Bowman's capsule is often lost (Selachii : Borcéa, **991**), and in Anura (Spengel, **1064** ; Nussbaum, **1045**) they come to open secondarily into the veins, Fig. 671 (cf. pronephros of Myxinoids, p. 676).

Characteristic of the mesonephros of the lower Gnathostomes is the tendency of its more anterior region to become reduced, acquire but few secondary tubules, and even degenerate more or less completely. In the Amniota, of course, it disappears (excepting for its duct and the tubules leading from the testis and corresponding vestiges in the female) completely as a functional excretory organ in embryonic or very early adult

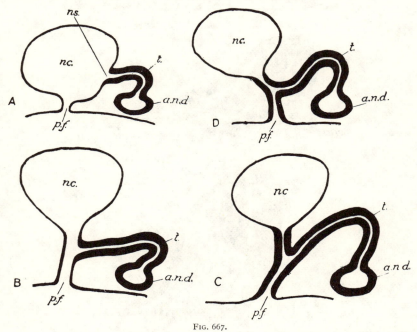

FIG. 667.

Illustrating variations in the relations of nephrocoele, tubule, and peritoneal canal in the pronephros of *Hypogeophis*. *a.n.d*, Archinephric duct ; *nc*, nephrocoele ; *ns*, nephrocoelostome ; *p.f*, peritoneal funnel ; *t*, tubule. (From Kerr, *Embryology*, 1919.)

life ; but even in Amphibia and Pisces the more posterior tubules tend to take on the chief excretory function and form the bulk of the adult organ, the hinder part of which may be developed from segments behind the cloaca (caudal kidney of Elasmobranchs and Teleosts). There is a general tendency for the openings of the tubules into the duct to shift backwards and even to combine, and these tubules may form a special duct opening at the base of the mesonephric duct (Selachii, Teleostei), and acquire a special blood supply from the aorta not belonging to the renal portal system. So marked is the tendency towards the differentiation of a special posterior kidney that this region has sometimes been called a metanephros. It is better, however, to reserve the name metanephros for the kidney of the adult Amniote with its special characteristics. Kerr has suggested the term 'opisthonephros' to include the whole series of tubules behind the pronephros in the lower Gnathostomes where a true metanephros is not present, Fig. 684.

The tubules of the anterior region of the mesonephros in Pisces tend to degenerate and become converted into a 'lymphoid' organ. This is most conspicuous in the Teleostei, where a large anterior mass is formed, composed chiefly of a network of blood-vessels, lymphatic bodies, and suprarenal elements (Audigé, 982).

The Metanephros and its Duct.—The definitive excretory organ of the Amniota may be defined as the true metanephros. This kidney takes on

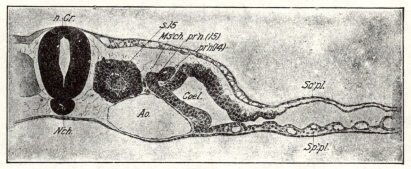

FIG. 668.

Transverse section through the fifteenth somite of a 16 s. embryo (from F. R. Lillie, *Devel t. Chick*, 1919). *Ao*, Aorta; *Coel*, coelom; *Nch*, notochord; *Ms'ch*, mesenchyne; *n.Cr*, neural crest; *pr'n* (14), (15), pronephric tubules of the fourteenth and fifteenth somites; *S.15*, fifteenth somite; *So'pl*, somatopleure; *Sp'pl*, splanchnopleure.

the excretory function in Reptiles, Birds, and Mammals, as the mesonephros degenerates; it has no connexion with the gonads. Developing relatively late, when the differentiation of other tissues is far advanced, it is found to arise from paired bands of 'nephrogenous tissue' following immediately behind that which gave rise to the mesonephros and like it originally derived from nephrotomes. These have long since separated off from their somites, practically all trace of segmentation is lost, and it is not possible to say how many segments may have contributed to the metanephrogenous mass—probably few and possibly only one. The ureter first appears as a diverticulum of the base of the mesonephric duct; it grows forwards and dorsally into the metanephrogenous tissue and with it extends dorsally to the mesonephros, Fig. 676. From the blind end of the ureter grow out numerous slender diverticula round which the nephrogenous tissue becomes grouped. From the latter develop a multitude of tubules each with a Bowman's capsule into which penetrates a glomerulus, and a coiled glandular section; these open into the diverticula which become the collecting tubules leading to a central expansion of the ureter, the pelvis of the kidney. Since all the tubules are of a secondary nature, no peritoneal funnels occur. The ureter comes to open separately later into the cloaca when the base of the mesonephric duct is merged into its wall, Figs. 677, 699-705.

It is clear that the metanephros does not differ in important essentials either in structure or in development from the mesonephros. Indeed,

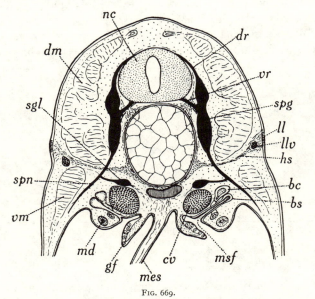

FIG. 669.

Thick transverse section of trunk of *Scyllium canicula* embryo 33 mm. long, showing spinal nerve and mesonephros. *bc,* Bowman's capsule ; *bs,* blind sac at end of funnel tube ; *cv,* posterior cardinal vein ; *dm,* dorsal part of myomere ; *dr,* dorsal root ; *gf,* genital fold ; *hs,* horizontal septum ; *ll,* lateral line ; *llv,* lateral-line vein ; *md,* Müllerian duct (Wolffian duct just above it) ; *mes,* mesentery ; *msf,* mesonephric peritoneal funnel ; *nc,* nerve cord ; *sgl,* sympathetic lateral ganglion ; *spg,* spinal ganglion ; *spn,* mixed spinal nerve (r. ventralis) ; *vm,* ventral part of myomere ; *vr,* ventral root.

the posterior region of the latter often shows the same specialisations, though less pronounced. While the earlier observers (Rathke, 1833; Remak, 1855; Koelliker, 1861) believed the whole metanephros to be derived from branches of the ureter, the modern view of its double origin from the ureter on the one hand and nephrogenous blastema on the other was initiated by v. Kupffer in 1865, upheld by many authors since, and finally established by Schreiner (**1053**).

Excretory Organs of the Cyclostomata.—The kidney of adult Petromyzontia consists on each side of a longitudinal fold extending for about half the length of the splanchnocoele into which it hangs, and containing convoluted tubules more numerous than the segments. The tubules open into a longitudinal duct, situated at the free edge of the fold, and the right and left ducts join posteriorly to a median sinus leading to a pore at the tip of a papilla behind the anus, Fig. 98. Each tubule has a glandular

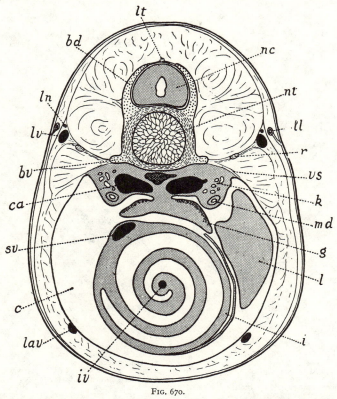

Fig. 670.

Scyllium canicula, late embryo ; transverse section of trunk. *bd*, Basidorsal ; *bv*, basiventral ; *c*, splanchnocoele ; *ca*, posterior cardinal ; *g*, genital ridge ; *i*, intestine ; *iv*, intestinal vein at edge of spiral valve ; *k*, mesonephros ; *l*, liver ; *lav*, lateral vein ; *ll*, lateral-line canal, and *ln*, its nerve ; *lt*, longitudinal ligament ; *lv*, lateral-line vein ; *md*, Müllerian duct ; *nc*, nerve-cord ; *nt*, notochord *r*, rib in dorsal septum ; *sv*, supraintestinal vein ; *vs*, ventral septum.

portion and a typical Malpighian body with capsule and glomerulus ; but it is noteworthy that there are no peritoneal funnels. This kidney is a mesonephros similar to that of the Gnathostomes, Fig. 678.

The development of the renal organs has been studied by many observers since Rathke (1827) first described it in the Ammocoete larva (W. Müller, 1875 ; Schneider, 1879 ; Scott, **1054** ; Goette, **825** ;

Fürbringer, 1878 ; Vialleton, 1890 ; Bujor, 1891). Recently a detailed account of the development of the pronephros and mesonephros has been given by Wheeler (**1072**), and by Hatta of the pronephros (**1020a**).

From six nephrotomes belonging to metaotic segments 4-9 arise six pronephric tubule rudiments as outgrowths of their parietal wall. In

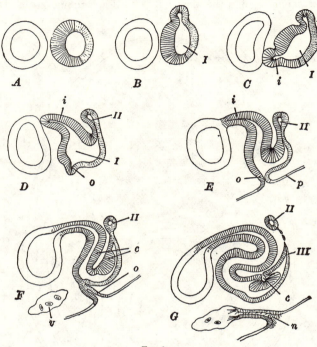

FIG. 671.

Series of diagrams illustrating development of primary mesonephric tubules in *R. sylvatica* (after Hall, from W. E. Kellicott, *Chordate Dev[*]Devel.*, 1913). The Wolffian duct is drawn in outline simply ; mesonephric vesicles are shaded ; somatic part of tubule is shaded by continuous lines, splanchnic part by dotted lines. A, Wolffian duct and simple mesonephric vesicle. B, Mesonephric vesicle dividing into large primary mesonephric unit and small dorsal chamber. The latter elongates antero-posteriorly and represents rudiment of secondary and later mesonephric units. C, Formation of rudiment of inner tubule. D, Inner tubule extending upward and toward mesonephric duct ; formation of rudiment of outer tubule. E, Outer tubule fused with peritoneum and rudiment of funnel thus established ; Bowman's capsule forming ; commencement of differentiation of secondary meso-nephric tubules. F, Separation of nephrostomal rudiment from remainder of tubule. G, Connexion of funnel with branch of posterior cardinal vein ; separation of secondary tubule, and beginning of tertiary tubule indicated. *c*, Bowman's capsule ; *i*, inner tubule ; *n*, peritoneal funnel ; *o*, outer tubule ; *p*, peritoneum ; *v*, branch of posterior cardinal vein ; *I*, primary mesonephric tubule ; *II*, secondary mesonephric tubule ; *III*, tertiary mesonephric tubule.

P. planeri, according to Wheeler, the five posterior rudiments develop into functional tubules ; but only the 3rd, 4th, and 5th do so in the species studied by Hatta. The outer ends of these tubules bend backwards and join to form a longitudinal collecting pronephric duct. Behind this region the duct is prolonged by the addition and fusion of segmental rudiments which separate off from the successive nephrotomes (and appear to represent vestigial tubules) until it reaches the level of the cloaca. The duct then joins the wall of the cloaca and opens into its cavity at the junction between ectoderm and endoderm (p. 707).

It is to be noticed that the pronephric tubules arise from segments anterior to the ductus Cuvieri, and so are related to the anterior cardinal veins ; also that the more anterior tubule rudiments extend into the branchial region. The larval functional organ begins just behind the last gill ; its tubules become somewhat crowded backwards losing their segmental disposition, and project into the pericardial coelom where they open by conspicuous ciliated funnels. These are the nephrocoelostomes

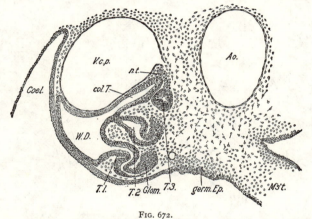

FIG. 672.

Transverse section through middle of Wolffian body of chick embryo of 96 hours (after F. R. Lillie, *Devolt. Chick*, 1919). *Ao*, Aorta ; *Coel*, coelom ; *col.T*, collecting tubule ; *Glom*, glomerulus ; *germ.Ep*, germinal epithelium ; *M's't*, mesentery ; *n.t*, nephrogenous tissue ; *T.*1, 2, 3, primary, secondary, and tertiary mesonephric tubules ; *V.c.p*, posterior cardinal vein ; *W.D*, Wolffian duct.

of segments whose nephrocoeles have opened out and become confluent with the pericardial cavity, and whose blood supply forms a glomus projecting into it, Figs. 522, 678. The pronephros atrophies at metamorphosis, its duct remaining in the mesonephric region.

Long before this atrophy the mesonephros has made its appearance further back in the nephric fold carrying the pronephric duct (Fürbringer, 1875 ; Vialleton, 1890 ; and especially Wheeler, 1072). The mesonephric tubules arise on the medial wall of this fold as thickenings of the peritoneum, which proliferate inwards, join and penetrate the wall of the duct. These cords of cells or tubule rudiments lengthen, separate off from the peritoneum, acquire a lumen, and expand at their inner ends into capsules into which blood-vessels are pushed, while their opposite ends open into the longitudinal duct. Thus each tubule has a Malpighian body, but no peritoneal funnel. Doubtless the cells of the rudiments were originally derived from the nephrotomes, which, however, at this late stage have become broken up and scattered. The rudiments are not strictly segmental, in most of the mesonephric region number about six to a segment, and several may join to open into the duct.

More interesting and instructive are the structure and development

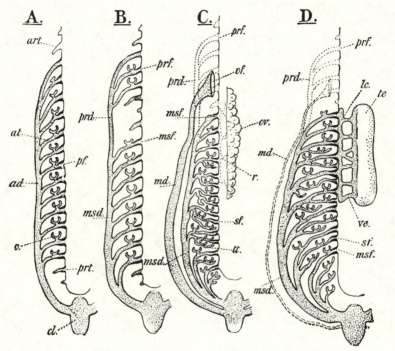

FIG. 673.

Diagrams of the urinogenital system in the Craniata. A, Hypothetical ancestral stage with continuous archinephros; B, Cyclostome with anterior pronephros; C, female Gnathostome (adult); D, male Gnathostome (adult). *ad*, Archinephric duct; *art*, anterior vestigial tubule; *at*, archinephric tubule; *c*, Malpighian capsule; *cl*, cloaca; *lc*, longitudinal canal; *md*, Müllerian duct; *msd*, mesonephric duct; *msf*, mesonephric funnel; *of*, coelomic funnel; *ov*, ovary; *pf*, coelomostome (funnel); *prd*, pronephric duct; *prf*, pronephric funnel; *prt*, posterior vestigial tubule; *r*, vestigial network of vasa efferentia; *sf*, secondary funnel; *te*, testis; *tt*, tertiary tubule; *ve*, vas efferens. The vestigial oviduct and the embryonic pronephros are represented by dotted lines in C and D. (From Goodrich, *Vert. Craniata*, 1909.)

of the renal organs of the Myxinoidea (J. Müller, 1836; Weldon, 1070; Maas, 1036; Spengel, 1065; Semon, 1061; Kirkaldy, 1034; Dean, 1003; Price, 1047; Conel, 1001). It is a remarkable fact that these Cyclostomes preserve the pronephros in adult life as an organ of considerable size projecting on each side of the oesophagus into the pericardial coelom (which retains a wide communication with the coelom of the trunk on the right side). Each pronephros consists of a number of tubules opening on the one hand by very numerous branches ending in funnels (nephrocoelostomes) into the cavity of the pericardium and on the other hand into the remains of the longitudinal pronephric duct, Figs. 679, 680. The latter is degenerate and discontinuous, its communication with the more posterior mesonephric region being interrupted. The tubules are surrounded by blood-spaces and the latter may secondarily open into the duct (Conel, 1001 ; compare Anura, p. 669). Posteriorly is a glomus projecting into a chamber almost closed off. The function of this modified adult pronephros is probably in the main phagocytal.

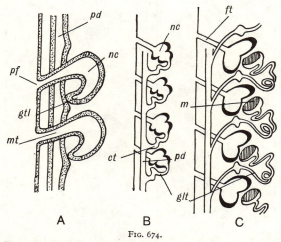

FIG. 674.

Diagrams illustrating development of mesonephros of *Squalus acanthias* (after J. Borcéa, 1905, from Ihle, *Vergl. Anat.*, 1927). A, Young stage showing blind end of primary tubule, *mt*, joining primary longitudinal duct, *pd*. B, Growth of nephrocoele chamber, *nc*, and of excretory collecting canal, *ct*. C, Separation of Malpighian capsule, *m*, from chamber. *ft*, Funnel canal ; *glt*, glandular tubule ; *pf*, peritoneal funnel.

The mesonephros extending along the whole length of the trunk coelom is no less remarkable, for it consists on each side of a longitudinal duct (former pronephric duct) provided with short tubules segmentally disposed, one pair to each segment. Each tubule ends blindly in a Malpighian body with capsule and glomerulus. The ducts open behind into the cloaca. The loss of peritoneal funnels is a sign of specialisation, but in other respects the mesonephros of Myxinoids is the most primitive known among Craniata, since it retains in the adult the presumably original strictly segmental disposition of the tubules and Malpighian bodies.

The early development is of great interest and has been well described by Price (1047) in *Bdellostoma*. The nephrotomes develop nephrocoeles

from about the 10th to the 80th segment of the trunk, and become somewhat constricted, the middle portion forming the rudiment of a tubule
and the outer thickened wall the rudiment of the longitudinal duct in
each segment. From about segment 30 backwards the nephrocoeles
become shut off from the splanchnocoele and give rise to Malpighian
bodies ; but from about the 30th segment forwards the nephrocoeles, on

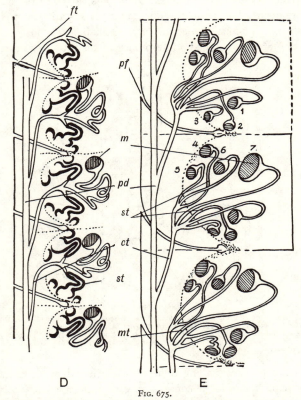

D   E

FIG. 675.

Diagrams illustrating late development of mesonephros of *Squalus acanthias* (after J. Borcéa,
1905, from Ihle, *Vergl. Anat.*, 1927). D, Growth from primary chamber of secondary chambers and
canals. E, Opening of secondary capsules, derived from secondary chambers, into base of primary
tubule, now collecting tubule (for lettering see Fig. 674).

the contrary, expand, fuse, and open out, merging with the general coelom,
and in this region no glomeruli are formed. The limit between pronephros
and mesonephros corresponds not to that between open and closed tubules,
but occurs at about the 33rd segment where develops the pericardial wall.
In front of this is formed the adult pronephros, composed of some three
posterior closed and eighteen more anterior tubules all collected together,

and preserving open nephrocoelostomes.   For, as the gill-sacs and heart
move backward in development (Dean, 1003) the more anterior fifteen
tubules are drawn backwards and crowded together, losing their original
segmental disposition, and the duct is correspondingly shortened up.

As indicated above, the longitudinal duct is formed *in situ* by the
coalescence of segmental contributions from each nephrotome along its
course, its lumen being an extension of the nephrocoeles.   Towards the
posterior end where the duct opens into the cloaca (p. 710) the tubules
are vestigial, and in about the last four segments the whole nephrotome is

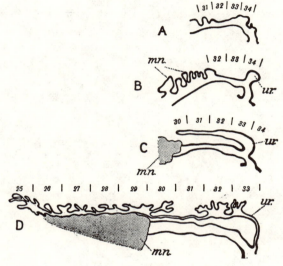

Fig. 676.

Reconstructed outlines of hind end of mesonephric duct and ureter in Bird embryos as seen
from the left side. (After Schreiner, 1902.) A, Duck embryo with 48 segments ; B, duck embryo
with 50 segments; C, duck embryo, 10·75 mm. ; D, fowl embryo, 13·5 mm. *mn*, Mesonephros ;
ur, ureter.   The Arabic numerals indicate the position of the mesoderm segments. (From Kerr,
Embryology, 1919.)

included in the formation of the duct.   There are several points of importance to be noticed in the development of the excretory organs of *Bdellostoma* : (*a*) The tubules at first form a continuous uninterrupted segmental
series ; (*b*) no secondary tubules are formed ; (*c*) the nephrocoeles of the
more anterior tubules become confluent with the splanchnocoele (pericardial
cavity), and the more posterior and greater number become closed
capsules surrounding glomeruli ; (*d*) no peritoneal funnels persist ; (*e*) the
limit between pro- and mesonephros does not coincide with that between
open and closed tubules ; (*f*) only one longitudinal duct is developed on
each side, and it is formed by the coalescence of segmental rudiments

from the nephrocoeles along its whole length ; (g) at first continuous, the duct becomes later interrupted between the pro- and mesonephric regions.

Distinction between Pro-, Meso-, and Metanephros.—Three different views have been held with regard to their general morphology. Gegenbaur, W. Müller, Koelliker, Fürbringer, held that they are three different, not homologous organs which replace each other completely. Sedgwick

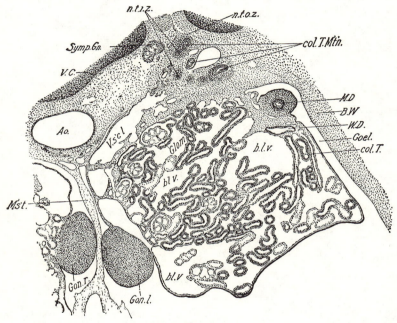

FIG. 677.

Transverse section through metanephros, mesonephros, gonads, and neighbouring parts of an 8-day chick (from F. R. Lillie, *Develt. Chick*, 1919). *b.l.v*, Blood-vessels (sinusoids) ; *B.W*, body-wall ; *col.T.M't'n*, collecting tubules of metanephros ; *M.D*, Müllerian duct ; *M's't*, mesentery ; *n.t.i.z*, inner zone of nephrogenous tissue (metanephric) ; *n.t.o.z*, outer zone of nephrogenous tissue ; *Symp.Gn*, sympathetic ganglion of twenty-first spinal ganglion ; *V.C*, centrum of vertebra. Other letter as in Fig. 691.

and Balfour, on the contrary, looked upon them as merely parts of one continuous organ, which develop successively from before backwards ; while Rückert held a somewhat intermediate view regarding them as three generations of similar tubules each more dorsal than its predecessor.

When the development of the mesonephros and metanephros was accurately worked out it had to be admitted that they are essentially of the same nature, and that the peculiarities of the metanephros are related to its late appearance and more specialised structure. The comparison

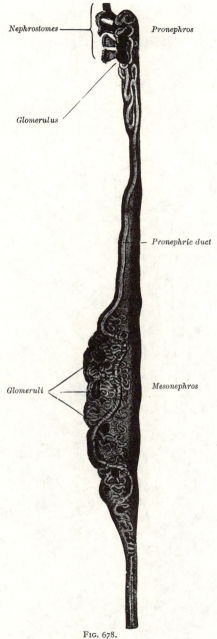

Nephrostomes — ⎧ — Pronephros

Glomerulus

Pronephric duct

Glomeruli — Mesonephros

FIG. 678.

The excretory system of a *Petromyzon fluviatilis*, 22 mm. in length, from the inner side. (After Wheeler.) About half the entire length of the primary urinary duct is represented, and behind the pronephros it is greatly coiled. Four pronephric funnels and a folded glomerulus are present, and between the pronephros and mesonephros is a portion wanting in tubules. (From Kerr, *Embryology*, 1919.)

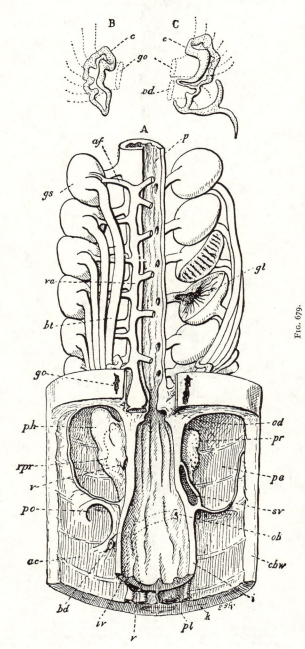

FIG. 679.

Myxine glutinosa, L. A, Ventral view of a dissection; B, cartilage near the opening of the right gill-pouches (indicated by dotted line); C, similar cartilage on the left side. *ac*, Abdominal coelom, exposed by cutting away ventral body-wall; *af*, afferent vessel; *bd*, bile-duct cut short; *c*, cartilage; *cbw*, cut body-wall; *gl*, gill-lamellae exposed in gill-sac; *go*, branchial opening; *gs*, gill-sac; *i*, intestine opened up; *k*, kidney-duct; *ob*, bile-duct opening; *od*, oesophageal duct; *p*, pharynx opened; *pe*, pericardium opened—the heart has been removed; *ph*, portal heart; *pl*, left posterior cardinal vein; *po*, opening from pericardial into abdominal coelom; *pr*, pronephros, with funnels on its surface; *rpr*, dotted line indicating hidden right pronephros; *sv*, sinus venosus cut across; *v*, portal vein; *va*, ventral aorta—the afferent vessels have been cut short on the left side, and the heart cut off behind the conus. (From Goodrich, *Vert. Craniata*, 1909.)

of the mesonephros with the pronephros is, however, not so easily made, and there is still much doubt as to the exact relation they bear to each other. The distinction once drawn between them as to structure (absence of peritoneal funnels, capsules, and glomeruli in pronephros) has broken down since Balfour described the pronephros of *Lepidosteus* and Brauer of *Hypogeophis*. It is also established that in both organs the tubules develop in essentially the same way from nephrotomes. Moreover, if it is true that secondary tubules are characteristic of the mesonephros and have never been proved to occur in the pronephric region, it must be remembered that they are totally absent in Myxinoidea, and that in any case this distinction is not fundamental. In many forms secondary

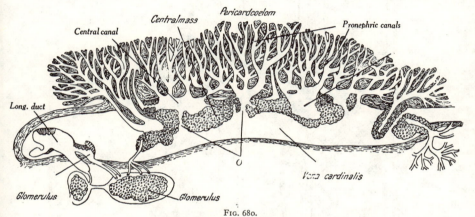

FIG. 680.

Reconstruction of 'head-kidney' of *Bdellostoma stouti* (after Conel, 1917; from Ihle, *Vergl. Anat.*, 1927). Pronephric tubules open by numerous branches into pericardial coelom; primary longitudinal duct interrupted, leaving short anterior 'central canal' disconnected from posterior longitudinal (mesonephric) duct; *O*, points at which pronephric tubules and central canal open secondarily into cardinal view.

tubules do not appear in several of the anterior segments of the mesonephros, Fig. 673.

Kerens has no doubt rightly insisted that there are no constant and fundamental differences in the development of the posterior pronephric and anterior mesonephric tubules of Amniota. Even the distinction, that, whereas the pronephric tubules are from the first continuous with the duct they help to form, the mesonephric tubules only secondarily fuse with it, breaks down; since a few of the posterior pronephric tubules may become connected with the duct in just the same fashion. The history of the tubules in *Bdellostoma* clearly shows that the anterior (pronephric) and posterior (mesonephric) are serially homologous organs which have diverged in structure in adaptation to different functions.

Moreover, the longitudinal duct is here derived all along its course from their outer ends. There can be little doubt that in Myxinoids pronephros

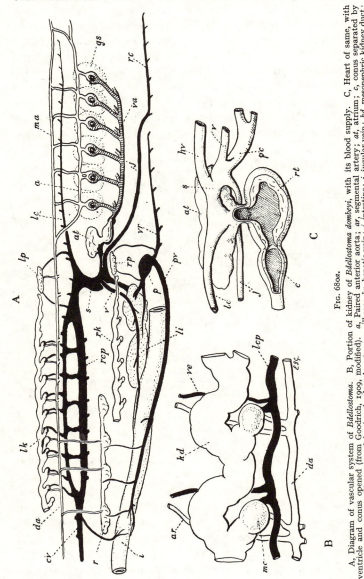

Fig. 680a.

A, Diagram of vascular system of *Bdellostoma*. B, Portion of kidney of *Bdellostoma dombeyi*, with its blood supply. C, Heart of same, with ventricle and conus opened (from Goodrich, 1909, modified). *a*, Paired anterior aorta; *ar*, segmental artery; *at*, atrium; *c*, conus separated by two valves from ventricle; *cv*, caudal vein; *da*, dorsal aorta; *gs*, gill-sac; *hv*, hepatic vein; *i*, intestine; *j*, jugular vein; *kd*, mesonephric kidney duct; *lc*, left anterior cardinal; *lcp*, left posterior cardinal; *li*, liver; *lk*, left mesonephros; *lp*, left pronephros; *ma*, median anterior aorta; *mc*, Malpighian capsule; *p*, portal heart; *pc*, united posterior cardinals; *pv*, portal vein; *r*, right ovary; *rc*, front end of right anterior cardinal; *rcp*, right posterior cardinal; *rk*, right mesonephros; *rp*, right pronephros; *rt*, cavity of ventricle separated by two valves from atrium; *s*, sinus venosus passing into left ductus Cuvieri; *sv*, intestinal vein; *va*, ventral aorta; *vr*, hinder end of right anterior cardinal.

and mesonephros represent two portions of an originally continuous and uniform archinephros with its archinephric duct; and it seems reasonable to suppose that the same conclusion holds for other Craniates in which, however, a secondary multiplication of tubules has taken place posteriorly. The only serious difficulty attending this interpretation is presented by the mesonephric duct. For if it is really separated off from the nephrotomes all along its course (Petromyzontia, Urodela, Anura), it might be objected that the first generation of tubules had therefore been exhausted in its formation, that the usual development of the duct by free backward growth (Selachii, Amniota) represents the same process shortened up in ontogeny, and that the mesonephric tubules must consequently belong to a second series. But it seems more probable that the material which goes to form the duct in these forms only represents the outer ends of the tubules, temporarily separated off from their inner ends, which two regions later on again become connected together. We may suppose, then, that the original mode of formation of the duct is seen in Myxinoids and less clearly in *Petromyzon* and some Amphibia; but that in other Craniates there has been an increasing tendency for the duct to be precociously developed while the tubules are delayed, and that backward free growth is merely a developmental device for the purpose.

Conclusion.—It may be concluded that the Craniata were originally provided throughout the trunk with a continuous series of segmental excretory tubules opening by peritoneal funnels into the coelom, and that by the growth backwards and coalescence of their outer ends they formed a longitudinal duct leading to the cloaca. Further, that such a primitive uniform archinephros no longer exists in any living form, but that owing to specialisation the series became differentiated into pronephric and mesonephric (opisthonephric) regions. Such a stage is represented in Myxinoids. The tendency for the pronephric tubules to develop early and to be replaced by later and more posterior tubules led to further specialisation of the pronephros, to the early completion of the duct before the mesonephric tubules developed, and their consequent secondary union with the duct. Instead of the duct being formed by the fusion of a succession of segmental rudiments, it tended more and more to be produced by the anterior tubules which grew freely back to the cloaca. Meanwhile, owing to the increasing importance of the mesonephros, the tubule-forming tissue (nephrotome) produced numbers of secondary tubules, and the original segmentation was lost.

THE GENITAL DUCTS OF GNATHOSTOMATA

Since the original function of the segmental tubules of the Vertebrata was probably to convey the spermatozoa and ova to the exterior (see p. 718), we might expect to find them still serving this purpose even in the highest forms.   Such indeed is the nature of the genital ducts in the male sex of all the Gnathostomes, though secondary specialisation may lead to considerable modification of their primitive structure.

The gonads develop in both sexes along paired genital folds, extending along nearly the whole length of the splanchnic coelom in primitive forms.   The germ-cells develop chiefly on the outer (lateral) surface of each fold, but tend in higher forms to become more and more restricted to a short fertile region, while the anterior and posterior parts of the fold become sterile and degenerate.   Now, it is an important fact that, while in the female sex the ova when ripe fall into the general splanchnic coelom (except in certain Actinopterygii, see p. 701), and are carried thence to the exterior usually by specialised ducts (p. 694), this primitive condition is never preserved in the male Craniata, except

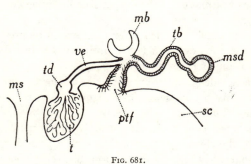

Fig. 681.

Diagram showing relation of mesonephric tubule to testis in *Gnathostomes* (except Selachian ?).   *mb*, Bowman's capsule; *ms*, mesentery; *msd*, mesonephric duct; *ptf*, peritoneal funnel; *sc*, splanchnic coelom; *t*, testis with seminiferous tubules; *tb*, glandular tube; *td*, testis duct or marginal canal; *ve*, vas efferens.

in Cyclostomes (p. 707). The testis of the male Gnathostome is always completely shut off from the general coelom (splanchnocoele), Fig. 681. The spermatozoa are never freely shed into this cavity, but are conveyed by a system of vasa efferentia to the primary mesonephric tubules and thence down the mesonephric duct to the exterior (see, however, secondary modifications in certain Osteichthyes described below).

In the more primitive condition, seen in *Acipenser, Lepidosteus, Amia,* and many Amphibia, the vasa efferentia extend across from testis to mesonephros along the whole length of the gonad.   They usually join to form a longitudinal marginal canal (Nierenrandkanal of Felix), and while the transverse canals running from the marginal canal to the kidney are necessarily segmental since they open into the anterior primary mesonephric tubules, those passing from the testis to the marginal canal are usually more numerous, less regular, and frequently anastomose. There is a tendency for this anterior genital region of the mesonephros to lose its renal function, for its secondary tubules to degenerate, and for the posterior region to enlarge and take on the chief function of excretion.

In the various groups there is also a tendency for the vasa efferentia to become restricted to either the anterior or the posterior region of the testis, and for the tubules into which they open to become simplified by the reduction of the coiling and disappearance of the Malpighian body, Fig. 682.

The whole system of channels leading the spermatozoa into the kidney tubules, and known as the testicular network, may be considered as of originally coelomic nature and has probably arisen by

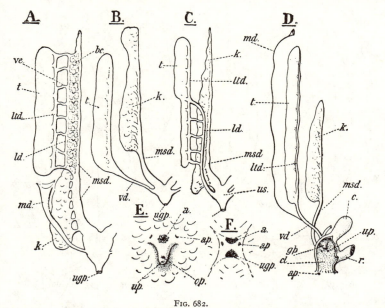

FIG. 682.

A, B, C, and D, diagrams of the urinogenital organs in male Dipnoi and Teleostomi. A, *Acipenser* (*Lepidosteus* and *Amia* are similar, but without the funnel, *md*) ; B, Teleostei ; and C, *Polypterus* (from Budgett's figures) ; D, *Protopterus* (from W. N. Parker's figures) ; E, urinogenital papilla of a female *Salmo*, ventral view ; F, similar view of a male *Polypterus* (after Budgett). *a*, Anus ; *ap*, abdominal pore ; *bc*, renal capsule ; *c*, cloacal bladder ; *gp*, genital papilla ; *k*, mesonephros ; *ld*, longitudinal duct ; *ltd*, longitudinal testis duct ; *md*, Müllerian duct ; *msd*, mesonephric duct ; *op*, oviducal pore ; *r*, rectum ; *t*, testis ; *ugp*, urinogenital pore ; *up*, urinary pore ; *us*, urinogenital sinus ; *vd*, vas deferens ; *ve*, vas efferens. (From Goodrich, *Vert. Craniata*, 1909.)

folds of the coelomic epithelium closing off ciliated grooves extending from the peritoneal funnels of the mesonephros to the genital fold. But the detailed homology of the various parts of the network is by no means thoroughly understood. The whole surface of the testis becomes covered by coelomic epithelium, while seminiferous tubules develop in its thickness ; these open into a collecting or central longitudinal testis canal from which start the vasa efferentia, Figs. 673, 681.

In Elasmobranchs the testicular network is restricted to the anterior region of the testis, Fig. 684. The number of vasa efferentia may still be considerable in sharks, but in skates may become reduced to the most anterior one. The renal tubules of the genital region degenerate, and the twisted anterior part of the mesonephric duct (Wolffian duct,

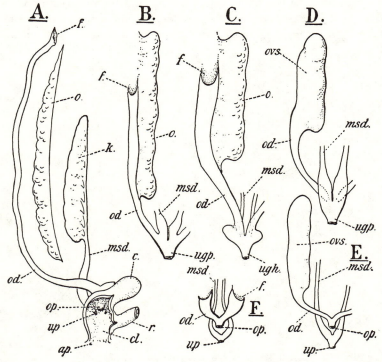

FIG. 683.

Diagrams of the female urinogenital ducts in the Dipnoi and Teleostomi derived from the figures of various authors. A, *Protopterus* (after Ayers and W. N. Parker) ; B, *Polypterus* (after Budgett) ; C, *Amia* (after Hyrtl and Huxley) ; D, *Lepidosteus* (after Balfour and Parker) ; E, a Teleost with closed ovisac ; F, a Salmonid (after Weber). *ap*, Abdominal pore ; *c*, cloacal bladder ; *cl*, cloaca ; *f*, open funnel of oviduct ; *k*, mesonephros ; *msd*, mesonephric duct ; *o*, ovary ; *od*, oviduct ; *op*, genital papilla and pore ; *ovs*, closed ovisac ; *r*, rectum ; *ugp*, urinogenital papilla ; *up*, urinary pore. In all the figures, except F, only the right oviduct is completely drawn. (From Goodrich, *Vert. Craniata*, 1909.)

duct of Leydig), no longer renal in function, becomes glandular and secretes a white fluid (Semper, **1063** ; Balfour, **985** ; Borcéa, **991**). The mesonephric peritoneal funnels which connect with the testis apparently reach its central canal into which they open (Balfour). Where it exists the longitudinal anastomosis or marginal canal appears to be derived from the capsules of the simplified Malpighian bodies (Borcéa). This is not the case in other fishes, such as Dipnoi, and lower Teleostomes, where the funnels appear to reach no further than the marginal canal.[1]

The extension of the testicular network along the whole testis seen in primitive forms is modified in *Ceratodus*[2] and *Lepidosiren* where only

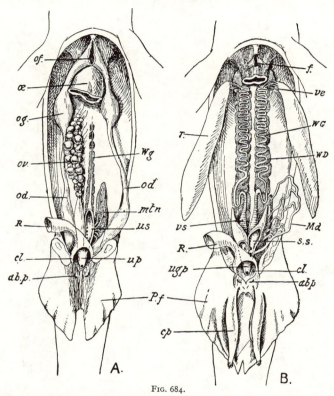

FIG. 684.

Scyllium canicula: urinogenital organs of female, A, and male, B, ventral view. *abp*, Abdominal pore; *cl*, cloaca; *cp*, clasper; *f*, vestige of oviduct in male; *Md*, ducts of posterior region of kidney (ureters); *mtn*, posterior excretory region of mesonephros (opisthonephros); *od*, Müllerian oviduct; *oe*, cut end of oesophagus; *og*, oviducal gland; *ov*, ovary; *P.f*, pelvic fin; *R*, rectum; *s.s*, sperm-sac; *T*, testis; *up*, urinary papilla in ♀; *ugp*, urinogenital papilla in ♂; *us*, urinary sinus; *ve*, vasa efferentia; *vs*, vesicula seminalis; *WD*, Wolffian mesonephric duct; *WG*, Wolffian gland or mesonephros. (From G. C. Bourne, *Comp. Anatomy*, 1902.)

some half-dozen vasa efferentia persist at the hind end of the testis (Kerr, 1032; Semon, 1062; Ballantyne, 986). In *Protopterus* they are still further reduced to a single canal leading from the degenerate sterile

[1] The exact relation of the testicular canals and marginal canal to the mesonephric tubules and peritoneal funnels has not yet been satisfactorily described in the various groups of Pisces.

[2] A detailed description of the urinogenital organs of Ceratodus is urgently needed.

tubular posterior region of the testis through the posterior mesonephric tubules to the base of the mesonephric duct (Kerr, 1032).

The Teleostomes show similar specialisations. While the more primitive forms, as mentioned above, have vasa efferentia all along the testis,

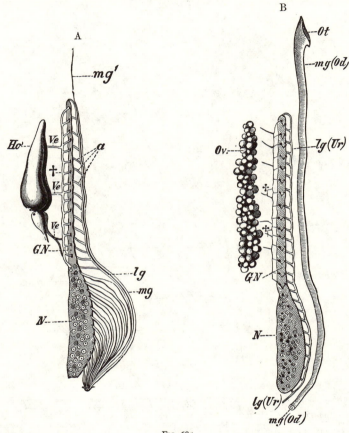

FIG. 685.

Diagram of the urinogenital system of (A) a male and (B) a female Urodele; founded on a preparation of *Triton taeniatus*. (After J. W. Spengel.) *a*, Collecting tubes of the mesonephros, which open into the Wolffian (urinogenital) duct (*lg*); in the female the latter serves simply as the urinary duct (*Ur*), and the system of the vasa efferentia (testicular network) is vestigial; *GN*, anterior portion of kidney (epididymis of the male); *Ho*, testis; *mg*, *mg¹*, Müllerian duct; *N*, posterior non-sexual portion of kidney; *Ot*, coelomic aperture of Müllerian duct (oviduct, *Od*); *Ov*, ovary; *Ve*, vasa efferentia of testis which open into the longitudinal canal of the mesonephros, †. (From Wiedersheim, *Comp. Anatomy*.)

in *Polypterus* (Budgett, 10; Kerr, 1033) the sterile posterior end of this organ consists of a collection of testicular canals leading to a single duct opening into the urinogenital sinus (base of mesonephric duct). The sperm thus no longer passes through the kidney at all, Fig. 682.

A structure of apparently similar origin is seen in the Teleostei (Kerr, 840). For in this group the testis is always quite separate from the kidney, and discharges its products usually into the base of the mesonephric ducts by a tubular prolongation which in some cases has been shown to be made up of testicular canals (Jungersen, 1028). The spermduct of the Polypterini and Teleostei is, then, to be interpreted as formed chiefly of the marginal canal which has grown backwards and come to open into the Wolffian duct separately from the kidney. Its opening into the duct may represent a single posterior tubule.

In Amphibia (Spengel, 1064) is also found an extensive testicular network in the more primitive Apoda and Urodela, with usually a well-developed marginal canal (Nierenrandkanal). But in Anura the network tends to become shortened, Figs. 681, 685-6.

The Amphibian network, however, is apparently of somewhat different structure from that of the fishes described above, since the primary peritoneal funnels open not into the vasa efferentia or marginal canal, but as usual into the splanchnic coelom. The transverse canals from the marginal canal communicate with the tubules at or near the capsules of the Malpighian bodies.

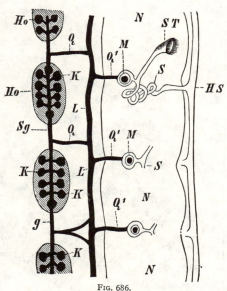

Fig. 686.

Diagram of a portion of the male generative apparatus in the Gymnophiona. *Ho*, Testis; *HS*, urinogenital duct; *K*, testicular capsules; *M*, Malpighian capsules; *N*, kidney; *Q*, transverse canals connecting the collecting duct with the longitudinal canal (*L, L*); *Q¹*, second series of transverse canals; *S*, convoluted portion of urinary tubule; *Sg*, collecting duct of testis; *ST*, nephrocoelostome. (From Wiedersheim, *Comp. Anatomy.*)

In ontogeny the whole network seems to arise from solid outgrowths of the capsules (Hoffmann, 1022; Semon, 1058-9; Gemmill, 1013). Except in Apoda the peritoneal funnels degenerate in the adult, and in Anura the Malpighian bodies as well.

The testicular network of the Amniota is doubtless built on the same plan as that of the Amphibia ; but no open funnels persist, and they are usually only vestigial and closed even in development. The anterior region of the mesonephric duct, and the group of simplified tubules derived from this end of the mesonephros and opening into the testicular canals, here constitute the epididymis of the adult.

There has been much difference of opinion concerning the homology of the network and the mode of development of its various parts in Amniotes. It is generally derived from a combination of ' rete-cords ' with ' sex-cords '; the former often appearing as solid rods of cells growing out of the capsules of Malpighian bodies, or directly from the coelomic

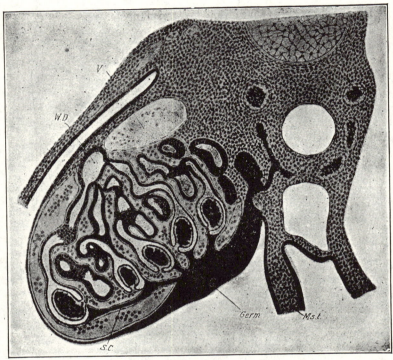

FIG. 687.

Cross-section through genital primordium of *Limosa aegocephala* (after Hoffmann, from F. R. Lillie, *Develt. Chick*, 1919). The stage is similar to that of a chick embryo of 4½ days. *Germ*, Germinal epithelium; *Ms.t*, mesentery; *S.C*, rete-cords; *V*, posterior cardinal vein; *W.D*, Wolffian duct.

epithelium, Fig. 687. The sex-cords are derived directly or indirectly from the epithelium of the genital ridge. As a rule these structures extend over more segments in young stages than in older (only those towards the middle persist to form the vasa efferentia), indicating that in Amniota also the genital region has been shortened and chiefly from behind. Both rete-cords and sex-cords are said to be derived from out-growths of the capsules by some authors (Braun, 1877–8, Weldon, **1071**,

Hoffmann, **1022**, in Reptiles; Semon, **1057**; Hoffmann, **1022**, in the Chick; Saimont, 1903, von Winiwarter, **1076**, in Mammals); while Janosik (**1027**) derived both from the coelomic epithelium.   Most observers

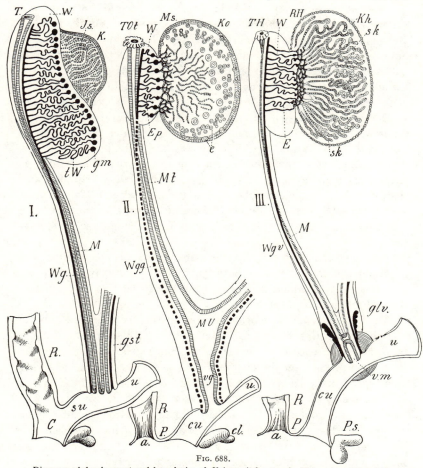

FIG. 688.

Diagrams of development and homologies of *Urinogenital system* in *Mammal*, omitting kidneys (after v. Milhalkovics, from M. Weber, 1927).   I., So-called indifferent stage, II., female, III., male. *a*, Anus; *C*, cloaca; *cl*, clitoris; *cu*, urinogenital canal; *e*, ova; *E*, epididymis; *Ep*, epoophoron; *gm*, glomeruli; *glv*, glandula vesicularis; *gst*, genital cord; *Js*, sex-cords; *K*, germinal epithelium, *Kh*, of testis, *Ko*, of ovary; *M*, *M.t*, Müllerian duct; *MU*, uterus; *P*, perineum; *Ps*, penis; *R*, rectum; *RH*, rete Halleri; *sk*, seminiferous canals of testis; *su*, sinus urinogenitalis; *T*, Müllerian funnel; *Th*, hydatid of Morgan; *Tt*, ostium abdominate; *t.W*, mesonephric or Wolffian tubule; *u*, urethra; *vg*, vagina; *vm*, vagina masculina (uterus masculinus); *W*, Wolffian body or mesonephros; *Wg*, Wolffian or mesonephric duct = *Wgg*, duct of Gartner in ♀, and *Wgv*, vas deferens in ♂.

now agree that the rete-cords come from the capsules or the peritoneum just lateral to the genital ridge, and the sex-cords from the germinal epithelium (Milhalkovics, **1041**).   This conclusion is probably correct and agrees best with the development of these parts in lower Gnathostomes, if we take the rete-cords to represent peritoneal funnel-canals leading to the capsules.[1] The observation of transient funnel-like structures joining the rete to the coelomic epithelium in the embryo of reptiles and mammals supports this interpretation (Allen, 981, in *Chrysemys* ; Fraser, 1010, in Marsupials ; Brambell, 994, in the mouse). In the higher mammals, however, these vestiges of funnels appear to be no longer recognisable (Allen, in pig, 980 ; Felix, in Man, 1006) ; and the rete-cords, then, have the appearance of prolongations of the sex-cords reaching the blind ends of the mesonephric tubules into which they eventually open, Figs. 688-9.

Our general conclusion with regard to the Tetrapoda is that the testicular network is formed by the combination of seminiferous tubules derived from the testis with mesonephric tubules originally provided with a peritoneal funnel opening into the splanchnocoele ; that the junction of the two takes place in the region of the funnel-canal or capsule ; and that, especially in the Amniota, the funnels themselves disappear more or less completely even in ontogeny. Anastomosis between the transverse vasa efferentia (Bowman's capsules ?) gives rise to a longitudinal marginal canal.

It is important to notice that the system of vasa efferentia, marginal canals, etc., in fact, the whole testicular network, is usually more or less completely developed in the female sex, though its vestiges may be much reduced in the adult, Figs. 685, 688.

The Müllerian Duct and Oviduct.—Besides the ducts described above, there exist in both sexes of Gnathostomes paired Müllerian ducts leading from the splanchnocoele to the exterior. They extend along the nephric folds, passing back on the outer side of the mesonephric duct (earlier archinephric duct), and primitively open in front close behind the septum transversum by a wide funnel into the coelom, and behind into the cloaca. They reach the front end of the cloaca by passing inwards ventrally to the mesonephric ducts. In the female sex the ova are shed into the splanchnocoele and are carried to the exterior by the Müllerian funnels and ducts ; in the male the ducts are vestigial, apparently functionless, and rarely open into the cloaca. In spite of the general resemblance of these ducts throughout the Gnathostomes their phylogenetic origin is still obscure, and owing to their apparent absence in some forms (certain Teleostomes) and discrepancies in their development, their very homology has been doubted.

---

[1] Such funnels represent the original communication of the nephrocoele with the splanchnocoele ; but, owing to modifications in development, might easily appear as ingrowths from the coelomic epithelium towards or as outgrowths from the capsules.

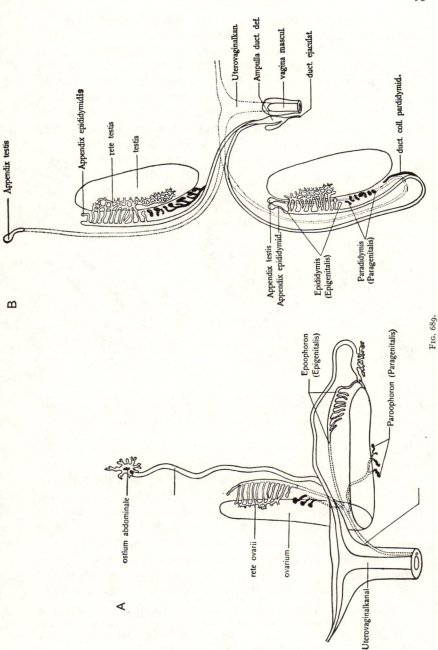

FIG. 689.

Diagrams illustrating fate of Mesonephros, Primary longitudinal duct, and Müllerian duct in human development of : A, female ; and B, male. Vestiges of posterior mesonephric excretory tubules in black (paroophoron, and paradidymis) ; more anterior tubules contribute to epigenital network ; while greater number of tubules still more anterior disappear. A, Ovary, though remaining in body-cavity, rotates through 90°. B, Testis wanders out of the body-cavity into scrotal sac.

All Tetrapoda are provided in the female with Müllerian ducts serving as oviducts,[1] Figs. 700-1. Their glandular wall secretes the albuminous covering and protective shell of the egg present except in some Marsupials and the placental Mammalia. Each duct opens in front by a wide 'ostium abdominale' or Müllerian funnel often provided with extensive ciliated lips drawn out in Mammals into fimbriated lobes. In the lower forms (Amphibia) the funnel retains its position far forward, Figs. 625, 685; but in others there is a progressive tendency for the duct to become shortened in the adult, and for the funnel to be drawn back to nearer the posterior region of the coelom of the trunk, especially in Mammals. In the Placental Mammals the two ducts fuse

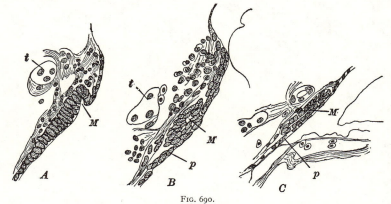

FIG. 690.

Sections through developing Müllerian duct of a 34 mm. tadpole of *R. sylvatica* (after Hall, from W. E. Kellicott, *Chordate Develt.*). A, Section passing through beginning of Müllerian evagination. B, Section posterior to A; duct established but still connected with peritoneum. C, Section still farther posterior, showing separation of duct from peritoneum. *M*, Müllerian duct; *p*, peritoneum; *t*, third pronephric tubule.

to a median vagina at their base, and in the higher groups their swollen uterine regions may also fuse to a median uterus.

Concerning the development of the Müllerian duct in the Tetrapoda there has been much controversy and the question cannot yet be considered as definitely answered. Balfour and Sedgwick maintained that in the chick the funnel is derived from the pronephros and the duct itself is split off from the archinephric (primary or Wolffian) duct. Most of the earlier observers believed this to be the usual mode of development of the Müllerian duct in Amniotes. In adopting this interpretation they were doubtless influenced by the fact that such had already been shown

[1] It has long been known that as a rule in Birds only the left ovary and oviduct are fully developed and functional in the adult. The left oviduct early ceases to grow and the left ovary likewise becomes atrophied. Occasionally these organs may be developed on both sides (Gadow, 1912).

to be the origin of the Müllerian duct in Selachians (Semper, **1063**; Balfour, **317**).

But it seems now to have been satisfactorily established by later workers that, at all events in Amniotes, the Müllerian funnel is developed by the closing over of a groove on an area of thickened epithelium (funnel area) situated at the edge of the anterior end of the nephric folds, and passing back along the lateral surface of the mesonephros towards the cloacal region as a longitudinal band of thickened coelomic epithelium

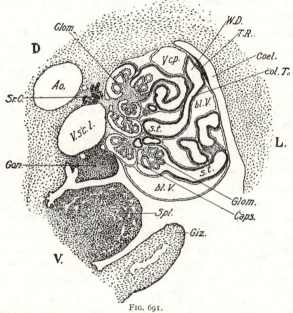

FIG. 691.

Transverse section through mesonephros and neighbouring parts of a 6-day chick, in region of spleen (from F. R. Lillie, *Devel. Chick*, 1919). *Ao*, Aorta; *bl.V*, blood-vessels (sinusoids); *Caps*, capsule of renal corpuscle; *Coel*, coelom; *col.T*, collecting tubule; *D*, dorsal; *Giz*, gizzard; *Glom*, glomerulus; *Gon*, gonad; *L*, left; *Spl*, spleen; *Sr.C*, cortical substance of suprarenal; *s.t*, secreting tubule; *T.R*, tubal ridge; *V*, ventral; *V.c.p*, posterior cardinal vein; *V.s'c.l*, left subcardinal vein; *W.D*, Wolffian duct.

often called the 'tubal ridge', Figs. 690-91. The groove forms a pit (sometimes two or three pits) which remains open in front and projects as a blind end into the fold behind (see further, p. 698). The opening becomes the ostium abdominale; the blind end gives rise to the duct itself by growing backwards independently between the outer 'tubal ridge' and the inner Wolffian duct, receiving contributions from neither of these structures (Braun, **1877–8**; v. Milhalkovics, **1041**, in Reptilia; Lillie, **845**, and others, in *Gallus*; Fraser, **1010**, in Marsupialia; Felix, **1005**; Brambell, **994**, and others, in Man and other Placentalia).

In the Amphibia also there extends back from the funnel area a band of specialised coelomic epithelium. According to Semon (1058-9) and Brauer (995) in Apoda the duct grows backwards from the funnel independently of both this band and the Wolffian duct. But according to some it may be derived anteriorly from the epithelium (MacBride, 1037, in *Rana* ; Wilson, 1075, in Urodela), and according to others from the wall of the Wolffian duct (Gemmill, 1013, in Urodela ; Hall, 1019, in Anura and Urodela). On the whole, the evidence seems to point to the Müllerian duct developing either independently, or possibly from the Wolffian duct at its posterior end only.

Scarcely less difficult to determine is the relation of the Müllerian funnel to the segmental tubules in Tetrapods. Clearly it is not merely an enlarged pronephric or mesonephric funnel ; yet it seems generally to be related to the pronephric or anterior mesonephric tubules. Although Brauer (995) states that in the Apodan *Hypogeophis* the funnel area has no connexion with pronephric peritoneal funnels, it has been shown by H. Rabl in *Salamandra* (1049) and by Hall in Urodela and Anura that the Müllerian funnel is developed from an area at first continuous with the lips of one or more pronephric nephrocoelostomes. According to Hall's careful description of *Amblystoma*, patches of thickened coelomic epithelium extend ventrally in the pronephric chamber from two nephro-coelostomes ; and a groove or funnel-like depression occurs in each patch. The patches join to a longitudinal funnel area, the blind ends of the pits grow back, fuse, and continue posteriorly as the rudiment of the duct. The funnel area passes forwards and downwards below the developing floor of the pronephric chamber, its edges close over, carrying the persisting opening of the posterior pit to its definitive position at the side of the liver. The ostium abdominale is thus originally derived from a patch of epithelium probably representing the remains of one or more pronephric peritoneal funnels. Although no such definite connexion with the pro-nephros has been described by recent observers in the Amniota except in the Crocodile (Wilson, 1074), yet the funnel area is usually situated very close to the vestigial pronephric or anterior mesonephric tubules. Certain structures in Mammals have been claimed by de Winiwarter (1076) and Wickmann to be remains of a connexion with the segmental tubules, and the latter describes pronephric funnels opening on the funnel area. Both in Birds and Mammals several depressions may occur on the surface of the area, and these may perhaps be interpreted as the last vestiges of peritoneal funnels ; sometimes they give rise to accessory Müllerian funnels. Recently Brambell (994) has described the origin of the Müllerian funnel in the mouse from an invagination of the coelomic

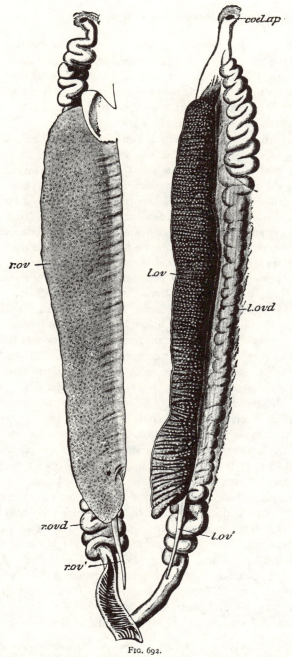

Fig. 692.

Ceratodus forsteri. Reproductive organs of female; the inner surface of the right and the outer surface of the left ovary shown. *coel.ap,* Coelomic aperture of oviduct; *l.ov,* left ovary; *l.ov',* its posterior termination; *l.ovd,* left oviduct; *r.ov,* right ovary; *r.ov',* its posterior termination; *r.ovd,* right oviduct. (After Günther, from Parker and Haswell, *Zoology*.)

epithelium in front of the mesonephros and occupying the same position on the urinogenital ridge as the peritoneal funnels further back.

The Dipnoi are provided with typical Müllerian ducts, which function as oviducts in the female. It is in this group that we might hope to find a clue to their morphology, but unfortunately they appear late, and their development is quite unknown, Fig. 692.

In Elasmobranchs the Müllerian ducts are both well developed and arise early in ontogeny, Figs. 684, 618-20. The funnels combine in the adult to a median ventral ostium situated behind the transverse septum and below the oesophagus and attachment of the liver. The ducts in the female acquire a large size, have a definite region specialised for secreting the egg-case, and open usually independently into the cloaca.[1] Although typical Müllerian ducts in function and anatomical relationships, they have been shown to develop from the pronephros and its duct (Semper, 1063; Balfour, 317). The funnel area is derived directly from the degenerate pronephric funnels, which combine to a single posterior opening (Balfour, Rabl, and others); it leads into the persisting last pronephric tubule and so to the longitudinal duct. The Müllerian duct itself develops by the gradual splitting off from before backwards of a second duct from the original primary duct. This progressive constriction of the primary duct into two gives rise to an upper true Wolffian duct and a lower Müllerian duct carrying the funnel at its anterior end. Posteriorly the two ducts come to open separately into the cloaca, Figs. 669, 670, 684.

Obviously, there is a striking difference between the development of the Müllerian ducts in Selachii and Tetrapoda; indeed, many have doubted its homology in the two groups. Yet so similar are the ducts in the adult condition both in function and in anatomical relationship that it can scarcely be doubted that they are homologous throughout the Gnathostomes (leaving the Teleostomes aside for the present; see below). Their constant presence in both sexes is a distinctive feature, and an important point is that in early stages the Müllerian funnel always occupies the same position—just behind the septum transversum and extending along the free edge of the anterior end of the nephric fold where it stretches down to the liver, Figs. 618-20, 625, 628, 649. Were it not for differences in its mode of development the homology of the Müllerian duct in Selachians and Tetrapods would scarcely have been questioned. But it is possible that these differences may be reconciled when its development becomes known in Holocephali and Dipnoi; and it is not

[1] Vestigial funnels remain in the male where the greater part of the duct is usually obliterated, but the posterior ends give rise to the 'seminal vesicles' (uterus masculinus).

impossible, even now, to devise a provisional reconciliation. Further knowledge of the development in other groups may enable us to solve this problem.[1]

Turning, now, to the conditions found in Teleostomes, the first fact that emerges is that in none of them have typical Müllerian funnels and ducts been found. This is one of the many features in their structure which exclude the Teleostomi from the direct line of ancestry of the Tetrapoda. When oviducts occur they are relatively short and their openings always somewhat far back, and, except in Acipenseridae, nothing resembling a Müllerian duct appears in the male, Fig. 682. Moreover, the oviduct extends along the outer side of the genital ridge close to its base,[2] instead of running along the outer side of the mesonephros, Fig. 683.

The Acipenseridae show, perhaps, the most primitive condition (Hyrtl, 1026). What appears to be a Müllerian duct is present in both sexes; in the female it is a wide tube lying close to the base of the elongated genital fold and opening about half-way up the ovary by a wide funnel. Posteriorly the oviduct opens into the urinogenital sinus leading to a median external pore. In the male a similar but smaller duct occurs in a corresponding position on the outer side of the genital fold; it has an anterior funnel and ends blindly in the wall of the urinogenital sinus.

[1] We have seen that in Tetrapods the funnel area is frequently connected with the pronephros, and may be derived from its peritoneal funnels; similarly in Selachians, since the pronephros is very degenerate and undergoes considerable remodelling at the time, it is possible that the funnel area also represents reduced pronephric peritoneal funnels. Even if the duct always develops by free backward growth in Tetrapods, it lies close to the latero-ventral side of the Wolffian duct—the splitting off, seen in Selachians, might have been omitted in the ontogeny of the Tetrapods, or on the contrary the continuity of the rudiments of the two ducts might have been secondarily established in the Selachian.

But another interpretation may prove to be nearer the truth. It is difficult to account for the presence of a 'tubal ridge' in Tetrapods if it has nothing to do with the development of the Müllerian duct, and it is possible that this duct may yet be shown to have arisen from the coelomic epithelium as a groove which became closed to form a tube opening behind into a mesonephric tubule and remained open in front. The peculiar mode of development of the duct in Selachians would then probably be secondary.

Since Müllerian duct and marginal canal often coexist in both sexes as separate structures they cannot be homologous, but the same objection does not apply with regard to the funnel area on which peritoneal funnels may open. It is not impossible that the funnel area may represent in front a special region of coelomic epithelium which further back becomes folded over to form the marginal canal (see further, p. 706).

[2] The statement frequently made in text-books that the oviduct lies on the inner or medial side of the genital fold in Teleostomes is erroneous. It extends between the genital and the nephric folds.

Very important is the condition seen in *Polypterus*, that isolated and in some respects primitive Actinopterygian (J. Müller, 1846; Hyrtl, 1026; Budgett, 10; Kerr, 1033). The oviduct of *Polypterus* has a wide opening about half-way up the ovary and passes back beyond it to open into the urinogenital sinus (near the place where the two mesonephric ducts join to a median sinus probably derived from a cloaca). The sinus opens on a urinogenital papilla behind the anus. *Amia* has very similar oviducts, which join and open between the anus in front and the urinary pore behind, Figs. 683, 693.

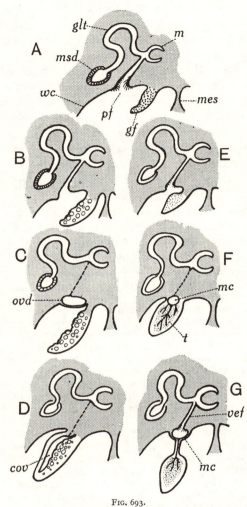

Lepidosteus presents a new type of structure in the female (Balfour and Parker, 2) since each ovary projects into a closed sac continuous with a duct behind, which opens into the base of the mesonephric duct. The two mesonephric ducts join to a sinus opening to the exterior by a median pore. In this fish, then, the ripe ova are carried out directly, and are not first shed into the general coelom, Figs. 683, 693.

Fig. 693.

Diagrams illustrating relation of *genital fold* and *mesonephric tubule* to oviduct on left and to marginal canal of testis on right in Teleostomes. A, Primitive indifferent condition; B, *Polypterus* ♀, young stage; C, *Polypterus* ♀, adult; D, *Lepidosteus* ♀, adult; E. *Polypterus* ♂, young stage; F, *Polypterus* ♂, adult; G, *Lepidosteus* ♂, adult. *cov*, Cavity of ovary; *gf*, genital fold; *glt*, glandular tube; *m*, capsule of Malpighian body; *mc*, marginal canal; *mes*, mesentery; *msd*, mesonephric duct; *ovd*, oviduct = marginal canal; *pf*, peritoneal funnel; *t*, testis; *vef*, vas efferens (funnel canal); *wc*, wall of splanchnocoele.

The Teleostei present various interesting types of structure in the female, some approaching that seen in *Amia* and others that seen in *Lepidosteus* (Rathke, 1820–25 ; Hyrtl, 1025). The morphology of the Teleostean ducts is very difficult to interpret, and the homology of the

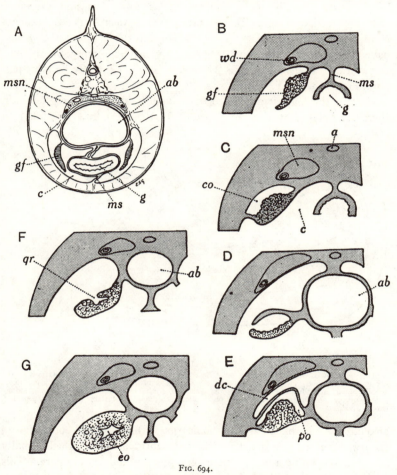

FIG. 694.

A, Transverse section of *Anguilla vulgaris*, showing free ovaries hanging in coelom. B-G, Diagrammatic sections showing development of closed ovaries. B and C, *Cobitis* ; early stage with genital fold free, and later stage with fold fused to coelomic wall (after Schneider, 1895). D and E, *Cyprinus* ; early stage with ovary still open, and later stage with fold fused and cavity closed (after Haller, 1905). F and G, *Acerina* ; early stage free and open, and late stage closed (after Jungersen, 1889). *ab*, Airbladder ; *c*, splanchnocoele ; *co*, ovarian cavity cut off from coelom ; *dc*, dorsal coelomic chamber ; *eo*, endovarial cavity ; *g*, gut ; *gf*, genital fold ; *ms*, mesentery ; *msn*, mesonephros ; *po*, parovarial cavity ; *qr*, parovarial groove ; *wd*, mesonephric duct.

parts by no means yet established. A few Teleosts have so-called ' free ovaries ', which shed the ova into the general coelom in the usual manner. In such forms the ova may be carried to the exterior by oviducts resem-

bling those of *Amia* or may pass out by mere pores. *Osmerus*, among the
Salmonidae, has oviducts almost as well developed as those of *Amia*;
but in *Salmo* they are reduced to short funnels behind the ovary and
leading to the median pore, while *Mallotus, Coregonus*, and *Argentina*
show intermediate stages (Huxley, **1024**; Weber, **1069**). The Galaxiidae,
Notopteridae, Hyodontidae, Osteoglossidae, and the Cyprinid *Misgurnus*,
have similar free ovaries and very short oviducts. Finally, in Anguilliformes the ova when shed into the coelom pass out directly by a median
pore, Figs. 683, 694-7.

But in the majority of Teleostei, including all the highest families,
there are ' closed ovaries ' or ovisacs. In these the ripe ova fall into

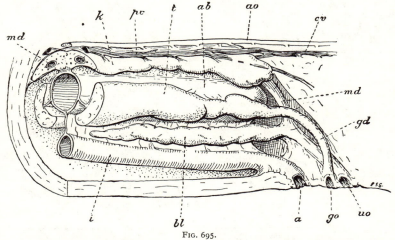

FIG. 695.

Left-side view of a dissection of a male *Esox lucius*, L., showing the median apertures of the
rectum, genital ducts, and kidney ducts. *a*, Anus; *ab*, air-bladder, blind hinder end; *ao*, dorsal
aorta; *bl*, urinary bladder; *cv*, *gd*, vas deferens; *go*, genital opening; *i*, intestine; *k*, kidney
(mesonephros); *md*, mesonephric duct; *pc*, posterior cardinal; *t*, testis; *uo*, urinary opening.
(From Goodrich, *Vert. Craniata*, 1909.)

an ovarian cavity having no communication with the general body-cavity;
the two ovarian sacs narrow behind to form oviducts which usually
combine to open by a median pore between anus and urinary pore. This
genital pore, situated behind the anus and in front of the urinary pore, is
doubtless homologous throughout the Teleosts (see further, p. 710).

Several questions arise in connexion with the female organs of the
Teleostei: Which is the primitive condition, the free ovary or the
ovarian sac? the long oviduct of *Osmerus* or the pore of *Anguilla*?
Are the oviducts homologous with Müllerian ducts or with the longitudinal
marginal canal occupying much the same position in the male of primitive
Teleostomes?

The cavity of the ovarian sac may be distinguished into two parts : the first or ovarial part related to the ovary itself, and the second part forming the lumen of the duct leading to the pore, Figs. 694, 696-7. While in such a form as *Anguilla* the genital fold remains as a simple band on the outer side of which the ova develop, in others a chamber is formed by the grooving of this ovarian surface, or the folding of the band on itself, or the fusion of its free edge with the coelomic wall (Macleod, 1038; Jungersen, 1028; Schneider, 1052; Haller, 1020). When the edge of the genital fold bends outwards and fuses with the outer wall of the coelom (*Rhodeus, Gobio, Esox,* etc.), or bends up to fuse close to its attachment, the

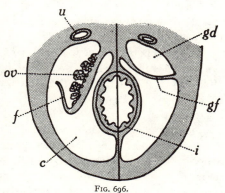

FIG. 696.

Diagrammatic transverse sections of a Teleost; on left through ovary, *ov*, on right more posteriorly through genital duct, *gd*. *c*, Splanchnocoele; *f*, free edge of genital fold; *gf*, genital fold forming behind ventral wall of duct; *i*, intestine; *u*, ureter (mesonephric duct).

chamber lies at the side of the ovary, and is called a parovarial canal. An endovarial canal occurs when the outer surface of the band becomes grooved and the two edges of the groove meet and fuse (*Acerina, Perca,* etc.). There is no fundamental distinction between the two types, and in *Salmo* the front end of the ovary is folded so as to

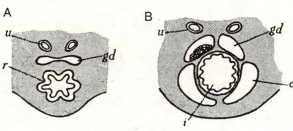

FIG. 697.

Diagrammatic transverse sections showing genital duct and coelom in ♀ *Teleost* : A, More posterior section just in front of anus ; B, section through hinder region of trunk coelom, *c*. *gd*, Oviduct or dorsal chamber of coelom containing hind end of ovary on left ; *i*, intestine ; *r*, rectum ; *u*, mesonephric duct.

form a short parovarial canal, while a short endovarial canal is developed behind, both remaining open (Felix u. Bühler, 1007). All Teleosts pass through a stage with free ovary. Clearly the closed sac is derived from a more primitive simple band-like genital fold. In answer to the question whether the ancestral Teleosts had free or closed ovaries, we have the view of Brock that the free ovary seen in *Anguilla* is primary, and that of Balfour that it is more probably secondary due to a return to an apparently primitive condition from the closed structure already established in *Lepidosteus*. The sporadic occurrence of the ' free ' condition in various unrelated Teleostean families is strong evidence in favour of Balfour's suggestion.

As to the homology of the Teleostean oviducts and genital pores the evidence is still uncertain. Waldeyer held that the duct is a reduced Müllerian duct, and that in forms with closed ovarian sacs the ovaries have been overgrown and enclosed by the Müllerian funnel (1067). Balfour suggested that in *Lepidosteus* the closed sac has probably been formed by the junction of an ovarian chamber of the coelom (parovarial canal) with a posterior short Müllerian duct opening into it (2). The ' free ' ovary in Teleosts would, then, be due to the failure of these two structures to unite. On the whole this suggestion agrees best with the facts. Jungersen (1028) has shown that the duct proper may develop by the hollowing out of a thickening of the coelomic epithelium before (*Acerina*) or after the formation of the ovarian cavity (*Esox, Zoarces*, etc.), and recently Essenberg has described the separate origin of the posterior duct and the parovarial canal in *Xiphophorus* (1004).

On the other hand, Brock (998-9), Felix (1007), and Lickteig (1035) maintain that the Teleostean oviduct is not a Müllerian duct, but merely a prolongation backward of the paired dorsal regions of the coelom (ovarian sacs in forms with closed ovaries) which join and open by a common pore to the exterior. Such a view necessitates the assumption that the Müllerian ducts and their primitive openings have disappeared altogether, and been replaced by new formations. Considering how universally Müllerian ducts occur in other groups, it seems more reasonable to suppose that the oviducts of Teleostomes are, at all events in part, homologous with them,[1] and that their original openings have persisted in phylogeny and are represented in Teleostei by the genital pore (pp. 704, 711).

Obscure as is the origin of the Müllerian duct in Gnathostomes generally, the oviduct of Teleostomes can scarcely be its complete homologue. For the researches of Balfour and Parker on *Lepidosteus* (2), of Budgett (10) and Kerr (1033) on *Polypterus* (and apparently of Maschkovzeff on

---

[1] It should not be forgotten that the oviducts in Acipenseridae are represented in the male (p. 701), and it is not impossible that vestiges of Müllerian ducts may yet be discovered in other Teleostomes where they appear to be absent when their structure and development come to be better known.

Acipenser (1040)), clearly prove that whereas the duct posterior to the ovary may represent a Müllerian duct, the more anterior part alongside the ovary is of different nature. This latter part is a longitudinal chamber of the coelom closed off at the base of the genital folds ; into it, at first, open the mesonephric peritoneal funnels, Fig. 693. This canal in the female has in fact much the same relations as the marginal canal in male Gnathostomes generally (see also female, p. 694). Felix, indeed, compares the whole oviduct of Teleostomes to a marginal canal ; but it would seem more probable that in these fish a short Müllerian duct has been prolonged by the closing over of a canal, carrying the ostium abdominale forward. This would give rise to the elongated oviducts seen in *Amia* or *Polypterus* ; by reduction from in front the condition seen in Salmonids and lastly in Eels would be brought about. The widening of the canal so as to enclose the whole of the fertile region of the genital fold and closure of the ostium might give rise to the ovarian sac of *Lepidosteus* and Teleostei.

THE GENITAL DUCTS OF THE CYCLOSTOMATA

There remain to be considered the genital pores or funnels of the Lampreys and Hag-fishes. It has already been mentioned that the Cyclostomes differ from the Gnathostomes in that their genital products do not pass out through the segmental tubules in the male. The gonads are built on the same general plan, and develop from paired genital folds ; but in both sexes the ripe germ-cells are shed into the general coelom and escape to the exterior through genital pores of paired origin. Thus, not only are there no vasa efferentia, but also no Müllerian ducts; unless, indeed, these pores represent them. In the Petromyzontia the pores open on either side into a urinogenital sinus provided with a median pore on a papilla situated behind the anus, Fig. 98. This sinus appears to be formed by the subdivision of an original endodermal cloaca into which open the primary archinephric ducts in the larva. The cloaca in the Myxinoidea is less completely divided, and the two genital pores here unite to open into it just behind the anus and in front of the urinary pore, Fig. 100.

The morphological significance of these genital pores is difficult to determine. It has been claimed that they are mere ' abdominal pores ' (Lickteig, 1035) ; but their anatomical relations do not support this view. The possibility that they represent a posterior pair of segmental tubules (coelomoducts), which have retained their original function of conveying the genital products to the exterior, must be kept in mind.

THE URINARY AND GENITAL PORES, AND THE CLOACA

Before leaving the subject of the excretory system of the Craniata something must be said about the external pores of the ducts, and the development and fate of the cloaca. A cloaca into which open rectum, urinary, and genital ducts is found in primitive Craniates. That of the Cyclostomes has been described above. All primitive Gnathostomes retain a cloaca (Selachii, Fig. 684, Dipnoi, Figs. 682-3, Amphibia, Reptilia,

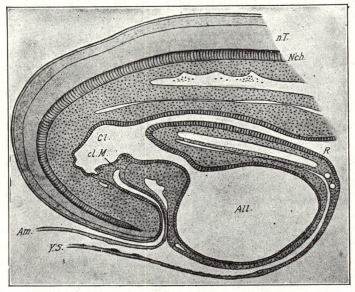

FIG. 698.

Median sagittal section of hind end of chick embryo on fourth day of incubation (after Gasser, from F. R. Lillie, *Develt. Chick*, 1919). *All*, Allantois ; *Am*, tail fold of amnion ; *cl.M*, cloacal membrane ; *Cl*, cloaca ; *N'ch*, notochord ; *n.T*, neural tube ; *R*, rectum ; *Y.S*, wall of yolk-sac.

Fig. 700, Aves, Fig. 701, Monotremata) ; but it is more or less completely lost in specialised groups (Holocephali, Fig. 167 c, Teleostomi, Figs. 682-3, 695, Mammalia Ditremata, Fig. 702). In the Gnathostomes the cloaca is formed partly from the endodermal gut and partly from an ectodermal invagination or proctodaeum. Endoderm and ectoderm coming into contact form a cloacal plate or membrane ; the primitive anus arises by the breaking through of this membrane generally late in embryonic life.[1]

---

[1] It is frequently held that the primitive anus in Vertebrates has been derived from the blastopore ; but the interpretation of its development has been much influenced by theories of more than doubtful value. Certainly in

Into the embryonic cloaca lined by endoderm come to open the primary archinephric and later the Müllerian ducts, while the posterior region of the cloaca lined by ectoderm opens to the exterior by the median cloacal aperture. The share taken by the endoderm and ectoderm in the formation of the definitive cloaca varies considerably in different groups.

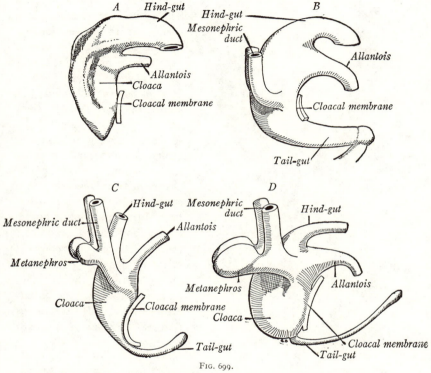

FIG. 699.

Four stages showing differentiation of cloaca into rectum, urethra, and bladder (after reconstructions by Pohlman). ×about 50. A, Human embryo, 3·5 mm.; B, about 4 mm.; C, 5 mm.; D, 7 mm. (From Prentiss and Arey, *Text-book of Embryology*, 1917.)

The cloaca of adult Selachians is shallow and widely open. Into it open the rectum, the mesonephric ducts (urinary sinus) at the end of a dorsal papilla, and the Müllerian ducts. The cloaca of the Holocephali is so shallow as to be lost, and the rectum and ducts come to open separately

---

Amphioxus the blastopore becomes the opening into the neurenteric canal, and has nothing to do with the anus which appears independently and later. The same is probably true of all vertebrates. Although in some animals, such as the Frog, the two apertures may appear closely related, it is difficult to see how this can be anything but secondary. The case of *Petromyzon*, in which the blastopore is said to persist as the anus, needs impartial reinvestigation.

to the exterior. The Dipnoi retain a primitive cloaca, and from the base of the united mesonephric ducts arises a bladder-like caecum lying dorsally to the rectum. On the other hand, the Teleostomi have lost the cloaca owing to its subdivision into ventral, rectal, and dorsal urinogenital portions. In the lower forms, such as *Polypterus*, *Acipenser*, *Lepidosteus*, and *Amia*, there is a urinogenital sinus opening by a median pore behind the anus; but in Teleosts this sinus is usually subdivided, so that a separate median genital pore occurs between the anus and urinary pore. Occasionally, however, the sperm-ducts open into the base of the mesonephric ducts (Anguilliformes, *Anablebs*, *Perca*, *Zoarces*, *Cyclopterus*, etc.), or together with the anus (*Lota*), or with both the anus and the kidney as in Lophobranchii (Hyrtl, **1025**; Stannius, 1854–6). A median mesodermal bladder-like diverticulum is often developed from the united bases of the mesonephric ducts which contribute to the formation of the urinogenital sinus, Fig. 695.

A typical cloaca is found in Amphibia, and from its mid-ventral endodermal wall is developed the true urinary bladder characteristic of the Tetrapoda. A similar cloaca occurs in Reptiles and Birds, but here its ectodermal ventral wall becomes strengthened by an upper corpus spongiosum and

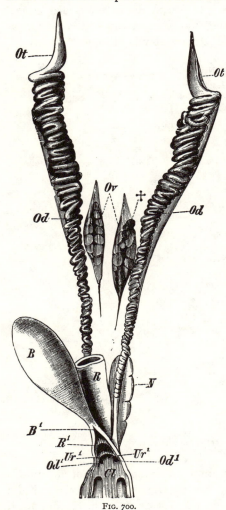

Fig. 700.

Female urinogenital apparatus of *Lacerta muralis*. *B*, Urinary bladder; *B¹*, neck of the bladder (cut open); *N*, kidneys; *Od*, oviducts, which open into the cloaca at *Od¹*; *Ot*, abdominal openings of oviducts; *Ov*, ovaries; *R*, rectum; *R¹*, opening of rectum into the cloaca (*Cl*); *Ur¹*, apertures of the ureters into the cloaca; †, remains of mesonephros. (From Wiedersheim, *Comp. Anatomy*.)

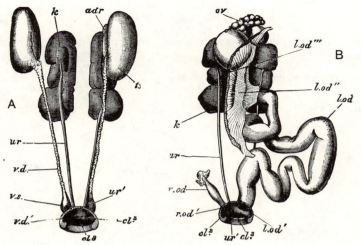

FIG. 701.

Columba livia. A, Male urinogenital organs. *adr*, Adrenal ; *cl.* 2, urodaeum ; *cl.* 3, proctodaeum ; *k*, kidney ; *ts*, testis, that of the right side displaced ; *ur*, ureter ; *ur′*, aperture of ureter ; *v.d*, vas deferens ; *v.d′*, its cloacal aperture ; *v.s*, vesicula seminalis. B, Female urinogenital organs. *cl.* 2, Urodaeum ; *cl.* 3, proctodaeum ; *k*, kidney ; *l.od*, left oviduct ; *l.od′*, its cloacal aperture ; *l.od″*, its coelomic funnel ; *l.od‴*, its coelomic aperture ; *ov*, ovary ; *r.od*, right oviduct ; *r.od′*, its cloacal aperture ; *ur*, ureter ; *ur′*, its cloacal aperture. (From Parker's *Zootomy*.)

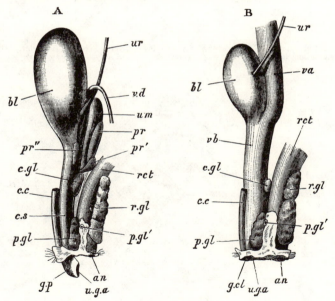

FIG. 702.

Lepus cuniculus. The urinogenital organs ; A, of male ; B, of female ; from the left side (half nat. size). The kidneys and proximal ends of the ureters, in A the testes, and in B the ovaries, Fallopian tubes and uteri are not shown. *an*, Anus ; *bl*, urinary bladder ; *c.c*, corpus cavernosum ; *c.s*, corpus spongiosum ; *c.gl*, Cowper's gland ; *g.cl*, apex of clitoris ; *g.p*, apex of penis ; *p.gl*, perineal gland ; *p.gl′*, aperture of its duct on the perineal space ; *pr*, anterior, *pr′*, posterior, and *pr″*, lateral lobes of prostate ; *rct*, rectum ; *r.gl*, rectal gland ; *u.g.a*, urinogenital aperture ; *u.m*, uterus masculinus ; *ur*, ureter ; *va*, vagina ; *vb*, vestibule ; *v.d*, vas deferens. (From Parker's *Zootomy*.)

a lower corpus fibrosum, which become converted into the copulatory organ of the male (Fleischmann, A., 1009; Pomayer, 1046; v.

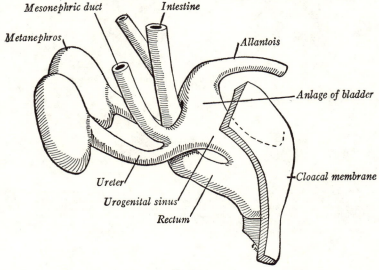

Fig. 703.

Reconstruction from 12 mm. human embryo showing partial subdivision of cloaca into rectum and urinogenital sinus (after Pohlman). × 65. (From Prentiss and Arey, *Text-book of Embryology*, 1917.)

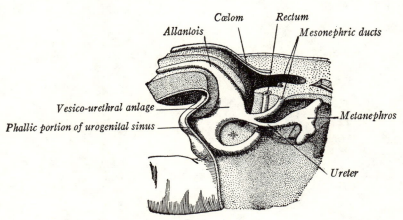

Fig. 704.

Reconstruction of caudal portion of 11·5 mm. human embryo showing differentiation of rectum, bladder, and urethra (after Keibel's model). × 25. (From Prentiss and Arey, *Text-book of Embryology*, 1917.)

Oordt, 1922; Boyden, 992; Hellmuth, 1021). The ventral endodermal diverticulum is enlarged to give rise in the embryo to the allantoic

sac characteristic of the Amniota, and the adult bladder is formed
from its enlarged base only,[1] Figs. 698-9. In the Mammalia
the cloaca tends to disappear as such in the adult (Gerhardt, 1014-
1015; Brock, 997). Its endodermal region becomes subdivided by
a backwardly growing fold into an upper or postero-dorsal rectum,
and a lower or antero-ventral urinogenital canal into which open the
bladder, ureters, and genital ducts. Its ectodermal region also becomes
subdivided in the male. For, while in Reptiles the spermatozoa pass along
open grooves on the copulatory organ, in the Mammal the median groove

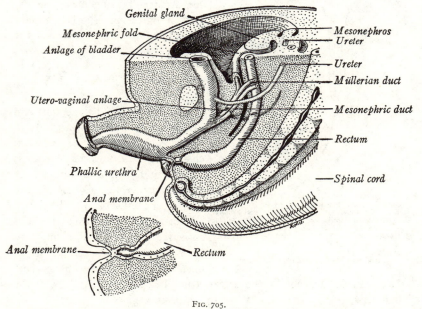

FIG. 705.

Reconstruction of caudal end of 29 mm. human embryo showing complete separation of rectum
and urinogenital sinus and relations of urinogenital ducts (after Keibel's model). ×15. (From
Prentiss and Arey, *Text-book of Embryology*, 1917.)

becomes closed over to form a tube, the penial urethra (phallic urethra).
The very primitive character of these structures in Monotremes is shown
in the persistence of the compound cloaca, and its single opening, and
the incomplete closure of the urethra anteriorly, so that the urinogenital

[1] In Birds there is a considerable ectodermal cloaca into which comes to
open a median dorsal bursa Fabricii, developed from the wall of the endodermal
cloaca. It usually atrophies in the adult and its function is doubtful; but it
may produce an internal secretion at the onset of sexual maturity (Wenckebach,
1888; Pomayer, 1046; Boyden, 992). Dorsal paired lateral bladder-like
pouches are developed in some Reptiles (Chelonia).

canal still remains in communication with the dorsal cloacal chamber by a narrow ' urinary canal ' (Keibel, 1030). In the Ditremata, on the other hand, the fold separating the rectal from the urinogenital chambers extends as far as the cloacal membrane, so that the anus and urinogenital apertures come to open separately (Keibel, 1030 ; Felix, 1006 ; Brock, 997). At the same time on the phallic thickening anterior to the latter is formed a groove which closes to a urethral canal in the male carrying the urinogenital opening to the extremity of the penis. The urinary canal (communication between the urinogenital and rectal chambers) is closed in the Ditremata (except in *Perameles* according to v. d. Brock), Figs. 703-5.

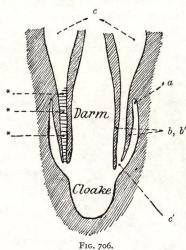

FIG. 706.

Diagrammatic horizontal section through the cloacal region of a Selachian. (After E. J. Bles.) *a*, Blind ectodermal invagination (cloacal pouch) ; *b*, *b'*, cloacal papilla ; *c*, peritoneal cavity, which opens by the abdominal pore at *c'* ; *Cloake*, cloaca ; *Darm*, rectum ; ***, points along the transversely striped section of the cloacal papilla at which the abdominal pore may break through, in which case the distal part of the papilla is solid (Raja).

In the Marsupialia the anus and urinogenital aperture lead into a considerable shallow ectodermal cloaca the opening of which is closed by a sphincter muscle. A small ectodermal cloacal region may occur in some of the lower Placentalia (some Rodents and Insectivores), but in the majority the anus and urinogenital apertures open quite separately on the surface. In the Placentalia the ureters come to open into the base of the bladder and the two oviducts join to a median vagina.

ABDOMINAL PORES

The problem of the homology of the genital ducts in Cyclostomes and Teleostomes has been complicated by the confusion by the older authors of ' genital pores ' and ' abdominal pores '. The latter name belongs to certain paired openings in the ventral body-wall in both sexes which lead from the coelom to the exterior (Bridge, 996 ; Weber, 1069 ; Ayers, 984 ; Bles, 990). They occur somewhat irregularly in Elasmobranchs, being as a rule present only when the fish is full-grown or sexually mature, Figs. 684, 706 (Notidani, Scyllioidei, Spinacidae, Rhinobatidae, Rajidae, Torpedinidae, etc., and Holocephali). They rarely open outside (Carcharias), but more usually inside the cloaca on projecting papillae. Similar paired abdominal pores opening to the exterior on either side of the anus are found in Chondrostei, Polypterini, *Amia*, and *Lepidosteus*, Fig. 682 E, F. The Dipnoi also possess such pores, but they may open by a single aperture (Owen, 1839 ; Günther, 1871). Among Teleostei the Salmonidae and Mormyridae possess them. They are unknown in Amphibia, but occur in Chelonia and Crocodilia among Amniota (Moens, 1042). In no Gnathostome do the abdominal pores ever serve for the exit of genital products. It has been clearly shown by Weber that in Teleosts they may coexist with ' genital pores ' (reduced genital ducts, see p. 704).[1] Unlike genital ducts and pores the abdominal pores are not related to the dorsal region of the coelom, but are openings pierced through the body-wall at the posterior extremity of the paired ventral prolongations of the coelom passing on either side of the ventral mesentery and often called peritoneal canals. They appear to be special openings of no great morphological significance and of doubtful function. Apparently they allow fluid to escape from the coelom and perhaps to enter it ; Bles has pointed out that they usually occur in those forms which have no open peritoneal funnels on the kidney.

THE PHYLOGENY OF THE SEGMENTAL TUBULES

The phylogenetic origin of the segmental tubules from which the various genital and excretory ducts of the Craniata have been derived must now be discussed. From what has been said above about the structure and development of these organs it may safely be concluded that they are all directly or indirectly developed from segmental out-growths of the wall of the coelom; that a pair of these funnel-like outgrowths was originally present in every segment of the body at least potentially ; that they failed to develop or have been secondarily lost in the head and tail regions ; that possibly they originally all opened separately to the exterior.

Without entering into a detailed discussion of the various theories of the origin of the coelom itself, it may be said that the only hypothesis which gives an explanation of the phylogenetic evolution of the coelom consistent with the facts revealed by a broad survey of the Triploblastic Metazoa (Coelomata) in general is the so-called ' Gonocoel Theory '. Founded on the conclusions of Hatschek and Meyer with regard to

---

[1] Lickteig has recently revived the view that the genital pores of Cyclo-stomes are abdominal pores (1035) ; but from what has been said above it appears that there is no good evidence for this interpretation.

Platyhelmia and Annelida, this theory has since been applied to all the higher Coelomata (Goodrich, 1016). It may briefly be stated as follows : In the ancestral form the genital cells tended to accumulate between the primary germ-layers, ectoderm and endoderm, and at maturity sought an escape to the exterior ; such accumulations, acquiring a definite wall from their more superficial cells, formed a pair of bilaterally symmetrical sacs ; and from the wall of each sac developed a ciliated funnel-like outgrowth which, fusing with the ectoderm (or the junction of ectoderm with endoderm), acquired an opening to the exterior. At first these sacs and funnels functioned merely as gonadial sacs and genital ducts (Platyhelminth stage) ; but soon the sacs tended to become regularly repeated along the elongated body in a bilateral series, to enlarge prematurely, and become solidly packed with genital cells only at the reproductive season (Nemertine stage). The restriction of the proliferation of genital cells to a definite region of the wall (henceforward known as the gonad proper) leads to the condition found in Annelida. Here the genital sacs become enlarged and arranged as a paired series of segmental coelomic chambers. Primitively each sac contains a testis or ovary (a proliferation from its wall), and from its wall develops a ciliated funnel (coelomostome) opening to the exterior. The coelomostomes, and the elongated coelomoducts which may develop from them, function as genital ducts.

In the majority of the segments the cavity becomes packed with reproductive cells at maturity; but, owing to increasing specialisation and differentiation between segments, some of the more anterior and posterior usually become sterile, failing to develop gonads. Thus, not only do the sacs tend to develop prematurely as chambers filled with fluid, but some of them remain in this condition. This is due to the acquisition of new functions. The coelom becoming a spacious cavity serves to distend the body-wall and afford a place in which the viscera can expand, spread, and indulge in muscular movements. Moreover, it also acquires an excretory function. In the metamerically segmented Annelida and Arthropoda this leads to a division of labour between the segments. In some the coelomoducts are devoted to excretion, in others they retain their primitive rôle of genital ducts. Moreover, in Arthropoda each sac, in the majority of segments, becomes subdivided into a dorsal genital and a ventral excretory portion. The genital portions usually combine to form the adult 'ovary' or 'testis'. A varying number of the segmental ventral portions persist as the excretory organs in most Arthropods. In the unsegmented Mollusca the history of the paired coelomic cavities, coelomostomes and coelomoducts, is essentially similar ; but here the differentiation arises between different parts of the same coelomic sac, which tends to become subdivided into genital, perivisceral (pericardial), and excretory compartments. The same general development of the coelom and coelomostome can be traced in all the groups of Triploblastic Invertebrata.

Turning now to the Vertebrata we find similar paired coelomic sacs developed from the mesoblast. Primitively they arise as separate segmentally disposed sacs from whose walls develop the gonads. Primitively also they were in all probability all of them fertile (as they still are to a great extent in *Amphioxus*) ; but anteriorly and posteriorly they become sterile. When, as in Craniates, the cavities of the segmental sacs became confluent ventrally even in the embryo (lateral plate region) the gonads combined to longitudinal paired bands showing little or no trace of segmentation. But, excepting in the head and tail regions, every segment still produced a pair of funnel-like outgrowths or coelomostomes, and these may still convey the genital products to the exterior. In the male Gnathostome this primitive function is performed by coelomostomes belonging to quite a large number of segments ; in the female the ducts are apparently of the same ultimate origin, but their metameric composition is obscure, and they represent perhaps the coelomoducts of only one segment.

Unfortunately theories of the morphology of the segmental tubules of the Vertebrates and of their origin from comparable organs in the Invertebrates were put forward and generally accepted before the structure and development of these organs had been correctly described or was understood. It was Semper who first definitely maintained that the Vertebrate tubules are homologous with the nephridia of Annelida. This attractive view was almost universally accepted at a time when almost any tube leading to the exterior was called by the convenient name ' nephridium '. But Semper's theory seems to be based on deceptive resemblances and erroneous interpretations. It is now well established that the Vertebrate tubules are coelomic in origin, of centrifugal growth, and that the ectoderm takes no share in their formation. On the other hand, it is now held that the nephridia are never of coelomic origin, are of centripetal growth from the surface, and probably always derived from the ectoderm, or at least from superficial cells. At first the nephridia have blind inner ends provided with flame-cells or solenocytes (protonephridial stage), and have nothing to do with the gonadial sacs (Platyhelminth and Nemertine stages).

But when the latter become large coelomic sacs the nephridia perforce are pushed into them, become related to them in excretion, and remain in the adult as the main excretory organs (Annelid stage). In some rare cases they may open into the coelom by nephridiostomes and lose the solenocytes (Oligochaeta, some Polychaeta).[1] But in the great majority of the Invertebrate Coelomata the nephridia appear only in early stages of development and are lost in the adult (Mollusca, Phoronidea, etc.), or do not appear at all (Arthropoda, Echinoderma). Thus, in the Invertebrates the nephridium is but rarely preserved as an adult excretory organ ; while the coelomostome is constantly present in all groups as a genital and often as an excretory duct as well.

It is clear that, if the vertebrate tubules are to be compared with any of the organs of an Invertebrate, it is not with the nephridia but with the coelomostomes that they must be homologised.

The argument would be clinched could we point to a Vertebrate possessing both well-developed nephridia and coelomic tubules. Unfortunately no such form is known to exist at the present day. The Craniates have preserved the coelomostomes, but lost the nephridia,[2] no longer necessary since the former have taken on the function of excreting. On the other hand, the Cephalochorda have preserved the nephridia (Boveri, Weiss), which are now known to be of typical protonephridial structure without internal openings and provided with well-developed solenocytes (Goodrich, 1017). But the genital ducts of the Cephalochorda appear to have been lost, the genital cells escaping by bursting through the wall of the genital sacs into the atrium.[3]

One more point remains to be considered : If the segmental tubules of the Vertebrates are homologous with the genital ducts of other Coelomata, is it the peritoneal funnel or the nephrocoelostome which represents the primitive coelomostome ? Many embryologists consider that the cavity

[1] In the vast majority of the Coelomate Invertebrates the nephridia and the coelomostomes have nothing to do with each other. In one class only, the Polychaeta, a connexion may be established between them, leading to the formation of a complex organ (nephromixium) serving for the exit of both excretory waste and genital products (Goodrich, 1899–1900). But even in this class there are forms where these organs retain their primitive independence (Capitellidae, Nereidae).

[2] Unless, indeed, these are represented by the thymus, as held by van Wijhe.

[3] A vestige of the primitive coelomostome may, however, be represented by the small knob or ' hilum ' formed by the fusion of the mesodermal wall of the genital sac with the atrial epithelium at the point where the rupture takes place. A somewhat similar reduction of the genital ducts to mere pores has been described above (pp. 704-7) in Teleostei and Cyclostomes. Among Invertebrates also the coelomoducts may be much reduced to quite short funnels or pores (oviducts of many Oligochaetes, genital funnels of many Polychaetes). Or, just as in *Amphioxus*, these vestigial ducts may cease to open to the exterior, the genital products bursting through the body-wall; some Polychaeta (such as epitokous forms among Nereidae, etc., and *Clistomastus* among Capitellidae).

of the intermediate cell-mass or stalk of the somite (which becomes the cavity of Bowman's capsule, and is often called the nephrocoele) is a chamber of the coelom itself.   In this case the nephrocoelostome would be the coelomostome opening from it, and the peritoneal funnel a specialised narrow channel of communication between this still segmental chamber and the general ventral splanchnocoele.   On the other hand, the peritoneal funnel may be the coelomostome and the chamber a specialised enlargement of the coelomoduct into which penetrates the glomerular blood-vessel.   Or, again, if the nephrocoelostome is the original funnel, it might by extension ventrally give rise to the peritoneal funnel as well.   There is something to be said for each of these interpretations, and at present it seems scarcely possible to decide between them.

THE PERIPHERAL NERVOUS SYSTEM

THE nervous system has been evolved to conduct nerve-impulses. Sensory
impulses are due to the action of stimuli from the external and internal
environment on receptor sensory cells on the surface of the body or in its
deeper tissues and organs; they are conveyed to the central nervous
system, where they set up excitor or motor impulses which are transmitted
to effector cells and organs, the muscles, and glands. Thus appropriate
responses are called forth, and the various functions of the organism are
integrated to its advantage. The complex system of interconnecting
peripheral nerves convey the sensory impulses to and excito-motor
impulses away from the central nervous system where these impulses are
co-ordinated. The central nervous system consists of the brain and spinal
cord situated dorsally, while the peripheral nervous system is made up of

ganglia and nerves distributed to all parts of the body. The essential elements of which nervous tissue is composed are the neurons or ganglion cells, with their branching conducting processes or fibres, of which the longest is called the axis-cylinder, axon, or neurite, and the others the dendrites. Both axon and dendrite may end in fine twigs for receiving or transmitting impulses. The neuron receives impulses by its dendrites and transmits them by its axon. Usually the axon does not divide till at or near its destination. The nerve-impulse passes from neuron to neuron, where the tips of the processes of one neuron meet the cell-body or processes of another. At this point of junction, known as the synapse, there would appear to be mere contact and not actual permanent continuity of the conducting fibril.[1] According to the ' Neuron theory ' of His and Waldeyer, now generally accepted, the whole nervous system, with the possible exceptions mentioned below, consists of chains of such neurons along which the impulses can pass when they overcome the resistance offered by the synapse. While it seems certain that the bulk of the nervous system, at all events of the Craniata, is built on this plan, yet there remain certain delicate nerve-plexuses on the blood-vessels, under the mucous membrane of the buccal cavity (of Amphibia and probably other forms), and in the wall of the gut, which apparently do not conform to it (Bethe, 1903 ; Prentiss, 1904 ; E. Müller, 1151-2). In these cases it seems possible that a true nerve net exists of anastomosing fibrils continuous from cell to cell, such as commonly occurs in the lower Invertebrata. In addition to the true nervous elements there are in the central nervous system packing neuroglia cells, and sheath cells on the fibres of the peripheral nerves. The nerves of anatomy are made up of bundles of such axons and their sheaths, the cell-bodies of the neurons being usually gathered either in the central nervous system or in ganglia outside it.

The first generalisation of importance, then, is that the peripheral nerves consist of two sets of nerve-fibres : one set of sensory or afferent fibres whose function is to carry impulses centripetally to the central nervous system, and another set of excito-motor or efferent fibres whose function is to carry impulses centrifugally from the central nervous system to muscles and glands. A nerve may be composed of either or of both kinds of fibre, Fig. 744.

It has already been explained (Chapter V.) that every segment of the body of a Vertebrate is primitively provided with a pair of dorsal and of ventral nerve-roots by which the nerve-fibres pass to or from the central nervous system. Moreover, every segment contains a pair of mesoblastic

---

[1] Whether the conducting fibril or substance is or is not ever continuous from one neuron to another is, however, still a matter of dispute.

segments, from each of which are developed a dorsal myotome and ventral lateral plate (p. 4). From the myotomes, which retain more or less their original segmentation, are derived the segmental muscles or myomeres of the adult (also the hypoglossal musculature, and the muscles of median fins and paired limbs). From the lateral plates, which lose their segmentation by longitudinal fusion either very early (*Amphioxus*) or from the very first (most Craniates), is derived the 'unsegmented' lateral plate mesoblast; this (together with mesoblast derived from sclerotome outgrowths of the dorsal region) gives rise to all the mesoblastic tissues of the body, excepting the myomeres. From these tissues, generally denoted as the unsegmented mesoblast, are therefore developed the contractile elements of the vascular system, certain muscles of the skin, the whole of the musculature of the alimentary canal and its appendages including the visceral muscles of the jaws and gill arches, and of the urinary and genital organs.

Returning to the description of the peripheral nerves, we find that primitively, as in *Amphioxus* and the Petromyzontidae, the dorsal and ventral roots are independent nerves; that in all Gnathostomes and in the Myxinoidea the dorsal and ventral nerves of a segment combine on each side in the spinal region to form compound spinal nerves each with a dorsal and a ventral root;[1] that in the cranial region the dorsal and ventral nerves remain separate.[2] Further important generalisations may be made: that the afferent fibres all reach the central nervous system by the dorsal roots, and that the cell-bodies of these sensory neurons are all situated in the segmental, cranial, and spinal ganglia outside the neural canal on these roots (except in *Amphioxus*); that the efferent fibres may pass out by both the dorsal and the ventral roots, and that the cell-bodies of these excito-motor neurons are all situated inside the central nervous system, with the exception of those belonging to the 'sympathetic' system as explained below (p. 772). It follows that while the ventral roots are, so far as we know, purely efferent, the dorsal roots may be of mixed character, Figs. 707, 747.

Although it must be supposed that originally the dorsal roots all along the body contained both afferent and efferent fibres, as they do in *Amphioxus*, a divergence has come about in the Craniata between the segmental

---

[1] These primary segmental roots must not be confused with the secondary rootlets into which they may be differentiated; for in the higher Craniata the fibres of the various components of the cranial nerves may become gathered together and more or less separated into distinct rootlets. Thus the rootlet of the visceral motor component is often confused with a primary ventral motor root.

[2] Except in so far as they may be connected by sympathetic fibres.

roots of the head and of the rest of the body ; so that while the dorsal cranial nerves of Craniates contain efferent fibres, they are stated to be totally absent from the dorsal roots of their spinal nerves, Figs. 745, 750. From time to time, however, it has been maintained that efferent

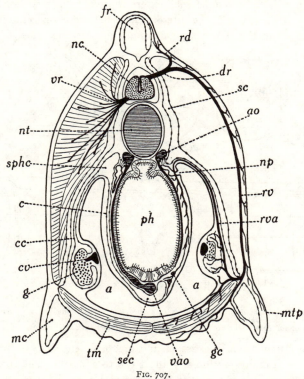

FIG. 707.

Diagrammatic transverse section of *Amphioxus* in pharyngeal region, showing peripheral nerves, coelomic cavities, etc.   Primary gill-bar on left, secondary on right.   *a*, Atrium ; *ao*, lateral dorsal aorta ; *c*, coelomic canal in primary bar ; *cc*, coelomic canal ; *cv*, posterior cardinal vein ; *dr*, dorsal root nerve ; *fr*, fin-ray ; *g*, gonad ; *gc*, ganglion cell ; *mc*, metapleural cavity ; *mtp*, metapleure ; *nc*, nerve-cord ; *np*, nephridiopore ; *nt*, notochord ; *ph*, pharynx ; *rd*, ramus dorsalis ; *rv*, ramus ventralis ; *rva*, ramus ventralis ascendens ; *sc*, sclerocoele ; *sec*, subendostylar coelom ; *sphc*, lateral suprapharyngeal coelomic chamber ; *tm*, transverse subatrial muscle ; *vao*, ventral aorta ; *vr*, ventral root nerve.

fibres (vasomotor ?) issue through the spinal dorsal roots ; future research may perhaps show that the primitive efferent component remains, at all events, in some of the lower forms (*Amphioxus*: Willey, **94** ; Heymans and v. d. Stricht, **1122** ; Dogiel, **1104** ; Kutchin, **1141** ; Franz, **1109**).

Amphioxus has no well-defined ganglia on its dorsal nerves, for the nucleated cell-bodies of the afferent fibres are still inside the spinal cord or scattered along the root of the nerve to near its bifurcation (Rohde, **1169** ; Hatschek, 1882 ; Johnston, **1127**).[1]

Concerning the general function and distribution of the sensory fibres it may here be mentioned that Sherrington distinguishes three main reception fields: the superficial extero-ceptive field receiving stimuli from the external world, the intero-ceptive field receiving stimuli at the inner surface of the alimentary canal, and the deep proprio-ceptive field

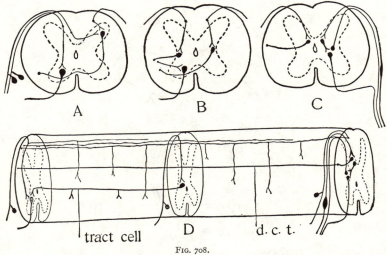

FIG. 708.

Diagrams illustrating several forms of reflex chains (from J. B. Johnston, *Nerv. Syst. of Vert.*, 1908). A, Somatic sensory and motor ; B, by way of tract cells ; C, visceral sensory and motor ; D, spinal cord and nerve roots from side. *d.c.t*, Direct cerebellar tract.

receiving stimuli from the tissues and organs in the body. In each of these fields stimuli may directly affect the terminal branches of sensory fibres, free nerve-endings ; or be received by special sensory cells or characteristic sense organs containing sensory cells specialised to respond to particular stimuli.

The afferent and efferent fibres, then, serve as paths for 'reflexes' or reflex arcs completed through the brain or spinal cord. The nerve-impulse passes along a chain of neurons ; entering by the sensory neuron, whose body is in the segmental ganglion, and emerging by the excito-

[1] It should also be noticed that in *Amphioxus* the sensory cells in the epidermis are said to send conducting fibres directly to the central nervous ·system, as in the Invertebrata ; whereas in the Craniata the sensory impulses, except in the case of the nasal olfactory epithelium, are transmitted by the axons of afferent neurons whose distal branching extremities are closely related to the sensory cells. In these respects *Amphioxus* seems to be, as in so many of its characters, more primitive than the Craniata.

motor neuron, whose body is in the central nervous system, Figs. 708, 747. The connexion between these two is made by one or more intermediate neurons situated entirely in the central nervous system (except in the case of the 'sympathetic' system, see below, p. 772). The behaviour of vertebrates is to a great extent made up of the co-ordination of such reflexes.

Lastly, an important point remains with regard to the function of the dorsal and ventral motor fibres (leaving out of account those belonging to the 'sympathetic' system); those of ventral roots supply only the muscles derived from the dorsal segmented myotomes (including, of course, the eye-muscles, hypoglossal muscles, and limb-muscles); while the motor fibres issuing by the dorsal roots innervate only the musculature derived from the ventral unsegmented mesoblast derived from the lateral plate.

The segmental value of the cranial and spinal nerves has been already dealt with above (Ch. V.). While comparative anatomists and embryologists were working out this interesting problem, other observers were studying the peripheral nerves more from the point of view of the function of the various neurons, and the peripheral distribution and central connexions of the fibres. Great advances have been made since the introduction of the silver impregnation technique of Golgi, the methylene blue staining of Ehrlich, and Waller's method of tracing the path of degenerating fibres severed from their cell-bodies. By these new methods, together with that of the artificial stimulation of nerve-fibres in the hands of many skilful experimentalists, the course of most of the important sets of fibres has been made out. Gaskell in 1886 also traced their origin and distribution by their histological characters, distinguishing somatic from splanchnic or visceral fibres (Merritt, 1147).

A new impetus was given to the comparative study of the functional components of the nervous system by the work, more especially, of Strong, who in 1895 made a complete analysis of the cranial nerves of the larval amphibian by the reconstruction of serial sections. Since then Strong (1176), Herrick (1120-21), Johnston (1126-30), Norris (1155-8), Willard (1184), and others have successfully applied the method to various Vertebrates from *Amphioxus* upwards. These detailed researches combined with the more purely anatomical studies of the older anatomists, such as Stannius (1849) and Fischer, 1843–54, and the modern work of Ewart and Mitchell on Selachians (1107), of Allis on *Amia* and other fishes (402, 404-5, 1081-5), and of Fürbringer (340), have helped to build up the important doctrine of Functional Components, of which the chief conclusions may now be summarised as follows :

The peripheral nervous system contains four chief components, two

sensory and two excito-motor, each subserving special functions, related
to special ' end-organs ', and connected to four corresponding longitudinal
regions of the central nervous system.   These regions appear in the spinal
cord in the form of ' columns ', of which the two sensory are dorsal and
the two motor ventral, and extend forward into the brain, where, however, they become much modified and complicated in higher forms.
The four component systems are known as the Somatic Sensory, the

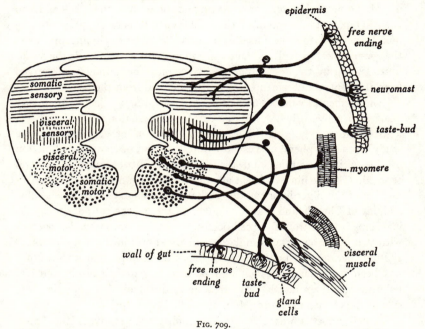

FIG. 709.

Diagram showing central origin from medulla and peripheral distribution of systems of *nerve-
components*.

Visceral Sensory, the Visceral Excito-motor, and the Somatic Motor,
Fig. 709.

Speaking generally, the somatic components are chiefly concerned with
responses to the animal's external environment, while the visceral components are chiefly concerned with responses to its internal environment ;
that is to say, with the internal processes of digestion, respiration, circulation, excretion, and reproduction.   But no hard-and-fast line can be
drawn between them on this account.

The primary sensory components may become subdivided owing
to each developing characteristic sense organs, such as the lateral-
line organs related to the Somatic Sensory component, and the taste-

buds related to the Visceral Sensory. The fibres belonging to each system present in a nerve are known as the components of that nerve, Fig. 728.

The functional divisions of the nervous system and components of the Craniata may be further described as follows :

A. **Somatic Sensory System.**—1. *General cutaneous* : Afferent fibres from the whole epiblastic surface of the body, the extero-ceptive field ;

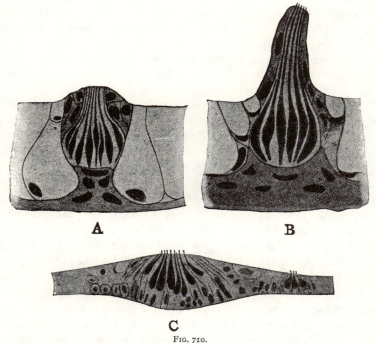

FIG. 710.

A, Taste-bud from oesophagus of *Catostomus* at time of hatching; B, taste-bud from pharynx of same ; C, two neuromasts from skin of same. (From J. B. Johnston, *Nerv. Syst. of Vert.*, 1908.)

with ' free nerve-endings ' among the epidermal cells, and more specialised end-organs, such as the corpuscles of Meissner and of Merkel, the end-bulbs of Krause, etc. These receptors are stimulated by pressure and vibrations (touch, radiant heat, light, cold, and pain). The cell-bodies of the neurons are in the cranial and spinal ganglia, and the fibres distributed primitively in every segment of the body by the dorsal cranial, and the spinal nerves (nervi profundus, trigeminus, facialis, glossopharyngeus, vagus, et spinales). The central connexions are with the dorsal horn or column of the spinal cord and its continuation in the brain, with associated

' nuclei ' (tractus spinalis trigemini of medulla oblongata, nucleus funiculi, n. acusticum, cerebellum, etc.).

2. The special visual organs, paired lateral eyes and dorsal pineal organs, were probably evolved from this system ; but were differentiated and have remained within the original wall of the brain.

3. *Special cutaneous or acustico-lateral system* : Afferent fibres related to neuromasts or organs of the ' lateral line ' (also pit-organs and ampullae), receptors in aquatic forms of relatively slow vibrations. Neuromasts possess superficial pear-shaped sensory cells which do not extend through the whole depth of the epithelium and bear sensory hairs on their outer surface, Figs. 710, 713.

The auditory organ represents a specialised region of the Special Cutaneous system, often therefore called the acustico-lateral or more shortly the lateralis system.

The acustico-lateral fibres are distributed typically by the facial, auditory, glossopharyngeal, and vagus nerves ; and their central connexions are chiefly with the nucleus acusticum and n. funiculis, and cerebellum.

This acustico-lateral system is dealt with more in detail below (p. 732).

4. With the General Cutaneous system may be associated the related afferent fibres of the ' muscular sense ' of the proprio-ceptive field. They have fine nerve-endings among the mesoblastic tissues, bones, tendons, and muscles, also Golgi organs and Pacinian bodies. The ' muscle spindles ' of striated muscle fibres derived from myotomes and from the lateral plate are end-organs of this system. The course of these fibres and their central connexions are similar to those of the General Cutaneous system.

B. **Visceral Sensory System.**—1. *General visceral* : Afferent fibres from the lining of the alimentary canal, the intero-ceptive field ; with free nerve-endings responding to mechanical stimuli. The cell-bodies of the neurons are in the dorsal cranial ganglia of the facial, glossopharyngeal, and vagus nerves, and the fibres are typically distributed in their visceral branches. The afferent fibres enter the central nervous system by the morphological dorsal roots and connect with the fasciculus communis or solitarius, lobus vagi, l. fascialis, and associated centres, and Clarke's column in the spinal cord. Sometimes known as the splanchnic or the communis system.

2. The special paired olfactory organ, whose sensory cells send conducting fibres to the bulbus and tractus olfactorius, should probably be considered as belonging to this system. Affected by chemical stimuli, the olfactory organ may have been primitively intero-ceptive ; but in most Craniates it acts chiefly as an extero-ceptor.

3. *Special visceral* : Afferent fibres related to taste-buds (end-buds),

typically situated on the mucous surface of the buccal cavity and pharynx, Fig. 710. Groups of special sensory cells whose bodies extend through the whole depth of the epithelium form these taste-buds which are receptors of chemical stimuli (from sweet, sour, salt, and bitter substances). The fibres enter the brain by visceral branches of dorsal cranial nerves (facial, glossopharyngeal, and vagus), the cell-bodies of the neurons are in their ganglia, and the central connexions are with the same centres as those of the General Visceral system (p. 730).

4. Associated with the General Visceral sensory system are afferent fibres travelling in the ' sympathetic ' nerves (p. 770). They have free nerve-endings among the smooth muscle-cells and glands supplied by the ' sympathetic ' system.

C. **Somatic Motor System.**—Efferent fibres issuing by ventral roots and supplying all the musculature derived from the segmental myotomes. These are the myomeres or body-wall muscles of the tail and trunk, and their derivatives the muscles of median and paired fins and limbs ; the muscles of the Mammalian diaphragm ; the dorsal epibranchial muscles of Elasmobranchs ; the hypoglossal musculature connected with the branchial arches and tongue and derived from certain myotomes of the gill region (ventral hypobranchial muscles, etc.) ; the external muscles of the eye-ball (modified anterior myotomes). The end‑organs are the ' motor end plates ' on the muscle fibres which are all striated. The cells giving origin to these somatic motor fibres are in the ventral horn of the spinal cord, and the corresponding nuclei of the hypoglossal, abducens, oculomotor, and trochlear cranial nerves. Their action is under the control of the will.

D. **Visceral Motor System.**—1. *Special visceral motor* : Efferent fibres, issuing by dorsal roots, and supplying muscles derived from the lateral plate mesoblast. These are the visceral constrictor muscles and their derivatives, including the muscles of the jaws, various muscles of the hyoid and branchial arches, and trapezius. The end-organs are ' motor end plates ' on the muscle fibres which are striated. They are under the control of the will, and the cell-bodies are situated in the lateral horn or intermediate zone of the medulla. The fibres issue by the trigeminal, facial, glossopharyngeal, and vagus (including the spinal accessory) cranial nerves.

2. ' Sympathetic ' or Autonomic excito-motor system of efferent fibres supplying glands and muscles (mostly unstriated and derived from the lateral plate mesoblast), and issuing by both dorsal and ventral roots. This system is described below.

This analysis of the peripheral nerve-fibres into components according to the function and distribution of their receptor and effector end-organs, and to their connexion with longitudinal zones of the central nervous system, has thrown much light on the structure and evolution of the nervous system in Vertebrates.   Passing from *Amphioxus* to Man we can trace the rise or the loss of components.   It offers an intelligible explanation of the great variation in the number and size of their nerve-branches within even small groups.   For a nerve varies according to the number of fibres of the components contained in it, and these again according to the abundance of the end-organs they supply.   Not only the size but the very presence or absence of a nerve in a particular animal depends on whether these end-organs are developed and needed for its life.   Nevertheless, there are weak points in the classification of the components in four main systems, and some of them fit with difficulty into the classification adopted by physiologists.   Especially is this so with some of the visceral components, which are better treated separately (see 'Sympathetic' System, p. 770).

Taste-bud System.—There are points of interest concerning the Special Visceral or Taste-bud system.   In the terrestrial Vertebrates the taste-buds are restricted to the internal lining of the anterior region of the alimentary canal.   For instance, in Mammalia they are scattered over the tongue, especially near the circumvallate and foliate papillae, on the soft palate, and sparsely on the larynx and epiglottis.   Of these the more abundant anterior buds are supplied by the facial nerve, the most posterior by the vagus, and some intermediate ones by the glossopharyngeal.   In most fishes they occur distributed over the roof, sides, and floor of the buccal cavity and pharynx (in some Teleosts even in the oesophagus).   But in *Petromyzon* they may also be found in the adult on the outer surface of the head and branchial region (Johnston, 1128).   In adult Gnathostomes they also tend to spread over the external surface (Herrick, 1120 ; Johnston, 1129).   *Amia*, many Teleostei, and some Amphibia have taste-buds not only internally, but also scattered over the head ; in some Teleosts (such as Siluriformes and Gadiformes) they extend to the base of the fins, and finally over the whole body including the tail.   This great increase in the area of distribution of taste-buds is, of course, accompanied by a corresponding increase in this component of the facial nerve, Fig. 711, and of its centres in the brain (Herrick, 1121).   In these aquatic forms the external taste-buds become extero-ceptive organs for the search of food.

Now the question arises whether these sense organs are really of endodermal origin and have spread outwards through the mouth by an outgrowth of endodermal tissue ; or are ectodermal organs which developed originally in that part of the buccal cavity derived from the stomodaeum and consequently lined by ectoderm. Thence, by growth of ectodermal tissue (possibly also through the gill-slits), they might have spread backwards into the pharynx and outwards over the body.[1] In this connexion it is interesting to note that excepting for taste-buds no receptor cells are known to occur in the endodermal lining of any Vertebrate ; and, since it is generally held on good evidence that all sensory and nervous cells are ultimately derived from the ectoderm (p. 757), it would require very rigid proof to establish that the taste-buds are an exception to this rule. Unfortunately, it is not possible accurately to determine the limit between ectoderm and endoderm in the buccal cavity of Craniates in the adult or

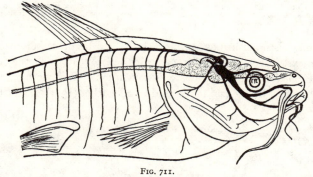

Fig. 711.

Cutaneous gustatory branches (special visceral component) of facial nerve of *Ameiurus* (after C. J. Herrick, from J. B. Johnston, *Nerv. Syst. of Vert.*, 1908).

even soon after the mouth has broken through, as all trace of the oral separating membrane usually disappears. No doubt the position of the hypophysial invagination indicates an ectodermal area, but with rare exceptions (*Polypterus*, p. 235) the connexion does not persist and the hypophysis is indeed separated off at a quite early stage of development. Dorsally, however, the limit is probably at the pituitary region. The same difficulty applies to the limit between the ectoderm and endoderm in the gill-slits. It may therefore be argued that ectodermal cells migrate into the endodermal lining of the pharynx even perhaps before the breaking through of the mouth or slits. Nevertheless, it seems highly probable that the taste-buds first arose in the endoderm and spread through the mouth on to the surface of the body. This is the view maintained by Johnston (**1129**), who finds that in *Petromyzon*, and the Teleosts *Catostomus* and *Coregonus*, they occur in the young only in the pharynx, and appear

[1] Such a spreading of the area supplied by a sensory component commonly occurs (see pp. 743 and 747).

afterwards on the lips, then on the head and body. Landacre (1142) believes the internal taste-buds to be of endodermal and the external

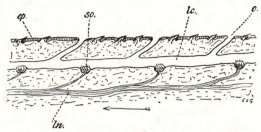

FIG. 712.

Diagram of the lateral-line canal of a Selachian seen in a section vertical to the surface. *ep*, Epidermis; *lc*, longitudinal canal; *ln*, lateral-line nerve; *o*, opening of branch canal on surface; *so*, sense organ. (From Goodrich, *Vert. Craniata*, 1909.)

(sometimes distinguished as ' terminal buds ') to be of ectodermal origin in *Ameiurus*. Cook and Neal (1102), after a careful study of the development of *Squalus*, support Johnston's conclusion. This view is easier to adopt if, as held by Botezat (1910) and G. H. Parker (1912), the taste-buds

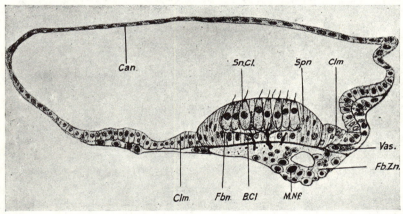

FIG. 713.

Transverse section of lateral-line canal of *Mustelus canis* (from S. E. Johnson, *J. Comp. Neurology*, 1917). *Can*, Canal wall; *Clm*, columnar supporting cell; *Fbn*, terminal fibrillae; *Fb.Zn*, longitudinal fibre-zone; *M.N.F*, medullated nerve-fibre; *Sn.Cl*, secondary sense cell; *Spn*, spindle-shaped supporting cell; *Vas*, blood-cell.

contain not true sensory cells, but specialised gland cells which on stimulation secrete a substance acting on the afferent nerve-endings.

Acustico-lateral System.—The acustico-lateral or neuromast system is of peculiar interest. It is present in all Craniates from Cyclostomes to Man, and is found in the earliest fossil fish of the Silurian. While typically developed only in aquatic forms, it survives in others as the inner ear; for it becomes differentiated into two parts : (1) the neuromasts of the lateral-line system proper, and related superficial organs, and (2) the auditory labyrinth and its contained sensory organs. The first is stimulated by slow vibrations of the watery environment, and serves for the orientation of the body in relation to waves and currents (Séde de Lièoux, 1884 ; Parker, 1160). The second has two distinct functions—equilibrium in space or balance, and hearing. The semicircular canals of the labyrinth are chiefly concerned with balance, while vibrations of high frequency (sound-waves) affect the sensory cells in the sacculus and lagena of lower forms

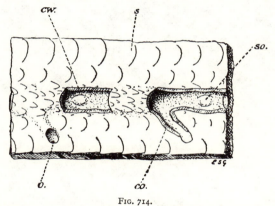

FIG. 714.

Gadus morrhua, L. Surface view of the skin, enlarged, showing the lateral line. *co*, Cut wall of canal to opening ; *cw*, cut wall of main canal ; *o*, opening ; *s*, scale ; *so*, sense organ (neuromast). (From Goodrich, *Vert. Craniata*, 1909.)

and the spirally coiled cochlea of Mammalia. The view that the sense organs of the internal ear are specialised deeply sunk neuromasts was put forth by Beard (1088) and Ayers (1086). It is supported by much weighty evidence, though no sufficiently primitive Craniate is yet known to show intermediate steps between the two kinds of organ. They both develop from similar dorso-lateral placodes (p. 765), both have sense cells provided with sensory hairs receiving stimuli from a liquid medium (water or endolymph), both tend to sink away from the surface, and both are innervated by fibres from corresponding and related centres in the brain.

The special organs of the lateral-line system proper, neuromasts of Wright (1186), are distributed over the body, typically along lines forming a definite pattern on the head and extending along each side of the body to the tip of the tail. The neuromasts of Cyclostomes are exposed on the surface, but in primitive Gnathostomes they are usually sunk in a closed canal embedded in the dermis and opening at intervals by tubes to the exterior (Leydig, 1851–68 ; Allis, **402** ; Herrick, **1121**). As a rule the tube opens into the canal between successive neuromasts in Selachians and opposite them in Teleostomes, Figs. 712-15. Sometimes the primary external pores become subdivided into secondary pores, as in *Amia*, *Lepidosteus*, and some Teleosts (Allis, **402**, **404-5**). In *Chlamydoselachus*, however, the main line of the trunk is in the form of an open groove, Fig. 27 (Garman, **1111**) ; also in *Chimaera*, where the edges of the groove do not

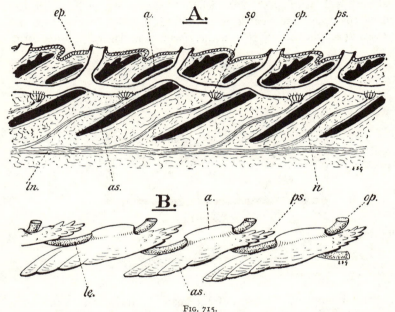

FIG. 715.

Diagrams showing the relation of the lateral-line canal to the scales on the trunk of *Perca fluviatilis*, L. A, Longitudinal section ; B, the scales and the canal seen from the side. *a*, Arch covering the canal ; *as*, anterior region of scale ; *ep*, epidermis ; *lc*, lateral-line canal ; *ln*, lateral-line nerve ; *n*, nerve to sense organ ; *op*, external opening of canal ; *ps*, posterior ctenoid edge of scale ; *so*, sense organ in canal. (From Goodrich, *Vert. Craniata*, 1909.)

completely close even on the head. Many of the more specialised Teleosts, and all modern aquatic Amphibia (Urodela and larval Anura and Apoda), have the neuromasts naked on the surface ; likewise the Dipnoi except on the head. This condition is no doubt secondary, at all events in Osteichthyes, for canals are found in their early fossil representatives, and even in Ostracodermi of the Silurian.

In the canals on the head the neuromasts may vary greatly in number, but along the main trunk line they are usually segmentally disposed.in Pisces. Now, in the Osteichthyes, where bony dermal plates are developed on the head and shoulder girdle, and scales on the remainder of the body,

the canals enclosing the neuromasts necessarily become more or less completely enclosed in these plates and scales, Fig. 715. There is thus
established a definite relation between the lateral-line system and
the exoskeleton (p. 285). In all primitive Osteichthyes, the main lateral-
line canal of the body pierces every scale along its course, passing from
its upper surface in front to its lower surface behind. A neuromast is
placed between each pair of scales, and a tube leads to a pore outside
each scale. The same structure may be seen in the secondary longitudinal lateral lines often developed dorsally and ventrally as branches
of the main line. Anteriorly the main line passes forwards through the
dorsal elements of the shoulder girdle (post-temporal, etc.) on to the head.

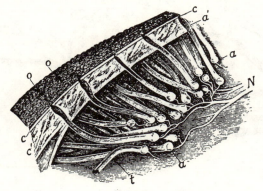

FIG. 716.

Portion of the snout of *Scyllium* in section, showing ampullary tubes. (After Gegenbaur, from
Sedgwick's *Zoology*.) *a*, Ampulla; *a′*, passage of a tube through the dermis; *c*, epidermis; *c′*, dermis;
N, nerve; *o*, external openings of tubes; *t*, tube.

This burial of the canals in the superficial skeleton occurs not only in
the earliest Teleostomes and Dipnoi, but also in the Palaeozoic Ostracodermi (Pteraspidomorphi, Cephalaspidomorphi, Pterichthyomorphi) and
Coccosteomorphi. In the Stegocephalia it was apparently less complete,
as the course of the canals in these extinct Amphibia is at most marked by
superficial grooves on the skull, Fig. 321. When, as in the later more
specialised forms (Dipnoi, modern Amphibia, and many Teleostei), the
dermal bones tend to sink far below the skin, the canals become again free
in the superficial dermis, or, as already mentioned, the neuromasts come
to lie naked on the surface.

There are present in Pisces other surface sense organs besides typical
neuromasts; pit-organs, ampullae, vesicles, nerve sacs. As Herrick
(1903) and Johnston (1129) have shown, all these organs belong to the
acustico-lateral system, have essentially the same structure as neuromasts,

are innervated by the same component as the lateral-line organs, and are connected with the same centres in the brain, Figs. 717-18, 720.

Pit-organs generally appear in Selachians scattered or in rows on the dorsal surface of the head, along the hyoid arch, and dorsally along a line extending from the head backwards along the trunk (Allis, 412 ; Norris and Hughes, 1158). Teleostomes have more or less complete lines of pit-organs in approximately the same positions (Herrick, 1120 ; Allis, 1081). They are well developed in *Amia* and occur in addition on the cheek, mandible, and median gular. Lines on the cranial bones of fossil Dipnoi and Teleostomes seem to indicate their pre-

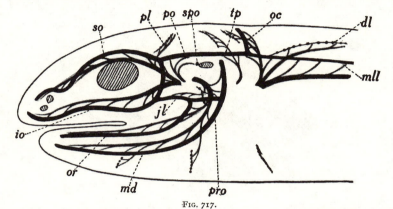

FIG. 717.

Left-side view of head of generalised Fish showing distribution of *lateral-line organs* and canals (black), and their *nerve-supply* (grey). *dl*, Dorsal line of pit-organs ; *io*, infraorbital canal ; *jl*, jugal canal ; *md*, mandibular canal ; *mll*, main canal of trunk ; *oc*, transverse occipital canal ; *or*, oral canal ; *pl*, anterior of three lines of pit-organs—similar lines occur ventrally ; *po*, post-orbital canal ; *pro*, preopercular or hyomandibular canal ; *so*, supraorbital canal ; *spo*, spiracular neuromast ; *tp*, temporal canal. For nerve-supply see Figs. 720 and 728.

sence in early forms. Sometimes these lines of neuromasts sunk in separate pits appear to represent in one form a true lateral-line canal in another.

The ampullae of Lorenzini of Selachians and Holocephali, on the other hand, are more specialised neuromasts sunk far below the surface in groups, above and below the snout, on the upper and lower jaws, and on the hyoid arch, Figs. 716, 718. A long tube leads from the swollen base of each to the external pore, which marks the point at which the organ was first developed (Sappey, 1880 ; Allis, 1082a). Doubtless the vesicles of Savi found in *Torpedo* are similar organs which have become separated off from the epidermis.

No ampullae occur in Osteichthyes ; but Herrick (1903) has described, besides the lateral-line neuromasts and pit-organs, a third set of ' small 'pit organs' scattered over the body of Siluroids, and 'nerve-sacs' are

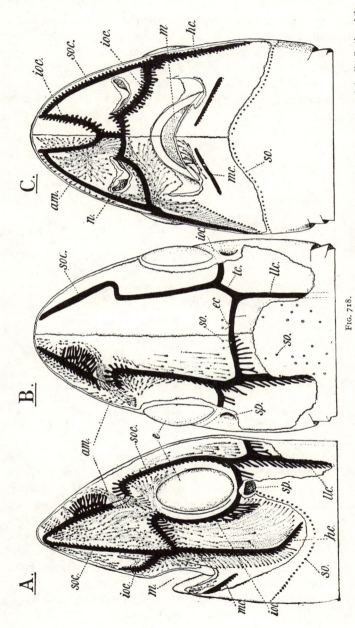

Fig. 718.

Mustelus laevis, Risso. A, side view, B, dorsal view, and C, ventral view, of the head, showing the course of the lateral-line canals, and the distribution of the ampullae (shown on the left side only). *Am*, Ampullae of Lorenzini; *e*, eye; *ec*, endolymphatic canal opening to the exterior; *hc*, hyomandibular canal; *ioc*, infraorbital canal; *llc*, lateral-line canal of the trunk; *mc*, mandibular canal; *m*, mouth; *mc*, mandibular canal; *n*, nostril; *so*, sense-organ; *soc*, supraorbital canal; *sp*, spiracle; *tc*, postorbital canal. (After Allis, from Goodrich, *Vert. Craniata*, 1909).

3 B

found below the skin in some forms. These may represent the ampullae of Selachians.

A complex pattern is formed by the lateral-line system on the head, remarkably constant, except for minor modifications, throughout the Gnathostomes. It is subdivided into regions innervated by the facial, glossopharyngeal, and vagus nerves by means of which three nerves (leaving aside the auditory) the fibres of this component are distributed not only over the head, but over the rest of the body as well. The course

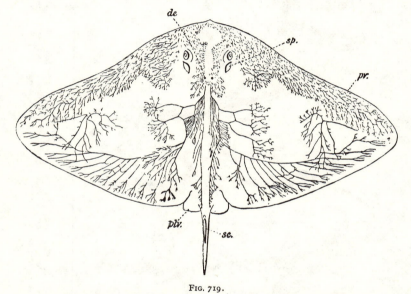

FIG. 719.

Dorsal view of *Pteroplatea Valenciennii*, Dum. (after Garman), showing the extensive development of the lateral-line organs. *de*, endolymphatic openings ; *pv*, pectoral fin ; *plv*, pelvic fin ; *sc*, spine ; *sp*, spiracle. (From Goodrich, *Vert. Craniata*, 1909.)

and nerve-supply of the canals or lines have been admirably described by many authors, among whom may be specially mentioned the works of Ewart on Selachians, Allis on *Amia*, and Herrick on *Menidia*. (Canals and innervation : Ewart and Mitchell, 1107, Allis, 1082a, Norris and Hughes, 1158, on Selachians ; Cole, 1099, on *Chimaera* ; Norris and Collinge, 476, on Chondrostei and *Lepidosteus* ; Pollard, 575, Allis, 1082, on *Polypterus* ; Herrick, 1120, Pollard, 1164, Cole, 1099a, Guitel, 1891, Allis, 404-5, in Teleostei ; Pinkus, 1161, in Dipnoi ; Strong, 1176, Coghill, 1098, Norris, 1155, Escher, 1106, in Amphibia. Courses of lateral line : Sappey, 1880, Garman, 1111, Reese, 1167, in Elasmobranchs ; Collinge, 477-8, Goodrich, 35, 518, Stensiö, 218, 606, Watson, 646, in fossil Teleostomes ; Malbranc, 1146, in Amphibia ; Moodie, 548-50, in Stegocephalia, Figs. 717-21, 725.

It will be readily seen from the figures that the general plan in Pisces is for the main canal or line of the trunk to run forward on to the head where it is continued to near the orbit. Here it divides one branch running above and the other below the eye to the snout. A canal may run across the cheek, and be continued along the lower jaw ; while another runs down the hyoid arch, and below the lower jaw. A dorsal

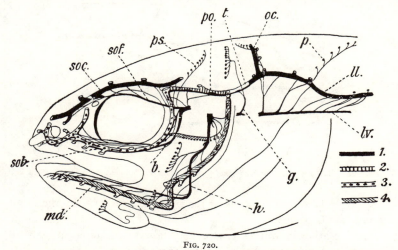

FIG. 720.

Diagram of the head of *Amia calva*, L., showing the system of lateral-line canals and pit-organs, and their nerve-supply (from Allis). *b*, Buccal branch of facial nerve ; *g*, dorsal branch of glosso-pharyngeal ; *h*, hyomandibular branch of facial ; *ll*, lateral line of trunk ; *lv*, lateral-line branch of vagus ; *md*, mandibular canal ; *oc*, occipital or supratemporal canal ; *p*, pit-organs on trunk ; *po*, postorbital canal ; *ps*, pit-organs on head ; *sob*, suborbital canal ; *soc*, supraorbital canal ; *sof*, superior ophthalmic branch of facial ; *t*, temporal canal. The system of distinguishing the canals is indicated on the right of the figure ; 1, supraorbital, and main canal of trunk ; 2, postorbital ; 3, suborbital ; 4, hyomandibular. (From Goodrich, *Vert. Craniata*, 1909.)

branch runs up across the occipital region. The chief parts of the head system have received names, the transverse occipital being known as the supratemporal, the horizontal and suborbital as the infraorbital, the part running above the orbit as the supraorbital, and the ventral branch as the hyomandibular, its continuation on the lower jaw being sometimes called the mandibular (in Osteichthyes these are often named preopercular and mandibular). But a more satisfactory nomenclature can be given, taking the nerve supply into account. The supraorbital line is supplied by the superior ophthalmic branch of the facial nerve ; the infraorbital by its buccal branch ; the hyomandibular and mandibular by the truncus hyomandibularis of the facial ; the jugal and oral by branches of the latter nerve ; the transverse occipital by the ramus supratemporalis of the vagus. There remains the horizontal line from the orbit to the main line of the trunk. This region is generally included in the ' infraorbital ' ; but, since between the anterior part supplied by the otic branch of the facial and the beginning of the main canal there is generally intercalated a short region innervated by the supratemporal branch of the glossopharyngeal, it would seem better to name these two horizontal regions the postorbital and temporal respectively (Goodrich, 35) as indicated in

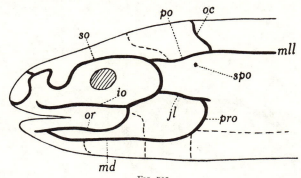

FIG. 721.

Diagram of lateral-line system of *Dipnoi* ; canals thick black lines ; pit organs broken lines. Lettering as in Fig. 717.

Figs. 717, 720.[1] The main line of the trunk is innervated by the ramus lateralis of the vagus.

Specialisation in various groups may lead to the interruption, subdivision, branching, and fusion of these primary lines.

The supraorbital canal of Selachians becomes much developed on the rostrum, and joins the infraorbital, which may pass forward between the mouth and the nostril (*Squalus, Mustelus*). A transverse commissure may be formed by the junction of the right and left occipital canals, while the hyomandibular fails to meet the postorbital but runs forward longitudinally (jugal canal) to join the infraorbital, the mandibular being separate. Missing parts of the hyomandibular canal seem to be represented by pit-organs.

The occipital transverse commissure is usually completed in Osteichthyes across the middle line. While in Actinopterygii, even in the fossil Palaeoniscoidei (Stensiö, 218 ; Watson, 646), the hyomandibular canal joins the postorbital dorsally, and does not run across the cheek to the infraorbital (being represented here by pit-organs in *Amia*, Allis, 1081). In Dipnoi, Osteolepidoti, and Coelacanthini (Stensiö, 605-6 ; Goodrich,

---

[1] The temporal canal is often very short and sometimes eliminated by the backward extension of the postorbital line, as in *Menidia* (Herrick, 1120).

518-19; Watson 644) there is usually a jugal canal across the cheek as in Selachians, Figs. 721-4.

The distribution of the neuromasts in modern Amphibia (Malbranc, 1146; Kingsbury, 1133; Escher, 1106) agrees in general with that of primitive fishes and is remarkably like that of Dipnoi.[1] The lateral-line grooves on the skull of Stegocephalia often clearly show the jugal connexion of the hyomandibular canal with the infraorbital mentioned above, Fig. 725 (Moodie 548-50).

On the other hand, the Cyclostomes differ considerably from the Gnathostomes in the pattern of the lateral-line system, which always remains superficial (Alcock, 1080; Johnston, 1128; Stensiö, 1926). It is better developed in Petromyzontia than in Myxinoidea, where it was discovered by Ayers and Worthington (1087).

In *Petromyzon* there is a main line running down the side of the body to the end of the tail, and a more dorsal line above. These pass forwards above the gill openings to the hind region of the head. The main line reaches to the orbit and another line extends beyond

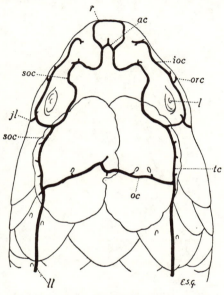

Fig. 722.

Dorsal view of head of *Ceratodus forsteri*, showing lateral line canals seen through skin and over-lying scales. *ac*, Anterior commissural; *ioc*, infraorbital; *jl*, jugal; *ll*, main trunk; *oc*, transverse occipital; *orc*, oral; *r*, rostral; *soc*, supraorbital, and *tc*, temporal canals; *l*, eye.

it; the former represents the postorbital and the latter probably the infraorbital canal. Further forward still is a line representing the supraorbital. Two dorsal transverse lines, one just behind the orbit and the other further back, may represent the dorsal pit-organs and occipital lines of Gnathostomes. A line runs round each side of the oral sucker, and from these extend two longitudinal ventral lines along the gill region. There are also groups of neuromasts between each pair of gill-slits, and

[1] The lateral-line organs and their nerves are eliminated in those Amphibia which adopt a permanent terrestrial life; but persist in those Urodela which retain the aquatic habit, and also, among Anura, in the aquatic Aglossa, and partially in *Bombinator* (Escher, 1106).

between the ventral lines are groups or short transverse lines corresponding to each gill opening. All these neuromasts appear to be innervated by branches of 7th, 9th, and 10th cranial nerves as usual. Stensiö (1926) has recently pointed out that the pattern of the lateral-line canals, first recognised by Lankester in the Pteraspidomorphi, closely resembles that of the lines of neuromasts in modern Cyclostomes; an important piece of evidence in favour of the view advocated by Cope, Woodward, and Stensiö, that these two groups are related. The lateral-line organs of the Pteraspidae are, however, in canals deeply embedded in the bony shields (A. S. Woodward, 663; Goodrich, 35).

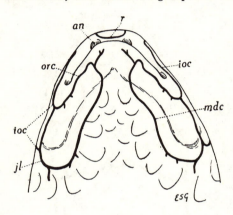

FIG. 723.

Ventral view of head of *Ceratodus forsteri* showing course of lateral-line canals running below skin and scales. *an*, Anterior nostril; *ioc*, infraorbital, *jl*, jugal, *mdc*, mandibular, *orc*, oral, and *r*, rostral canals.

Whether the acustico-lateral system was originally segmented or not is an important question we may now consider. It is a striking fact that in all Craniata the whole system is innervated by cranial nerves. That the main lateral line extending to the tip of the tail is supplied by a branch of a cranial nerve, the vagus, requires explanation. Some authors, laying stress on the fact that in many Selachians and especially in Teleostomes the neuromasts of the main lateral line are strictly segmental, consider that this was their original disposition (Solger, 1879–82; Bodenstein, 1882; Hoffmann, 1883; Beard, 1088), and it has been suggested that the ramus lateralis vagi is a collector nerve (Eisig, 1887), presumably formed by the gathering forwards of all the posterior lateralis components of the spinals into the root of the vagus; a similar gathering of other components may account for the longitudinal branchial branch of the same cranial nerve (p. 767). Others, on the contrary, would consider that the acustico-lateral system was not originally segmental at all, belonged to the head region, stood as it were apart, and has only secondarily been included in and distributed by the 7th, 9th, and 10th cranial nerves. The majority of modern authors adopt a somewhat intermediate view (Strong, 1176; Herrick, 1120; Johnston, 1129, and others): that it is a special neuromast component differen-

tiated from the general visceral sensory system of these three cranial
nerves which has secondarily extended backwards on to the trunk and
tail.   As evidence that the main line has phylogenetically grown

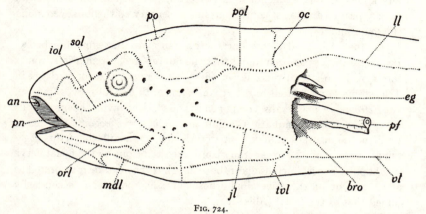

FIG. 724.

Protopterus annectens, left-side view of anterior region showing lateral line system; pectoral fin
cut short.   *an*, Anterior nostril;   *bro*, branchial opening;   *eg*, external gill;   *iol*, infraorbital line;
jl jugular line;   *ll*, main line of trunk;   *mdl*, mandibular line;   *oc*, transverse occipital line;   *orl*,
oral line;   *pf*, pectoral fin;   *po*, transverse postorbital line;   *pol*, postorbital line;   *pn*, posterior
nostril;   *sol*, supraorbital line;   *tvl*, transverse ventral line;   *vl*, longitudinal ventro-lateral line.

backwards from the head it is pointed out that such invariably is its
development in the embryo.   For it is always derived from an ectodermal thickening or dorso-lateral placode, in the region of the vagus,

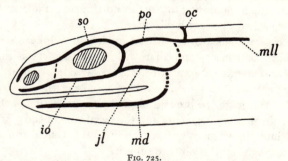

FIG. 725.

Diagram of lateral-line system in *Stegocephalia*.   Lettering as in Fig. 717.

which later extends over the trunk, growing at its tip and burrowing its
way through the epidermis until it reaches the end of the tail (Balfour, **317**;
Dohrn, **1105**; Johnston, **1129**; Wilson, **1185**; Allis, **1081**).   Harrison by
an ingenious experiment has shown that, if the tail region of one tadpole
be grafted on to the head of another, the lateral-line rudiment of the

latter will grow backwards on to the tail in a most independent manner (1903).[1]

Moreover the neuromasts are not segmentally arranged on the head of any known Craniate, and in some Selachians (*Mustelus*, Johnson, 1125a) they appear as an almost continuous line of sensory epithelium even along the main line.

On the other hand, certain facts seem to support the segmental theory. In most forms it has been found that the lateral-line system arises from a series of dorso-lateral placodes, at first independent (except perhaps in Teleostomes) and related to the 7th, 9th, and 10th cranial nerves (p. 765). In *Petromyzon* a lateralis component is associated with the profundus nerve, and in other aquatic Craniates lateralis fibres become associated with the ramus ophthalmicus superficialis and r. maxillaris of the trigeminal. These fibres are generally considered to belong to the facial nerve (r. of ophth. superf. facialis and r. buccalis), but they can be interpreted as a component of the trigeminal whose lateralis rootlet has secondarily joined that of the facial (Allis, 402).

Similarly with the metaotic segments, the vagus may have had a lateralis component in each of its segmental branchial branches. According to Alcock, indeed, the branchiomeric groups of neuromasts in the young Ammocaete are supplied by fibres passing down each gill arch (this important point has, however, not been confirmed by Johnston (1128)), the first two groups being supplied from the facial and glossopharyngeal. Also lateralis fibres to the ventral group of neuromasts near the yolk stalk of Selachians according to Norris and Hughes (1150), pass down the arch between the 4th and 5th gill-slit.

These observations all suggest that originally each head segment had its lateralis component, and that with increasing specialisation they became more and more concentrated into the facial, glossopharyngeal, and vagus segments, and in some cases survive only in the facial and vagus segments. This might be brought about by the collecting of the lateralis rootlets of the first two segments into the facial, and of the posterior body segments into the vagus. Further study of the system in Cyclostomes might throw light on this obscure subject.

The development of the lateral-line system has been described by many embryologists since Balfour (von Kupffer, 363, and Koltzoff, 361, in *Petromyzon*; Mitrophanow, 1148, Klinkhardt, 1134, Dohrn, 1105,

---

[1] These observations and experiments are not quite conclusive. It may still be argued that the building material for the lateral line, although phylogenetically derived from successive body segments, has been in ontogeny precociously gathered near the root of the vagus and grows backwards later.

in Selachians ; Allis, **1081**, in *Amia* ; Wilson, **1185**, in Teleosts ; Platt, **1162**, Mitrophanow, 1887, Brauer, **1094**, Brachet, **993**, in Amphibia). Good accounts have recently been given of the origin of the lateral-line system in Selachians by Johnson (**1125a**) and Ruud (**1171**). The ectoderm early becomes thickened over wide lateral and ventral regions on the head, and in these 'fields' appear special thickenings or placodes (p. 765), connected by strands with the underlying rudiments of the cranial ganglia. There can soon be distinguished supraorbital, infraorbital, and postorbital placodes in front of the auditory placode, and a row of post-auditory placodes overlying the glossopharyngeus and four branches of the vagus. The backward growth of the coalesced two posterior vagal placodes gives rise to the main lateral line of the trunk. The placode of the second vagal segment grows dorsally and caudally to form the more dorsal row of neuromasts, while that of the

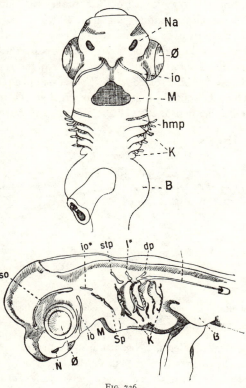

Fig. 726.

Ventral and left-side views of anterior region of embryo of *Spinax niger*, showing rudiments of lateral line system (from G. Ruud, *Zool. Jahrb.*, 1920). *B*, Pectoral fin ; *dp*, dorsal trunk line ; *hmp*, mandibular line ; *io*, infraorbital line ; *K*, gills ; *l*, main trunk line ; *l**, transverse occipital line ; *M*, mouth ; *N* and *Na*, nasal organ ; *Ø*, orbit ; *so*, supraorbital line ; *Sp*, spiracle ; *stp*, temporal line.

first vagal segment forms the transverse occipital line. The glosso-pharyngeal placode contributes to the main line and dorsally gives rise to the supratemporal pit-organs. A ventral placode yields the pit-organs near the base of the pectoral fin, Figs. 726-7.

From the lengthened postorbital placode arises the postorbital line supplied by the otic branch of the facial. The infraorbital line is developed from the corresponding placode overlying the maxillary branch of the trigeminal, and the hyomandibular line from a placode extending along the hyoid bar. The mandibular line appears in Torpedo to be developed from a separate placode related to the mandibular branch of the trigeminal. The exact history of the supraorbital line is not yet clear. In Selachians it appears to develop from a supraorbital placode at first related to the developing profundus ganglion. But in *Petromyzon* two placodes are described in this region, one at first connected with the profundus ganglion and the other with the facial. Eventually the supraorbital placodes give rise to the supraorbital line supplied by the r. ophthalmicus facialis, and the infraorbital placode to the line supplied by the r. buccalis facialis. These various lines grow in length, like the main line of the trunk, by multiplication of the cells at the tip, and burrow their way through the

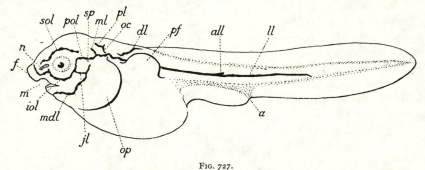

FIG. 727.

Larva of *Amia calva* (after Allis, 1889). *a*, Anus ; *all*, accessory lateral line bud ; *dl*, dorsal line of trunk ; *iol*, infraorbital, *ll*, main trunk, *mdl*, mandibular, *oc*, occipital, *pol*, postorbital, *sol*, supraorbital lateral-line canal rudiments ; *f*, fixing organ ; *m*, mouth ; *n*, undivided nasal opening ; *jl*, *ml*, and *pl*, pit lines ; *op*, opercular fold ; *pf*, pectoral fin ; *sp*, position of closed spiracle.

general ectoderm. *Amia* and Amphibia resemble very closely the Selachii in the distribution of placodes and development of the lateral lines, Fig. 727. In *Necturus*, however, the ventral neuromasts are more developed and the lines pass below all the gill-slits (Platt, 1162). The true relation of the rudiments of the lateral lines to the three preotic segments is still obscure. Although in the adult all the lines developed from them appear to be innervated from the facial nerve, yet at their first appearance placodes seem to belong to each of these three segments. This observation, together with that of an originally separate placode on the glossopharyngeal and each of the vagal segments, lend support to the view discussed above that the lateral-line system was originally distributed segmentally on the head.

General Cutaneous and other Components.—The history of the other components is easier to deal with, Fig. 728. It is clear that the general cutaneous was originally present in every segment of the body as it still is in *Amphioxus* (Johnston, 1127). In the Craniate it is the dominant component in the first two segments of the head, the profundus being apparently composed of general cutaneous fibres only, and they are always abundant in the trigeminal. General cutaneous fibres are retained in all the other segments of Cyclostomes (*Petromyzon*, Johnston, 1128); but in Gnathostomes, although well represented in all the spinal nerves, this component tends to be reduced in the intermediate segments. The sensory area supplied by the profundus and trigeminal covers the greater part of the head and meets that supplied by the first complete spinal in Amniota, excepting for the area innervated by the r. auricularis of the vagus.

General visceral sensory fibres are probably present in every segment of the body in *Amphioxus*, except perhaps the first and last few. They are greatly developed in the 7th, 9th and 10th cranial nerves of Craniata. When present, as in Teleosts, in the maxillary and mandibular branches of the trigeminal, they are generally supposed to have been borrowed from the facial. This component is peculiarly well developed in the branchial region of Cyclostomes and Fishes.

The general occurrence and distribution of the components in the various groups of vertebrates may be gathered from Fig. 728.

Branchial and Spinal Nerve.—Before describing the cranial nerves the structure of a complete dorsal nerve of the branchial region of a fish may be explained and contrasted with that of a typical spinal nerve. A complete branchial nerve would have lateralis, general cutaneous, general visceral, special visceral, and visceral excito-motor rootlets, and a dorsal ganglion. The lateralis fibres would be given off as dorsal and lateral branches (ramus supratemporalis, etc.); the general cutaneous fibres as a dorsal branch (r. auricularis, etc.). The general visceral sensory fibres would pass inwards as a ramus pharyngeus, also containing special visceral fibres to taste-buds, and on either side of the gill-slit as pre- and post-trematic branches. Of these, the r. pretrematicus internus and externus are composed of visceral sensory fibres, as are the two external and one internal post-trematic rami. The visceral motor fibres pass down the r. post-trematicus posticus. Such a theoretically complete branchial nerve is rarely if ever found in fishes, since two or more branches may combine and some of the components may be absent (Sewertzoff, 1173; Allis, 404-5; Norris, 1156-7). Usually, however, there are three distinct branches, a visceral sensory pharyngeal, a visceral sensory pre-trematic, and a mixed visceral sensory and motor post-trematic; there may also be a small dorsal general cutaneous branch and lateralis branches. Certain visceral sensory and excito-motor fibres may pass into ' sympathetic ' nerves. The somatic motor-root of the segment is, of course,

separate and contributes to the hypoglossal nerve in metaotic segments.

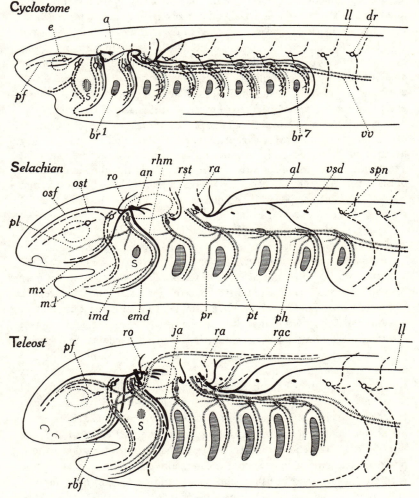

FIG. 728.

Diagrams of *Components* of dorsal root cranial nerves of *Gnathostomes*; ventral root nerves
(somatic motor component) to eye-muscles and hypoglossal muscles omitted. *a*, Auditory capsule;
al, accessory line; *an*, auditory nerve; *br¹⁻⁷*, branchial slits; *cht*, chorda tympani; *clbr*, closed
branchial slit; *dr*, dorsal root nerve of trunk; *e*, eye; *emd*, external mandibular; *fa*, facial and
auditory rootlets; *gl*, glossopharyngeal; *imd*, internal mandibular; *ja*, Jacobson's anastomosis;
ll, main lateral line nerve; *md*, mandibular; *mx*, maxillary; *osf*, superior ophthalmic of facial;

A typical spinal nerve in a fish has a dorsal root chiefly of general
cutaneous fibres and a ganglion; and this root joins a ventral root

chiefly of somatic motor fibres. The mixed nerve gives off a sensory anterior dorsal branch, and a motor posterior dorsal branch which joins

Aquatic Amphibian

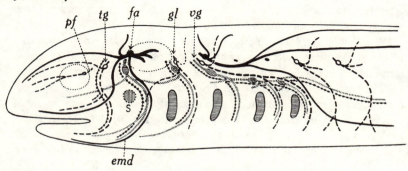

Terrestrial Amphibian

Amniote

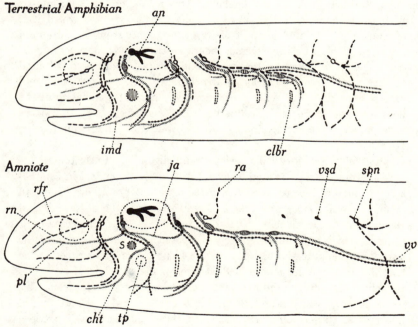

FIG. 728.

ost, sup. ophthalmic of trigeminal ; *pf*, profundus ; *ph*, pharyngeal ; *pl*, palatine ; *pr*, pretrematic ; *pt*, postrematic ; *ra*, dorsal ramus ; *rac*, recurrent accessory ramus of facial and of vagus ; *rbf*, buccal ; *rfr*, ramus frontalis ; *rhm*, hyomandibular ; *rn*, orbitonasal ; *ro*, ramus oticus ; *rst*, r. supratemporalis ; *s*, spiracular slit ; *spn*, dorsal root of spinal nerve ; *tg*, trigeminal ; *tp*, tympanic membrane ; *vsd*, vestigial dorsal root ; *vg*, vagus ; *vv*, visceral ramus. Components : black line=lateral line ; broken line=general cutaneous ; beaded line=visceral motor ; cross-hatched line=visceral sensory.

the sensory branch of the segment behind. These dorsal rami supply the skin of the dorsal region and the muscles of the median fin (p. 116). A mixed ramus medialis innervates the dorsal part of the myomere above the horizontal septum and the skin of that region, and a mixed ramus ventralis innervates the ventral part of the myomere and the skin of the latero-ventral region. Visceral motor and sensory fibres pass to the ' sympathetic ' by the ramus communicans (p. 775).

The Cranial Nerves :—We may now briefly review the nerves of the head in order from before backwards, Fig. 728. The first is the nervus terminalis composed of general cutaneous fibres.

The **nervus terminalis,** first clearly described by Pinkus in *Protopterus* (1161), has since been found in all Gnathostomes from sharks to man (Locy, 1145, in Selachii ; Brookover, 1910, Sheldon, 1909, in Teleostomi ; Pinkus, 1894, Bing and Burkhardt, 1905, in Dipnoi ; Herrick, 1909, McKibben, 1911, in Amphibia ; Johnston, 1913, Larsell, 1919, in Reptilia ; Brookover, 1914, 1917, Johnston, 1914, Huber and Guild, 1913, Larsell, 1918, Stewart, 1175, in Mammalia). This nerve, not to be confused with the vomero-nasal division of the olfactory nerve with which it is often closely associated for part of its course, issues from the fore-brain (telencephalon) near the recessus neuroporicus and lamina terminalis, has a ganglion (ganglion terminale) and peripheral fibres distributed to free nerve-endings chiefly in the region of the nasal septum and external nostril. Van Wijhe, 1918, considers the n. terminalis to be the homologue of the first or apical nerve of *Amphioxus*, also a sensory nerve with ganglion cells on its course and distributed to the rostrum (Ayers, 1919).

Profundus Nerve.—Well developed and distinct in Cyclostomes this nerve becomes closely related or even fused to the trigeminal in Gnathostomes, and for a long time was considered to be a part of it,[1] until Marshall (366), van Wijhe (396), and others showed that it is the dorsal nerve of the premandibular segment with its own ganglion. The general cutaneous fibres of which it is almost entirely composed innervate the skin of the anterior region of the head, especially the snout and neighbourhood of the nostril. It is variable in size in Pisces, consisting typically of two chief branches : a main longitudinal ramus ophthalmicus profundus (r. nasalis, or nasociliaris) which crosses the orbit between the dorsal and ventral divisions of the oculomotor, dorsally to the optic nerve, and between the superior and inferior oblique muscles of the eye, passing through the nasal capsule to the snout (Chapter VI.) ; and a more dorsal ramus frontalis or portio ophthalmicus profundi, which in Pisces usually joins the superior ophthalmic branches of the facial and trigeminal nerves. The

---

[1] Consequently the profundus is often called V[1], the r. maxillaris V[2], and the r. mandibularis V[3].

r. frontalis may be much reduced in some Selachians and Teleostomes (being then apparently replaced by the r. ophth. superf. trigemini), or well developed and separate as in *Polypterus* (Allis, 410). The r. ophth. profundus may also disappear in Selachians (*Scyllium*) and Teleosts, leaving only the radix longa to the ciliary ganglion and long ciliary nerves which always arise from it (p. 774). The r. frontalis remains an important branch in Amniota, as well as the r. nasalis.

The second nerve of *Amphioxus*, a sensory root-nerve with ganglion cells on its course, is considered by van Wijhe to be the homologue of the profundus nerve of Craniates (1918).

Trigeminal Nerve.—From the trigeminal or Gasserian ganglion come two main branches : a general cutaneous ramus maxillaris to the region of the upper jaw, and a mixed visceral motor and general cutaneous r. mandibularis. The former becomes closely connected with the buccal branch of the facial. The r. mandibularis supplies sensory fibres to the region of the lower jaw, and motor fibres to constrictor muscles and their derivatives (constrictor superficialis dorsalis of this segment, adductor mandibulae, levator labialis superioris, levator maxillae sup. of Pisces, also protractor hyomandibularis, dilator operculi, levator arcus palatini of Teleostomi ; masseter or temporalis or capiti-mandibularis, pterygoideus anterior and posterior of Amphibia, Reptilia, and Aves ; masseter, temporalis pterygoideus internus and externus, digastricus (pars anterior), tensor veli palatini, and tensor tympani of Mammalia).

The ramus ophthalmicus superficialis forms a general cutaneous third branch more dorsal and so closely associated with the r. frontalis profundi and r. ophth. superf. facialis that it is doubtful how far these sensory fibres really belong to the trigeminal. In Amniota it is no longer distinguishable from the r. frontalis.

Facial Nerve.—The facialis bears the Geniculate ganglion and contains all the dorsal components in Teleostomes, but loses the general cutaneous almost if not entirely in Elasmobranchs, Amphibia, and Amniota. The latter, of course, have no lateralis component left in the facial nerve. Lateralis fibres pass ventrally into the postspiracular truncus hyomandibularis, and are distributed more dorsally in rami oticus, ophthalmicus superficialis, and buccalis (certain neuromasts on the trunk are supplied by a recurrent r. lateralis in Amphibia) ; these lateral-line branches disappear in terrestrial forms. In Teleosts visceral sensory fibres pass into rr. maxillaris and mandibularis trigemini to supply the inner surface of the jaws and teeth, and in those forms with external taste-buds visceral fibres run to them in these nerves and the r. ophth. superficialis. The external taste-buds farther back are supplied by a special dorsal recurrent branch of the facialis, with which a similar branch from the vagus combines to form the ramus 'lateralis accessorius'; lateralis fibres may enter this nerve, and the spinal nerves may contribute general cutaneous fibres to it. The well-developed pharyngeal branch of visceral sensory fibres in all Gnathostomes runs forwards as the palatine (great superficial petrosal) nerve to the roof of the buccal cavity. A ventral branch from it supplies the roof of the buccal cavity posteriorly and in Pisces the anterior wall of the spiracle and spiracular pseudobranch; it represents the pre-trematic branch of the facialis nerve.

In most Osteichthyes and Tetrapoda (but not in Elasmobranchii) the pharyngeal branch of the glossopharyngeal joins the r. palatini facialis forming 'Jacobson's anastomosis'.

The post-trematic branch or truncus hyomandibularis carries lateralis fibres to the preoperculo-mandibular, oral, and jugal canals in aquatic forms, and supplies general cutaneous fibres to the region of the hyoid and lower jaw in a ramus mandibularis externus in Osteichthyes; a branch composed of these fibres also innervates the operculum. But in Selachians the general cutaneous component is small as in Notidani (Kappers, 1132), or said to be altogether absent as in *Squalus* (Norris and Hughes, 1158). Visceral sensory fibres form a r. mandibularis internus to the mucous surface of lower jaw and buccal cavity, including taste-buds (chorda tympani).

The distribution of these components of the tr. hyomandibularis in Amphibia closely resembles that of fishes; in Urodela it may divide into a lateralis r. mentalis, a visceral r. mandibularis internus (r. alveolaris), and a motor r. jugularis receiving general cutaneous fibres from the glosso-pharyngeal. In Amniota the r. mandibularis internus (combined with 'sympathetic' fibres) is known as the 'chorda tympani' (p. 462).

The motor fibres supply the constrictor muscles and their derivatives (adductor hyomandibularis, adductor and levator operculi of Osteichthyes; depressor mandibulae, part of mylohyoideus, sphincter colli and stapedial of Reptilia and Aves; pars posterior of digastric occipito-frontalis, stylo-hyoideus, stapedial, platysma, and facial muscles of Mammalia).

The auditory nerve with its ganglion is a special development of the acustico-lateral component of the facial segment with various branches supplying the sensory epithelium of the labyrinth. Though the rootlets of the auditory and facial nerves are separated in higher Gnathostomes, they are closely connected in Amphibia and Pisces, especially in early stages of ontogeny.

Glossopharyngeal Nerve.—This closely approaches in Pisces the ideal 'branchial nerve' described above, and contains usually all the dorsal components excepting the general cutaneous.   Fibres of this component

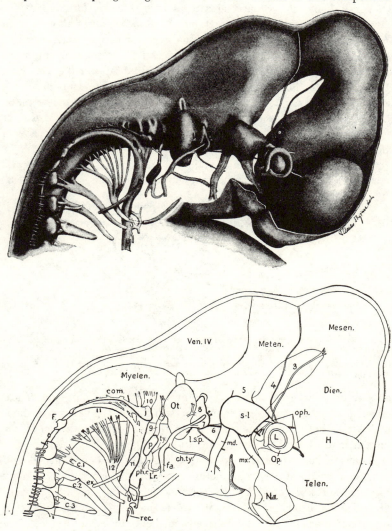

FIG. 729.

Reconstruction of brain and cerebral nerves of a 12 mm. pig embryo (from J. B. Johnston, *Nerv. Syst. of Vert.*, 1908, after Minot and Lewis).   3-12, Cranial nerves ;   *s.l*, semilunar ganglion of trigeminal ;   *oph*, profundus branches ;   *l.s.p*, large superficial petrosal ;   *ch.ty*, chorda tympani ;   *fa*, main (hyomandibular) facial ;   *s*, superior, and *p*, petrosal ganglion ;   *ty*, tympanic ;   *lr*, lingual, and *phr*, pharyngeal branches of glossopharyngeal ;   *j*, jugular, and *n*, nodosal ganglion ;   *rec*, recurrent n ;   *ex*, spinal accessory branch to trapezius ;   *F*, Froriep's vestigial ganglia ;   *c*$^{1-3}$, cervical spinal nerves ;   12, hypoglossal.

are still present in some Elasmobranchs (Notidani, Kappers, **1132**; *Mus-*

telus, Hauser ; *Laemargus*, Ewart and Cole, 1108 ; *Chimaera*, Cole, 1099);
they issue by the ramus supratemporalis, chiefly formed of the
lateralis fibres of this segment.   General cutaneous fibres also occur in
Amphibia, pass to the facial by Jacobson's anastomosis or other branches,
and into the post-trematic branch ; the dorsal fibres probably enter the
r. auricularis of the vagus.   This component has disappeared in Amniota.
Visceral sensory fibres make up the r. pharyngeus (r. communicans ix
ad vii, Jacobson's anastomosis, tympanic branch) and pre-trematic
branches, and enter the post-trematic branch to supply the posterior
ventral region of the buccal cavity and tongue by the r. lingualis in
higher forms.

The post-trematic branch contains visceral motor fibres to visceral
muscles of the 1st branchial arch (levator arcus branchialis, ceratohyoideus,

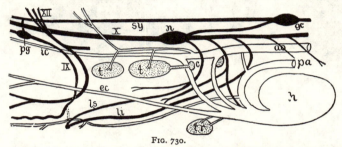

FIG. 730.

Diagrammatic left-side view showing relation of nerves and arteries in neck of *Sphenodon* (after
van Bemmelen, from J. S. Kingsley, *Comp. Anat. of Vertebrates*, 1926).   Nerves black, arteries white.
ao, Dorsal aorta ; *c,* carotid gland ; *ec,* external carotid ; *gc,* cervical sympathetic ganglion ; *ic,*
internal carotid ; *li, ls,* inferior and superior laryngeal nerves ; *n,* ganglion nodosum ; *pa,* pulmonary
artery ; *pg,* petronal ganglion ; *sy,* sympathetic trunk ; *t,* thymus ; *tr,* thyroid.

and certain laryngeal muscles in Tetrapoda).   The glossopharyngeal
ganglion may be very closely connected with that of the vagus ; in Mammalia it becomes subdivided into a superior or jugular ganglion in the
jugular foramen, and an inferior or petrous ganglion.

Vagus or Pneumogastric Nerve.—The largest of the cranial nerves is of
compound origin (its general structure is discussed below, p. 767).   Many
rootlets lead to its ganglion jugulare or nodosum, Fig. 729.   The lateralis
fibres have already been dealt with (p. 742) ; they issue in a dorsal r.
supratemporalis, a main r. lateralis, and smaller branches to the secondary
dorsal and ventral lines of neuromasts.

Few general cutaneous fibres survive ; they pass dorsally in a r. auricularis, better developed in Teleostomes and Tetrapods than in Selachians.
Visceral sensory fibres are distributed in Pisces, in pre- and post-trematic
branches at each branchial slit, and to the dorsal wall of the pharynx by
corresponding pharyngeal branches ; also by the large r. intestinalis to the

endodermal region of the alimentary canal and its glandular appendages and lungs, Fig. 730.

The visceral excito-motor fibres also pass to the alimentary canal by the r. intestinalis (p. 774), to visceral branchial muscles in Pisces, and corresponding muscles of hyoid apparatus and larynx in Tetrapoda. Some posterior motor rootlets of fibres supplying the trapezius muscle become separated off in Mammalia, and form a nerve called the eleventh cranial or spinal accessory nerve.

SPIRACULAR SENSE-ORGAN

The first or spiracular gill-slit is always specialised in Gnathostomes. Its opening is reduced in the course of ontogeny from below to a relatively

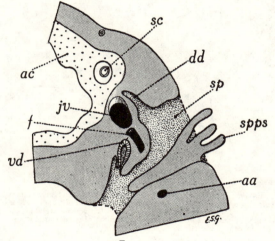

FIG. 731.

Scyllium canicula, advanced embryo. View from in front of transverse section through spiracle, diagrammatic reconstruction. *aa*, Afferent artery ; *ac*, auditory capsule ; *dd*, dorsal diverticulum ; *f*, hyomandibular branch of facial nerve ; *jv*, jugular vein (v. cap. lat.) ; *sc*, semicircular canal ; *sp*, external opening of spiracular slit ; *spps*, spiracular pseudobranch ; *vd*, ventral diverticulum.

small spiracle situated behind the eye ; and even this aperture is closed in the adult of many Pisces (Holocephali, Dipnoi, most Teleostomi) and all Tetrapoda (p. 755).

It was long ago noticed by J. Müller (1841) that in many Selachians the spiracular slit gives off a dorsal diverticulum which becomes applied to the ventro-lateral wall of the auditory capsule below the prominence formed by the horizontal semi-circular canal, and morphologically ventrally to the articulation of the hyomandibula and jugular vein, Figs. 731-2. This 'auditory diverticulum' possibly conveys vibrations to the auditory labyrinth (J. Müller, 1833–43 ; v. Bemmelen, 1091 ; Ridewood, 1168).

Nearer the internal opening of the spiracular cleft there is another diverticulum from its anterior wall, as shown by Wright ; the blind end of this diverticulum contains a neuromast sense organ (supplied by the otic branch of the facial) and in late stages becomes constricted off as a closed vesicle (Hoffmann, 354 ; van Wijhe, 1183 ; Norris and Hughes, 1158). A similar neuromast was discovered by Wright (1186) in *Acipenser*, *Lepidosteus*, and *Amia*, where it is lodged in a dorsal diverticulum of the spiracle, Fig. 733. This diverticulum passes up outside the wall of the

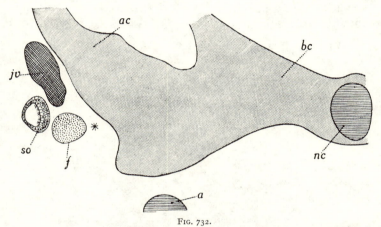

FIG. 732.

Portion of transverse section through auditory region of head of late embryo of *Heterodontus* (*Cestracion*) *Philippi*, 70 mm. long. *a*, Lateral aorta ; *ac*, auditory capsule ; *bc*, basal plate ; *f*, hyomandibular branch of facial nerve ; *jv*, jugular vein (v. capitis lateralis) ; *nc*, notochord ; *so*, spiracular sense organ in vesicle closed off from diverticulum of spiracular slit, a star marks position of more posterior and ventral diverticulum (cp. Fig. 731).

trigemino-facialis chamber in front of the hyomandibula, its blind end projecting dorsally through a canal piercing the postorbital process, and the neuromast being there supplied by a twig of the otic branch of the facial nerve. The canal is formed by overgrowth of cartilage from the capsule in front and wall of the trigemino-facialis chamber behind. The Dipnoi also possess a spiracular sense-organ discovered by Pinkus (1611) in *Protopterus*, where it is in the form of a closed vesicle lodged in the cartilaginous postorbital process. Agar (1079) has shown that in *Lepidosiren* it arises as an offshoot from the ectodermal region of the spiracle. There can be no doubt that the spiracular sense organ of all these fishes is a special neuromast, derived from that part of the acustico-lateral system supplied by the otic nerve, which has sunk into the spiracle. The interesting sense-organ recently described by Vitalli (1179-80) in a vesicle between the tympanic cavity and the auditory capsule in Birds is possibly of the same nature (Ranzi, 1165).

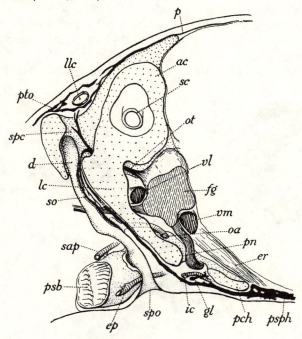

FIG. 733.

Reconstructed thick transverse section of head of larval *Amia* showing spiracular sense-organ in dorsal diverticulum of spiracular pouch. *ac*, Auditory capsule; *d*, dorsal diverticulum of spiracular pouch penetrating into spiracular canal; *ep*, efferent pseudobranchial artery; *er*, posterior or external rectus; *fg*, portion of facial ganglion in trigemino-facial chamber; *gl*, visceral branch of glosso-pharyngeal; *ic*, internal carotid in parabasal canal; *lc*, lateral cartil. commissure, outer wall of trigemino-facialis chamber; *llc*, postorbital lateral-line canal; *oa*, orbital artery; *ot*, otic branch of facial supplying spiracular neuromast and lateral-line; *pch*, anterior parachordal cartilage; *pn*, palatine nerve passing forwards; *psb*, spiracular pseudobranch; *psph*, parasphenoid with lateral ascending wing; *sap*, secondary afferent pseudobranchial artery; *sc*, semicircular canal; *so*, spiracular neuromast; *spc*, spiracular canal in cartilage; *spo*, opening of spiracular pouch; *vl*, vena lateralis; *vm*, vena medialis. Posterior view.

THE DEVELOPMENT OF THE PERIPHERAL NERVOUS SYSTEM

Considerable light is thrown on the structure of the peripheral nervous system by a study of its development, of which only a brief sketch can here be given.

The central nervous system (brain and spinal cord) is developed in all Vertebrates from a single essentially unsegmented dorsal neural plate of thickened ectoderm which sinks inwards and folds to form a hollow tube. The two edges of the plate or neural folds coming together dorsally and fusing become separated off from the superficial general ectoderm. The tube closes first in the anterior trunk region and then progressively from that point backwards and forwards. At the hind end the blastopore becomes enclosed by the neural folds and persists as a rule for a time as a neurenteric canal leading from the neural canal above to the cavity of the enteron below. Growth in length of all the germ-layers takes place here, more especially at the dorsal lip of the blastopore. The last point to remain open anteriorly is the temporary neuropore, which finally closes in

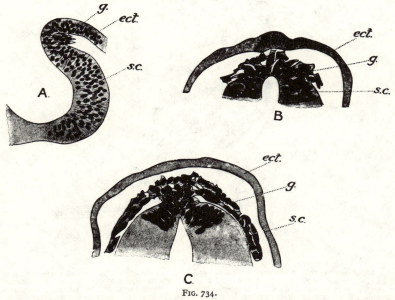

FIG. 734.

Transverse sections illustrating the mode of origin of the spinal ganglia. A, Fowl embryo with four mesoderm segments (after Neumayr, 1906); B and C, *Torpedo* 4 mm. embryo (after Dohrn, 1902). *ect*, Ectoderm; *g*, rudiment of ganglion; *s.c*, spinal cord. (From Kerr, *Embryology*, 1919).

front of the neural plate, and may be marked in later stages by a small depression on the inner face of the lamina terminalis of the brain, the recessus neuroporicus. The whole central nervous system thus soon forms a closed tube entirely surrounded by mesoblastic tissue; but while the neural plate is separating from the external ectoderm there appears along each side a longitudinal ridge or thickening of ectodermal cells, the neural crest of Marshall (366) and Beard (1089), Figs. 734-5. The exact time at which the crest appears varies in different regions and in different animals; it may appear when the neural folds meet, or a considerable time before as in the cerebral region of most forms. When the separation of the external ectoderm from the neural tube has taken place the crests remain attached to the latter and may meet in the middle line temporarily. The crests may be considered as proliferations of a narrow zone differentiated in the ectoderm along the edges of the neural plate, or perhaps as a differentiation of the margin of the plate itself. Probably the crest is primitively continuous from neuropore to blastopore. Segmental proliferations of the cells of the neural crest soon appear in the spinal region and develop from behind the head to near the tip of the tail ; as these segmental rudiments of the spinal ganglia enlarge and grow downwards between the neural tube and the outer mesoblastic somites the neural crest between them disappears. In the head region the neural

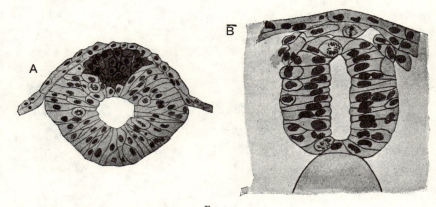

FIG. 735.

Transverse section of neural tube of embryo *Siredon* (*Amblystoma*) *punctatum*. Neural crest cells darkly shaded (from J. B. Johnston, *Nerv. Syst. of Vert.*, 1908). A, Just after closing ; B, later stage.

crest becomes very early interrupted, and proliferations from it give rise to the rudiments of the vagus, glossopharyngeal, facial and auditory, trigeminal and profundus ganglia, Figs. 736-9 (Balfour, 317 ; Marshall, 366 ; van Wijhe, 396 ; Beard, 1089 ; His, 1879–93, and others ; more recent general accounts will be found in text-books, more especially in those of Neumayer, 1164 ; Keibel and Mall, 1910–12 ; Brachet, 993 ; Johnston, 359). The nervus terminalis probably develops from its extreme anterior end.

The subsequent history of the development of the spinal nerves is comparatively simple. As the rudiments of the ganglia enlarge and grow downwards they lose their primitive connexion with the neural tube. Some of the indifferent crest-cells become ganglion cells or neurons, others become neuroglia and sheath-cells. The ganglion cells send a fibre growing inwards centripetally into the central nerve tube to form the definitive dorsal afferent root, and a fibre growing outwards centrifugally to form the afferent nerve. The ventral root is formed by the outgrowth from cells in the neural tube of efferent fibres which pass outwards to the corresponding myotome. At first in Selachians (Balfour) this ventral root is separate, but it soon meets the dorsal root beyond the ganglion to form a mixed nerve which runs in the septum posterior to its myotome (the

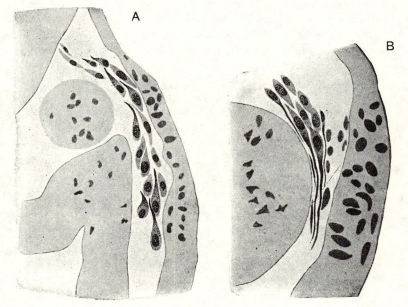

A

B

FIG. 736.

Transverse sections of embryo *Siredon* (*Amblystoma*) *punctatum* (from J. B. Johnston, *Nerv. Syst. of Vert.*, 1908). A, ganglion of glossopharyngeal at time of formation of central processes ; B, ganglion of trigeminal with outgrowing axons of ramus mandibularis.

segmental relationships of nerves and myotomes are discussed below ; see also p. 218).

Particular interest attaches to the development of the cranial nerves and ganglia (Marshall, 366 ; v. Wijhe, 396 ; Beard, 1088-9 ; Froriep, 498 ; v. Kupffer, 363 : Neal, 368 ; Kolzoff, 361 ; Dohrn, 1105 ; Landacre, 1142a-43a ; Belogolowy, 1090 ; Brachet, 993 ; Chiarugi, 1097 ; Goronowitsch, 1115-16 ; Klinkhardt, 1134 ; Neumayer, 1154 ; Guthke, 1117 ; Knouff, 1135).

These cranial ganglia and nerves seem to arise from three separate sources : (1) the neural crest ; (2) dorso-lateral placodes ; (3) epibranchial placodes. The neural crest, the chief source of the ganglionic cells, develops much as in the trunk, becoming early broken up into sections which give rise by proliferation to segmental downgrowths. These rudiments of the ganglia, however, remain near the surface on the outer

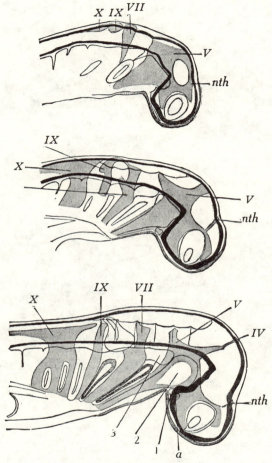

FIG. 737.

Diagrams of head-region of embryos of *Squalus acanthias* showing contribution from neural crest to development of cranial nerves (after H. V. Neal, from J. B. Johnston, *Nerv. Syst. of Vert.*, 1908). *a*, Anterior mesoblast ('head cavity'); 1, 2, 3, prootic somites; cranial nerves *IV-X*; *nth*, temporary nervous thalamicus.

side of the mesoblastic somites, here little developed. They develop at first intersegmentally (Neal, 368), that is posteriorly to the somite to which they are assigned (p. 219). The first proliferation gives rise to the profundus ganglion [1] (ganglion ophthalmicum, or mesocephalicum), which

---

[1] Considerable confusion has arisen in the literature owing to this rudiment being named by several embryologists the ciliary ganglion (v. Wijhe, **396**;

is usually closely associated with the next trigeminus or Gasserian ganglion. From the second rudiment, extending down into the hyoid arch, develops

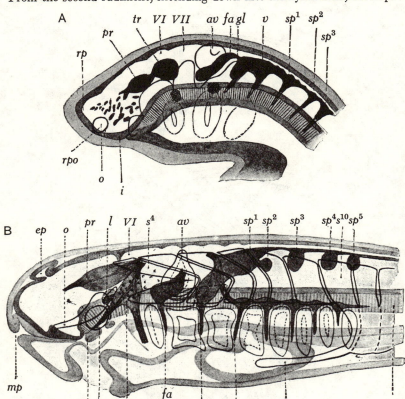

FIG. 738.

Reconstructions of head region of embryo *Petromyzon*, left-side view (after N. K. Koltzoff, from J. B. Johnston, *Nerv. Syst. of Vert.*, 1908). A, Young stage, with 2 gill-slits open and showing 3 prootic and 5 metaotic somites; B, older stage, 4 mm. long. *av*, Auditory vesicle; *ep*, epiphysis; *i*, infundibulum; *l*, lens; *mp*, median nasal pore; *nt*, notochord; *o*, optic cup; *pg*, preoral endoderm; *rp*, recessus neuroporicus; *rpo*, recessus preopticus; *rv*, post-trematic branch of vagus; s^1, pre-mandibular somite; s^2, mandibular somite is just posterior to it; s^4, first metaotic somite; sp^{1-5}, dorsal root nerves of spinal region; *t*, hypoglossal muscle; *VI*, *VII*, neuromeres. Other letters as in Fig. 741.

the facialis or geniculate and the acusticus ganglia; and from the third, extending into the first branchial arch, arises the glossopharyngeus ganglion. A more extensive region of the neural crest gives rise to the compound vagus ganglion (composed of ganglia, each corresponding to

Klinkhardt, **1134**, and others). The true ciliary ganglion belongs to the sympathetic system (p. 774).

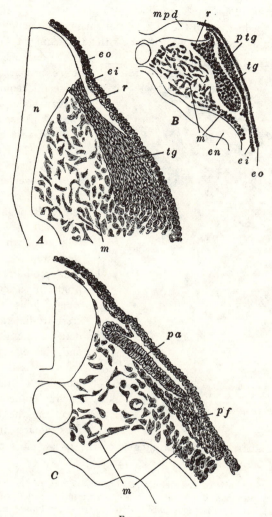

FIG. 739.

Portions of sections through head of frog (*Rana fusca*), illustrating formation of placodes and history of crest ganglia after Brachet, from W. E. Kellicott, *Chordate Develt.*, 1913). A, Transverse section through trigeminal ganglion of embryo, 3 mm. B, Transverse section through acustico-facialis ganglion of embryo with three or four pairs of mesodermal somites. C, Transverse section through facial ganglion and auditory placode of embryo, 2·8 mm. *ei*, Inner or nervous layer of ectoderm; *en*, endoderm; *eo*, outer layer of ectoderm; *m*, mesoderm; *mpd*, definite medullary plate; *n*, nerve cord; *pa*, auditory placode; *pf*, facial placode; *ptg*, trigeminal placode; *r*, spinal prolongation of ganglion; *tg*, trigeminal ganglion.

one branchial segment) and sends downgrowths into the remaining branchial arches, Figs. 736-9, 740.[1]

In the branchial region the rudiment of each cranial ganglion derived from the neural crest comes into contact with a thickening of the ectoderm at the dorsal edge of the corresponding gill-slit. These thickenings are the epibranchial placodes or ' branchial sense organs ' of Beard (1088) and Froriep (1887). There is no sufficient evidence that they are or ever have been branchiomeric organs of sense, although it is possible that they give rise to cells from which develop the taste-buds. Each placode proliferates inwards producing a mass of cells which joins the neural crest rudiment, becomes detached from the surface, and, sinking inwards, contributes to the definitive ganglion of its segment. Such epibranchial placodes occur in the embryo of all Craniates from Cyclostomes to Man related to the ganglia of the facial, glossopharyngeal, and each of the vagal segments, Fig. 741.

Fig. 740.

Transverse section immediately behind the first visceral pouch of chick embryo of thirteen somites (after Goronowitsch, from F. R. Lillie, *Develt. Chick*, 1919). *Ad*, Aorta descendens ; *c*, rounded mesenchyme cells ; *f*, proliferation of ectoderm from placode ; *g*, place where cells derived from neural crest unite with mesenchyme cells of periaxial mass ; *ms*, mesoderm ; *p*, spindle-shaped peripheral mesenchyme cells.

[1] As the rudiments of the definitive nervous system of the head become differentiated portions of the neural crest not directly used up may become scattered and mixed indistinguishably with the true mesoblastic mesenchyme. There is reason to believe that such scattered ectoderm cells go to build up the peripheral nerves, probably in the form of sheath-cells. It has frequently been stated that they give rise to ' mesectoderm ', contributing to the formation of the connective tissues or even the endo-skeleton of the head, with the help of other cells said to be proliferated from the inner surface of the covering ectoderm (Platt, 1162 ; v. Kupffer, 363 ; Dohrn, 1105, and others). This doctrine of the formation of special " mesectoderm " in the head is, however, almost certainly founded on misinterpretations and erroneous observations on unsuitable material (Brachet, 993 ; Adelmann, 1078).

Another contact of the neural crest ganglia takes place more dorsally with thickenings of the ectoderm known as dorso-lateral placodes. There is a series of such thickenings of which the auditory placode giving rise by invagination to the auditory sac (developing later into the membranous labyrinth and its sensory patches) is the largest and most constant. The dorso-lateral placodes are best developed in aquatic Craniates, but may appear also in Amniotes in a reduced condition. In spite of statements to the contrary there is good reason to believe that they are merely the rudiments of the lateral-line organs and their nerves, and it is doubtful whether even in fishes they make any important contribution to the definitive nervous system other than the neurons concerned with the lateral-line system. (See further development of lateral-line system, p. 744.)

There are thus two longitudinal rows of segmentally arranged placodes: a dorso-lateral series and a ventro-lateral or epibranchial series—the latter related to each gill-slit and branchial nerve ; the former possibly represented in addition in the trigeminal and profundus segments, Figs. 728, 741.

It is tempting to associate each of these three possible sources of ganglionic cells with components (Strong, 1176), and it has been concluded that the neurons of the general cutaneous and general visceral components are derived from the neural crest cells, those of the special somatic (lateral-line system) component from the dorso-lateral placodes, and those of the special visceral (taste-bud system) from the epibranchial placodes (Landacre, 1142-43a). In the lower Craniates these three main elements of the ganglia may remain fairly distinct even in the adult and have each its own rootlet. The above conclusion is in accordance with the observation that the dorsolateral placodes (excepting the auditory) are reduced in terrestrial forms, and that the epibranchial placodes are preserved and correspond to those segmental nerves which supply taste-buds ; but it is difficult to reconcile with the view of Herrick and Johnston that the two special systems have been phylogenetically differentiated from the general systems. Much further evidence concerning the precise derivation of the functional neurons making up the peripheral ganglia is required before safe conclusions can be reached.

Observations about the origin of the cranial ganglia are still very contradictory. It should be remembered that in at all events many of the lower forms the appearance of definite proliferations is preceded in early stages by a transient thickening of the epidermis of the head over extensive areas of which the placodes and rudiments of the lateral-line organs are local developments, and that these are not always clearly marked off from each other. Also in higher forms vestiges of the placodes and rudiments may apparently occur which contribute little or nothing to the definitive nervous system.

It seems fairly well established in the case of the vagus and glosso-pharyngeal ganglia that the neural crest supplies the general cutaneous and general visceral portions, and that the special visceral portions come from the epibranchial placodes and lateralis portions from the dorso-lateral placodes. More difficult is the interpretation of the acustico-facialis complex. Here the special visceral is doubtless derived from the first epibranchial placode; and the general visceral, and what general

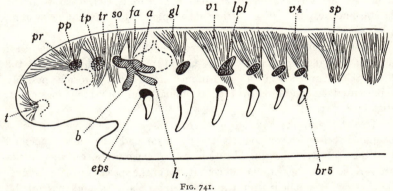

Fig. 741.

Diagram showing origin of segmental ganglia in Gnathostome. Contribution from neural crest is striated, from dorso-lateral placodes obliquely hatched, from epibranchial placodes black. *a*, Auditory branch; *b*, buccal branch; *br*5, fifth branchial gill-slit; *eps*, epibranchial placode of spiracular slit; *fa*, facial; *gl*, glossopharyngeal; *h*, hyoid branch; *lpl*, lateral-line placode of vagus; *pp*, profundus placode; *pr*, profundus; *so*, superior ophthalmic branch; *sp*, spina ganglion; *t*, terminal; *tp*, placode of trigeminal; *tr*, trigeminal; *v*$^{1-4}$, first to fourth segment of vagus.

cutaneous component there may be, from the neural crest. There appear also in the lower forms a dorso-lateral and ventro-lateral placode in addition to the auditory placode, and all three should probably be con-sidered as differentiations from an extensive dorso-lateral placodal area from which are derived the auditory ganglion and the two lateralis portions of the facial ganglion. But opinions differ as to the origin of the auditory ganglion in Amniotes, some deriving it more or less completely from the neural crest. Concerning the origin of the trigeminal and profundus ganglia there is still less agreement. Some would derive them entirely from neural crest cells (Goette, Corning, in Amphibia), others from both neural crest and dorso-lateral placodes or from the latter chiefly. Most authors, however, describe a contact with the epidermis in the case of both these ganglia not only in the lower forms (*Petromyzon*: Kupffer, 363; Koltzoff, 361. Selachians: Beard, 1089; Hoffmann, 354; Goette, 1114;

Klinkhardt, 1134; Gast, 1113; Guthke, 1117; de Beer, 1924. Amphibia: Brachet, 993; Knouff, 1135), but also in Birds (Goronowitsch, 1892–3; Neumayer, 1154) and Mammals (Chiarugi, 1097; Giglio-Tos, 1902: Adelmann, 1078).

Such a contact with the epidermis, even if temporary and not involving any contribution of cells, may indicate a vestigial lateralis ganglion belonging to the trigeminal segment. The profundus connexion is usually more pronounced and the ganglion seems to be derived, in considerable part at all events, from an epidermal placode. Now the profundus nerve is composed of general cutaneous fibres and has no lateralis component in Gnathostomes ; we should expect it, therefore, to develop like a spinal ganglion from the neural crest. To explain this anomaly it has been suggested that the placode in this case belongs not to the dorso-lateral series but to the neural crest, which in the anterior region may spread outwards and so not become involved in the involution and sinking of the neural tube as it does farther back. The same explanation may apply to that part of the trigeminal ganglion which appears to develop from a dorso-lateral placode (Knouff, 1135).

Morphology of the Vagus.—There remains to be discussed the question of the morphology of the vagus nerve, which differs so markedly from the other nerves of the branchial region and supplies all the branchial arches behind the first. The known facts of its development and comparative anatomy do not allow us to explain this distribution of its branchial nerves as due to the branching of an originally single segmental nerve ; nor do they support Gegenbaur's view that the vagus is a compound nerve formed by the gathering together of segmental dorsal nerves and ganglia equal in number to the gill-slits it supplies. This conception of the vagus as compounded of a number of complete dorsal nerves necessitates the further supposition that a corresponding number of myomeres have been compressed and more or less completely obliterated behind that corresponding to the first vagal segment, a supposition for which there is no evidence. If myomeres are suppressed at all it is immediately behind the auditory capsule (p. 226). Not only do the segments of the vagal region yield myomeres persistent in the adult, but there is developed in each of the vagal segments behind the first a pair of vestigial and usually transitory dorsal roots and ganglia, Figs. 240, 742. These were long ago described by Froriep in mammalian embryos (339), and are clearly seen in such lower forms as the Selachii (v. Wijhe, 396 ; Goodrich, 349 ; de Beer, 320). Whatever may be the explanation of the structure of the vagus, these vestigial segmental ganglia show that it has not been formed by their coalescence.

But the best evidence comes from a study of the anatomy and development of *Petromyzon* (Hatschek, 352 ; Koltzoff, 361; Johnston, 359, 1128). In this Cyclostome every segment in the region occupied by the vagus is provided with a myomere innervated by a ventral root nerve (coalesced in the first three metaotic segments), and a ganglionated dorsal root nerve giving off a dorsal general cutaneous ramus and a ventral branch connecting with the longitudinal ' epibranchial ' vagus. There has been

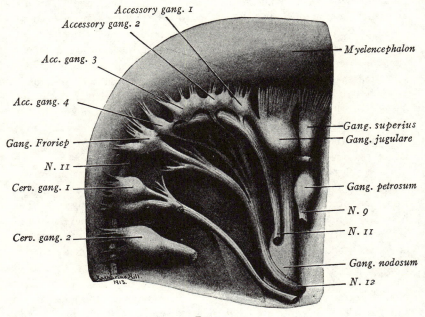

Accessory gang. 1
Accessory gang. 2
Acc. gang. 3
Acc. gang. 4
Gang. Froriep
N. 11
Cerv. gang. 1
Cerv. gang. 2

Myelencephalon
Gang. superius
Gang. jugulare
Gang. petrosum
N. 9
N. 11
Gang. nodosum
N. 12

Fig. 742.

Dissection of head of 15 mm. pig embryo from right side to show accessory vagus ganglia with peripheral roots passing to hypoglossal nerve. × 25. (From Prentiss and Arey, *Textbook of Embryology*, 1917.)

here no general ' collecting ' of dorsal nerves and no suppression of segments, Figs. 728, 743.

Some other explanation must therefore be sought, and the best is based on the theory of partial polymerisation suggested in a general way by Hatschek. Adopting the doctrine of nerve components we may suppose that the rootlets of certain components of the segmental dorsal nerves supplying the branchial segments have been completely, as in the case of the special somatic and special and general visceral, or incompletely, as in the case of the general cutaneous components, gathered forwards into the anterior root of the vagus, some of the last component being left behind in its original position (vestigial ganglia of Gnathostomes, and more complete dorsal nerves of Cyclostome). The longitudinal 'epibranchial'

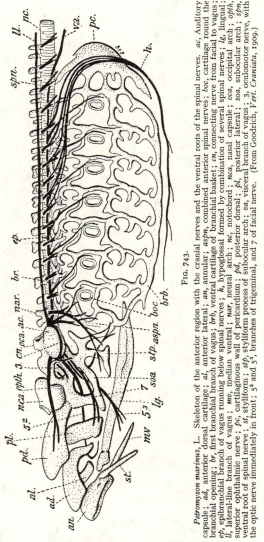

FIG. 743.

Petromyzon marinus, L. Skeleton of the anterior region with the cranial nerves and the ventral roots of the spinal nerves. *ac*, Auditory capsule; *ad*, anterior dorsal cartilage; *al*, anterior lateral; *an*, annular; *aspn*, combined anterior spinal nerves; *boc*, cartilage round the branchial opening; *br*, first branchial branch of vagus; *brb*, ventral cartilage of branchial basket; *cn*, connecting nerve from facial to vagus; *ep*, epibranchial branch of vagus running below spinal nerves; *h*, hypoglossal formed by combination of several spinal nerves; *lg*, lingual; *ll*, lateral-line branch of vagus; *mv*, median ventral; *nar*, neural arch; *nc*, notochord; *nca*, nasal capsule; *oca*, occipital arch; *opth*, superior ophthalmic nerve; *pc*, cartilaginous wall of pericardium; *pd*, posterior dorsal; *pl*, posterior lateral; *soa*, subocular arch; *spn*, ventral root of spinal nerve; *st*, styliform; *stp*, styliform process of subocular arch; *va*, visceral branch of vagus; *3*, oculomotor nerve, with the optic nerve immediately in front; *5²* and *5³*, branches of trigeminal, and *7* of facial nerve.   (From Goodrich, *Vert. Craniata*, 1909.)

branch of the vagus would then be a partial collector nerve. This explanation avoids the assumption that new central connexions have been established with the brain (Johnston, 359). Moreover, the fact, otherwise difficult to account for, that the number of branchial branches of the vagus corresponds to and varies with the number of branchial slits is naturally explained on the supposition that the process of 'collecting' extends over all the segments of the region where branchial slits are developed ; neither excalation nor intercalation of nerves between the vagus and the first spinal need have taken place.

THE 'SYMPATHETIC' NERVOUS SYSTEM

All the parts of the peripheral nervous system of a Vertebrate are so connected with each other and the central nervous system that it is not

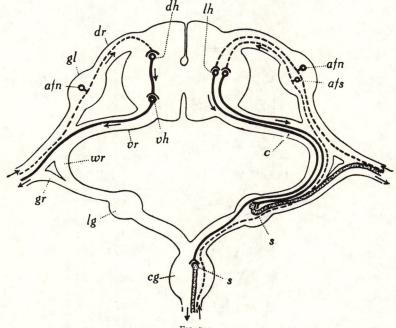

FIG. 744.

Reflex paths in peripheral nervous system. Diagram illustrating origin and distribution of nerve-fibres in trunk of Gnathostome (Mammal). Somatic reflex arc on left ; autonomic reflex arc on right. Afferent fibres represented as broken lines, efferent fibres as black lines ; sympathetic fibres red. Arrows indicate course of nerve impulses. *afn*, Afferent sensory neuron ; *afs*, afferent neuron related to viscera ; *cg*, collateral sympathetic ganglion ; *dh*, intermediate neuron of dorsal horn ; *dr*, dorsal root ; *gl*, spinal ganglion ; *gr*, grey ramus ; *lg*, lateral sympathetic ganglion ; *lh*, preganglionic neuron of lateral horn related to sympathetic ; *s*, sympathetic neuron ; *vh*, motor neuron of ventral horn ; *vr*, ventral root ; *wr*, white ramus.
*Instead of red, stippled.

possible satisfactorily to subdivide it into anatomically separate regions. We may distinguish the central nervous system from the cranial and spinal nerves ; but if, with the anatomist, we define what remains when these have been removed as the 'sympathetic system', consisting of nerves, ganglia, and plexuses supplying chiefly unstriated muscles of the vascular system, alimentary canal, and other viscera, and glands, we are met at once with the difficulty that many of the fibres coming from the sympathetic ganglia are distributed in the branches of the spinal and cranial nerves

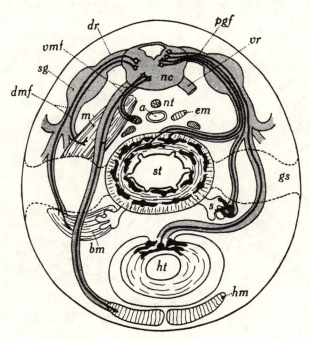

Fig. 745.

Efferent nervous system of head. Diagram of transverse section of branchial region of Gnathostome, showing distribution of excito-motor neurons in cranial nerves (black), and 'sympathetic' neurons (red)*. *a*, Dorsal aorta ; *bm*, ventral branchial muscle ; *dr*, dorsal root ; *dmf*, dorsal motor fibres ; *em*, epibranchial muscle ; *gs*, position of gill-slit ; *hm*, hypobranchial muscle ; *ht*, heart ; *m*, dorsal branchial muscle ; *nc*, medulla ; *nt*, notochord ; *pgf*, connector preganglionic fibres ; *s*, salivary gland ; *sg*, segmental ganglion ; *st*, stomach ; *vmf*, ventral motor fibres ; *vr*, ventral root. Dorsal root nerve represents vagus, and ventral root nerve hypoglossal. *em* and *hm*, derived from segmented somites. *bm* and *m* from unsegmented lateral plate. 'Sympathetic' on right side.

*Instead of red, **thick black** lines.

to the skin, and indeed perhaps to all the muscles of the body both striated and smooth.[1] The time-honoured name sympathetic, then, used by the anatomist, indicates an arbitrarily distinguished portion of the peripheral

[1] A great deal of controversy has arisen of late years concerning the alleged double innervation of voluntary striated muscle fibres by both ordinary motor and autonomic nerve fibres. Thoracico-lumbar sympathetic fibres undoubtedly penetrate the muscles and are distributed in them, but whether they supply their own muscle fibres, fibres already provided with motor end plates, the blood-vessels, or a combination of these elements, is still uncertain (Peroncito, 1901 ; Boeke, 1913 ; Agduhr, 1920 ; Botezat, 1910 ; Kulchitsky, 1924 ; Hunter, 1925 ; also Wilson, 1921, for general review of the subject).

nervous system connected here and there with the cranial and spinal nerves. A better subdivision of the peripheral nerves can be made on functional grounds, taking into account the course of the nerve fibres serving as paths of impulses to and from the central nervous system (see p. 725).

The sympathetic of the anatomist contains many sensory afferent fibres from the viscera and internal organs ; but, since these differ in no essential from other afferent fibres, and since their cell-bodies are situated

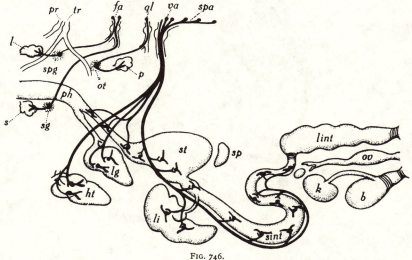

FIG. 746.

Diagram of Autonomic *dorsal root system* of *Mammal* (modified from W. H. Gaskell, 1916). *b*, Urinary bladder ; *fa*, facial nerve ; *gl*, glossopharyngeal ; *ht*, heart ; *k*, kidney ; *l*, lacrimal gland ; *lg*, lung ; *li*, liver ; *lint*, large intestine ; *ot*, otic ganglion ; *ov*, oviduct ; *p*, parotid gland ; *ph*, pharynx ; *pr*, profundus ; *s*, submaxillary gland ; *sg*, submaxillary ganglion ; *sint*, small intestine ; *sp*, spleen ; *spa*, spinal accessory ; *st*, stomach ; *tr*, trigeminal ; *va*, vagus.

in the dorsal root ganglia of the cranial and spinal nerves, they require no special treatment and can be set aside as the visceral sensory component of the peripheral nerves (p. 729).

To the whole efferent sympathetic system Gaskell gave the name ' involuntary ' since its action is usually not under the control of the will ; but Langley substituted the less objectionable name ' autonomic ', which is now generally used by physiologists (Gaskell, 1112 ; Langley, 1144 ; Müller, 1153 ; Huber, 1125). This autonomic system consists essentially of peripheral neurons, generally grouped in ganglia, receiving impulses from the central nervous system by branches of the cranial or spinal nerves, and carrying impulses by fibres distributed to practically all regions of the body, including the viscera. But the complete system includes cells in the brain and spinal cord whose axons pass to the outer ganglia or plexuses where they come into synaptic relation with other cells whose axons supply the effector end organs (pigment cells, contractile cells of blood-vessels and muscles, and gland cells). The peripheral cell may be known as the sympathetic neuron, the central cell as the connector neuron. The connector axon is known as the preganglionic fibre, the sympathetic axon as the postganglionic fibre, Figs. 744-5, 750. In Gnathostomes, whereas the preganglionic fibres are provided with the usual medullary sheath, the postganglionic fibres remain non-myelenated, and can thus be

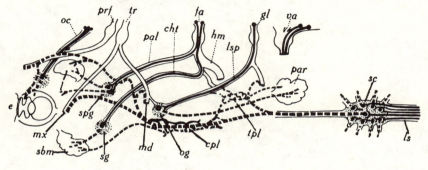

FIG. 747.

Diagram of *Autonomic supply in head* of Mammal ; left-side view of cranial nerves, etc. (modified from Gaskell, 1916). Preganglionic neurons black ; postganglionic neurons red.* *cht*, Chorda tympani ; *cpl*, carotid plexus ; *e*, eye ; *fa*, facial ; *gl*, glossopharyngeal ; *hm*, hyomandibular branch of facial ; *l*, lacrimal gland ; *ls*, longitudinal sympathetic cord ; *lsp*, lesser superficial petrosal ; *md*, mandibular branch ; *mx*, maxillary branch ; *oc*, oculomotor ; *og*, otic ganglion ; *pal*, palatine branch ; *par*, parotid gland ; *prf*, profundus ; *sbm*, submaxillary gland ; *sc*, superior cervical ganglion ; *sg*, submaxillary ganglion ; *spg*, sphenopalatine ganglion ; *tpl*, tympanic plexus ; *tr*, trigeminal ; *va*, vagus.

*Instead of red, broken line.

distinguished ; moreover, except in the heart, the muscle fibres supplied by sympathetic neurons are of the ' smooth ' unstriated kind.[1]

In the Gnathostomes there are paired ganglia in the head connected to some of the cranial nerves ; paired segmental ganglia in the trunk and anterior region of the tail, extending below the vertebral column on each side, joined together by a longitudinal nerve trunk and connected to the corresponding spinal nerves by rami communicantes ; also median ganglia in the dorsal mesentery, and smaller peripheral ganglia distributed in the viscera.

Except possibly in the enteric plexus, there are no short cuts peripherally to sympathetic ganglia, and (neglecting possible ' axon reflexes ') sympathetic reflex arcs can only be completed through the central nervous

[1] These distinctions, however, are not absolute, especially in lower forms. For instance, in birds the postganglionic fibres to the iris muscles are medullated, and these muscles are striated (Gaskell, 1112).

system to which all the sensory impulses are carried by afferent fibres whether by spinal, cranial, or visceral nerves.

Our modern knowledge of the detailed structure and function of the sympathetic nervous system is due chiefly to Gaskell and Langley ; and since it is best known in the higher vertebrates it will be well to begin by describing it in a Mammal. The sympathetic neurons are connected

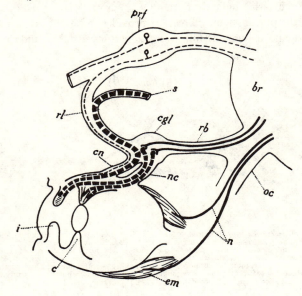

FIG. 748.

Diagram illustrating relation of dorsal cranial nerve (*prf*, profundus) and ventral cranial nerve (*oc*, oculomotor) with ciliary ganglion, *cgl*, and 'sympathetic'. Afferent fibres broken lines ; efferent fibres black ; 'sympathetic' fibres red. *br*, Brain ; *c*, ciliary muscle ; *cn*, nerve, with sympathetic fibres to iris ; *em*, external eye-muscle ; *i*, iris muscle ; *n*, motor fibres ; *nc*, ciliary nerve, with sympathetic fibres to iris and ciliary muscle ; *rb*, radix brevis ; *rl*, radix longa ; *s*, sympathetic nerve from cervical ganglion.

*Instead of red, **thick** broken lines.

with the central nervous system by four 'outflows' of preganglionic fibres :

(1) **The Mid-brain Outflow**, through the oculomotor nerve to the ciliary ganglion, supplies the iris sphincter and the ciliary muscle, Figs. 747-8-9.

(2) **The Bulbar** or **Hind-brain Outflow** through the facial, glosso-pharyngeal, and vagus nerves : through the 7th nerve to the spheno-palatine ganglion by the palatine or great superficial petrosal nerve supply-ing the lacrimal gland, and by the chorda tympani to the submaxillary gland ; through the 9th to the otic ganglion supplying the parotid ; through the 10th to sympathetic neurons in the heart (inhibitory), in the lungs, and

in the wall of the alimentary canal and its glands (including the liver) as far as the posterior end of the small intestine. It is these neurons of the alimentary canal which form the plexus of Auerbach (plexus myentericus) between the muscle layers, and the plexus of Meissner (plexus submucosus) near the mucous membrane of the stomach and small intestine, Figs. 745, 750 (see p. 782). Outflows (1) and (2) are related to cranial nerves and issue in front of the pectoral limb-plexus, Figs. 745-7.

(3) **The Thoracico-lumbar Outflow** is related to the spinal nerves between the pectoral and the pelvic limb-plexuses. The connector neurons, situated in the lateral column of the spinal cord, have axons which issue by the ventral spinal roots as medullated fibres which leave the mixed spinal nerves forming the white rami communicantes passing ventrally to a corresponding segmental series of vertebral or lateral ganglia on either side of the dorsal aorta. The vertebral ganglia are joined together by a longitudinal nerve trunk formed by the connector fibres and their collateral branches, and were originally no doubt distributed segmentally all along the body ; but they dwindle away in the tail, and in the cervical region run together to coalesce into a large superior and inferior ganglion still connected to the corresponding spinal nerves by segmental grey rami. Moreover the longitudinal nerve runs forward to connect with the cranial nerves and successive cranial sympathetic ganglia of outflows (1) and (2). From sympathetic neurons constituting all these lateral ganglia issue non-medullated fibres which pass as grey rami to the corresponding spinal nerves to supply the sweat-glands and smooth muscles of the skin (erector muscles of the hair, etc.) over the entire body. Sympathetic fibres also pass from the superior cervical ganglion to the radial ciliary muscle, the submaxillary gland (inhibitory), the buccal region, the larynx, the heart ; while the inferior cervical ganglion supplies the heart (accelerator) and larynx, Figs. 749 and 750.

More connector fibres pass out through the ventral thoracic roots and lateral ganglia into the median mesentery where they form the splanchnic nerves running to the unpaired prevertebral or collateral ganglia. Of these the most anterior, the coeliac, supplies the spleen, liver, and stomach; the superior mesenteric ganglion sends inhibitor fibres to the small intestine, and motor fibres to the ileo-caecal valve; from the most posterior ganglion, the inferior mesenteric, run fibres to the large intestine, the sphincter ani and sphincters of the urinary bladder and urethra, and muscles of the copulatory organs.

Lastly, a set of preganglionic connector fibres pass out still farther by the lumbar ventral roots and, running through the inferior mesenteric ganglion, join to a hypogastric nerve which branches to a plexus of sympathetic neurons supplying the muscular and glandular walls of the excretory and genital ducts.

(4) **The Sacral Outflow** takes place behind the pelvic plexus by the ventral roots of the first three sacral nerves, where connector fibres pass out and join to a pelvic nerve (nervus erigens) which branches to a peripheral plexus in the wall of the large intestine, rectum, and bladder, and supplies the blood-vessels of the external copulatory organs, Fig. 750.

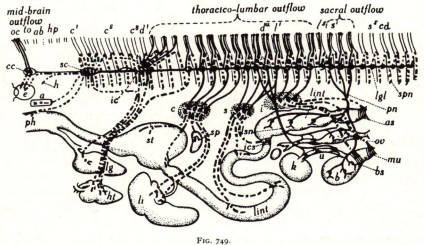

FIG. 749.

Diagram of Autonomic *ventral root system* of *Mammal* (modified from W. H. Gaskell, 1916). In black, preganglionic neurons of right side whose axons issue by ventral roots ; in red, postganglionic neurons. *a*, Aorta ; *ab*, abducens ; *as*, anal sphincter ; *bs*, bladder sphincter ; *c*, semilunar ganglion ; *cc*, ciliary ganglion ; *e*, eye ; *h*, branch to head ; *i*, inferior mesenteric ganglion ; *ic*, stellate ganglion ; *lgl*, chain of lateral ganglia ; *lsn*, lumbar splanchnic ; *mu*, muscle of urethra ; *pn*, pelvic nerve ; *s*, superior mesenteric ganglion ; *spn*, spinal nerve receiving sympathetic fibre from lateral chain by ramus communicans. Above are indicated ventral nerve roots from *oc*, oculomotor, anteriorly to fourth caudal posteriorly ; *to*, trochlear ; *ab*, abducens ; *c*$^{1-8}$, cervical ; *d*$^{1-12}$, dorsal or thoracic ; *l*$^{1-5}$, lumbar ; *s*$^{1-5}$, sacral ; *c*^{1}, caudal. Other letters as in Fig. 746.

*Instead of red, broken lines.

From the above description it appears that in the Mammal the autonomic system has been subdivided by physiologists into four divisions or outflows from the central nervous system ; but from the point of view of the comparative anatomist we may group these into two sets : (1) the dorsal root system comprising the outflow through the 7th, 9th, and 10th cranial nerves (and perhaps the nervus terminalis), and (2) the ventral root system comprising the outflows through the 3rd cranial nerve, the thoracico-lumbar and sacral nerves.

It is important to notice that an organ frequently receives a double autonomic supply, usually antagonistic. For instance the striated musculature of the heart receives accelerator impulses from the thoracic outflow and inhibitory impulses from the bulbar outflow, while the reverse is the case with the unstriated musculature. The bulbar outflow stimulates in the intestine peristaltic action which is inhibited by the thoracic outflow. The secretion of the submaxillary gland is increased by impulses through the chorda tympani branch of the 7th nerve, but decreased by impulses from the superior cervical ganglion. In these instances one supply comes from the dorsal and the other from the ventral root system ; but in others the double supply may issue through the ventral

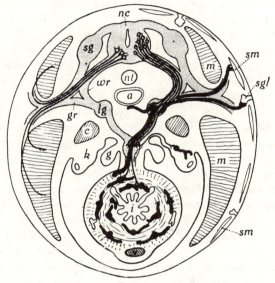

FIG. 750.

Efferent nervous system of trunk. Diagram of transverse section of trunk region of Gnathostome, showing distribution of excito-motor neurons in spinal nerve (black), and ' sympathetic ' neurons (red). Compare Fig. 745. *c*, Cardinal vein ; *g*, gonad ; *gr*, grey ramus ; *i*, intestine ; *lg*, lateral ganglion (median collateral ganglion shown dorsal to intestine) ; *m*, myomere ; *nc*, spinal cord ; *sgl*, skin gland ; *sm*, smooth muscle ; *wr*, white ramus.

*Instead of red, thick black lines.

roots, as in the case of the large intestine, though here the antagonism may possibly be not of quite the same nature.

Another remarkable fact is that excitor fibres of the skin (erector muscles and glands) are distributed over segmental areas corresponding to the sensory areas of the mixed spinal nerves in the branches of which they run. But those supplying the greater part of the skin of the head pass forwards from the superior cervical ganglion into branches of the fifth nerve. Similarly vasomotor fibres may pass forwards and be distributed in the branches of the sensory cranial nerves.

Although the sympathetic system is by no means so well known in the lower Vertebrates as in Mammals, yet enough has been made out to

establish that it is built on the same general principle in all the Gnathostomes. That of Birds and Reptiles differs only in detail from that of
the Mammal. Even in the Selachian (Chevrel, 1096 ; E. Müller, 1149) are
found essentially the same chain of paired lateral ganglia, splanchnic
nerves to unpaired collateral ganglia, and peripheral sympathetic neurons
and plexuses in the alimentary canal and other viscera, all belonging to the
ventral root system. A ciliary ganglion supplies the iris muscles. The
dorsal root system is represented by nerves and ganglia connected with
the 7th and 9th cranial nerves, as well as an extensive supply from the
vagus to the stomach, small intestine, and other parts. Some Teleosts
possess not only segmental lateral ganglia in the trunk, but also a
ganglion to each of the sensory cranial nerves, all connected by a
longitudinal cord (Allis, 402, 404), a condition possibly indicating a
primitive stage in which segmental ganglia were regularly distributed
along the whole body. In many Teleostomes, however, the system
seems to be less definitely differentiated, the lateral ganglia less
regular, and the sympathetic neurons scattered along the cardinal veins
(Kuntz, 1137).

In the Cyclostomes it is undoubtedly less well differentiated than in the
typical Gnathostome. The sympathetic fibres are here no longer distinguishable by the absence of a medullary sheath, since no fibres in these
lower Craniates have yet acquired such a sheath. The whole system is
difficult to make out, and is but very imperfectly known, especially in the
Myxinoids. The chain of lateral ganglia described by Julin in the larva
has not been found by subsequent observers, nor are collateral ganglia
known to occur ; but the thoracico-lumbar outflow seems to be represented by cells scattered along the cardinal veins, grouped near the segmental vessels and dorsal roots of the spinal nerves, where they are
associated with chromafine cells distributed in almost all the segments of
the body, and forming a mass on the sinus venosus (Giacomini, 1902-4 ;
Fusari, 1889 ; Favaro, 1924). Moreover, V. Kupffer and Johnson have
described sympathetic branches from the 7th and 10th cranial nerves to
the pharyngeal region, and the vagus further supplies the whole of the
alimentary canal. Here it is related to a plexus apparently representing Auerbach's plexus in the Gnathostomes (Dogiel, 1103 ; Smirnow ;
Brandt, 1093 ; Tretjakoff, 1178).

No definite sympathetic system has been found in *Amphioxus*.
Scattered ganglion cells have been described in the wall of its pharynx ;
and such fibres as pass to the vascular system and alimentary canal
come from visceral branches of the dorsal spinal roots, Fig. 707. In a
general way these may represent the vagal supply of the Gnathostomes

(Fusari, 1889 ; Dogiel, **1104** ; Heymans and Stricht, v. d., **1122** ; Kutchin, **1141** ; Johnston, **1127** (see p. 782)).

Development and Phylogeny.—If we turn to embryology for a clue to the origin of the sympathetic system, we find that, whereas Remak and the early observers believed that it developed *in situ* from mesoblastic

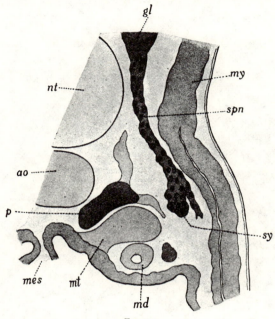

FIG. 751.

Scyllium canicula, embryo stage, L. Portion of transverse section showing developing mesonephros and sympathetic system. *mt*, Rudimentary mesonephric tubule not yet joined to duct, *md.* ; *sy*, rudiment of sympathetic lateral ganglion from spinal nerve.

tissues, the modern and almost universally adopted view is that it is entirely of epiblastic origin.

Balfour in 1877 first established the epiblastic origin of the lateral ganglia in Selachians, showing that they develop from the mixed spinal nerves below the spinal ganglia as outgrowths which pass towards the aorta, Figs. 751-2. In each segment the drawn-out connecting strand remains as the ramus communicans, while the longitudinal cord only develops later. These observations were confirmed and extended by Onodi in 1886, who traced from the same source both lateral and collateral ganglia in Selachians and Lizards. Similar views have been upheld by van Wijhe, 1889, Rabl, 1889, and Hoffmann, 1899, with regard to the Selachii, and since by others with regard to other groups of Gnathostomes,

Fig. 753 (Aves: Abel, 1077; Ganfini, 1110; Rau and Johnson, 1166). According to Schenk and Birdsall, 1878, the whole system in the body of the duck and mammal develops from the spinal ganglia. But it was His junior who first described in birds and mammals the whole system, including even the peripheral plexuses, as formed of epiblastic cells which actively

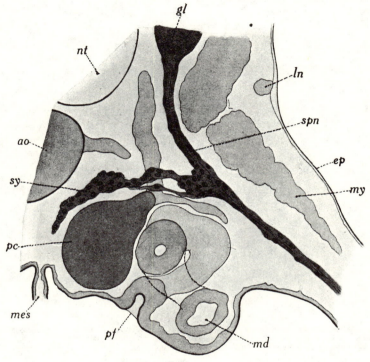

FIG. 752.

Scyllium canicula, embryo 23 mm. long, stage *M*. Portion of transverse section of anterior trunk segment showing developing mesonephros and sympathetic system. *ao*, Dorsal aorta; *ep*, epidermis; *gl*, spinal ganglion; *ln*, lateral-line nerve; *md*, mesonephric duct; *mes*, mesentery; *my*, myomere; *nt*, notochord; *pc*, posterior cardinal vein; *pf*, peritoneal funnel; *spn*, spinal nerve; *sy*, rudiment of lateral 'vertebral' sympathetic ganglion.

migrate from the neural tube to the spinal ganglia and thence farther and farther until they reach the remotest parts of the viscera. This active migration of independent cells into the mesenchyme towards the aorta seems to occur more in the higher than in the lower forms. Now, while many, like Held (1119), Marcus (1909), Kohn (1136), and E. Müller (1149, 1151), believe that the sympathetic cells migrate out through the dorsal root or are more directly produced from cells in the spinal ganglion, Froriep working on *Lepus* and *Torpedo,* and Cajal on the chick, maintain that these motor neurons pass out from the neural tube by the ventral roots (Froriep; also Harrison in *Salmo*, 'o1, and Carpenter and Main, 'o7). Hoffmann and Neumayer, on the other hand, hold that both roots are concerned; Kuntz in a series of important studies on the development of the sympathetic in Pisces, Amphibia, Reptilia, Aves, and Mammalia (**1137**), maintains that in all these cells migrate from the neural tube by both the dorsal and the ventral roots, a conclusion also supported by Ganfini (**1110**). Kuntz made the important observation that the cells of the peripheral plexuses supplied by the vagus, including the myenteric and submucous plexuses, all pass out along the branches of the vagus which supply them. Certainly cells do migrate out by the ventral roots, and the only doubt is as to whether they are destined to develop into sympathetic neurons and not merely sheath-cells; but Kuntz seems to have set this doubt at rest by showing that, if the ventral region of the neural tube be removed in the chick embryo, the sympathetic fails to develop completely. By similar experiments on the frog, however, E. Müller claims to have shown that no neurons come from the ventral roots (**1151**).

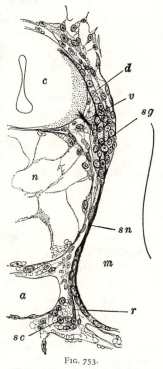

FIG. 753.

Transverse section through 8·6 mm. larva of *Rana esculenta*, illustrating relations of sympathetic cord and spinal nerve (after Held, from W. E. Kellicott, *Chordate Develt.*, 1913.) *a*, Dorsal aorta; *c*, spinal cord; *d*, dorsal (sensory afferent) root of spinal nerve; *m*, myotome; *n*, notochord; *r*, ramus communicans; *sc*, sympathetic cord; *sg*, spinal ganglion; *sn*, spinal nerve trunk; *v*, ventral (motor, efferent) root of spinal nerve.

The development of the cranial ganglia has been studied by Kuntz in man (**1139**) and by Stewart in the rat (**1174**). The otic ganglion is formed by cells migrating by the glossopharyngeal nerve, the submaxillary and sublingual ganglia of cells migrating by the chorda tympani, and the spheno-palatine of cells migrating by the palatine branch of the facial. These observations agree with the known course of the connector fibres. Kuntz, however, also describes a contribution to all three from the trigeminal ganglion. Moreover, the ciliary ganglion is said to be formed by cells migrating not only by the oculomotor nerve, but also by the profundus branch of the fifth nerve. Yet no connector fibres have been proved to be contained in this nerve.

The general result of these modern researches is that the sympathetic neurons are of epiblastic origin, and pass outwards in two main streams : one of cells of neural-tube or neural-crest origin which migrate through the dorsal roots and ganglia of the cranial and spinal nerves ; the other of cells from the lateral column of the central nervous system which migrate along the third cranial nerve and the ventral spinal roots. If these conclusions are correct it would follow that the emigrating cells may take the same course as the connector fibres in the adult. It is clear that this whole autonomic or ' sympathetic ' system might be defined as that part of the peripheral excitor nervous system whose special neurons are situated outside the central nervous system and segmental dorsal ganglia, and receive impulses through preganglionic fibres.

Gaskell long ago represented the whole efferent sympathetic system as formed by the outflowing of cells farther and farther from the central nervous system, with which they remain in connexion by correspondingly lengthening preganglionic connector fibres. Such indeed appears to be its ontogenetic development ; but it can scarcely be its phylogenetic history. For it cannot be held that the early Vertebrate had no efferent supply to its viscera, vascular system, skin muscles and glands. Rather may it be supposed that this peripheral nervous system, such as it occurs in Gnathostomes, has been differentiated out of a more diffuse general network, like that found in primitive Invertebrates, where the various efferent paths are imperfectly distinguished, and where the nerve fibres anastomose to a true network, perhaps not yet subdivided into neurons separated by synapses.

It may be suggested that in Vertebrates the central nervous system arose by the concentration of neurons originally distributed in such a diffuse network throughout the body ; and that the sympathetic system is composed of those excitor neurons which have not yet been so concentrated, but remain distributed at various points along the nerves and in the viscera. The history in phylogeny would then have been the reverse of what is believed to occur in ontogeny. But it must be confessed that the evidence for such a history is still far from satisfactory. A better knowledge of the structure, function, and development of the peripheral nervous system of the lower forms, more especially of the Cyclostomes, is urgently needed. A study of the peripheral nervous system in the Enteropneusta might also throw light on the early history of this system in the Vertebrata.

It is tempting to see the remains of a primitive network in the enteric plexus of Gnathostomes. The exact nature of this plexus is, however, by no means yet determined. A variety of cells have been described in it

(Dogiel, 1103; Cajal, 1893–4; Michailow; E. Müller, 1150, 1152; Kuntz, 1138; Hill, 1123), some of which have been held to be afferent, thus giving rise to reflex arcs within the wall of the gut; but this view has not been thoroughly established. Some authors describe the plexuses as composed of true networks (Bethe, 1903; E. Müller, 1149-50, 1152), while others describe them as made up of neurons separated by synapses (Cajal, 1911; Kuntz, 1138; Hill, 1123). According to E. Müller, in birds and mammals the myenteric plexus is formed of neurons related to vagal connector fibres, and the submucous plexus of a network related to fibres from the prevertebral ganglia; while in Selachians both would still be true networks.

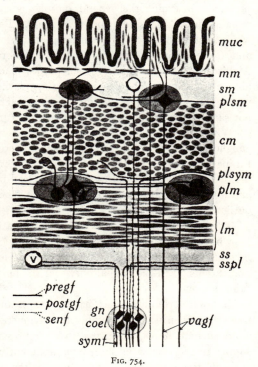

Fig. 754.

The complex structure of the nerve plexuses of the intestine of Mammals may be briefly described as follows according to the latest work (C. J. Hill, 1123). The ganglia of the myenteric plexus of Auerbach and the submucous plexus of Meissner contain sympathetic neurons connected with preganglionic fibres of the vagus nerve. These sympathetic neurons may not only supply the muscles, but also send fibres to other neurons in neighbouring ganglia and from one plexus to the other. Both the outer coat of longitudinal muscles and the inner coat of circular muscles are supplied from the myenteric plexus. The submucous plexus supplies the thin innermost coat of muscles belonging to the mucosa and the muscles of the villi.

Diagram illustrating relations of nervous plexuses in wall of intestine of Mammal, seen in longitudinal section (from C. J. Hill, *Tr. Roy. Soc.*, 1927). *cm*, Circular muscle; *gn.coel*, coeliac ganglion; *lm*, longitudinal muscle; *mm*, muscularis mucosae; *muc*, mucosa with villi projecting into lumen; *plm*, myenteric plexus of Auerbach; *plsm*, submucous plexus of Meissner; *plsym*, sympathetic plexus; *sm*, submucosa; *ss*, subserosa; *sspl*, subserous plexus; *symf*, sympathetic fibres; *v*, blood-vessel; *vagf*, vagal fibres; *pregf*, preganglionic fibre; *postgf*, postganglionic fibre; *senf*, sensory fibre.

Postganglionic sympathetic fibres pass down the mesentery from the collateral ganglia to the intestine and are distributed chiefly to the blood-vessels and possibly to the muscle cells, Fig. 754.

Dorsal- and Ventral-root Systems.—An interesting and very important problem concerning the general morphology of the autonomic system remains to be discussed (Goodrich, 1927). Physiologists subdivide the whole system into (*a*) sympathetic proper or thoracico-lumbar outflow, and (*b*) parasympathetic (mid-brain, hind-brain, and sacral outflows). A third division, (*c*) enteric, is sometimes made for the plexuses of the alimentary canal. The division into sympathetic and parasympathetic may be justified on physiological grounds (action of drugs, etc.), but is not satisfactory from the point of view of morphology (p. 776).

What strikes the morphologist is that some of the outflows of pre-ganglionic fibres pass out through dorsal (posterior) nerve-roots, and others through ventral (anterior) nerve-roots. The whole system may be sub-divided into (*a*) the dorsal-root system (Hind-brain outflow, Figs. 745-6), and (*b*) the ventral-root system (Mid-brain, Thoracico-lumbar, and Sacral outflows, Figs. 749, 750).

The problem then arises : How has the latter system been evolved ? How have 'sympathetic' neurons come to be connected with the central nervous system through ventral roots ? For the connexion through dorsal roots (as in the cranial nerves of the hind-brain) appears to be the primitive one.

In lower vertebrates, such as *Amphioxus* and *Petromyzon*, the ventral roots throughout the body supply the myomeres only, so far as we know. Any efferent fibres, belonging to or representing the 'sympathetic' supply to the viscera or skin, must apparently in these forms pass out through the dorsal roots, not only in the head but throughout the whole length of the body. This condition persists in the hind-brain region of the higher forms, where the facial, glossopharyngeal, and vagus nerves represent the dorsal roots (of three segments), which remain separate from their corresponding ventral roots (abducens and hypoglossal; see above, pp. 216, 221, and 226).

Moreover, embryology seems to show that the hind-brain outflow of sympathetic cells is through dorsal-root cranial nerves, and although the evidence is still somewhat conflicting with regard to the thoracico-lumbar and sacral outflows these seem to take place also to some extent through the dorsal roots of the spinal nerves.

Clearly, the evidence as a whole points to the autonomic system having been originally related to the dorsal roots, and it remains to be explained how in the Gnathostomes sympathetic neurons belonging to the mid-brain, thoracico-lumbar, and sacral outflows acquired a connexion with the central nervous system by means of preganglionic fibres issuing through the ventral roots. It can hardly be supposed that the whole 'ventral-root system' is peculiar to and has arisen in the Gnathostome Vertebrates.

BIBLIOGRAPHY

CHAPTER I

VERTEBRAL COLUMN AND RIBS AND STERNUM

1. **Albrecht, P.** Über d. Proatlas. *Zool. Anz.*, v. 3, 1880.
2. **Balfour, F. M.**, and **Parker, W. K.** On the Structure and Devel. of Lepidosteus. *Phil. Trans. R.S. London*, v. 173, 1882.
3. **Bardeen, C. R.** Devel. of Thoracic Vertebrae in Man. *Am. Jour. Anat.*, v. 4, 1905, and *Human Embryology*, 1910.
4. **Barge, J. A. J.** Entwickl. der Craniovertebralgrenze beim Schaf. *Anat. Hefte*, 1 Abt., v. 55, 1917.
5. **Baur, G.** Morph. der Wirbelsäule. *Biol. Centralbl.*, v. 6, 1886.
6. —— Proatlas, Atlas, and Axis of Crocodilia, *Amer. Nat.*, v. 20, 1886 ; Proatlas e. Schildkröte, *Anat. Anz.*, v. 10, 1895.
7. **Beer, G. R. de.** Devel. of Head in Heterodontus. *Quart. J. Micr. Sci.*, v. 68, 1924.
8. **Boeke, J.** Geldrollenstadium d. vertebraten Chorda. *Anat. Anz.*, v. 33, 1908.
9. **Brünauer, E.** Entwickl. d. Wirbelsäule b. d. Ringelnatter. *Arb. Zool. Inst. Wien*, v. 18, 1910.
10. **Budgett, J. S.** Structure of Larval Polypterus. *Trans. Zool. Soc.*, v. 16, 1902.
11. **Burlet, H. M. de.** Rudim. Wirbelkörper a. d. Spitze d. Dens epistrophei b. Bradypus. *Morph. Jahrb.*, v. 45, 1913 (Bibl.).
12. **Cope, E. D.** Rhachitomous Stegocephali and other papers. *Am. Nat.*, 1882–87.
13. —— Intercentrum of Terrestrial Vertebrata. *Trans. Amer. Phil. Soc.*, Philadelphia, v. 16, 1888.
14. **Cornet, Y.** Le prétendu proatlas des mammifères et de hatteria. *Bull. Ac. R. Belg.*, v. 15, 1888.
15. **Corning, H. K.** Neugliederung der Wirbelsäule, etc. *Morph. Jahrb.*, v. 17, 1891.
16. **Davison, A.** Devel. of Vert. Column and its Appendages (Amphiuma). *Anat. Anz.*, v. 14, 1897.
17. **Dean, B.** Chimaeroid Fishes. Carnegie Instit., Washington, 1906.
18. **Dollo.** Sur le Proatlas. *Zool. Jahrb.*, Abt. Anat., v. 3, 1889.
19. **Ebner, V. v.** Urwirbel u. Neugliederung der Wirbelsäule. *Sitz. Ak. Wiss. Wien*, v. 97, 1888 ; and v. 101, 1892.
20. —— Wirbel der Knochenfische u. die Chorda dorsalis der Fische u. Amphibien. *Sitz. Ak. Wiss. Wien*, v. 105, 1896.
21. **Field, H. H.** Bemerkungen über die Entwicklung der Wirbelsäule bei

der Amphibien ; nebst Schilderung eines abnormen Wirbelsegmentes. *Morph. Jahrb.*, v. 22, 1895, p. 340.

22. **Fortman, J. P. de G.** Entwickl. der Wirbelsäule von Megalobatrachus. *Tijdschr. Ned. Dierk. Ver.*, v. 16, 1918.

23. **Fritsch, A.** *Fauna der Gaskohle* . . . 4 vols. Prag., 1883–1901.

24. **Froriep, A.** Entwickl. der Wirbelsäule, insbes. des Atlas, Epistropheus, u. der Occipitalregion. *Arch. Anat. u. Phys.*, Anat. Abt., 1883 and 1886.

25. **Fürbringer, K.** Morph. d. Skeletes der Dipnoer, etc. *Denk. Med. Nat. Ges. Jena*, vol. 4, 1904.

26. **Gadow, H.** Evolution of Vertebral Column of Amphibia and Amniota. *Phil. Trans. R.S. London*, v. 187, 1896.

27. **Gadow, H.,** and **Abbott, E. C.** Evol. of Vertebral Column in Fishes. *Phil. Trans. R.S. London*, v. 186, 1895.

28. **Gaudry, A.** *Les Enchaînements du monde animal.* Paris, 1883.

29. **Gegenbaur, C.** *Vergl. Anat. der Wirbelsäule bei Amphibien u. Reptilien.* Leipzig, 1862.

30. —— Entw. der Wirbelsäule des Lepidosteus. *Jen. Zeitschr.*, v. 3, 1867.

31. **Goette, A.** Beitr. z. vergl. Morph. d. Skeletsystems. *Arch. f. mikr. Anat.*, v. 14, 1877 ; v. 15, 1878; v. 16, 1879.

32. —— *Entw. der Unke (Bombinator).* Leipzig, 1875.

33. —— Wirbelsystem. *Zool. Anz.*, v. 1, 1878 ; and v. 17, 1894.

34. —— Wirbelbau bei d. Reptilien. *Zeitschr. wiss. Zool.*, v. 62, 1897.

35. **Goodrich, E. S.** Vertebrata Craniata, *Treatise on Zoology*, Pt. 9. London, 1909.

36. —— On Scales of Fish. *Proc. Zool. Soc. London*, 1908 ; and Bone of Fishes, *ibid.*, 1913.

37. **Grassi, B.** Entw. der Wirbelsäule der Teleostier. *Morph. Jahrb.*, v. 18, 1882.

38. **Hasse, C.** *Das naturl. Syst. d. Elasmobranchier.* 2 Parts and supplement. Jena, 1882–85.

39. —— Entwickl. d. Wirbelsäule : Triton. *Zeitschr. wiss. Zool.*, v. 53, 1892.

40. —— Elasmobranchier, Dipnoi. *Ibid.*, v. 55, 1893.

41. —— Ganoiden. *Ibid.*, v. 57, 1894.

42. —— Cyclostomen. *Ibid.*, v. 57, 1894.

43. **Hatschek, B.** Studien z. Segmenttheorie des Wirbeltierkopfes. *Morph. Jahrb.*, v. 40, 1910.

44. **Hatta, S.** Segmentation of Mesoblast in Petromyzon, etc. *Annot. Zool. Japan*, v. 4, 1901.

45. **Hay, O. P.** Vertebral Column of Amia. *Field Columb. Mus. Publ. Zool.*, v. 6, 1895.

46. **Hayek, H.** Proatlas u. Kopfgelenke b. Säugetieren. *Sitz. Ak. Wiss. Wien*, v. 130, 1923.

47. —— Schicksal d. Proatlas u. Entw. d. Kopfgelenke b. Reptilien u. Vögeln. *Morph. Jahrb.*, v. 53, 1924 (Bibl.).

48. **Higgins, G. M.** Devel. of Primitive Reptilian Vertebral Column (Alligator). *Am. J. Anat.*, v. 31, 1923.

49. **Howes, G. B.,** and **Swinnerton, H. H.** Devel. of Skeleton of Tuatara, Sphenodon punctatus. *Tr. Zool. Soc. London*, v. 16, 1901.

50. **Huene, F. v.** Phylog. Bedeutung des Wirbelbau der Tetrapoden. *Paläont. Zeitschr.*, v. 7, 1926.

51. **Jaekel, O.** Organisation von *Archegosaurus*. *Zeit. d. Deutsch. Geol. Ges.*, v. 48, 1896.

52. Kapelkin, W. Entwickl. des axialen Skeletts der Amphibien. *Bull. Soc. Nat. Moscou*, v. 14, 1900.

53. Klaatsch, H. Vergl. Anat. der Wirbelsäule: I. Urzustand der Fishwirbelsäule; II. Bildung knorpeliger Wirbelkörper bei Fischen; III. Phylogenese der Chordascheiden. *Morph. Jahrb.*, v. 19, 1893; v. 20, 1893; v. 22, 1895.

54. Kölliker, A. Unters. ü. d. Scheiden der Chorda dorsalis. *Verh. Phys.-med. Ges. Würzburg*, v. 9, 1859; v. 10, 1860; N.F., v. 3, 1872.

54a. —— *Ende d. Wirbelsäule d. Ganoiden*, etc., 1860.

55. Kuppfer, C. von. *Vergl. Entw. d. Kranioten.* I. Acipenser, 1893. II. and III. Petromyzon, 1894. München u. Leipzig.

56. Lankester, E. R. Contrib. to Knowledge of Amphioxus. *Quart. J. Micr. Sci.*, v. 29, 1889.

57. Lotz, Th. Bau der Schwanzwirbelsäule der Salmoniden, etc. *Zeitschr. f. wiss. Zool.*, v. 14, 1864.

58. Lwoff, B. Studien ü. d. Chorda u. Chordascheide. *Bull. Soc. Nat. Moscou*, 1887.

59. Macallister, A. Devel. and Variation of Atlas. *J. of Anat. and Physiol.*, v. 27, 1893; Devel. and Variation of Second Cervical Vertebra. *Ibid.*, v. 28, 1894.

60. Männer, H. Entw. d. Wirbelsäule b. Reptilien. *Zeitschr. f. wiss. Zool.*, v. 66, 1899.

61. Männich, H. Entwickl. der Wirbelsäule v. Eudyptes. *Jen. Zeitschr.*, v. 37, 1903.

62. Marcus, H., u. Blume, W. Wirbel u. Rippen bei Hypogeophis. *Zeitschr. f. Anat. u. Entw.*, v. 80, 1926.

63. Marshall, W. Bau der Vögel. *Naturw. Bibl. Leipzig*, 1895.

64. Murray, J. A. Vert. Column of Primitive Urodela. *Anat. Anz.*, v. 13, 1897.

65. Osborn, H. F. Intercentra and Hypapophyses in Mosasaurs, Lizards, and Sphenodon. *Amer. Nat.*, v. 34, 1900.

66. Parker, T. J. Anat. and Devel. of Apteryx. *Phil. Trans. R.S. London*, v. 182, 1891; v. 183, 1892.

67. Parker, W. K. Skeleton of the Marsipobranch Fishes. *Phil. Trans. R.S.*, v. 174, 1883.

68. Peter, K. Wirbelsäule der Gymnophionen. *Ber. Nat. Ges. Freiburg*, v. 9, 1895.

69. —— Bedeutung des Atlas der Amphibien. *Anat. Anz.*, v. 10, 1895 (Bibl.).

70. Piiper, J. Evolution of Vertebral Column in Birds. *Phil. Trans. R.S.* B. v. 216, 1928.

71, 72. Rabl, C. Theorie d. Mesoderms. *Morph. Jahrb.*, v. 15, 1889; v. 19, 1892.

73. Rabinerson, A. Vergl. Anat. der Wirbelsäule der Knorpelfische. *Anat. Anz.*, 59, 1925.

74. Ridewood, W. G. Devel. of Vertebral Column in Pipa and Xenopus. *Anat. Anz.*, v. 13, 1897, p. 359.

75. —— Caudal Diplospondyly of Sharks. *J. Linn. Soc. Zool.*, v. 27, 1899.

76. —— Calcification of Vertebral Centra in Sharks and Rays. *Phil. Trans. R.S. London*, v. 210, B., 1921.

77. Rosenberg, E. Entw. der Wirbelsäule des Menschen. *Morph. Jahrb.*, v. 1, 1875.

78. **Schauinsland, H.** Entw. d. Wirbelsäule nebst Rippen u. Brustbein. *Hertwigs Handb. Entw. Wirbeltiere*, v. 3, Jena, 1906 (Bibl.).
79. **Scheel, C.** Entw. der Teleostierwirbelsäule. *Morph. Jahrb.*, v. 20, 1893.
80. **Schmidt, L.** Wirbelbaues von Amia Calva. *Zeitschr. f. wiss. Zool.*, v. 54, 1892.
81. **Schneider, A.** *Beitr. z. vergl. Anat. der Wirbeltiere*, Berlin, 1879.
82. **Schultze, O.** Embryonale u. bleibende Segmentierung. *Verh. Anat. Ges.*, Berlin, 1896.
83. **Schwarz, H.** Morphogenie der Wirbelsäule. *Sitzb. Ges. Nat. Fr.*, Berlin, 1908 ; *Paläontogr.* v. 21, 1908.
84. **Sečerov, S.** Entstehung der Diplospondylie der Selachier. *Arb. Zool. Inst. Wien*, v. 19, 1911.
85. **Shufeldt, R. W.** Osteology of Amia Calva. *Rep. U.S. Com. of Fish and Fisheries* (1883), 1885.
86. **Sunier, A. L. J.** Différentiation interne du Myotome, etc. *Tijdschr. Nederl. Dierk. Ver.*, v. 12, 1911.
87. **Tretjakoff, D.** Wirbelsäule des Neunauges. *Anat. Anz.*, v. 61, 1914.
88. —— Chordascheide u. Wirbel bei Zyklostomen u. Fische. *Zeitschr. Zellf. u. mikr. Anat.*, v. 4, 1924.
89. —— Chordascheiden der Urodelen. *Z. Zellforsch. u. mikr. Anat.*, v. 5, 1927.
90. **Ussow, S.** Anat. u. Entwickl. der Wirbelsäule der Teleostier. *Bull. Soc. Imp. Nat. Moscou*, 1900.
91. **Weiss, A.** Entwickl. d. Wirbelsäule d. Ratte. *Z. wiss. Zool.*, v. 69, 1901.
92. **Wieland, G. R.** Terminology of Vertebral Centra. *Am. Journ. of Sci.*, v. 8, 1899.
93. **Wijhe, J. W. v.** Entwickl. des Kopf- und Rumpfskeletts von Acanthias. *Bijdr. tot de Dierk.*, Amsterdam, 1922.
94. **Willey, A.** *Amphioxus and Ancestry of Vertebrates*, 1894.
95. **Williston, S. W.** *Osteology of Reptiles.* Harvard U.P., 1925.
96. **Welcker, H.** Bau u. Entw. der Wirbelsäule. *Zool. Anz.*, v. 1, 1878.
97. **Woodland, W. N. F.** Caudal Anatomy and Regeneration in Gecko. *Quart. J. Micr. Sci.*, v. 65, 1920.
98. **Zittel, K. A.** *Handbuch den Paläontologie*, v. 3 (Pisces, Amphibia, Reptilia, Aves), v. 4 (Mammalia), Munich, 1887, and later translations (Bibl.).
99. **Zykoff, W.** Knorpels der Chorda bei Siredon. *Bull. Soc. Nat. Moscou*, 1893.

RIBS AND STERNUM

100. **Baur, G.** Morphology of Ribs. *Amer. Nat.*, v. 21, 1887.
101. —— Morph. of Ribs and Fate of Actinosts in Fishes. *J. of Morph.*, v. 3, 1889.
102. —— Rippen u. ähnliche Gebilde. *Anat. Anz.*, v. 9, 1894.
103. **Boulenger, G. A.** On the Nature of ' Haematophyses '. *Ann. Mag. Nat. Hist.*, v. 12, 1893.
104. **Dollo, L.** Sur la morph. des côtes. *Bull. sc. de la France et Belgique*, v. 24, 1892 ; Sur la morph. de la colonne vertébrale. *Ibid.*, v. 25, 1893.
105. **Gamble, D. L.** Morph. of Ribs and Transv. Processes in Necturus. *J. Morph.*, v. 36, 1922.
106. **Göppert, E.** Kenntniss der Amphibienrippen. *Morph. Jahrb.*, v. 22, 1895.
107. —— Unters. z. Morph. der Fischrippen. *Morph. Jahrb.*, v. 23, 1895.

108. **Göppert, E.** Morph. der Amphibienrippen. *Gegenbaur Festschr.*, 1, 1896.
109. **Hatschek, B.** Rippen der Wirbeltiere. *Verh. Anat. Ges.*, Berlin, 1889.
110. **Hoffmann, C. K.** Morphologie der Rippen. *Niederl. Archiv. Zool.*, v. 4, 1878.
111. **Knickmeyer, C.** *Entwickl. der Rippen bei Triton*, Munich, 1891.
112. **Mayerhofer, F.** Morph. u. Entwickl. der Rippensystems der Urodelen. *Arb. Zool. Inst. Wien*, v. 17, 1909.
112a. **Paterson, A. M.** Sternum, devel. etc. *J. Anat. and Phys.*, v. 35, 1901.
112b. **Ruge, G.** Entw. a. Brustbein. *Morph. Jahrb.*, v. 18, 1880.
113 **Parker, T. J.** Origin of the Sternum. *Trans. New Zeal. Inst.*, v. 23, 1891.
114. **Schöne, G.** Befestigungen der Rippen an der Wirbelsäule. *Morph. Jahrb.*, v. 31, 1902.
115. **Whitehead, R. H.,** and **Waddell, J. A.** Early Devel. of Mammalian Sternum. *Am. Jour. Anat.*, v. 12, 1911.

CHAPTER II

MEDIAN FINS

116. **Balfour, F. M.,** and **Parker, W. K.** Structure and Devel. of Lepidosteus. *Phil. Trans. R.S. London*, v. 173, 1882.
117. **Breder, C. M.** Locomotion of Fishes. *Zoologica, N.Y. Zool. Soc.*, v. 4, 1926.
118. **Bridge, T. W.** Mesial Fins of Ganoids and Teleosts. *J. Linn. Soc. Zool.*, v. 25, 1896.
119. **Dollo, L.** Sur la phylogénie des dipneustes. *Bull. Soc. Belge Géol.*, v. 9, 1895.
120. —— Éthologie paléont. relative aux poissons. *Bull. Soc. Belge Géol. et Pal.*, v. 20.
121. **Emery, C.** Fierasfer. *Fauna u. Flora d. Golfes v. Neapel*, Monogr. 2, 1880.
122. **Goodrich, E. S.** Dermal Fin-rays of Fishes. *Q. Journ. Micr. Sci.*, v. 47, 1903.
123. **Gregory, W. K.** Orders of Teleostomous Fishes. *Ann. N.Y. Acad. Sci.*, v. 17, 1907.
124. **Hatschek, B.** Schichtenbau von Amphioxus. *Anat. Anz.*, v. 3, 1888.
125. **Huxley, T. H.** Devel. of some Parts of Skeleton of Fishes. *Q. Journ. Micr. Sci.*, v. 7, 1859.
126. **Kiaer, J.** Downtonian Fauna of Norway : I. Anaspida. *Vidensk. Skr. Mat. Naturw. Kl.*, 1924.
127. **Kirkaldy, J. W.** Genera and Species of Branchiostomidae. *Quart. J. Micr. Sci.*, v. 37, 1895.
128. **Mayer, P.** Unpaaren Flossen der Selachier. *Mitth. Zool. Sta. Neapel*, v. 6, 1885.
129. **Regan, C. T.** Caudal Fin of Elopidae. *Ann. Mag. Nat. Hist.*, ser. 8, v. 5, 1910.
130. **Ryder, J. A.** Origin of Heterocercy and Evol. of Fins and Fin-rays of Fishes. *U.S. Com. of Fish. Report for 1884*, 1886.
131. **Salensky, W.** Entw. der unpaaren Flossen der störartigen Fische. *Ann. Mus. Zool. Ac. Imp. St-Pétersbourg*, 1899.
132. **Schaffer, J.** Bau u. Entw. des Schwanzflossenknorpels von Petromyzon. *Anat. Anz.*, 19, 1901.

133. **Schmalhausen, J. J.** Morph. der unpaaren Flossen : I. Entwickl. *Zeitschr. wiss. Zool.*, vol. 100, 1912 ; II. Bau u. Phylogenese. *Ibid.*, vol. 104, 1913.

134. **Stensiö, E. A. S.** Triassic Fishes from Spitzbergen : II. Saurichthyidae. *K. Svenska Vet.-Akad. Handlingar*, v. 2, 1925.

135. —— Downtonian and Devonian Vertebrates of Spitzbergen (Cephalaspidae). *Norske Videnskaps-Akad. Oslo*, No. 12, 1927.

136. **Totton, A. K.** Struct. and Devel. of Caudal Skel. of Pleurogramma. *Proc. Zool. Soc. London*, 1914.

137, 138. **Traquair, R. H.** Palaeospondylus Gunni. *Proc. Roy. Phys. Soc. Edinburgh*, v. 12, 1893 ; *Proc. Zool. Soc.*, 1897.

139. **Whitehouse, R. H.** Caudal Fin of Teleostomi. *Proc. Zool. Soc. London*, 1910.

139. —— Caudal Fin of Fishes. *Proc. Roy. Soc.*, London, B, v. 82, 1910.

140. —— Evol. of Caudal Fin. *Rec. Ind. Mus.*, v. 15, 1918.

CHAPTER III

PAIRED FINS

141. **Agar, W. E.** Devel. of Ant. Mesoderm and Paired Fins in Lepidosiren and Protopterus. *Trans. Roy. Soc. Edinburgh*, vol. xlv., 1907.

142. **Balfour, F. M.** Devel. of Skeleton of Paired Fins of Elasmobranchs. *Proc. Zool. Soc.*, 1881.

143. **Bateson, W.** *Materials for Study of Variation*, London, 1894.

144. **Braus, H.** Innervation der paarigen Extremitäten bei Selachiern. *Jen. Zeitschr.*, v. 29, 1895.

145. —— Entwickl. der Muskulatur u. des periph. Nervensyst. d. Selachier : I. Teil, Die metotischen Urwirbel ; II. Teil, Die paarigen Gliedmassen. *Morph. Jahrb.*, v. 27, 1899.

146. —— Versteinerter Gliedmassenknorpel von Selachiern. *Verh. Phys. Med. Ges. Würzburg*, 34, 1901.

147. —— Muskeln u. Nerven der Ceratodusflosse. *Zool. Forsch. Denkschr. Med. Nat. Ges.*, Jena, 1901.

148. —— Entw. d. Form der Extremitäten u. d. Extremitätenskelets. *Hertwigs Handb. Entw. Wirbeltiere*, v. 3, Jena, 1906.

148a. —— Segmentalstruct. d. mot. Nervenplexus. *Anat. Anz.*, v. 34.

149. **Brongniart, C.** Pleuracanthus Gaudryi. *Bull. Soc. Géol. de France*, v. 16, 1888.

150. **Broom, R.** Origin of Cheiropterygium. *Bull. Am. Mus. N.H.*, v. 32, 1913.

151. **Bunge, A.** Biserial Archipterygium bei Selachiern u. Dipnoern. *Jen. Zeitschr. f. Naturw.*, v. 8, 1874.

152. **Cope, E. D.** Homologies of Fins of Fishes. *Am. Nat.*, v. 24, 1890.

152a. —— On Symmorium. *Amer. Natur.*, v. 27, 1893.

153. **Davidoff, M. v.** Vergl. Anatomie der hinteren Gliedmasse der Fische : Pt. I. *Morph. Jahrb.*, 5, 1879 ; Pt. II. *Morph. Jahrb.*, 6, 1880 ; Pt. III. *Morph. Jahrb.*, 9, 1883.

154. **Dean, B.** Morph. of Cladoselache. *Journ. Morph.*, v. 9, 1894 ; *Trans. N. York Acad. Sci.*, v. 13, 1894.

155. —— *Fishes Living and Fossil*, New York, 1895 (Bibl.).

156. **Dean, B.** Fin-fold Origin of Paired Limbs, in Light of Ptychopterygia of Palaeozoic Sharks. *Anat. Anz.*, v. 11, 1896.

157. **Derjugin, K.** Bau u. Entwickl. d. Schultergürtels u. d. Brustflossen b. d. Teleostiern. *Zeitschr. wiss. Zool.*, v. 96, 1910.

158. **Döderlein, L.** Skelet v. Pleuracanthus. *Zool. Anz.*, v. 12, 1889.

159. **Dohrn, A.** Paarigen u. unpaaren Flossen der Selachier. *Mitt. Zool. Sta. Neapel*, v. 5,, 1884.

159a. —— Unpaare Flosse, u. d. Reste der Beckenflosse bei Petromyzon, *Mitt. Zool. Sta. Neapel*, v. 6, 1885.

160. **Ducret, E.** Dével. des membres pairs et impairs des téleostéens. *Trutta Inaug. Diss.*, Lausanne, 1894.

161. **Frechkop, S.** Struct. et dével. de l'organe copulateur des raies. *Arch. de Biol.*, v. 35, 1925.

162. **Fürbringer, M.** Lehre v. den Umbildungen der Nervenplexus. *Morph. Jahrb.*, v. 5, 1879.

163. —— *Unters. z. Morph. u. System der Vögel*, Jena, 1888.

164. —— Über d. spino-occipitalen Nerven. *Festschr. z. Geb. C. Gegenbaur*, v. 3, 1897.

165. **Gadow, H.** Myologie d. hinteren Extremitäten der Reptilien. *Morph. Jahrb.*, v. 7, 1881.

166. **Gegenbaur, C.** *Unters. z. vergl. Anat. d. Wirbeltiere*, Leipzig, Heft 1, 1864 ; Heft 2, 1865 ; Heft 3, 1872.

167. —— Skelet der Gliedmassen in allgemein und der Hintergliedmassen der Selachier insbesondere ; Skelet der Hintergl. bei der Männchen der *Selachier* und *Chimären*. *Jen. Zeit. Med. u. Naturw.*, v. 5, 1870.

168. —— Ueber das Archipterygium. *Jen. Zeit. Med. Nat.*, v. 7, 1873. Flossenskelet der Crossopterygier u. Archipterygium. *Morph. Jahrb.*, v. 22, 1895.

169. —— Morph. der Gliedmaassen der Wirbelthiere. *Morph. Jahrb.*, v. 2, 1876.

170. —— *Vergl. Anatomie der Wirbeltiere.* 2 vols. Leipzig, 1898.

171. **Goodrich, E. S.** On the Pelvic Girdle and Fin of Eusthenopteron. *Quart. J. Micr. Sci.*, v. 45, 1901.

172. —— Devel., etc., of the Fins of Fish. *Quart. J. Micr. Sci.*, v. 50, 1906.

173. —— Segmental Struct. of Motor Nerve-plexus. *Anat. Anz.*, v. 36, 1910.

173a. —— Metameric Segmentation and Homology. *Quart. J. Micr. Sci.*, v. 59, 1913.

174. **Gregory, W. K., Miner, R. W.,** and **Noble, G. K.** Carpus of Eryops and Primitive Cheiropterygium. *Bull. Am. Mus. N.H.*, v. 48, 1923.

175. **Guitel, F.** Dével. des nageoires paires du Cyclopterus lumpus. *Arch. Zool. Expér.*, v. 4, 1896.

176. **Haller, B.** Schultergürtel der Teleostier. *Arch. mikr. Anat.*, v. 67, 1905.

177. **Hamburger, R.** Paarigen Extremitäten von Squalus, Trigla, Periophthalmus u. Lophius. *Rev. Suisse de Zool.*, 12, 1904.

178. **Hammarsten, O. D.** Innervation d. Bauchflossen b. d. Teleostiern. *Morph. Jahrb.*, v. 42, 1911.

179. **Harrison, R. G.** Entw. der Flossen der Teleostier. *Arch. f. mikr. Anat.*, v. 46, 1895.

180. **Haswell, W. A.** Paired Fins of Ceratodus. *Proc. Linn. Soc. N.S. Wales*, v. 7, 1883.

181. —— Elasmobranch Skeleton. *Ibid.*, v. 9, 1885.

182. **Herringham, W. P.** Minute Anat. of Brachial Plexus. *Proc. Roy. Soc. London*, v. 41, 1886.

183. **Howes, G. B.** Skeleton and Affinities of Paired Fins of Ceratodus. *Proc. Zool. Soc. London*, 1887.

184. —— Pectoral Fin-Skeleton of Living Batoid Fishes and of Extinct Squaloraja. *Proc. Zool. Soc. London*, 1890.

185. **Huber, O.** Kopulationsglieder der Selachier. *Z. wiss. Zool.*, v. 70, 1901.

186. **Jaekel, O.** Ü. d. Organisation der Pleuracanthiden. *Sitz. Ges. naturf. Fr. Berlin*, 1895.

187. **Jhering, H. v.** *Das periphere Nervensystem der Wirbelthiere*, Leipzig, 1878.

188. **Jungersen, H. F. E.** Bauchflossenanhänge der Selachiermännchen. *Anat. Anz.*, v. 14, 1898.

189. —— Appendices Genit. in Greenland Shark, Somniosus, etc. *Danish Ingolf Exped.*, v. 2, Copenhagen, 1899.

190. **Kiaer, J.** Downtonian Fauna of Norway : I. Anaspida. *Videnskaps. Skr. Mat. naturw. Kl.*, 1924.

191. —— Mouth of oldest known Vertebrates. *Palaeobiologica*, v. 1, 1928.

192. **Klaatsch, H.** Brustflosse der Crossopterygier. *Festschr. f. C. Gegenbaur*, Leipzig, 1896.

193. **Kryžanovsky, S.** Entwickl. d. paarigen Flossen bei Acipenser, Amia u. Lepidosteus. *Acta Zool.*, v. 8, 1927.

194. **Lankester, E. R.** The Cephalaspidae (and Pteraspidae). *Mon. Palaeontogr. Soc.*, 1868–70.

195. **Lebedinsky, N. G.** Entwickl.-mech. Unters. an Amphibien, II. *Arb. Vergl. Anat. Inst. Lettl. Univ.*, Riga, 1925.

195a. **Leigh-Sharpe, W. H.** Sec. Sexual Char. of Elasmobranchs. *J. of Morph. and Phys.*, v. 34, 1920 ; v. 35, 1921 ; v. 39, 1924.

196. **Mivart, St. G.** On Fins of Elasmobranchs. *Trans. Zool. Soc. London*, v. 10, 1879.

197. **Mollier, S.** Die paarigen Extremitäten der Wirbeltiere : I. Das Ichthyopterygium. *Anat. Hefte*, Abt. 1, v. 3, 1893.

197a. —— Die paarigen Extremitäten der Wirbeltiere : II. Das Cheiropterygium. *Anat. Hefte*, Abt. 1, v. 5. 1894–95.

198. —— Entw. d. paar. Flossen des Störs. *Anat. Hefte*, Abt. 1, v. 8, 1897.

199. **Müller, E.** Brustflosse der Selachier. *Anat. Hefte*, v. 39, 1909.

200. —— Muskeln u. Nerven der Brustflosse bei Acanthias. *Ibid.*, v. 43, 1911.

201. —— Anat. u. Entwickl. des peripheren Nervensystems b. d. Selachiern. *Arch. mikr. Anat.*, v. 81, 1913.

202. **Osburn, R. C.** Origin of the Paired Limbs of Vertebrates. *Am. J. Anatomy*, v. 7, 1907.

203. **Paterson, A. M.** Position of Mammalian Limb. *J. Anat. and Phys.*, v. 23, 1889.

204. —— Origin and Distrib. of Nerves of Lower Limb. *Ibid.*, v. 28, 1893.

205. **Petronievics, B.** Pectoral Fin of Eusthenopteron. *Ann. Mag. Nat. Hist.*, v. 2, 1918.

206. **Rabl, C.** Ursprung der Extremitäten. *Zeitschr. wiss. Zool.*, v. 70, 1901.

207. **Rautenfeld, E. v.** Skelett der hinteren Gliedmassen von Ganoiden und Teleostiern. *Inaug. Diss.*, Dorpat, 1882 ; and *Morph. Jahrb.*, v. 9, 1884.

208. **Regan, C. T.** Phylogeny of Teleostomi. *Ann. Mag. Nat. Hist.*, v. 13, 1904.

209. **Rosenberg, E.** Entwickl. der Wirbelsäule, etc. *Morph. Jahrb.*, v. 1, 1875.

210. **Rosenberg, E.** Modus des Zustandekommens der Regionen a. d. Wirbelsäule des Menschen. *Ibid.*, v. 36, 1907.

211. **Ruge, E.** Entw. d. Skelettes d. Extremität v. *Spinax niger. Morph. Jahrb.*, vol. 30, 1902.

212. **Ruge, G.** Verschiebungen i. d. Endgebieten der Nerven des Plexus lumbalis. *Morph. Jahrb.*, v. 20, 1893.

213. **Salensky, W.** Dével. de l'ichthyoptérygie des poissons ganoides et dipnoides. *Ann. Mus. Zool. Ac. Imp. St-Pétersbourg*, v. 3, 1898.

214. **Semon, R.** Entw. d. paarigen Flossen der Ceratodus. *Denkschr. Med. Nat. Ges. Jena*, v. 4, 1898.

215. **Sewertzoff, A. N.** Entwickl. d. Muskeln, etc., d. Extremitäten d. niederen Tetrapoda. *Bull. Soc. Imp. Nat. Moscou*, 1907.

215. —— Dével. des extrémités de Chamaeleo. *Jour. Russe de Zool.*, v. 1, 1916.

216. —— Devel. of Pelvic Fins of Acipenser. *J. Morph. and Physiol.*, v. 41, 1926.

217. —— Morph. d. Brustflossen d. Fische. *Jen. Zeitschr. f. Naturw.*, v. 62, 1926.

218. **Stensiö, E. A. S.** *Triassic Fishes from Spitzbergen*, Vienna, 1921.

219. **Thacher, J. K.** Median and Paired Fins. *Trans. Connecticut Acad.*, v. 3, 1877 ; Ventral Fins of Ganoids. *Ibid.*, v. 4, 1878.

220. **Traquair, R. H.** Structure and Affinities of Tristichopterus alatus. *Trans. Roy. Soc. Edinburgh*, v. 27, 1874.

221. —— Lower Devonian Fishes of Gemünden. *Trans. Roy. Soc. Edinburgh*, v. 40, 1893 ; Supplement. *Ibid.*, v. 41, 1895.

222. —— On Cladodus Neilsoni. *Trans. Geol. Soc. Glasgow*, v. 11, 1897.

223. —— On Thelodus Pagei, from Old Red Sandstone of Forfarshire. *Trans. Roy. Soc. Edinburgh*, v. 39, 1898.

223a. —— Fossil Fishes . . . in Silurian Rocks of South of Scotland. *Trans. Roy. Soc. Edinburgh*, v. 39, 1898.

224. —— Devonian Fishes of Campbeltown and Scaumenac Bay. *Geol. Mag.*, v. 7, 1890 ; v. 10, 1893.

225. —— Drepanaspis Gemündenensis. *Geol. Mag.*, v. 7, 1900 ; v. 9, 1902.

226. **Vogel, K.** Entw. d. Schultergürtels u. d. Brustflossen Skelettes der Forelle (Trutta). *Jena. Zeitschr. Naturw.*, v. 45, 1909.

227. **Watson, D. M. S.** Cheiropterygium. *Geol. Mag.*, D. 6, v. 1, 1914.

228. **Welker, H.** Bau u. Entwickl. der Wirbelsäule. *Zool. Anz.*, v. 1, 1878.

229. **Whiteaves, J. F.** Fossil Fishes of Devonian Rocks of Canada. *Trans. Roy. Soc. Canada*, v. 4, 1887 ; and v. 6, 1888.

230. **Wiedersheim, R.** *Gliedmassenskelet der Wirbeltiere*, Jena, 1892.

231. **Woodward, A. Smith.** *Outlines of Vertebrate Palaeontology* Cambridge, 1898 (Bibl.),

CHAPTER IV

Pectoral and Pelvic Girdles

232. **Andrews, C. W.** Devel. of Shoulder Girdle in Plesiosaur. *Ann. Mag. Nat. Hist.*, v. 15, 1895.

233. —— *Cat. of Marine Reptiles of Oxford Clay.* 2 Pts. London, 1910–13.

234. **Anthony, R.** Morph. of the Shoulder Girdle. 17th Internat. Cong. Med., 1913.

235. **Anthony, R., et Vallois H.** Ceinture scapulaire chez les Batraciens. *Bibl. Anat.*, v. 24, 1914.

236. **Baur, G.** Pelvis of Testudinata, etc. *Jour. Morph.*, v. 4, 1891.

237. **Bogoljubsky, S.** Brustbein etc. b. e. Lacertiliern. *Zeitschr. wiss. Zool.*, v. 110, 1914.

238. **Braus, H.** Brustschulterapparat der Froschlurche. *Sitzb. d. Heidelberger Akad. d. Wiss.*, 1919.

239. **Broom, R.** Sterno-Coracoidal Articulation in a Foetal Marsupial. *J. Anat. and Phys.*, v. 31, 1899. Devel. and Morph. of Marsupial Shoulder Girdle. *Trans. Roy. Soc. Edinb.*, v. 39, 1900.

239a. —— Structure of Mesosaurus. *Trans. S. Afr. Phil. Soc.*, v. 15, 1904.

240. —— Lacertilian Shoulder Girdle. Devel. of App. Skel. of Ostrich. *Trans. S. Afr. Philos. Soc.*, v. 16, 1906.

241. —— Morphology of Coracoid. *Anat. Anz.*, v. 41, 1912.

242. **Bunge, A. v.** *Entwickl. des Beckens der Amphibien, Reptilien u. Vögel.* Dorpat, 1880.

243. **Cope, E. D.** Shoulder Girdle and Extremities of Eryops. *Trans. Am. Phil. Soc.*, v. 16, 1890.

244. —— Degenerate Types of Scapular and Pelvic Arches in Lacertilia. *J. Morph.*, v. 7, 1892.

245. **Eggeling, H.** Morph. des Manubrium Sterni. *Festschr. z. E. Haeckel*, 1904.

245. —— Clavicula, etc. *Anat. Anz.*, v. 29, 1906.

246. **Eisler, P.** Homologie der Extremitäten. *Abh. naturf. Ges. Halle*, v. 19, 1893–95.

247. **Fraas, E.** *Ichthyosaurier d. süddeutschen Trias u. Jura.* Tübingen, 1891.

248. **Fraas, O.** Pterodactylus suevicus. *Palaeontographica*, v. 25, 1878.

249. **Fuchs, H.** Entw. u. vergl. Anat. des Brustschultergürtels der Wirbeltiere: 1. Clavicula bei Talpa etc. *Zeitschr. f. Morph. u. Anthrop.*, 1912. 2. Schultergürtel der Amphibia anura. *Ibid.*, v. 22, 1922. 3. *Ibid.*, v. 24, 1924. 5. (Cartilago procoracoidea, etc.). *Anat. Anz.*, 61, 1926. 6. (Praezonales etc.). *Verh. Anat. Ges. Freiburg*, 1926. 7. Amphibien, Sauriern u. Testudinaten. *Anat. Anz.*, 64, 1927.

250. **Fürbringer, M.** Vergl. Anat. des Brustschulterapparates. *Jen. Zeitschr.*, v. 7, 1873; v. 8, 1874; v. 34, 1900; v. 36, 1902.

250. —— *Morph. Jahrb.*, v. 1, 1875.

251. **Gegenbaur, C.** Clavicula u. Cleithrum. *Morph. Jahrb.*, v. 23, 1895.

252. **Gelderen, C. van.** Devel. of Shoulder Girdle and Episternum in Reptiles. *Proc. Kon. Acad. v. Wet. Amsterdam*, v. 26, 1923.

253. —— Entw. des Brustschulterapparats bei Sauriern. *Anat. Anz.*, v. 59, 1925.

254. **Goette, A.** Morph. des Skelettsystems der Wirbeltiere: Brustbein u. Schultergürtel. *Arch. mikr. Anat.*, v. 14, 1877.

255. **Goodrich, E. S.** Pectoral Girdle in young Clupeids. *J. Linn. Soc. London, Zool.*, v. 34, 1922.

256. **Gregory, W. K.** Origin of Birds. *Ann. N. York Ac. Sci.*, v. 27, 1916.

256a. **Gregory, W. K., and Camp, C. L.** Comp. Myology and Osteology. *Bull. Am. Mus. N.H.*, v. 38, 1918.

257. **Hanson, F. B.** Ontogeny and Phylogeny of the Sternum. *Am. Jour. Anat.*, v. 26, 1919.

258. —— Problem of the Coracoid. *Anat. Rec.*, v. 19, 1920.

259. **Hart, P. C.** Entw. d. Schultergürtels der Schildkröten. *Tijdschr. Niederl. Dierk. Ver.*, v. 19, 1925.
260. **Haswell, W. A.** Elasmobranch Skeleton. *Proc. Linn. Soc. N.S.W.*, v. 9, 1884.
261. **Hoffmann, C. K.** Vergl. Anat. der Wirbeltiere : Schultergürtel u. Brustbein. *Niederl. Arch. Zool.*, v. 5, 1879.
262. **Hommes, J. H.** Devel. of Clavicula and Sternum in Birds and Mammals. *Tijdschr. Niederl. Dierk. Ver.*, v. 19, 1924.
263. **Howes, G. B.** Morph. of Mammalian Coracoid. *J. Anat. and Phys.*, v. 21, 1887.
264. —— Coracoid of Terrestrial Vertebrata. *P.Z.S.*, 1893.
265. **v. Huene, Fr.** Übersicht über der Reptilien der Trias. *Geol. u. Paläont. Abh.*, v. 6, 1902.
266. —— Praepubisfrage bei Dinosaurien u. anderen Reptilien. *Anat. Anz.*, v. 33, 1908.
267. —— Erythrosuchus (Pelycosimia). *Geol. Paläontol. Abh.*, v. 10, 1911.
268. **Huxley, T. H.** Animals intermediate between Reptiles and Birds. *Proc. Roy. Soc. London*, 1868.
269. —— Affinity between Dinosaurian Reptiles and Birds. *Quart. J. Geol. Soc. London*, v. 26, 1869.
270. **Johnson, A.** Devel. of Pelvic Girdle of Chick. *Quart. Jour. Micr. Sci.*, v. 23, 1883.
271. **Juhn, M.** Entw. d. Sternums bei Lacerta. *Acta Zool.*, v. 4, 1923.
272. **Knopfli, W.** Entw. des Brustschulterskeletts b. d. Vögeln. *Jena. Z. f. Naturw.*, 55, 1919 (Bibl.).
273. **Kravetz, L. P.** Entw. des Sternum- u. des Episternalapparats der Säugetiere. *Bull. Soc. Imp. Nat. Moscou*, v. 19, 1905 (1906).
274. **Lebedinsky, N. G.** Morph. u. Entw. des Vogelbeckens. *Jena. Zeitschr.*, v. 50, 1913.
275. —— Entw.-mech. Unters. an Amphibien : II. Die Umformungen des Grenzwirbels bei Triton, etc. *Biol. Zentralbl.*, v. 45, 1925.
276. **Levin, B.** Topogr. d. n. obturatorius in vorknorpeligen Vogelbecken. *Arb. Zool. Inst. Lettland. Univ.*, Riga, 1926.
277. **Lindsay, B.** Avian Sternum. *Proc. Zool. Soc. London*, 1885.
278. **Lydekker, R.** Coracoidal Element in Adult Sloths. *Proc. Zool. Soc.*, 1893.
279. **McGregor, J. H..** The Phytosauria. *Mem. Am. Mus. Nat. Hist.*, v. 9, 1906.
279a. —— Mesosaurus. *Comm. de Estudos d. Minas do Brazil*, 1908.
280. **Mehnert, E.** Entwickl. des os pelvis der Vögel. *Morph. Jahrb.*, v. 13, 1888.
281. —— Entwickl. des Beckengürtels bei Säugetieren. *Ibid.*, v. 15. 1889.
282. ——Entwickl. des Beckengürtels der Emys. *Ibid.*, v. 16, 1890.
283. **Merriam, J. C.** Triassic Ichthyosauria. *Mem. Univ. California*, v. 1, 1908.
284. **Meyer, H. v.** Ueber Belodon. *Palaeontogr.*, v. 10, 1863.
285. **Miner, R. W.** Pectoral Limb of Eryops, etc. *Bull. Am. Mus. N.H.*, v. 51, 1925.
286. **Nauck, E. Th.** Vergl. Morph. des Beutelknochens. *Morph. Jahrb.*, v. 55, 1925.
287. —— Entw. d. ventralen Schultergürtelabschnittes b. Alytes obstetricans. *Morph. Jahrb.*, v. 60, 1928.

287a. **Nauck, E. Th.** Episternum von Echidna. *Anat. Anz.*, v. 67, 1929.
288. **Parker, W. K.** Shoulder Girdle and Sternum in Vertebrata. *Ray Society*, 1868.
289. **Petronievics, B.,** and **Woodward, A. S.** Pectoral and Pelvic Arches of Archaeopteryx. *Proc. Zool. Soc. London*, 1917.
290. **Romer, A. S.** Comparison of Mammalian and Reptilian Coracoids. *Anat. Rec.*, v. 24, 1922.
291. ——— Ilium in Dinosaurs and Birds. *Bull. Am. Mus. N.H.*, v. 46, 1922.
292. ——— Locomotor Apparatus of Mammal-like Reptiles. *Ibid.*
293. ——— Pelvic Musculature of Saurischian Dinosaurs. *Bull. Am. Mus. N.H.*, v. 48, 1923.
294. ——— Pectoral Limb and Shoulder Girdle in Fish and Tetrapods. *Anat. Rec.*, 1924.
295. ——— P. m. of Ornithischian Dinosaurs. *Acta Zool.*, v. 8, 1927.
296. **Schmalhausen, I.** Dermal Bones of Shoulder Girdle of Amphibia. *Rev. Zool. Russe*, v. 2, 1917.
297. **Seeley, H. G.** *Ornithosauria.* Cambridge, 1870.
297a. ——— Mesosauria. *Quart. J. Geol. Soc.*, v. 48, 1892.
298. ——— Nature of Shoulder Girdle in Sauropterygia. *Proc. Roy. Soc. London*, v. 51 and v. 54, 1892–93.
299. **Siebenrock, F.** Verbindungsweise des Schultergürtels mit den Schädel bei den Teleosteern. *Ann. K.K. Nat. Hofmuseums*, Wien, v. 16, 1901.
300. **Swinnerton, H. H.** Pectoral Skeleton of Teleosteans. *Quart. J. Micr. Sci.*, v. 49, 1905.
301. **Swirski, G.** Entw. des Schultergürtels u. d. Skel. d. Brustflosse des Hechtes. *Inaug. Diss.*, Dorpat, 1880.
302. **Vialleton, L** Ceinture pectorale et thorax des tétrapodes. Basin des tétrapodes. *Bull. Ac. Sc. Montpellier*, 1917.
303. **Villiers, C. G. S. de.** Brustschulterapp. b. d. Anuren. *Acta Zool.*, 3, 1922.
304. **Vogel, R.** Entw. d. Schultergürtels u. d. Brustflossenskelettes der Forelle. *Jena. Zeitschr. f. Naturw.*, v. 45, 1909.
305. **Wasnetzoff, W. W.** Function des Mesocoracoid. *Rev. Zool. Russe*, v. 3, 1922.
306. **Watson, D. M. S.** Batrachiderpeton lineatum. *Proc. Zool. Soc. London*, v. 2, 1913.
307. ——— Evol. of Tetrapod Shoulder Girdle. *J. of Anat.*, v. 5, 1917.
308. ——— Elasmosaurid Shoulder Girdle. *Proc. Zool. Soc. London*, 1924.
309. **Wiedersheim, R.** Becken der Fische. *Morph. Jahrb.*, v. 7, 1881.
310. ——— Phylogenie der Beutelknochen. *Zeitschr. wiss. Zool.*, v. 53, 1892.
310a. ——— Comp. Anatomy of Vertebrates, transl. by W. N. Parker, 3rd ed., 1907.
311. **Whipple, J. L.** Ypsiloid Apparatus of Urodeles. *Biol. Bull.*, v. 10, 1906.

CHAPTER V

Morphology of Head Region

312. **Addens, J. L.** Eye-muscle Nerves of Petromyzonts. *Proc. R. Acad. Sci. Amsterdam*, v. 31, 1928.
313. **Adelmann, H. B.** Devel. of Premandibular Head Cavities, etc. *J. of Morph. and Phys.*, v. 42, 1926.
314. ——— Development of Eye-muscles of Chick. *J. Morph.*, v. 44, 1927.

315. **Ayers, H.** Unity of Gnathostome Type. *Am. Nat.*, v. 40, 1906.
316. —— Vertebrate Cephalogenesis : V. Origin of Jaw Apparatus and Trigeminus Complex, Amphioxus, etc. *J. Comp. Neur.*, v. 33, 1921.
317. **Balfour, F. M.** The Devel. of Elasmobranch Fishes. *Journ. Anat. and Phys.*, 1876–78.
318.. —— *Comparative Embryology,* v. 2, London, 1881.
319. **Beccari, N.** Scheletro i miotomi di Trota, etc. *Arch. Ital. di Anat. ed Embriol.*, v. 19, 1922.
320. **Beer, G. R. de.** Segmentation of Head in Squalus. *Quart. Jour. Micr. Sci.*, v. 66, 1922.
321. —— Devel. of Head in Heterodontus. *Quart. J. Micr. Sci.*, v. 68, 1924.
322. —— Prootic Somites of Heterodontus and Amia. *Quart. J. Micr. Sci.*, v. 68, 1924.
323. —— Orbitotemporal Region of Skull. *Quart. Jour. Micr. Sci.*, v. 70, 1926,
324. —— Devel. of Skull in Torpedo. *Ibid.*
325. **Born, G.** Über d. Nasenhöhlen u. d. Thränennasengang der Amphibien. *Morph. Jahrb.*, 2, 1876.
326. **Brock, G. T.** Devel. of Skull of Leptodeira. *Quart. Jour. Micr. Sci.*, v. 73, 1929.
327. **Brohmer, P.** Der Kopf e. Embryo v. Chlamydoselachus. *Jena Zeitschr.*, v. 44, 1909.
327a. **Broom, R.** Homology of Alisphenoid. *Rep. S. Afr. Assoc.*, 1907.
327b. —— Devel. of Marsupial Skull. *Proc. Linn. Soc. N.S. Wales*, v. 34, 1909.
328. **Burlet, H. M. de.** Entwickl. des Walsschädels (Balaenoptera). *Morph. Jahrb.*, v. 49, 1914, and v. 50, 1916.
329. **Chiarugi, G.** Dével. des nerfs vague, etc., chez les sauropsides et mammifères. Myotomes et nerfs, etc., des Amphibies anoures. *Arch. Ital. Biol.*, v. 13, 1890, and 15, 1891.
330. **Cords, E.** Nervensorgung d. Augenm. v. Petromyzon. *Anat. Anz.*, v. 66, 1928.
331. **Corning, H. K.** Entwickl. d. Kopf- und Extremitätenmuskulatur b. Reptilien. *Morph. Jahrb.*, v. 28, 1900.
332. —— Vergl. Anat. der Augenmuskulatur. *Morph. Jahrb.*, v. 29, 1900.
333. **Dohrn, A.** Stud. z. Urgeschichte d. Wirbeltierkörpers. Visceralbogen b. Petromyzon. *Mitt. Zool. Sta. Neapel*, v. 5, 1884. Zungenbein u. Kiefer d. Selachier. *Ibid.*, v. 6, 1885.
334. —— Mandibularhöhle der Selachier. Praemandibularhöhle. *Mitt. Zool. Sta. Neapel*, v. 17, 1904.
335. **Dücker, M.** Augen der Zyklostomen. *Jen. Zeitschr. f. Naturw.*, 60, 1924.
336. **Elliot, A. I. M.** Devel. of Frog. Part 1 : Segments of Occipital Region of Skull. *Quart. Jour. Micr. Sci.*, v. 51, 1907.
337. **Filatoff, D.** Metamerie der Kopfes v. Emys. *Morph. Jahrb.*, v. 37, 1907.
338. **Fraser, E. A.** Head-cavities and Devel. of Eye-muscles in Trichosurus. *Proc. Zool. Soc. London*, 1915.
339. **Froriep, A.** Occipitalen Urwirbel der Amnioten in Vergleich mit d. Selachier. *Verh. Anat. Ges.*, 1905.
340. **Fürbringer, M.** Über die Spino-occipitalen Nerven. *Festschr. v. C. Gegenbaur*, v. 3, Leipzig, 1897.
340a. **Fürbringer, P.** Muskulatur d. Kopfsk. d. Cyclostomen. *Jen. Zeitschr.*, v. 9, 1875.
341 **Gast, R.** Entwickl. des Oculomotorius, etc. *Mitt. Zool. Sta.*, v. 19, 1909.

342. **Gaupp, E.**   Metamerie des Schädels.   *Anat. Hefte Ergeb.*, v. 7, 1897 (Bibl.).

343. —— Entwicklung des Kopfskelettes.   *Hertwigs Handb. Entw. Wirbeltiere*, vol. iii., Jena, 1905 (Bibl.).

344. **Gegenbaur, C.**   Kopfnerven v. Hexanchus.   *Jen. Zeitschr.*, v. 6, 1871.

345. —— Über d. Occipitalregion . . . der Fische.   *Festschr. z. A. v. Kölliker*, Leipzig, 1887.

346. —— Metamerie des Kopfes.   *Morph. Jahrb.*, v. 13, 1888.

347. **Goodrich, E. S.**   Segmentation of Occipital Region in Urodela.   *Proc. Zool. Soc. London*, 1911.

348. —— ' Proboscis Pores ' in Craniate Vertebrates.   *Quart. Jour. Micr. Sci.*, v. 62, 1917.

349. —— Devel. of Segments of Head in Scyllium.   *Quart. Jour. Micr. Sci.*, v. 63, 1918.

350. **Hatschek, B.**   Studien ü. Entwickl. von Amphioxus.   *Arb. Zool. Inst. Wien*, v. 4, 1881.

351. —— Mitth. ü. Amphioxus.   *Zool. Anz.*, v. 7, 1884.

352. —— Metamerie des Amphioxus u. des Ammocoetes.   *Verh. Anat. Ges. Wien*, 1892.   Corrected *Anat. Anz.*, 8, 1893.

353. **Higgins, G. M.**   Nasal Organs in Amphibia.   *Illinois Biol. Monogr.*, v. 6, 1920 (Bibl.).

354. **Hoffmann, C. K.**   Entw. der Selachii.   *Morph. Jahrb.*, v. 27, 1899.

355. **Huxley, T. H.**   On the Theory of the Vertebrate Skull.   *Proc. Royal Soc.*, v. 9, 1857–59.

356. **Jackson, W. H.**, and **Clarke, W. B.**   The Brain and Cranial Nerves of Echinorhinus spinosus, etc.   *J. Anat. and Phys.*, v. 10, 1876.

357. **Jager, J.**   Segmentierung der Hinterhauptregion.   *Bez. der Cartilago Acrochordalis zur Mesodermcommissur*, Gröningen, 1924.

358. **Johnson, C. E.**   Devel. of Prootic Somites, etc. (Chelydra).   *Amer. J. Anat.*, v. 14, 1913.

359. **Johnston, J. B.**   Morph. of Vertebrate Head.   *Jour. Comp. Neurology*, v. 15, 1905.

360. **Kingsbury, B. F.**, and **Adelmann, H. B.**   Morph. Plan of Head.   *Quart. J. Micr. Sci.*, v. 68, 1924.

361. **Koltzoff, N. K.**   Entwickl. d. Kopfes v. Petromyzon Planeri.   *Anat. Anz.*, v. 16, 1899 ; *Bull. Soc. Imp. Nat. Moscou*, v. 15, 1901.

362. **Krawetz, L.**   Entw. d. Knorpelschädels v. Ceratodus.   *Bull. Soc. Nat. Moscou*, v. 24, 1911.

363. **Kupffer, C. von.**   *Entwickl. des Kopfes von Ammocoetes.*   München u. Leipzig, 1894.

363a. **Lamb, A. B.**   Devel. of Eye-Muscles in Acanthias.   *Am. J. Anat.*, v. 1, 1902.

364. **Levi, G.**   Contrib. a. conosc. d. condrocranio cer. d. Mammiferi.   *Monit. Z. Ital.*, Anno 20, 1909.

365. **Marshall, A. M.**   On the Head Cavities, etc.   *Q. Jour. Micr. Sci.*, v. 21, 1881.

366. —— Segmental Value of Cranial Nerves.   *Jour. Anat. and Phys.*, v. 16, 1882.

367. **Matys, W.**   Entwickl. der Muskulatur der Orbita b. Vögeln.   *Arch. Anat. u. Phys.*, Anat. Abt. 1908.

368. **Neal, H. V.**   Segmentation of Nervous Syst. in Squalus acanthias.   *Bull. Mus. Comp. Zool. Harvard*, v. 31, 1898.

369. **Neal, H. V.** History of Eye Muscles. *J. of Morph.*, v. 30, 1918.
370. **Noordenbos, W.** Entw. des Chondrocraniums der Säugetiere. *Petrus Camper*, v. 3, 1905.
371. **Pedaschenko, D.** Entw. der Augenmuskelnerven. *Anat. Anz.*, v. 47, 1914.
372. **Pehrson, T.** Devel. of Teleostomian Fishes. *Acta Zool.*, v. 3, 1922.
373. **Platt, J. B.** Morph. of Vertebrate Head, based on a Study of Acanthias. *J. Morph.*, v. 5, 1891.
374. —— Devel. of Cartil. Skull in Necturus. *Morph. Jahrb.*, v. 25, 1897.
375. **Rex, H.** Entwickl. d. Augenmuskulatur der Ente. *Arch. mikr. Anat.*, v. 57, 1900.
376. **Rice, E. L.** Devel. of Skull in Skink, Eumeces. *J. of Morph.*, v. 34, 1920 (Bibl.).
377. **Rosenberg, E.** *Die Occipitalregion des Cranium* . . . *einiger Selachier*, Dorpat, 1884.
378. **Sagemehl, M.** Beitr. z. vergl. Anat. der Fische. I. Das Cranium von Amia calva. *Morph. Jahrb.*, v. 9, 1884. III. Das Cranium der Characiniden. *Ibid.*, v. 10, 1884. IV. Das Cranium der Cyprinoiden. *Ibid.*, v. 17, 1891.
379. **Salvi, G.** Cavités prémandibulaires . . . chez les Sauriens. *Bibl. Anat.*, v. 10, 1902.
380. **Scammon, R. E.** Devel. of Squalus. *Norm. Tafeln*, 1911.
381. **Schreiner, K. E.** Occipitalregion von Amia und Lepidosteus. *Zeitschr. f. wiss. Zool.*, vol. 72, 1902.
382. **Seters, W. H. van.** Dével. du chondrocrâne *d'Alytes obstetricans*. *Arch. de Biol.*, 32, 1922.
383. **Sewertzoff, A.** Entw. der Occipitalregion der niederen Vertebraten. *Bull. Soc. Imp. Nat. Moscou*, 1895.
384. **Sewertzoff, A. N.** Entwickl. des Selachierschädels. *Festschr. C. v. Kuppfer*, Jena, 1899.
385. **Sonies, F.** Entwickl. d. Chondrocraniums u. d. Knorp. Wirbelsäule b. d. Vögeln. *Petrus Camper*, v. 4, 1907.
386. **Stöhr, P.** Entw. d. Urodelenschädels. *Zeitschr. f. wiss. Zool.*, v. 33, 1879.
387. —— Entw. d. Anurenschädels. *Ibid.*, v. 36, 1881.
388. —— Entwickl. des Kopfskelettes der Teleostier. *Festschr. der J. Max, Univ. Würzburg*, v. 2, 1882.
389. **Swinnerton, H. H.** Morph. of Teleostean Head Skeleton (Gasterosteus aculeatus). *Quart. Jour. Micr. Sci.*, v. 45, 1902.
390. **Terry, R. J.** Primordial Cranium of Cat. *J. of Morph.*, v. 29, 1917.
391. **Toeplitz, Ch.** Bau u. Entwickl. der Knorpelschädels v. Didelphys. *Zoologica*, v. 27, 1920.
392. **Veit, O.** Besonderheiten am Primordialcranium v. Lepidosteus. *Anat. Hefte*, v. 33, 1907.
393. —— Entw. d. Primordialcranium von *Lepidosteus osseus*. *Anat. Hefte*, 1ste Abt. 44, 1911.
394. **Voit, M.** Primordialkranium des Kaninchens. *Anat. Hefte*, v. 38, 1909.
395. —— Abducensbrücke b. Menschen. *Anat. Anz.*, v. 52, 1919.
396. **Wijhe, J. W. van.** Mesodermsegmente d. Selachierkopfes. *Verh. K. Akad. Wet. Amsterdam*, v. 22, 1882 ; reprinted 1915.
397. —— Entwickl. des Kopf- und Rumpfskeletts von Acanthias. *Bijdr. t. d. Dierkunde*, v. 22, 1922.

3 F

398. **Wilson, J. T.** Taenia clino-orbitalis in Echidna and Ornithorhynchus. *J. Anat. und Physiol.*, v. 40, 1906.

399. **Ziegler, H. E.** Phylog. Entstehung des Kopfes. *Jen. Zeitschr. Naturw.*, v. 43, 1908.

CHAPTER VI

Skull

400. **Adelmann, H. B.** Prechordal Plate. *Am. J. Anat.*, v. 21, 1922.

401. **Allis, E. P.** Petrosal Bone and Sphenoidal Region of Skull of *Amia calva. Zool. Bull.*, v. 1, 1897.

402. —— Cranial Muscles and Cranial and First Spinal Nerves in *Amia calva. J. Morph.*, v. 12, 1897.

403. —— Squamosal Intercalar Exoccipital Extrascapular Bones in *Amia calva. Anat. Anz.*, v. 16, 1899.

404. —— Skull and Cranial Muscles, etc., in Scomber. *J. Morph.*, v. 18, 1903.

405. —— Cranial Anat. of Mail-cheeked Fishes. *Zoologica*, v. 22, 1909.

406. —— Pituitary Fossa and Trigemino-facialis Chamber in Selachians. *Anat. Anz.*, v. 46, 1914.

407. —— in Ceratodus. *Ibid.*, v. 46, 1914.

408. —— Trigemino-facialis Chamber in Amphibians and Reptiles. *Ibid.*, v. 47, 1914.

409. —— Myodome, etc., in Fishes. *J. of Morph.*, v. 32, 1919.

410. —— Cranial Anatomy of Polypterus. *J. of Anat.*, v. 56, 1922.

410a. —— Myodome, etc., in Coelacanthidae, etc. *Ibid.*

411. —— Palatoquadrate in Coelacanthid Fishes. *Proc. Zool. Soc. London*, 1923.

412. —— Cranial Anatomy of Chlamydoselachus. *Acta Zool.*, v. 4, 1923.

413. —— Pituitary Fossa, Myodome, etc. *J. of Anat.*, v. 63, 1928 (Bibl.).

414. **Baur, G.** Phylog. Arrangement of Sauropsida. *J. of Morph.*, v. 1, 1887.

415. —— Morph. of the Vertebrate Skull. *J. of Morph.*, v. 3, 1889.

416. —— Temporal Part of the Skull. *Am. Nat.*, v. 29, 1895.

417. —— Osteol. d. Schläfergegend d. höheren Wirbeltiere. *Anat. Anz.*, v. 10, 1895.

418. —— Morphologie des Unterkiefers der Reptilien. *Anat. Anz.*, 11, 1895.

419 —— The Stegocephali. *Anat. Anz.*, v. 11, 1896.

420. —— Systematische Stellung der Microsaurier. *Anat. Anz.*, v. 14, 1898.

421. **Beer, G. R. de.** Studies on Vertebrate Head: Part 1: Fish. *Quart. J. Micr. Sci.*, v. 68, 1924. Part 2: Orbito-temporal region. *Ibid.*, v. 70, 1926.

421a. —— Devel. of Skull in Sturgeon, *Ibid.*, v. 69, 1925.

422. —— Early devel. of Chondrocranium of Salmo fario. *Quart. J. Micr. Sci.*, v. 71, 1927.

422a. —— Skull of Shrew. *Phil. Trans. R.S.*, v. 217, 1929.

423. **Bemmelen, J. F. van.** Schädelbau der Monotremen. *Denkschr. Med. Nat. Ges. Jena*, v. 6, 1901.

424. **Boas, J. E. V.** Schläfenüberdachung, etc. *Morph. Jahrb.*, v. 49, 1914.

425. **Böker, H.** Schädel v. *Salmo salars. Anat. Hefte*, v. 49, 1913.

426. **Boulenger, G. A.** Fishes (Teleostei). *Cambridge Nat. Hist.*, London, 1904.

426a. **Bradley, O. C.** Muscles, etc., of Skull in Lacertilia. *Zool. Jahrb.*, Abt. Anat., v. 18, 1903.

427. **Branson, E. B.** Structure and Relationships of American Labyrinthodontidae. *J. of Geol.*, v. 13, 1905.

428. —— Skull of Pariotichus. *J. of Geol.*, v. 19, 1911.

429. **Bridge, T. W.** Cranial Osteology of Amia calva. *J. of Anat. and Phys.*, v. 11, 1877.

430. —— Osteology of Polyodon folium. *Phil. Trans. R.S.*, v. 169, 1879.

431. —— Cranial Anatomy of Polypterus. *Proc. Phil. Soc. Birmingham*, v. 6, 1887–89.

432. —— Skull of Osteoglossum formosum. *Proc. Zool. Soc.*, 1895.

433. —— Skull of Lepidosiren. *Trans. Zool. Soc.*, v. 14, 1897.

434. —— Fishes. *Cambridge Nat. Hist.*, v. 7, London, 1904.

435. **Broili, F.** Eryops megacephalus. *Palaeontogr.*, v. 46, 1899.

436. —— Pelycosaurierreste v. Texas. *Z. d. Deutsch. Geol. Ges.*, v. 56, 1904.

437. —— Stammreptilien. *Anat. Anz.*, v. 25, 1904.

438. —— Unpaare Elemente im Schädel von Tetrapoden. *Anat. Anz.*, v. 49, 1917.

439. **Broom, R.** Nasal-floor Bone in the Hairy Armadillo. *J. Anat. and Phys.*, v. 31, 1897.

440. —— Prevomer in Gomphognathus. *J. Anat. and Phys.*, v. 31, 1897.

441. —— Palate in Dicynodon. *Tr. S. Af. Phil. Soc.*, v. 11, 1901.

441a. —— Vomerine Bones. *Proc. Linn. Soc. N.S.W.*, 1904.

442. —— S. Afr. Pseudosuchian Euparkeria. *Ann. S. Afr. Mus.*, v. 3, 1906 ; *Proc. Zool. Soc. London*, 1913. Youngina. *Ibid.*, 1914.

443. —— Origin of Mammal-like Reptiles. *Proc. Zool. Soc. London*, 1908.

444. —— Permian Reptiles. *Bull. Am. Mus. Nat. Hist.*, v. 28, 1910.

445. —— Skull of Cynodonts. *Proc. Zool. Soc. London*, 1911.

446. —— Mesosuchus browni, etc. *Rec. Albany Mus.*, v. 2, 1913.

447. —— Squamosal in Mosasaurs and Lizards. *Ibid.*, v. 32, 1913.

448. —— Permian Temnospondylous Stegocephalians. *Ibid.*, v. 32, 1913.

449. —— Cotylosaurian, Pantylus. *Ibid.*, v. 32, 1913.

450. —— Structure of Mandible in Stegocephalia. *Anat. Anz.*, 45, 1913.

451. —— Origin of Mammals. *Phil. Trans. R.S.*, v. 206, B, 1914.

452. —— Diadectid Skull. *Bull. Am. Mus. Nat. Hist.*, v. 33, 1914.

453. —— S. Af. Dinocephalians and Am. Pelycosaurs. *Ibid.*, v. 33, 1914.

454. —— Permian Triassic and Jurassic Reptiles of S. Af. *Ibid.*, v. 25, 1915.

455. —— Persist. of Mesopterygoid in Reptilia. *Proc. Zool. Soc. London*, 1922.

456. —— Temporal Arches of Reptilia. *Ibid.*, 1922.

457. —— Skeleton of Youngina. *Ann. Transv. Mus.*, 1922.

458. —— Skull in Carnivorous Dinocephalian Reptiles. *Proc. Zool. Soc.*, 2, 1923.

459. —— Classification of Reptilia. *Bull. Am. Mus. Nat. Hist.*, v. 51, 1924.

460. —— Pareiasaurian Skull. *Proc. Zool. Soc. London*, 1924.

461. —— Origin of Lizards. *Proc. Zool. Soc.*, 1925.

462. —— Skeleton of Eosuchian (Palaeagama). *Ibid.*, 1926.

462a. —— Mammalian Presphenoid and Mesethmoid. *Ibid.*, 1926.

462b. —— Mammalian Basicranial Axis. *Ibid.*, 1927.

463. —— Sphenosuchus and Origin of Crocodiles. *Ibid.*, 1927.

464. —— Anningia. *Ibid.*, 1927.

465. **Bryant, W. L.** Structure of Eusthenopteron. *Bull. Buffalo Soc. Nat. Sci.*, v. 13, 1919.

465a. **Bulman, O. M.,** and **Whittard, W. F.** Branchiosaurus and Allied Genera. *Proc. Zool. Soc.,* 1926.

466. **Case, E. C.** Pariotichus incisivus. *Zool. Bull.*, v. 2, 1898–99.

467. —— Palaeontological Notes. *Contrib. fr. Walker Mus.*, Chicago, v. 1, 1902.

468. —— Osteology of Embolophorus. *Jour. Geol.*, 11, 1903.

469. —— Structure and Relationship of American Pelycosauria. *Am. Nat.*, 37, 1903.

470. —— Osteology of Diadectidae. *J. of Geol.*, v. 13, 1905.

471. —— Skull of Pelycosaurian Dimetrodon. *Trans. Am. Phil. Soc.*, v. 21, 1905.

472. —— Skull of Lysorophus. *Bull. Am. Mus. Nat. Hist.*, v. 24, 1908.

473. —— Revision of Cotylosauria. *Ibid.*, publ. 145, 1911 (Bibl.).

474. —— Amphibia and Pisces of Permian of N. Am. *Carnegie Inst.* publ. 146, 1911.

475. **Cole, F. J.** Cranial Nerves and Lateral Sense Organs of Fishes, genus *Gadus. Tr. Linn. Soc.*, s. 2, v. 7, 1896–1900.

476. **Collinge, W. E.** Lat. Canal System of Polypterus—of Lepidosteus. *Pr. Birmingham Phil. Soc.*, v. 8, 1891–93.

477. —— Morph. of Sensory Canal System in some Fossil Fishes. *Pr. Birmingham Phil. Soc.*, 9, 1895.

478. —— Sensory Canal System of Fishes, Teleostei, Suborder A, Physostomi. *Proc. Zool. Soc. London,* 1895.

479. **Cope, E. D.** Posterior Cranial Arches in Reptilia. *Trans. Am. Phil. Soc.*, v. 17, 1892.

480. —— Baur on the Temporal part of Skull, etc. *Am. Nat.*, v. 29, 1895.

481. —— Palaeozoic Order Cotylosauria. *Am. Nat.*, v. 30, 1896.

482. —— Reptilian Order Cotylosauria. *Proc. Am. Phil. Soc.*, v. 34, 1896.

483. **Cords, E.** Primordialcranium v. Perameles. *Anat. Hefte*, v. 52, 1915.

484. **Credner, H.** Stegocephalen und Saurier aus den Rothliegenden. *Zeit. der Deutsch. Geol. Ges.,* 1881–93.

485. —— Hylonomus and Petrobates and Discosaurus. *Zeit. d. Deut. Geol. Ges.*, v. 42, 1890.

486. **Dames, W.** Archaeopteryx. *Palaeont. Abh.*, v. 2, 1884.

487. **Daniel, J. F.** *Elasmobranch Fishes,* Berkeley, 1922; 2nd edit., 1928 (Bibl.).

488. **Fawcett, E.** Devel. of Human Sphenoid. *J. of Anat. and Physiol.*, v. 44, 1910.

489. —— Primordial Cranium of Microtus. *J. of Anat.*, v. 51, 1917.

490. —— P. C. of Erinaceus. *Ibid.*, v. 52, 1918.

491. —— P. C. of Poecilophoca. *Ibid.*, v. 52, 1918.

492. —— P. C. of Miniopterus. *Ibid.*, v. 53, 1919.

493. —— Septo-maxillare in Tatusia. *J. of Anat.*, v. 53, 1919.

494. **Fejérváry, G. J.** Dermal Bones of Skull. *Ann. Mus. Nat. Hungar.*, v. 16, 1918.

495. **Fischer, E.** Primordialcranium u. Talpa europaea. *Anat. Hefte*, v. 17, 1901.

496. —— Entwickl. der Affenschädels. *Z. Morph. u. Anthrop.*, v. 5, 1903.

497. **Fraas, E.** Labyrinthodonten d. swäb. Trias. *Palaeontogr.*, v. 36, 1889.

498. **Froriep, A.** Ganglion des Hypoglossus und Wirbelanlagen in der Occipitalregion. *Arch. f. Anat. u. Phys.*, 1882. Chorda Tympani. *Anat. Anz.*, v. 2, 1887.

499. —— Kraniovertebralgrenze b. d. Amphibien. *Arch. f. Anat. u. Phys.* Anat. Abt., 1917.

500. **Frost, G. A.** Int. Cranial Elements of Dapedius. *Q. J. Geol. Soc.*, v. 69, 1913.

501. **Fuchs, H.** Schläfengegend a. Schädel der Quadrupeden. *Anat. Anz.*, v. 35, 1909.

502. —— Septomaxillare e. rec. Säugetieres (Dasypus). *Ibid.*, v. 38, 1911.

503. —— Bau u. Entw. d. Schädels d. Chelone. Voeltzkow, *Reise in Ostafr.*, 1915.

504. **Gaupp, E.** Primordialcranium u. Kieferbogen v. Rana. *Morph. Arb.*, v. 2, 1893.

505. —— Schläfengegend am Knöch. Wirbelt.-Schädel. *Morph. Arb.*, v. 4, 1895.

506. —— Chondrocranium v. Lacerta. *Anat. Hefte*, v. 15, 1900.

507. —— Alte Probleme u. neure Arbeiten ü. d. Wirbeltierschädel. *Ergeb. Anat. u. Entw.*, v. 10, 1900 (Bibl.).

508. —— Über d. Ala temporalis des Säugerschädels u. d. Regio orbitalis. *Anat. Hefte*, v. 19, 1902.

509. —— Neue Deutungen a. d. Gebiete der Lehre vom Säugetierschädel. *Anat. Anz.*, v. 27, 1905.

510. —— Fragen a. d. Lehre von Kopfskelett. *Verh. Anat. Ges.*, 1906.

511. —— Morph. des Schädels von Echidna. *Denkschr. Med. Nat. Ges. Jena*, v. 6, 1907–8.

512. —— Ersten Wirbel u. d. Kopfgelenke v. Echidna. *Denkschr. Med. Nat. Ges. Jena*, v. 6, 1907.

513. —— Säugepterygoid u. Echidnapterygoid. *Anat. Hefte*, v. 42, 1910.

514. —— N. trochlearis der Urodelen. *Anat. Anz.*, v. 38, 1911.

515. —— Unterkiefer der Wirbeltiere. *Anat. Anz.*, v. 39, 1911.

516. **Gegenbaur, C.** Kopfskelet von Alepocephalus. *Morph. Jahrb.*, v. 4, suppl., 1878.

517. **Goodrich, E. S.** Classification of Reptilia. *Proc. Roy. Soc. London*, B, v. 89, 1916.

518. —— Head of Osteolepis. *J. Linn. Soc. London, Zool.*, v. 34, 1919.

519. —— Cranial Roofing-bones in Dipnoi. *J. Linn. Soc. London, Zool.*, v. 36, 1925.

520. —— Polypterus a Palaeoniscid ? *Palaeobiologica*, v. 1, 1928.

521. **Gregory, W. K.** Orders of Mammalia. *Bull. Am. Mus. N.H.*, v. 27, 1910.

522. —— Critique . . . on Morph. of Vertebrate Skull. *J. of Morph.*, v. 24, 1913.

523. —— Origin of Tetrapoda. *Ann. N. York Ac. Sci.*, v. 26, 1915.

524. —— Lacrymal Bone. *Bull. Am. Mus. N.H.*, v. 42, 1920.

524a. **Gregory, W. K.,** and **Noble, G. K.** Origin of Mammalian Alisphenoid. *J. Morph.*, v. 39, 1924.

525. **Greil, A.** Entw. d. Kopfes, etc., v. Ceratodus. *Denkschr. Med. Nat. Ges. Jena*, v. 4, 1913.

526. **Hafferl, A.** Neurocranium des Gecko. *Zeitschr. Anat. u. Entw.*, v. 62, 1921.

527. **Heilmann, G.** *Origin of Birds.* London, 1926.

528. Howes, G. B., and **Swinnerton, H. H.** Devel. of Skeleton of Sphenodon. *Trans. Zool. Soc. London*, v. 16, 1901.

529. Huene, F. v. Os interparietale d. Mammalia. *Anat. Anz.*, v. 42, 1912.

530. —— Cotylosauria der Trias. *Palæontogr.*, v. 59, 1912.

531. —— Skull Elements of Permian Tetrapoda. *Bull. Am. Mus. Nat. Hist.*, v. 32, 1913.

531a. —— Lysorophus. *Anat. Anz.*, v. 43, 1913.

532. —— Osteol. d. Dicynodon-Schädels. *Palaeont. Zeitschr.*, v. 5, 1922.

533. Huxley, T. H. Fishes of Devonian, etc. *Mem. Geol. Surv.*, 1861. Crossopterygian Ganoids (Coelacanthini). *Ibid.*, 1866 and 1872.

533a. —— Classification of Birds. *Proc. Zool. Soc. London*, 1867.

534. —— Skull and Heart of Menobranchus. *Proc. Zool. Soc.*, 1874.

535. —— On Ceratodus Forsteri. *Proc. Zool. Soc.*, 1876.

536. Jaekel, O. Wirbeltierfunde a. d. Keuper Stegochelys (Triassochelys) dux. *Palaeont. Zeitschr.*, 2, 1918.

537. Kindred, J. E. Skull of Amiurus. *Illin. Biol. Monogr.*, v. 5, 1919.

538. —— Chondrocranium of Syngnathus fuscus. *J. Morph.*, v. 35, 1921. Devel. of Skull of Syngnathus. *Am. J. Anat.*, v. 33, 1924.

539. Kingsley, J. S. Bones of Reptilian Lower Jaw. *Am. Nat.*, v. 39, 1905.

540. Kunkel, B. W. Devel. of Skull of Emys. *J. of Morph.*, v. 23, 1912.

540a. Lakjer, T. Gaumenregion bei Sauriern. *Zool. Jahrb.*, Abt. Anat., v. 49, 1927.

541. Lapage, E. O. Septomaxillary. I. In Amphibia Urodela ; II. A. Anura and Reptilia. *J. Morph. and Phys.*, v. 45, 1928 ; and v. 46, 1928.

542. Lehn, Ch. Primordialschädels v. Polypterus. *Zeit. f. angew. Anat. u. Konstitut.*, v. 2, 1918.

543. Lubosch, W. Kiefergelenk der Edentaten u. Marsupialier. *Denkschr. Med. Nat. Ges. Jena*, v. 7, 1897–1912.

544. Matthes, E. Neuere Arbeiten ü. d. Primordialkranium der Säugetiere. *Ergeb. d. Anat. u. Entw.*, v. 23, 1921 (Bibl.).

545. Mayrew, R. L. Skull of Lepidosteus platostomus. *J. Morph.*, v. 38, 1924.

546. Merriam, J. C. Thalattosauria. *Mem. Calif. Ac. Sci.*, v. 5, 1905.

547. ——Triassic Ichthyosauria. *Mem. Univ. Calif.*, v. 1, 1908.

548. Moodie, R. T. Lateral Line System in Extinct Amphibia. *J. Morph.*, v. 19, 1908 ; and *J. Comp. Neurol.*, v. 25, 1916.

549. —— Coal Measure Amphibia of N. Am. *Carnegie Inst.* publ. 238, 1916.

550. —— Influence of Lat. Line Syst. on Osseous Elements. *J. Comp. Neurol.*, v. 34, 1922.

551. Norman, J. R. Devel. of Chondrocranium of Eel (Anguilla). *Phil. Trans.*, v. B, 214, 1926.

552. Osawa, G. Anat. der Hatteria. *Arch. f. mikr. Anat.*, v. 51, 1898.

553. Osborn, H. F. Occipital Condyles. *Am. Nat.*, v. 34, 1900.

554. —— Reptilian subclasses Diapsida and Synapsida. *Mem. Am. Mus. Nat. Hist.*, v. 1, 1903.

555. Parker, T. J. Cranial Osteology of Dinornithidae. *Tr. Zool. Soc.*, v. 25, 1895.

Parker, W. K.
Struct. and devel. of Skull of—

556. —— Ostrich. *Phil. Trans. R.S.*, v. 156, 1866.

557. —— Fowl. *Ibid.*, v. 159, 1869.

558. —— Rana. *Ibid.*, v. 161, 1871.

Parker, W. K.
Struct. and devel. of Skull of—
559. —— Batrachia. *Ibid.*, v. 161, 1871.
560. —— Salmon. *Ibid.*, v. 163, 1873.
561. —— Pig. *Ibid.*, v. 164, 1874.
562. —— Urodela. *Ibid.*, v. 167, 1876.
563. —— Tropidonotus. *Ibid.*, v. 169, 1879.
564. —— Lacertilia. *Ibid.*, v. 170, 1880.
565. —— Acipenser. *Ibid.*, v. 173, 1882.
566. —— Lepidosteus. *Ibid.*, v. 173, 1882.
567. —— Edentata and Insectivora. *Ibid.*, v. 176, 1885.
568. —— Birds. *Trans. Zool. Soc.*, v. 9 and 10, 1875–76.
569. —— Sharks and Skates. *Ibid.*, v. 10, 1879.
570. —— Crocodilia. *Ibid.*, v. 11, 1883.
571. —— Opisthocomus. *Ibid.*, v. 13, 1891.
572. —— Birds. *Tr. Linn. Soc. Zool.*, v. 1, 1875.
573. —— Urodela. *Ibid.*, v. 2, 1879–88.
574. —— Devel. of Green Turtle (Chelone viridis). *'Challenger' Reports*, v. 1, 1880.
575. **Pollard, H. B.** Anat. and Phylog. Position of Polypterus. *Zool. Jahrb.* Abt. Anat., v. 5, 1892.
576. **Pycraft, W. P.** Palaeognathae and Neognathae. *Tr. Zool. Soc.*, v. 15, 1900.
577. **Regan, C. T.** Skeleton of Lepidosteus. *Proc. Zool. Soc. London*, 1923.
578. **Ridewood, W. G.** Cranial Osteology of Elopidae and Albulidae, etc. *Proc. Zool. Soc.*, v. 2, 1904.
579. —— Cranial Osteology of Clupeoid Fishes. *Proc. Zool. Soc.*, v. 2, 1904.
580. —— Cranial Osteol. of Osteoglossidae, Pantodontidae and Phractolaemidae. *Jour. Linnean Soc.*, v. 29, 1904.
581. —— Cranial Osteol. of Mormyridae, Notopterydae, and Hyodontidae. *Jour. Linnean. Soc.*, v. 29, 1905.
582. **Riese, H.** Anatomie des Tylototriton verrucosus. *Zool. Jahrb.* Abt. Anat., v. 5, 1891.
583. **Schauinsland, H.** Entw. u. Anat. der Wirbeltiere : Sphenodon, Chamäleo, Callorhynchus. *Zoologica*, 16, 1903.
584. **Schleip, W.** Entw. der Kopfknochen b. d. Lachs u. d. Forelle. *Anat. Hefte*, v. 23, 1904.
Seeley, H. G.
Structure, etc., of Fossil Reptilia—
585. —— Protorosaurus. *Phil. Trans. R.S.*, v. 178, B, 1887.
586. —— Pareiasaurus. *Ibid.*, v. 179, B, 1888.
587. —— Keirognathus. *Ibid.*
588. —— Anomodont Reptilia. *Ibid.*, v. 180, B, 1889.
589. —— Diademodon. *Ibid.*, v. 185, B, 1894.
590. —— Gomphodontia. *Ibid.*, v. 186, B, 1895.
591. —— Cynodontia. *Ibid.*, v.
592. **Sewertzoff, A. N.** Entwickl. des *Ceratodus Forsteri*. *Anat. Anz.*, v. 21, 1902.
593. —— Morph. d. Schädels v. *Polypterus delhesi*. *Anat. Anz.*, v. 59, 1925.
594. —— Head Skeleton and Muscles of Acipenser ruthenus. *Acta Zoologica*, v. 9, 1928.
595. **Seydel, O.** Nasenhöhle u. Mundhöhlendache von Echidna u. das Gaumen

der Wirbeltiere. *Denk. Med. Nat. Ges. Jena*, v. 3, Zool. Forsch., 1899 (Bibl.).
596. **Shaner, R. F.** Devel. of Skull of Turtle. *Anat. Rec.*, v. 32, 1925.
597. **Shiino, K.** Chondrocranium v. Crocodilus. *Anat. Hefte*, v. 50, 1914.
598. **Siebenrock, F.** Osteologie d. Hatteria-kopfes. *Sitz. Ak. Wiss. Wien*, v. 102, 1893.
599. —— Skelet d. Lacerta. *Ibid.*, v. 103, 1894.
600. —— Skelet d. Agamidae. *Ibid.*, v. 104, 1895.
601. —— Kopfskelet der Schildkroten. *Ibid.*, v. 106, 1897.
602. **Sollas, J. B. J.**, and **Sollas, W. J.** Skull of Dicynodon. *Phil. Trans. R.S.*, v. 204, B, 1913.
603. **Sollas, W. J.** Lysorophus. *Phil. Trans. R.S.*, v. 209, 1920.
604. **Stadtmüller, F.** Entw. d. Kopfskeletts d. Salamandra. *Zeit. f. Anat. u. Entw.*, v. 75, 1924.
605. **Stensiö, E. A. S.** Coelacanthiolen a. d. Oberdevon. *Palaeont. Zeitschr.*, v. 4, 1922.
606. —— Notes on Crossopterygians. *Proc. Zool. Soc.*, 1922.
607. —— Downtonian and Devonian Vert. of Spitzbergen : Cephalaspidae. *Akad. Oslo*, 1927.
608. **Sushkin, P. P.** Schädel v. Tinunculus. *Nouv. Mém. Soc. Imp. Nat. Moscou*, v. 16, 1899.
608a. —— Cranial Morph. of Captorhinus. *Palaeobiologica*, v. 1, 1928.
609. **Sutton, J. B.** Parasphenoid, Vomer, and Palatopterygoid Arcade. *Pr. Zool. Soc.*, 1884.
610. **Thévenin, A.** Anciens quadrupèdes de France. *Ann. de. Paléont.*, v. 5, 1910 (Bibl.).
611. **Thyng, F. W.** Squamosal Bone in Tetrapodous Vertebrata. *Proc. Bost. Soc. N.H.*, v. 32, 1906 (Bibl.).
612. **Tonkoff, W.** Entwickl. des Hühnerschädels. *Anat. Anz.*, v. 18, 1900.
613. **Traquair, R. H.** Cranial Osteology of Polypterus. *Jour. Anat. and Phys.*, v. 5, 1871.
614. —— Struct., etc., of Cheirolepis. *Ann. Mag. Nat. Hist.*, v. 15, 1875.
615. —— Structure of Lower Jaw in Rhizodopsis and Rhizodus. *Ann. Mag. Nat. Hist.*, v. 19, 1877.
616. —— Palaeoniscidae. *Monogr. Palaeontogr. Soc.*, 1877.
617. —— Genera Dipterus, Palaedaphus, Holodus and Cheirodus. *Ann. Mag. Nat. Hist.*, v. 2, 1878.
618. —— Struct. and Affinities of Platysomidae. *Trans. Roy. Soc. Edinburgh*, v. 29, 1879.
619. —— Cranial Osteology of Rhizodopsis *Trans. Roy. Soc. Edinburgh*, v. 30, 1883.
620. —— Chondrosteus acipenseroides. *Geol. Mag.*, v. 4, 1887.
621. —— Struct. and Classif. of Asterolepida. *Ann. Mag. Nat. Hist.*, v. 2, 1888.
622. —— Fossil Dipnoi and Ganoids of Fife and Lothians. *Proc. Roy. Soc. Edinburgh*, v. 17, 1890.
623. **Versluys, J.** Parasphenoid bei Dermochelys. *Zool. Jahrb.*, v. 28, 1909.
624. —— Streptostylie bei Dinosaurien. *Ibid.*, v. 30, 1910.
625. —— Das Streptostylie-Problem. *Ibid.*, suppl. 15, v. 2, 1912.
626. —— Phylogenie der Schläfengruben u. Jochbogen. *Sitz. Heidelb. Ak. Wiss*, 1919.
627. —— Schädel d. Trachodon. *Abh. Senckenb. Nat. Ges.*, v. 38, 1923.

628. **Vrolik, A. J.** Verknöcherung d. Schläfenbeins der Säugetiere. *Niederl. Arch. f. Zool.*, v. 1, 1873.
629. —— Studien ü. d. Verknöcherung . . . der Teleostier. *Niederl. Arch. f. Zool.*, v. 1, 1873.
630. **Walther, J.** Entw. d. Deckknochen des Hechtes. *Jen. Zeitschr.*, v. 16, 1883.
631. **Watson, D. M. S.** Larger Coal Measure Amphibia. *Mem. Manchester Lit. Phil. Soc.*, v. 57, 1912.
631a. —— Skull of Diademodon. *Ann. Mag. Nat. Hist.*, v. 8, 1911.
632. —— Reptilian Lower Jaws. *Ibid.*, v. 10, 1912.
633. —— Micropholis stowi. *Geol. Mag.*, v. 10, 1913.
634. —— Skull, Brain, etc., of Diademodon. *Ibid.*, v. 12, 1913.
635. —— Anomodont Brain-case. *Anat. Anz.*, v. 44, 1913.
636. —— Varanosaurus. *Ann. Mag. Nat. Hist.*, v. 13, 1914.
637. —— Skull of Pareiasaurian. *Proc. Zool. Soc. London*, 1914.
638. —— Pleurosaurus and Lizard's Skull. *Ann. Mag. Nat. Hist.*, v. 14, 1914.
639. —— Procolophon. *Proc. Zool. Soc. London*, 1914.
640. —— Deinocephalia. *Ibid.*, 1914. Cynodontia. *Ibid.*, 1920.
641. —— Carnivorous Therapsids. *Ibid.*, 1914.
642. —— Monotreme Skull. *Phil. Trans. R.S.*, v. 207, 1916.
643. —— Seymouria. *Proc. Zool. Soc. London*, 1919.
644. —— Evol. and Origin of Amphibia. 1. *Phil. Trans. Roy. Soc.*, v. 209, B, 1919 ; 2. *Ibid.*, v. 214, B, 1925 (Bibl.).
645. —— On Coelacanth Fish. *Ann. Mag. N.H.*, v. 8, 1921.
646. —— Structure of Palaeoniscids. *Proc. Zool. Soc. London*, 1925.
647. **Watson, D. M. S., and Day, H.** Notes on Palaeozoic Fishes. *Mem. Manchester Lit. Phil. Soc.*, 1916.
648. **Watson, D. M. S., and Gill, E. L.** Structure of Palaeozoic Dipnoi. *J. Linn. Soc. Zool.*, v. 35, 1923.
649. **Wellburn, E.** Genus Coelacanthus, etc. *Proc. Yorks. Geol. and Pal. Soc.*, v. 14, 1902.
650. **Wells, Grace A.** Skull of Acanthias vulgaris. *J. of Morph.*, v. 28, 1917.
651. **Wiedersheim, R.** Kopfskelet der Urodelen. *Morph. Jahrb.*, v. 3, 1877.
652. —— *Anat. der Gymnophionen.* Jena, 1879.
653. —— Skelet, etc., v. Lepidosiren annectens. *Morph. Studien*, 1, Jena, 1880.
654. **Wijhe, J. W. v.** Visceralskelet u. Nerven des Kopfes der Ganoiden u. von Ceratodus. *Niederl. Arch. f. Zool.*, v. 5, 1879–82.
655. **Williston, S. W.** Temporal Arches of Reptilia. *Biol. Bull.*, v. 7, 1904.
656. —— Lysorophus a Permian Urodele. *Biol. Bull.*, v. 15, 1908.
657. —— Permian Vertebrates : Trematops. *J. of Geol.*, v. 17, 1909.
658. —— Cacops, Desmospondylus. *Bull. Geol. Soc. Am.*, v. 21, 1910.
659. —— Mandible in Amphibians and Reptiles. *J. of Geol.*, v. 21, 1913.
660. —— Trimerorhachis. *J. of Geol.*, v. 23, 1915.
661. —— Synopsis of Am. Permocarboniferous Tetrapoda. *Contrib. Walker Mus.*, v. 1, 1916.
662. **Winslow, C. M.** Chondrocranium in Ichthyopsida. *Tufts Coll. Studies*, No. 5, 1898.
663. **Woodward, A. Smith.** *Catalogue of Fossil Fishes in Brit. Mus.* 4 vols. London, 1889–1901.

664. **Woodward, A. Smith.** Cranial Osteology of . . . Lepidotus and Dapedius. *Proc. Zool. Soc.*, 1893.

665. —— Fossil Fishes of English Chalk. *Palaeontogr. Society*, London, 1902.

666. **Zimmermann, S.** Chondrocr. v. *Angius fragilis*. *Anat. Anz.*, v. 44, 1913.

CHAPTER VII

VISCERAL ARCHES AND LABIAL CARTILAGES

667. **Allis, E. P.** Hyomandibular of Gnathostome Fish. *J. of Morph.*, v. 26, 1915.

668. —— Labial Cartilages of Raja. *Quart. Jour. Micr. Sci.*, v. 62, 1917.

669. —— Origin of Hyomandibular. *Anat. Rec.*, v. 15, 1918.

670. —— Polar and Trabecular Cartilages. *J. of Anat.*, v. 57, 1922 ; v. 58, 1923 ; and v. 59, 1925.

671. —— Hyomandibula and Preoperculum. *Ibid.*, v. 62, 1928.

672. **Berrill, N. J.** Devel. of Skull of Sole and Plaice. *Quart. Jour. Micr. Sci.*, v. 69, 1925.

673. **Braus, H.** Kiemenapparat von *Heptanchus*. *Anat. Anz.*, v. 29, 1906.

674. **Daniel, J. F.** Anat. of Heterodontus : II. Endoskeleton. *J. Morph.*, v. 26, 1915.

675. **Drüner, L.** Zungenbein, Kiemenbogen- u. Kehlkopfmuskeln der Urodelen. 1 Th., *Zool. Jahrb.*, Abt. Anat., v. 15, 1902 ; 2 Th., *Zool. Jahrb.*, Abt. Anat., v. 19, 1904.

676. **Edgeworth, F. H.** Quadrate in Cryptobranchus, Menopoma, and Hynobius. *J. of Anat.*, v. 57, 1923.

677. —— Larval Hyobranchial Skeleton and Musculature of Cryptobranchus, etc. *Ibid.*, v. 57, 1923.

678. —— Autostylism of Dipnoi and Amphibia. *J. of Anat.*, v. 59, 1925.

679. —— Hyomandibula of Selachii, Teleostomi, and Ceratodus. *Ibid.*, v. 60, 1926.

680. **Foote, E.** Extra-branchial Cartilages. *Anat. Anz.*, v. 13, 1897.

681. **Fürbringer, K.** Beitr. z. K. d. Visceralskeletts d. Selachier. *Morph. Jahrb.*, v. 31, 1903.

682. —— Morph. d. Skeletes der Dipnoer, etc. *Denk. Med. Nat. Ges. Jena*, v. 4, 1904.

683. **Gaupp, E.** Das Hyobranchialskelet. *Anat. Hefte, Ergeb.*, v. 14, 1904.

684. **Gibian, A.** Hyobranchialskelett der Haie. *Morph. Jahrb.*, v. 45, 1913 (Bibl.).

685. **Goodey, T.** Skel. Anat. of Chlamydoselachus. *Proc. Zool. Soc. London*, 1910.

686. **Göppert, E.** Kehlkopf der Amphibien u. Reptilien. *Morph. Jahrb.*, v. 26, 1898 ; v. 28, 1899.

687. —— Vergl. Anat. des Kehlkopfes. *Denkschr. Med. Nat. Ges. Jena*, v. 6, 1901 ; v. 11, 1904 (Bibl.).

688. **Gregory, W. K.** Relations of Anterior Visceral Arches to Chondrocranium. *Biol. Bull.*, v. 7, 1904.

689. **Hawkes, O. A. M.** Vestigial Sixth Branch. Arch in Heterodontidae. *J. Anat. and Phys.*, v. 40, 1905.

690. **Jaekel, O.** Das Mundskelett der Wirbeltiere. *Morph. Jahrb.*, v. 55, 1925.

691. **Kallius, E.** Entwickl. des Kehlkopfes. *Anat. Hefte*, v. 9, 1897.

692. **Kesteven, H. L.** Palate and Upper Jaw of Fishes. *J. of Anat.*, v. 56, 1922.
693. **Kravetz.** Entw. d. Knorpelschädels v. Ceratodus. *Bull. Soc. Nat. Moscou*, 1911.
694. **Krivetski, A.** Morph. de l'arc hyoide chez les Sélaciens. *Rev. Zool. Russe*, v. 2, 1917.
695. **Luther, A.** N. trigeminus innervierte Muskulatur der Selachier. *Acta Soc. Sc. Fennicae*, v. 36-7, 1909. Ganoiden u. Dipneusten. *Ibid.*, v. 41, 1913.
696. **Pollard, H. B.** Oral Cirri of Siluroids. *Zool. Jahrb.*, Abt. Morph., v. 8, 1895.
697. —— Suspension of Jaws in Fish. *Anat. Anz.*, v. 10, 1895.
698. **Ridewood, W. G.** Hyoid Arch of Ceratodus. *Proc. Zool. Soc.*, 1894.
699. **Schmalhausen, J. J.** Suspensorialapparat d. Fische. *Anat. Anz.*, v. 56, 1923.
700. —— Autostylie d. Dipnoi. *Ibid.*
701. **Sewertzoff, A. N.** Évolution des Vertébrés: Cyclostomes, etc. *Arch. russes d'Anat., d'Hist. et d'Embr.*, v. 1, 1916.
701a. —— Appareil viscéral des Elasmobranches. *Publ. Staz. Zool. Napoli*, v. 18, 1927.
702. **White, P. J.** Skull and Visceral Skeleton of Laemargus. *Tr. R.S. Edin.*, v. 37, 1891–95.
703. **Woodward, A. S.** Mandib. and Hyoid Arches in Hybodus. *Proc. Zool. Soc. London*, 1886.

CHAPTER VIII

MIDDLE EAR AND EAR OSSICLES

704. **Baur, G.** Quadrate of Mammalia. *Quart. J. Micr. Sci.*, v. 28, 1887; *Biol. Centralbl.*, v. 6.
705. **Bender, O.** Homologie des Spritzloches der Selachier u. d. Paukenhöhlen, etc. *Anat. Anz. Ergh.*, v. 30, 1907.
706. —— Entwickl. d. Visceralsk. bei Testudo. *Abh. bayer. Akad. Wiss., Math.-phys. Kl.*, v. 25, 1912.
707. **Bondy, G.** Vergl. Anat. d. Gehörorgans d. Säuger. *Anat. Hefte*, 1 Abt., v. 35, 1907.
708. **Broman, I.** Die Entwickl. der Gehörknöchelchen, p. 507. *Anat. Hefte*, 1 Abt., v. 11, 1899.
709. **Broom, R.** Theriodont Mandible, etc. *Proc. Zool. Soc. London*, 1904.
710. —— Internal Ear in Dicynodon and Homology of Auditory Ossicles. *Ibid.*, 1912.
711. **Cope, E. D.** Hyoid and Otic Elements in Batrachia. *J. Morph.*, v. 2, 1889.
712. **Cords, E.** Entwickl. d. Paukenhöhle v. Lacerta. *Anat. Hefte*, Abt. 1, v. 52, 1915.
713. **Dollo, J.** Malleus of Lacertilia. *Quart. J. Micr. Sci.*, v. 23, 1883.
714. **Doran, A. H. G.** Morph. of Mammalian Ossicula auditus. *Trans. Linn. Soc. Zool.*, v. 1, 1878.
715. **Dreyfuss, R.** Entw. des Mittelohres u. d. Trommelfels. *Morph. Arb.*, v. 2, 1893.

716. Drüner, L. Anat. u. Entw. des Mittelohres beim Menschen u. bei der Maus. *Anat. Anz.*, v. 24, 1904.

717. Dunn, E. R. Sound-transmitting App. of Salamanders and Phylog. of Caudata. *Am. Nat.*, v. 56, 1922.

718. Edgeworth, F. H. Devel. of Mandibular and Hyoid Muscles. *Quart. J. Micr. Sci.*, v. 59, 1914.

719. Emmel, V. E. Relation of Chorda Tympani in Microtus. *J. Comp. Neur.*, v. 14, 1904.

720. Eschweiler, R. Entw. d. schalleit. Apparates. *Arch. mikr. Anat. u. Entw.*, v. 63, 1904.

721. —— Entw. d. M. stapedius u. d. Stapes. *Ibid.*, v. 77, 1911.

722. Esdaile, Ph. C. Struct. and Devel. of Skull of Perameles. *Phil. Trans. Roy. Soc.*, B, v. 207, 1916.

723. Fox, H. Devel. of Tympano-Eustachian Passage, etc. *Proc. Acad. Nat. Sci. Philad.*, v. 53, 1901.

724. Frey, H. Hammer-Amboss Verbindung der Säuger. *Anat. Hefte*, v. 44, 1911.

725. Froriep, A. Homologon der Chorda tympani bei niederen Wirbelthieren. *Anat. Anz.*, v. 2, 1887.

726. —— Anlagen v. Sinnesorganen am Facialis. *Arch. Anat. u. Phys.*, Abt. Anat., 1885.

727. Fuchs, Hugo. Entw. des Operculums der Urodelen u. d. Distelidiums (' Columella ' auris) einiger Reptilien. *Anat. Anz. Ergh.*, v. 30, 1907.

728. Fuchs, H. Über Knorpelbildung in Decknochen u. Unters. u. Betr. ü. Gehörknöchelchen. *Arch. Anat. Phys.*, Abt. Anat., Suppl., 1909.

729. Gadow, H. Evolution of Auditory Ossicles. *Anat. Anz.*, 19, 1901.

730. —— Modifications of Visceral Arches. *Phil. Trans. R.S.*, v. 186, 1895.

731. Gaupp, E. Ontog. u. Phylog. des schalleitenden Apparates. *Ergeb. Anat. u. Entw.*, v. 8, 1899 (Bibl.).

732. —— Die Reichertsche Theorie. *Arch. Anat. u. Physiol.*, Abt. Anat., Suppl. v., 1913 (Bibl.).

733. Goldby, F. Devel. of Columella auris in Crocodilia. *J. Anat.*, v. 59, 1925.

734. Goodrich, E. S. Chorda tympani and Middle Ear in Reptiles. *Quart. J. Micr. Sci.*, v. 61, 1915.

735. Gregory, W. K. Orders of Mammals. *Bull. Am. Mus. Nat. Hist.*, v. 27, 1910.

736. Hammar, J. A. Entw. des Vorderdarms, 1 Abt. Morph. d. Schlundspalten —Mittelohrraumes, etc. *Arch. mikr. Anat.*, v. 59, 1902.

737. Hoffmann, C. K. Ontw. v. h. Gehoororgaan b. de Reptilien. *Natuurk. Verh. Kon. Ac. Amsterdam*, v. 28, 1889.

738. Huxley, T. H. Devel. of Columella auris in Amphibia. *Nature*, v. 11, 1875.

739. —— Malleus and Incus of Mammalia. *Proc. Zool. Soc. London*, 1869.

740. Jenkinson, J. W. Devel. of Ear-bones in Mouse. *J. Anat. and Phys.*, v. 45, 1911.

741. Kampen, P. N. van. Tympanalgegend des Säugetierschädels. *Morph. Jahrb.*, v. 34, 1905.

742. Killian, G. Anat. u. Entw. d. Ohrmuskeln. *Anat. Anz.*, v. 5, 1890.

743. Kingsbury, B. F. Columella auris and Nervus facialis in Urodela. *J. Comp. Neur.*, v. 13, 1903.

744. Kingsbury, B. F., and Reed, H. D. Columella auris in Amphibia. *Anat. Rec.*, v. 2, 1908 ; *J. of Morph.*, v. 20, 1909.

745. **Kingsley, J. S.** Ossicula auditus. *Tuft's Call. Studies,* 1900.
746. **Kingsley, J. S., and Ruddick, W. H.** Ossicula auditus and Mammalian Ancestry. *Am. Nat.,* 33, 1899.
747. **Klaauw, C. J. van d.** Entw. des Entotympanicus. *Tydschr. Nederl. dierk. Ver.,* v. 18, 1922.
748. —— Skelettstückchen i. d. Sehne d. Musculus stapedius. *Zeitschr. Anat. u. Entw.,* v. 69, 1923.
748a. —— Skelettst. von Paauw u. d. Verlauf d. Chorda tympani. *Anat. Anz.,* v. 57, 1923.
748b. —— Bau u. Entw. der Gehörknöchelchen. *Ergeb. d. Anat. Entw.,* v. 25, 1924 (Bibl.).
749. **Krause, G.** Die Columella der Vögel. Berlin, 1901.
750. **Litzelmann, E.** Visceralapparat der Amphibien. *Zeitsch. f. Anat. u. Entw.,* v. 67, 1923.
751. **Magnien, L.** Anat. comp. de la corde du tympan des oiseaux. *C.R. Acad. Sc. Paris,* v. 101, 1885.
752. **Marcus, H.** Gymnophionen : Entwickl. des Kopfes. *Morph. Jahrb.,* v. 40, 1910.
753. **Möller, W.** Entwickl. d. Gehörknöchelchens b. d. Kreuzotter. *Arch. mikr. Anat.,* v. 65, 1905.
754. **Noack.** Entwickl. d. Mittelohres v. Emys. *Arch. mikr. Anat.,* v. 69, 1907.
755. **Okajima, K.** Entwickl. d. Gehörknöchelchens b. d. Schlangen. *Anat. Hefte,* v. 53, 1915.
756. **Osawa, G.** Anat. d. Hatteria. *Arch. mikr. Anat.,* v. 51, 1898.
757. **Palmer, W. R.** Lower Jaw and Ear Ossicles of Foetal Perameles. *Anat. Anz.,* v. 43, 1913.
758. **Peter, K.** Entwickl. des Schädels von *Ichthyophis glutinosus. Morph. Jahrb.,* v. 25, 1896–98.
759. **Peyer, B.** Entwickl. d. Schädelskeletts v. Vipera. *Morph. Jahrb.,* v. 44, 1912.
760. **Reed, H. D.** Sound-transmitting Apparatus in Necturus. *Anat. Rec.,* v. 9, 1915.
761. —— Morph. of Sound-transmitting Apparatus in Caudate Amphibia, etc. *J. of Morph.,* v. 33, 1920.
762. **Salensky, W.** Entwickl. der knorpeligen Gehörknöchelchen bei Säugetieren. *Morph. Jahrb.,* 6, 1880.
763. **Sarasin, P. and F.** Entwickl. u. Anat. d. Ichthyophis. *Ergeb. Nat. Forsch. Ceylon,* v. 2, 1890.
764. **Shiino, K.** Chondrocranium v. Crocodilus. *Anat. Hefte,* v. 50, 1914.
765. **Smith, G. W.** Middle Ear and Columella of Birds. *Quart. J. Micr. Sci.,* v. 48, 1905.
766. **Smith, L. W.** Columella auris in Chrysemis. *Anat. Anz.,* v. 46, 1914.
767. **Spemann, H.** Entw. d. Tuba Eustachii u. d. Kopfskeletts v. *Rana temporaria. Zool. Jahrb.,* Abt. Anat., v. 11, 1898.
768. **Sushkin, P.** Mandibular and Hyoid Arches in early Tetrapoda. *Paläont. Zeitschr.,* v. 8, 1927.
769. **Versluys, J.** Mittl. u. äussere Ohrsphäre d. Lacertilia u. Rhynchocephalia. *Zool. Jahrb.,* Abt. Anat. u. Entw., v. 12, 1898.
770. —— Entwickl. der Columella auris b. d. Lacertilien. *Zool. Jahrb.,* Abt. Anat. u. Ont., v. 19, 1903.
771. **Villy, F.** Devel. of Ear and Accessory Organs in Common Frog. *Quart. J. Micr. Sci.,* v. 30, 1890.

772. **Witebsky, M.** Entwickl. d. schalleit. Apparates d. Axolotl. *Inaug. Diss. Berlin*, 1896.
773. **Wyeth, F. J.** Devel. of Audit. Apparatus in Sphenodon. *Phil. Trans. Roy. Soc.*, B, 212, 1924.

CHAPTER IX

VISCERAL CLEFTS AND GILLS

774. **Assheton, R.** The Devel. of Gymnarchus niloticus. *Budgett Mem. Vol.*, London, 1907.
775. **Biétrix, E.** *Morph. du système circulatoire . . . des poissons.* Paris, 1895.
776. **Budgett, J. S.** Breeding Habits of some West African Fishes, etc. *Trans. Zool. Soc.*, v. 16, 1901.
777. **Clemens, P.** Äusseren Kiemen der Wirbeltiere. *Anat. Hefte*, v. 5, 1894.
778. **Dean, B.** Embryology of Bdellostoma. *Festschr. f. C. v. Kupffer*, 1899.
779. —— Notes on Japanese Myxinoids. *J. Coll. Sci. Tokyo*, v. 19, 1904.
780. **Dohrn, A.** Spritzlochkieme der Selachier, etc. *Mitth. Zool. Sta. Neapel*, v. 7, 1886.
781. **Faussek, V.** Histologie der Kiemen bei Fischen u. Amphibien. *Arch. mikr. Anat.*, v. 60, 1902.
782. **Goette, A.** Entwick. der Teleostierkieme. *Zool. Anz.*, v. 1, 1878.
783. —— Über die Kiemen der Fische. *Zeitschr. f. wiss. Zool.*, v. 69, 1901.
784. **Greil, A.** Homologie der Anamnierkiemen. *Anat. Anz.*, 28, 1906.
785. **Harrison, R. G.** Devel. of Balancer in Amblystoma. *J. Exper. Zool.*, v. 41, 1925 (Bibl.).
786. **Jacobshagen, E.** Homologie der Wirbeltierkiemen. *Jen. Zeitschr. f. Naturw.*, v. 57, 1920.
787. **Kastschenko, N.** Schicksal d. Schlundspalten b. Säugetieren. *Arch. mikr. Anat.*, v. 30, 1887.
788. —— Schlundspaltsystem des Hühnchens. *Arch. Anat. u. Phys.*, 1887.
789. **Kellicott, W. E.** Devel. of Vascular System of Ceratodus. *Mem. Acad. Sci. New York*, 1905.
790. **Kerr, J. G.** Devel. of Lepidosiren paradoxa. *Phil. Trans.*, v. 192, 1900 ; *Quart. J. Micr. Sci.*, v. 46, 1902.
791. —— Devel. of Polypterus senegalus. *Budgett Mem. Vol.*, London, 1907.
792. **Lankester, E. R., and Willey, A.** Devel. of Atrial Chamber of Amphioxus. *Quart. J. Micr. Sci.*, v. 31, 1890.
793. **Moroff, T.** Entw. der Kiemen bei Knochenfischen. *Arch. mikr. Anat.*, v. 60, 1902.
794. —— Entw. der Kiemen bei Fischen. *Arch. mikr. Anat.*, v. 64, 1904.
795. **Oppel, A.** Lehrb. d. vergl. mikr. Anat., v. 6, 1905.
796. **Orton, J. H.** Ciliary Mechanisms in Amphioxus, etc. *J. Mar. Biol. Ass.*, v. 10, 1913–15.
797. **Plehn, M.** Zum feineren Bau der Fishkeime. *Zool. Anz.*, v. 24, 1901.
798. **Regan, C. T.** A Classification of the Selachian Fishes. *Proc. Zool. Soc.*, 1906.
799. **Reiss, J. A.** Bau der Kiemblätter b. d. Knochenfischen. *Arch. f. Naturgesch.*, 47 Jahrg., 1881.

800. **Schulze, F. E.** Die inneren Kiemen der Batrachierlarven. *Abh. K. Preuss. Ac. Berlin*, 1888 and 1892.

801. **Stadtmüller, F.** Entw. u. Bau d. papillenf. Erhebungen a. d. Branchialbogen d. Salamandridenlarven. *Zeit. f. Morph. u. Anthr.*, v. 24, 1924.

CHAPTER X

VASCULAR SYSTEM, HEART

802. **Allis, E. P.** Pseudobranchial Circulation in Amia. *Zool. Jahrb.*, Abt. Anat., v. 14, 1900.

803. —— Pseudobr. and Carotid Arteries in Ameiurus. *Anat. Anz.*, v. 33, 1908.

804. —— In Chlamydoselachus. *Ibid.*, v. 39, 1911.

805. —— In Polyodon. *Ibid.*

806. —— In Esox, Salmo, Gadus, and Amia. *Ibid.*, v. 41, 1912.

807. **Ayers, H.** Morph. of Carotids (Chlamydoselachus). *Bull. Mus. Harvard*, v. 17, 1889.

808. **Beddard, F. E.,** and **Mitchell, P. C.** Heart of Alligator. *Proc. Zool. Soc.*, 1895.

809. **Bemmelen, J. F. van.** Halsgegend b. Reptilien. *Zool. Anz.*, v. 10, 1887.

810. **Benninghoff, A.** Anat. u. Entw. des Amphibienherzens. *Morph. Jahrb.*, v. 51, 1921.

811. **Boas, J. E. V.** Conus arteriosus b. Butirinus, etc. *Morph. Jahrb.*, v. 6, 1880.

812. —— Hertz und Arterienbogen bei Ceratodus und Protopterus. *Morph. Jahrb.*, 6, 1880.

813. —— Conus art. u. Arterienbogen d. Amphibien. *Ibid.*, 7, 1882.

814. **Boveri, T.** Nierencanälchen des Amphioxus. *Zool. Jahrb.*, v. 5, 1892.

815. **Bremer, J. L.** I. Devel. of Heart: II. Left Aorta of Reptiles. *Am. J. Anat.*, v. 42, 1928.

816. **Bruner, H. S.** Heart of Lungless Salamanders. *J. Morph.*, v. 16, 1900.

817. —— Cephalic Veins and Sinuses of Reptiles. *Am. J. Anat.*, v. 7, 1907.

818. **Carazzi, O.** Sist. arterioso di Selache. *Anat. Anz.*, v. 26, 1905.

819. **Erdmann, B.** Entw. Atrioventrikularklappen bei den Anuren. *Morph. Jahrb.*, v. 51, 1921.

820. **Fuchs, F.** Entw. des Kiebitzherzens (Vanellus). *Z. f. Anat. u. Entw.*, v. 75, 1925.

821. **Gaupp, E.** Anat. des Frosches, 1896–1901.

822. **Gegenbaur, C.** Vergl. Anat. des Herzens. *Jen. Zeitschr. Naturw.*, v. 2, 1866.

823. —— Conus arteriosus der Fische. *Morph. Jahrb.*, v. 17, 1891.

824. **Gelderen, C. van.** Morph. d. Sinus durae matris. *Zeitschr. f. Anat. u. Entw.*, v. 73 and v. 74, 1924; v. 75, 1925.

825. **Goette, A.** *Entw. des Flussneunauges (Petromyzon)*, 1890.

826. **Goodrich, E. S.** On Reptilian Heart. *J. of Anat.*, v. 53, 1919.

827. **Granel, F.** Structure et dével. de la Pseudobranchie des Téléostéens. *C.R. Ac. Sci. Paris*, v. 175, 1922.

828. —— Signification morphologique de la pseudobranchie des Téléostéens. *Ibid.*, v. 175, 1922.

829. —— Étude histologique et embr. sur la pseudobranchie des Téléostéens. *Arch. d'Anat. Histol. et Embr.*, v. 2, 1923.

830. **Granel, F.** Sur la pseudobranchie de Chrysophrys. *C.R. Ass. des Anatomistes*, 1923.
831. **Greil, A.** Vergl. Anat. u. Entw. des Herzens u. d. Truncus art. der Wirbeltiere. *Morph. Jahrb.*, v. 31, 1903.
832. **Grosser, O., u. Brezina, E.** Entw. d. Venen des Kopfes bei Reptilien. *Morph. Jahrb.*, v. 23, 1895.
833. **Hafferl, A.** Entwicklungsgech. der Kopfgefässe des Gecko (Platydactylus). *Anat. Hefte*, v. 59, 1921.
834. —— Entw. der Kopfarterien b. Kiebitz (Vanellus). *Ibid.*
835. **Hochstetter, F.** Entw. d. Blutgefässsystems. Hertwigs *Handb. Entw. Wirbeltiere*, v. 3, Jena, 1906 (Bibl.).
836. —— Blutgefässsyst. der Krocodile. Voeltzkow, *Reise in Ostafrika*, v. 4, 1906.
837. **Hopkins, G. S.** Heart of some Lungless Salamanders. *Am. Nat.*, 30, 1896.
838. **Hoyer, H.** Morphologie des Fischherzens. *Bull. intern. Acad. Sc. Cracovie*, 1900.
839. **Kern, A.** Das Vogelherz. *Morph. Jahrb.*, v. 56, 1926.
840. **Kerr, J. G.** Text-book of Embryology, vol. 2, London, 1919.
841. **Langer, A.** Entw. der Bulbus cordis bei Amphibien u. Reptilien. *Morph. Jahrb.*, v. 21, 1894.
842. —— bei Vögeln u. Säugetieren. *Ibid.*, v. 22, 1894.
843. **Lankester, E. R.** Hearts of Ceratodus, Protopterus, and Chimaera. *Trans. Zool. Soc.*, v. 10, 1878.
844. **Legros, R.** Appareil vasculaire de l'amphioxus. *Mitth. Zool. Sta. Neapel*, vol. 15, 1902.
845. **Lillie, F. R.** Devel. of Chick. New York, 1919.
846. **Marshall, A. M., and Bles, E. J.** Devel. of Blood-vessels in Frog. Stud. Owens College, 1890.
847. **Masius, J.** Dével. du cœur chez le poulet. *Arch. de Biol.*, v. 9, 1889.
848. **Maurer, F.** Pseudobranchie der Knochenfische. *Morph. Jahrb.*, v. 9, 1884.
849. —— Kiemen u. i. Gefässe b. Amphibien u. Teleostiern. *Morph. Jahrb.*, v. 14, 1888.
850. **Mayer, P.** Entw. des Herzens, etc., bei Selachier. *Mitth. Zool. Sta. Neapel*, vol. 7, 1887.
851. **Müller, F. W.** Entw. u. morph. Bedeutung der Pseudobranchie bei Lepidosteus. *Arch. mikr. Anat.*, v. 49, 1897.
852. **O'Donoghue, C. H.** Blood Vascular Syst. of Sphenodon. *Phil. Trans. Roy. Soc.*, B, v. 210, 1920 (Bibl.).
853. **Parker, T. J.** Blood-vessels of Mustelus. *Phil. Trans. Roy. Soc.*, B, v. 177, 1886.
854. **Parker, W. N.** Anat. and Phys. of Protopterus annectens. *Trans. Irish Acad.*, v. 30, 1892.
855. **Platt, J. R.** Morph. of Vertebrate Head. *J. of Morph.*, v. 5, 1891.
856. **Raffaele, D. F.** Sviluppo d. sist. vascolare n. Selacei. *Mitt. Zool. Sta. Neapel*, v. 10, 1892.
857. **Rao, C. R. N., and Ramana, R. S.** Conus arteriosus of Engystomatidae, etc. *Proc. Zool. Soc. London*, 1926.
858. **Rau, A. S.** Heart, etc., of Ceratophrys. *J. of Anat.*, v. 58, 1924.
859. —— Heart of Tiliqua and Eunectes. *J. of Anat.*, v. 59, 1924.
860. **Ridewood, W. G.** Circulus Cephalicus in Teleostean Fishes. *Proc. Zool. Soc.*, 1899.

861. **Robertson, J. I.** Devel. of Heart of Lepidosiren. *Quart. J. Micr. Sci.*, v. 59, 1913.
862. **Röse, C.** Beitr. z. vergl. Anat. d. Herzens. *Morph. Jahrb.*, v. 16, 1890.
863. **Rückert, J.** Anlage d. Herzens, etc., b. Selachierembryonen. *Biol. Centrabl.*, v. 8, 1888.
864. **Sabatier, A.** Syst. aortique d. la série des Vertébrés. *Ann. Sc. Nat. Zool.*, v. 19, 1874.
865. —— Études sur le cœur, etc., des Vertébrés. Montpellier, 1873.
866. **Salzer, H.** Entw. d. Kopfvenen der Meerschweinchens. *Morph. Jahrb.*, v. 23, 1895.
867. **Sappey, Ph. C.** L'appareil mucipare et sur le syst. lymphatique des poissons. Paris, 1880.
868. **Senior, H. D.** Conus arteriosus in Tarpon. *Biol. Bull.*, v. 12, 1907.
869. **Sewertzoff, A. N.** Entwickl. der Kiemen u. Kiemenbogengefässe. *Zeitschr. wiss. Zool.*, v. 121, 1923.
870. **Shiino, T.** Arteriellen Kopfgefässe d. Reptilien. *Anat. Hefte*, v. 51, 1914.
871. **Smith, W. C.** Conus arteriosus in Teleosts. *Anat. Rec.*, v. 15, 1918.
872. **Spencer, W. B.** Ceratodus. The Blood-vessels. *Macleay Memorial Vol.*, 1892.
873. **Stöhr, Ph.** Klappenapparat in Conus arteriosus der Selachier u. Ganoiden. *Morph. Jahrb.*, v. 2, 1876.
874. **Takahashi, So.** Formation of Cardiac Septa in Chick. *J. of Anat.*, v. 57, 1923.
875. **Tandler, J.** Vergl. Anat. d. Kopfarterien b. d. Mammalia. *Denkschr. k. Akad. d. Wiss.*, 67, Wien, 1899.
876. —— Entw. d. Kopfarterien b. d. Mammalia. *Morph. Jahrb.*, 30, 1902.
877. —— Vergl. Anat. d. Kopfarterien b. d. Mammalia. *Anat. Hefte*, v. 18, 1902.
878. **Twining, G. H.** Embryonic History of Carotid Arteries in the Chick. *Anat. Anz.*, v. 29, 1906.
879. **Vialleton, L.** Cœur des lamproies. *Arch. d'anat. Micr.*, v. 6, 1903.
880. **Vialli, M.** Branchie e pseudobranchie d. Sturione. *Publ. Staz. Zool. Napoli*, v. 5, 1925.
881. —— Le pseudobranchie dei pesci. *Arch. Ital. Anat. e Embr.*, v. 23, 1926.
882. **Wright, R. R.** On the Hyomandibular Clefts and Pseudobranchs of Lepidosteus and Amia. *J. Anat. and Phys.*, v. 19, 1885.

CHAPTER XI

Air-bladder and Lungs

883. **Aeby, Ch.** Bronchialbaum d. Säugetiere. Leipzig, 1880.
884. **Anton, W.** Nasenhöhle der Perennibranchiaten. *Morph. Jahrb.*, v. 44, 1922.
885. **Assheton, R.** Devel. of Gymnarchus. *Budgett Mem. Vol.*, 1907.
886. **Ballantyne, F. M.** Air-bladder and Lungs. *Tr. Roy. Soc. Edin.*, v. 55, 1927.
887. **Beaufort, L. F. de.** Schwimmblase der Malacopterygii. *Morph. Jahrb.*, v. 39, 1909.
888. **Bloch, L.** Schwimmblase, Knochenkapsel and Weber'scher Apparat von Nemachilus. *Jen. Zeit. f. Naturw.*, 34, 1900.

889. **Bohr, C.** Infl. of Section of Vagus n. on Gases in Air-bladder. *J. of Physiol.*, v. 15, 1894.

890. **Bremer, J. L.** Lung of Opossum. *Am. J. of Anat.*, v. 3, 1904.

891. **Bridge, T. W.**, and **Haddon, A. C.** Air-bladder and Weberian Ossicles of Siluroid Fishes. *Phil. Trans. Roy. Soc.*, B, v. 184, 1893 (Bibl.).

892. **Bruner, H. L.** Pulmonary Respir. in Amphibians. *Morph. Jahrb.*, v. 48, 1914.

893. **Camerano, L.** Ric. anat.-fisiol. i. a. Salamandrini apneumoni. *Anat. Anz.*, v. 9, 1894 ; v. 12, 1896.

894. **Campana, J. C.** Phys. de la respiration chez les oiseaux. Paris, 1875.

895. **Coggi, A.** Blutdrüsen i. d. Schwimmblase d. Hechtes. *Morph. Jahrb.*, v. 15, 1889.

896. **Corning, H. K.** Wundernetzbild. i. d. Schwimmblasen. *Morph. Jahrb.*, v. 14, 1888.

897. **Evans, H. M.** Anat. and Phys. of Air-bladder and Weberian Ossicles in Cyprinidae. *Proc. Roy. Soc.*, B, v. 97, 1925.

898. **Evans, H. M.**, and **Damant, G. C. C.** Physiol. of Swim-bladder in Cyprinoid Fishes. *Brit. J. Exper. Biol.*, v. 6, 1928.

899. **Fischer, G.** Bronchialbaum d. Vögel. *Zoologica*, v. 19, 1905 (Bibl.).

900. **Flint, J. M.** Devel. of Lungs. *Am. J. of Anat.*, v. 6, 1906.

901. **Goette, A.** Ursprung der Lungen. *Zool. Jahrb.*, v. 21, Anat., 1905.

902. **Göppert, E.** Schwimmblase, Lunge u. Kehlkopf. Hertwig's *Handb. Vergl. Entw. d. Wirbeltiere*, v. 2, 1902 (Bibl.).

903. **Greil, A.** Frage n. d. Ursprunge der Lungen. *Anat. Anz.*, 26, 1905.

904. **Guillot, N.** App. de la respiration dans les oiseaux. *Ann. Sc. Nat.*, v. 5 (3), 1846.

905. **Guyénot, E.** Étude de la vessie natatoire des Cyprinides. *C.R. Soc. Biol. Paris*, 1905.

906. **Haldane, J. S.** Secret. and Absorpt. of Gas in Swim-bladder. *Science Progr.*, v. 7, 1898.

907. **Hall, F. G.** Functions of Swim-bladder. *Biol. Bull.*, v. 47, 1924 (Bibl.).

908. **Hardiviller, A. d'.** Ramif. bronchique chez le lapin. *Bibliogr. Anat.*, v. 4, 1896 ; v. 5, 1897.

909. **Hesser, K.** Entw. d. Reptilienlunge. *Anat. Hefte*, v. 29, 1905.

910. **Huntington, G. S.** Pulmonary Evol. of Mammalia. *Am. J. of Anat.*, v. 27, 1920 (Bibl.).

911. **Huxley, T. H.** Respiratory Organs of Apteryx. *Proc. Zool. Soc. London*, 1882.

912. **Juillet, A.** Poumon des oiseaux. *Arch. Zool. Exp. et Gén.*, v. 9, 1912 (Bibl.).

913. **Justesen, P. Th.** Entw. u. Verz. d. Bronchialbaumes. *Arch. mikr. Anat. u. Entw.*, v. 56, 1900.

914. **Kerr, J. G.** Devel. of Alim. Canal in Lepidosiren and Protopterus. *Quart. J. Micr. Sci.*, v. 54, 1910.

915. **Larsell, O.** Recurrent Bronchi of Air-sacs of Chick. *Anat. Anz.*, v. 47, 1914.

916. **Locy, W. A.**, and **Larsell, O.** Embryology of the Bird's Lung. Pt. 1, *Am. J. Anat.*, v. 19, 1916; Pt. 2, *Am. J. Anat.*, v. 20, 1916.

917. **Lühe, M.** Lungenlose Urodelen. *Zool. Zentralbl.*, v. 7, 1900.

918. **Makuschok, M.** Phylog. Entw. der Lungen. Pt. 1, *Anat. Anz.*, v. 39, 1911 ; Pt. 2, *Anat. Anz.*, v. 42, 1912.

919. **Makuschok, M.** Genetische Beziehung zwischen Schwimmblase u. Lungen. *Anat. Anz.*, v. 44, 1913.

920. **Milani, A.** Reptilienlunge. *Zool. Jahrb.*, Abt. Anat., v. 7, 1894; v. 10, 1897.

921. **Moser, F.** Entw. d. Wirbeltierlunge. *Arch. mikr. Anat. u. Entw.*, v. 60, 1902.

922. —— Entwickl. der Schwimmblase. *Anat. Anz.*, 23, 1903.

923. **Müller, B.** Air-sacs of Pigeon. *Smithsonian Misc. Publ.*, 1724, 1908.

924. **Narath, A.** Entw. d. Lunge v. Echidna. *Denkschr. Med. Nat. Ges. Jena*, v. 5, 1896.

925. —— Bronchialbaum d. Säugetiere. *Bibl. Medica*, 1901.

926. **Neumayer, L.** Entw. des Darmkanales, etc., b. Ceratodus. *Denkschr. Med. Nat. Ges. Jena*, v. 4, 1904.

927. **Nusbaum, J.** Gehörorgan u. Schwimmblase b. d. Cyprinoiden. *Zool. Anz.*, v. 4, 1881.

928. —— Gasdrüse u. Oval. *Anat. Anz.*, v. 31, 1907.

929. **Parker, T. J.** Air-bladder and Audit-organ in Red Cod. *Tr. New Z. Inst.*, v. 15, 1882.

930. **Piper, H.** Entw. u. Leber, Pankreas, Swimmblase u. Milz von Amia. *Anat. Anz.*, 21, 1902.

931. **Potter, G. E.** Respiratory Function of Swim-bladder in Lepidosteus. *J. Exp. Zool.*, v. 49, 1927.

932. **Rauther, M.** Vergl. Anat. der Swimmblase der Fische. *Ergeb. u. Fortschr. der Zool.*, v. 5, 1923 (Bibl.).

933. **Reiss, K.,** and **Nusbaum, J.** Histol. d. Gasdrüse i. d. Schwimmblase. *Anat. Anz.*, v. 27, 1905; v. 28, 1906.

934. **Ridewood, W. G.** The Air-bladder and Ear of Brit. Clupeoid Fishes. *Journ. Anat. and Phys.*, v. 26, 1891.

935. **Robinson, A.** Devel. of Lungs of Rats and Mice. *J. of Anat. and Physiol.*, v. 23, 1889.

936. **Roché, G.** Anat. comp. des réservoirs aériens chez les oiseaux. *Ann. d. Sc. Nat. Zool.*, v. 11 (7), 1891.

937. **Rösler, H.** Erste Anlage der Lungen u. d. Nebengekröse einiger Vogelarten. *Anat. Hefte*, 1ᵗᵉ Abt., v. 44, 1911.

938. **Rowntree, W. S.** Visceral Anatomy of Characinidae, etc. *Trans. Linn. Soc.*, v. 9, 1903.

939. **Sörensen, W.** Air-bladder and Weberian Ossicles. *J. Anat. and Phys.*, v. 29, 1895.

940. **Spencer, W. B.** Bau d. Lungen v. Ceratodus u. Protopterus. *Denkschr. Med. Nat. Ges. Jena*, v. 4, 1898.

941. **Spengel, J. W.** Schwimmblasen, Lungen u. Kiementaschen. *Zool. Jahrb.*, Suppl. 7, 1904.

942. **Tracy, H. C.** Morph. of Swim-bladder in Teleosts. *Anat. Anz.*, v. 38, 1911.

943. —— Clupeoid Cranium in its Relation to the Swim-bladder, etc. *J. of Morph.*, 33, 1920.

944. **Vincent, S.,** and **Barnes, S. A.** Red Glands in Swim-bladder. *J. Anat. and Phys.*, v. 30, 1896.

945. **Weber, A.,** et **Buvignier, A.** Dével. de l'app. pulmonaire chez Miniopterus. *Bibl. Anat.*, v. 12, 1903. Chez le canard, etc. *C.R. Soc. Biol.*, v. 55, 1903.

946. **Wiedersheim, R.** Kehlkopf der Ganoiden u. Dipnoer. *Anat. Anz.,* v. 22, 1903.
947. **Wilder, H. H.** Amphibian Larynx. *Zool. Jahrb.,* Abt. Anat., v. 9, 1896.
948. —— Lungless Salamanders. *Anat. Anz.,* v. 9, 1894 ; v. 12, 1896.
949. **Woodland, W. N. F.** Structure and Function of Gas Glands, etc. *Proc. Zool. Soc. London,* 1911.
950. —— Physiol. of Gas Production. *Anat. Anz.,* v. 40, 1912.
951. **Wright, R. R.** Skull and Auditory Organ of Siluroid Hypophthalmus. *Trans. R. Soc. Canada,* Sect. 4, 1885.

CHAPTER XII

Subdivision of Coelom, and Diaphragm

952. **Beddard, F. E.** Diaphragm and Muscular Anat. of Xenopus, etc., Anat. of Pipa. *Proc. Zool. Soc.,* 1895.
953. —— Struct. and Classif. of Birds. London, 1898.
954. **Bertelli, D.** Pieghe d. reni primitivi. Pisa, 1897.
955. —— Rich. di embriol. e di anat. comp. sul diaframma e sull' apparechio respiratorio. *Arch. di Anat. e di Embr.,* v. 4, 1905.
956. **Brachet, A.** Dével. de la cavité hépato-entérique, etc. *Arch. de Biol.,* v. 13, 1895 ; *Anat. Anz.,* v. 11, 1896.
957. —— Dével. du diaphragme et du foie. *J. de l' Anat. et Physiol.,* 1895 and 1897.
958. —— Entwickl. d. g. Körperhöhlen u. Entwickl. d. Zwerchfells. *Ergeb. Anat. u. Entw.,* v. 7, 1897 (Bibl.).
959. **Broman, I.** Entw. der Bursa omentalis. Wiesbaden, 1904 (Bibl.).
960. —— Entw. d. Membr. peric. u. d. Zwerchfells. *Ergeb. Anat. u. Entw.,* v. 20, 1911 (Bibl.).
961. **Brouha, M.** Dével. du foie, etc., chez les oiseaux. *J. de l'Anat. et Physiol.,* 1898.
962. **Butler, G. W.** Subdiv. of Body-cavity in Lizards, Crocodiles, and Birds. *Proc. Zool. Soc. London,* 1889. In Snakes. *Ibid.,* 1892.
963. —— Suppression of the Right Lung in *Amphisbaenidae* and of Left Lung in Snakes, Lizards, and Amphibians. *Proc. Zool. Soc.,* 1895.
964. **Goette, A.** Entwick. des Flussneunauges (Petromyzon). *Abh. z. Entw. der Tiere,* 5tes Heft, 1890.
965. **Goodrich, E. S.** Devel. of Peric.-peritoneal Canals in Selachians. *J. of Anat.,* v. 53, 1918.
966. **Gössnitz, W. v.** Diaphragmafrage. *Denkschr. Med. Nat. Ges. Jena,* v. 4, 1901.
967. **Hochstetter, F.** Scheidewandbild., etc., b. Sauriern. *Morph. Jahrb.,* v. 27, 1899.
968. —— Scheidewand zw. peric. u. Peritoneal-höhle v. Acanthias. *Morph. Jahrb.,* v. 29, 1900.
969. —— Scheidewandbild. i. d. Krocodile. Voeltzkow, *Reise Ostafrika,* v. 4, 1906.
970. **Keith, A.** Mammalian Diaphragm and Pleural Cavities. *J. Anat. and Phys.,* v. 39, 1905.
971. **Lockwood, C. B.** Devel. of Pericardium, Diaphragm, etc. *Phil. Trans. R.S.,* v. 179, 1888.

972. **Mall, F. P.** Devel. of Human Diaphragm. *Proc. Assoc. Am. Anatomists*, 1900.

973. **Mathes, P.** Morph. d. Mesenterialbildungen b. Amphibien. *Morph. Jahrb.*, v. 53, 1895.

974. **Poole, M.** Subdiv. of Pleuro-perit. Cav. in Birds. *Proc. Zool. Soc.*, 1909.

975. **Ravn, E.** Bild. d. Scheidewand in Säugetier. *Arch. f. Anat. u. Entw.*, 1889.

976. —— Entw. d. Zwerchfells, etc. *Ibid.*, Suppl. 1889.

977. —— Bild. d. Septum transversum. *Ibid.*, 1896.

978. **Swaen, A.** Dével. du foie, etc. *J. de l' Anat. et Physiol.*, 1896–97.

979. **Uskow, N.** Entw. d. Zwerchfelles, etc. *Arch. f. mikr. Anat.*, v. 22, 1883.

CHAPTER XIII

EXCRETORY ORGANS AND GENITAL DUCTS

980. **Allen, B. M.** Devel. of Ovary and Testis of Mammals. *Am. J. Anat.*, v. 3, 1904.

981. —— Rete-cords and Sex-cords of Chrysemys. *Ibid.*, v. 5, 1905.

982. **Audigé, J.** Reins des Téléostéens. *Arch. Zool. Expér.*, v. 4, 1910.

983. **Ayers, H.** Beitr. z. Anat. u. Phys. der Dipnoër. *Jen. Zeitschr.*, v. 18, 1884.

984. —— Pori abdominales. *Morph. Jahrb.*, v. 10, 1885.

985. **Balfour, F. M.** Comparative Embryology. v. 2, London, 1881.

986. **Ballantyne, F. M.** Male Genito-urinary Organs of Ceratodus. *Proc. Zool. Soc. London*, 1928.

987. **Bates, G. E.** Pronephric Duct in Elasmobranchs. *J. Morph.*, v. 25, 1914.

988. **Beard, J.** Origin of Segmental Duct. *Anat. Anz.*, v. 2, 1887.

989. —— Pronephros of Lepidosteus. *Anat. Anz.*, v. 10, 1894.

990. **Bles, E. S.** Openings in Wall of Body-cavity. *Proc. Roy. Soc.*, 1897.

991. **Borcéa, J.** Syst. uro-génital des Élasmobranches. *Arch. Zool. exp. et gén.*, v. 4, 1905.

992. **Boyden, E. A.** Devel. of Cloaca in Ostrich. *Anat. Rec.*, v. 24, 1922. Devel. of Avian Cloaca. *J. Exper. Zool.*, v. 46, 1924.

993. **Brachet, A.** Embryologie des vertébrés. Paris, 1921.

994. **Brambell, F. W. R.** Devel. and Morph. of Gonads of Mouse. *Proc. Roy. Soc. London*, B, v. 102, 1927.

995. **Brauer, A.** Entw. u. Anat. der Gymnophionen III. : Entw. der Excretionsorgane. *Zool. Jahrb.*, Abt. Anat., v. 16, 1902.

996. **Bridge, T. W.** Pori abdominales. *Jour. Anat. and Phys.*, v. 14, 1879.

997. **Brock, A. J. P. v. d.** Urogenitalapp. der Beutler, etc. *Morph. Jahrb.*, v. 41, 1910.

998. **Brock, J.** Geschlechtsorgane der Knochenfische. *Morph. Jahrb.*, 4, 1878.

999. —— Geschlechtsorgane e. Muraenoiden. *Mitth. Zool. Sta. Neapel*, v. 2, 1881.

1000. **Burlend, T.** Pronephros of Chrysemys. *Zool. Jahrb.*, Abt. Anat., v. 36, 1913 ; v. 37, 1914.

1001. **Conel, J. Lek.** Urogenital System of Myxinoids. *J. Morph.*, v. 29, 1917.

1002. **Dean, B.** Larval Devel. of Amia. *Zool. Jahrb.*, Abt. Syst., v. 9, 1897.

1003. Dean, B. Embryology of Bdellostoma. *Festschr. f. C. v. Kupffer*, 1899.
1004. Essenberg, J. M. Sex-differentiation in Xiphophorus. *Biol. Bull.*, v. 45, 1923.
1005. Felix, W. Entw. des Harnapparates. Hertwig's *Handb. Entw. d. Wirbeltiere*, v. 3, Jena, 1906.
1006. —— Devel. of Urinogenital Organs. V. 2, Keibel and Mall, *Human Embryology*, 1912.
1007. Felix, W., u. Bühler, A. Entw. d. Keimdr. u. i. Ausführgänge. Hertwig's *Handb. Entw. d. Wirbeltiere*, v. 3, Jena, 1906..
1008. Field, H. H. Devel. of Pronephros and Segmental Duct in Amphibia. *Bull. Harv. Mus. Comp. Zool.*, v. 21, 1891.
1009. Fleischmann, A. Morph. Stud. ü. Kloake, etc. (Mammalia). *Morph. Jahrb.*, v. 32, 1903.
1010. Fraser, E. A. Devel. of Urogenital Syst. in Marsupialia. *J. of Anat.*, v. 53, 1918. Pronephros and Mesonephros in Cat. *J. of Anat.*, v. 54, 1920.
1011. —— Devel. of Pronephros of Acipenser. *Quart. J. Micr. Sci.*, v. 71, 1928 (Bibl.).
1012. Fürbringer, M. Entw. d. Amphibienniere, Heidelberg, 1877. Anat. u. Entw. d. Exkretionsorg. *Morph. Jahrb.*, v. 4, 1878.
1013. Gemmill, J. F. Entsteh. d. Müller's Ganges b. Amphibien. *Arch. Anat. u. Embr.*, 1896.
1014. Gerhardt, U. Entw. d. bleib. Niere. *Arch. mikr. Anat.*, v. 57, 1901.
1015. —— Kopulationsorgane d. Säugetiere. *Jen. Zeitschr.*, v. 39, 1905.
1016. Goodrich, E. S. Coelom, Genital Ducts, and Nephridia. *Quart. J. Micr. Sci.*, v. 37, 1895.
1017. —— Excretory Organs of Amphioxus. *Quart. J. Micr. Sci.*, v. 45, 1902.
1018. Guitel, F. Anat. des reins des Gobiésocidés. *Arch. Zool. Expér.*, v. 5, 1906.
1019. Hall, R. W. Devel. of Mesonephros and Müllerian Ducts in Amphibia. *Bull. Mus. Comp. Zool. Harvard*, vol. 45, 1904.
1020. Haller, B. Ovarialsack der Knochenfische. *Anat. Anz.*, v. 27, 1905.
1020a. Hatta, S. Devel. of Pronephros, etc., in Petromyzon. *J. Coll. Sc. Imp. Univ. Tokyo*, v. 13, 1900.
1021. Helmuth, K. Morph. Stud. ü. Kloake, etc. (Chelonia, Crocodilia). *Morph. Jahrb.*, v. 30, 1902.
1022. Hoffmann, C. K. Entw. d. Urogen.-Organe b. Anamniern. *Zeitschr. wiss. Zool.*, v. 44, 1886 ; b. Reptilien. *Ibid.*, v. 48, 1889.
1023. Holzbach, E. Die Hemmungsbildungen der Müllerschen Gänge. *Beitr. Geburtsb. Gynäkol.*, v. 14, 1909.
1024. Huxley, T. H. Oviducts of Osmerus, etc. *Proc. Zool. Soc.*, 1883.
1025. Hyrtl, J. Morph. der Urogenital-Organe der Fische. *Denk. Ak. Wien*, v. 1, 1850.
1026. —— Geschlechts- und Harnwerkzeuge b. d. Ganoiden. *Denk. Ak. Wien*, v. 8, 1854.
1027. Janosik, J. Embr. d. Urogenitalsyst. *Sitz. Ber. Akad. Wiss. Wien*, v. 91, 1885 ; v. 109, 1890.
1028. Jungersen, H. F. E. Entw. der Geschlechtsorgane bei der Knochenfischen. *Arb. Zool. Inst. Würzburg*, v. 9, 1889.

1029. **Jungersen, H. F. E.** Embryonalniere des Störs. *Zool. Anz.*, v. 16, 1893. Embryonalniere von *Amia calva*. *Ibid.*, v. 17, 1894. Urogenital-organe v. Polypterus u. Amia. *Ibid.*, v. 23, 1900.

1030. **Keibel, F.** Entw. des Urogenitalapparates von Echidna. *Denkschr. Med. Nat. Ges. Jena*, v. 6, 1904.

1031. **Kerens, B.** Dével. de l'app. excréteur des Amniotes. *Arch. de Biol.*, v. 22, 1907.

1032. **Kerr, J. G.** Male Genito-urinary Organs of Lepidosiren and Protopterus. *Proc. Zool. Soc.*, 1901 ; *Proc. Philos. Soc. Cambridge*, v. 11, 1902.

1033. —— Devel. of Polypterus. *Budgett Mem. Vol.*, London, 1907.

1034. **Kirkaldy, J. W.** Head Kidney of Myxine. *Quart. J. Micr. Sci.*, v. 35, 1894.

1035. **Lickteig, A.** Geschlechtsorgane der Knochenfische. *Zeitschr. wiss. Zool.*, v. 106, 1913.

1036. **Maas, O.** Vorniere u. Urniere bei Myxine. *Zool. Jahrb.*, Abt. Anat., v. 10, 1897.

1037. **MacBride, E. W.** Devel. of Oviduct in Frog. *Quart. J. Micr. Sci.*, v. 33, 1892.

1038. **MacLeod.** Appareil reprod. femelle des Téléostéens. *Arch. Biol.*, v. 2, 1881.

1039. **Martin, A.** Anlage d. Urniere b. Kaninchen. *Arch. mikr. Anat.*, v. 32, 1888.

1040. **Maschkowzeff, A.** Zur Phylogenie d. Urogenitalsystems der Wirbeltiere (Acipenser). *Zool. Jahrb.*, Abt. Anat., v. 48, 1926.

1041. **Milhalkovics, v.** Entw. d. Harn- u. Geschlechtsapp. d. Amnioten. *Intern. Monatsschr. Anat. u. Phys.*, v. 2, 1885.

1042. **Moens, N. W. I.** Peritonealkanäle der Schildkröton u. Krokodile. *Morph. Jahrb.*, v. 44, 1911.

1043. **Mollier, S.** Vornierensyst. bei Amphibien. *Arch. Anat. u. Entw.*, 1890.

1044. **Nussbaum, J.** Entw. samenbl. Wege b. Anuren. *Zool. Anz.*, v. 3, 1880.

1045. —— Wimpertricher i. d. Niere d. Anuren. *Ibid.*

1046. **Pomayer, C.** Morph. Stud. ü. Kloake, etc., Vögel. *Morph. Jahrb.*, v. 30, 1902.

1047. **Price, G. C.** Devel. of Excretory Organs in Bdellostoma. *Zool. Jahrb.*, Abt. Anat., v. 10, 1897 ; *Amer. Journ. Anat.*, v. 4, 1904.

1048. **Rabl, C.** Entw. d. Urogenitalsystems der Selachier. *Morph. Jahrb.*, v. 24, 1896.

1049. **Rabl, H.** Vorniere u. Bildung des Müller'schen Ganges bei Salamandra, etc. *Arch. mikr. Anat.*, v. 64, 1904.

1050. **Rückert, J.** Über d. Entsteh. d. Exkretionsorgane bei Selachiern. *Arch. Anat. u. Entw.*, 1888.

1051. —— Entw. d. Exkretionsorgane. *Anat. Hefte, Ergeb.*, 1892.

1052. **Schneider, G.** Entw. der Genitalcanäle bei Cobitis u. Phoxinus. *Mém. Acad. Imp. St-Pétersb.*, v. 11, 1895.

1053. **Schreiner, K. E.** Entw. der Amniotenniere. *Zeitschr. f. wiss. Zool.*, vol. 71, 1902.

1054. **Scott, W. B.** Beitr. z. Entw. Gesch. der Petromyzonten. *Morph. Jahrb.*, v. 7, 1881. Development of Petromyzon. *Jour. Morph.*, v. 1, 1887.

1055. **Sedgwick, A.** Head Kidney in Chick. *Quart. J. Micr. Sci.*, v. 19, 1879 ; v. 20, 1880.

1056. —— Early Devel. of Wolffian Duct, etc. *Quart. J. Micr. Sci.*, v. 21, 1881.

1057. **Semon, R.** Keimdrüse b. Hühnchen, etc. *Jen. Zeitschr.*, v. 21, 1887.
1058. —— Morph. Bedeut. d. Urniere. *Anat. Anz.*, v. 5, 1890.
1059. —— Bauplan d. Urogenitalsyst. *Jen. Zeitschr. Nat.*, v. 19, 1891.
1060. —— N. ü. d. Zusammenhang der Harn- u. Geschlechtsorgane b. d. Ganoiden. *Morph. Jahrb.*, v. 17, 1891.
1061. —— Exkretionsystem der Myxinoiden. *Festschr. f. C. Gegenbaur*, v. 3, Leipzig, 1896.
1062. —— Urogenitalsyst. d. Dipnoer. *Zool. Anz.*, v. 24, 1901.
1063. **Semper, C.** Urogenitalsystem der Selachier. *Arb. Inst. Würzburg*, v. 2, 1875.
1064. **Spengel, J. W.** Urogenitalsyst. der Amphibien. *Arb. Zool. Inst. Würzburg*, v. 3, 1876.
1065. —— Exkretionsorg. b. Myxine. *Anat. Anz.*, v. 13, 1897.
1066. **Swaen, A., et Brachet, A.** Dével. du mésoblaste du Téléostéens. *Arch. de Biol.*, v. 16, 1899 ; v. 18, 1901.
1067. **Waldeyer, W.** Eierstock u. Ei. Leipzig, 1870.
1068. **Walsche, L. de.** Dével. du pro- et mésonéphros des Chéloniens. *Arch. de Biol.*, v. 39, 1929.
1069. **Weber, M.** Abdominalporen der Salmoniden. *Morph. Jahrb.*, v. 21, 1887.
1070. **Weldon, W. F. K.** Head-kidney of Bdellostoma. *Quart. J. Micr. Sci.*, v. 24, 1884.
1071. —— Early Devel. of Lacerta. *Quart. J. Micr. Sci.*, v. 31, 1890.
1072. **Wheeler, W. M.** Devel. of Urogenitalorgans of Lamprey. *Zool. Jahrb. Anat.*, v. 13, 1899.
1073. **Wijhe, J. W. v.** Mesodermsegmente des Rumpfes u. d. Entw. d. Exkretionssyst. *Arch. f. mikr. Anat.*, v. 23, 1889.
1074. **Wilson, G.** Devel. of Ost. Abdom. in Crocodile. *Trans. Roy. Soc. Edin.*, v. 38, 1897.
1075. —— Devel. Müll. Duct of Amphibians. *Ibid.*
1076. **Winiwarter, H. v.** Corps de Wolff et dével. du canal de Müller. *Arch. de Biol.*, v. 25, 1910.

CHAPTER XIV

PERIPHERAL NERVOUS SYSTEM AND SENSE ORGANS

1077. **Abel, W.** Devel. of Symp. n. s. in Chick. *J. of Anat. and Phys.*, v. 46, 1912 ; v. 47, 1913.
1078. **Adelmann, H. B.** Devel. of Neural Folds and Cranial Ganglia of Rat. *J. Comp. Neur.*, v. 39, 1925.
1079. **Agar, W. E.** Spiracular Gill Cleft in Lepidosiren and Protopterus. *Anat. Anz.*, v. 28, 1906.
1080. **Alcock, R.** Peripheral Distribution of Cranial Nerves of Ammocaetes. *J. Anat. and Phys.*, v. 33, 1898–99.
1081. **Allis, E. P., Jr.** Anat. and Devel. of the Lateral Line System in Amia. *Jour. Morph.*, ii., 1889.
1082. —— Lateral Sensory Canals of *Polypterus*. *Anat. Anz.*, 17, 1900.
1082a. —— Lat. Sensory Canals, etc., of Mustelus. *Quart. J. Micr. Sci.*, v. 45, 1901.
1083. —— Lateral Canals and Cranial Bones of Polyodon folium. *Zool. Jahrb.*, Abt. Anat., 17, 1903.

1084. **Allis, E. P., Jr.** Lat. Sensory System in *Muraenidae*. *Int. Monat. Anat. u. Phys.*, v. 20, 1903.

1085. —— Latero-sensory Canals and related Bones in Fishes. *Intern. Mon. An. Phys.*, v. 21, 1905.

1085a. —— Lat. Sens. Canals of Plagiostomi. *J. Comp. Neur.*, v. 35, 1922.

1086. **Ayers, H.** Morph. of Vertebrate Ear. *Jour. Morph.*, v. 6, 1892.

1087. **Ayers, H.,** and **Worthington, J.** Skin End-organs of Trigeminus and Lateralis Nerves of Bdellostoma. *Am. J. of Anatomy*, v. 7, 1907.

1088. **Beard, J.** Sense Organs of Lateral Line and Morph. of Vert. Auditory Organ. *Zool. Anz.*, v. 7, 1884.

1089. —— Branchial Sense Organs and their Associated Ganglia in Ichthyopsida. *Quart. J. Micr. Sci.*, v. 26, 1885.

1090. **Belogolowy, J.** Entw. d. Kopfnerven der Vögel. *Bull. Soc. Imp. Nat. Moscou*, v. 22, (1908) 1910.

1091. **Bemmelen, J. F. van.** Rudimentäre Kiemenspalten bei Elasmobranchiern. *Mitth. Zool. Sta. Neapel*, v. 6, 1885.

1092. **Bender, O.** Schleimhautnerven d. Facialis Glossopharyngeus u. Vagus. *Denk. Med. Nat. Ges. Jena*, v. 7, 1907.

1093. **Brandt, W.** Darmnervensyst. v. Myxine. *Zeitschr. f. Anat. u. Entw.*, v. 65, 1922.

1094. **Brauer, A.** Entw. u. Anat. d. Gymnophionen. *Zool. Jahrb.*, v. 7, Suppl., 1904.

1095. **Camus, R.** Entw. d. symp. Nervensyst. b. Frosch. *Arch. mikr. Anat.*, v. 81, 1913.

1096. **Chevrel, R.** Syst. nerveux grand sympathique des Élasmobranches et poissons osseux. *Arch. Zool. Exp. et Gén.*, v. V., Suppl., 1887.

1097. **Chiarugi, G.** Dével. des nerfs vague, etc. *Arch. Ital. Biol.*, v. 13, 1890.

1098. **Coghill, G. E.** Afferent Syst. of Head of Amblystoma. *J. Comp. Neur.*, 25, 1916.

1099. **Cole, F. J.** Cranial Nerves of Chimaera monstrosa. *Trans. R. Soc. Edin.*, v. 38, 1896.

1099a. —— Of Gadus. *Trans. Linn. Soc. London, Zool.*, v. 7, 1893.

1100. **Collinge, W. E.** Sensory Canal System of Fishes. Pt. I. Ganoidei. *Quart. J. Micr. Sci.*, v. 36, 1894.

1101. —— Sensory and Ampullary Canals of Chimaera. *Proc. Zool. Soc.*, 1895.

1102. **Cook, M. H.,** and **Neal, H. V.** Are Taste-buds of Elasmobr. of Endod. Origin ? *J. Comp. Neur.*, v. 33, 1921.

1103. **Dogiel, A. S.** Darmgeflechte b. d. Säugetiere. *Anat. Anz.*, v. 10, 1895. Sympathischer Nervenzellen. *Ibid.*, v. 11, 1896. Geflechten des Darmes. *Arch. Anat. u. Phys.*, Anat. Abt., 1899.

1104. —— Periphere Nervensystem des Amphioxus. *Anat. Hefte*, 1. Abt., v. 21, 1903.

1105. **Dohrn, A.** Stud. z. Urgesch. d. Wirbeltierkörpers. 9, Unpaaren Flosse. *Mitt. Zool. Sta. Neapel*, v. 6, 1885. 13, Ammocaetes u. Petromyzon. 14, Rückenmarksnerven b. Selachiern. *Ibid.*, v. 7, 1888. 15, Metamerie des Kopfes. *Ibid.*, v. 9, 1890. 16, Entw. d. Augenmuskelnerven, v. 10, 1891. 18-22, Occipitalsomite, etc. *Ibid.*, v. 15, 1901–2.

1106. **Escher, K.** Seitenorgane u. i. Nerven C. Übergang z. Landleben. *Acta Zool.*, v. 6, 1927.

1107. **Ewart, J. C.**, and **Mitchell, J. C.** Lat. Sense Organs of Elasmobranchs. I. Laemargus. II. Common Skate Raja batis. *Tr. R.S. Edin.*, v. 37, 1891–95.

1108. **Ewart, J. C.**, and **Cole, F. J.** Dorsal Branches of Cranial and Spinal Nerves of Elasmobranchs. *Tr. Roy. Soc. Edin.*, v. 20, 1895.

1109. **Franz, V.** Nervensyst. d. Akranier. *Jena. Zeitschr.*, v. 59, 1923.

1110. **Ganfini, C.** Sviluppo d. s. n. simpatico n. ucelli. *Arch. Ital. d. Anat.*, 1916.

1111. **Garman, L.** Lat. Canal Syst. of Selachia and Holocephala. *Bull. Mus. Comp. Zool.*, v. 17, 1888.

1112. **Gaskell, W. H.** Involuntary Nervous System. London, 1916 (Bibl.).

1113. **Gast, R.** Entw. des Oculomotorius u. seiner Ganglien bei Selachier. *Mitth. Zool. Sta. Neapel*, v. 19, 1909.

1114. **Goette, A.** Entw. d. Kopfnerven b. Fischen u. Amphibien. *Arch. mikr. Anat.*, v. 85, 1914.

1115. **Goronowitsch, N.** Gehirn u. Cranialnerven v. Acipenser. *Morph. Jahrb.*, v. 13, 1888.

1116. —— Der Trigemino-facialis complex v. *Lota vulgaris. Gegenbaurs Festschr.*, v. 3, 1896.

1117. **Guthke, E.** Ganglien, etc., v. Torpedo. *Jena. Zeitschr.*, v. 42, 1907.

1118. **Harrison, R. G.** Histog. d. periph. Nervensyst. b. Salmo. *Arch. mikr. Anat.*, v. 57, 1901 ; *Biol. Bull.*, v. 2, 1901.

1119. **Held, H.** *Entw. des Nervengewebes b. d. Wirbeltieren.* Leipzig, 1909.

1120. **Herrick, C. J.** Cranial and First Spinal Nerves of Menidia. *J. Comp. Neur.*, v. 9, 1899.

1121. —— Doctrine of Nerve Components. *J. Comp. Neur.*, v. 14, 1904.

1122. **Heymans, J. F.**, and **Stricht, O. van d.** Syst. nerveux de l'Amphioxus. *Mém. Cour. Ac. Roy. Belgique*, v. 56, 1898.

1123. **Hill, C. J.** Enteric Plexuses. *Phil. Trans. R.S. London*, v. 215, 1927 (Bibl.).

1124. **His, W., Jr.** Entw. d. Bauchsympathicus. *Arch. Anat. u. Phys.*, Abt. Anat., Suppl., 1897.

1125. **Huber, G. C.** Morph. of Sympathetic System, Folia Neuro-biol., v. 7, 1913 ; and Intern. Congress of Medicine, London, 1913.

1125a. **Johnson, S. E.** Sense Organs of Lat. Canal Syst. of Selachians. *J. Comp. Neur.*, v. 28, 1917.

1126. **Johnston, J. B.** Gehirn u. Cranialnerven der Anamnier. *Ergeb. Anat. u. Entw.*, v. 11, 1902.

1127. —— Cranial and Spinal Ganglia and Viscero-motor Roots in Amphioxus. *Biol. Bull.*, v. 9, 1905.

1128. —— Cranial Nerve Components of Petromyzon. *Morph. Jahrb.*, v. 34, 1905. Add. Notes on Cranial Nerves of Petromyzonts. *J. Comp. Neur.*, v. 18, 1908.

1129. —— Nervous Syst. of Vertebrates. London, 1908 (Bibl.).

1130. —— Central Nervous System of Vertebrates. *Ergeb. u. Fortschr. der Zoologie*, v. 2, 1910.

1131. **Julin, C.** Grand sympathique de l'Ammocoetes. *Anat. Anz.*, v. 2, 1887.

1132. **Kapers, C. A.** Der Geschmack, etc. *Psychiatr. neur. Bladen*, 1914.

1133. **Kingsbury, F.** Lateral Line System in American Amphibia and comparison in Dipnoans. *Tr. Am. Micr. Soc.*, v. 17, 1896.

1134. **Klinkhardt, W.** Beitr. z. Entw. d. Kopfganglien u. Sinneslinien der Selachier. *Jenaisch. Zeitschr.*, v. 40, 1905.

1135. **Knouff, H. W.** Origin of Cranial Ganglia of Rana. *J. Comp. Neur.*, v. 44, 1927.

1136. **Kohn, A.** Entw. d. symp. Nervensyst. d. Säugetiere. *Arch. mikr. Anat.*, v. 70, 1907.

1137. **Kuntz, A.** Devel. of Sympathetic (Mammals, Birds, and Fishes). *J. Comp. Neur.*, v. 20-21, 1910–11.

1138. —— Innervation of Digestive Tube. *J. Comp. Neur.*, v. 23, 1913.

1139. —— Devel. of Symp. Nerv. Syst. in Man. *Ibid.*, v. 32, 1920.

1140. **Kuntz, A.,** and **Batson, O. V.** Experim. Observations on Histogenesis of Sympathetic Trunks in Chick. *J. Comp. Neur.*, v. 32, 1920.

1141. **Kutchin, H. L.** Periph. Nervous Syst. of Amphioxus. *Proc. Am. Acad.*, v. 49, 1913.

1142. **Landacre, F. L.** Origin and Dist. of Taste-buds of Ameiurus. *J. Comp. Neur.*, v. 17, 1907.

1142a. —— Origin of Cranial Ganglia of Ameiurus. *J. Comp. Neur.*, v. 20, 1910.

1143. —— Epibr. Placodes of Lepidosteus. *J. Comp. Neur.*, v. 22, 1912.

1143a. —— Fate of Neural Crest in Urodeles. *Ibid.*, v. 33, 1921.

1144. **Langley, J. N.** Sympathetic System. Schaefer's *Text-book of Physiology*, v. 2, 1900. Autonomic Nervous System. Cambridge, 1921.

1145. **Locy, W. A.** On a Newly Recognised Nerve. *Anat. Anz.*, v. 26, 1905.

1146. **Malbranc, M.** Seitenlinie u. ihren Sinnesorgane bei Amphibien. *Z.W.Z.*, 26, 1875.

1147. **Merritt, O. A.** Theory of Components. *J. of Anat. and Phys.*, v. 39, 1904.

1148. **Mitrophanow, P.** Étude embryogénique sur les Sélaciens. *Arch. Zool. Exp. et Gén.*, S. 3, v. 1, 1893.

1149. **Müller, E.** Autonome Nervensystems b. Selachiern. *Arch. mikr. Anat.*, v. 94, 1920.

1150. —— Darmnervensystem. *Upsala Läk. förhandl.*, v. 26, 1921.

1151. **Müller, E.,** and **Ingvar, S.** Sympathicus b. d. Amphibien. *Upsala Läk. förh.*, v. 26, 1921.

1152. **Müller, E.,** and **Liljestrand, G.** Autonome Nervensyst. der Elasmobranchier, etc. *Arch. Anat. u. Phys.*, Abt. Anat., 1918.

1153. **Müller, L. R.** Lebensnerven. Berlin, 1924 (Bibl.).

1154. **Neumayer, L.** Histog. u. Morphog. d. periph. Nervensystems. Hertwig's *Handb. Entwickl.*, v. 2, 1906 (Bibl.).

1155. **Norris, H. W.** Cranial Nerves of Siren lacertina. *J. of Morph.*, v. 24, 1913.

1156. —— Branchial Nerve Homologies. *Zeit. f. Morph. u. Anthr.*, v. 24, 1924.

1157. —— Cranial Nerves of Certain Ganoid Fishes. *J. Comp. Neur.*, v. 39, 1925.

1158. **Norris, H. W.,** and **Hughes, S. P.** Nerves of Squalus acanthias. *J. Comp. Neur.*, v. 31, 1920.

1158a. —— Spiracular Sense Organ in Elasmobranchs, Ganoids, and Dipnoans. *Anat. Rec.*, v. 18, 1920.

1159. **Onodi, A.** Entw. d. symp. Nervensyst. *Arch. mikr. Anat.*, v. 26, 1885.

1160. **Parker, G. H.** Functions of Lat. Line Organs. *Bull. Bur. Fish.*, v. 24, 1904.

1161. **Pinkus, F.** Hirnnerven des Protopterus. *Morph. Arb.*, v. 4, 1895.
1162. **Platt, J. B.** Devel. of Periph. Nervous Syst. (Necturus). *Quart. J. Micr. Sci.*, v. 38, 1896.
1163. **Plessen, J. v., u. Rabinovicz, J.** *Kopfnerven v. Salamandra.* 1891.
1164. **Pollard, H. B.** Lateral Line System of Siluroids. *Zool. Jahrb.*, Abt. Anat., 5, 1892.
1165. **Ranzi, S.** Organo di senso d. pr. placode epibr. d. Selaci. *Rend. Acad. Lincei Roma*, v. 1, 1925.
1166. **Rau, A. S., and Johnson, P. H.** Devel. of Symp. n. s., etc., in Sparrows. *Proc. Zool. Soc. London*, 1923.
1167. **Reese, A. M.** Lateral Line System of Chimaera colliei. *J. Exp. Zool.*, v. 9, 1910.
1168. **Ridewood, W. G.** Spiracle, etc., in Elasmobranch Fishes *Anat. Anz.*, v. 11, 1896.
1169. **Rohde, E.** Nervensyst. v. Amphioxus. *Zool. Beitr.*, v. 2, 1888.
1170. **Ruge, G.** Gebiet d. Nervus facialis. *Festschr. C. Gegenbaur*, v. 3, Leipzig, 1897.
1171. **Ruud, G.** Hautsinnesorgane bei *Spinax niger*. *Zool. Jahrb.*, Abt. Anat., v. 41, 1920.
1172. **Rynberk, G. van.** Versuch einer Segmentalanatomie. *Anat. Hefte*, 2te Abt., v. 18, 1908 (Bibl.).
1173. **Sewertzoff, A. N.** Kiemenbogennerven der Fische. *Anat. Anz.*, v. 38, 1911.
1174. **Stewart, F. W.** Devel. of Cranial Symp. Ganglia in Rat. *J. Comp. Neur.*, v. 31, 1920.
1175. —— Origin of Nervus terminalis of Albino Rat. *J. Comp. Neur.*, v. 32, 1920.
1176. **Strong, O. S.** Cranial Nerves of Amphibia. *J. of Morph.*, v. 10, 1895.
1177. **Tretjakoff, D.** Skelett u. Muskulatur d. Flussneunauges. *Zeit. wiss. Zool.*, v. 128, 1926.
1178. —— Periph. Nervensystem d. Fl. *Ibid.*, v. 129, 1927.
1179. **Vitali, G.** Organo n. d. senso nell' orechio medio d. ucelli. *Intern. Monats. f. Anat. u. Physiol.*, 30, 1914.
1180. —— *Richerche di morfologia*, v. 3, 1923 ; v. 4, 1924.
1181. **Watkinson, G. B.** Cranial Nerves of Varanus. *Morph. Jahrb.*, v. 35, 1906.
1182. **Wijhe, J. W. van.** Nervus terminalis fr. Man to Amphioxus. *Verh. d. k. Ak. v. Wet. Amsterdam*, v. 21, 1918.
1183. —— Thymus, Spiracular Sense Organ, etc. *K. Akad. v. Wetensch. Amsterdam*, v. 26, 1923.
1184. **Willard, W. A.** Cranial Nerves of Anolis. *Bull. Mus. Comp. Zool. Harvard*, v. 59, 1915 (Bibl.).
1185. **Wilson, H. V.** Embryology of Sea Bass (Serranus). *Bull. U.S. Fish. Comm.*, v. 9, 1889.
1186. **Wright, R. R.** Hyomand. Cleft and Pseudobranchs of Lepidosteus and Amia. *J. Anat. and Physiol.*, v. 19, 1885.

INDEX

All the numbers refer to pages, and the numbers followed by f. *refer to figures on the pages indicated.*

THE END